ROAD TRAFFIC: SAFETY, MODELING AND IMPACTS

ROAD TRAFFIC: SAFETY, MODELING AND IMPACTS

SOPHIE E. PATERSON
AND
LUCY K. ALLAN
EDITORS

Nova Science Publishers, Inc.
New York

For permission to use material from this book please contact us:
Telephone 631-231-7269; Fax 631-231-8175
Web Site: http://www.novapublishers.com

LIBRARY OF CONGRESS CATALOGING-IN-PUBLICATION DATA

Road traffic : safety, modeling & impacts / [edited by] Sophie E. Paterson and Lucy K. Allan.
p. cm.
ISBN 978-1-60456-884-4 (hardcover)
1. Traffic engineering--Developing countries. 2. Traffic engineering--Social aspects--Developing countries. 3. Traffic safety--Developing countries. 4. Traffic accidents--Developing countries. I. Paterson, Sophie E. II. Allan, Lucy K.
HE368.7.R63 2008
388.3'1091724--dc22

2008032041

Published by Nova Science Publishers, Inc. ✣ New York

CONTENTS

PREFACE

Chapter 1 - Traffic flow is a kind of self-driven many-particle system of strongly interacting vehicles. Various models are presented to understand the rich variety of physical phenomena exhibited by vehicular traffic. Jams and stop-and go-waves are a typical signature of the complex behavior of vehicular traffic. This chapter reviews the dynamical models and simulation for vehicular traffic. The car-following and cellular automaton models are in a general modeling framework for self-driven many-particle systems. The car-following model on a single-lane highway is extended to that on a multi-lane highway. The authors present the fundamental diagram (current-density profile) for the traffic flow of mixed vehicles on a multi-lane highway. The traffic jams induced by slowdown sections are studied on a two-lane and multi-lane highway. The authors clarify the characteristics of the traffic breakdown induced by speed limit. The nonlinear-map models are presented to study city traffic controlled by traffic signal. They clarify the dependence of traffic flow on signal characteristics. The traffic signal is controlled by cycle time, split, and offset time. Also, the authors study the effect of signal interval and its variation on vehicular behavior. They discuss the chaos induced by the interaction between the speedup and signals

Chapter 2 - In motor vehicle deaths, autopsies are performed to determine the cause and manner of death, detect any disease or factor that may precipitate or contribute to the death, document all findings for subsequent use in either criminal or civil actions and establish positive identification of the body, especially if it is burnt or severely mutilated (Saukko and Knight, 2004).

When victims of traffic accidents are autopsied, the standard autopsy procedure should be followed with detailed documentation of the injuries. It is very important that the body be seen clothed, if brought dead to the mortuary or hospital, so that injuries can be matched against soiling and damage to the garments. If this is not possible, as when survival allowed admission to a hospital or accident department, the clothing should be preserved and examined by the pathologist. In any event, the clothes should be retained by the police for submission to the forensic science laboratory, usually when criminal proceedings are likely. Toxicological analysis for alcohol is imperative, whether search for drugs depends on the information available. Blood samples should be retained for blood grouping and perhaps even 'DNA fingerprinting' (Saukko & Knight, 2004, DiMaio & DiMaio, 2001). All types of trace evidence may be found by a pathologist, from paint flakes and glass debris to parts of the vehicle structure and any of the above foreign bodies or particles, either in the clothing, hair, on the skin or in the wounds, must be carefully retained for forensic science examination (Smock et al., 1989, Curtin and Langlois, 2007).

Chapter 3 - *Background:* Head injury is a common cause of accidental deaths and of severe disabilities. It usually results from road traffic accident especially in developing countries. The presence of skull fracture is an important indicator of the nature and severity of the impact and risk of an operable intracranial lesion. Whereas in developed countries patients with skull fractures are managed by neurosurgeons, such patients are usually managed by general surgeons in developing countries. The authors present the experience of general surgeons on the management of depressed skull fractures in Nigeria.

Patients and Method: This study was conducted in the department of surgery Ahmadu Bello university teaching hospital Zaria, Nigeria. Adult patients seen between 1995 and 2005 with clinical and radiological diagnosis of depressed skull fracture were retrospectively reviewed. Patient evaluation included assessment of level of consciousness using the Glasgow coma scale, and other neurologic findings. All patients had skull x-rays. CT scan was done in some patients. Clinical and radiological features were used to select patients that required operative intervention. The records of these patients were analysed in respect of age and sex distribution and mechanism of injury. The type and location of skull fracture, clinical course, operative findings and neurologic outcome were also reviewed.

Results: There were 235 patients with depressed skull fractures which represent 3.6% of head injured patients. Their ages ranged 15 to 68 years, mean of 30 ± 5.7SD. Male to female ratio was 3.8:1. Road traffic accident caused fractures in 155 (66.0%) patients. Blows and missiles accounted for 15.7% and 11.5% respectively. Road traffic accident caused the most severe skull fractures. Of the 235 depressed skull fractures 152 (64.7%) were compound. The frontal bones were fractured in 115 (48.9%) patients while the parietal bone was involved in 61 (26.0%) patients. In 13 (5.5%) patients the fractures were located on cranial venous sinuses. The admission Glasgow coma score was ≤ 8 in 28 (11.9%) patients and 9-12 in 67 (28.5%). Elevation of depressed skull fracture was performed in 128 (54.5%) patients of which 80 (62.5%) had additional treatment of intracranial pathology. At discharge from hospital, 185 (78.7%) patients had complete recovery while additional 32(13.6%) had residual neurologic deficit but live an independent life. Overall, mortality was 10.2%.

Conclusion: Skull fracture is common among head injured patients and is usually caused by road traffic accident. The complications and sequelae of depressed skull fracture can be minimised by early diagnosis and treatment. With careful selection many of these patients can be safely manage non-operatively.

Chapter 4 – This chapter outlines a theory-based approach to the construction and psychometric evaluation of more behavioural based measures of willingness to take risks in road traffic. Based on the risk homeostasis theory (Wilde, 1994) several traffic situations varying on the degree of objective danger where filmed. Respondents were asked to indicate at which point the action that is contingent to the described situation will become too dangerous for them to carry out. Latencies at the item level were obtained. The first study deals with the dimensionality of this newly developed measure which is referred to as the Vienna Risk Taking Test Traffic. The second study investigates the generalizability of these results to a sample of professional driver applicants. The third study locates the newly developed measure within the context of the Giant Three model of personality. Confirmatory factor analyses indicated that the Vienna Risk Taking Test Traffic loads on the Giant Three factor 'Adventurousness', thereby confirming the construct validity of the behavioural based measures of willingness to take risks. Two additional studies investigate the criterion validity of the newly constructed measure using accident rates as well as respondents' performance in a standardized driving test as criterion measure. Despite the wealth of literature supporting the predictive validity of personality-related and ability-related determents of fitness to drive these two kinds of predictor variables have previously been studied in isolation. In order to overcome this methodological shortcoming the incremental validity of the Vienna Risk Taking Test over and above a rather comprehensive set of ability and personality tests was investigated. In line with theoretical considerations the newly developed objective personality test contributed incrementally to the prediction of respondents' performance in a standardized driving test as well as to the prediction of their accident rates. The theoretical implications of the results presented in this chapter are discussed in the light of current theoretical model on fitness to drive.

Chapter 5 - The increased availability of traveller information and onboard navigation systems is dramatically changing the characteristics of urban traffic. Traveller information (e.g., real time traffic information) helps individual travellers in minimising their travel costs through pre-trip scheduling and en-route dynamic route choice. On the other hand, traveller information systems provide traffic managers with an alternative and powerful means to implement transport policy and handle congestion by influencing travellers' dynamic route choice. This research investigates travellers' dynamic responses to real time traffic information and proposes a cost-effective way to support traveller information system development. In this research, the authors have developed a portable application program to implement the variable message sign (VMS) based route guidance system in a simulated urban city environment. Expected travel delay on the targeted road links is the primary information distributed by the VMS. Individual drivers' inherent characteristics (i.e., awareness and aggression), their minimum acceptable travel delay and perceived travel cost variations determine their dynamic route choice decisions. The program has been tested using the Adelaide CBD microscopic traffic simulation model which was developed in our previous research. The preliminary results demonstrate the capability of the program in mimicking travellers' dynamic response to VMS (and traffic information in general). Meanwhile, the tests reveal that the impact of a traveller information system on urban traffic should not be overestimated, and empirical results from travel behaviour studies are essential for any traveller information system development.

Chapter 6 - People aged over 65 represent the most rapidly growing segment of the population in most industrialized countries. Compared with previous generations, these people will, in future, demand greater transport mobility. As the general population ages, the percentage of elderly drivers on the road will increase. It is well established that age per se is not the best predictor of driving performance. However, it is recognized that aging has an effect on visuo-attentional and cognitive abilities, which can have an impact on driving ability and could be potentialy compensated by metacognitive abilities. The purpose of this chapter is to highlight the main trends in traffic safety research concerning senior drivers with health conditions, as a cross-disciplinary issue.

The first part of the chapter is concerned with advances in research related to neurocognitive abilities and driving. In past decades, research in this area has evolved from a classical medical approach to a more neuropsychological one. Numerous research protocols including neuropsychological tests (paper-and-pencil or computer-based), specific scenarios implemented in driving simulators and field observations have been used to study the impact of neurocognitive impairments on the driving performance of older drivers. Most of these studies have been conducted to explore the value of these tests in predicting future driving performance. A review of the various research protocols developed in these studies provides an opportunity to discuss and comment on some neuropsychological models that are capable of explaining some of the underlying mechanisms.

This chapter also addresses visuo-attentional issues in aging, focusing mainly on the useful visual field concept. Several experimental approaches have been used to study this concept, which has a number of clinical and practical applications. Different tests developed in these studies are presented, allowing a better understanding of numerous scientific approaches that highlight two main points: how is the useful visual field affected by aging?; and what are the consequences of this deterioration for driving activity?

In the final section of the chapter, the authors suggest an explanation on the potential adaptation of the driving behavior of some elderly drivers to their visual or cognitive deficiencies. The increasing body of research on metacognition is presented. Here, the objective is to emphasize the potential benefit for the development of more efficient training procedures for elderly drivers.

Simultaneous consideration of these three aspects provides an overarching framework for the conduct of future research on older drivers' fitness to drive.

Chapter 7 - This study creates a sequential approach to represent the dynamic update forecast of accident duration. This procedure includes two Artificial Neural Network-based models. Model A is used to forecast the duration time at the instant of accident notification while Model B provides multi-period updates of duration time after the moment of accident notification. These two models together provide a sequential forecast of accident duration from the accident notification to the accident site clearance. With these two models, the estimated duration time can be provided by plugging in relevant traffic data as soon as an incident is reported. This study shows very promising practical applicability of the proposed models in the Intelligent Transportation Systems context.

Chapter 8 - A signalized road network design problem is to find the optimal signal settings and link capacity expansions while taking into account the route choice of users. This problem can be formulated as a bi-level optimization problem by taking the user equilibrium traffic assignment as a constraint. At the upper level an optimization problem of signal timings and link capacity expansions for road networks is determined, among which the signal timing plan for coordinated fixed time control is defined by the common cycle time, the start and duration of greens. At the lower level the user equilibrium traffic assignment obeying Wardrop's first principle can be formulated as a minimization problem where the link travel time function is defined as the sum of the undelayed travel time on the link and the average delay incurred by vehicles at the downstream end of the link. In this chapter, the methods solving the signalized road network design problem are investigated via numerical calculations on real data road network. Improvement on a locally optimal search by combining the technique of parallel tangents with conjugate gradient projections is particularly investigated. A Hybrid Search heuRistic (HSR) is therefore proposed by combining the technique of parallel tangents with conjugate gradient projections. As it shows, the proposed HSR method achieved substantially better performance than did conventional approaches when solving the signalized road network design problem with equilibrium flows.

Chapter 9 - Social cognitive factors involved in automobile driving are generally understudied. Typically, human factors researchers who conduct research on driving examine the interaction of the driver and the vehicle, with emphasis on either the solitary driver or vehicle design. Social psychologists rarely look at the human factors of driving. However, as people must drive on roads together, social cognition plays a role in how people interact. Consequently, social cognition plays a central role in driving. Mather and DeLucia (2007) recently examined the interaction between social psychology and human factors of driving. Following their empirical study of implicit attitudes and pedestrian-vehicle collisions, the current chapter proposes various potential research topics on social cognitive human factors of automobile traffic safety. Basic and applied research in social psychology have much to contribute to research on the human factors of driving. Social interaction inside of the car can lead to distraction (e.g., cell phones; passenger interactions). Social interaction outside of the car can lead to death (e.g., teenagers tossing an item from one moving vehicle to another), injury (e.g., waving a car through when the other car is not clear), and saving lives (e.g., pointing to another driver's flat tire). Some possible areas of social psychological research that could contribute to research on the human factors of automobile driving include: motivation (e.g., need for closure), expectancies (e.g., second guessing another driver at a four-way stop), aggression (e.g., road rage), social facilitation (e.g., speeding up to pass another car or slowing down to keep from passing another car; general driving performance), attitudes and persuasion (e.g., increasing compliance with seat belt laws), and implicit racial attitudes (e.g., pedestrian-vehicle collisions). Potential contributions from these areas can be helpful to a single driver trying to drive safely and to anticipate danger from the road, obstacles, and other drivers. In summary, as automobile driving itself is an inherently social phenomenon, social psychological research is centrally relevant to research on driving. The current chapter examines in detail various social psychological research that is relevant to the human factors of automobile driving and traffic safety.

Chapter 10 - Road traffic accidents are a major source of mortality and morbidity in the United Kingdom. Speed is a known risk factor for accident risk, and speed cameras are being increasingly used to control vehicle speeds. However, if cameras are to be effective in modifying behaviour, it is important to understand how drivers respond to them. This research was undertaken using a postal questionnaire of drivers observed passing a speed camera in the county of Norfolk, England. Respondents were classified into a four category typology of conformers, deterred drivers, manipulators or defiers. Views and attitudes towards speed cameras and speed related behaviour were compared between groups. Differing perceptions and knowledge of the four types partially explained driving styles. Conformers and deterred drivers were least likely to exceed speed limits and were most favourable towards camera enforcement. Manipulators and defiers were younger, less experienced motorists who felt that they were unlikely to be prosecuted. This view means that manipulators and defiers are most difficult drivers to target for behavioural modification. In conclusion, the driver groups studied clearly showed divergent views and opinions. In order to improve the deterrent effect of speed cameras, specifically designed strategies should be used for the different groups.

Chapter 11 - Avalanches pose a significant natural hazard in many parts of the world. Avalanche hazard is managed using a range of methods in different mountain areas. A first step for hazard management is a risk assessment to evaluate the hazard posed to a road. This assessment can then be used as a tool to determine the appropriate mitigation measures and techniques to be employed. Care must be taken when using any risk assessment equation as the sensitivity of the assumptions made are rarely fully understood. This chapter will examine the application of two key risk assessment methods to examine avalanche risk on highways, the Avalanche Hazard Index (AHI) and the probability of death to individuals (PDI). First, this chapter will review the historical development and modifications to the AHI and PDI methods. Second, this chapter will examine the sensitivity of these two methods to the various assumptions made in the analysis and results will be compared to other commonly accepted levels of risk for other hazards. Finally, using both methods, examples of risk assessments for several avalanche prone roads from around the world will be presented and compared to data from the Milford Road, New Zealand.

Chapter 12 - This study aims to explore the effects of different road environments and their changes on driving behaviors and cognitive task performance of fatigued drivers. Twenty-four participants volunteered in a 2 (road environment) x 3 (fatigue level) within-subject factorial design simulated driving experiment. Participants were asked to perform basic numerical calculation and distance estimation of traffic signs when driving normally, and provide answers to a questionnaire on fatigue rating. Results show that fatigued drivers faced greater attention demand, were less alert, and tended to overestimate the distance to roadside traffic signs. Fatigue caused by driving in complex road environment had the greatest negative impact on driving behavior and visual distance estimation, and the fatigue transfer effect worsened significantly both driving behavior and performance of fatigued drivers when switching from a complex to a monotonous road environment and vice versa. Notably, this study shows that fatigued drivers performed relatively better in arithmetic tasks than non-fatigued ones. In addition, when switching from a monotonous to a complex road environment, drivers' performance in visual distance estimation and arithmetic tasks improved though driving behavior deteriorated, revealing that the fatigue effect upon drivers might be explained to some extent by the driver's alertness and arousal levels.

Chapter 13 - Injuries in traffic accidents are important causes of mortality in industrialized countries. Head trauma represents one of the most severe damages in road traffic accidents. Pedestrians, bicyclists and motor vehicle occupants have often suffered fatal intracranial injuries. The incidence of fatal head trauma depends on the role of victim, the speed, the protecting facilities of motor vehicles, and the available medical care. In this chapter the purpose of our examination was to calculate the rate of lethal intracranial injuries in different pathomechanisms of traffic accidents, and to evaluate the severity and outcome of head trauma.

Chapter 14 - Road trauma is an important problem in the developing world, with serious implications particularly in resource poor areas. 90% of road traffic accident related deaths worldwide occur in low- and middle-income countries, accounting for approximately 1.1 million deaths each year. Road trauma mortality is expected to increase by 80% over the next twenty years in the developing world and become the second leading cause of death. This is in contrast to developed nations, where road traffic mortality has steadily declined.

The greater population density in developing countries means pedestrians are relatively more likely to be involved in road traffic accidents, with a lower incidence of multiple car involvement. Injury patterns reflect this. Aside from soft tissue injuries, the most frequent presentations involve limb fractures, followed by blunt abdominal trauma. Developing countries have lower rates of seatbelt usage causing different patterns of abdominal trauma, whilst a high number of two-wheelers also lead to patients presenting with more serious and varied patterns of injury. Oro-facial trauma is particularly common with two-wheelers.

The high incidence of road trauma in developing countries has been attributed to increasing number of vehicles and poorly developed infrastructure unable to accommodate different types of vehicles as well as large pedestrian populations. Studies have also demonstrated associations with speed and alcohol usage.

Trauma management in the clinical setting contrasts sharply to that in the developed world. This is mainly due to a lack of specialized care in some areas, with poor access to diagnostic radiology and operating theatre resources. In blunt abdominal trauma, for example, this necessitates a greater reliance on techniques like diagnostic peritoneal lavage for diagnosis, with a lower threshold for laparotomy. Conversely, the poorest nations can only manage patients conservatively due to a lack of facilities and expertise. Patient management is also compromised by late patient presentation caused by lack of education, geographic barriers, and transport constraints.

Improving trauma outcomes in the developing world remains a significant challenge. While the advent of specialist trauma centers and improved access to public hospitals is going some way to achieve this in certain parts of the world, the poorest nations still suffer. In order to effect change, advancements in infrastructure, and road legislation and its enforcement are required, as well as a commitment to public health and education projects. This must be instigated at a governmental and organizational level.

Chapter 15 - Asian Development Bank (ADB) showed that road crashes killed more than 75,000 people in ASEAN in 2003 and it cost more than 2.2 percent of the region's annual gross domestic product (GDP).

Malaysia as one of the Association of South East Asian Nations (ASEAN) countries is a developing nation encompassing approximately 25 million population comprising Malays, Chinese, Indians and other races. Road traffic accident (RTA) is a major issue in this country as it is the predominant cause of mortality and morbidity in young people. In 2006, RTA resulted in 6287 fatalities and 58% or 3683 were fatalities involving motorcyclists and the leading cause of death was head injuries as a result of non-wearing or not securing the helmets properly.

Safety Planning Matrix which was based from the Haddon Matrix was introduced comprising pre-crash, crash and post-crash strategies involving human, vehicle and environment factors. Human and environmental safety programs are among the most extensively addressed and implemented. Human prevention program involves behavior modification and police enforcement program. Behavior modification program is directed largely at young road users especially motorcyclists and this mainly involves the media and television commercials. Safety and speed limit awareness are the key features in the police enforcement program.

Under the environmental safety program, exposure control program involves encouraging commuters to use other modes of transport to reduce them from being physically exposed to bodily impact during motorcycle rides; conspicuity program to increase visibility of a motorcyclist and road engineering program which involves setting-up exclusive motorcycle lanes, paved road shoulder and end treatment of non-exclusive motorcycle lanes.

It is targeted by the Malaysian Government that by the year 2010, the accident rate should reduce to more than 50 percent, i.e., deaths due to RTA should decline from the current 4.2 deaths per 10,000 vehicles to 2 deaths per 10,000 vehicles by 2010.

Chapter 16 - This chapter provides a descriptive case study to demonstrate the use of a process-based approach to change regarding the implementation of an information system for road traffic accident reporting in a UK police force. The supporting tools of process mapping and business process simulation are used in the change process and assist in communicating the current process design and people's roles in the overall performance of that design. The simulation model is also used to predict the performance of new designs incorporating the use of information technology. The case study examines both the steps in developing the simulation model and the organisational context in which the simulation study takes place. Change in this public sector organisation is discussed in terms of process, policy, legal and staffing issues and the simulation method is assessed in relation to these factors. The benefits of the simulation method in the case study are also outlined. For example, the ability of simulation to proof new designs was seen as particularly important in a government agency were past failures of information technology investments had contributed to a risk averse approach to their implementation.

Chapter 17 - Road traffic becomes a dominant source of air pollution in urban areas. The emissions of inorganic compounds and volatile organic compounds caused by motor vehicles have increased trends, particularly in urban areas, which together with other emitted compounds can cause human health problems in a long-time period. Procedures for estimating of road traffic emissions in the USA and in the European region are based on guidelines and recommendations. In order to implement the guidelines, decision-making tools are necessary to provide information in an easy understandable form. Thus, the spatial information system is needed to manage spatio-temporal data, provide analysis, solve numerical models and visualize the results. Generally, spatial data include street networks and monitoring networks. Temporal data are represented by changes of traffic intensity and by time series of measured pollutants. Geographic information systems (GISs) extended by environmental modeling tools offer advance estimates of traffic emissions improved by the digital terrain data and by the wind flow effects. The estimates of air pollution in the surrounded areas can be based on spatial interpolation. In a wide range of spatial interpolations, the deterministic techniques or the geostatistical methods are used in dependence on the input data and the results provided by the exploratory spatial data analysis (ESDA). Each spatial interpolation attached to the map layer defines concentration levels at a specified time period. The time series of map layers can demonstrate the variability of the air pollution distribution, which brings a new insight into this research.

In addition to mapping of the urban air pollution in dependence on the traffic intensity and the measured concentrations, numerical modeling and simulation in wind tunnels are presented as important tools for exploration of the dispersion and the transport of pollutants caused by wind flows and other effects. A number of software tools based on Gaussian dispersion principles are mentioned in the framework of US EPA guidelines.

As a case study, spatial modeling of air pollution is focused on the urban area of the city of Prague, because a number of motor vehicles registered on the Prague territory is growing. In addition to the registered vehicles, the specific phenomenon in the central Europe is represented by the abrupt increase in traffic of trucks. In spite of new methods for reduction of emissions from motor vehicles, urban development and successive reconstructions of existing roads cause other emissions. Thus, the street network also becomes a set of line sources of stirred up suspended particulates generated by the passing vehicles. These dust emissions, so-called secondary dust, are estimated and partially validated by measurements of the fraction PM_{10}. As an example, estimates of nitrogen oxides are carried out by spatial interpolation based on data from the surface monitoring network and the DIAL-LIDAR measurements.

Integration of spatio-temporal data together with environmental modeling tools brings new possibilities to compare spatial interpolations created for individual compounds at the high temporal and spatial resolution. In the framework of the GIS, mapping of air pollution and human exposures in streets, at workplace locations and residence addresses can help to reduce traffic emissions by optimization of transport scenarios.

In: Road Traffic: Safety, Modeling and Impacts
Editors: S. E. Paterson and L. K. Allan, pp. 1-13
ISBN 978-1-60456-884-4

Expert Commentary

DRIVING UNDER THE INFLUENCE OF CANNABIS

S. Athanaselis, S. Papadodima, C. Maravelias, C. Spiliopoulou
Dept. of Forensic Medicine and Toxicology, School of Medicine,
University of Athens, Greece

Cannabis is probably the most popular psychoactive substance used for recreational purposes in western nations, second only to alcohol (WHO, 2005, Mura et al., 2006, Gonzalez-Wilhelm, 2007, Hall and Degenhardt, 2007, Jones et al., 2008).

As it can be expected, with a high prevalence of use in the general population, cannabis is often involved in traffic accidents and other mishaps of operation that require skill.

The rising prevalence of driving under the influence of illegal and medicinal drugs (DUID) and its potential impact on traffic safety have raised awareness among media, scientists,and policy makers all over the world and prompted calls for more effective control.

The facts that the active constituents of cannabis are dependence producing substances and that the use of cannabis products leads to reckless behavior, especially after smoking a joint, have led many countries to consider the growing of cannabis plants, as well as purchase, possession, and selling of the different cannabis products as criminal offences, classifying cannabis and its various preparations as illicit drugs (Kalant, 2004, Raes and Verstraete, 2006, Jones et al., 2008).

Despite that, these psychoactive substances are used widely for recreational purposes and thereby represent a well recognizable problem for traffic safety, although serious questions (and sometimes doubts) about the extent of the risk appear in the literature (Robbe and O' Hanlon, 1993, Bates and Blakeley, 1999, Movig et al., 2004, Walsh et al., 2004, Asbridge, 2005, Mura et al., 2006, Alvarez at al., 2007, Bedardt et al., 2007). Cannabis is used for the primary purpose of "getting high" and escaping from reality, a behavior that is not compatible with performing skilled tasks such as driving (Grotenhermen, 2007). Indeed, it is generally accepted that the use of cannabis negatively alters a person's behavior, deteriorating its cognitive, sensory and motor functioning, faculties necessary for driving a vehicle (Grotenhermen, 2003).

Cannabis and its various products (marijuana, hashish, and hashish oil) are derived from the plant *Cannabis sativa*. The main psychoactive constituent of all cannabis products is delta-9-tetrahydrocannabinol (THC). Marijuana contains 0.5-5% THC while hashish (the resin of the plant) contains 2-20% and hashish oil 15-50% THC (Hall and Solowij, 1998).

The dose absorbed after smoking a cannabis joint ranges between 5 and 30 mg. The plasma concentration of THC reaches a peak after 3 to 8 minutes and then decreases quickly (half-life ≈ 30 min). THC is normally detectable (limit of detection (LOD) 1 ng/ml) in plasma for approximately 5 hours (UNIDCP, 1993, Verstraete, 2004).

Due to the complex pharmacokinetic profile, the high lipid solubility, the protein binding, and the large volume of distribution that THC displays in some cases, it can be detected in blood for up to 48 hours, depending of the amount of cannabis smoked and its potency. Its metabolites are detectable in blood or urine for days or weeks after the last cannabis use, depending mainly on the frequency of its use (UNIDCP, 1993, Skopp et al., 2003, Grotenhermen, 2003, Grotenhermen, 2007).

The undisputable evidence that verifies the use of cannabis by an individual is the detection of THC or its main metabolites 6-hydroxy-THC (THC-OH) and carboxy-THC (THC-COOH) in blood, urine or other body fluid (Grotenhermen, 2003, Goulle and Lacroix, 2006). It has to be mentioned here that THC-OH is possibly more potent than THC itself and may be responsible for some of the effects of cannabis (Ashton, 2001, Huestis, 2002, Grotenhermen, 2003) while THC-COOH is inactive. It has also to be mentioned that more than 20 other metabolites are known, some of which have long half-lifes of several days. These metabolites are partly excreted in the urine (25%) but mainly into the gut (65%), from which they are reabsorbed, further prolonging their pharmacologic actions (Ashton, 2001).

The most meaningful indicator of impairment is the concentration of THC in blood. It is generally accepted that even a high dose of THC typically causes acute impairment of driving skills for only 3-4 hours. This means that these users may still test positive for THC even when a long enough waiting time has elapsed between cannabis use and driving and impairment has dissipated (Grotenhermen, 2003, Grotenhermen, 2007).

Because of the pharmacokinetic characteristics of cannabinoids and especially the sequestration of fat and the presence of active metabolites, there is a very poor relationship between plasma or urine concentrations and degree of cannabis induced intoxication (Ashton, 2001).

The acute effects of cannabis on the user are well known and include mild euphoria, relaxation, increased sociability, heightened sensory perception, and increased appetite. Short-term psychomotor and cognitive effects related to the use of cannabis include impaired attention, memory, ability to process complex information, altered perception of the passage of time, and impaired performance in a wide variety of tasks, including handwriting and motor coordination tests. (Bates and Blakeley, 1999). Heavy, chronic cannabis users remain impaired even when they are not actually intoxicated (Hall et al., 1994) and some of these impairments can last for many weeks, months or even years after cessation of cannabis use (Solowij, 1998).

In order to determine whether a driver, involved in an accident or stopped at a roadside checkpoint, is impaired or under the influence of a certain substance there are three basic approaches used by the current DUID laws (Grotnhermen et al., 2007). One is the *effect-based* or *impairment* approach where the fitness of the drivers is observed and assessed. This requires that each suspect should be examined by a physician, who will look for signs and symptoms of drug influence and will also conduct various clinical tests of impairment. The other two approaches are versions of the *per se* approach. In one version of this approach, as in the case of alcohol, a science-based finite limit is used for the "tolerable" concentration of a drug or its metabolites in the driver's blood. The other version is the *zero tolerance* case where the detection of any detectable amount of the drug in the driver's blood is penalized as it constitutes an offence. In theory, the *impairment* approach is the one that meets the objectives of DUID laws. In fact, there are severe shortcomings to its adoption like the lack of standardized methods for estimating or measuring the impairment caused by consumption of a specific drug and the fact that impairment may arise from several other, synergistic, factors like fatigue or consumption of alcohol or CNS affecting drugs. Sobriety tests can reliably detect drug-induced impairment especially the one caused by CNS depressants. However, these tests are less sensitive to modest impairment (Papafotiou et al., 2005a , Papafotiou et al., 2005b). So, the assessment of driver's impairment in such cases is somewhat arbitrary and can easily raise legal disputes. All these shortcomings have led many jurisdictions to adopt *per se* limits for driving under the influence of cannabis (DUIC). Many of them use as *de facto* zero limits the limit of detection (LOD) of the analytical method used. It has to be mentioned here that the different forensic laboratories world wide do not use the same analytical methodology so, obviously, this *de facto* limit varies. It should also be mentioned that adoption of a zero tolerance policy penalizes the presence of THC or its metabolites in any biological fluid at any concentration, which does not necessarily correspond to actual impairment of driving skills (Huestis et al., 2005, Khabani et al.,2006, Jones et al., 2008). On the other hand, to put a "tolerable" cut-off concentration in the law is equivalent to condoning the use of cannabis, which remains illegal in almost all countries. In such a case, questions like: "how much cannabis must I consume without being positive during a road control or after an accident?" will arise (Walsh et al., 2004).

Despite all these, scientists attempt to establish concentration limits for driving after use of cannabis with THC concentration in blood serving as *per se* evidence of impairment (Grotenhermen et al., 2007). The scientific data for setting these limits are based on critical reviews of scientific literature, as well as on epidemiological studies of traffic accidents in which THC was identified in driver's blood, on laboratory studies of the effect of cannabis use on psychomotor and cognitive skills of individuals and on on-the-road-driving performance (Kurtzhaler et al., 1999, Ramaekers et al., 2000, Drummer et al., 2004, Ramaekers et al., 2004, Papafotiou et al., 2005, Wadsworth et al., 2006a, Jones et al., 2008).

Laboratory findings suggest clearly that acute cannabis use affects the attention, tracking and psychomotor skills used in driving (Moskowitz, 1985, Murray, 1986, Coambs and Mc Andrews 1994, Kurzthaler et al., 1999, Ramaekers at al., 2006).

Epidemiological and experimental studies provide conclusions that are heterogeneous and not robust enough to prove that such consumption represents a crash risk factor of significant magnitude (Brookoff, 1998, Bates and Blakeley, 1999, Fergusson and Horwood, 2001, Shope et al., 2001, Vingilis et al., 2002, Chipman et al., 2003, Drummers et al., 2004, Ramaekers et al., 2004, Asbridge et al., 2005, Blows et al., 2005).

It is extremely difficult to draw satisfactory conclusions from the different *epidemiological studies* mainly due to methodological problems like: selection bias, due particularly to low response rates; measurement bias relating to difficulty in accurately measuring cannabis use; failure to adjust for other significant confounders, like fatigue , speed, habitual or acute cannabis use, concurrent alcohol consumption; use of self-reported and non-injury crashes as outcome; and long time delays between measurement of exposure and outcome (Blows et al., 2005).

In most epidemiological studies, study subjects (mainly drivers but also pedestrians) are classified in terms of outcome (i.e., culpable or not culpable) or in terms of exposure status. (presence or absence of THC or THC-COOH in the biological fluids). In both cases misclassification is likely. The decision whether a driver should be considered responsible for an accident is in many cases subjective and is judged from detailed reconstructions of the individual accidents. On the other hand the detection of THC-COOH in the blood of a driver does not necessarily means that he or she was driving under the influence of cannabis. When THC itself is measurable, a causal role of cannabis is more strongly suggested by the fact that the probability of culpability increased progressively with increasing plasma levels of THC (Hunter et al., 1998, Ramaekers et al.,2004, Drummer et al., 2004). In any case, epidemiological studies measure the effect of drug use on driving performance and accident risk under real-life conditions and are thus suited to correlate the concentrations of a drug use indicator to an actual risk (Grotenhermen et al., 2007). However, studies of the effects of cannabis on driving under more realistic conditions on roads have shown much more modest impairments than the impairments of recreational doses of cannabis on driving performance in laboratory simulators and standardized driving courses (Hall and Solowij, 1998).

Experimental studies have also some disadvantages. Due to ethical considerations it is not possible to administer high enough cannabis doses to obtain the THC concentrations often found in real life situations. The studies in most of the cases are single dose experiments. The subjects included in experimental research are less often experienced users, excluding possible tolerance as a part of the studies. It could thus be argued that findings from such experimental studies would have limited relevance for real-life impairment of cannabis in experienced users and that concentration-effect relationship for THC would not be clear in a population with mixed cannabis experience (Khiabani et al., 2006).

The existing studies have demonstrated a connection between accidents and cannabis use but have not established causation by cannabis, despite the knowledge that individuals who use cannabis have impaired performance in driving simulator and on-the-road tests (Moskowitz, 1985, Smiley 1986, Kurtzthaler et al., 1999, Walsh and Mann, 1999, Ramaekers et al., 2000). In the driving studies, the strongest decrements were in the drivers' ability to concentrate and maintain attention, estimate time and distance, and demonstrate coordination on divided attention tasks, all important requirements for driving (Ogden and Moskowitz, 2004, Papafotiou et al., 2005).

Findings of experimental studies suggest that not all driving tasks are equally sensitive to the detrimental effects of THC. Performance was always worst in tests measuring driving skills at the operational level, i.e. tracking and speed adjustment as compared to performance in tests measuring driving performance at maneuvering level, i.e. distance keeping and braking, and the strategic level, i.e. observation and understanding of traffic, risk assessment and planning (Ramaekers et al., 2004).

Scientists find it virtually impossible to agree upon the concentration of a psychoactive substance in blood that leads to impairment in the vast majority of people, owing to individual differences in response, habituation, potency of the abused drug and differences related to dose, mode of administration as well as the pharmacokinetic profile of the abused drug.

Indeed, establishing a threshold concentration of THC in blood above which impairment of driving skills occurs is not an easy task. Unlike in case of alcohol, impairment of performance from cannabis use is not easily correlated to blood concentrations of THC or its metabolites due to a complex concentration-effect relationship (Huestis et al., 2005, Jones et al., 2008). It is interesting to notice that maximum effects from cannabis use may occur later than peak blood concentrations of THC or its active metabolites. Effects on the brain continue as blood concentrations of active drug decrease, a process termed *hysteresis* (Cone and Huestis, 1993, Huestis et al., 2005).

Experimental studies have shown clear but modest impairment of driving skills and actual driving performance in subjects smoking small or moderate doses of cannabis but that the drivers appeared to be less aggressive, more cautious and more aware of their impairment than subjects impaired to a similar degree by alcohol. However, it was noted that cannabis produced a significant decrease in attention and in ability to react to sudden unexpected emergencies, and that this posed a potential risk to driving safety (Kalant et al., 1999).

Cannabis users may be unaware of any impact on their cognitive performance as studies show that there is no association between cannabis use and cognitive failure (Block, 1996, Wadsworth et al., 2006a). This is consistent with the finding that measurable cognitive performance deficits among cannabis users are observed but they show relatively little awareness of any detrimental performance effects at work (Wadsworth et al., 2006b) and eventually at driving performance.

It is interesting that, as the results of observational and experimental (with driving simulators) studies and controlled driving situations show, cannabis, has no positive association with, and may even reduce, overall traffic crash fatality and serious injury risk despite the fact that impairment induced by cannabis still increase the overall risk of a crash. However, the fact that cannabis use is associated with reduced consequences of traffic crashes does not necessarily reflect a reduction in the risk of traffic crashes themselves (Bates and Blakeley, 1999).

A possible explanation is that cannabis-intoxicated drivers modify their driving behavior to compensate for their perceived impairment. This means that they seldom take risks and tend not to drive at speeds likely to result in fatalities or serious injuries. In fact, drivers under the influence of cannabis tend to overestimate the adverse effects of the drug on their driving ability and compensate accordingly (e.g. by slowing down) (Robbe and O' Hanlon 1993; Grotenhermen et al., 2007). It is also possible, that, despite their slower speeds, such drivers may be sufficiently impaired and despite the fact that their crash rate could be increased, the result is minor injuries and vehicle damage rather than deaths and serious injuries. Compensation might not be so successful in emergencies situations or during monotonous periods of driving (i.e on a highway) (Bates and Blakeley, 1999).

Drivers under the influence of cannabis tend to over-estimate the adverse effect of the drug on their driving quality and compensate when they can; e.g. by increasing effort to accomplish a task, increasing headway or slowing down, or a combination of these (Robbe and O'Hanlon,1993).

However, during the elimination phase of THC in blood, the subjective experience of the symptoms may have diminished and drivers may no longer feel it necessary to compensate for the effects of the drug, resulting in more impaired driving performance (Papafotiou et al., 2005).

Previous studies have shown that behavioral signs of intoxication, though small, outlast physiological and subjective reactions to THC (Reeve at al., 1983; Yesavage et al., 1985). It has also been noticed that cannabis can adversely affect complex human performance up to 24 hours after smoking (Heishman et al., 1990). So, if driving impairment still occurs after THC disappears from blood, it could mean that most epidemiological research has underestimated the proportion of drivers who were driving under the influence of cannabis at the times their accidents occurred. To examine this hypothesis, future research studies should extend actual driving performance measurements 4, 8, 16 and 24 hours after smoking (Robbe and O'Hanlon, 1993). According to this report of Robbe and O'Hanlon it is not possible to conclude anything about driver's impairment on the basis of his/her plasma concentration of THC determined in a single sample and that the THC's adverse effects on driving performance appear relatively small as the observed impairment manifests itself mainly in the ability to maintain a steady lateral position on the road, but its magnitude is not exceptional in comparison with changes produced by many medicinal drugs and alcohol (Hall and Homel, 2007).

A consensus panel of experts in 2004 reported that most behavioral and physiological effects return to baseline levels within 3-5 h after drug use (Couper et al., 2004) although some investigators have demonstrated residual effects in specific behaviors such as complex divided attention and performance tasks for up to 24 h (Leirer et al., 1991, O'Kane C et al., 2002, Huestis et al., 2005).

Plasma concentrations of THC have been shown to vary widely between 1 and 35 ng/ml in drivers suspected of driving under the influence (Augsburger et al., 2005) and between 1 and 100 ng/ml in fatally injured drivers (Drummer at al., 2004, Ramaekers et al., 2006). Apprehended impaired drivers in Norway showed a median THC concentration in blood of 2.2 ng/ml and a range from 0.3-45.3 ng/ml (Khiabani et al., 2006) while the respective concentrations of DUID suspects in Sweden were ranged from 0.3-67 ng/ml (median 1.0 ng/ml) (Jones et al., 2008).

It has to be mentioned here that the use of plasma THC concentrations to assess the impairment is recommended as plasma levels of THC are higher (almost double) than whole blood concentrations and plasma extracts GC-MS chromatograms are generally cleaner than those obtained with whole blood samples (Giroud et al., 2001).

It is interesting to notice that maximum impairment associated with cannabis consumption may occur once peak THC plasma levels have plateaued like it is observed in the case of benzodiazepines (Rush et al., 1996).

Previous studies demonstrated that recent exposure and possible measurable impairment have been linked to plasma THC concentrations in excess of 2-5 ng/ml (Mason and McBay, 1984; Tzambazis and Stough, 2002; Huestis et al.,1992) concentrations that correspond to 1-2.5 ng/ml in whole blood (Khiabani et al., 2006).

The use of cannabis or THC at low doses that do not cause psychotropic effects is not associated with an increased accident risk. Regular users of cannabis have no increased accident risk. Drivers with low THC blood concentrations may not have a higher accident risk than drug-free controls but THC blood concentrations above 5 ng/ml may be associated with an increased accident risk (Grotenhermen et al., 2007).

The findings of the study of Ramaekers et al., (2006) showed that slight and selective impairment of tracking performance was already notable at THC ranges between 2 and 5 ng/ml, but impairments became truly prominent across all performance domains at serum THC concentrations between 5 and 10 ng/ml.

These findings are in agreement with previous studies (Drummer et al., 2004; Laumon et al., 2005) according to which significant odds ratio (OR) of crash risk for THC blood concentrations ranging between 1 and 2 ng/ml and 2 and 5 ng/ml were 1.45 and 2.13 respectively. When the concentration was higher than 5 ng/ml the ORs ranged from 2.1 to 6.6.

All these findings suggest that THC serum concentrations between 2 and 5 ng/ml establish the lower and upper range of a *per se* limit for defining general performance impairment above which drivers are at risk. It should be stressed, however, that the predictive validity of such a *per se* limit is confined to the driving population at large and are not necessarily applicable to each and every driver as an individual. Individual drivers can widely differ in their sensitivity for THC induced impairment as evinced by the weak correlations between THC in serum and magnitude of performance impairment. Even at the higher limit of 5 ng/ml in 10-30% of the observations there was no impairment at all (Ramaekers et al., 2004). Drivers who claimed to be using cannabis on a regular basis were less often judged as impaired despite no difference in THC concentration between regular users and not regular users (Khiabani et al., 2006). This may support other studies that report that experienced users develop some degree of tolerance to acute effects of THC (Howlett et al., 2004).

The purpose of setting a *per se* limit is to indicate only the average THC concentration above which drivers are at risk and should be interpreted as such. A THC concentration in serum of between 7 and 10 ng/ml which corresponds to 3.5 – 5 ng/ml in blood was suggested by Grotenhermen et al., (2007) as a threshold *per se* limit for prosecution. These limits look high since if these THC concentrations will be adopted a very large percentage – higher than 75% - of people who had used cannabis a few hours before driving would not be liable to prosecution as the concentration of THC in their blood would drop below the *per se* limit for driving between the time of use and obtaining blood samples for toxicological analysis (Jones et al., 2008). The THC blood concentration at the time of blood sampling will be significantly less than at the time of arrest or the time of the accident, which can be up to one and a half hour earlier and driving which was still earlier (Jones et al., 2008). Back-extrapolation of the determined THC concentration from time of sampling blood to the time of driving (like in alcohol cases) is not possible mainly due to the complex pharmacokinetic profile of THC (Ashton, 2001, Grotenhermen 2003, Grotenhermen 2007, Jones et al., 2008).

In case that a zero tolerance law will be adopted many more people will be liable to prosecution with only a trace amount of THC in blood as the limit of quantitation (LOQ) of the different analytical methods range from 0.3 to 0.8 ng/ml (Giroud et al., 2001, Jones et al., 2006). In such cases the issue of residual THC after a previous use of cannabis has to be considered. There is evidence that THC can be detected (1.3-6.4 ng THC/ml serum) in moderate or heavy users 24 to 48 hours after smoking the last joint (Skopp et al., 2003). This is mainly explained by the high solubility of THC in fat and its slow washout into the circulation after termination of use (Grotenhermen et al., 2007).

In Sweden, LOQ for analysis of THC in blood (0.3 ng/ml) serves as the threshold concentration for prosecution (Jones, 2005, Jones et al.,2008). Germany have established a consensus limit of 1.0 ng/ml THC in serum which corresponds to 0.5 ng/ml in blood.

Belgium has adopted a punishable THC limit of 2.0 ng/ml in plasma (1.0 ng/ml in blood) and Switzerland enforces a limit of 1.5 ng/ml in blood. In the other countries where a zero-tolerance law operates, such as France, Finland, Poland and Australia the laboratory LOQ determines the threshold limit for a DUID prosecution (Walsh et al., 2004, Raes and Verstraete, 2006, Butler, 2007). A similar zero-tolerance law operates in Greece.

Passive exposure to cannabis smoke remains always an issue and is often used as an excuse by defendant drivers.

Controlled studies have shown that reaching measurable amounts of THC in blood after passive inhalation is virtually non-existent (Hayden, 1991, Busuttil et al., 1996). However, other studies support that in case of passive exposure to cannabis smoke, measurable THC concentrations can be observed but without causing concurrent impairment (Mason et al., 1983, Morland et al., 1985). The detection of THC-COOH in urine after passive smoking is more likely to occur, but it cannot be associated with impairment (Perez-Reyes et al., 1983, Morland et al., 1985).

Another important issue is that most of the conducted studies on cannabis and driving are experimental and concern living individuals. The application of the suggested limits to cases of fatal car accidents has a risk of significant errors mainly due to postmortem changes of blood composition and postmortem redistribution of THC. It is impossible to relate postmortem blood and antemortem plasma, serum or blood THC concentrations (Huestis et al.,2005).

Zero-limit DUIC based on analysis of THC-COOH in blood or urine lack scientific support and cannot be defended, as residual amounts of it are still measurable in biological fluids several days or weeks since last use (UNIDCP, 1993, Skopp et al., 2003, Verstraete, 2004).

CONCLUSION

There is accumulative evidence showing a strong link between driving under the influence of cannabis and risk of a motor vehicle crash. Whether these data imply causality is yet to be resolved, but even if the risks prove to be small, the public health outcomes could potentially be quite catastrophic. While epidemiological research can often identify causal linkages, well designed experimental studies offer the best approach for testing causality.

In order to establish a causal association between cannabis use and motor vehicle crashes

- more thorough elucidation of the behavioral and cognitive effects of cannabis and
- more thorough investigation of the relationship between cannabis biomarkers and intoxication or functional impairment
- are needed while, at the same time,
- more correctly designed experimental studies of the impact of cannabis on performance and particularly on driving behavior and
- more studies of the prevalence of cannabis (with and without alcohol or drugs affecting driving ability) in drivers

should be conducted.

Such studies will contribute to a science-based foundation for government policies and law enforcement practices on cannabis and driving that will help the policy makers in their efforts to mitigate the risk of DUIC by the establishment to engage more just and effective prevention efforts and cost-effective prevention strategies (Jones et al., 2006 Fergusson, 2005). In connection with appropriate legislative and enforcement measures (Asbridge, 2006, Alvarez et al., 2007) and possibly by the development of scientifically based "tolerable" THC blood concentration limits for DUIC this can be accomplished.

Identifying cannabis users who are most at risk of DUIC is of contemporary relevance to road accident prevention policy. As there is a lack of uniformity in the way in which the different countries approach the DUIC driver problem the standardization and the harmonization of laws through the development of "model" legislation should be encouraged (Walsh et al., 2004).

Data to demonstrate the effectiveness of the *impairment* approach over the *per se* approach (the adoption of which is the recent trend) is still in development.

Using the laboratory LOQ to establish the threshold limit for prosecution (zero- tolerance policy) seems that is the most realistic way to enforce DUIC legislation although establishing limit concentrations of THC in blood, like alcohol, depend more on political decisions and the specific issues of the narcotics policy in the countries concerned rather than scientific research.

In any case, it must be kept in mind that the use of illicit drugs, in general and not only in connection with driving, should not be tolerated by healthy societies and the fight for a drug free world must be continued.

References

Alvarez, FJ; Fierro, I; Del Rio, MC. (2007). Cannabis and driving: Results from a general population survey. *Forensic Sci Int,* 170: 111-116.

Asbridge, M; Poulin; C, Donato, A. (2005). Motor vehicle collision risk and driving under the influence of cannabis: Evidence from adolescents in Atlantic Canada. *Accid Anal Prev,* 37(6): 1025-1034.

Asbridge; M. (2005). Letter to the editor and reply on: "Drugs and driving" Traffic Injury Prevention, 5:241-253, 2004. *Traffic Inj Prev* 6:197.

Asbridge, M. (2006). Drugs and driving: when science and policy don't mix. *Can J PublicHealth,* 97:283-285.

Ashton, CH. (2001). Pharmacology and effects of cannabis: a brief review. *Br J Psychiatry,* 178:101-106.

Augsburger, M; Donze, N; Menterey, A; Brossard, C; Sporkert, F; Giroud, C; Mangin, P. (2005). Concentrations of drugs in blood of suspected impaired drivers. *Forensic Sci Int;* 153(1):11-15.

Bates, MN; Blakely, TA. (1999). Role of cannabis in motor vehicle crashes. *Epidemiol Rev,* 21(2): 222-232.

Bedard, M; Dubois, S; Weaver, B. (2007). The impact of cannabis on driving. *Can J Public Health,* 98(1):6-11.

Block, RI. (1996). Does heavy marihuana use impairs human cognition and brain function? *JAMA,* 275:560-562.

Blows, S; Ivers, R; Connor, J; Ameratunga, S; Woodward, M; Norton, R. (2005). Marijuana use and car crash injury. *Addiction,* 100: 605-611.

Brookoff, D. (1998). Marijuana and injury: is there a connection? *Ann Emerg Med,* 32: 361-363.

Busuttil, A; Obafunwa, J; Bulgin, S. (1996). Passive inhalation of cannabis smoke: a novel defence strategy? *J Clin Forensic Med,* 3:99-104.

Butler, M. (2007). Australia's approach to drugs and driving. *Of Subst, 5*:24-26.

Chipman, M; Macdonald, S; Mann, R. (2003). Being 'at fault' in traffic crashes: does alcohol, cannabis, cocaine, or polydrug abuse makes a difference? Inj Prev, 9:343-348.

Coambs, R; Mc Andrews, M. (1994). The effects of psychoactive substances on workplace performance. In Macdonald, S and Roman, P. (eds), *Drug testing in workplace.* New York, Plenum Press, pp 77-102.

Cone, EJ; Huestis, MA. (1993). Relating blood concentrations of tetrahydrocannabinol and metabolites to pharmacologic effects and time of marijuana usage. *Ther Drug Monit,* 15:527-532.

Couper, F; Logan, B; Corbett, M; Farell, L; Huestis M; Jeffrey, W; et al. *Drugs and human performance fact sheets.* National Highway Traffic Safety Administration, Washington DC, 2004.

Drummer, OH; Gerostamoulos, J; Batziris, H; Chu, M; Caplehorn, J; Robertson, MD; Swann, P. (2004). The involvement of drugs in drivers of motor vehicles killed in Australian road traffic crashes. *Accid Anal Prev,* 36:239-248.

Fergusson, DM. (2005). *Marijuana use and driver risks: the role of epidemiology and experimentation,* Editorial, Addiction, 100: 577-578, 2005.

Fergusson, D; Horwood, L. (2001). Cannabis use and traffic accidents in a birth cohort of young adults. *Accid Anal Prev:* 33, 703-711.

Giroud, C; Mentrey, A; Augsburger, M; Buclin, T; Sanchez-Mazas, P; Mangin, P. (2001). Δ^9-THC, 11-OH-Δ^9-THC and Δ^9-THCCOOH plasma or serum to whole blood concentrations distribution ratios in blood samples taken from living and dead people. *Forensic Sci Int,* 123: 159-164.

Gonzalez- Wilhelm L. (2007). Prevalence of alcohol and illicit drugs in blood specimens from drivers involved in traffic law offences. Systematic review of cross-sectional studies. *Traffic Inj Prev,* 8:189-198.

Goulle. JP; Lacroix. C. (2006). Mise en evidence des cannabinoides: quell milieu biologique? *Ann Pharm Fr,* 64:181-191, 2006.

Grotenhermen, F; (2003). Pharmacokinetics and pharmacodynamics of cannabinoids. *Clin Pharmacokinet,* 42 (4):327-360.

Grotenhermen, F; (2007). The toxicology of cannabis and cannabis prohibition. *Chem Biodivers,* 4(8):1744-1769.

Grotenhermen, F; Leson, G; Berghaus, G; Drummer, O; Kruger, H; Longo, M; Moskowitz, H; Perrine, B; Ramaekers, J; Smiley, A; Turnbridge, R. (2007). *Developing limits for driving under cannabis. Addiction,* 102: 1910-1917.

Hall, W; Degenhardt, L. (2007). Prevalence and correlates of cannabis use in developed and developing countries. *Curr Opin Psychiatry;* 20(4):393-397.

Hall, W; Homel, R. (2007). Reducing cannabis-impaired driving: is there sufficient evidence for drug testing of drivers. *Addiction,* 102: 1918-1919.

Hall, W; Solowij, N. (1998). Adverse effects of cannabis. *Lancet,* 352: 1611-1615.

Hall, W; Solowij, N; Lemon, J. (1994). The health and social consequences of cannabis use. *National Drug Strategy Monograph Series No 25*. Canberra: Australian Government Publishing Service.

Hayden, JW. (1991). Passive inhalation of marijuana smoke: a critical review. *J Subst Abuse;* 3:85-90.

Heishman, SJ; Huestis, MA; Henningfield, JE; Cone, EJ. (1990). Acute and residual effects of Marijuana: Profiles of plasma THC levels, physiological, subjective, and performance measures. *Pharmacol Biochem Behav,* 37(3): 561-565.

Howlett, AC; Breivogel, CS; Chidren, SR; Deadwyler, SA; Hampson, RE; Porino, LJ. (2004). Cannabinoid physiology and pharmacology: 30 years of progress. *Neuropharmacology*, 47(Suppl 1):345-358.

Huestis, MA. (2002). Cannabis (marijuana)-effects on human performance and behaviour. *Forensic Sci Rev,* 14:15-60.

Huestis, MA; Barnes, A; Smith, ML. (2005). Estimating the time of last cannabis use from plasma Δ^9-Tetrahydrocannabinol and 11-nor-9-carboxy-Δ^9-Tetrahydrocannabinol concentrations. *Clin Chem,* 51(12): 2289-2295.

Huestis MA, Henningfield JE, Cone EJ. (1992). Blood cannabinoids.I Absorption of THC and formation of 11-OH-THC and THC-COOH during and after smoking of marijuana. J Anal Toxicol, 16:276-282.

Hunter, C; Lokan, R; Longo, M. (1998). The prevalence and role of alcohol, cannabinoids, benzodiazepines and stimulants in non fatal crashes, *Forensic science,* Department of Administration and Information Services, Adelaide, South Australia.

Jones, AW. (2005). Driving under the influence of drugs in Sweden with zero concentration limits in blood for controlled substances. T*raffic Inj Prev,* 6:317-322.

Jones, AW; Holmgren, A; Kugelberg, FC. (2008). Driving under the influence of cannabis: a 10-year study of age and gender differences in the concentration of tetrahydrocannabinol in blood. *Addiction,* 103: 452-461.

Jones, C; Donelly, N; Swift, W; Weatherburn, D. (2006). Preventing cannabis users from driving under the influence of cannabis. *Accid Anal Prev,* 38: 854-861.

Kalant, H; Corrigall, W.A; Hall, W; Smart, R.G. Editors, (1999). *The Health Effects of Cannabis,* Toronto, ARF Books (CAMH), pp. 1–526.

Kalant H., Adverse effects of cannabis on health: an update of the literature since 1996. (2004). *Prog Neuropsychopharmacol Biol Psychiatry,* 28(5): 849-863.

Khiabani, HZ; Bramness, JG; Bjorneboe, A; Morland, J. (2006). Relationship between THC concentration in blood and impairment in apprehended drivers. *Traffic Inj Prev,* 7: 111-116.

Kurzthaler, I; Hummer, M; Miller, C; Sperner-Uterweger, B; Gunther, V; Wechdorn, H; Battista, HJ; Fleischhacker, WW. (1999). Effects of cannabis use on cognitive functions and driving ability. *J Clin Psychiatry,* 60(6): 395-399.

Laumon, B; Gadegbeku, B; Martin, JL; Biecheler, MB; the SAM group. (2005). Cannabis intoxication and fatal road crashes in France: population based control study. *BMJ,* doi: 10.1136/bmj38648.617986.1F (published 2 December 2005).

Leirer, V; Yesavage, J; Morrow, D. (1991). Marijuana carryover effects on aircraft pilot performance. *Aviat Space Environ Med,* 62:221-227.

Mason, A; McBay, A. (1984). Ethanol, marijuana, and other drug use in 600 drivers killed in single-vehicle crashes in North Carolina. 1978-1981. *J Forensic Sci,* 29, 987-1026.

Moskowitz, H. (1985). Marijuana and driving. *Accid Anal Prev* !7: 323-345.

Morland, J; Bugge, A; Skuterud, B; Steen, A; Wethe, G; Kjeldsen, T. (1985). Cannabinoids in blood and urine after passive inhalation of cannabis smoke. *J Forensic Sci,* 30:997-1002.

Movig, KL; Mathijssen, MP; Nagel, PH; van Egmond, T; de Gier, JJ; Leufkens, HG; Egberts, AC. (2004). Psychoactive substance use and the risk of motor vehicle accidents. *Accid Anal Prev.* 36(4):631-636.

Mura, P; Brunet, B; Favreau, F; Hauet, T. (2006). Cannabis et accidents de la voie publique: resultants des dernieres etudes francaises. *Ann Pharm Fr,* 64: 192-196.

Murray, J. (1986). Marijuana's effects on human cognitive functions, psychomotor functions and personality. *J Gen Psychol;* 113(1):23-55.

Ogden, E; Moskowitz, H. (2004). Effects of alcohol and other drugs on driver performance. *Traffic Inj Prev,* 5:185-198.

O'Kane, C; Tutt, D; Bauer, L. (2002). Cannabis and driving: a new perspective. *Emerg Med;* 14: 296-303.

Papafotiou, K; Carter, JD; Stough, C. (2005a). The relationship between performance on the standardized field sobriety tests, driving performance and the level of Δ9-tetrahydrocannabinol (THC) in blood. *Forensic Sci Int,* 155: 172-178.

Papafotiou, K; Stough, C; Carter, JD.(2005b). An evaluation of the sensitivity of the standardized field sobriety tests (SFSTs) to detect impairment due to marihuana intoxication. *Psychopharmacology;* 180(1):107-114.

Perez-Reyes, M; Di Giuseppi, S; Mason, A; Davis, K; (1983). Passive inhalation of marihuana smoke and urinary excretion of cannabinoids. *Clin Pharmacol Ther;* 34:36-41.

Raes, E; Verstraete, AG; (2006). Cannabis et conduite automobile: la situation en Europe. *Ann Pharm Fr,* 64(3):197-203.

Ramaekers, JG; Robbe, HW; O'Hanlon, JF. (2000). Marijuana, alcohol and actual driving performance. *Hum Psychopharmacol,* 15(7):551-558.

Ramaekers, JG; Berghaus, G; van Laar, M; Drummer OH. (2004). Dose related risk of motor vehicle crashes after cannabis use. *Drug Alcohol Depend,* 73(2):109-119.

Ramaekers, JG; Moeller, MR; van Ruitenbeek, P; Theunissen, EL; Schneider, E; Kauert, G. (2006). Cognition of motor control as a function of Δ^9-THC concentration in serum and oral fluid: Limits of impairment. *Drug Alcohol Depend,* 85:114-122.

Reeve, VC; Robertson, WB; Grant, J; Soares, JR; Zimmerman, EG; Gillespie, HK; Hollister, LE. (1983). Hemolyzed blood and serum levels of Δ^9- THC: Effects on the performance of roadside sobriety tests. *J Forensic Sci,* 28: 963-971.

Robbe, H; O'Hanlon, J. (1993). *Marijuana and actual driving performance.* Report DOT HS 808 78, Washington U.S. Department of Transportation, National Highway Traffic Safety Administration.

Rush, C; Griffiths, R. (1996). Zolpidem, triazolam and temazepam: behavioural and subject-rated effects in normal volunteers. *J Clin Psychopharmacol,* 16(2): 146-157.

Shope, J; Waller, P; Raghunathan, T; Patil, S. (2001). Adolescent antecedents of high-risk driving behavior into young adulthood: substance use and parental influences. *Accid Anal Prev;* 33: 649-658.

Skopp, G; Richter, B; Potsch, L. (2003). Cannanoidbefunde im Serum 24 bis 48 Stunden nach Rauchkonsum [Serum cannabinoid levels 24 to 48 hours after cannabis smoking]. *Arch Kriminol,* 212(3-4): 83-95.

Smiley, A. (1999). Marijuana: on road and driving simulator studies. In: Kalant H, Corrigal W, Hall W, Smart R, editors. The Health effects of Cannabis. Toronto, Canada, *Addiction Research Foundation,* p. 173-191.

Solowij, N. (1998). *Cannabis and cognitive functioning. Cambridge,* Cambridge University Press.

Tzambazis, K; Stough, C. (2002). *The SFST and driving ability. Are they related?* (Ed. Dussault, C.), Montreal, Quebec, Canada, ICADTS.

UNIDCP. (1993). Recommended methods for the detection and assay of Heroin and Cannabinoids in biological specimens, *Manual for use by national laboratories.* New York, United Nations, p. 31-43.

Verstraete, AG. (2004). Detection times of drugs of abuse in blood, urine, and oral fluid. *Ther Drug Monit,* 26(2):200-205.

Vingilis, E; Macdonald, S. (2002). Drugs and traffic collisions. *Traffic Inj Prev* 3:1-11.

Wadsworth, EJK; Moss, SC; Simpson, SA; Smith, AP. (2006a). A community based investigation of the association between cannabis use, injuries and accidents. *J Psychopharm,* 20(1):5-13.

Wadsworth, EJK; Moss, SC; Simpson, SA; Smith, AP. (2006b). Cannabis use, cognitive performance and mood in a sample of workers. *J Psychopharmacol,* 20(1):14-23.

Walsh, JM; DeGier, JJ; Christopherson, AS; Verstraete, AG. (2004). Drugs and Driving. *Traffic Inj Prev,* 5:241-253.

Walsh, GW; Mann, RE. (1999). On the high road: driving under the influence of cannabis in Ontario. *Can J Public Health,* 90(4):260-3.

World Health Organization. (2005). Management of substance abuse: facts and figures at http://www.who.int/substance_abuse/facts

Yesavage, JA; Leirer, VO; Denari, M; Holister, LE. (1985). Carry-over effects of marijuana intoxication on aircraft pilot performance: a preliminary report. *Am J Psychiatry,* 142:1325-1329.

In: Road Traffic: Safety, Modeling and Impacts
Editors: S. E. Paterson and L. K. Allan, pp. 15-224
ISBN 978-1-60456-884-4

Chapter 1

MODELS AND SIMULATION FOR TRAFFIC JAM AND SIGNAL CONTROL

Takashi Nagatani
Department of Mechanical Engineering, Division of Thermal Science, Shizuoka University, Hamamatsu 432-8561, Japan

ABSTRACT

Traffic flow is a kind of self-driven many-particle system of strongly interacting vehicles. Various models are presented to understand the rich variety of physical phenomena exhibited by vehicular traffic. Jams and stop-and go-waves are a typical signature of the complex behavior of vehicular traffic. This chapter reviews the dynamical models and simulation for vehicular traffic. The car-following and cellular automaton models are in a general modeling framework for self-driven many-particle systems. The car-following model on a single-lane highway is extended to that on a multi-lane highway. We present the fundamental diagram (current-density profile) for the traffic flow of mixed vehicles on a multi-lane highway. The traffic jams induced by slowdown sections are studied on a two-lane and multi-lane highway. We clarify the characteristics of the traffic breakdown induced by speed limit. The nonlinear-map models are presented to study city traffic controlled by traffic signal. We clarify the dependence of traffic flow on signal characteristics. The traffic signal is controlled by cycle time, split, and offset time. Also, we study the effect of signal interval and its variation on vehicular behavior. We discuss the chaos induced by the interaction between the speedup and signals

1. INTRODUCTION

Physics, other sciences and technologies meet at the frontier area of interdisciplinary research. The concepts and techniques of physics are being applied to such complex systems as transportation systems [1-5]. The scientific studies for traffic problems have started about

1935 [6]. In 1955, Lighthill and Whitham have presented the oldest and most popular macroscopic traffic model based on the fluid-dynamic theory [7]. They have studied the traffic jam as a shock wave by treating traffic as an effectively one-dimensional compressible fluid. Prigogine and co-workers [8,9] have presented the gas-kinetic model based on the Boltzmann equation. In 1961, Newell has proposed the microscopic, optimal velocity model based on the assumption of a delayed adaptation of velocity [10]. In 1976 and 1978, Musha and Higuchi have studied the noisy behavior of traffic flow and have conjectured that the fluctuations of traffic current exhibit the so-called *1/f* noise [11,12].

Although there were already some early pioneer's works like Lighthill and Whitham [7] and Prigogine [8], the papers of Biham *et al* [13], Nagel and Schreckenberg [14] and Kerner and Kohnhauser [15] triggered the main activities in traffic physics. Then, an avalanche of publications started in various international physics journal. The development of the modern traffic theories is due to the availability of computer and the concepts and techniques of modern physics. Traffic is modeled as a system of interacting vehicles driven far from equilibrium. The traffic models exhibit a rich variety of physical phenomena such as the dynamical jamming transition, critical phenomena, metastability, self-organized criticality and nonlinear waves (soliton), etc. The physical research works have been reviewed in the references [1-5].

For decades, the functional relations between the vehicle current and the vehicle density (so-called fundamental diagram) have highly attracted attention of traffic researchers. At low densities, the traffic shows the linear dependence of the traffic current on the density. In contrast, at high densities, the traffic current decreases with increasing density. There are strong fluctuations of the current at high densities. The vehicles move freely at low densities, while the vehicles are in a congested state at high densities. Thus, with increasing vehicle density, the traffic state changes from the free traffic at low densities to the congested traffic at high densities. Near the density of the maximal current, the jamming transition from the free traffic to the congested traffic occurs. The jamming transition exhibits the hysteresis and metastability. The modern traffic theory has developed to clarify the fundamental aspects of the jamming transition.

The traffic science aims to discover the fundamental properties and laws in transportation systems. On the other-hand, traffic engineering aims at making the planning and implementation of the road network and control systems. In the traffic engineering area, the very complex traffic models have been proposed to forecast or estimate the traffic current in real transportation systems [16-19]. The models include so many factors that it is difficult to discover the essential factors affecting on the traffic behavior. Physicists have proposed the simplified traffic models including a few factors at most to clarify the cause and effect. Despite the complexity of traffic and complications of human behavior, physical traffic theory is an example of a highly quantitative description for a living system.

In this review, we try to give the modeling and simulation for the single-lane and multi-lane traffic flows. Attention is paid to the formulation of the traffic dynamics and its application. We discuss the methods and results for the traffic flow under typical situations of real traffic in detail after explaining the traffic models. We focus on the traffic phenomena induced by the slowdown and signal control.

The outline of this review is as follows. In Section 2, we present the various traffic models where there are different conceptual frameworks for modeling traffic. In the macroscopic description, traffic is viewed as a compressible fluid. In the microscopic description, traffic is treated as a system of interacting particles driven far from equilibrium. In Section 3, we discuss the fundamental diagrams and traffic characteristics obtained from the modern traffic models. In Section 4, we study the traffic jams induced by slowdown sections. In Section 5, we extend the single-lane traffic simulation to that on the multi-lane traffic flow. We present the simulation result obtained from the traffic flow on the multi-lane highway. In Section 6, we present the nonlinear-map models for the vehicular traffic through a sequence of traffic signals. We study the vehicular traffic controlled by traffic signals and clarify the dependence of vehicular behavior on signal characteristics. In Section 7, we discuss the maximal flow and clustering controlled by traffic lights. We clarify the dependence of the maximal flow on the signal characteristics. In section 8, we study the chaotic behavior of vehicles induced by the speedup and interaction with other vehicles. Section 9 presents the summary.

2. Models of Vehicular Traffic

This section presents the various traffic models with different conceptual frameworks for modeling traffic. In the microscopic models, the traffic is treated as a system of interacting self-driven particles. In contrast, in the so-called macroscopic models, the traffic is viewed as a compressible fluid.

2.1. Microscopic Car-following Models

A vehicle is represented by a Newtonian particle of a mass with a friction [4]. The equation of motion of a mass with a simple friction is given by

$$m\frac{d^2x_j(t)}{dt^2}+\gamma\frac{dx_j(t)}{dt}=F(x_1(t),x_2(t),--x_j(t),--,x_N(t)),\tag{1}$$

where $x_j(t)$ is the position of vehicle j at time t, m is the mass of a vehicle, γ is the friction coefficient of a vehicle, and $F(x_1(t),x_2(t),--,x_N(t))$ is the driving force to accelerate or decelerate the vehicle. The driving force depends on the configuration of other vehicles.

The car-following model is a typical one of microscopic traffic models. The non-integer car-following models are called *follow-the-leader* models. The vehicle j is affected only by the vehicle ahead j+1, called the *leading* vehicle. Then, the driving force is described by

$$F(x_1(t),x_2(t),--,x_N(t))=F(\Delta x_j(t)),\tag{2}$$

where $\Delta x_j(t)\ (=x_{j+1}(t)-x_j(t))$ is the headway of vehicle j at time t.

If one defines new parameter $\tau\equiv m/\gamma$ and variable $F(\Delta x_j(t))\equiv\gamma V(\Delta x_j(t))$, one has the equation of vehicular motion

$$\frac{d^2x_j(t)}{dt^2}=\left(\frac{1}{\tau}\right)\left(V(\Delta x_j(t))-\frac{dx_j(t)}{dt}\right). \tag{3}$$

Eq. (3) has been presented by Bando *et al* [20]. The model is called as the optimal velocity model. The inverse $(1/\tau)$ of delay time is called the sensitivity a.

Newell has proposed the original *optimal velocity* model [10]. The equation of motion for vehicle *j* is described

$$\frac{dx_j(t+\tau)}{dt}=V(\Delta x_j(t)), \tag{4}$$

where $x_j(t)$ is the position of vehicle j at time t, τ is the delay time, $\Delta x_j(t)\ (=x_{j+1}(t)-x_j(t))$ is the headway of vehicle j at time t and $V(\Delta x_j(t))$ is the optimal velocity. The idea is that a driver adjusts the vehicle velocity according to the observed headway $\Delta x_j(t)$. The delay time τ allows for the time lag that it takes for the vehicle velocity to reach the optimal velocity $V(\Delta x_j(t))$ when the traffic flow is varying.

By Taylor-expanding Eq. (4), one obtains the optimal velocity model (3).

By transforming the time derivative to the forward difference in Eq. (4), one obtains the difference equation model [21, 22]

$$x_j(t+2\tau)=x_j(t+\tau)+\tau V(\Delta x_j(t)). \tag{5}$$

It is useful to convert Eq. (5) to the headway equation:

$$\Delta x_j(t+2\tau)=\Delta x_j(t+\tau)+\tau[V(\Delta x_{j+1}(t))-V(\Delta x_j(t))]. \tag{6}$$

Generally, it is necessary that the optimal velocity function has the following properties: it is a monotonically increasing function and it has an upper bound (maximal velocity). Bando *et al* [20] propose the relation

$$V(\Delta x_j(t))=\frac{v_{\max}}{2}\{\tanh(\Delta x_j(t)-x_c)+\tanh(x_c)\}, \tag{7}$$

where x_c is a constant representing the safety distance. When $\Delta x_j\to\infty$ and $x_c>0$, $V(\infty)\cong v_{\max}$. Equation (7) has a turning point (inflection point) at $\Delta x_j=x_c$. It is important that the optimal velocity function has the turning point. Otherwise, one cannot obtain a robust density wave representing a traffic jam [4].

The optimal velocity model (3) is simple and convenient for computer simulation and theoretical analysis. As the optimal velocity model does not take into account a driver response to the relative velocity with respect to the vehicle ahead, it produces a collision accident with increasing delay τ. To avoid this collision, Treiber *et al* [23, 24] have proposed the intelligent driver model taking into account the relative velocity. Also, the relative velocity has been taken into account in some models [25-31].

The optimal velocity model is extended to take into account the vehicle interaction before the next vehicle ahead (the next-nearest-neighbor interaction) [32]. If the headway Δx_{j+1} of the next vehicle *j*+1 ahead is larger than Δx_j of vehicle *j*, the driver of vehicle *j* may wish to proceed with larger velocity than the optimal velocity $V(\Delta x_j)$. The motion equation of the next-nearest-neighbor model is given by

$$\frac{d^2 x_j(t)}{dt^2} = \left(\frac{1}{\tau}\right)\left(V(\Delta x_j(t)) + \gamma\left(V(\Delta x_{j+1}(t) - V(\Delta x_j(t))\right) - \frac{dx_j(t)}{dt}\right). \tag{8}$$

Here parameter γ represents the strength of the next-nearest-neighbor interaction and $0 \leq \gamma \leq 1$. The second term on the right-hand side is the increase of the desired velocity by the next-nearest-neighbor interaction. The next-nearest-neighbor interaction stabilizes the traffic flow and enhances the traffic current.

Mason and Woods [33] have generalized the model (3) to take into account the two different types of vehicles, say, cars and trucks:

$$\frac{d^2 x_j(t)}{dt^2} = \left(\frac{1}{\tau_j}\right)\left(V_j(\Delta x_j(t)) - \frac{dx_j(t)}{dt}\right), \tag{9}$$

where τ_j is the delay time of vehicle *j*, depending on the type of vehicles.

The optimal velocity model is extended to take into account the backward vehicle *j-1* [34, 35]. A driver looks at the following vehicle *j-1* as well as the preceding vehicle *j+1*. The motion equation of the backward looking model is given by

$$\frac{d^2 x_j(t)}{dt^2} = \left(\frac{1}{\tau}\right)\left(V(\Delta x_j(t)) + V_B(x_j - x_{j-1}) - \frac{dx_j(t)}{dt}\right). \tag{10}$$

Here, $V_B(x_j - x_{j-1})$ is the optimal velocity function for backward looking. The backward interaction stabilizes the traffic flow and enhances the traffic current similarly to the next-nearest interaction.

Treiber *et al* [23, 24] have presented the intelligent driver model to take into account the relative velocity. The model has the following advantage: the vehicle behaves as if accident-free. Lubashevsky *et al* [36] have proposed the generalization of the optimal velocity model similar to the intelligent driver model by the use of a variational principle.

A coupled-map model based on optimal-velocity function is introduced by discretizing the time variable of Eq. (3) [37, 38].

2.2. Cellular Automata (CA)

Cellular automata models have been used for simulating various physical systems because of the simplifications. The simplest traffic model is the CA 184 [13, 39, 40]. The model has been investigated as the totally asymmetric simple exclusion model on one-dimensional lattice for the prototype of interacting systems far from equilibrium [41, 42]. The dynamics is described by

$$x_j(t+1) = x_j(t) + \min[1, x_{j+1}(t) - x_j(t) - 1]. \tag{11}$$

In this model, a particle (vehicle) moves by one lattice spacing if the site ahead is not occupied by other particles. Otherwise, it stops on the site. All particles are updated in parallel (simultaneously). The velocity is either one or zero. Later, Fukui and Ishibashi [43] have proposed the extended CA model

$$x_j(t+1) = x_j(t) + \min[v_{\max}, x_{j+1}(t) - x_j(t) - 1]. \tag{12}$$

The velocity takes the integer value ranging from 0 to $v_{\max}$. The velocity depends on the headway $\Delta x_j(t)$. If $\Delta x_j(t)$ is larger than maximal velocity $v_{\max}$, the vehicle j moves with the maximal velocity. If $\Delta x_j(t)$ is less than maximal velocity $v_{\max}$, the vehicle moves with velocity $\Delta x_j - 1$. This model is rewritten

$$x_j(t+1) - 2x_j(t) + x_j(t-1) = \min[v_{\max}, \Delta x_j(t) - 1] - \{x_j(t) - x_j(t-1)\}. \tag{13}$$

By taking a continuous limit $\Delta t \to 0$, one obtains

$$\frac{d^2 x_j}{dt^2} = \min[v_{\max}, \Delta x_j(t) - 1] - \frac{dx_j}{dt}.$$

The first term on the right-hand side represents the optimal velocity function. This equation is equivalent to the optimal velocity model (3) with delay time $\tau = 1$ by replacing $V(\Delta x_j(t))$ with $\min[v_{\max}, \Delta x_j(t) - 1]$ [44, 45].

The extended CA model (13) is a simplified version of the Nagel-Schreckenberg (NaSch) model [14]. The NaSch model has been introduced by Nagel and Schreckenberg (1992). The CA model has been recognized as the pioneer work for simulating the real traffic flow. The dynamics is formulated as follows:

$$x_j(t+1) = x_j(t) + \max\lfloor 0, \min\{v_{\max}, x_{j+1}(t) - x_j(t) - 1, x_j(t) - x_j(t-1) + 1\} - \xi_j(t) \rfloor, \quad (14)$$

where the Boolean random variable $\xi_j(t) = 1$ with probability p and 0 with probability 1-p. The vehicles are updated in parallel according to the fourth steps: motion, acceleration, deceleration, and randomization. The NaSch model has been extended by some researchers [46, 47]. Schadschneider and Schreckenberg [48] and Schrekenberg *et al* [49] have presented the mean-field theory for the model.

Takayasu and Takayasu [50] have proposed the CA model of the slow-start rule to take into account the inertia of vehicle. The dynamics is given by

$$x_j(t+1) = x_j(t) + \min[1, x_{j+1}(t) - x_j(t) - 1, x_{j+1}(t-1) - x_j(t-1) - 1]. \quad (15)$$

The motion of vehicle depends not only on the headway at time t but also on the headway at t-1. This model exhibits the metastability. Schadscheider and Schreckenberg [51] have also extended the NaSch model to include the slow-start rule. Very recently, Nishinari [44] has proposed the generalized CA model taking into account the inertia effect and has shown that the model reproduces the real fundamental diagram. Krauss and coworkers [52, 53] have developed and have investigated the Gipps model [54]. The model exhibits the three distinct behaviors: the free traffic, the metastability and the congested traffic.

Helbing and Schreckenberg [55] have presented the discrete and noisy optimal velocity model to construct a link between the NaSch model and the optimal velocity model. Various CA models have been proposed and investigated (see the review of Chowdhury *et al* [2], therein references). Cheybami *et al* [56] have studied the effect of the stochastic boundary conditions on the traffic flow in the deterministic NaSch model.

The CA model has been extended to take into account quenched random hopping probabilities of the individual cars [57-59]. It has been found that Bose-Einstein-like condensation occurs for platoon formation.

2.3. Gas-kinetic Models

The kinetic theory treats vehicles as a gas of interacting particles. The various versions have been developed to extend and modify the gas-kinetic theory for traffic flow [9, 60-68]. Prigogine and Herman have proposed the Boltzmann equation for the traffic

$$\frac{\partial f(x,v,t)}{\partial t} + v\frac{\partial f(x,v,t)}{\partial x} = -\frac{f(x,v,t) - \rho(x,t)F_{des}(v)}{\tau_{rel}} + \left(\frac{\partial f(x,v,t)}{\partial t}\right)_{\text{int}}, \quad (16)$$

where the first term on the right-hand side represents the relaxation of the velocity distribution function $f(x,v,t)$ to the desired velocity distribution $\rho(x,t)F_{des}(v)$ with the relaxation time τ_{rel} in the absence of the interactions of vehicles and the second term on the right-hand side takes into account the change arising from the interactions among vehicles.

Lehmann [69] has presented the simplest model which uses the desired velocity distribution determined from the empirical data:

$$\frac{\partial f(x,v,t)}{\partial t}+v\frac{\partial f(x,v,t)}{\partial x}=-\frac{f(x,v,t)-f_{des}(v,\rho)}{\tau_{rel}}. \tag{17}$$

By taking into account the different personalities of drivers, Faveri-Fontana [60] has proposed the generalized gas-kinetic model:

$$\frac{\partial g(x,v,v_{des},t)}{\partial t}+v\frac{\partial g(x,v,v_{des},t)}{\partial x}+\frac{\partial}{\partial v}\left[\left(\frac{v_{des}-v}{\tau}\right)g(x,v,v_{des},t)\right]=\left(\frac{\partial g(x,v,v_{des},t)}{\partial t}\right)_{\text{int}}, \tag{18}$$

$$\left(\frac{\partial g(x,v,v_{des},t)}{\partial t}\right)_{\text{int}}=f(x,v,t)\int_v^{\infty}dv'(1-P_{pass})(v'-v)g(x,v',v_{des},t)$$
$$-g(x,v,v_{des},t)\int_0^{v}dv'(1-P_{pass})(v-v')f(x,v',t), \tag{19}$$

where $f(x,v,t)=\int_0^{\infty}dv_{des}g(x,v,v_{des},t)$ and P_{pass} is the probability of passing. The velocity distribution function $g(x,v,v_{des},t)dxdvdv_{des}$ represents the number of vehicles at time t between x and $x+dx$, having acutual velocity between v and $v+dv$ and desired velocity between v_{des} and $v_{des}+dv_{des}$. This model accounts for the distribution of desired velocity inherent for the driver-vehicle system, i.e. the difference among the individual vehicles.

By starting with the master equation, Helbing [28] has presented the following kinetic equation

$$\frac{\partial f(x,v,t)}{\partial t}+\frac{\partial(vf(x,v,t))}{\partial x}+\frac{\partial}{\partial v}\left[\left(\frac{v_{des}-v}{\tau}\right)f(x,v,t)\right]=\frac{1}{2}\frac{\partial^2(D_{vf}f(x,v,t))}{\partial v^2}+\left(\frac{\partial f(x,v,t)}{\partial t}\right)_{\text{int}}, \tag{20}$$

where D_{vf} is a velocity diffusion constant.

Helbing and coworkers [70, 71] have derived the macroscopic traffic model from (20). Thus, they have constructed a micro-macro link. Henneke *et al* [72] have performed the simultaneous micro- and macro-simulation.

Ben-Naim and Krapivsky [62] have studied the power-law platoon formation (bunching of cars) as aggregation phenomena by using the kinetic theory.

Mahnke and Pieret [73] have presented the simple model of master equation approach to the jam growth. The evolution equation of jam size n is given by:

$$\frac{dP(n,t)}{dt} = W_+(n-1)P(n-1,t) + W_-(n+1)P(n+1,t) - \{W_+(n) + W_-(n)\}P(n,t), \tag{21}$$

where $P(n,t)$ is the probability distribution of jam size n at time t and $W_+(n)$ ($W_-(n)$) is the transition rate from jam size n to n+1 (n-1).

The kinetic theories of a single-lane highway have been extended to the two-dimensional flow for city traffic and the multi-lane traffic.

2.4. Macroscopic Traffic Models

The macroscopic traffic theory treats traffic as an effectively one-dimensional compressible fluid. The traffic states at position x and time t is described in terms of the spatial vehicle density $\rho(x,t)$ and the average velocity $v(x,t)$. Lighthill and Whitham [7] have proposed the oldest continuum model. The model is described by the continuity equation of fluids

$$\frac{\partial \rho(x,t)}{\partial t} + \frac{\partial q(x,t)}{\partial x} = 0, \tag{22}$$

where $q(x,t) = \rho(x,t)v(x,t)$ is the traffic current (or flow). Lighthill and Whitham assume that the traffic current is determined by the fundamental (flow-density) diagram: $q(x,t) = Q_0(\rho(x,t))$. The nonlinear equation describes the propagation of kinematic waves. To avoid an instability of shock front, a small diffusion term is added

$$q(x,t) = Q_0(\rho(x,t)) - D\frac{\partial \rho(x,t)}{\partial x}. \tag{23}$$

Assuming the simple fundamental diagram $Q_0 = v_{\max}\rho(x,t)(1-\rho(x,t))$, the Burgers equation is obtained. Until now, various macroscopic traffic models have been proposed [74-76]. Finally, the complete continuum model of the highway traffic flow is given by

$$\frac{\partial \rho}{\partial t} + \frac{\partial(\rho v)}{\partial x} = 0, \tag{24}$$

$$\rho\frac{\partial v}{\partial t} + \rho v\frac{\partial v}{\partial x} = \frac{\rho}{\tau}[V(\rho) - v] - c_0{}^2\frac{\partial \rho}{\partial x} + \mu\frac{\partial^2 v}{\partial x^2}, \tag{25}$$

where τ, c_0^2 and μ are phenomenological constants. The phenomenological function $V(\rho)$ represents the desired velocity achieved in the steady state. The constant τ is the relaxation time to the steady state. The desired velocity $V(\rho)$ corresponds to the optimal velocity in the microscopic model. The relaxation time τ corresponds to the delay time in the optimal velocity model.

Kerner and Konhauser [77, 78] have investigated the continuum model and have shown that the jamming transition occurs at high density and the density waves appear as the autosolitons. Later, Lee *et al* [79] have extended the model to take into account the inflow on-ramp. Some attempts have been made for deriving the macroscopic traffic model from the microscopic model. Berg *et al* [80] have derived the continuum model from the optimal velocity model (2). By using a series expansion of the headway in terms of the density, the following expression is obtained

$$\frac{\partial v}{\partial t}+v\frac{\partial v}{\partial x}=\frac{1}{\tau}[V(\rho)-v]+\frac{V'(\rho)}{\tau}\left[\frac{1}{2\rho}\frac{\partial\rho}{\partial x}+\frac{1}{6\rho^2}\frac{\partial^2\rho}{\partial x^2}-\frac{1}{2\rho^3}\left(\frac{\partial\rho}{\partial x}\right)^2\right]. \tag{26}$$

Equation (26) is analogous to the Kerner-Konhausser model (25). However, an important difference between (25) and (26) lies in the coefficients. The coefficients of (25) are the phenomenological parameters, while the coefficients of (26) depend on the parameters of the microscopic model. It is easy to identify the parameters of (26) though it is difficult to determine the phenomenological constants τ, c_0^2 and μ in (25).

Nelson [81] has derived the modified Lighthill-Whitham model by the use of the different method. He assumes that drivers compensate for the delay τ by adjusting to the density seen at some anticipation distance L_a ahead of their current position. The actual speed at position x and time t is given by

$$v(x,t)=V(\rho(x+L_a-V\tau,t-\tau)), \tag{27}$$

where $V(\rho)$ is the desired velocity at density ρ. By expanding the right-hand side to first order of τ and L_a, the following expression of traffic current is obtained instead of (23)

$$q(x,t)=\rho(x,t)v(x,t)=Q_0(\rho(x,t))+\rho(x,t)\left[L_aV'(\rho(x,t))+\tau\rho(x,t)\{V'(\rho(x,t))\}^2\right]\frac{\partial\rho}{\partial x}. \tag{28}$$

It is easy to identify the parameters τ and L_a since the model is connected to the microscopic model (4).

A lattice hydrodynamic model has been proposed to have the same mathematical properties as the optimal velocity model [82, 83]. The model has been extended to the two-dimensional lattice for the city traffic. The two types of vehicles are considered: the first type moves only to the positive x-direction and the second type only to the positive y-direction. The continuity equations of x and y vehicles are given respectively

$$\frac{\partial \rho_x(x,y,t)}{\partial t} + \frac{\partial \rho_x(x,y,t)u(x,y,t)}{\partial x} = 0, \tag{29}$$

$$\frac{\partial \rho_y(x,y,t)}{\partial t} + \frac{\partial \rho_y(x,y,t)v(x,y,t)}{\partial y} = 0, \tag{30}$$

where $\rho_x(x,y,t)$ and $\rho_y(x,y,t)$ are the local densities of x and y vehicles at position (x,y) at time t and $u(x,y,t)$ and $v(x,y,t)$ are the local speeds of x and y vehicles at position (x,y) at time t. The traffic currents of x and y vehicles are given respectively

$$\rho_x(x,y,t)u(x,y,t) = \rho_{x,0}V(\rho(x+L_a,y,t-\tau)), \tag{31}$$

$$\rho_y(x,y,t)u(x,y,t) = \rho_{y,0}V(\rho(x,y+L_a,t-\tau)), \tag{32}$$

where $\rho(x,y,t)(=\rho_x(x,y,t)+\rho_y(x,y,t))$ is the local total density at position (x,y) at time t and $\rho_{x,0}$ and $\rho_{y,0}$ are the average densities of x and y vehicles.

3. Fundamental Diagram and Characteristics

3.1. Fundamental Diagram

The current (flow)-density diagram is called the fundamental diagram. It is most important one in the field of traffic science and engineering. Figure 1 shows the plot of the traffic current q against density ρ for model (5) where $v_{\max} = 2.0$, $\tau = 0.5$ and $x_c = 5.0$. The traffic current increases with density in the low-density region, reaches the maximal value, decreases discontinuously at the gap and then decreases continuously with increasing density. At low density, vehicles move freely and result in the free traffic. The critical density at the gap agrees with the jamming transition point (neutral stability point). At intermediate density after the maximal point, the stop- and go-waves occur. At high density, vehicles move slowly and result in the homogeneous congested traffic. Thus, traffic states change with density. The fundamental diagram is one of characteristics of vehicular traffic [1-5].

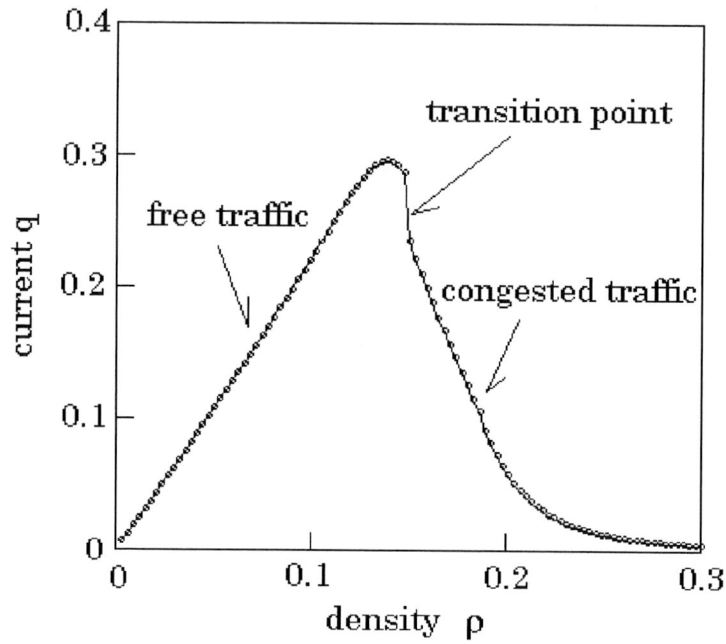

Figure 1. Fundamental diagram for model (5). Plot of the traffic current q against density ρ.

3.2. Linear Stability

We consider the stability of the uniform traffic flow. In the optimal velocity models, all the vehicles move with the same headway h and the optimal velocity $V(h)$ at a low density of vehicles. When the density is higher than the critical value, the traffic jam appears as density waves propagating backward. The uniform traffic flow is defined by such a state that all vehicles move with constant headway h and the optimal velocity $V(h)$. The uniform traffic flow is a solution of the optimal velocity model. The solution is given by

$$x_{j,0}(t) = hj + V(h)t \text{ with } h = L/N, \tag{33}$$

where N is the number of vehicles, L is the road length and density ρ is $1/(h+1)$.

We apply the linear stability theory to the optimal velocity model (4). By adding a small fluctuation to the steady-state solution, one can study whether or not fluctuations amplify. If fluctuations added to the steady-state solution decay in time, the steady state is stable. Otherwise, fluctuations amplify in time and the uniform traffic flow changes the different dynamical state. Let $y_j(t)$ be small deviation from the uniform solution $x_{j,0}(t)$: $x_j(t) = x_{j,0}(t) + y_j(t)$. Then, the linear equation is obtained from Eq.(4)

$$\frac{dy_j(t+\tau)}{dt} = V'(h)\Delta y_j(t), \tag{34}$$

where $V'(h)$ is the derivative of optimal velocity $V(\Delta x)$ at $\Delta x = h$. By expanding $y_j(t) = Y\exp(ikj + zt)$, one obtains

$$ze^{z\tau} = V'(h)(e^{ik} - 1). \tag{35}$$

By solving Eq. (35) with *z*, one finds that the leading term of *z* is order of *ik*. When $ik \to 0$, $z \to 0$. Let us derive the long wave expansion of *z*, which is determined order by order around $ik \approx 0$. By expanding $z = z_1(ik) + z_2(ik)^2 + \cdots$, the first- and second-order terms of *ik* are obtained

$$z_1 = V'(h) \text{ and } z_2 = -V'(h)\big(2V'(h)\tau - 1\big)/2. \tag{36}$$

If z_2 is a negative value, the uniform flow becomes unstable for long wavelength modes. When z_2 is a positive value, the uniform flow is stable. The neutral stability condition is given by $z_2 = 0$:

$$\tau = \frac{1}{2V'(h)}. \tag{37}$$

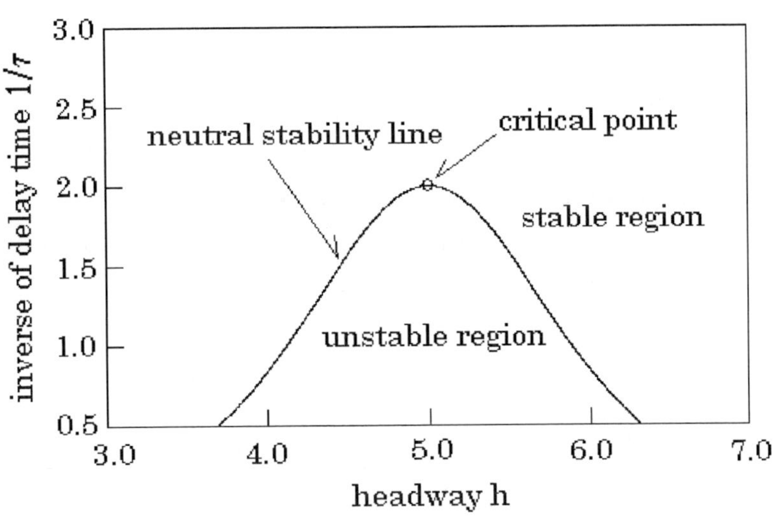

Figure 2. Region map in parameter space $(h, 1/\tau)$ for $x_c = 5$ and $v_{\max} = 2.0$. The solid curve indicates the neutral stability line. In the region above the neutral stability line, the traffic flow is stable. In the region below the neutral stability line, the traffic flow becomes unstable.

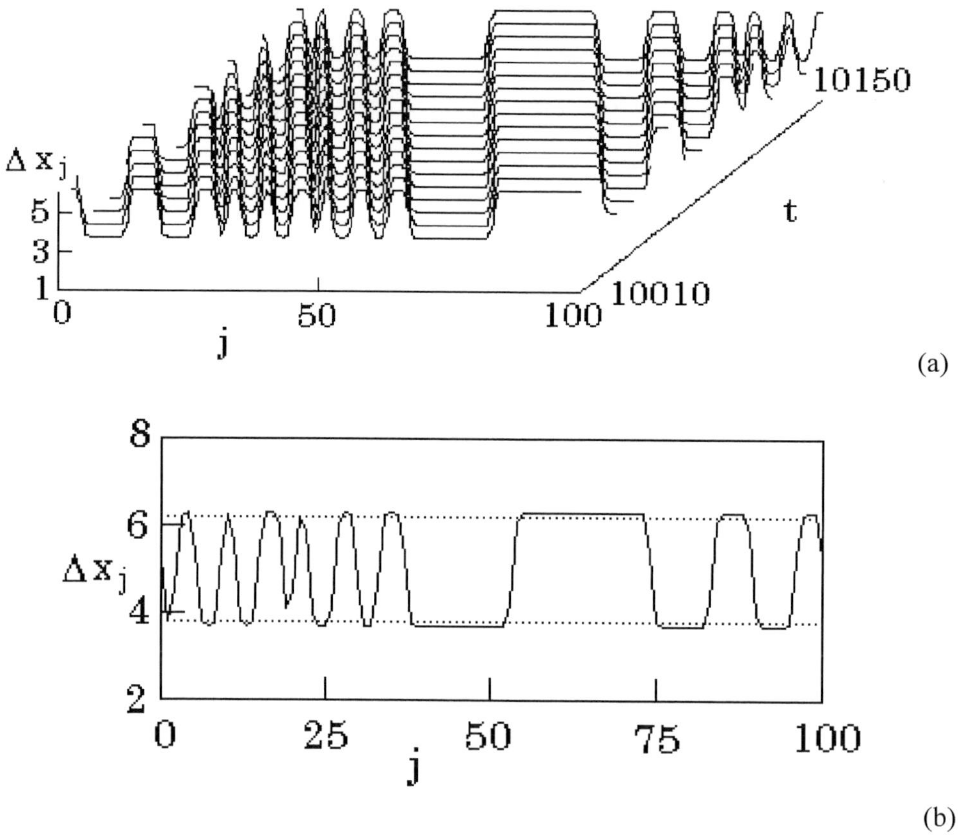

Figure 3. Typical profile of the inhomogeneous flow obtained from simulation of (5). The pattern (a) indicates the time evolution of the headway profile in the unstable region in Figure 2 The headway profile (b) is obtained after sufficiently large time.

For small disturbances of long wavelengths, the uniform traffic flow is unstable if delay τ is larger than $1/2V'(h)$: $\tau > 1/2V'(h)$. Otherwise, it is stable. Figure 2 shows the region map in parameter space $(h, 1/\tau)$ for $x_c = 5$ and $v_{\max} = 2.0$. The solid curve indicates the neutral stability line. In the region above the neutral stability line, the traffic flow with the uniform headway and velocity profiles is stable. In the region below the neutral stability line, the traffic flow becomes unstable. One finds that there is a critical point at $h(=\Delta x) = x_c$ and $\tau = \tau_c = 1/v_{\max}$. Therefore, if $\tau < \tau_c$, the uniform flow is always stable irrespective of density (headway).

The neutral stability condition of differential equation model (3) is also obtained and is consistent with Eq. (37). The neutral stability condition of difference equation model (5) is given by $\tau = 1/3V'(h)$ and is different from Eq. (37). The neutral stability line agrees with the jamming transition curve obtained from simulation. The neutral stability curves are obtained for the next-nearest-neighbor interaction model (8) and the backward looking model (10). The linear stability conditions for models (3), (4), (5), (8) and (10) are summarized as follow:

$$\tau < \frac{1}{2V'(h)} \text{ for Eqs. (3) and (4),}$$

$$\tau < \frac{1}{3V'(h)} \text{ for Eq. (5),}$$

$$\tau < \frac{(1+2\gamma)}{2V'(h)} \text{ for Eq. (8),}$$

$$\tau < \frac{\left(V'(h) - V'_B(h)\right)}{2\left[V'(h) + V'_B(h)\right]^2} \text{ for Eq. (10).} \quad (38)$$

The linear stability analysis has been applied to the macroscopic traffic models. The neutral stability conditions have been derived. Kerner and Konhauser [15] and Kurtze and Hong [84] have derived the linear stability condition for the fluiddynamic model of traffic described by Eqs. (24) and (25). Eqs. (24) and (25) admit a simple steady-state solution representing uniform flow: $\rho(x,t) = \rho_0$, $v(x,t) = V(\rho_0) \equiv v_0$. By assuming that traffic is initially in a state which differs infinitesimally from the uniform flow, one decomposes this flow into

$$\rho(x,t) = \rho_0 + \sum_k \delta\rho_k \exp(ikx + \sigma_k t),$$
$$v(x,t) = v_0 + \sum_k \delta v_k \exp(ikx + \sigma_k t). \quad (39)$$

By substituting the expressions into (24) and (25) and linearizing in $\delta\rho_k$ and δv_k, one obtains the linear equations. One finds that each linear growth rate σ_k must satisfy the quadratic equation

$$(\sigma_k + iv_0 k)^2 + \left[\frac{1}{\tau} + \mu k^2\right](\sigma_k + iv_0 k) + c_0^{\ 2} k^2 + i\frac{\rho_0 V'(\rho_0)}{\tau} k = 0, \quad (40)$$

where the prime represents derivative with respect to ρ_0. If both roots of this quadratic have negative real parts, the uniform traffic flow is stable against all infinitesimal perturbations. The linear stability condition is given by

$$\rho_0 V'(\rho_0)^2 < c_0^{\ 2} \rho_0^{\ -1}. \quad (41)$$

This linear stability condition for the macroscopic model (24) and (25) is compared with (38) for the optimal velocity models.

If one uses the expression $c_0^{\ 2} = V'(\rho_0^{\ -1}) / 2\tau\rho_0^{\ 2}$ in the macroscopic model derived by Lee *et al* [85], the following linear stability condition is obtained

$$\tau < \frac{1}{2V'(\rho_0^{-1})}, \tag{42}$$

where the prime represents derivative with respect to ρ_0^{-1}. Thus, the macroscopic linear stability condition (42) agrees with the microscopic stability condition (38).

When small noises are added to the uniform traffic flow, its flow is stable and robust for low density because the stability condition (38) is satisfied. When the density is higher than the critical density, the condition (38) is not satisfied. Then, the uniform flow including noises is unstable, the noises amplify in time and the uniform flow changes finally to the inhomogeneous traffic flow with propagating density waves. Figure 3 shows the typical profile of the inhomogeneous flow obtained from simulation of (5) where $v_{max} = 2.0$, $\tau = 0.5$, $x_c = 5.0$ and average headway $\Delta x_0 = 5.0$. The pattern (a) indicates the time evolution of the headway profile in the unstable region in Figure 2 The headway profile (b) is obtained after sufficiently large time. The initial uniform flow evolves in time to the inhomogeneous flow with the kink density waves. The density wave propagates backward with constant speed. The density wave is robust and keeps the kink form. The high-density region is the jammed state of vehicles and the low-density region is the free traffic state. Thus, the coexisting phase of both jammed and free states appears in the unstable region of Figure 2.

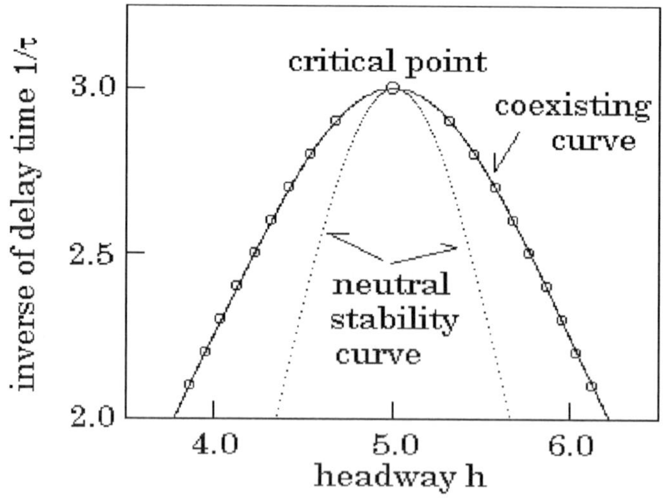

Figure 4. Plot of the headways within and outside the jam for various values of τ. The solid curve connecting their circles gives the coexisting curve in the phase space $(h, 1/\tau)$. The dotted line indicates the neutral stability curve.

The upper dotted line of headway profile (b) indicates the value of headway outside the jammed state. The bottom dotted line indicates the value of headway within the jam. These values keep constant values for constant τ and depend only on τ. By considering long-time evolution, only two distinct headways survive for the coexisting phase. Figure 4 shows the plot of the headways within and outside the jam for various values of τ. The solid curve connecting their circles gives the coexisting curve in the phase space $(h, 1/\tau)$. The dotted line indicates the neutral stability curve.

3.3. Multi-Phase Model

A multi-phase traffic model is presented to take into account the complex motion of vehicles [86, 87]. The multi-phase model is an extended one of the optimal velocity model. The optimal velocity (OV) function is modified to have multi turning points. The original OV model with a single turning point exhibits two phase traffic, while the model with n ($n \geq 2$) turning points displays $n+1$ phase traffic. The multiple phase transitions occur by varying the density. The phase transitions depend highly on the sensitivity (the inverse of delay time). The neutral stability line is obtained by using the linear stability analysis and is consistent with the jamming transition points. The phase diagrams are presented for the multiple jamming transitions.

We consider the vehicular traffic flowing on a single-lane highway where there are no bottlenecks, inflow, and outflow. We assume that vehicles do not pass over the other vehicles. It is known that the conventional optimal velocity model displays the two phases of traffic states. Real traffic flow shows more complex traffic behavior than that of two phases. For the complex traffic flow, we extend the two-phase traffic model to the multi-phase one. The OV model is described by the following equation of motion of vehicle i

$$\frac{d^2 x_i(t)}{dt} = a\left[V(\Delta x_i(t)) - \frac{dx_i(t)}{dt}\right], \tag{43}$$

where $V(\Delta x_i(t))$ is the optimal velocity function, $x_i(t)$ is the position of vehicle i at time t, $\Delta x_i(t)(= x_{i+1}(t) - x_i(t))$ is the headway of vehicle i, and $a(=1/\tau)$ is the sensitivity (τ : delay time).

The optimal velocity of vehicles has been given by

$$V(\Delta x_i) = \frac{v_{\max}}{2}\left[\tanh \alpha(\Delta x_i - x_c) + \tanh \alpha(x_c)\right], \tag{44}$$

where $v_{\max}$ is the maximal velocity of vehicles and x_c is the safety distance of vehicles. The OV function is shown by the solid curve in Figure 5 for $v_{\max} = 6.0$, $x_c = 6.0$, and $\alpha = 0.5$. It has a single turning point at x_c.

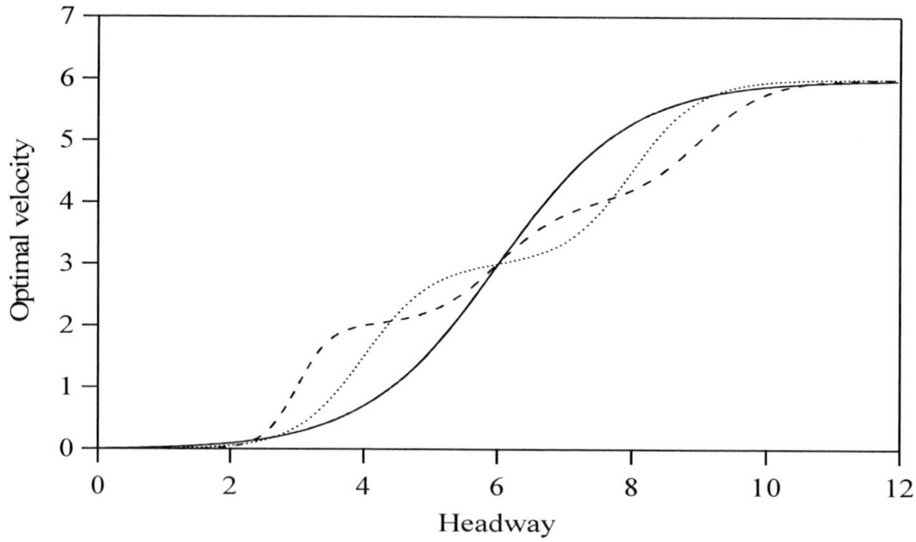

Figure 5. Optimal-velocity functions. The solid curve indicates the plot of Eq. (44) for $v_{max} = 6.0$, $x_c = 6.0$, and $\alpha = 0.5$. The dotted curve indicates the plot of Eq. (45) for $v_{max} = 6.0$, $x_{1,c} = 4.0$, $x_{2,c} = 8.0$, and $\alpha = \beta = 1.0$. The dashed curve indicates the plot of Eq. (46) for $v_{max} = 6.0$, $x_{1,c} = 3.0$, $x_{2,c} = 6.0$, $x_{3,c} = 9.0$, $\alpha = 2.0$, $\beta = \gamma = 1.0$.

We extend the OV function (44) to the following:

$$V(\Delta x_i) = \frac{v_{max}}{4}\left[\tanh\alpha(\Delta x_i - x_{1,c}) + \tanh\alpha(x_{1,c}) + \tanh\beta(\Delta x_i - x_{2,c}) + \tanh\beta(x_{2,c})\right] \tag{45}$$

The OV function (45) has two turning points at $x_{1,c}$ and $x_{2,c}$. The dotted curve in Figure 5 shows the plot of Eq. (45) for $v_{max} = 6.0$, $x_{1,c} = 4.0$, $x_{2,c} = 8.0$, and $\alpha = \beta = 1.0$.

Furthermore, we extend the OV function to the following:

$$V(\Delta x_i) = \frac{v_{max}}{6}\begin{bmatrix}\tanh\alpha(\Delta x_i - x_{1,c}) + \tanh\alpha(x_{1,c}) + \tanh\beta(\Delta x_i - x_{2,c}) + \tanh\beta(x_{2,c}) \\ + \tanh\gamma(\Delta x_i - x_{3,c}) + \tanh\gamma(x_{3,c})\end{bmatrix} \tag{46}$$

The OV function (46) has three turning points at $x_{1,c}$, $x_{2,c}$, and $x_{3,c}$. The dashed curve in Figure 5 shows the plot of Eq. (46) for $v_{max} = 6.0$, $x_{1,c} = 3.0$, $x_{2,c} = 6.0$, $x_{3,c} = 9.0$, $\alpha = 2.0$, $\beta = \gamma = 1.0$.

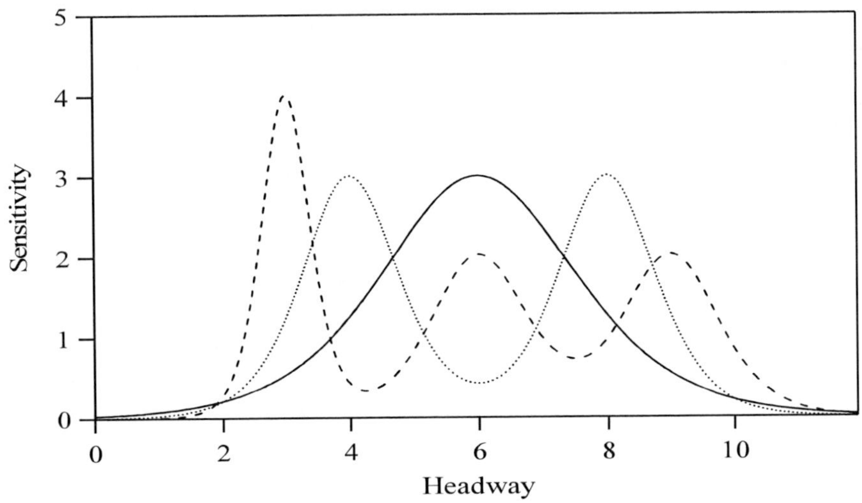

Figure 6. Plots of neutral stability points against the headway for Eqs. (44)- (46). The solid curve indicates the neutral stability line for the original two-phase model (44). The dotted curve represents the neutral stability line for the optimal velocity function (45) with two turning points. The dashed line indicates the neutral stability curve for the optimal velocity function (46) with three turning points.

Generally, we present the following OV function:

$$V(\Delta x_i) = \frac{v_{\max}}{2n}\left[\begin{array}{l}\tanh\alpha_1(\Delta x_i - x_{1,c}) + \tanh\alpha_1(x_{1,c}) + \tanh\alpha_2(\Delta x_i - x_{2,c}) + \tanh\alpha_2(x_{2,c}) \\ + - - - - + \tanh\alpha_n(\Delta x_i - x_{n,c}) + \tanh\alpha_n(x_{n,c})\end{array}\right]. \tag{47}$$

The OV function (47) has n turning points at $x_{1,c}$, $x_{2,c}$, ---, and $x_{n,c}$.

We calculate the neutral stability line. The neutral stability condition is given by

$$a = 2V'(\Delta x_0), \tag{48}$$

where $V'(\Delta x_0)$ is the derivative of the optimal velocity function at the mean headway Δx_0. Eq. (48) holds for the optimal velocity functions (44)-(46). Figure 6 shows the plots of neutral stability points against the headway for Eqs. (44)- (46). The solid curve indicates the neutral stability line for the original two-phase model (44). The dotted curve represents the neutral stability line for the optimal velocity function (45) with two turning points. Two local maximum points appear at the headways $\Delta x_0 = 4.0$ and $\Delta x_0 = 8.0$ of two turning points. The single local minimum point appears at $\Delta x_0 = 6.0$. The homogeneous traffic state appears above the neutral stability curve, while homogeneous traffic state becomes unstable and evolves in time to an inhomogeneous traffic state below the neutral stability curve. The dashed line indicates the neutral stability curve for the optimal velocity function (46) with three turning points. Three local maximum points appear at the headways $\Delta x_0 = 3.0$,

$\Delta x_0 = 6.0$, and $\Delta x_0 = 9.0$ of three turning points. Two local minimum points appear at $\Delta x_0 = 4.2$ and $\Delta x_0 = 7.5$. The homogeneous traffic state appears above the neutral stability curve, while homogeneous traffic state becomes unstable below the neutral stability curve.

Thus, the neutral stability line depends highly on the optimal velocity function. The number of local maximum points agrees with that of turning points.

We calculate the theoretical flow-density curve. The theoretical flow-density curve is defined as that of the homogeneous traffic flow with no jams. If the traffic jams do not occur, the traffic flow exhibits a homogeneous state. Then, the velocity of vehicles is consistent with the optimal velocity $V(\Delta x_0)$. The flow (current) is given by

$$J = \frac{V(\Delta x_0)}{\Delta x_0}, \tag{49}$$

where density ρ is defined as $\rho \equiv 1/\Delta x_0$. Figure 7 shows the plots of flow against density for optimal velocity functions (44)-(46). The solid, dotted, and dashed lines indicate, respectively, the theoretical flow-density curves for optimal velocity functions (44)-(46). There exists only a local maximum point in the original two-phase model (44), while there are two local maximum points and single local minimum point for optimal velocity function (45) and there exit three local maximum points and two local minimum points for optimal velocity function (46).

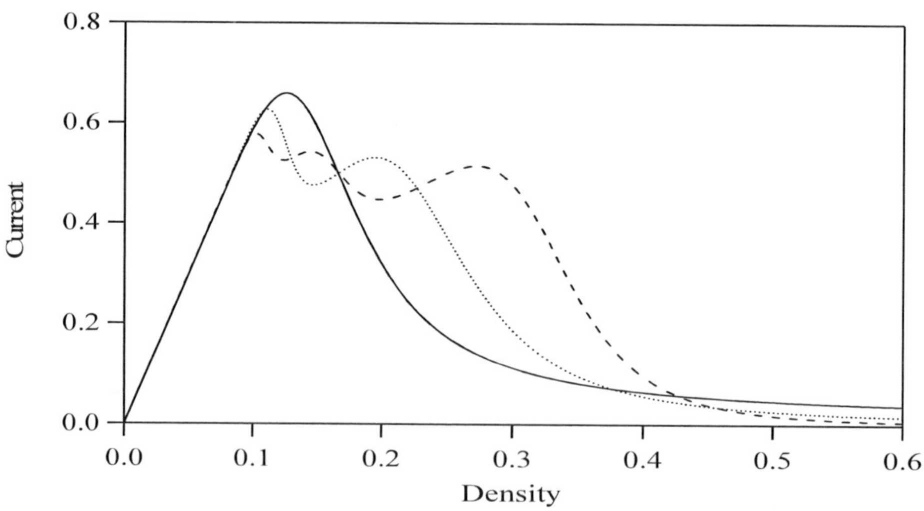

Figure 7. Plots of flow against density for optimal velocity functions (44)-(46). The solid, dotted, and dashed lines indicate, respectively, the theoretical flow-density curves for optimal velocity functions (44)-(46).

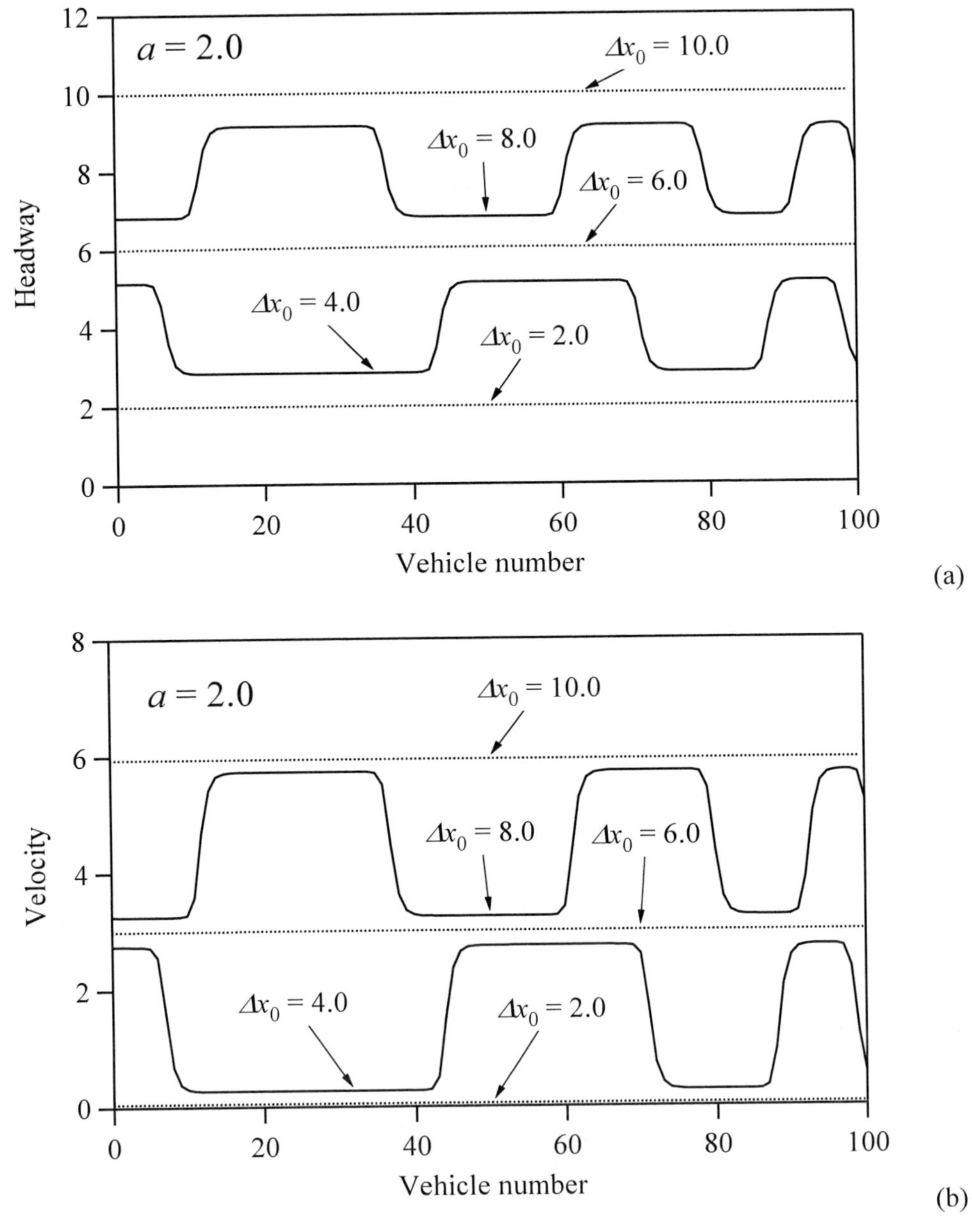

Figure 8. (a) Plots of headway against vehicle's number for sensitivity $a = 2.0$ where $v_{\max} = 6.0$, $x_{1,c} = 4.0$, and $x_{2,c} = 8.0$. Five headway profiles are shown for five mean densities $\Delta x_0 = 2.0, 4.0, 6.0, 8.0, 10.0$. (b) Plots of velocity against vehicle's number. Five headway profiles are shown for five mean densities $\Delta x_0 = 2.0, 4.0, 6.0, 8.0, 10.0$.

Thus, the theoretical flow-density curve depends highly on the optimal velocity function. The number of local maximum points agrees with that of turning points, corresponding to the neutral stability curve.

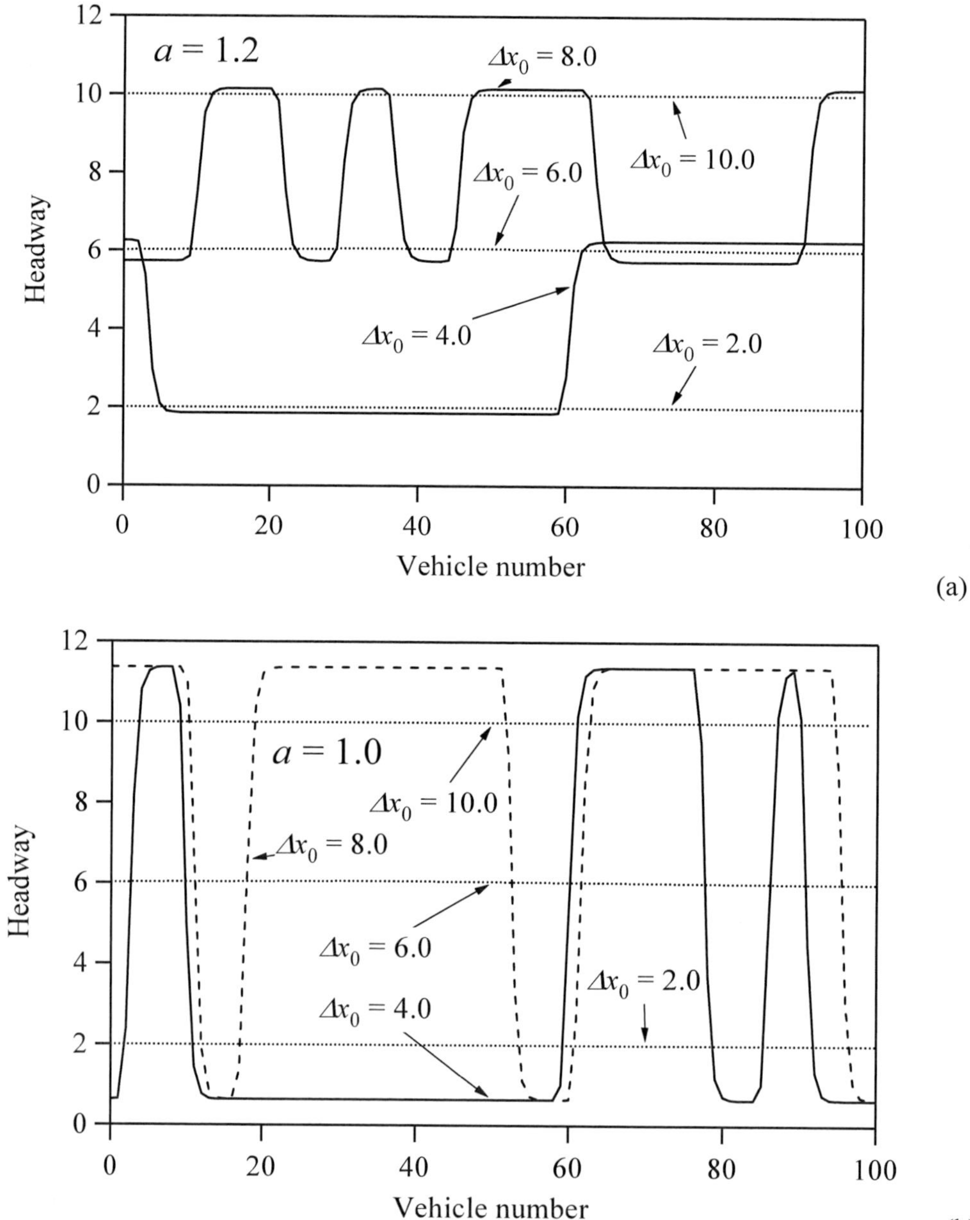

Figure 9. (a) Plots of headway against vehicle's number for sensitivity $a = 1.2$ where $v_{max} = 6.0$, $x_{1,c} = 4.0$, and $x_{2,c} = 8.0$. Five headway profiles are shown for five mean densities $\Delta x_0 = 2.0, 4.0, 6.0, 8.0, 10.0$. (b) Plots of headway against vehicle's number for sensitivity $a = 1.0$ where $v_{max} = 6.0$, $x_{1,c} = 4.0$, and $x_{2,c} = 8.0$. Five headway profiles are shown for five mean densities $\Delta x_0 = 2.0, 4.0, 6.0, 8.0, 10.0$.

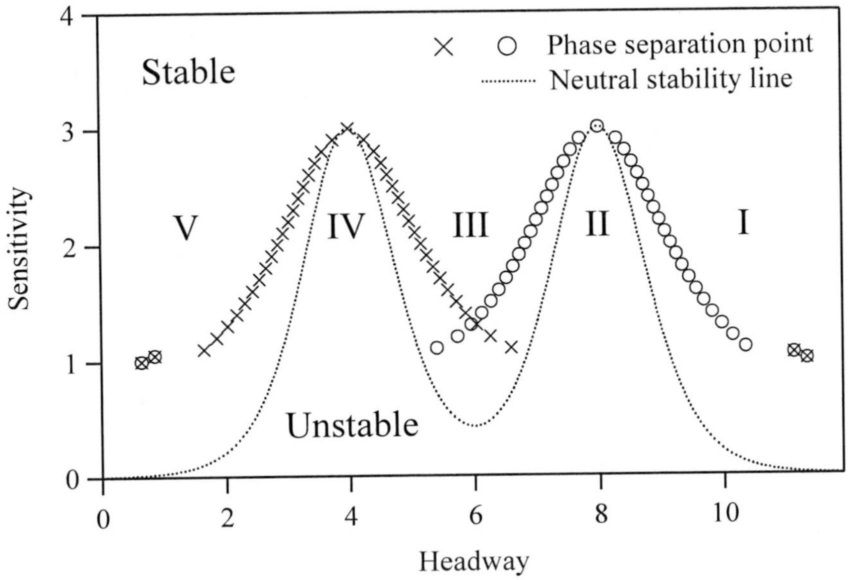

Figure 10. Phase diagram in the phase space $(\Delta x_0, a)$. The circles and crosses indicate the phase separation lines. The dotted line indicates the spinodal line.

We simulate the single-lane traffic flow under the periodic boundary condition. We study the dynamical behavior of traffic. We solve numerically Eq. (43) with optimal velocity functions (45) and (46) by using fourth-order Runge-Kutta method where the time interval is $\Delta t = 1/128$. The simulation is performed until the traffic flow results in a steady state.

First, we investigate the traffic states for three-phase model (45) by varying the density and sensitivity. When the sensitivity is higher than 3.0, the traffic flow is stable and evolves in time to a homogeneous traffic even if the initial headway has any profile. If the sensitivity is less than 3.0, the traffic flow displays the complex behavior. Figure 8(a) shows the plots of headway against vehicle's number for sensitivity $a = 2.0$ where $v_{\max} = 6.0$, $x_{1,c} = 4.0$, and $x_{2,c} = 8.0$. Five headway profiles are shown for five mean densities $\Delta x_0 = 2.0, 4.0, 6.0, 8.0, 10.0$. Figure 8(b) shows the plots of velocity against vehicle's number. Five velocity profiles are shown for five mean densities $\Delta x_0 = 2.0, 4.0, 6.0, 8.0, 10.0$. Each velocity profile corresponds to headway profile in Figure 8(a). At $\Delta x_0 = 2.0$, the traffic flow exhibits the homogeneous congested state (phase 1). At $\Delta x_0 = 4.0$, the traffic flow becomes unstable and results in the inhomogeneous coexisting phase (coexisting phase of phases 1 and 2). At $\Delta x_0 = 6.0$, the traffic flow exhibits again the homogeneous state (phase 2). At $\Delta x_0 = 8.0$, the traffic flow becomes again unstable and results in the inhomogeneous coexisting phase (coexisting phase of phases 2 and 3). At $\Delta x_0 = 10.0$, the traffic flow displays the free traffic state of homogeneous state (phase 3). Thus, when the density increases, the dynamical phase transitions occur from phase 1 to coexisting phase of phases 1 and 2 at first stage, from coexisting phase of phases 1 and 2 to phase 2 at second stage, from phase 2 to coexisting phase of phases 2 and 3 at third stage, and from coexisting phase of phases 2 and 3 to phase 3 at fourth stage. The traffic flow of model (45) has three phases 1-3 and two coexisting phases. Therefore, model (45) displays three-phase traffic.

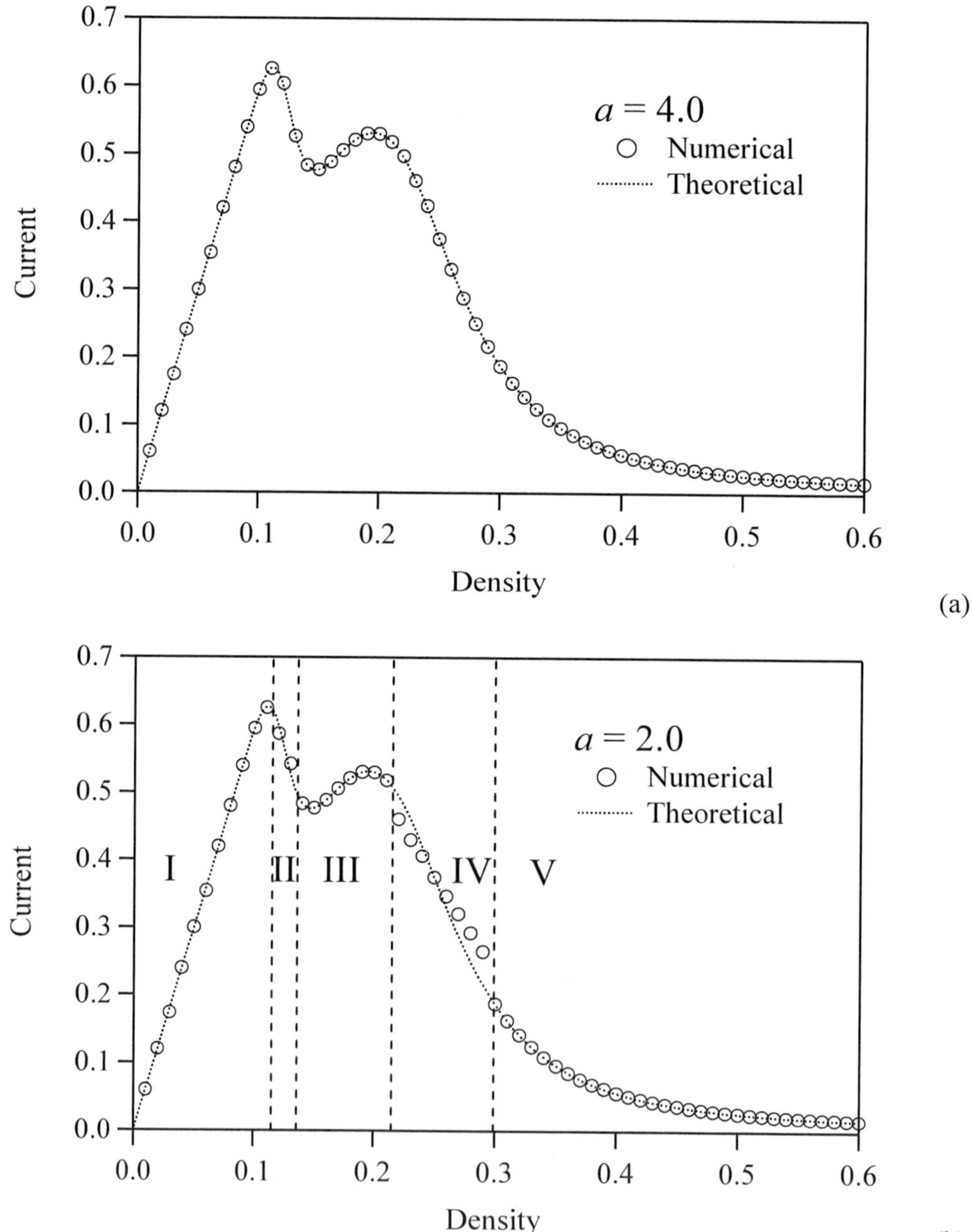

Figure 11. (Continued on next page)

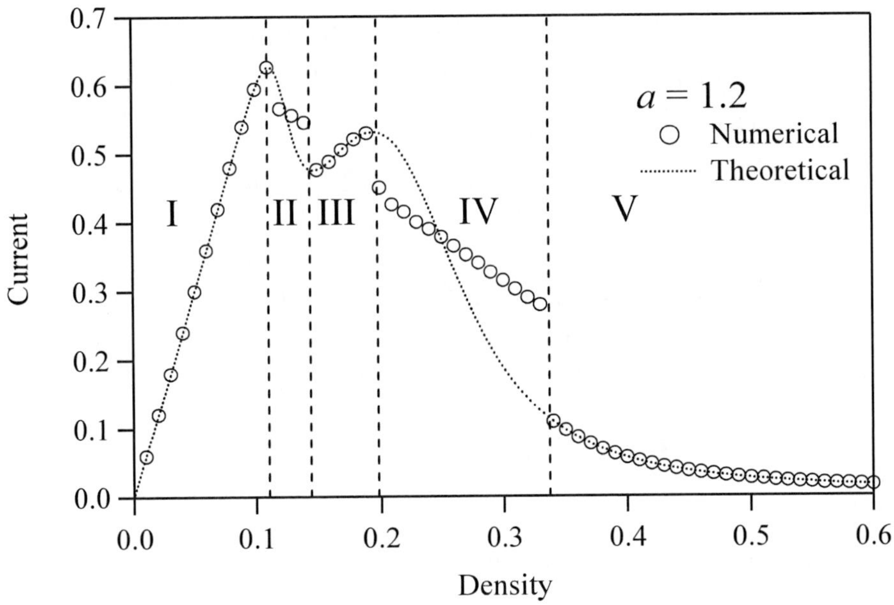

(c)

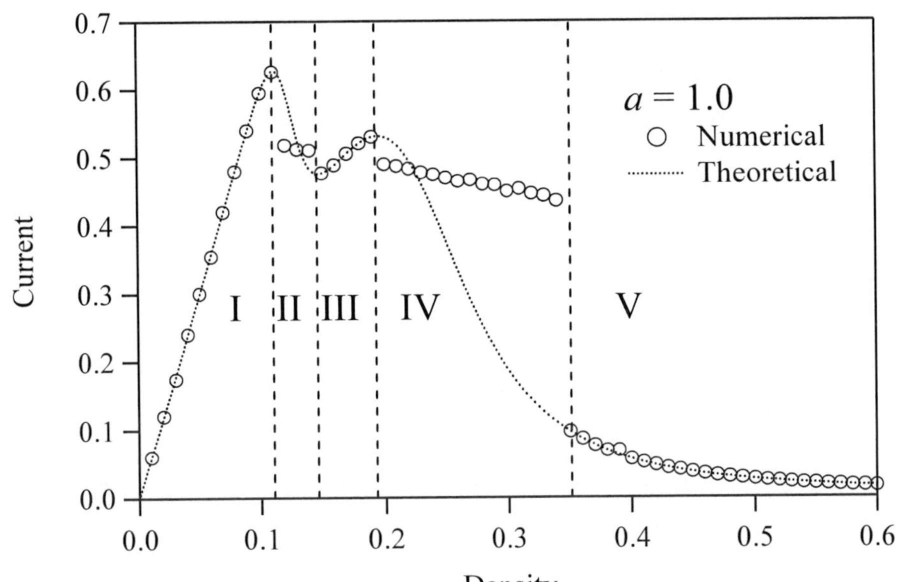

(d)

Figure 11. Plots of current against density. The dotted line indicates the theoretical current-density curve. The circles indicate the simulation result. Diagrams (a)-(d) represent, respectively, the fundamental diagrams for $a = 4.0, 2.0, 1.2, 1.0$.

We study the headway profile for different values of sensitivity. Figure 9(a) shows the plots of headway against vehicle's number for sensitivity $a = 1.2$ where $v_{max} = 6.0$, $x_{1,c} = 4.0$, and $x_{2,c} = 8.0$. Five headway profiles are shown for five mean densities $\Delta x_0 = 2.0, 4.0, 6.0, 8.0, 10.0$. At $\Delta x_0 = 2.0$, the traffic flow exhibits the homogeneous congested state (phase 1). At $\Delta x_0 = 4.0$, the traffic results in the inhomogeneous coexisting phase (coexisting phase of phases 1 and 2). At $\Delta x_0 = 6.0$, the traffic flow exhibits again the homogeneous state (phase 2). At $\Delta x_0 = 8.0$, the traffic flow results again in the inhomogeneous coexisting phase (coexisting phase of phases 2 and 3). At $\Delta x_0 = 10.0$, the traffic flow displays the free traffic state of homogeneous state (phase 3). When the density increases, the dynamical phase transitions similar to Figure 8 occur at four stages. However, there exists a difference between the headway profiles for $a = 2.0$ and $a = 1.2$. In the coexisting phase of phases 1 and 2, the headway of phase 1 is less than that of phase 1 at $\Delta x_0 = 2.0$ and the headway of phase 2 is higher than that of phase 2 at $\Delta x_0 = 6.0$. Furthermore, in the coexisting phase of phases 2 and 3, the headway of phase 2 is less than that of phase 2 at $\Delta x_0 = 4.0$ and the headway of phase 3 is higher than that of phase 3 at $\Delta x_0 = 10.0$.

Figure 9(b) shows the plots of headway against vehicle's number for sensitivity $a = 1.0$ where $v_{max} = 6.0$, $x_{1,c} = 4.0$, and $x_{2,c} = 8.0$. Five headway profiles are shown for five mean densities $\Delta x_0 = 2.0, 4.0, 6.0, 8.0, 10.0$. At $\Delta x_0 = 2.0$, the traffic flow exhibits the homogeneous congested state (phase 1). At $\Delta x_0 = 4.0$, the traffic results in the inhomogeneous coexisting phase (coexisting phase of phases 1 and 3). The inhomogeneity is very higher than that of $a = 2.0$ and $a = 1.2$ in Figures 8(a) and 9(a). The strong jam is formed and propagates backward. At $\Delta x_0 = 6.0$, the traffic flow exhibits again the homogeneous state (phase 2). At $\Delta x_0 = 8.0$, the traffic flow results again in the very inhomogeneous coexisting phase (coexisting phase of phases 1 and 3). At $\Delta x_0 = 10.0$, the traffic flow displays the free traffic state of homogeneous state (phase 3). The coexisting phase at $\Delta x_0 = 4.0$ agrees with that of $\Delta x_0 = 8.0$. Therefore, the coexisting phases at $\Delta x_0 = 4.0$ and $\Delta x_0 = 8.0$ are the mixture of phases 1 and 3. Thus, the coexisting phase of $a = 1.0$ is definitely different from that of $a = 2.0$ and $a = 1.2$. There exist three phases 1-3 and one coexisting phase between 1 and 3. When the density increases, the dynamical phase transitions similar to Figure 8 occur at four stages: (1) from phase 1 to coexisting phase between phases 1 and 3, (2) from coexisting phase between phases 1 and 3 to phase 2, (3) from phase 2 to coexisting phase between phases 1 and 3, and (4) from coexisting phase between phases 1 and 3 to phase 3.

We derive the phase diagram for three-phase model (45). The local minimum value of headway in the coexisting phase between 1 and 2 is the headway value of phase 1. The local maximum value of headway in the coexisting phase between 1 and 2 is the headway value of phase 2. The local minimum value of headway in the coexisting phase between 2 and 3 is the headway value of phase 2. The local maximum value of headway in the coexisting phase between 2 and 3 is the headway value of phase 3. The local minimum value of headway in the coexisting phase between 1 and 3 is the headway value of phase 1. The local maximum value of headway in the coexisting phase between 1 and 3 is the headway value of phase 3. The curve connecting these values represents the coexisting lines. The spinodal curve is given by the neutral stability line. Thus, one obtains the phase diagram. Figure 10 shows the phase diagram in the phase space $(\Delta x_0, a)$. The circles and crosses indicate the phase separation lines. The dotted line indicates the spinodal line. By comparing Figure 10 with Figure 4, the three-phase model exhibits two mountains and single valley, while two-phase model displays a single mountain. The phases 1, 2, and 3 appear, respectively, in regions V, III, I. The coexisting phase of phases 1 and 2 occurs in region IV. The coexisting phase of phases 2 and 3 occurs in region II. Thus, the three-phase model has the characteristics different from the two-phase model.

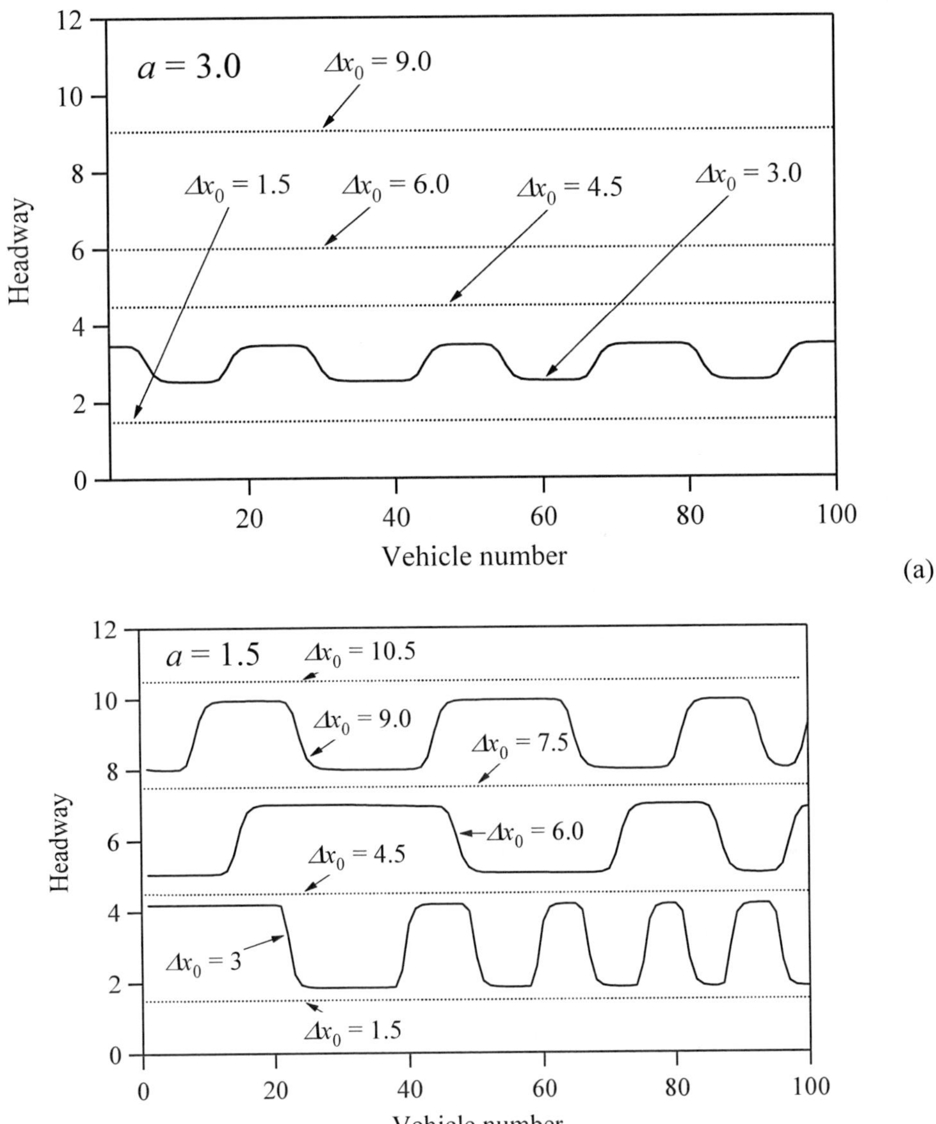

Figure 12. (Continued)

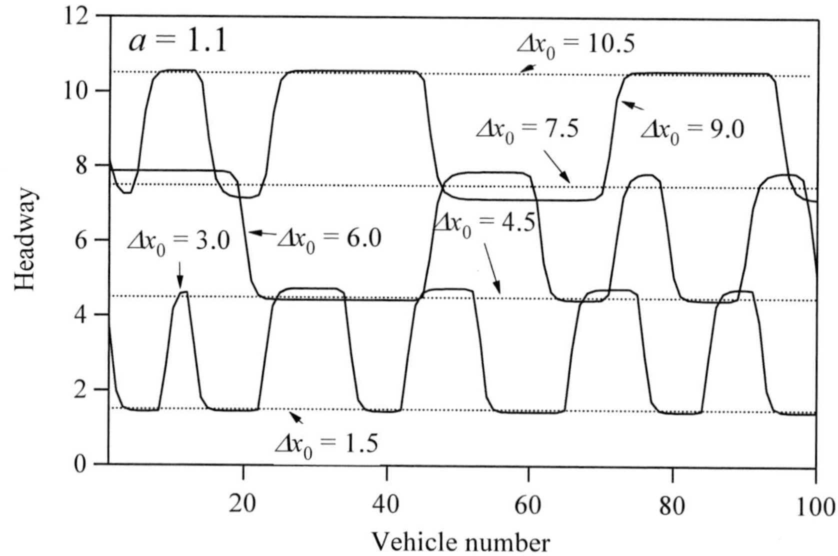

(c)

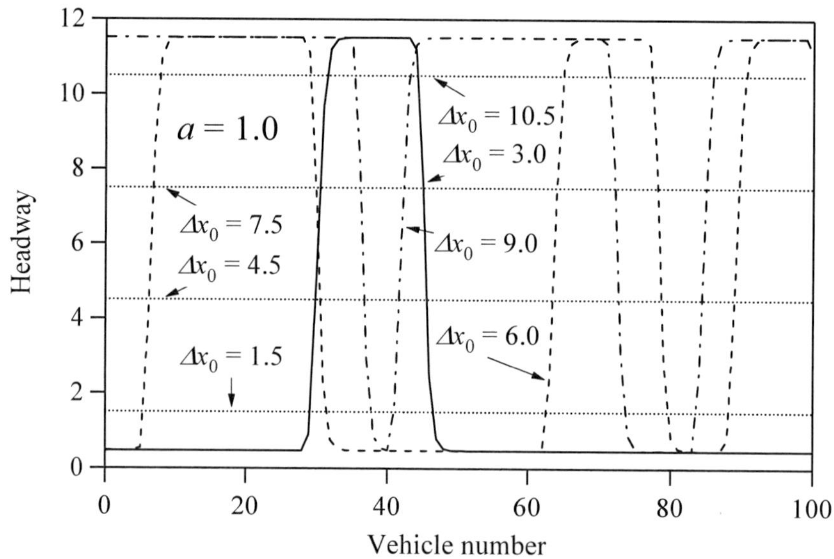

(d)

Figure 12. (a) Plots of headway against vehicle's number for sensitivity $a = 3.0$ where $v_{max} = 6.0$, $x_{1,c} = 3.0$, $x_{2,c} = 6.0$, and $x_{2,c} = 9.0$. Five headway profiles are shown for five mean densities $\Delta x_0 = 1.5, 3.0, 4.5, 6.0, 9.0$. (b) Plots of headway against vehicle's number for sensitivity $a = 1.5$. Seven headway profiles are shown for seven mean densities $\Delta x_0 = 1.5, 3.0, 4.5, 6.0, 7.5, 9.0, 10.5$. (c) Plots of headway against vehicle's number for sensitivity $a = 1.1$. Seven headway profiles are shown for seven mean densities $\Delta x_0 = 1.5, 3.0, 4.5, 6.0, 7.5, 9.0, 10.5$. (d) Plots of headway against vehicle's number for sensitivity $a = 1.0$. Seven headway profiles are shown for seven mean densities $\Delta x_0 = 1.5, 3.0, 4.5, 6.0, 7.5, 9.0, 10.5$.

We study the fundamental (current (flow)-density) diagram for various values of sensitivity. Figure 11 shows the plots of current against density. The current (flow) is obtained by averaging the number of vehicles passing a point over sufficiently large time. The dotted line indicates the theoretical current-density curve. The circles indicate the simulation result. Diagrams (a)-(d) represent, respectively, the fundamental diagrams for $a = 4.0, 2.0, 1.2, 1.0$. For (a) $a = 4.0$, the simulation points collapse on the theoretical current-density curve. This is due to no jams for $a > 3.0$.

For (b) $a = 2.0$, the traffic is classified into five states: (I) free traffic (phase 3), (II) coexisting phase between phases 3 and 2, (III) homogeneous traffic (phase 2) at intermediate density, (IV) coexisting phase between phases 2 and 1, and (V) homogeneous congested traffic (phase 1) at high density. The distinct traffic states I-V are shown in Figure 10. In regions II and IV, the density wave appears and propagates backward. The simulation result deviates slightly from the theoretical current-density curve in regions II and IV because the jams occur. For (c) $a = 1.2$ and (d) $a = 1.0$, the traffic is classified into five states. Below the neutral stability line, the traffic flow results in the inhomogeneous coexisting state. With decreasing sensitivity, the strong traffic jam is formed. As result, the current deviates highly from the theoretical current-density curve. In region IV of (d) $a = 1.0$, the current does not decreases highly with density. Irrespective of increasing density, the high current is maintained.

We investigate the traffic states for model (46) by varying the density and sensitivity. When the sensitivity is higher than 4.0, the traffic flow is stable and evolves in time to a homogeneous traffic even if the initial headway has any profile. If the sensitivity is less than 4.0, the traffic flow displays the complex behavior. Figure 12(a) shows the plots of headway against vehicle's number for sensitivity $a = 3.0$ where $v_{\max} = 6.0$, $x_{1,c} = 3.0$, $x_{2,c} = 6.0$, and $x_{2,c} = 9.0$. Five headway profiles are shown for five mean densities $\Delta x_0 = 1.5, 3.0, 4.5, 6.0, 9.0$. At $\Delta x_0 = 1.5$, the traffic flow exhibits the homogeneous congested state (phase 1). At $\Delta x_0 = 3.0$, the traffic flow becomes unstable and results in the inhomogeneous coexisting phase (coexisting phase of phases 1 and 2). At $\Delta x_0 = 4.5, 6.0, 9.0$, the traffic flow exhibits again the homogeneous state (phase 2). Figure 12(b) shows the plots of headway against vehicle's number for sensitivity $a = 1.5$. Seven headway profiles are shown for seven mean densities $\Delta x_0 = 1.5, 3.0, 4.5, 6.0, 7.5, 9.0, 10.5$. At $\Delta x_0 = 1.5$, the traffic flow exhibits the homogeneous congested state (phase 1). At $\Delta x_0 = 3.0$, the traffic flow becomes unstable and results in the inhomogeneous coexisting phase of phases 1 and 2. At $\Delta x_0 = 4.5$, the traffic flow exhibits again the homogeneous state (phase 2). At $\Delta x_0 = 6.0$, the traffic flow results again in the inhomogeneous coexisting phase of phases 2 and 3. At $\Delta x_0 = 7.5$, the traffic flow exhibits again the homogeneous state (phase 3). At $\Delta x_0 = 9.0$, the traffic flow results again in the inhomogeneous coexisting phase of phases 3 and 4. At $\Delta x_0 = 10.5$, the traffic flow exhibits again the homogeneous state (phase 4). Thus, with increasing headway, the dynamical transitions occur at the six stages. The traffic flow displays four stable states (phases 1-4) and three kink-antikink waves (coexisting phases 1 and 2, 2 and 3, and 3 and 4). This model (46) exhibits the four-phase traffic. Figure 12(c) shows the plots of headway against vehicle's number for sensitivity $a = 1.1$. Seven headway profiles are shown for seven mean densities $\Delta x_0 = 1.5, 3.0, 4.5, 6.0, 7.5, 9.0, 10.5$. The traffic flow shows the similar behavior to that in Figure 12(b). However, there exists such a difference that the maximum value of coexisting phase of phases 1 and 2 is higher than the minimum value of coexisting phases 2 and 3 and the maximum value of coexisting phase of phases 2 and 3 is higher than the minimum value of coexisting phases 3 and 4. Figure 12(d) shows the plots of headway against vehicle's number for sensitivity $a = 1.0$. Seven headway profiles are shown for seven mean densities $\Delta x_0 = 1.5, 3.0, 4.5, 6.0, 7.5, 9.0, 10.5$. For $\Delta x_0 = 1.5, 4.5, 7.5, 10.5$, the traffic flow exhibits the homogeneous congested state. For $\Delta x_0 = 3.0$, the traffic flow is unstable and results in the homogeneous coexisting phase. The strong density wave (stop-and go-wave) is formed and propagates backward. For $\Delta x_0 = 6.0, 9.0$, the traffic flow results in the inhomogeneous coexisting phase and the similar density wave is formed. The density waves of $\Delta x_0 = 6.0, 9.0$ agrees with that of $\Delta x_0 = 3.0$.

We derive the phase diagram for four-phase model (46). The local minimum value of headway in the coexisting phase between 1 and 2 is the headway value of phase 1. The local maximum value of headway in the coexisting phase between 1 and 2 is the headway value of phase 2. The local minimum value of headway in the coexisting phase between 2 and 3 is the headway value of phase 2. The local maximum value of headway in the coexisting phase between 2 and 3 is the headway value of phase 3. The local minimum value of headway in the coexisting phase between 3 and 4 is the headway value of phase 3. The local maximum value of headway in the coexisting phase between 3 and 4 is the headway value of phase 4. The curve connecting these values represents the coexisting lines. The spinodal curve is given by the neutral stability line. Thus, one obtains the phase diagram. Figure 13 shows the phase diagram in the phase space $(\Delta x_0, a)$. The circles, crosses, and triangles indicate the phase separation lines. The dotted line indicates the neutral stability line (spinodal line). By comparing Figure 13 with Figure 4, the four-phase model exhibits three mountains and two valleys, while two-phase model displays only single mountain.

For four-phase model (46), we study the fundamental (current(flow)-density) diagram for various values of sensitivity. Figure 14 shows the plots of current against density. The current (flow) is obtained by averaging the number of vehicles passing a point over sufficiently large time. The dotted line indicates the theoretical current-density curve. The circles indicate the simulation result. Diagrams (a)-(d) represent, respectively, the fundamental diagrams for $a = 3.0, 1.5, 1.1, 1.0$. Fundamental diagrams for four-phase model are definitely different from those for two-phase and three-phase models.

We have extended the conventional optimal velocity function to have n turning points. We have found that the model with n turning points displays $n+1$ phase traffic. We have clarified that the multi-phase model exhibits multiple phase transitions with increasing density.

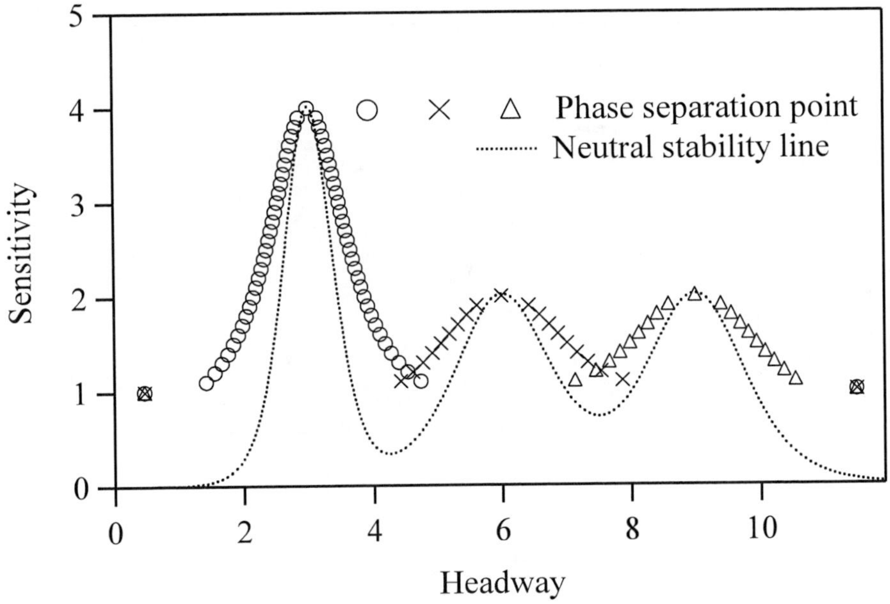

Figure 13. Phase diagram in the phase space $(\Delta x_0, a)$ for the four-phase model. The circles, crosses, and triangles indicate the phase separation lines. The dotted line indicates the neutral stability line (spinodal line). By comparing Figure 13 with Figure 4, the four-phase model exhibits three mountains and two valleys, while two-phase model displays only single mountain.

4. Traffic Jams Induced by Slowdown Sections

The city traffic is controlled by a speed limit for security and priority for a road. Such speed limit as slowdown often induces traffic jams when a density of vehicles is high. One is interested in the structure and formation of traffic jams induced by slowdown. When the sections of slowdown are introduced on a highway, where, when, and how are the traffic jams occur? In this section, we study the traffic states and jams induced by the slowdown sections on a highway [88-90].

4.1. Jams at a Slowdown Section

When a density of vehicles is low, vehicles move freely with no jams even if the slowdown sections exist on a highway. If the density is higher than a critical value, the traffic jam is formed just before the section of slowdown. The speed of vehicles within the jam becomes lower than the speed limit of slowdown. The traffic jam ends with forming a queue of slow vehicles. A discontinuous front appears at the end (edge) of traffic jam. Before and after the discontinuity, the traffic state changes abruptly.

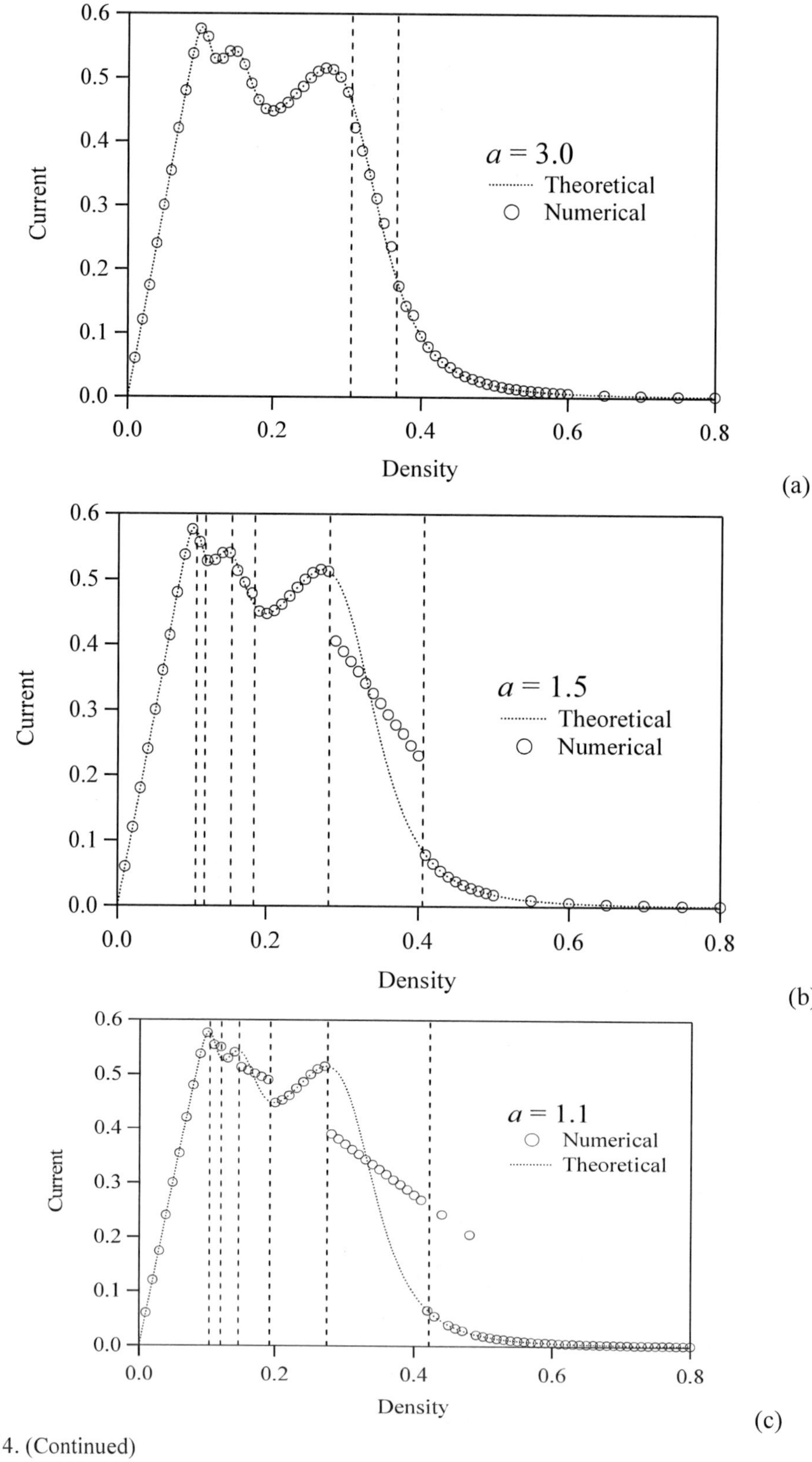

Figure 14. (Continued)

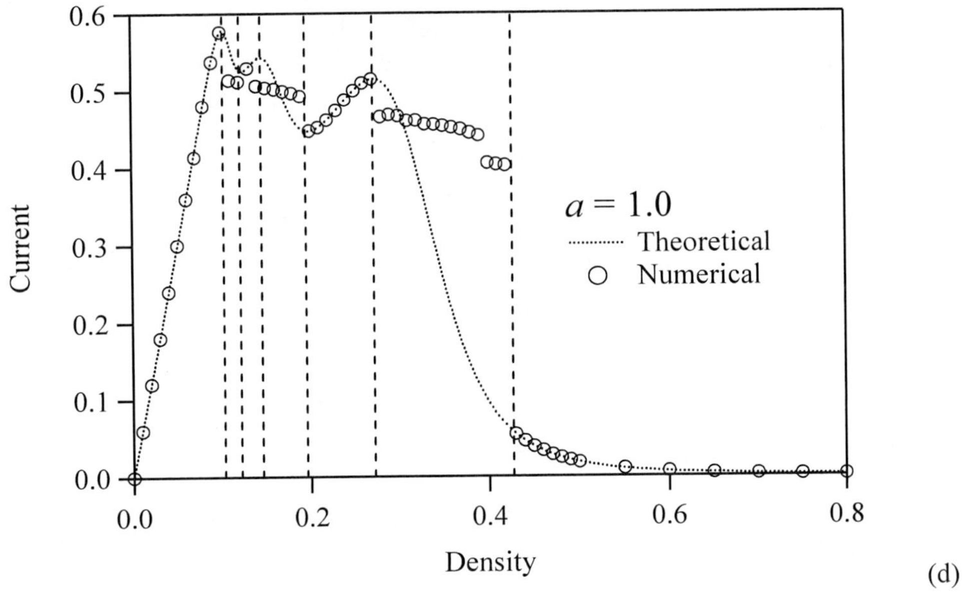

Figure 14. Plots of current against density for four-phase model. The dotted line indicates the theoretical current-density curve. The circles indicate the simulation result. Diagrams (a)-(d) represent, respectively, the fundamental diagrams for $a = 3.0, 1.5, 1.1, 1.0$.

However, it is little known about the discontinuity. How does the traffic state change near the discontinuous front? How does the traffic flow change by introducing a slowdown? Does such relationship as Rankine-Hugoniot equations of shock wave hold for the discontinuity? The discontinuity induced by the slowdown has little been investigated by the use of the dynamic models.

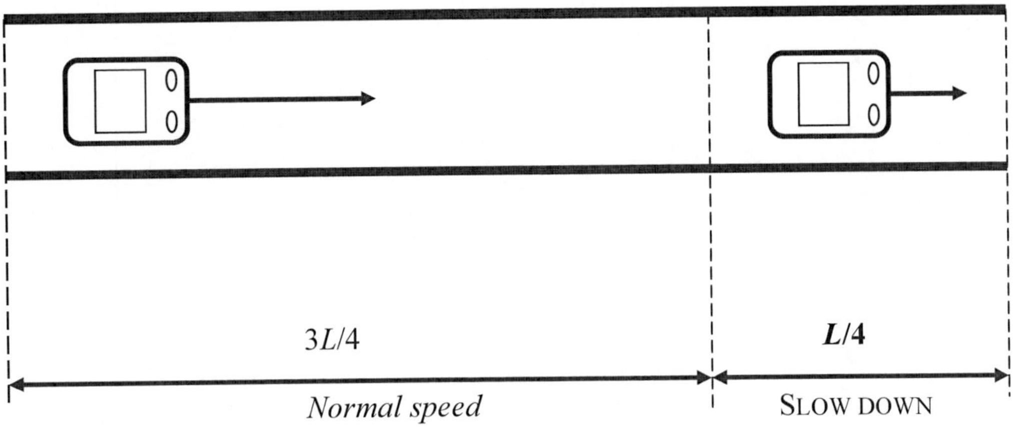

Figure 15. Schematic illustration of the traffic model for the single-lane highway with the section of slowdown. Vehicles move with low speed in the section of slowdown, while they move with the normal velocity except for the section of slowdown.

We extend the optimal velocity model to take into account the slowdown. We study the traffic states and jams by using the extended version. We clarify the dynamical states of traffic and the characteristic of discontinuous front. We present the fundamental diagram in the traffic flow including the slowdown. We show how the traffic state changes with increasing density of vehicles and with a degree of slowdown.

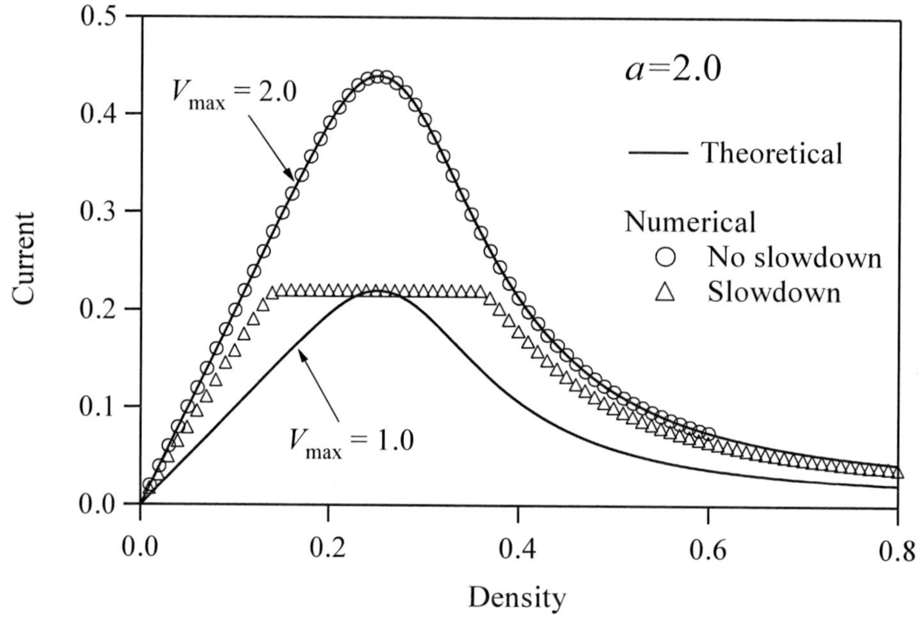

(a)

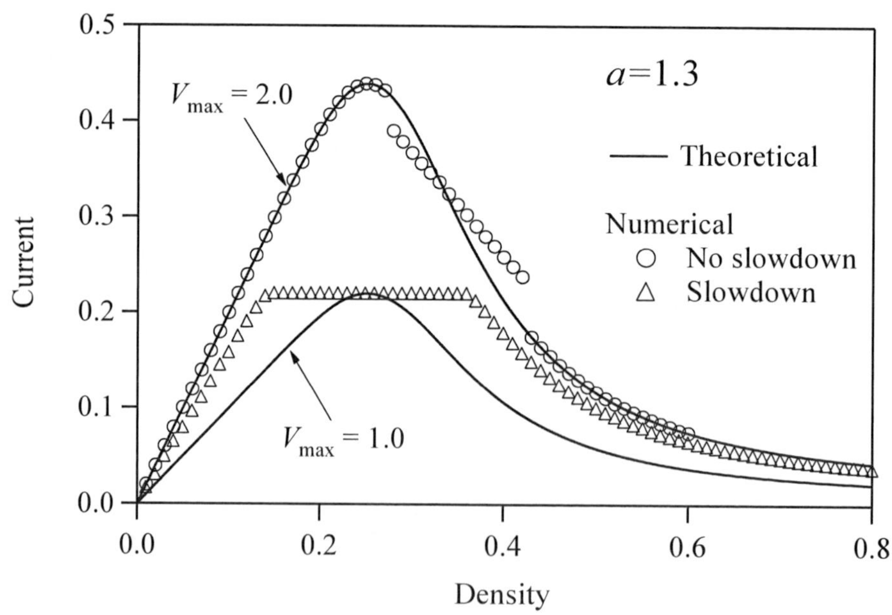

(b)

Figure 16. (Continued)

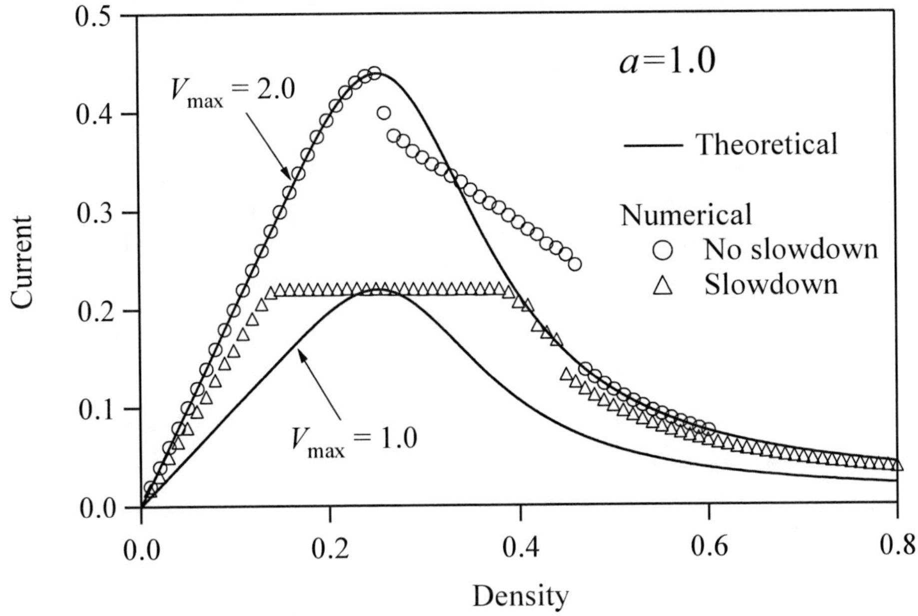

(c)

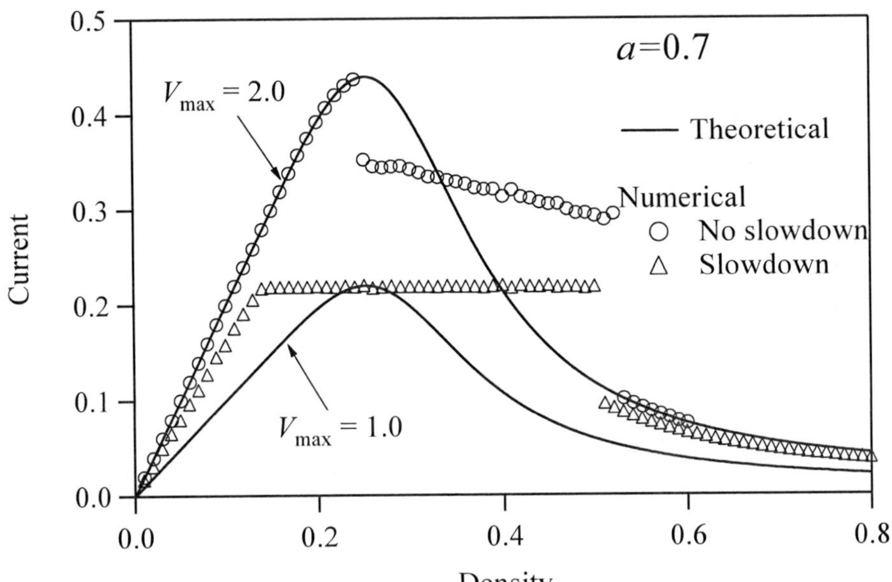

(d)

Figure 16. Plots of traffic currents against density for (a) sensitivity $a = 2.0$, (b) $a = 1.3$, (c) $a = 1.0$, and (d) $a = 0.7$, where $v_{max} = 2.0$, $v_s = 1.0$, and 200 vehicles. The open circles indicate the traffic current with no slowdown. The open triangles indicate the traffic current for velocity ratio $v_s / v_{max} = 0.5$ of the slowdown. For comparison, the traffic currents for two cases $v_{max} = 2.0$ and $v_{max} = 1.0$ without traffic jams are shown by two solid lines.

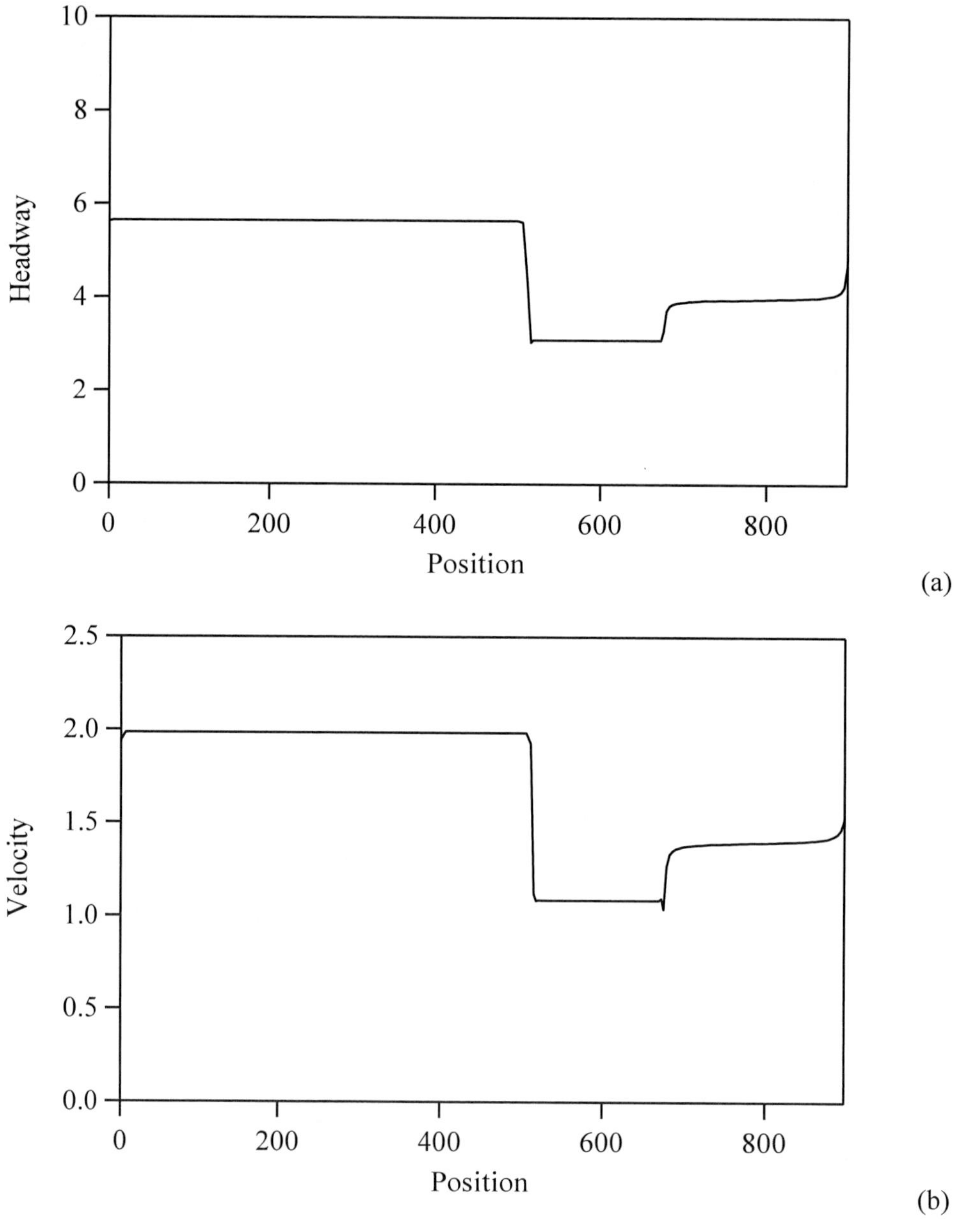

Figure 17. (a) Plot of headway against position of vehicles for sensitivity $a = 2.0$, average(initial) headway $\Delta x_0 = 4.5$, velocity ratio $v_s / v_{\max} = 0.8$, and 200 vehicles. (b) Plot of velocity against position of vehicles.

We consider the traffic of vehicles flowing on the single-lane roadway. Vehicles flow with no passing on the single-lane roadway under periodic boundary condition. We assume that vehicles are forced to slow down when they enter into the section of the slowdown. Figure 15 shows the schematic illustration of the traffic model for the single-lane highway with the section of slowdown. Vehicles move with low speed in the section of slowdown, while they move with the normal velocity except for the section of slowdown. We apply the optimal velocity model to the traffic flow. The optimal velocity model is described by Eq. (3) (or Eq. (43)) for the equation of motion of vehicle *i*. A driver adjusts the vehicle speed to approach the optimal velocity determined by the observed headway. In the region of normal speed, the optimal velocity of vehicles is given by

$$V(\Delta x_i) = \frac{v_{\max}}{2}\left[\tanh(\Delta x_i - x_c) + \tanh(x_c)\right], \tag{50}$$

where $v_{\max}$ is the maximal velocity of vehicles and x_c is the safety distance of vehicles.

In the section of slowdown, vehicles move with forced low speed. The speed should be lower than the speed limit of slowdown. When vehicles enter into the section of slowdown, the optimal velocity is given by

$$V(\Delta x_i) = \frac{v_s}{2}\left[\tanh(\Delta x_i - x_{cs}) + \tanh(x_{cs})\right], \tag{51}$$

where v_s is the speed limit of slowdown, $v_s < v_{\max}$, and x_{cs} is the safety distance in the section of slowdown.

Thus, when the density of vehicles is very low, vehicles move at the maximal velocity $v_{\max}$ except for the section of slowdown, while they move at the forced low speed v_s in the section of slowdown. With increasing density, vehicles interact with each other. The dynamics is determined by Eqs. (3), (50), and (51). Then, various dynamic states of traffic appear and traffic jams may occur. We study the dynamic states and traffic jams in the traffic flow described by model in Figure 15.

The simulation is performed until the traffic flow reaches a steady state. We solve numerically Eq. (3) with optimal velocity functions (50) and (51) by using fourth-order Runge-Kutta method where the time interval is $\Delta t = 1/128$.

We carry out simulation by varying the initial headway, sensitivity, and velocity ratio $v_s / v_{\max}$ for 200 and 2000 vehicles, safety distances $x_c = x_{cs} = 4.0$, and maximal velocity $v_{\max} = 2.0$. Initially, we put all vehicles on the single-lane highway with the same headway Δx_{int}. The density ρ is given by the inverse of the headway. The length *L* of highway varies with the initial headway. The section of slowdown is set on the downstream position $L/4$ of the highway.

Figure 16(a)-(d) show the plots of traffic currents against density for (a) sensitivity $a = 2.0$, (b) $a = 1.3$, (c) $a = 1.0$, and (d) $a = 0.7$, where $v_{\max} = 2.0$, $v_s = 1.0$, and 200 vehicles. The traffic current is obtained by averaging the current from $t = 2000$ to $t = 5000$. The open circles indicate the traffic current with no slowdown. The open triangles indicate the traffic current for velocity ratio $v_s / v_{\max} = 0.5$ of the slowdown. For comparison, the traffic currents for two cases $v_{\max} = 2.0$ and $v_{\max} = 1.0$ without traffic jams are shown by two solid lines. The current is called as the theoretical current curve. It is given by $J = V(\Delta x)/\Delta x$ where Δx is the average value of headway and $V(\Delta x)$ is the optimal velocity.

The case of no slowdown corresponds to the conventional traffic flow. When sensitivity is higher than critical value 2.0, no traffic jams occur and the current agrees with the theoretical current curve (see Figure 16(a)). If sensitivity is lower than the critical value, traffic jams occur and current deviates from the theoretical current curve in the region of the density at which traffic jams appear (see Figure 16(b)-(d)).

In the case of slowdown, the current is shown by triangles. The current increases linearly with density at low density, but is lower than the current of no slowdown. Then, the current saturates at the first critical density and keeps a constant value until the second critical density. When the density is higher than the second critical density, the current decreases with increasing density. The first critical density does not depend on the sensitivity but the second critical density depends highly on the sensitivity. The value of second critical density increases with decreasing sensitivity. Also, the saturated value of current is consistent with the maximal value of the theoretical current curve for $v_{max} = 1.0$.

In the region of saturated current, a traffic jam appears just before the section of slowdown. We study the headway and velocity profiles for the traffic flow with jam induced by the slowdown. Figure 17(a) shows the plot of headway against position of vehicles for sensitivity $a = 2.0$, average(initial) headway $\Delta x_0 = 4.5$, velocity ratio $v_s / v_{max} = 0.8$, and 200 vehicles. Figure 17(b) shows the plot of velocity against position of vehicles, corresponding to the headway profile in Figure 17(a). The section of slowdown begins at $x = 675$ and ends at $x = 900$. The traffic jam begins just before the section of slowdown and ends at $x = 510$. The discontinuous front appears just after the edge of traffic jam.

Figure 18(a) shows the plot of headway against position of vehicles for sensitivity $a = 1.0$, average(initial) headway $\Delta x_0 = 4.5$, velocity ratio $v_s / v_{max} = 0.8$, and 200 vehicles. Figure 18(b) shows the plot of velocity against position of vehicles, corresponding to the headway profile in Figure 18(a). The oscillatory jam begins just before the section of slowdown and ends at $x = 489$. The discontinuous front appears just after the edge of traffic jam. When the value of sensitivity is low, the oscillatory jam occurs.

We study the characteristic properties of the discontinuity. We derive, numerically, the headways before and after discontinuous front by varying velocity ratio v_s / v_{max}. Figure 19(a) shows the plot of headways before and after the discontinuous front against the velocity ratio for sensitivity $a = 2.0$ where the jam is not oscillatory but uniform. Open circle indicates the value of headway just before the continuous front. Open square indicates the value of headway just after the continuous front. Open triangle indicates the value of average headway within the section of slowdown.

Figure 19(b) shows the plot of headways before and after the discontinuous front against the velocity ratio for sensitivity $a = 1.0$ where the jam is uniform or oscillatory. Full square indicates the mean value of headway just after the discontinuous front when the oscillatory jam occurs. When the velocity ratio is not low, the oscillatory jam appears. The solid lines indicate the headways obtained from the theoretical analysis later.

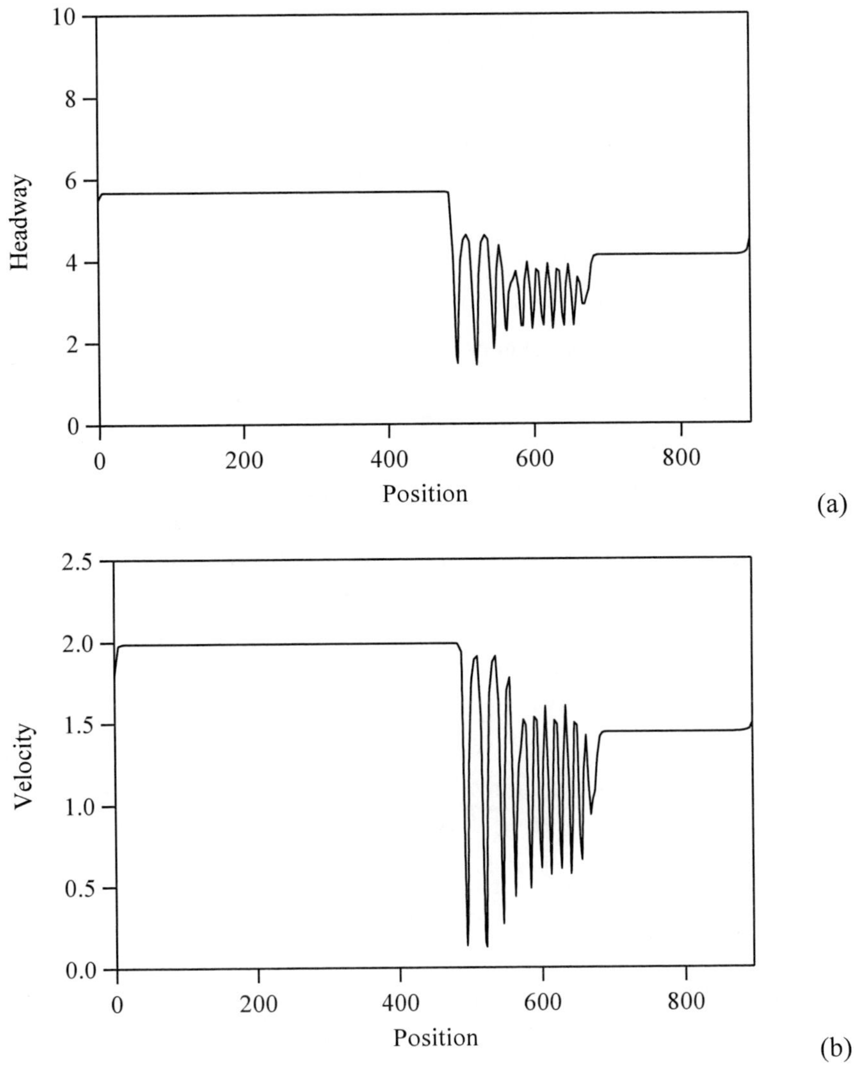

Figure 18. (a) Plot of headway against position of vehicles for sensitivity $a = 1.0$, average(initial) headway $\Delta x_0 = 4.5$, velocity ratio $v_s / v_{\max} = 0.8$, and 200 vehicles. (b) Plot of velocity against position of vehicles.

We derive the headways before and after discontinuous front analytically. Figure 20 shows two theoretical current curves for $v_{\max} = 2.0$ and $v_{\max} = 1.0$. The solid and dashed lines represent, respectively, the theoretical curves for $v_{\max} = 2.0$ and $v_{\max} = 1.0$. At the steady state, the traffic current is uniform over the highway irrespective of slowdown section. When the traffic current saturates, the traffic current throughout the section of slowdown becomes maximal value of the theoretical current curve of $v_{\max} = 1.0$. If no traffic jam occurs, the maximal current is given by the maximum point of the theoretical current curve with optimal velocity function (51). The maximum point is indicated by the full triangle in Figure 20. Because the traffic current is uniform over the highway, the crossing points of solid curve (theoretical current curve of $v_{max} = 2.0$) with the horizontal line on the maximal point of full triangle give such densities that are allowed to take on the highway except for the slowdown section. The values of densities are indicated by full circle and square. Thus, one obtains the headways (the inverse of densities) before and after the discontinuous front for various values of velocity ratio. The headways obtained from the theoretical analysis are shown by the solid lines in Figure 19. For $a = 2.0$ in Figure 19(a), the theoretical result agrees with the simulation result. For $a = 1.0$ in Figure 19(b), the theoretical result agrees with the simulation result when velocity ratio v_s / v_{max} is low. However, when velocity ratio is not low and traffic jam is oscillatory, the simulation value deviates a little from the theoretical value.

We derive the region map of distinct jams by studying the distinct states of traffic. We find three distinct jams: (1) the uniform jam, (2) the oscillatory jam, and (3) the propagating jam. The uniform jam is stationary in the highway and the headway and velocity take constant values within the jam. The headway and velocity profiles are shown in Figure 17. In the oscillatory jam, the headway and velocity oscillate. The jam propagates from the front of slowdown section to the discontinuous front. The headway and velocity profiles are shown in Figure 18. Propagating jam propagates backward over the whole highway as a single pulse jam.

Figure 21 shows the region map obtained for 2000 vehicles and $\Delta x_0 = 4.0$. Here, cross points represent uniform jam (1). Circles indicate oscillatory jam (2). Triangles represent propagating jam (3). The uniform jam appears for high values of sensitivity, while the oscillatory jam occurs for low values of sensitivity. When the sensitivity is low and the velocity ratio is high, the propagating jam appears only in the narrow region of Figure 21. The curve indicated by the dashed line represents the neutral stability line. We derived the neutral stability curve by studying the linear stability for the uniform traffic jam shown by the full square in Figure 20. The neutral stability line is almost consistent with the boundary between the uniform and oscillatory jams. The boundary obtained from the simulation of 200 vehicles gives low values of sensitivity, but the boundary approaches to the neutral stability line with increasing number of vehicles.

We study how the length of traffic jam varies with density. Figure 22 shows the plots of jam length ratio against density for sensitivities $a = 2.0, 1.3, 1.0, 0.7$ where the number of vehicles is 200. The jam length ratio is defined by the jam length divided by the length of normal speed section. When the jam reaches the position $x = 0$ (starting point of highway), the value of ratio takes one. Under a constant value of sensitivity, the jam length increases linearly with density. When the uniform jam appears ($a = 2.0, 1.3$), the jam length does not depend on the sensitivity. However, the jam length depends highly on the sensitivity for the oscillatory jam. Until the traffic jam reaches the position $x = 0$, the current saturates and takes a constant value. If the traffic jam passes over $x = 0$, the current decreases and becomes lower than the saturated current. Such density that the jam length becomes one is consistent with such value that the traffic current begins to decrease from the saturated current in Figure 16.

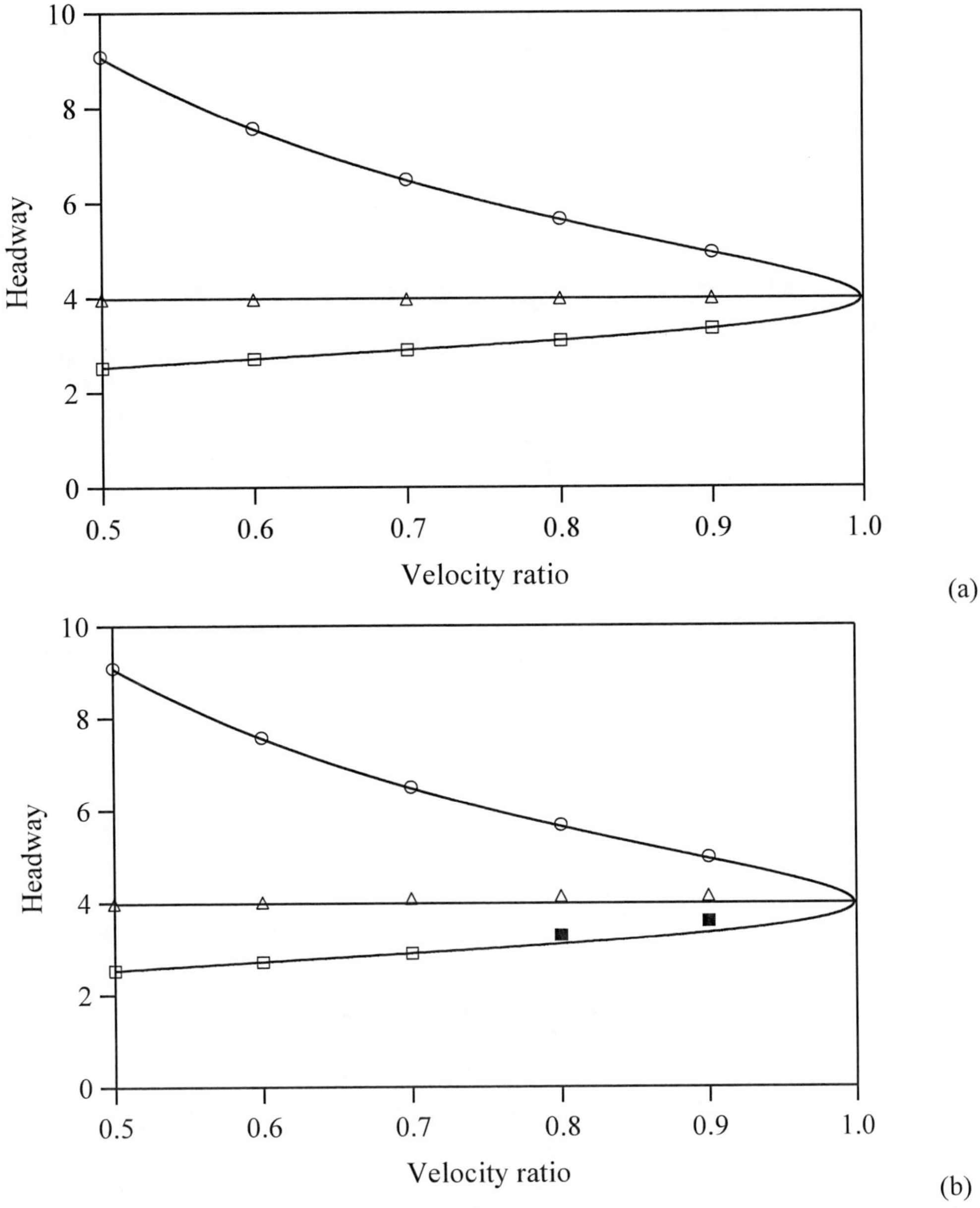

Figure 19. (a) Plot of headways before and after the discontinuous front against the velocity ratio for sensitivity $a = 2.0$. Open circle indicates the value of headway just before the continuous front. Open square indicates the value of headway just after the continuous front. Open triangle indicates the value of average headway within the section of slowdown. (b) Plot of headways before and after the discontinuous front against the velocity ratio for sensitivity $a = 1.0$. Full square indicates the mean value of headway just after the discontinuous front when the oscillatory jam occurs. The solid lines indicate the headways obtained from the theoretical analysis later.

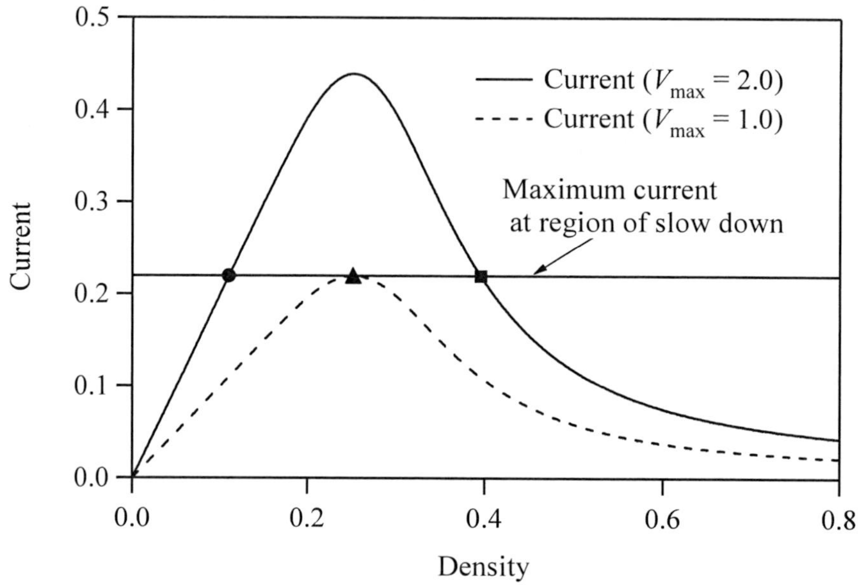

Figure 20. Two theoretical current curves for $v_{max} = 2.0$ and $v_{max} = 1.0$. The solid and dashed lines represent, respectively, the theoretical curves for $v_{max} = 2.0$ and $v_{max} = 1.0$.

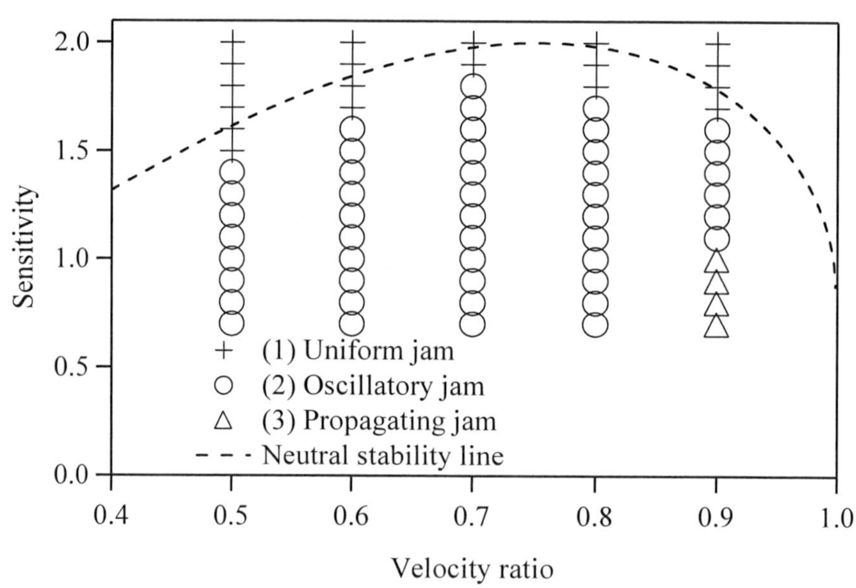

Figure 21. Region map obtained for 2000 vehicles and $\Delta x_0 = 4.0$. Here, cross points represent uniform jam (1). Circles indicate oscillatory jam (2). Triangles represent propagating jam (3). The curve indicated by the dashed line represents the neutral stability line.

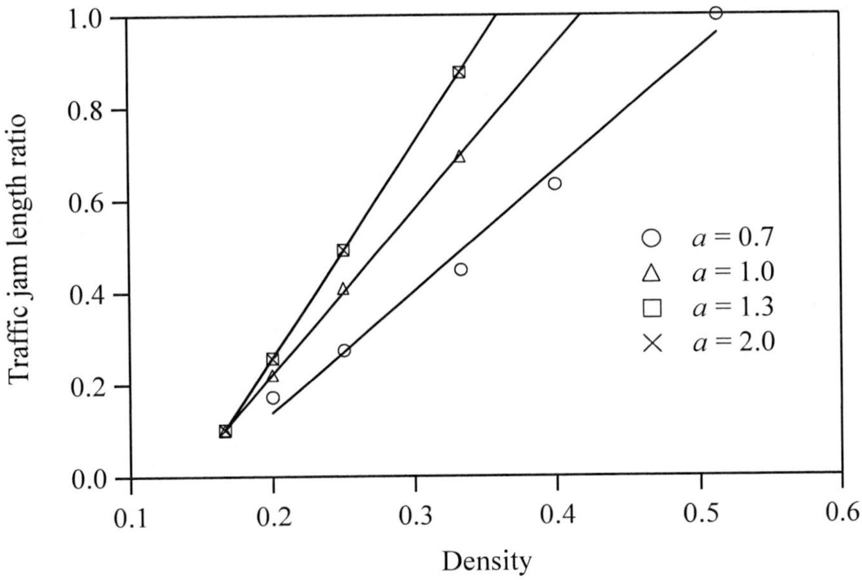

Figure 22. Plots of jam length ratio against density for sensitivities $a = 2.0, 1.3, 1.0, 0.7$ where the number of vehicles is 200.

4.2. Jams in Three-phase Model

Real traffic is very complex. The traffic jams induced by slowdown are also complex. We extend the conventional optimal-velocity model to the three phase (two-stage optimal-velocity) model described in Sec. 3.3. We study the traffic states and discontinuous front induced by the slowdown, by using the three-phase model. We clarify the dynamical states of traffic and the characteristic of discontinuous front. We present the fundamental diagram in the traffic flow including the slowdown. We show how the traffic state changes with increasing density of vehicles, with a degree of slowdown, and by two-stage optimal-velocity function.

Vehicles are forced to slow down when they enter into the section of the slowdown. Figure 15 shows the schematic illustration of the traffic model for the single-lane highway with the section of slowdown. Vehicles move with low speed in the section of slowdown, while they move with the normal velocity except for the section of slowdown. We apply the three-phase traffic model to the traffic flow [87]. In the region of normal speed, the two-stage optimal velocity function of vehicles is given by

$$V(\Delta x_i) = \frac{v_{f,\max}}{4}\left[\begin{array}{l}\{\tanh\alpha(\Delta x_i - x_{f,c,1}) + \tanh\alpha(x_{f,c,1})\} \\ \qquad + \{\tanh\alpha(\Delta x_i - x_{f,c,2}) + \tanh\alpha(x_{f,c,2})\}\end{array}\right], \tag{52}$$

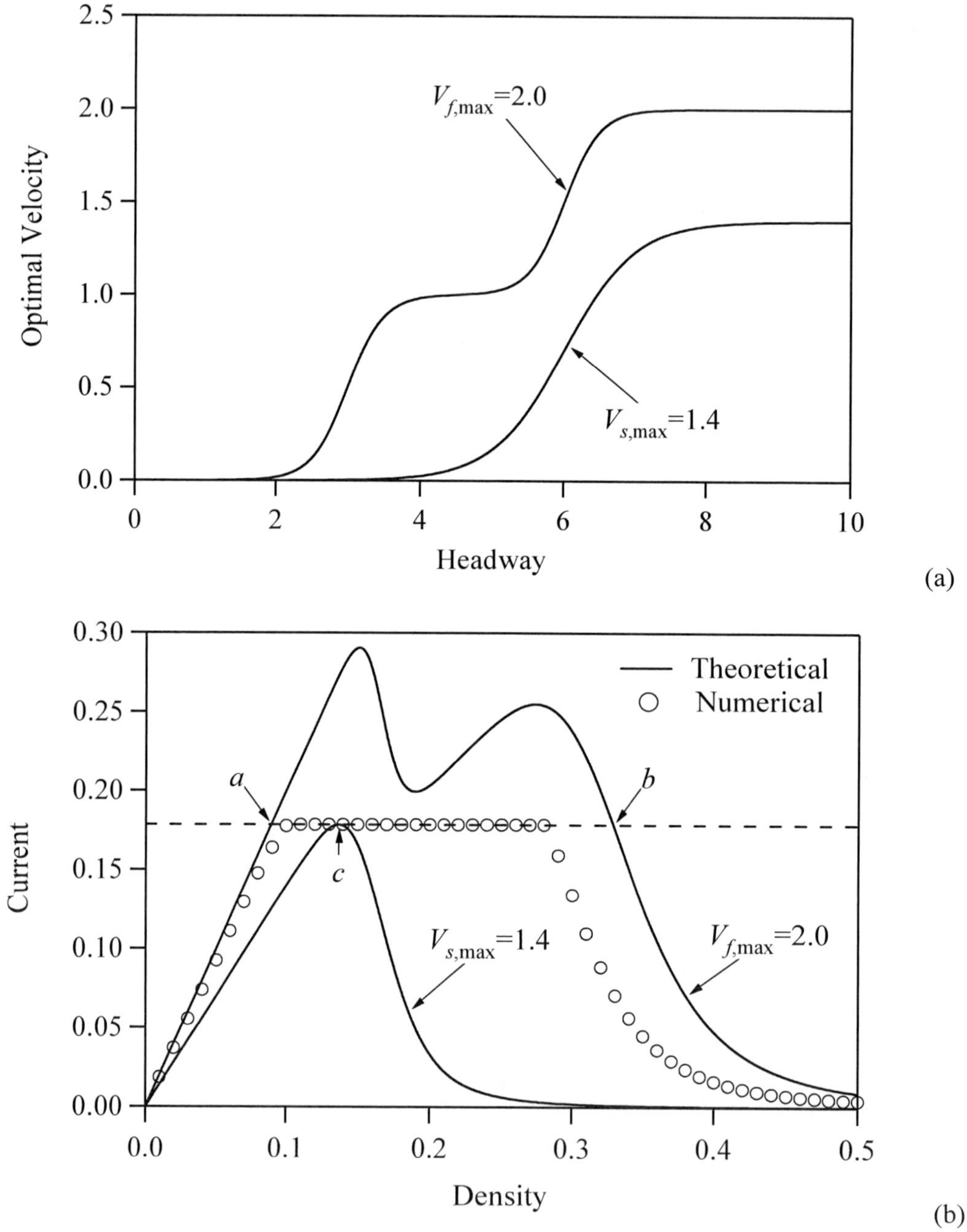

Figure 23. (a) Solid curves represent the optimal velocity functions. (b) Plots of traffic currents against density. The open circle indicates the traffic current obtained by simulation.

where $v_{f,\max}$ is the maximal velocity of vehicles on the roadway except for the section of slowdown, $x_{f,c,1}$ the position of first turning point, $x_{f,c,2}$ the position of second turning point, and α the slope at the turning point [87].

In the section of slowdown, vehicles move with forced low speed. The speed should be lower than the speed limit of slowdown. When vehicles enter into the section of slowdown, they obey the conventional or two-stage optimal-velocity functions:

$$V(\Delta x_i) = \frac{v_{s,\max}}{2}\left[\tanh\beta(\Delta x_i - x_{s,c}) + \tanh\beta(x_{s,c})\right], \tag{53}$$

or

$$V(\Delta x_i) = \frac{v_{s,\max}}{4}\left[\begin{array}{l}\{\tanh\beta(\Delta x_i - x_{s,c,1}) + \tanh\beta(x_{s,c,1})\} \\ \qquad + \{\tanh\beta(\Delta x_i - x_{s,c,2}) + \tanh\beta(x_{s,c,2})\}\end{array}\right], \tag{54}$$

where $v_{s,\max}$ is the speed limit of slowdown, $v_{s,\max} < v_{f,\max}$, $x_{s,c,1}$ the position of first turning point, $x_{s,c,2}$ the position of second turning point, and β the slope at the turning point.

Thus, when the density of vehicles is very low, vehicles move at the maximal velocity $v_{f,\max}$ except for the section of slowdown, while they move at a speed lower than the forced speed limit $v_{s,\max}$ in the section of slowdown. With increasing density, vehicles interact with each other. Then, various dynamic states of traffic appear and traffic jams may occur. We study the dynamic states and traffic jams in the traffic flow described by model in Figure 15. We restrict ourselves to such case that sensitivity is higher than the critical point, $a > a_c$, since we do not investigate the spontaneous jam but study the traffic jams induced by slowdown.

We carry out simulation by varying the initial headway, slowdown's velocity $v_{s,\max}$ for 500 vehicles, sensitivity $a = 2.5$ and maximal velocity $v_{f,\max} = 2.0$. Initially, we put all vehicles on the single-lane highway with the same headway Δx_{int}. The density ρ is given by the inverse of the headway. The length L of highway varies with the initial headway. The section of slowdown is set on the downstream position $L/4$ of the highway.

First, we study the traffic behavior for following optimal velocity functions: Eq. (52) with $v_{f,\max} = 2.0$, $\alpha = 2.0$, $x_{f,c,1} = 3.0$, and $x_{f,c,2} = 6.0$ for the section of normal velocity and Eq.(53) with $v_{s,\max} = 1.4$, $\beta = 1.0$ and $x_{s,c,1} = 6.0$ for the section of slowdown. The above optimal velocity functions are shown by solid curves in Figure 23(a). Figure 23(b) shows the plots of traffic currents against density. The open circle indicates the traffic current obtained by simulation. The traffic current is obtained by averaging the current over $t = 9000 - 10000$. The current increases linearly with density along the theoretical curve of

$v_{f,\max} = 2.0$. At $\rho = 0.1$, the current saturates and keeps a constant value which agrees with the maximal value of theoretical current curve of $v_{s,\max} = 1.4$. When density is higher than $\rho = 0.28$, the current decreases with increasing density.

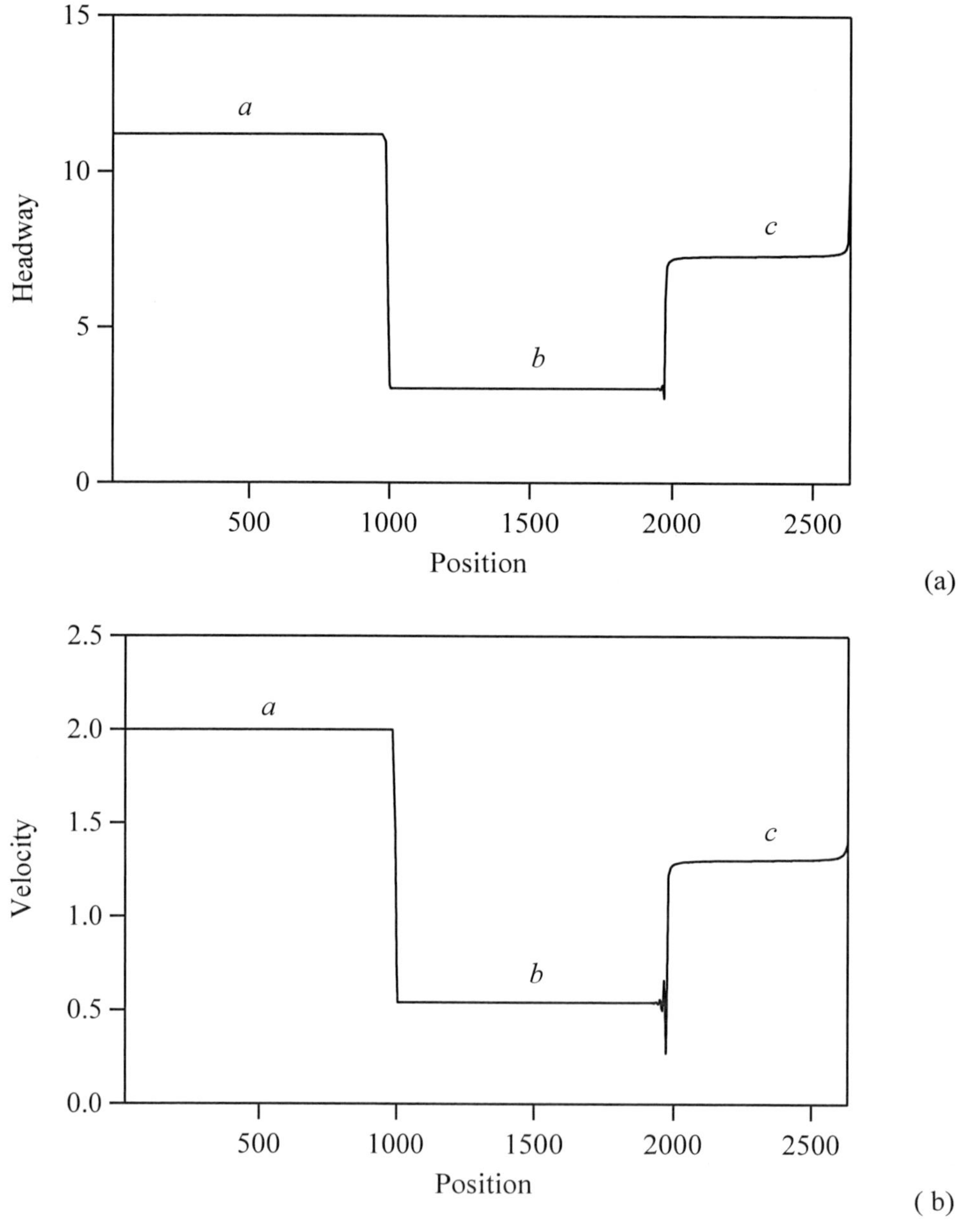

Figure 24. (a) Headway and (b) velocity profiles at $\rho = 0.19$.

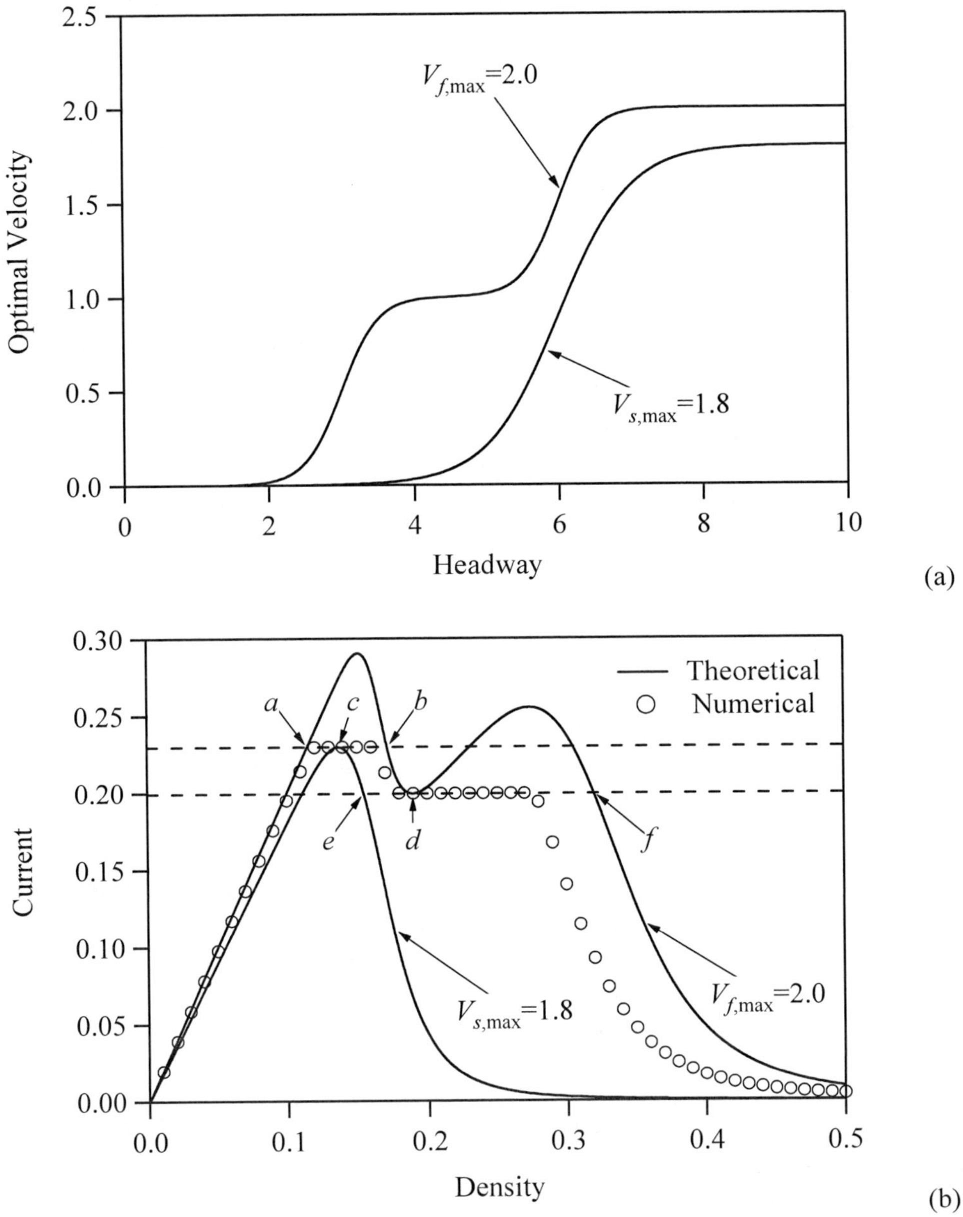

Figure 25. (a) The solid curves represent the optimal velocity functions. (b) Plots of traffic currents against density. The open circle indicates the traffic current obtained by simulation. The theoretical current curves for two optimal velocity functions are shown by two solid lines.

For comparison, the traffic currents for two optimal velocity functions without traffic jams are shown by two solid lines. The current is called as the theoretical current curve. It is given by $J = V(\Delta x)/\Delta x$ where Δx is the average value of headway and $V(\Delta x)$ is the optimal velocity. The case of no slowdown corresponds to the conventional traffic flow. When sensitivity is higher than critical value 2.0, no traffic jams occur and the current agrees with the theoretical current curve. If sensitivity is lower than the critical value, traffic jams occur and the current deviates from the theoretical current curve in the region of the density at which traffic jams appear.

Figure 24(a) and (b) show, respectively, the headway and velocity profiles at $\rho = 0.19$. In the region where the current saturates, the discontinuous front appears in the section of normal velocity. The traffic jam occurs just before the section of slowdown and ends at the discontinuous front. The densities (inverse of headway) before and after the discontinuous front are consistent with those at points *a* and *b* in Figure 23(b). The density at the section of slowdown agrees with that at point *c* of the maximal current of $v_{s,\max} = 1.4$. The velocities at regions *a*, *b*, and *c* are given, respectively, by the optimal velocities at densities *a*, *b*, and *c*.

We study the traffic behavior for $v_{s,\max} = 1.8$ where the other values of parameters are the same as those in Figure 23. Then, the optimal velocity functions are given by the solid curves in Figure 25(a). Figure 25(b) shows the plots of traffic currents against density. The open circle indicates the traffic current obtained by simulation. The theoretical current curves for two optimal velocity functions are shown by two solid lines. The current increases linearly with density along the theoretical curve of $v_{f,\max} = 2.0$. Between $\rho = 0.12 - 0.16$, the current saturates and keeps a constant value which agrees with the maximal current of theoretical current curve of $v_{s,\max} = 1.8$. When density is higher than $\rho = 0.16$, the current decreases until $\rho = 0.18$. Again, between $\rho = 0.18 - 0.27$, the current keeps a constant value which agrees with that of extreme value of theoretical current curve at point *d*. When density is higher than $\rho = 0.27$, the current decreases with increasing density. Figure 26(a) shows the headway profile at $\rho = 0.14$. In the region where the current saturates first, the discontinuous front appears in the section of normal velocity. The traffic jam occurs just before the section of slowdown and ends at the discontinuous front. The densities (inverse of headway) before and after the discontinuous front are consistent with those at points *a* and *b* in Figure 25(b). The density at the section of slowdown agrees with point *c* of the maximal current of $v_{s,\max} = 1.8$. The velocities at regions *a*, *b*, and *c* are given, respectively, by the optimal velocities at densities *a*, *b*, and *c*. Figure 26(b) shows the headway profile at $\rho = 0.23$. In the region where the current saturates second, the discontinuous front appears in the section of normal velocity. The traffic jam appears at two stages. The first traffic jam occurs just before the section of slowdown and ends at the discontinuous front. The second jam occurs just before the discontinuous front. The densities (inverse of headway) before and after the discontinuous front are consistent with those at points *d* and *f* in Figure 25(b). The density at the section of slowdown agrees with point *e* of the theoretical current curve of $v_{s,\max} = 1.8$. The velocities at regions *d*, *e*, and *f* are given, respectively, by the optimal velocities at densities *d*, *e*, and *f*. The velocity at region *d* is lower than that at the section of slowdown. If we compare Figure 26(b) with Figure 24(a), we find that two kinds of discontinuous fronts appear when the maximal point *c* is higher than the local minimal point *d*.

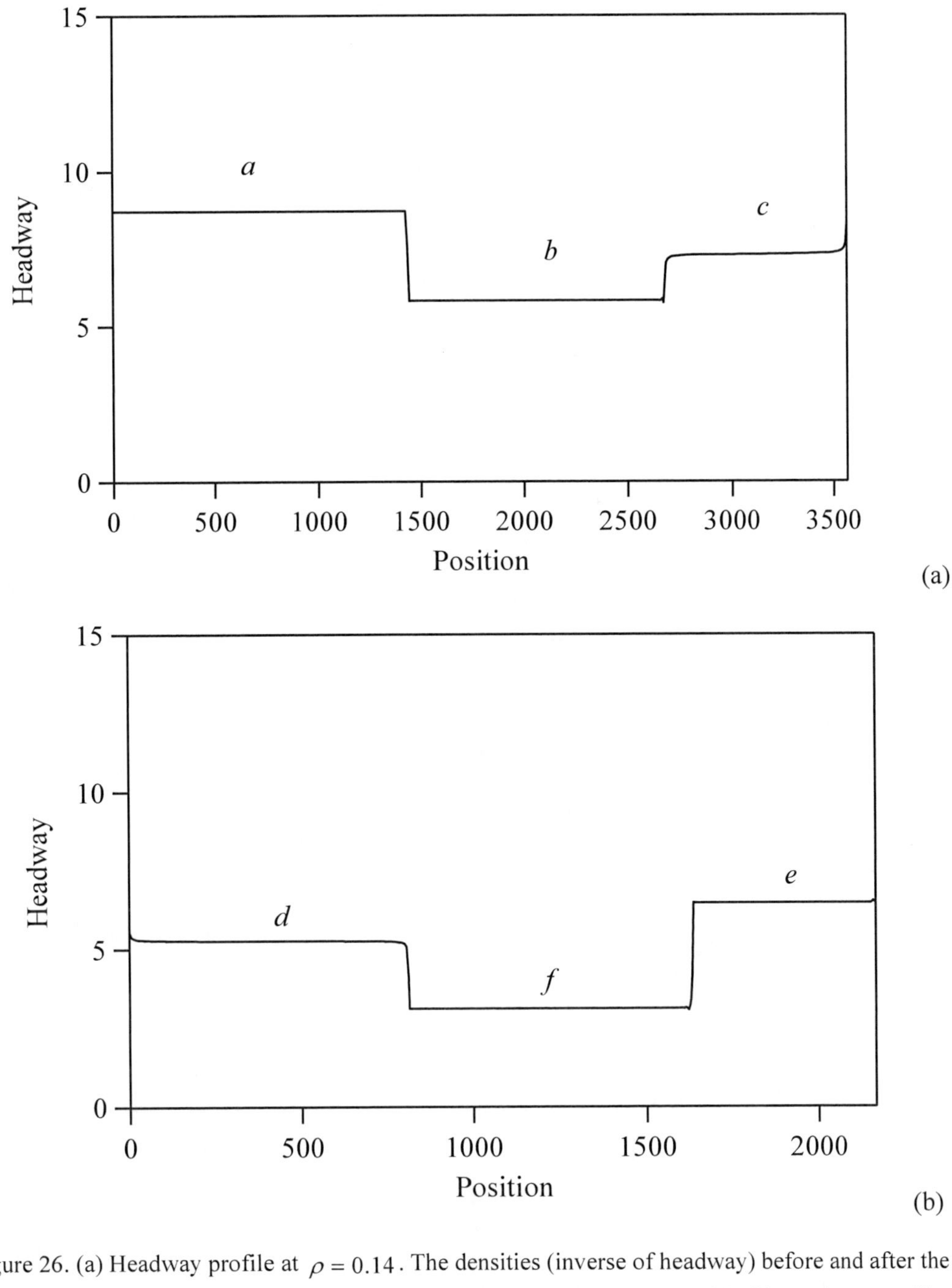

Figure 26. (a) Headway profile at $\rho = 0.14$. The densities (inverse of headway) before and after the discontinuous front are consistent with those at points *a* and *b* in Figure 25(b). (b) Headway profile at $\rho = 0.23$. The densities (inverse of headway) before and after the discontinuous front are consistent with those at points *d* and *f* in Figure 25(b). The density at the section of slowdown agrees with point *e* of the theoretical current curve of $v_{s,\max} = 1.8$.

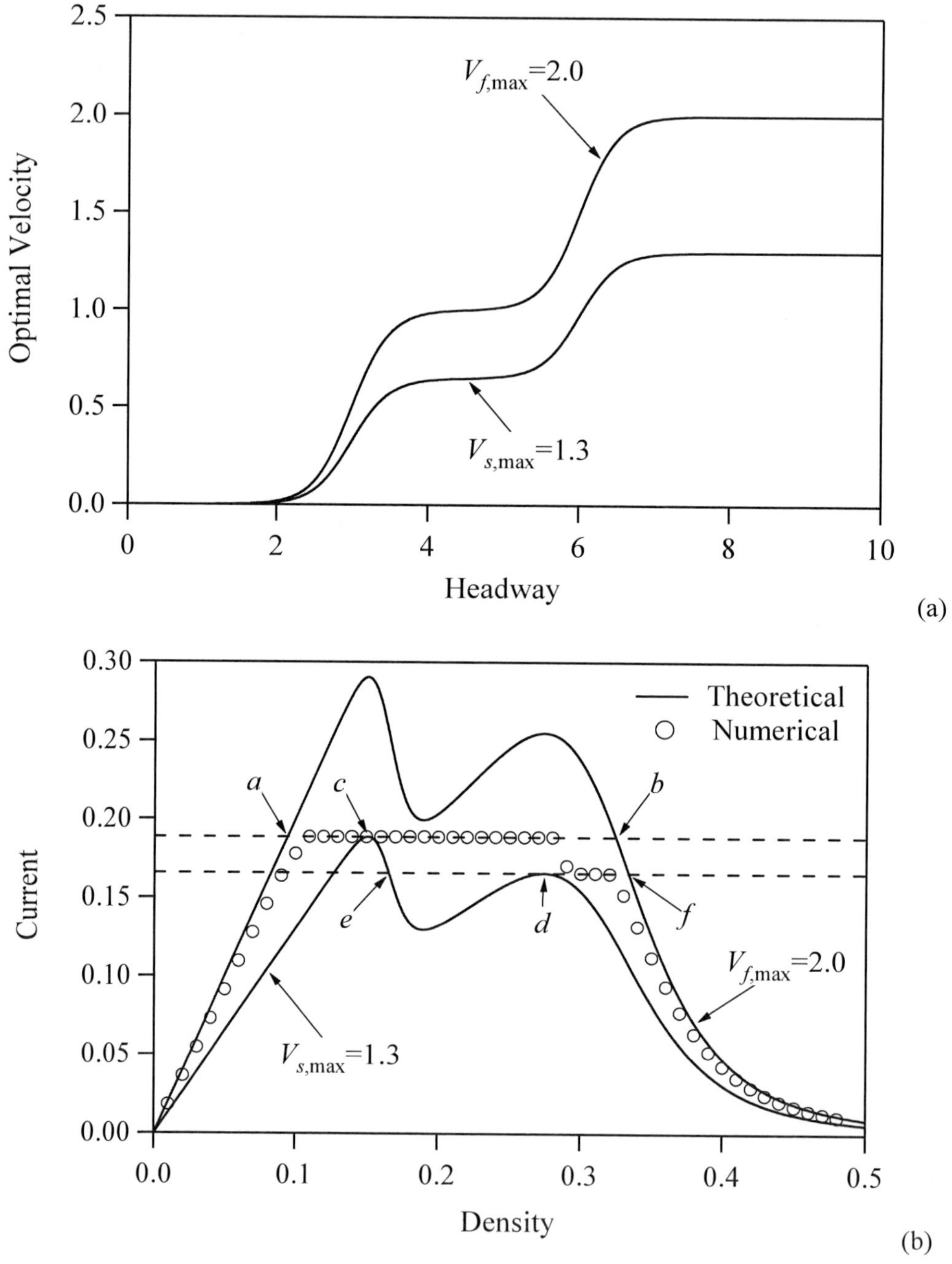

Figure 27. (a) The solid curves represent the optimal velocity functions. (b) Plots of traffic currents against density. The open circle indicates the traffic current obtained by simulation. For comparison, the theoretical currents for two optimal velocity functions without traffic jams are shown by two solid lines.

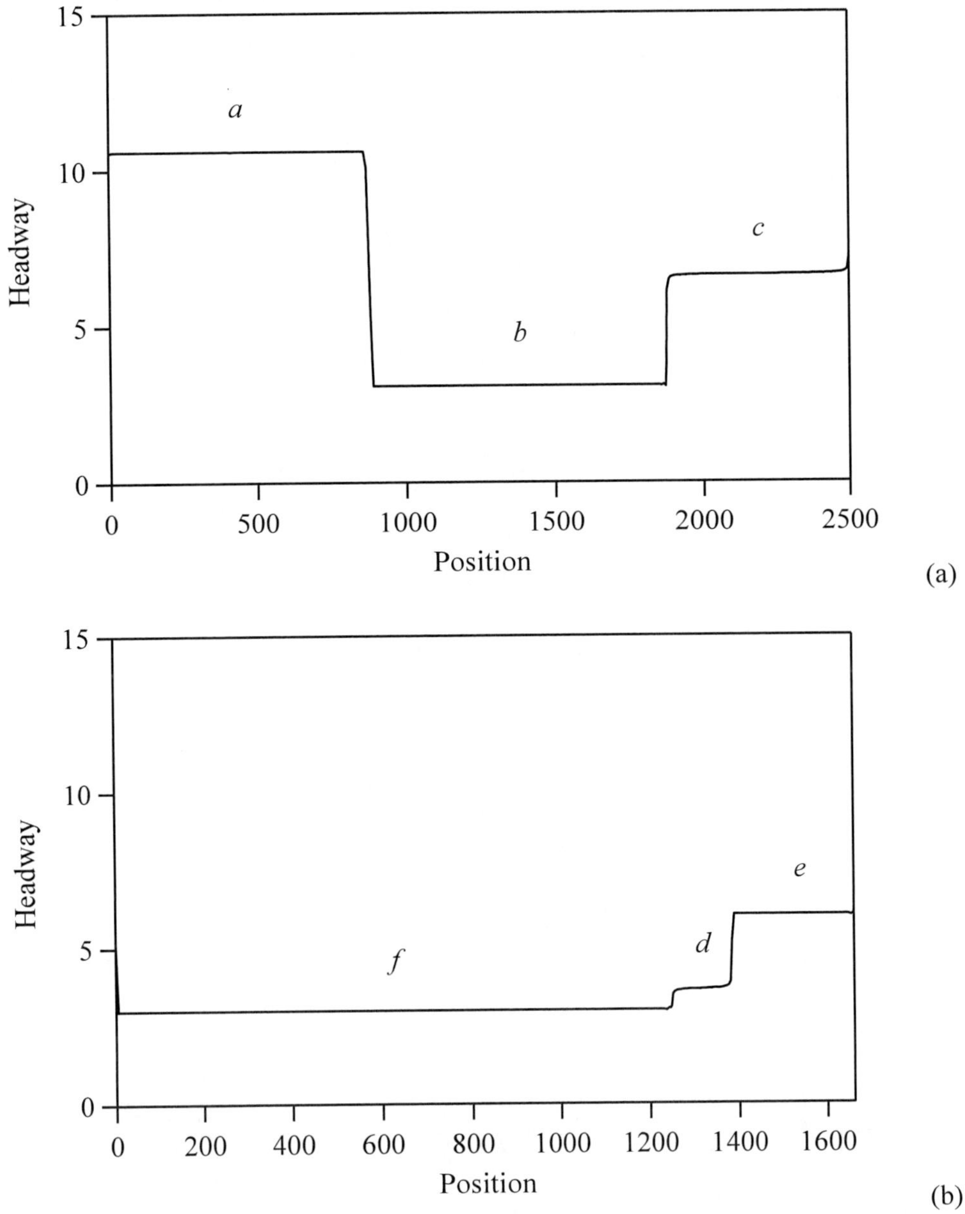

Figure 28. (a) Headway profile at $\rho = 0.20$. The densities (inverse of headway) before and after the discontinuous front are consistent with those at points *a* and *b* in Figure 27(b). (b) Headway profile at $\rho = 0.30$. The density at the section of normal velocity is given by that at point *f* in Figure 27(b).

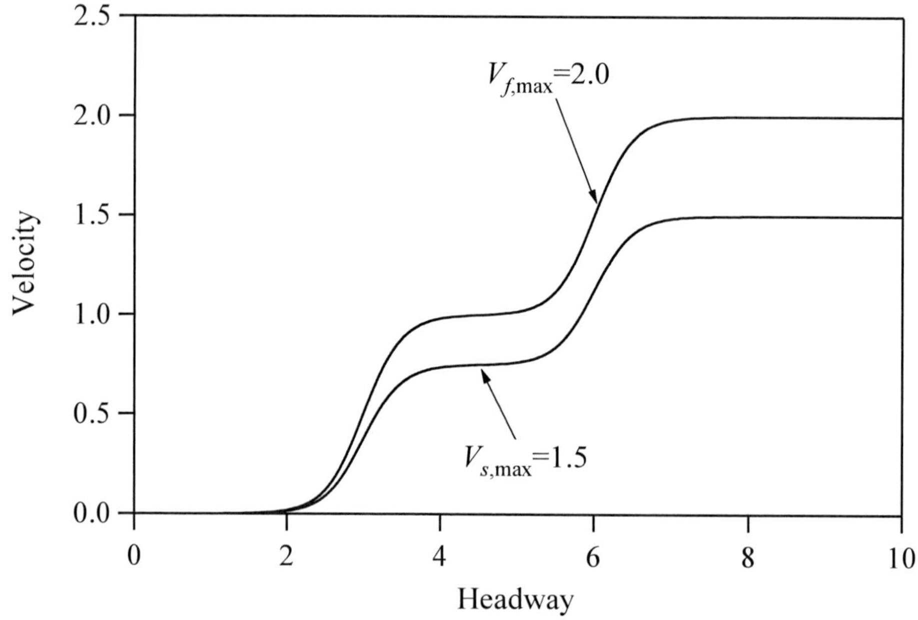

(a)

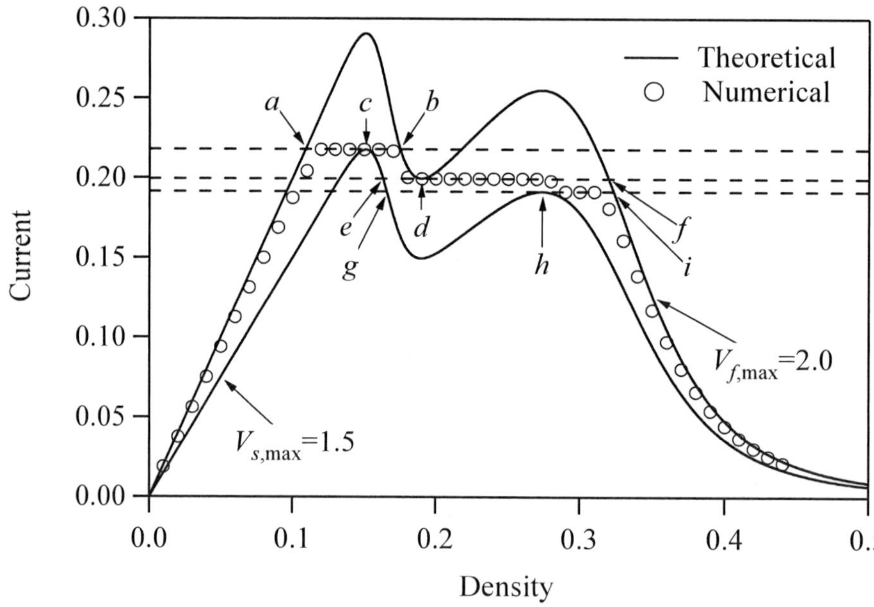

(b)

Figure 29. (a) The solid curves represent the optimal velocity functions. (b) Plots of traffic currents against density. The open circle indicates the traffic current obtained by simulation. Two solid curves indicate the theoretical currents for two optimal velocity functions.

We study the traffic behavior for such case that both optimal velocities at the sections of normal velocity and slowdown have the two-stage function with two turning points. The optimal velocity functions are given, respectively, by Eq.(52) with $v_{f,\max} = 2.0$, $\alpha = 2.0$, $x_{f,c,1} = 3.0$, and $x_{f,c,2} = 6.0$ for the section of normal velocity and Eq.(54) with

$v_{s,\max} = 1.3$, $\beta = 2.0$, $x_{f,c,1} = 3.0$, and $x_{s,c,1} = 6.0$ for the section of slowdown. The above optimal velocity functions are shown by solid curves in Figure 27(a). Figure 27(b) shows the plots of traffic currents against density. The open circle indicates the traffic current obtained by simulation. The traffic current is obtained by averaging the current over $t = 9000 - 10000$. For comparison, the theoretical currents for two optimal velocity functions without traffic jams are shown by two solid lines. The current increases linearly with density along the theoretical curve of $v_{f,\max} = 2.0$. At $\rho = 0.11$, the current saturates and keeps a constant value which agrees with the maximal current at point *c* of theoretical current curve of $v_{s,\max} = 1.3$. When density is higher than $\rho = 0.28$, the current decreases with increasing density. If density is higher than $\rho = 0.30$, the current again saturates and keeps a constant value until $\rho = 0.32$. The second saturated value of current is consistent with the current at point *d* of second local maximum of the theoretical current curve of $v_{s,\max} = 1.3$. Then, the current decreases with increasing density.

Figure 28(a) shows the headway profile at $\rho = 0.20$. In the region where the current saturates first, the discontinuous front appears in the section of normal velocity. The traffic jam occurs just before the section of slowdown and ends at the discontinuous front. The densities (inverse of headway) before and after the discontinuous front are consistent with those at points *a* and *b* in Figure 27(b). The density at the section of slowdown agrees with point *c* of the maximal current of $v_{s,\max} = 1.3$. The velocities at regions *a*, *b*, and *c* are given, respectively, by the optimal velocities at densities *a*, *b*, and *c*. Figure 28(b) shows the headway profile at $\rho = 0.30$. In the region where the current saturates second, the discontinuous front does not appear in the section of normal velocity but occurs in the section of slowdown. The traffic jam occurs over the whole section of normal velocity. The density at the section of normal velocity is given by that at point *f* in Figure 27(b). We note that the discontinuous front appears within the section of slowdown. The densities (inverse of headway) before and after the discontinuous front are consistent with those at points *e* and *d* in Figure 27(b). Thus, the position which the discontinuous front appears is definitely different with those in Figures 24 and 26.

We study the traffic behavior when the first local maximum of theoretical current curve at the slowdown section is higher than the local minimum of theoretical current curve at the section of normal velocity and the second local maximum of theoretical current curve at the slowdown section is lower than the local minimum of theoretical current curve at the section of normal velocity. The optimal velocity functions are given, respectively, by Eq.(52) with $v_{f,\max} = 2.0$, $\alpha = 2.0$, $x_{f,c,1} = 3.0$, and $x_{f,c,2} = 6.0$ for the section of normal velocity and Eq.(54) with $v_{s,\max} = 1.5$, $\beta = 2.0$, $x_{f,c,1} = 3.0$, and $x_{s,c,1} = 6.0$ for the slowdown section. The values of parameters are the same as those of Figure 27 except for $v_{s,\max} = 1.5$. The above optimal velocity functions are shown by solid curves in Figure 29(a). Figure 29(b) shows the plots of traffic currents against density. The open circle indicates the traffic current obtained by simulation. Two solid curves indicate the theoretical currents for two optimal velocity functions. The current increases linearly with density along the theoretical curve of

$v_{f,\max} = 2.0$. At $\rho = 0.12$, the current saturates and keeps a constant value which agrees with the maximal current at point *c* of theoretical current curve of $v_{s,\max} = 1.5$. When density is higher than $\rho = 0.17$, the current decreases with increasing density. If density is higher than $\rho = 0.18$, the current again saturates and keeps a constant value until $\rho = 0.27$. The second saturated value of current is consistent with the current at point *d* of the local minimum of the theoretical current curve of $v_{f,\max} = 2.0$. Then, the current decreases with increasing density. When density is higher than $\rho = 0.29$, the current again saturates and keeps a constant value until $\rho = 0.31$. The third saturated value of current is consistent with the current at point *h* of the local maximum of the theoretical current curve of $v_{s,\max} = 1.5$. Then, the current decreases with increasing density. Thus, the current saturates three times.

Figure 30(a) shows the headway profile at $\rho = 0.14$. In the region where the current saturates first, the discontinuous front appears in the section of normal velocity. The traffic jam occurs just before the section of slowdown and ends at the discontinuous front. The densities (inverse of headway) before and after the discontinuous front are consistent with those at points *a* and *b* in Figure 29(b). The density at the slowdown section agrees with point *c* of the maximal current of $v_{s,\max} = 1.5$. The velocities at regions *a*, *b*, and *c* are given, respectively, by the optimal velocities at densities *a*, *b*, and *c*. Figure 30(b) shows the headway profile at $\rho = 0.23$. In the region where the current saturates secondly, the discontinuous front appears in the section of normal velocity. The traffic jam occurs at two stages. The first jam begins just before the section of slowdown and ends at the discontinuous front. The second jam begins just before the discontinuous front. The densities (inverse of headway) before and after the discontinuous front are consistent with those at points *d* and *f* in Figure 29(b). The density at the slowdown section agrees with point *e* of the theoretical current curve of $v_{s,\max} = 1.5$. The velocities at regions *d*, *e*, and *f* are given, respectively, by the optimal velocities at densities *d*, *e*, and *f*.

Figure 30(c) shows the headway profile at $\rho = 0.30$. In the region where the current saturates thirdly, the discontinuous front does not appear in the section of normal velocity but occurs in the section of slowdown. The traffic jam occurs over the whole section of normal velocity. The density at the section of normal velocity is given by that at point *i* in Figure 29(b). We note that the discontinuous front appears within the section of slowdown. The densities (inverse of headway) before and after the discontinuous front are consistent with those at points *h* and *g* in Figure 29(b). Thus, three kinds of discontinuous fronts appears, corresponding to three kinds of saturations of current.

We study the traffic behavior when the first and second local maxima of theoretical current curve at the slowdown section are higher than the local minimum of theoretical current curve at section of normal velocity. The optimal velocity functions are given, respectively, by Eq.(52) with $v_{f,\max} = 2.0$, $\alpha = 2.0$, $x_{f,c,1} = 3.0$, and $x_{f,c,2} = 6.0$ for the section of normal velocity and Eq.(54) with $v_{s,\max} = 1.7$, $\beta = 2.0$, $x_{f,c,1} = 3.0$, and

$x_{s,c,1} = 6.0$ for the section of slowdown. The values of parameters are the same as those of Figure 29 except for $v_{s,\max} = 1.7$. The above optimal velocity functions are shown by solid curves in Figure 31(a). Figure 31(b) shows the plots of traffic currents against density. The open circle indicates the traffic current obtained by simulation. Two solid curves indicate the theoretical currents for two optimal velocity functions. The current increases linearly with density along the theoretical curve of $v_{f,\max} = 2.0$. At $\rho = 0.13$, the current saturates and keeps a constant value which agrees with the maximal current at point *c* of theoretical current curve of $v_{s,\max} = 1.7$. When density is higher than $\rho = 0.16$, the current decreases with increasing density. If density is higher than $\rho = 0.19$, the current again saturates and keeps a constant value until $\rho = 0.20$. The second saturated value of current is consistent with the current at point *d* of the local minimum of the theoretical current curve of $v_{f,\max} = 2.0$. Then, the current increases with density. When density is higher than $\rho = 0.23$, the current again saturates and keeps a constant value until $\rho = 0.30$. The third saturated value of current is consistent with the current at point *h* of the local maximum of the theoretical current curve of $v_{s,\max} = 1.7$. Then, the current decreases with increasing density. Thus, the current saturates three times.

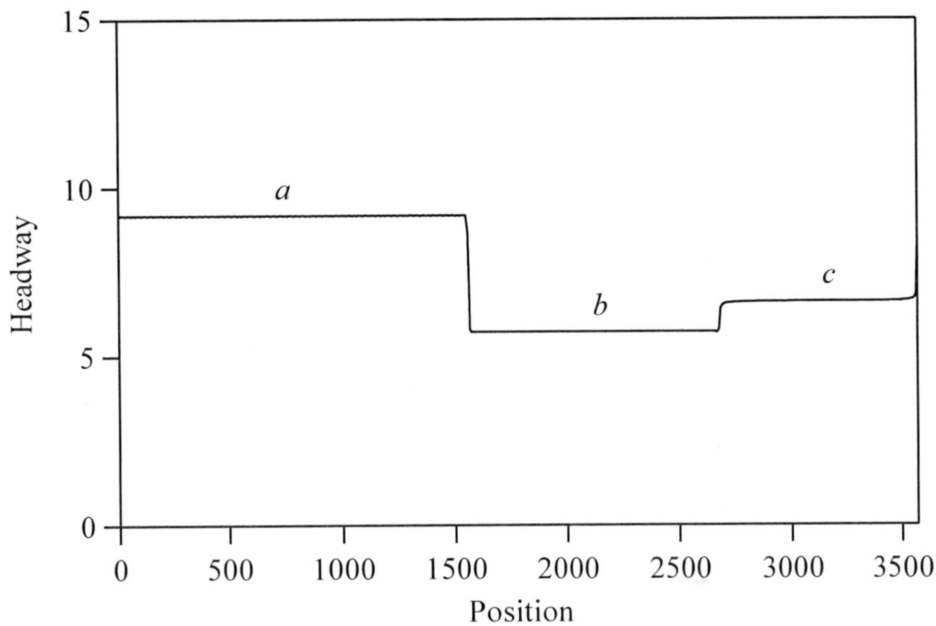

(a)

Figure 30. (Continued)

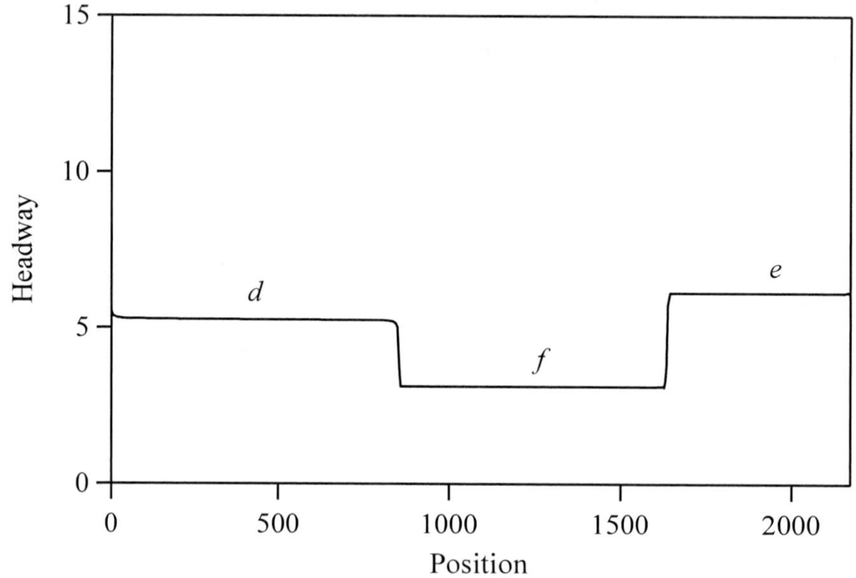

(b)

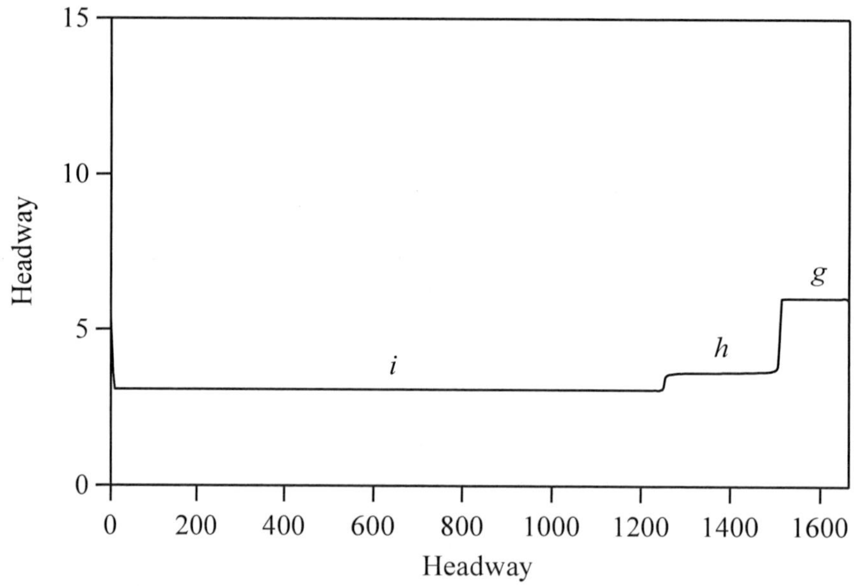

(c)

Figure 30. (a) Headway profile at $\rho = 0.14$. The densities (inverse of headway) before and after the discontinuous front are consistent with those at points *a* and *b* in Figure 29(b). The density at the slowdown section agrees with point *c* of the maximal current of $v_{s,\max} = 1.5$. (b) Headway profile at $\rho = 0.23$. The densities (inverse of headway) before and after the discontinuous front are consistent with those at points *d* and *f* in Figure 29(b). The density at the slowdown section agrees with point *e* of the theoretical current curve of $v_{s,\max} = 1.5$. (c) Headway profile at $\rho = 0.30$. The density at the section of normal velocity is given by that at point *i* in Figure 29(b). The densities (inverse of headway) before and after the discontinuous front are consistent with those at points *h* and *g* in Figure 29(b).

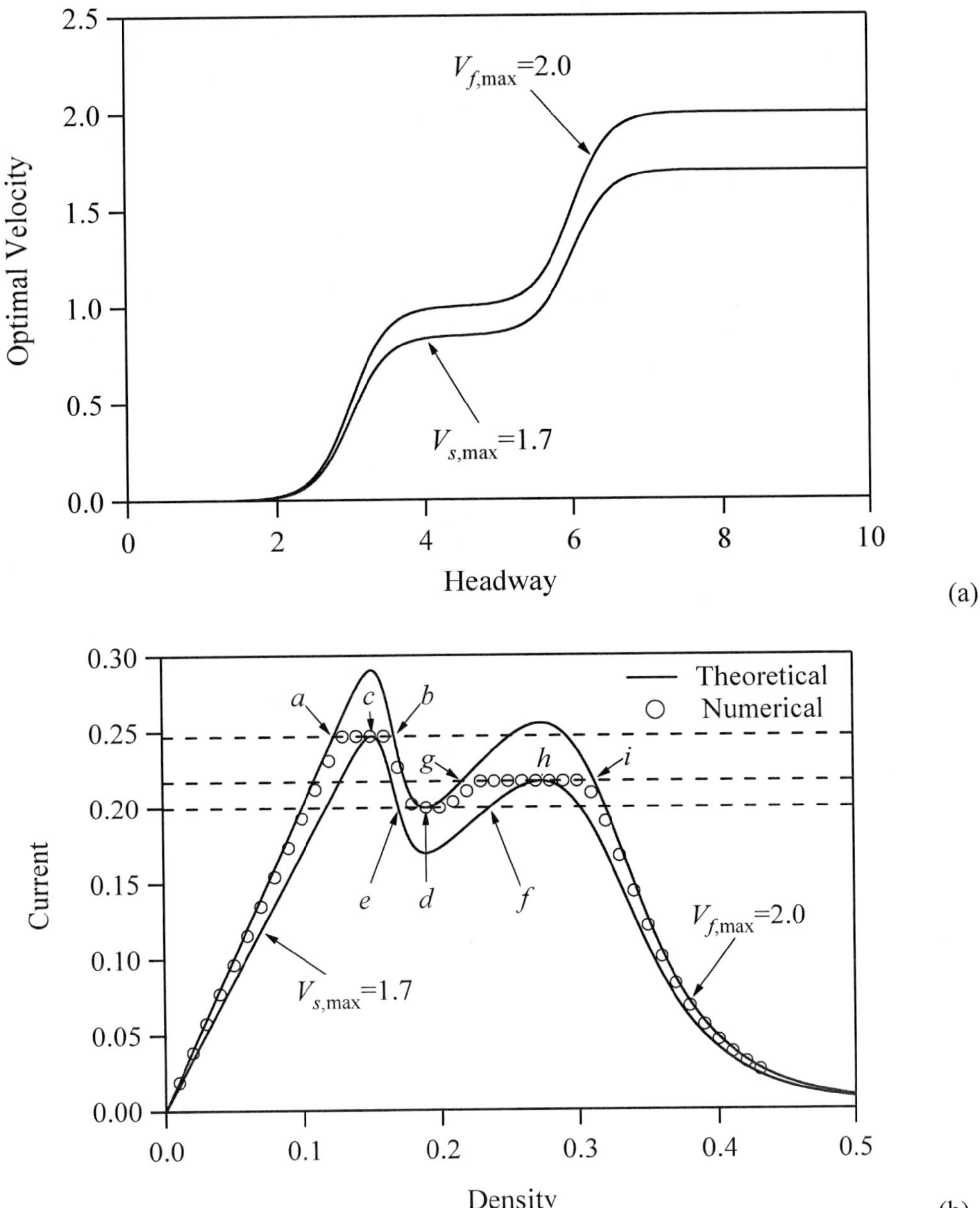

Figure 31. (a) The solid curves represent the optimal velocity functions. (b) Plots of traffic currents against density. The open circle indicates the traffic current obtained by simulation. Two solid curves indicate the theoretical currents for two optimal velocity functions.

Figure 32(a) shows the headway profile at $\rho = 0.15$. In the region where the current saturates first, the discontinuous front appears in the section of normal velocity. The traffic jam occurs just before the section of slowdown and ends at the discontinuous front. The densities (inverse of headway) before and after the discontinuous front are consistent with those at points a and b in Figure 31(b). The density at the slowdown section agrees with point

c of the maximal current of $v_{s,\max} = 1.7$. The velocities at regions a, b, and c are given, respectively, by the optimal velocities at densities a, b, and c. The property of discontinuous front in Figure 32(a) agrees with that in Figure 30(a). Figure 32(b) shows the headway profile at $\rho = 0.19$. In the region where the current saturates secondly, the discontinuous front appears in the slowdown section. One of two jams appears in the slowdown section and the other in the section of normal velocity. The first jam begins just before the discontinuous front in the slowdown section. The densities (inverse of headway) before and after the discontinuous front are consistent with those at points f and e in Figure 31(b). The density at the section of normal velocity agrees with point d of the theoretical current curve of $v_{f,\max} = 2.0$. The velocities at regions d, e, and f are given, respectively, by the optimal velocities at densities d, e, and f.

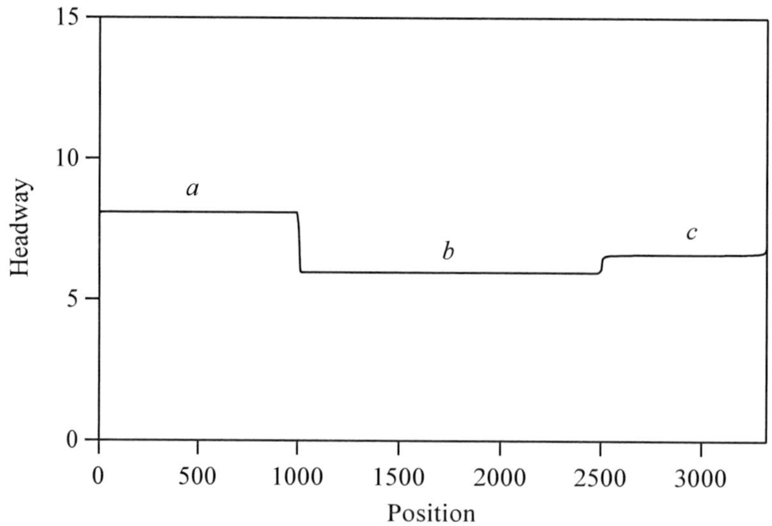

(a)

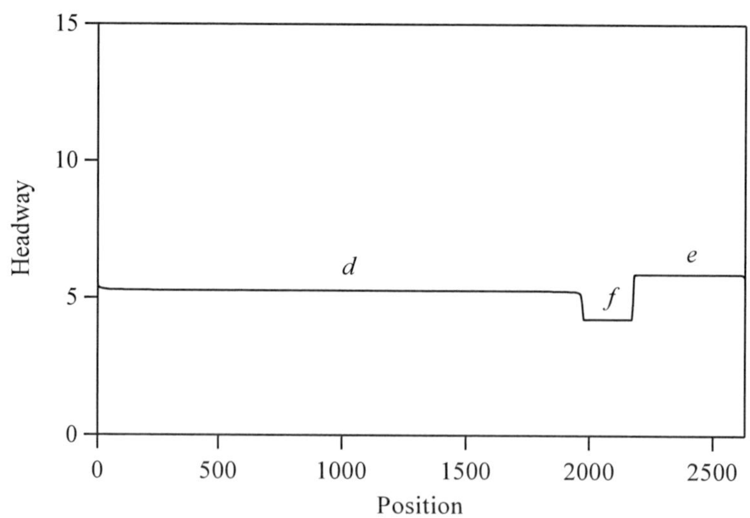

(b)

Figure 32. (Continued)

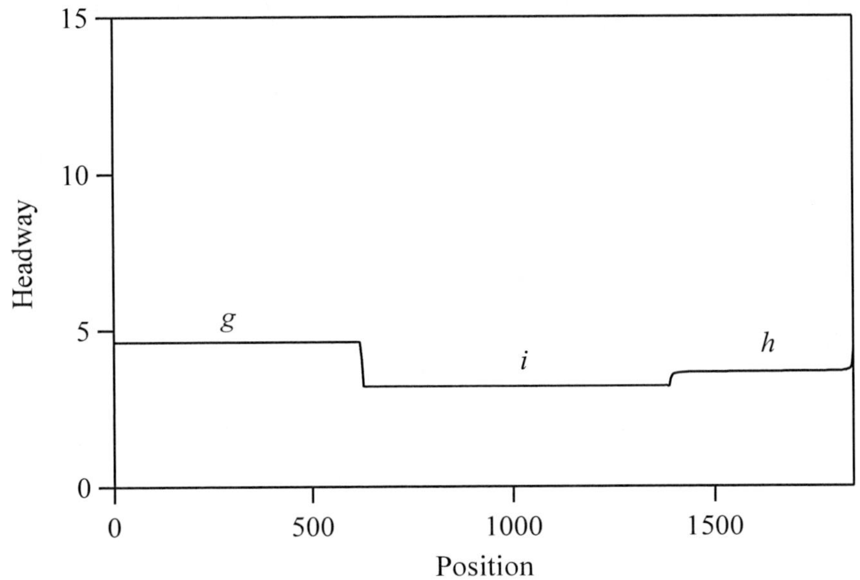

(c)

Figure 32. (a) Headway profile at $\rho = 0.15$. The densities (inverse of headway) before and after the discontinuous front are consistent with those at points *a* and *b* in Figure 31(b). The density at the slowdown section agrees with point *c* of the maximal current of $v_{s,\max} = 1.7$. (b) Headway profile at $\rho = 0.19$. The densities (inverse of headway) before and after the discontinuous front are consistent with those at points *f* and *e* in Figure 31(b). The density at the section of normal velocity agrees with point *d* of the theoretical current curve of $v_{f,\max} = 2.0$. (c) Headway profile at $\rho = 0.27$. The densities (inverse of headway) before and after the discontinuous front are consistent with those at points *g* and *i* in Figure 31(b). The density at the slowdown section agrees with that at point *h* in Figure 31(b).

Figure 32(c) shows the headway profile at $\rho = 0.27$. In the region where the current saturates thirdly, the discontinuous front again appears in the section of normal velocity. The densities (inverse of headway) before and after the discontinuous front are consistent with those at points *g* and *i* in Figure 31(b). The density at the slowdown section agrees with that at point *h* in Figure 31(b). Thus, three kinds of discontinuous fronts appear, corresponding to three kinds of saturations of current.

We study the traffic behavior for following optimal velocity functions: conventional function (Eq. (53)) with $v_{f,\max} = 2.0$, $\alpha = 2.0$, and $x_{f,c} = 3.5$ for the section of normal velocity and Eq.(54) with $v_{s,\max} = 1.6$, $\beta = 2.0$, $x_{s,c,1} = 3.0$ and $x_{s,c,2} = 6.0$ for the section of slowdown. The above optimal velocity functions are shown by solid curves in Figure 33(a). Figure 33(b) shows the plots of traffic currents against density. The open circle indicates the traffic current obtained by simulation. The theoretical currents for two optimal velocity functions without traffic jams are shown by two solid lines. The current increases linearly with density along the theoretical curve of $v_{f,\max} = 2.0$. At $\rho = 0.13$, the current saturates and keeps a constant value which agrees with the first local maximum of theoretical current curve of $v_{s,\max} = 1.6$. When density is higher than $\rho = 0.27$, the current decreases with increasing density. If density is higher than $\rho = 0.29$, the current again saturates until $\rho = 0.31$ and keeps a constant value which agrees with the second local maximum (point *d*) of $v_{s,\max} = 1.6$. When density is higher than $\rho = 0.31$, the current again decreases with increasing density.

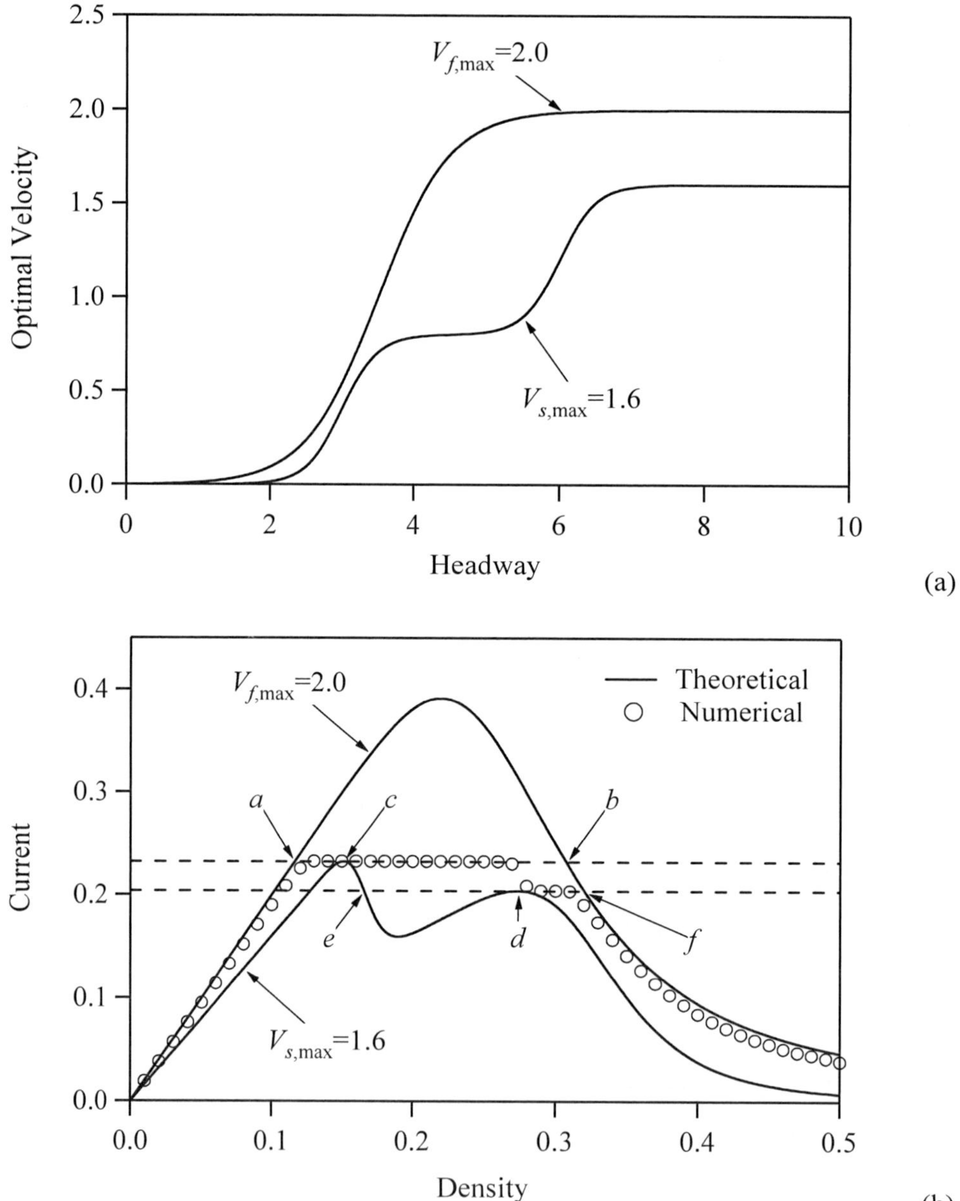

Figure 33. (a) The solid curves represent the optimal velocity functions. (b) Plots of traffic currents against density. The open circle indicates the traffic current obtained by simulation. The theoretical currents for two optimal velocity functions without traffic jams are shown by two solid lines.

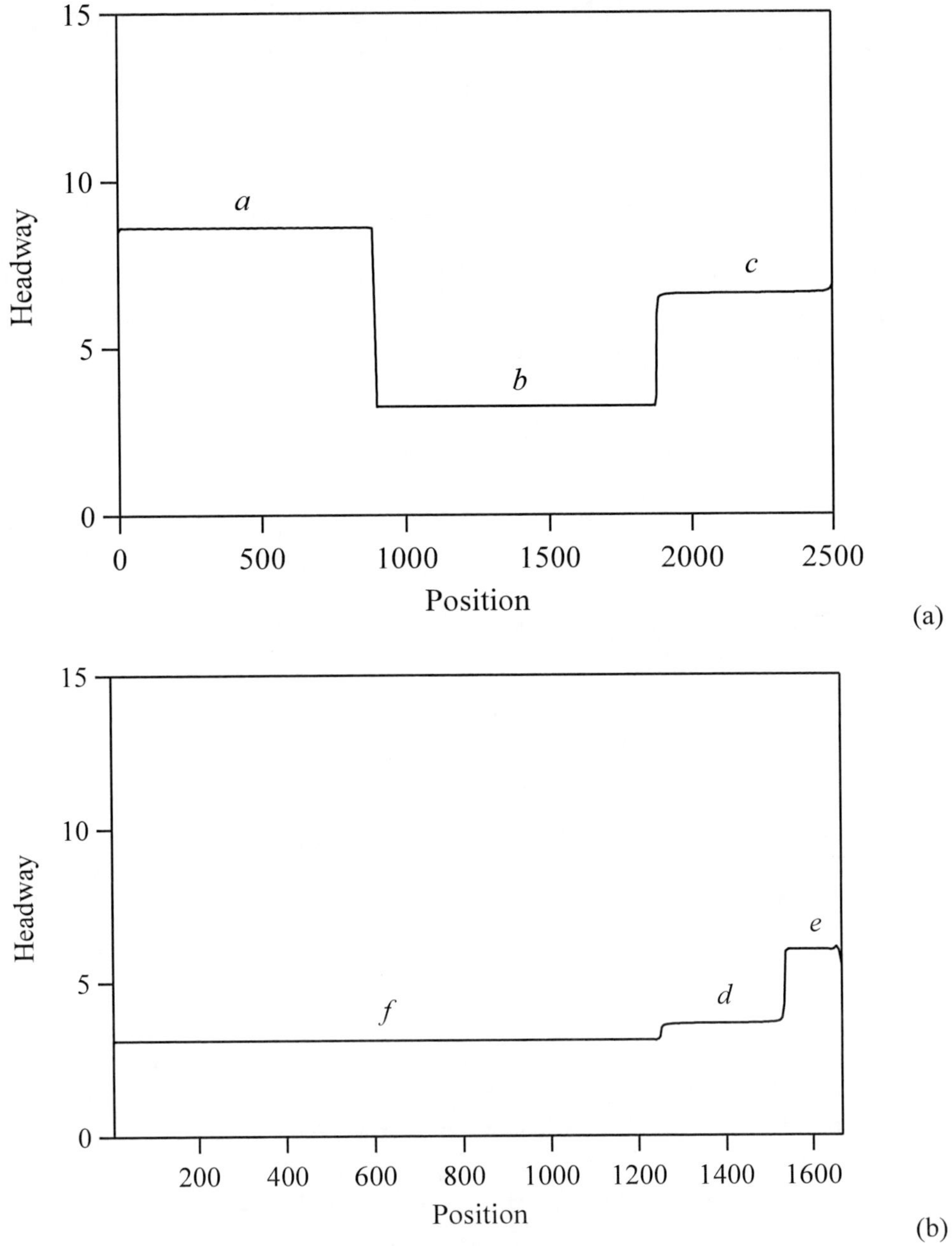

Figure 34. (a) Headway profile at $\rho = 0.20$. The densities (inverse of headway) before and after the discontinuous front are consistent with those at points *a* and *b* in Figure 33(b). The density at the section of slowdown agrees with point *c* of the first local maximal current of $v_{s,\max} = 1.6$. (b) Headway profile at $\rho = 0.30$. The densities (inverse of headway) before and after the discontinuous front are consistent with those at points *d* and *e* in Figure 33(b). The density at the section of normal velocity agrees with point *f* of the theoretical current of $v_{f,\max} = 2.0$.

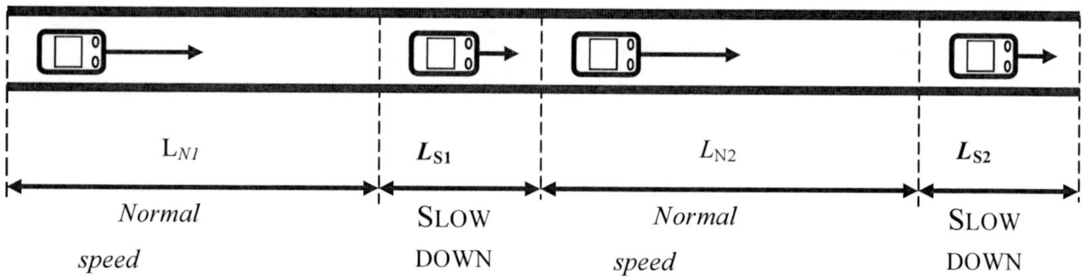

Figure 35. Schematic illustration of the traffic model for the single-lane highway with two slowdown sections. The lengths of slowdown sections are L_{S1} and L_{S2}. The lengths of normal-speed sections are L_{N1} and L_{N2} where the road length is $L = L_{N1} + L_{N2} + L_{S1} + L_{S2}$.

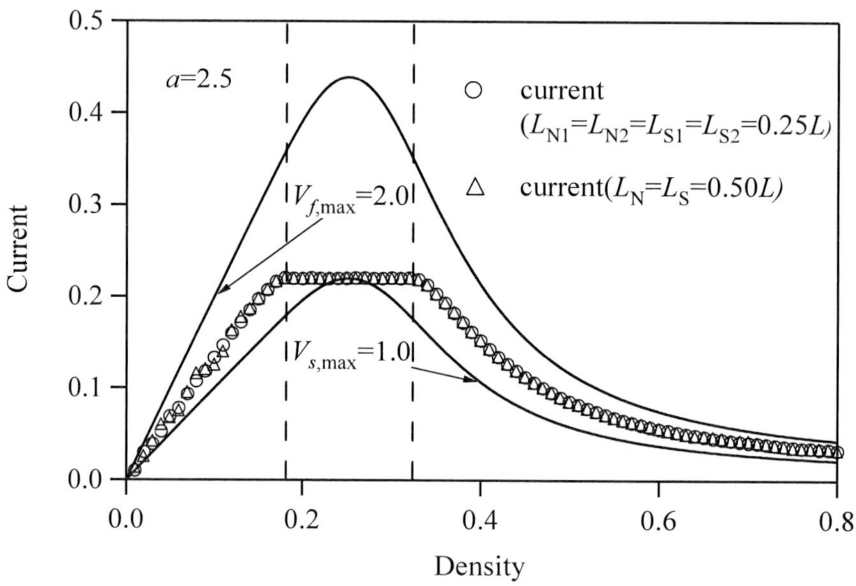

Figure 36. Plots of traffic currents against density. The open circle indicates the traffic current obtained by simulation. For comparison, the values of traffic current obtained by simulation for a single slowdown section ($L_N = L_S = L/2$) are indicated by open triangles.

Figure 34(a) shows the headway profile at $\rho = 0.20$. In the region where the current saturates firstly, the discontinuous front appears in the section of normal velocity. The traffic jam occurs just before the section of slowdown and ends at the discontinuous front. The densities (inverse of headway) before and after the discontinuous front are consistent with those at points *a* and *b* in Figure 33(b). The density at the section of slowdown agrees with point *c* of the first local maximal current of $v_{s,\max} = 1.6$. The velocities at regions *a*, *b*, and *c* are given, respectively, by the optimal velocities at densities *a*, *b*, and *c*. Figure 34(b) shows the headway profile at $\rho = 0.30$. In the region where the current saturates secondly, the discontinuous front appears in the slowdown section. The first jam occurs just before the discontinuous front. The densities (inverse of headway) before and after the discontinuous front are consistent with those at points *d* and *e* in Figure 33(b). The second jam occurs over the whole section of normal velocity. The density at the section of normal velocity agrees with point *f* of the theoretical current of $v_{f,\max} = 2.0$. The velocities at regions *d*, *e*, and *f* are given, respectively, by the optimal velocities at densities *d*, *e*, and *f*. Thus, the traffic jam induced by slowdown depends highly on the maximal velocities of slowdown section and the turning points of optimal velocity function.

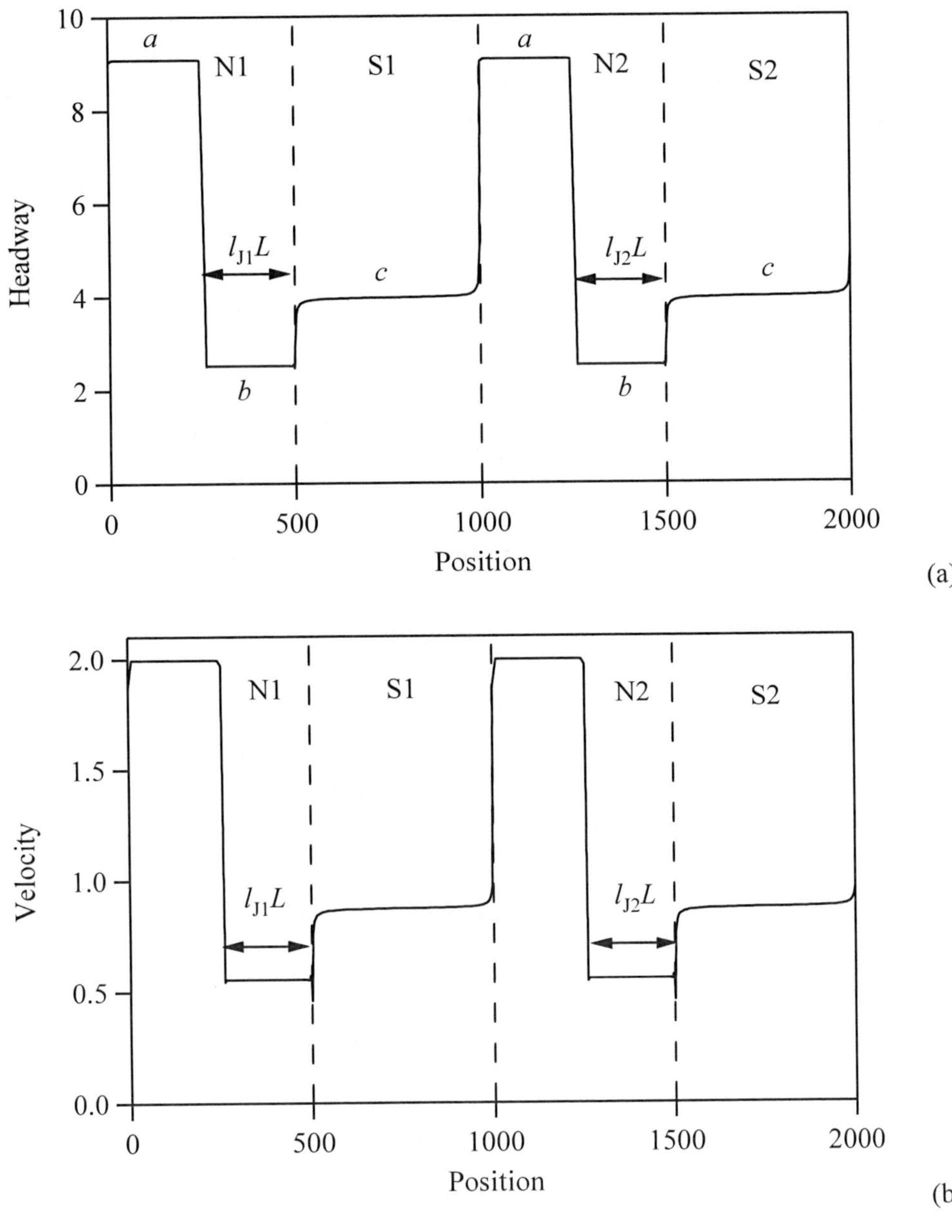

Figure 37. (a) and (b) Headway and velocity profiles at $\rho = 0.25$ and $t = 50000$ where $L_{N1} = L_{N2} = L_{S1} = L_{S2} = L/4$, $v_{f,\max} = 2.0$, $v_{s1,\max} = v_{s2,\max} = 1.0$, and $a = 2.5$.

We find that the characteristics of discontinuous front depends highly on the extreme values of theoretical current determined by the optimal velocity functions.

4.3. Jams on a Highway with Some Slowdown Sections

In real traffic, some sections of slowdown exist on a highway. When a few sections of distinct slowdowns exist on a highway, where, when, and how do traffic jams occur on a highway? The traffic jams induced by some slowdown sections have little been studied by using modern traffic models. We study the traffic states and jams induced by some slowdown sections. We clarify the dynamical states of traffic and the characteristic of traffic jams.

We consider the vehicular traffic flowing on the single-lane roadway with some sections of slowdown. Vehicles are forced to slow down when they enter into a section of the slowdowns. Figure 35 shows the schematic illustration of the traffic model for the single-lane highway with two slowdown sections. The lengths of slowdown sections are L_{S1} and L_{S2}. The lengths of normal-speed sections are L_{N1} and L_{N2} where the road length is $L = L_{N1} + L_{N2} + L_{S1} + L_{S2}$. Vehicles move with low speed in the slowdown sections, while they move with the normal velocity except for the slowdown sections. In the region of normal-speed sections, the optimal velocity function of vehicles is given by

$$V(\Delta x_i) = \frac{v_{f,\max}}{2}\left[\tanh\alpha(\Delta x_i - x_{f,c}) + \tanh\alpha(x_{f,c})\right], \tag{55}$$

where $v_{f,\max}$ is the maximal velocity of vehicles on the roadway except for the slowdown sections and $x_{f,c}$ the position of turning point.

In the slowdown sections, vehicles move with the forced speed and the vehicular velocity should be lower than the speed limits of slowdown. When vehicles enter into the first section of slowdown, they obey the conventional optimal-velocity function with speed limit $v_{s1,\max}$:

$$V(\Delta x_i) = \frac{v_{s1,\max}}{2}\left[\tanh(\Delta x_i - x_{s,c}) + \tanh(x_{s,c})\right], \tag{56}$$

where $v_{s1,\max}$ is the speed limit of first slowdown section, $v_{s1,\max} < v_{f,\max}$, and $x_{s,c}$ the position of turning point.

Similarly, when vehicles enter into the second slowdown section, they obey the following optimal-velocity function with speed limit $v_{s2,\max}$:

$$V(\Delta x_i) = \frac{v_{s2,\max}}{2}\left[\tanh(\Delta x_i - x_{s,c}) + \tanh(x_{s,c})\right], \tag{57}$$

where $v_{s2,\max}$ is the speed limit of second slowdown section and $v_{s2,\max} < v_{f,\max}$.

Thus, when the density of vehicles is very low, vehicles move at the maximal velocity $v_{f,\max}$ except for the sections of slowdown, while they move at a speed lower than the forced speed limits $v_{s1,\max}$ or $v_{s2,\max}$ in the slowdown sections.

We carry out simulation by varying the initial headway, slowdown-section lengths L_{S1} and L_{S2}, and slowdown's velocities $v_{s1,\max}$ and $v_{s2,\max}$ for 500 vehicles, sensitivity $a = 2.5$ and maximal velocity $v_{f,\max} = 2.0$. Initially, we put all vehicles on the single-lane highway with the same headway Δx_{int}. The density ρ is given by the inverse of the headway. The length L of highway varies with the initial headway.

First, we study the traffic behavior for such case that the length and speed limit of the first slowdown section agree with those of the second slowdown section: $L_{N1} = L_{N2} = L_{S1} = L_{S2} = L/4$ and $v_{s1,\max} = v_{s2,\max} = 1.0$. Figure 36 shows the plots of traffic currents against density. The open circle indicates the traffic current obtained by simulation. The current increases linearly with density, then the current saturates and keeps a constant value which agrees with the maximal value of theoretical current curve of $v_{s1,\max} = v_{s2,\max} = 1.0$. When density is higher, the current decreases with increasing density.

For comparison, the values of traffic current obtained by simulation for a single slowdown section ($L_N = L_S = L/2$) are indicated by open triangles. The traffic current of $L_N = L_S = L/2$ is consistent with that of $L_{N1} = L_{N2} = L_{S1} = L_{S2} = L/4$. Also, the traffic currents for two optimal velocity functions without traffic jams are shown by two solid lines. The current is called as the theoretical current curve. It is given by $J = V(\Delta x)/\Delta x$ where Δx is the average value of headway and $V(\Delta x)$ is the optimal velocity. The case of no slowdown corresponds to the conventional traffic flow. When sensitivity is higher than critical value 2.0 in traffic flow on a highway with no sections of slowdown, no traffic jams occur and the current agrees with the theoretical current curve.

Figure 37(a) and (b) show, respectively, the headway and velocity profiles at $\rho = 0.25$ and $t = 50000$ where $L_{N1} = L_{N2} = L_{S1} = L_{S2} = L/4$, $v_{f,\max} = 2.0$, $v_{s1,\max} = v_{s2,\max} = 1.0$, and $a = 2.5$. In the region where the current saturates, two discontinuous fronts appear simultaneously in two sections of normal velocity. Two traffic jams occur simultaneously just before two sections of slowdown and end at the discontinuous fronts. Figure 38 shows the crossing points at the theoretical current curves through the horizontal line on the maximal current of $v_{s,\max} = 1.0$. The crossing points are indicated by the full circle and the full square. The maximal current of $v_{s,\max} = 1.0$ is indicated by the full triangular point. The densities (inverse of headway) before and after two discontinuous fronts are consistent with those at points a and b in Figure 38. The density at the sections of slowdown agree with that at point c of the maximal current of $v_{s1,\max} = v_{s2,\max} = 1.0$. The velocities at regions *a*, *b*, and *c* are given, respectively, by the optimal velocities at densities *a*, *b*, and *c*.

The lengths of two jams in Figure 37 increase with density. Figure 39 shows the plot of jam-length ratio (jam length/*L*) against density for $L_{N1} = L_{N2} = L_{S1} = L_{S2} = L/4$. Jam-length ratio l_{J1} at section *N*1 and jam-length ratio l_{J2} at section *N*2 are indicated by open circles and crosses respectively. Jam-length ratio l_{J1} at section *N*1 agrees with jam-length ratio l_{J2} at section *N*2. Two jams occur simultaneously when the density is higher than $\rho = 0.18$. This value of density is consistent with the starting point of current saturation in Figure 36 and is shown by the vertical dashed line. Two discontinuous fronts of jams disappear simultaneously when the density is higher than $\rho = 0.325$. This value of density agrees with the ending point of current saturation in Figure 36 and is shown by the dashed line with the arrow. The total account of two jam-length ratios is plotted by open triangles. The total length of two jams agrees with that obtained by the theoretical analysis which is shown by the solid line in Figure 39.

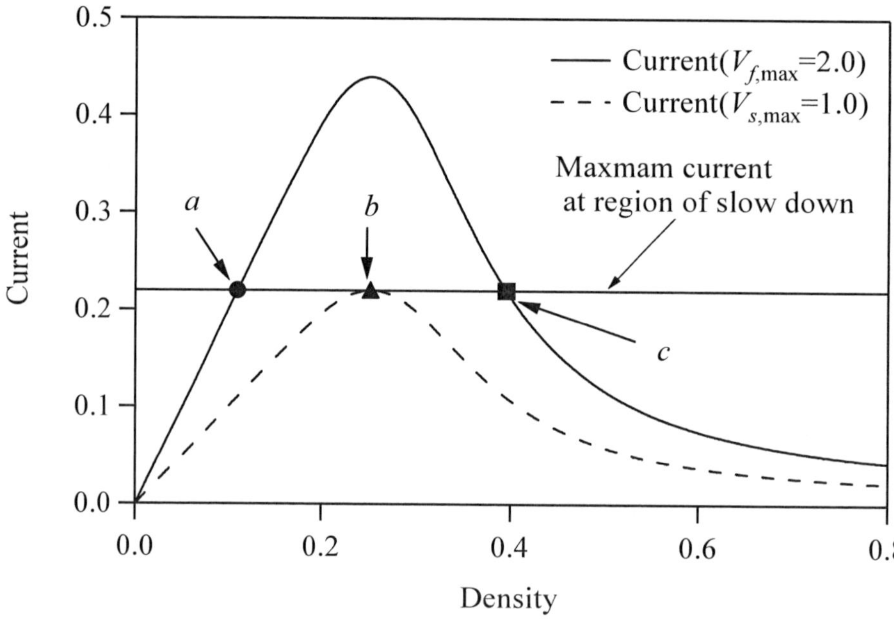

Figure 38. Crossing points at the theoretical current curves through the horizontal line on the maximal current of $v_{s,\max} = 1.0$. The crossing points are indicated by the full circle and the full square. The maximal current of $v_{s,\max} = 1.0$ is indicated by the full triangular point. The densities (inverse of headway) before and after two discontinuous fronts are consistent with those at points *a* and *b*.

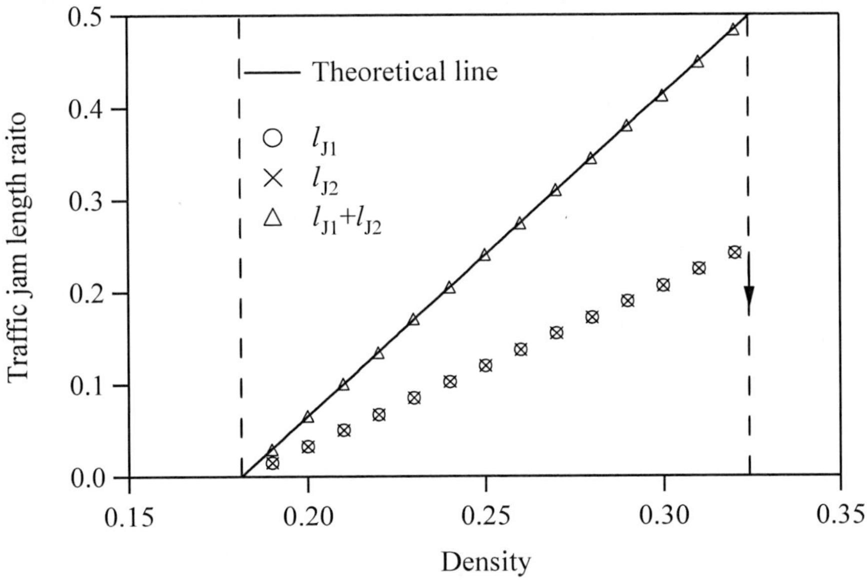

Figure 39. Plot of jam-length ratio (jam length/L) against density for $L_{N1} = L_{N2} = L_{S1} = L_{S2} = L/4$. Jam-length ratio l_{J1} at section $N1$ and jam-length ratio l_{J2} at section $N2$ are indicated by open circles and crosses respectively.

We study the traffic flow for such case that $L_{N1}(=0.35L) > L_{N2}(=0.15L)$, $L_{S1} = L_{S2} = 0.25L$, and $v_{s1,\max} = v_{s2,\max} = 1.0$. Figure 40(a) shows the headway profile at $\rho = 0.31$. Figure 40(b) shows the plot of jam-length ratio (jam length/L) against density. Jam-length ratio l_{J1} at section $N1$ and jam-length ratio l_{J2} at section $N2$ are indicated by open circles and crosses respectively. Jam-length ratio l_{J1} at section $N1$ is different from jam-length ratio l_{J2} at section $N2$. First, the jam occurs at the first section of normal speed before the first slowdown section when the density is higher than $\rho = 0.18$. This value of density is consistent with the starting point of current saturation in Figure 36 and is shown by the vertical dashed line. Then, the second jam occurs at the second section of normal speed before the second slowdown section. With increasing density, both jams grow. When the second jam reaches the boundary between the first slowdown section and the second normal-speed section, the second jam does not grow but the first jam continues to grow until the first jam reaches the boundary between the first normal-speed section and the second slowdown section. In Figure 40(b), the down arrow indicates the point at which the second jam reaches the boundary between S1 and N2. The second discontinuity stops to move when the density is higher than $\rho = 0.325$. This value of density is shown by the dashed line. The total account of two jam-length ratios is plotted by open triangles. The total length of two jams agrees with that obtained by the theoretical analysis which is shown by the solid line in Figure 40(b). Thus, the occurrence and disappearance of two jams are different each other when length

L_{N1} of the first normal-speed section is different from L_{N2} of the second normal-speed section.

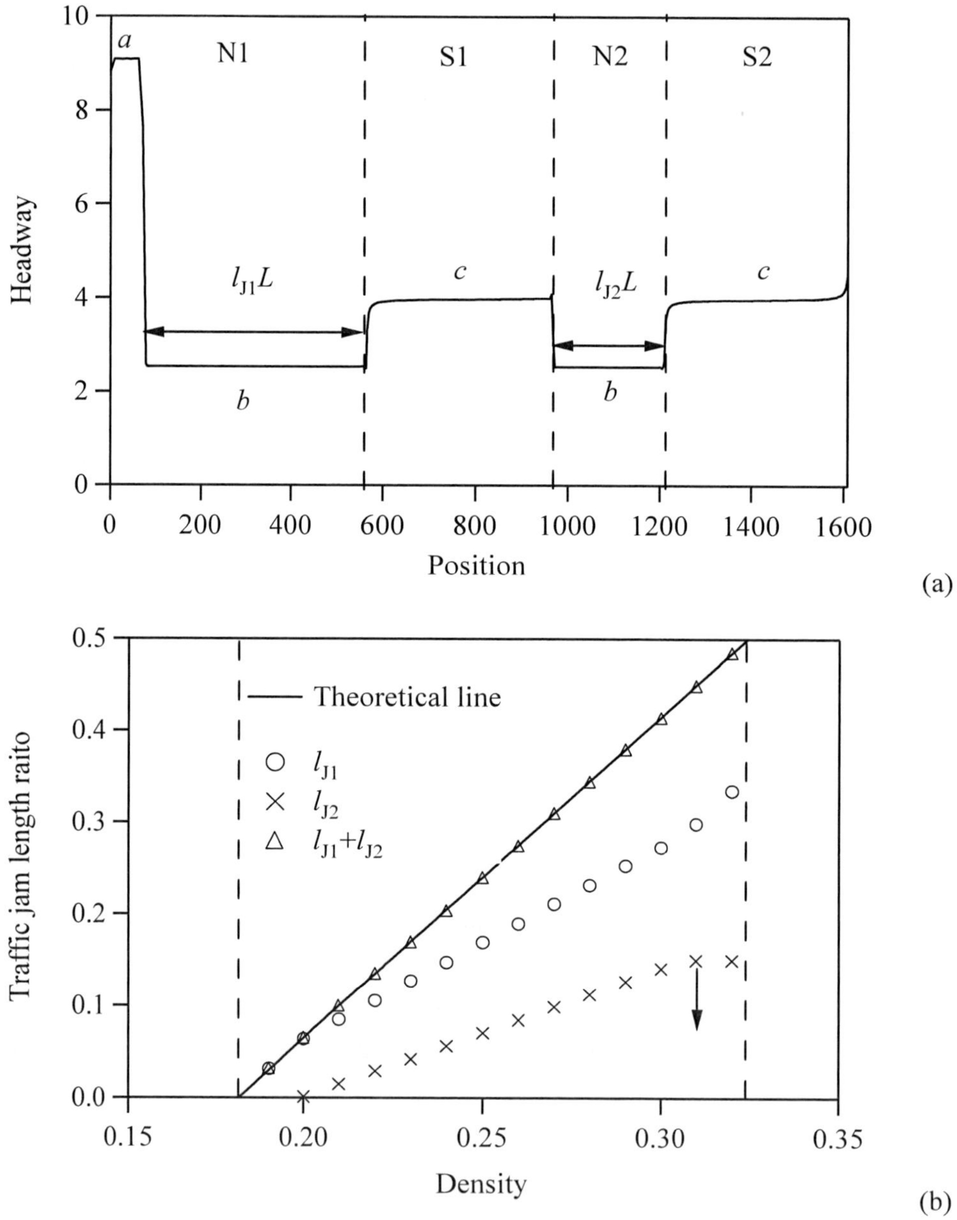

Figure 40. (a) Headway profile at $\rho = 0.31$. (b) Plot of jam-length ratio (jam length/L) against density. Jam-length ratio l_{J1} at section $N1$ and jam-length ratio l_{J2} at section $N2$ are indicated by open circles and crosses respectively.

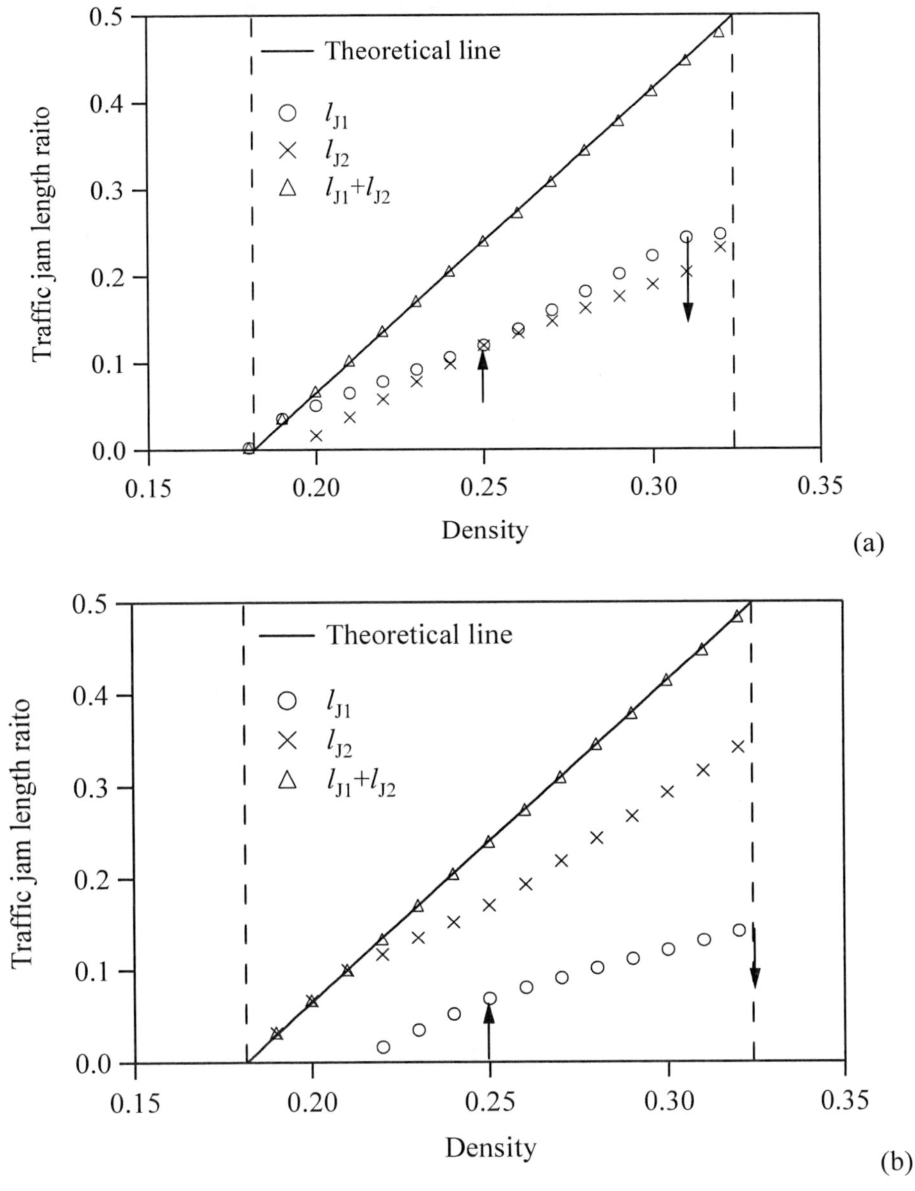

Figure 41. (a) Plots of jam-length ratio (jam length/L) against density for $L_{N1} = L_{N2} = 0.25L$, $L_{S1}(= 0.35L) > L_{S2}(= 0.15L)$, and $v_{s1,\max} = v_{s2,\max} = 1.0$. (b) Plots of jam-length ratio (jam length/L) against density for $L_{N1}(= 0.15L) < L_{N2}(= 0.35L)$, $L_{S1}(= 0.15L) < L_{S2}(= 0.35L)$, and $v_{s1,\max} = v_{s2,\max} = 1.0$.

We study the traffic flow for such case that $L_{N1} = L_{N2} = 0.25L$, $L_{S1}(= 0.35L) > L_{S2}(= 0.15L)$, and $v_{s1,\max} = v_{s2,\max} = 1.0$. Figure 41(a) shows the plots of jam-length ratio (jam length/L) against density. Jam-length ratio l_{J1} at section $N1$ is different from jam-length ratio l_{J2} at section *N*2. First, the jam occurs at first normal-speed section *N*1 when the density is higher than $\rho = 0.18$. Then, the second jam occurs at second normal-speed section *N*2. With increasing density, both jams grow. Length l_{J1} of the first jam agrees with length l_{J2} of the second jam at $\rho = 0.25$ indicated by the up arrow. At the density indicated by the down arrow, the first jam reaches the boundary between *N*1 and *S*2 and the first discontinuous front stops to move at this density. The total length of two jams agrees with that obtained by the theoretical analysis.

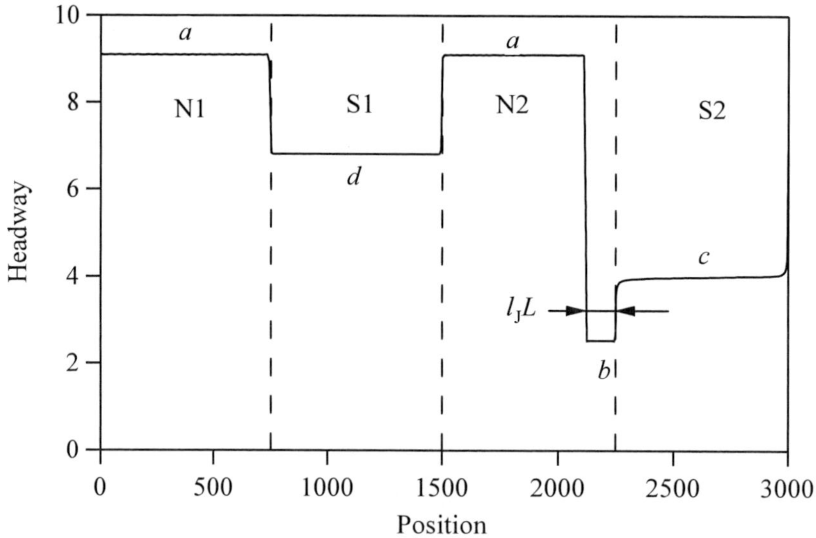

(a)

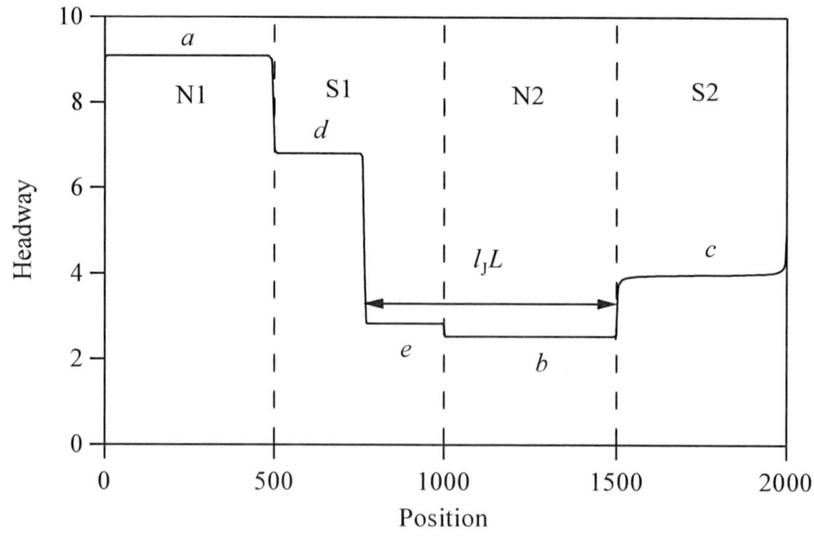

(b)

Figure 42. (Continued)

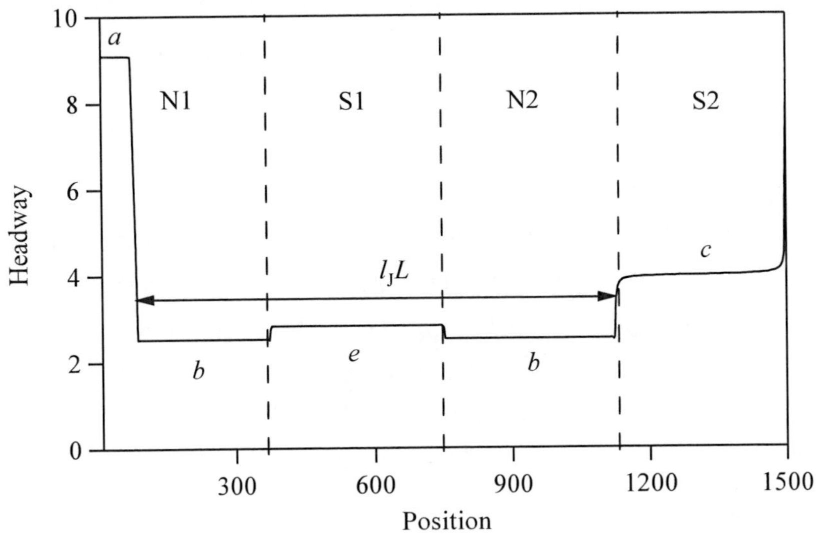

(c)

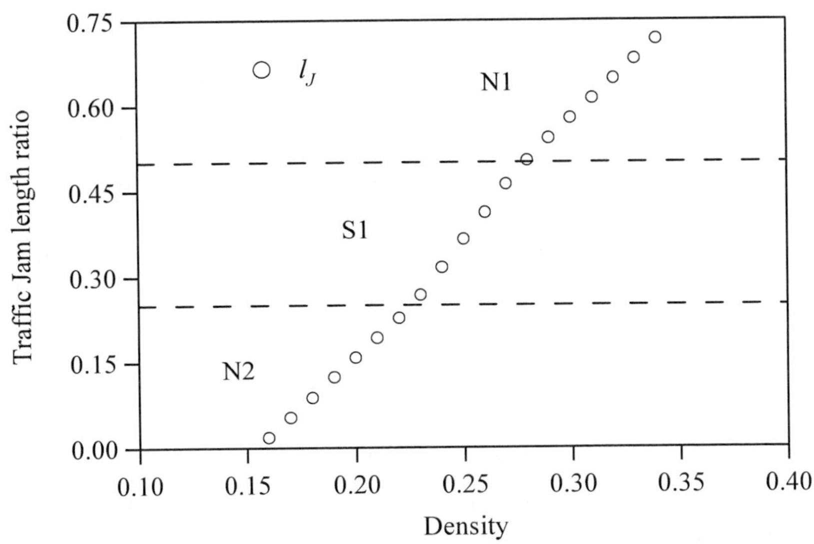

(d)

Figure 42. (a) Headway profiles at (a) $\rho = 0.16$, (b) $\rho = 0.25$, and (c) $\rho = 0.33$ for $L_{N1} = L_{N2} = L_{S1} = L_{S2} = L/4$ and $v_{s1,\max}(=1.5) > v_{s2,\max}(=1.0)$. (d) Plot of jam-length ratio (jam length/L) against density.

We study the traffic flow for such case that $L_{N1}(=0.15L) < L_{N2}(=0.35L)$, $L_{S1}(=0.15L) < L_{S2}(=0.35L)$, and $v_{s1,\max} = v_{s2,\max} = 1.0$. Figure 41(b) shows the plots of jam-length ratio (jam length/L) against density. Jam-length ratio l_{J1} at section $N1$ is different from jam-length ratio l_{J2} at section $N2$. First, the jam occurs at second normal-speed section $N2$ when the density is higher than $\rho = 0.18$. Then, the second jam occurs at first normal-speed section $N1$. With increasing density, both jams grow. Length l_{J2} of the first jam is longer length l_{J1} of the second jam. At $\rho = 0.25$ indicated by the up arrow, growth rates of both jams change. At the density indicated by the down arrow, both jams reach, simultaneously, the boundaries and both discontinuous fronts stop to move at density $\rho = 0.325$. The total length of two jams agrees with that obtained by the theoretical analysis.

We study the traffic behavior for such case that the length of all the sections is the same and the speed limit of the first slowdown section is higher than that of the second slowdown section: $L_{N1} = L_{N2} = L_{S1} = L_{S2} = L/4$ and $v_{s1,\max}(=1.5) > v_{s2,\max}(=1.0)$. Figure 42 shows the headway profiles at (a) $\rho = 0.16$, (b) $\rho = 0.25$, and (c) $\rho = 0.33$. Figure 42(d) shows the plot of jam-length ratio (jam length/L) against density. Figure 43 shows the theoretical current curves for three optimal-velocity functions. The solid, chain, and dotted curves represent, respectively, the theoretical current curves for $v_{f,\max} = 2.0$, $v_{s1,\max} = 1.5$, and $v_{s2,\max} = 1.0$. Point c indicates the maximal value on the current curve of $v_{s2,\max} = 1.0$. Points a, b, d, e indicate such points that the horizontal line on point c cross at the theoretical curves of $v_{f,\max} = 2.0$ and $v_{s1,\max} = 1.5$. First, the jam occurs at section $N2$ before section $S2$ of the strong speed limit. The headway profile at $\rho = 0.16$ is shown in Figure 42(a). The densities before and after the discontinuity are given by those at points a and b in Figure 43. With increasing density, the discontinuous front moves to the upstream. If the discontinuity reaches the boundary between $S1$ and $N2$, it goes through the boundary and the discontinuous front is formed at section $S1$. The headway profile at $\rho = 0.25$ is shown in Figure 42(b). The densities before and after the discontinuity are given by those at points d and e in Figure 43. When the density increases furthermore, the discontinuous front goes through the boundary between $N1$ and $S1$ and the discontinuity is formed at section $N1$. The densities before and after the discontinuity are given by those at points a and b in Figure 43. The jam continues to grow with increasing density. Jam length l_J is plotted by open circles against density in Figure 42(d). Thus, the jam occurrence and growth are definitely different from the cases of $v_{s1,\max} = v_{s2,\max} = 1.0$.

We study the traffic flow on a highway with three slowdown sections which are positioned alternately with three normal-speed sections. We take the lengths of sections as $L_{N1} = L_{N2} = L_{N3} = L_{S1} = L_{S2} = L_{S3} = L/6$. Three jams occur simultaneously at normal-speed sections $N1$, $N2$, and $N3$. Figure 44(a) shows the plots of jam-length ratio (jam length/L) against density. All the jam-length ratios at sections $N1$, $N2$, and $N3$ are the same. For $L_{N1} = L_{S1} = L/4$, $L_{N2} = L_{S2} = 0.15L$, and $L_{N3} = L_{S3} = L/10$, jam-length ratios are plotted against density in Figure 44(b). The jam occurs first at section $N1$. Then, two jams occur simultaneously at both sections $N2$ and $N3$. The lengths of three jams are different each other. Thus, the occurrence and growth of jams depend highly on the lengths of normal-speed and slowdown sections.

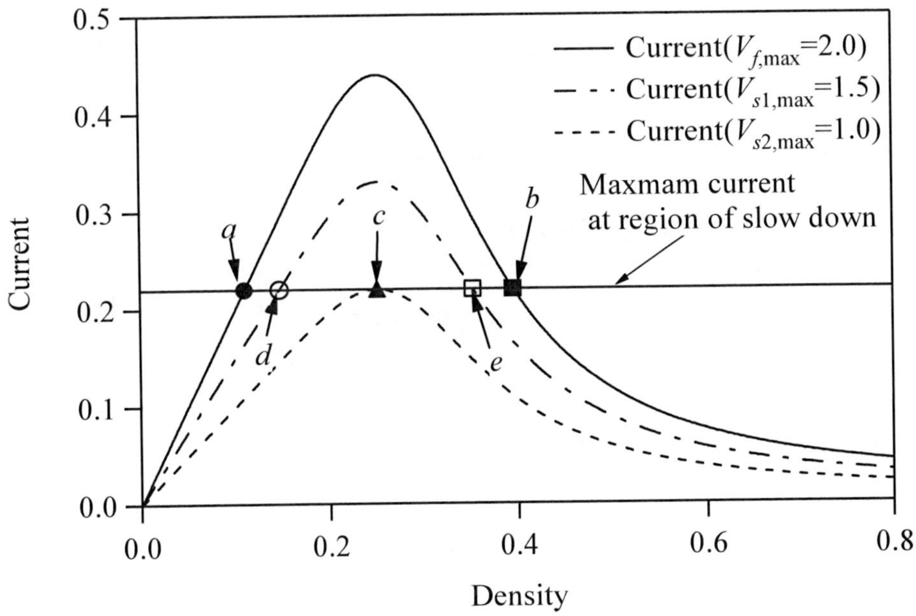

Figure 43. Theoretical current curves for three optimal-velocity functions. The solid, chain, and dotted curves represent, respectively, the theoretical current curves for $v_{f,\max} = 2.0$, $v_{s1,\max} = 1.5$, and $v_{s2,\max} = 1.0$.

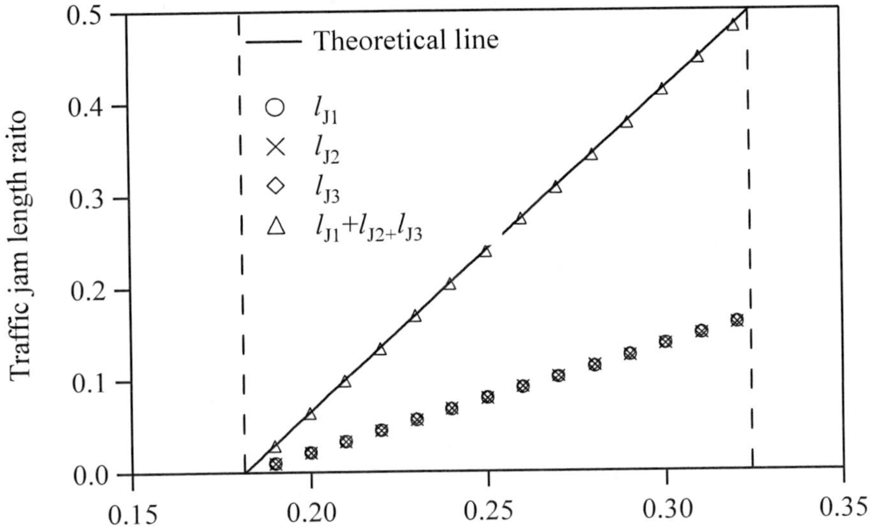

(a)

Figure 44. (Continued)

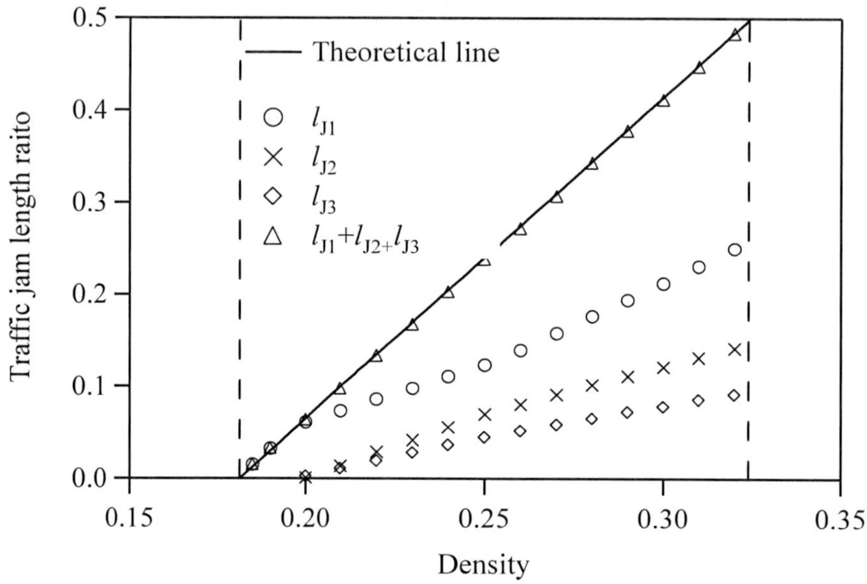

(b)

Figure 44. (a) Plots of jam-length ratio (jam length/L) against density for $L_{N1} = L_{N2} = L_{N3} = L_{S1} = L_{S2} = L_{S3} = L/6$. (b) Plot of jam-length ratios against density for $L_{N1} = L_{S1} = L/4$, $L_{N2} = L_{S2} = 0.15L$, and $L_{N3} = L_{S3} = L/10$.

We derive the relationship between the jam length and density analytically. We consider such case that there exist two slowdown sections on the highway and speed limit $v_{s1,\max}$ at first slowdown section $S1$ is consistent with $v_{s2,\max}$ at second slowdown section $S2$. Either a single jam or two jams are formed at normal-speed sections $N1$ or/and $N2$. Figure 45(a) shows the schematic illustration of the above case. Then, the amount of vehicular number is conserved and always equals total number N. One obtains the following

$$N = L \times \rho = (L_{S1} + L_{S2})\rho_c + (l_{J1} + l_{J2})L\rho_b + \{L - (l_{J1} + l_{J2})L - (L_{S1} + L_{S2})\}\rho_a, \tag{58}$$

where ρ_a and ρ_b are the densities before and after the discontinuous front and ρ_c is the density at the slowdown sections. Densities ρ_a and ρ_b are given by crossing points a and b in Figure 45 (b). Density ρ_c is given by point c at the maximal value of the current curve with $v_{s1(or2),\max}$.

By solving Eq. (58), one obtains the amount of jam lengths

$$(l_{J1} + l_{J2})L = \frac{1}{\rho_c - \rho_a}\{L\rho - (L_{N1} + L_{N2})\rho_a - (L_{S1} + L_{S2})\rho_b\}. \tag{59}$$

The jam length (59) obtained from the theory is shown by the solid line in Figure 39, Figure 40(b), Figure 41(a), and Figure 41(b). The theoretical lines are consistent with the simulation result.

We extend the above theory to such case that there exist more slowdown sections than two on the highway. For m sections of slowdown, one obtains the following

$$\sum_{i=1}^{m} l_{Ji} L = \frac{1}{\rho_c - \rho_a}\left\{L\rho - \left(\sum_{i=1}^{m} L_{Ni}\right)\rho_a - \left(\sum_{i=1}^{m} L_{Si}\right)\rho_b\right\}. \tag{60}$$

The theoretical line (60) for m=3 is shown by the solid line in Figure 44(a). The theoretical result agrees with the simulation result.

5. Traffic Flow on Multi-lane Highway

In real traffic, a highway has more lanes than one. We extend the conventional optimal-velocity model to the two- and multi-lane lane traffic flow.

5.1. Traffic States Induced by Slowdown Sections on a Two-lane Highway

In real traffic, some sections of slowdown exist on a two-lane highway. When a few slowdown sections with a distinct configuration exist on a two-lane highway, where, when, and how do traffic jams occur? The traffic jams induced by the slowdown sections on a two-lane highway have little been studied by using modern traffic models.

Here, we extend the conventional optimal-velocity model to the two-lane traffic flow on a highway with some slowdown sections. We study the traffic states and jams induced by the slowdown sections on the two-lane highway. We clarify the dynamical states of traffic and the fundamental diagrams on the two-lane highway with slowdown sections. We show where and how the traffic jams occur by increasing density of vehicles and by varying configuration of slowdown sections.

We consider such situation that many vehicles move ahead with changing lane on a two-lane highway. Some slowdown sections are positioned on the two-lane highway. Traffic flow is under the periodic boundary condition. We assume that vehicles are forced to slow down when they enter into a section of the slowdowns. Vehicles pass over other vehicles by changing lane if the criteria of lane changing are satisfied.

Lane changing is implemented as a pure sideways movement. We assume that the vehicular movement is divided into two parts: one is the forward movement and the other is the sideways movement. We apply the optimal velocity model to the forward movement. Figure 46 shows the schematic illustrations of the traffic models for the two-lane highway with various slowdown sections. Vehicles move with low speed in the slowdown sections, while they move with the normal velocity except for the slowdown sections. In Figure 46, the slowdown sections are illustrated by gray color. We consider four typical configurations (a)-(d) of slowdown sections. In configuration (a), the slowdown sections are set at the same position on the two lanes where the length of slowdown section is $L_S = L/2$ and the length of normal-speed section is $L_N = L/2$ (L: the road length). In configuration (b), the slowdown section is set only on the first lane. In configuration (c), two slowdown sections are set at different positions on first and second lanes where the length of slowdown section is $L/4$. In configuration (d), two slowdown sections are set on the first lane.

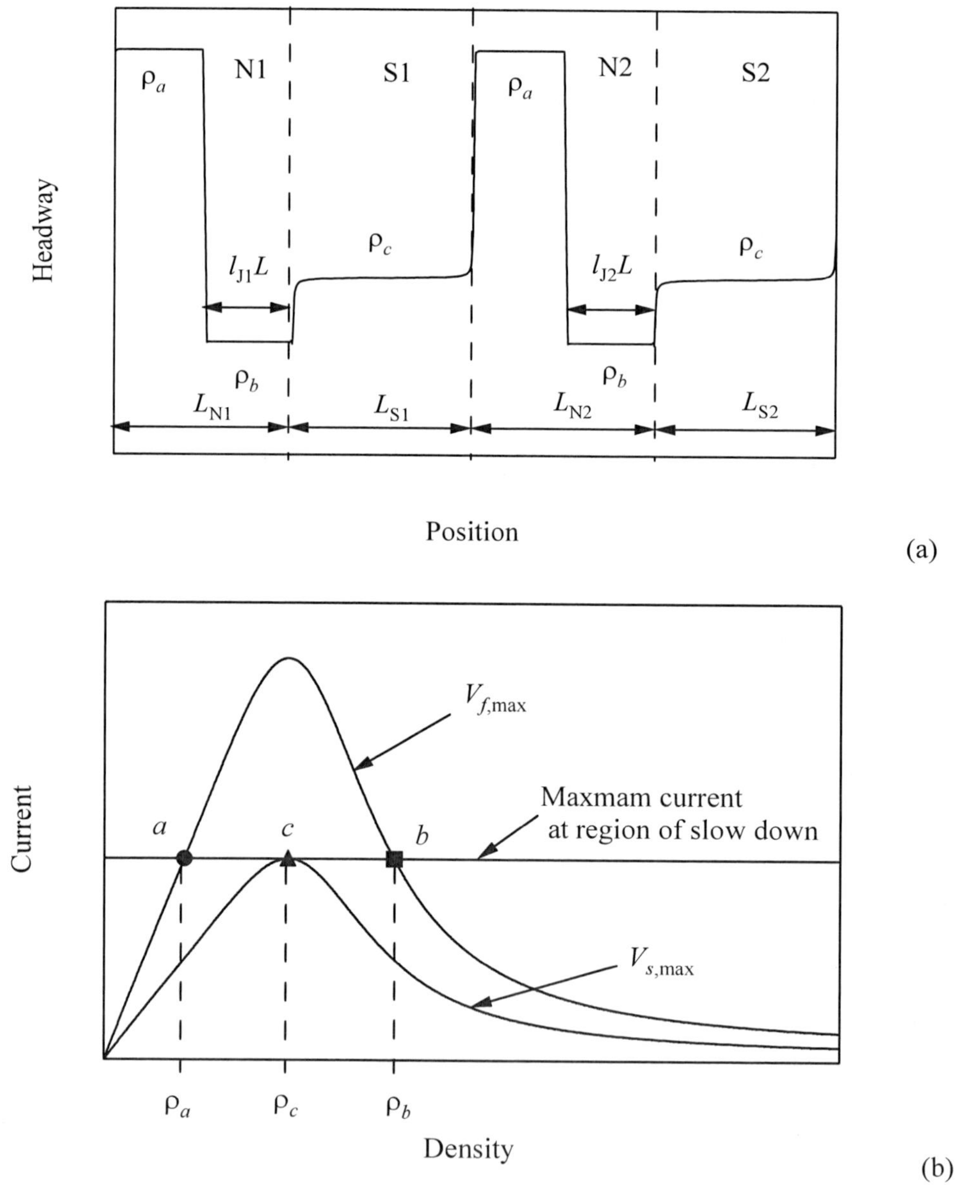

Figure 45. (a) Headway profile for such case that there exist two slowdown sections on the highway and speed limit $v_{s1,\max}$ at first slowdown section $S1$ is consistent with $v_{s2,\max}$ at second slowdown section $S2$. (b) Fundamental diagram.

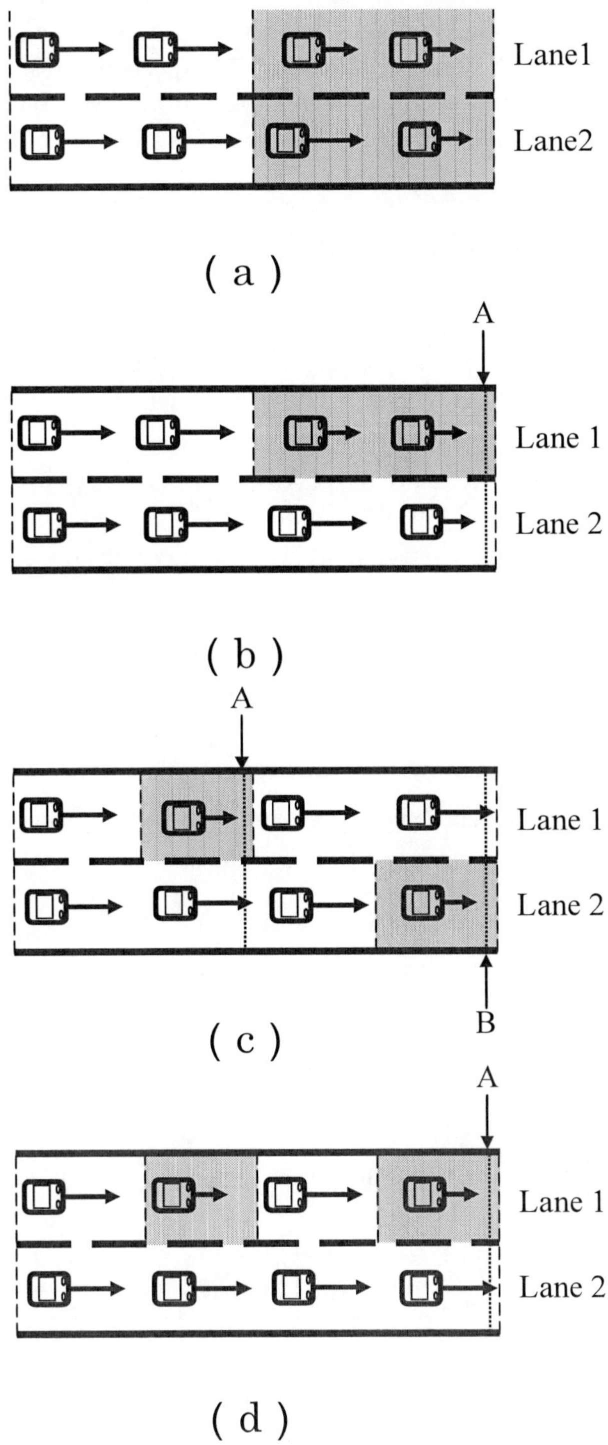

Figure 46. Schematic illustrations of the traffic models for the two-lane highway with various slowdown sections. The slowdown sections are illustrated by gray color. Four typical configurations (a)-(d) of slowdown sections.

We apply the optimal velocity model to the forward movement. The optimal velocity model is described by Eq. (3). Vehicles move with low speed in the slowdown sections, while they move with the normal velocity except for the slowdown sections. In the region of normal-speed sections, the optimal velocity function of vehicles is given by

$$V(\Delta x_i) = \frac{v_{f,\max}}{2}\left[\tanh \alpha(\Delta x_i - x_{f,c}) + \tanh \alpha(x_{f,c})\right], \tag{61}$$

where $v_{f,\max}$ is the maximal velocity of vehicles on the roadway except for the slowdown sections and $x_{f,c}$ the position of turning point.

In the slowdown sections, vehicles move with the forced speed and the vehicular velocity should be lower than the speed limits of slowdown. When vehicles enter into the first section of slowdowns, they obey the conventional optimal-velocity function with speed limit $v_{s1,\max}$:

$$V(\Delta x_i) = \frac{v_{s1,\max}}{2}\left[\tanh(\Delta x_i - x_{s,c}) + \tanh(x_{s,c})\right], \tag{62}$$

where $v_{s1,\max}$ is the speed limit of first slowdown section, $v_{s1,\max} < v_{f,\max}$, and $x_{s,c}$ the position of turning point. When vehicles enter into the second section of slowdowns, they obey the conventional optimal-velocity function with speed limit $v_{s2,\max}$:

$$V(\Delta x_i) = \frac{v_{s2,\max}}{2}\left[\tanh(\Delta x_i - x_{s,c}) + \tanh(x_{s,c})\right], \tag{63}$$

where $v_{s2,\max}$ is the speed limit of second slowdown section, $v_{s2,\max} < v_{f,\max}$. Thus, when the density of vehicles is very low, vehicles move at the maximal velocity $v_{f,\max}$ except for the sections of slowdown, while they move at a speed lower than the forced speed limits $v_{s1,\max}$ or $v_{s2,\max}$ in the slowdown sections.

If there are no specific public rules, it is natural to consider symmetric incentive criteria. We adopt symmetric lane changing rules. We apply the following lane changing rule for the two-lane highway:

$$\begin{aligned} &\Delta x_i < 2x_c \quad \text{for the incentive criterion,} \\ &\Delta x_{fi} > \Delta x_i \text{ and } \Delta x_{bi} > x_c \text{ for the security criterion,} \end{aligned} \tag{64}$$

where Δx_{fi} is the headway between vehicle *i* and the vehicle ahead on the target lane and Δx_{bi} is the headway between vehicle *i* and the vehicle behind on the target lane. A driver wants to change the lane when the headway is less than two times safety distance. In addition to the incentive criterion, when the headway between his vehicle and the front vehicle on the target lane is larger than his headway and the headway between his vehicle and the back vehicle on the target lane is larger than the safety distance, it is successful for his vehicle to change the lane.

With increasing density, vehicles interact with each other. Then, various dynamic states of traffic appear and traffic jams may occur. Here, we restrict ourselves to such case that sensitivity is higher than the critical point, $a > a_c$, since we do not investigate the spontaneous jam but study the traffic jams induced by slowdown sections.

We perform computer simulation for the traffic model shown in Figure 46. The simulation is performed until the traffic flow reaches a steady state. We solve numerically Eq. (3) with optimal velocity functions (61)-(63) by using fourth-order Runge-Kutta method where the time interval is $\Delta t = 1/128$.

We carry out simulation by varying the initial headway for 200 vehicles, sensitivity $a = 2.5$, maximal velocities $v_{f,\max} = 2.0$ and $v_{s,\max} = 1.0$, and safety distance $x_{f,c} = x_{s,c} = 3.0$. Initially, we put 100 vehicles on each lane with the same headway Δx_{int}. The density ρ is given by the inverse of the headway. The length *L* of highway varies with the initial headway.

First, we study the traffic behavior for the case in Figure 46(a) in which the slowdown sections are set at the same position on the two lanes. Figure 47(a) shows the plots of traffic currents against density. The open circles and triangles indicate the traffic currents on first and second lanes obtained by simulation. The upper solid curve represents the theoretical current curve for $v_{f,\max} = 2.0$ without slowdown sections. The lower solid curve represents the theoretical current curve for all slowdown sections of $v_{s,\max} = 1.0$. The current on the first lane is consistent with that on the second lane. The current increases linearly with density, then the current saturates and keeps a constant value which agrees with the maximal value of theoretical current curve of $v_{s,\max} = 1.0$. In the saturated-current traffic, the traffic jam is formed just before the slowdown section. When density is higher, the current decreases with increasing density. Thus, the traffic are classified into three states: the free traffic, the jammed traffic, and the congested traffic. In the jammed traffic where the stationary jam occurs, the discontinuous front is formed just after the jam. The densities before and after the discontinuity are given by those at points *a* and *b* in Figure 47(a) and the density at slowdown section is given by that at point *c*. The characteristics of jam is the same as that of the single-lane traffic flow.

We study the traffic flow for the case in Figure 46(b) where the slowdown section is positioned only at the first lane. Figure 47(b) shows the plots of traffic currents against density. The traffic current is calculated at section *A* in Figure 46(b). The open circles and triangles indicate the traffic currents on first and second lanes obtained by simulation. The upper solid curve represents the theoretical current curve for $v_{f,\max} = 2.0$ without slowdown sections. The lower solid curve represents the theoretical current curve for all slowdown sections of $v_{s,\max} = 1.0$. The current on the first lane is different from that on the second lane. The traffic jam occurs only at region Ⅳ.

Figure 48(a) shows the plot of vehicular occupancy on each lane against density. The occupancy means the fraction on first (second) lane that vehicles exist on the first (second) lane for all vehicles. In regions Ⅰ and Ⅵ, the fraction on first lane is consistent with that on the second lane. In region Ⅱ, the fraction on second lane is higher than that on first lane. In regions Ⅳ and Ⅴ, the fraction on the first lane is higher than that on the second lane. In region Ⅲ, the fraction on the first lane interchanges with that on the second lane. Figure 48(b) shows the plot of lane-changing vehicular number (per unit time) against density. In regions Ⅲ and Ⅳ, the lane changing occurs steadily. In regions Ⅱ and Ⅴ, the lane changing occurs at early stage but does not occur after sufficiently large time. The traffic is classified into six traffic states. In traffic state Ⅰ, vehicles are in the free traffic without changing lanes. In traffic state Ⅱ, vehicles change lanes at early stage, do not change lanes after sufficiently large time, and flow in free traffic with different occupancies on first and second lanes. In region Ⅲ, vehicles are in free traffic with changing lanes steadily. In region Ⅳ, the jam occurs on the first lane, the current on first lane saturates, and vehicles flow with changing lanes steadily. In region Ⅴ, vehicles change lanes at early stage but do not change lanes after sufficiently large time, and are in the congested traffic with different occupancies on first and second lanes. In region Ⅵ, vehicles do not change lanes and are in the congested traffic with the same occupancy on each lane.

Figure 49 shows the headway and velocity profiles at region Ⅳ in which the jam occurs on the first lane. Figure 49(a) shows the headway profile at density $\rho = 0.21$. The solid and dotted lines indicate the headway profiles on the first and second lanes. Figure 49(b) shows the velocity profile at density $\rho = 0.21$. The solid and dotted lines indicate the velocity profiles on the first and second lanes. Just before the slowdown section, the jam appears only on the first lane where the headway takes the lowest value, vehicles move at lowest speed, and the headway is given by the value at point *b* in Figure 47(b). Before the jam, vehicles move almost at maximal velocity and the maximal headway is given by the value at point *a* in Figure 47(b). The headway within the slowdown section is given by the value at point *c* in Figure 47(b). The headway values before and after the discontinuity just behind the jam is given by those derived by the theoretical analysis for the single-lane traffic flow.

Figure 50 shows the plot of jam-length ratio (jam length/*L*) on the first lane against density. The jam lengths increase with density. The jam appears only on the first lane.

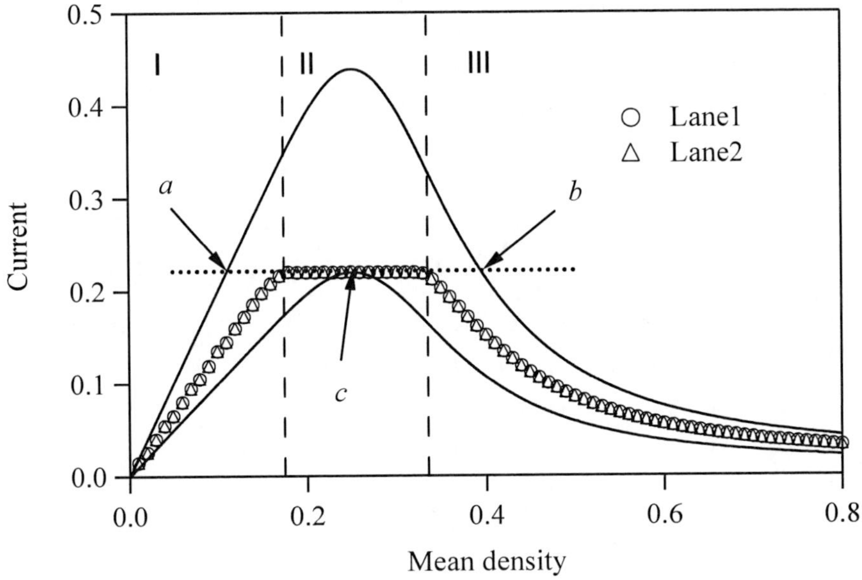

(a)

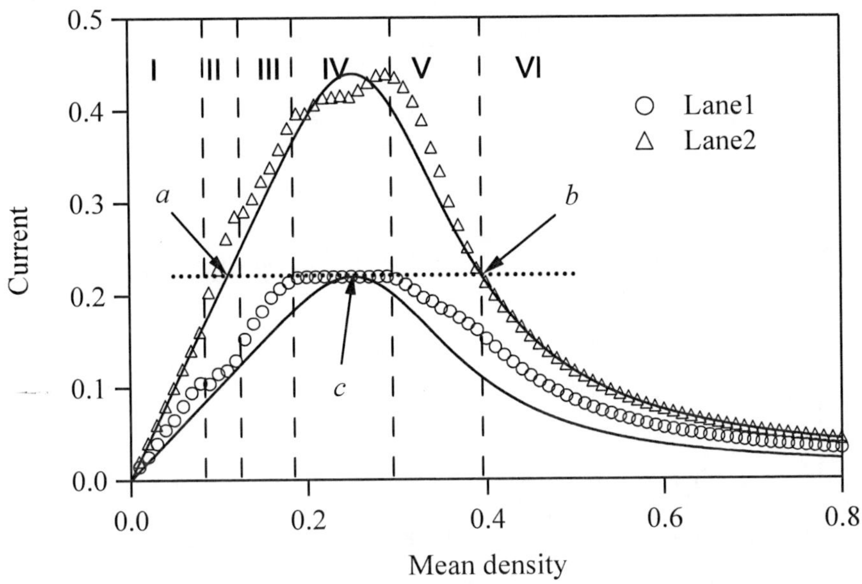

(b)

Figure 47. (a) Plots of traffic currents against density. The open circles and triangles indicate the traffic currents on first and second lanes obtained by simulation. The upper solid curve represents the theoretical current curve for $v_{f,\max} = 2.0$ without slowdown sections. The lower solid curve represents the theoretical current curve for all slowdown sections of $v_{s,\max} = 1.0$. (b) Plots of traffic currents against density for the case in Figure 46(b) where the slowdown section is positioned only at the first lane.

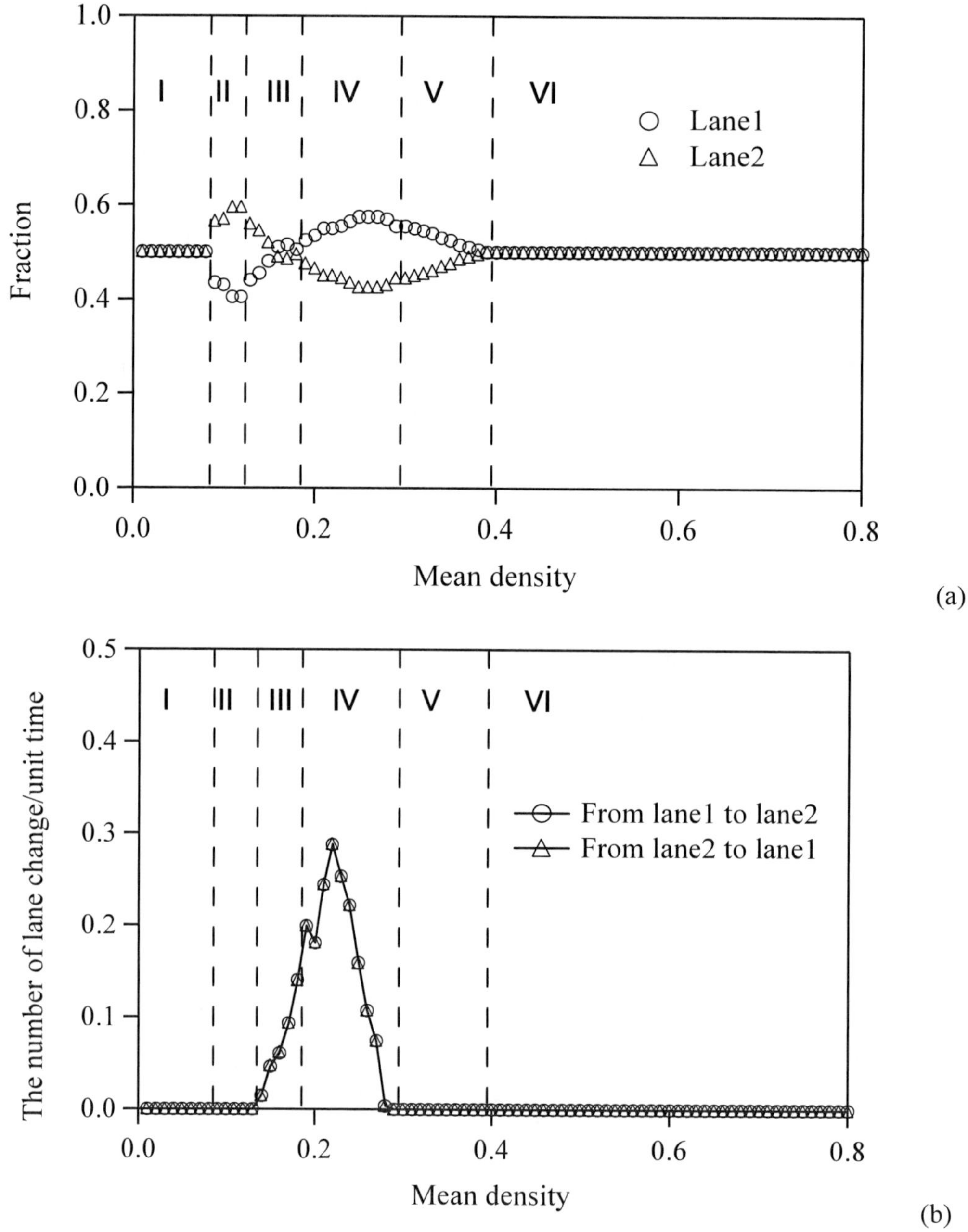

Figure 48. (a) Plot of vehicular occupancy on each lane against density for the case in Figure 46(b). (b) Plot of lane-changing vehicular number (per unit time) against density for the case in Figure 46(b).

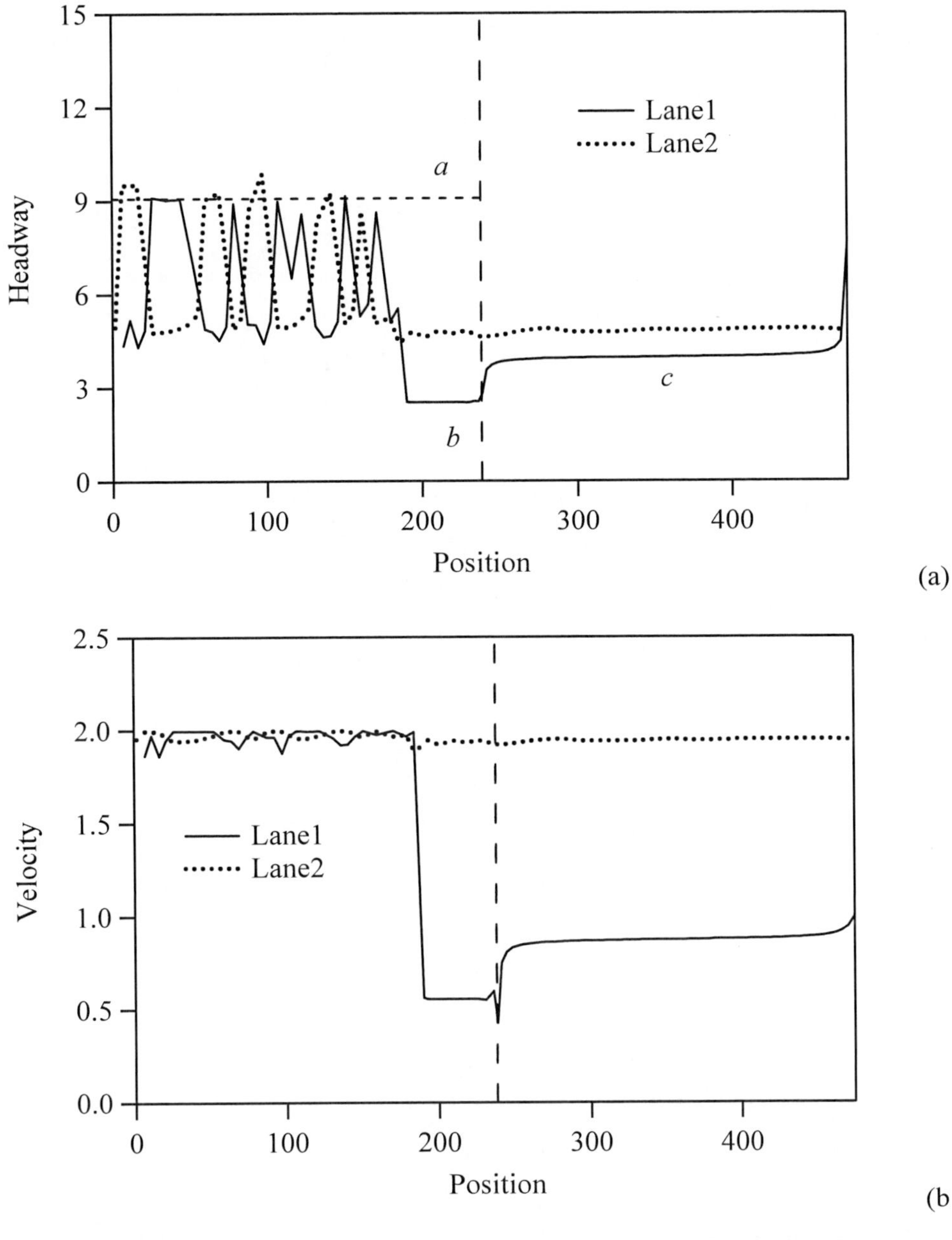

Figure 49. (a) Headway profile at density $\rho = 0.21$ for the case in Figure 46(b). The solid and dotted lines indicate the headway profiles on the first and second lanes. (b) Velocity profile at density $\rho = 0.21$ for the case in Figure 46(b). The solid and dotted lines indicate the velocity profiles on the first and second lanes.

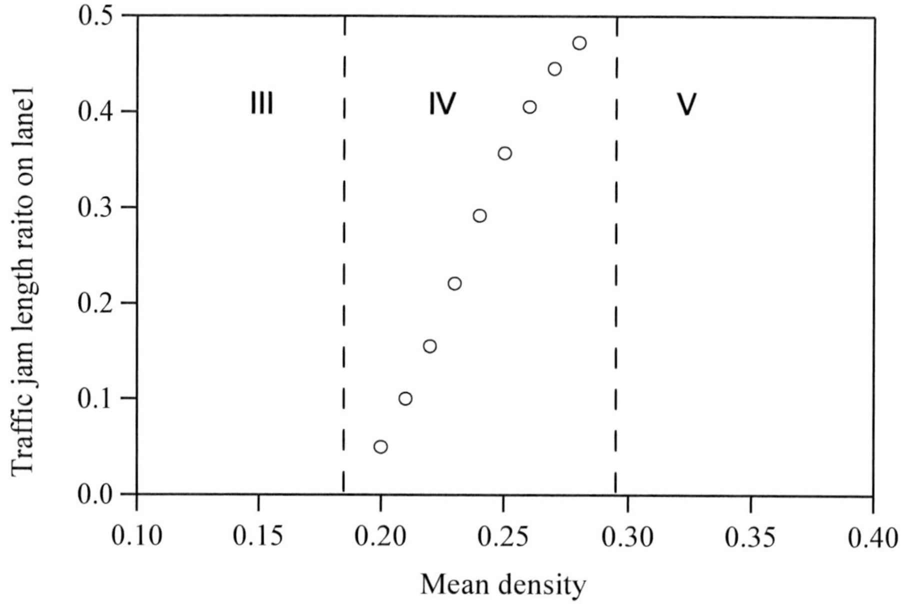

Figure 50. Plot of jam-length ratio (jam length/L) on the first lane against density for the case in Figure 46(b).

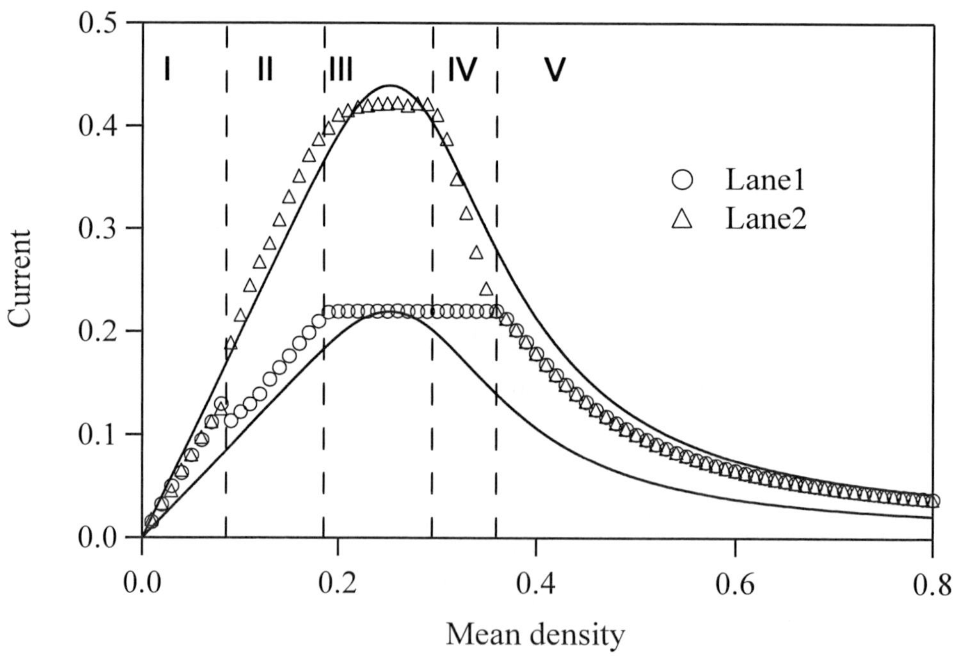

Figure 51. Plots of traffic currents against density at section B in Figure 46(c). The open circles and triangles indicate the traffic currents on first and second lanes obtained by simulation.

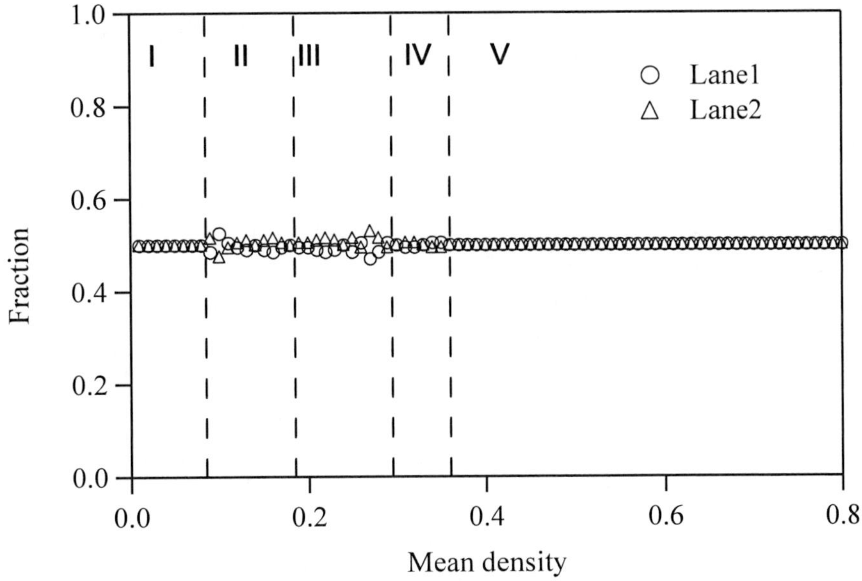

(a)

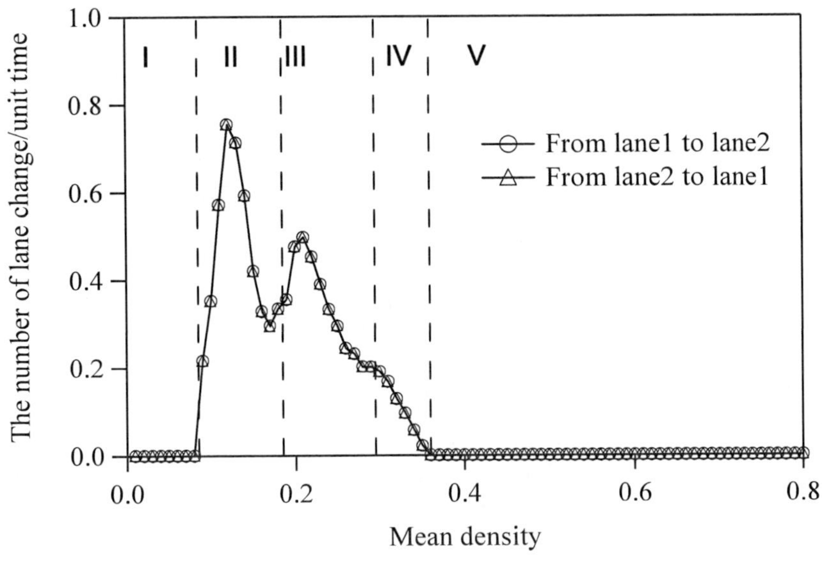

(b)

Figure 52. (a) Plot of vehicular occupancy (fraction) on each lane against density for the case in Figure 46(c). (b) Plot of lane-changing vehicular number (per unit time) against density for the case in Figure 46(c).

We study the traffic behavior for the case in Figure 46(c) in which the slowdown sections are set at the different position on the first and second lanes. Figure 51 shows the plots of traffic currents against density at section *B* in Figure 46(c). The open circles and triangles indicate the traffic currents on first and second lanes obtained by simulation. When the current is measured at section *A* in Figure 46(c), the current on the first lane interchanges with that on the second lane. The upper solid curve represents the theoretical current curve for $v_{f,\max} = 2.0$ without slowdown sections. The lower solid curve represents the theoretical current curve for all slowdown sections of $v_{s,\max} = 1.0$. The current on the first lane is different highly from that on the second lane. Figure 52(a) shows the plot of vehicular occupancy (fraction) on each lane against density. Figure 52(b) shows the plot of lane-changing vehicular number (per unit time) against density. Traffic are classified into five states I - V. In region I, vehicles do not change the lane, move freely, and the state results in the free traffic. In regions II-IV, vehicles change steadily the lane. In region V, vehicles do not change the lane and the state results in the congested traffic. In regions III and IV, the current on the first lane at section *B* in Figure 46(c) saturates and takes the maximal current on the lower theoretical curve.

Figure 53(a) shows the velocity profile at density $\rho = 0.24$ in region III. The solid and dotted lines indicate the velocity profiles on the first and second lanes. Jams appear on both lanes, the jam on the first lane occurs just before the slowdown section on the first lane, and the jam on the second lane appears just before the slowdown section on the second lane. Figure 53(b) shows the velocity profile at density $\rho = 0.33$ in region IV where both jam lengths saturate. The solid and dotted lines indicate the velocity profiles on the first and second lanes. Jams extend over the slowdown sections on both lanes.

Figure 54 shows the plot of jam-length ratio (jam length/*L*) on the first and second lanes against density. Open circles and triangles indicate, respectively, the jam-length ratios on the first and second lanes. Open squares represent the sum (total length ratio) of the jam-length ratio on the first and second lanes. The jam lengths increase with density in region III. The jam appears on both first and second lanes. In region IV, both jam lengths saturate and do not grow with density.

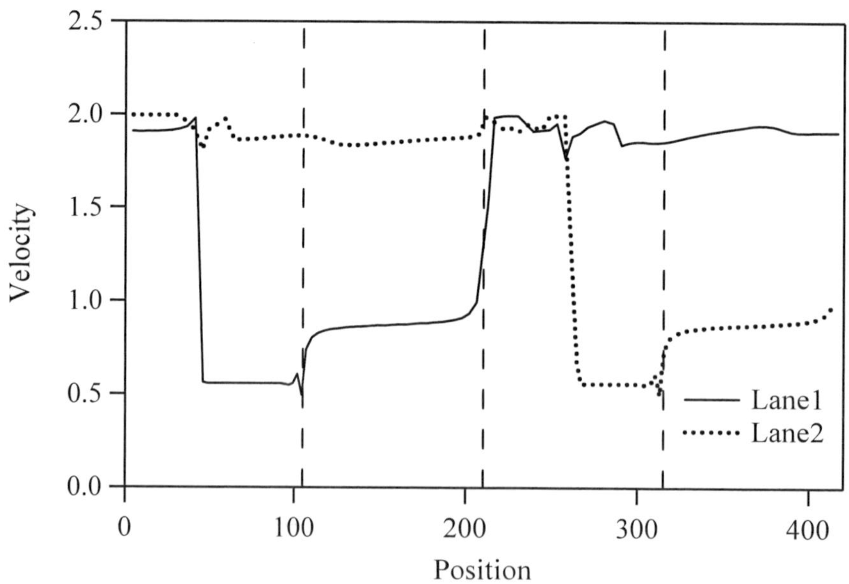

(a)

Figure 53. (Continued)

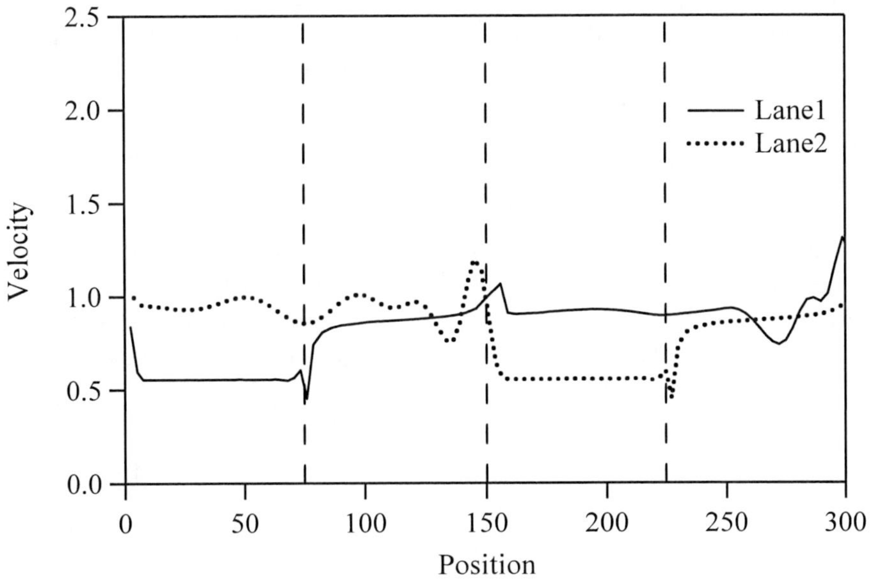

(b)

Figure 53. (a) Velocity profile at density $\rho = 0.24$ in region III for the case in Figure 46(c). The solid and dotted lines indicate the velocity profiles on the first and second lanes. (b) Velocity profile at density $\rho = 0.33$ in region IV for the case in Figure 46(c)

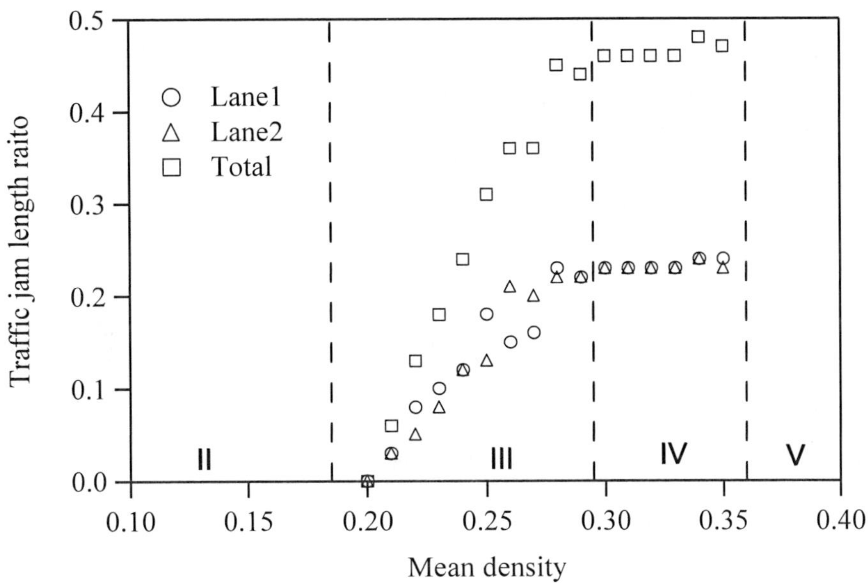

Figure 54. Plot of jam-length ratio (jam length/L) on the first and second lanes against density for the case in Figure 46(c). Open circles and triangles indicate, respectively, the jam-length ratios on the first and second lanes. Open squares represent the sum (total length ratio) of the jam-length ratio on the first and second lanes.

We study the traffic behavior for the case in Figure 46(d) in which the slowdown sections are set at the different position on the first lane. Figure 55 shows the plots of traffic currents against density at section A in Figure 46(d). The open circles and triangles indicate the traffic currents on first and second lanes obtained by simulation. The upper solid curve represents the theoretical current curve for $v_{f,\max} = 2.0$ without slowdown sections. The lower solid curve represents the theoretical current curve for all slowdown sections of $v_{s,\max} = 1.0$. The current on the first lane is different highly from that on the second lane. Figure 56(a) shows the plot of vehicular occupancy (fraction) on each lane against density. Figure 56(b) shows the plot of lane-changing vehicular number (per unit time) against density. Traffic are classified into five states I - VI. In region I, vehicles do not change the lane, move freely, and the state results in the free traffic. In region II, vehicles change the lane at early stage and do not change the lane in time. The fraction on the first lane is different from that on the second lane. In region III, vehicles change steadily the lane and the current saturates. In regions IV-VI, vehicles do not change the lane. In regions III- V, the current on the first lane in Figure 46(d) saturates and takes the maximal current on the lower theoretical curve.

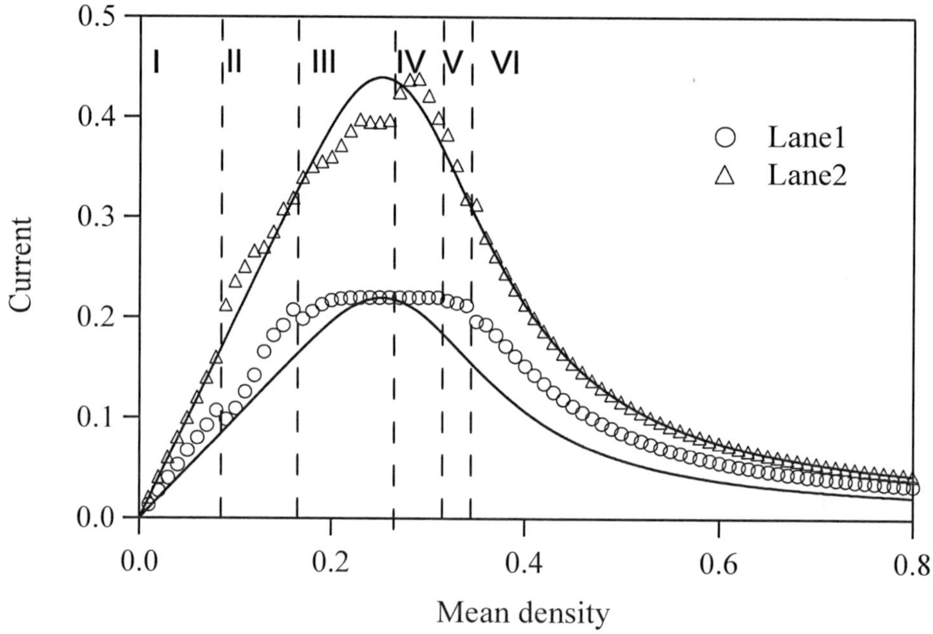

Figure 55. Plots of traffic currents against density at section A in Figure 46(d). The open circles and triangles indicate the traffic currents on first and second lanes obtained by simulation.

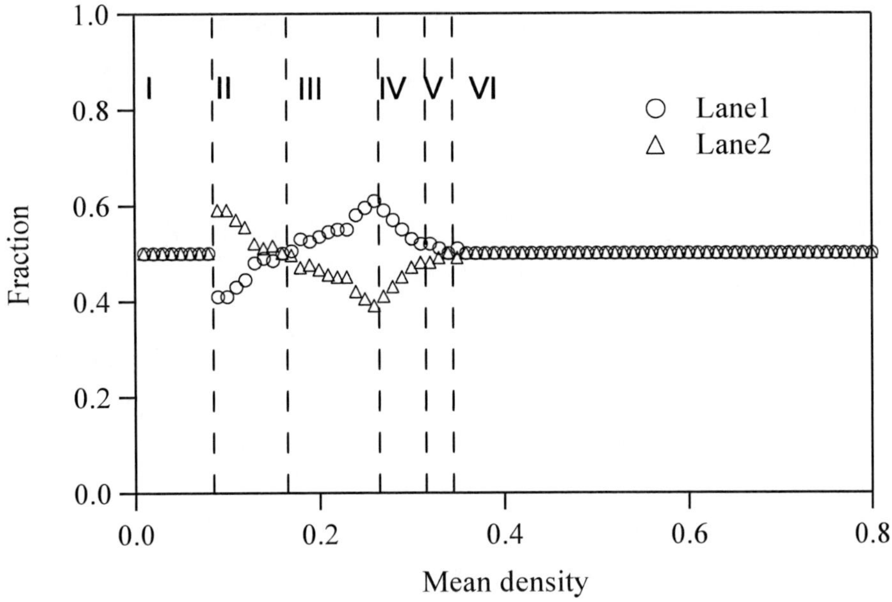

(a)

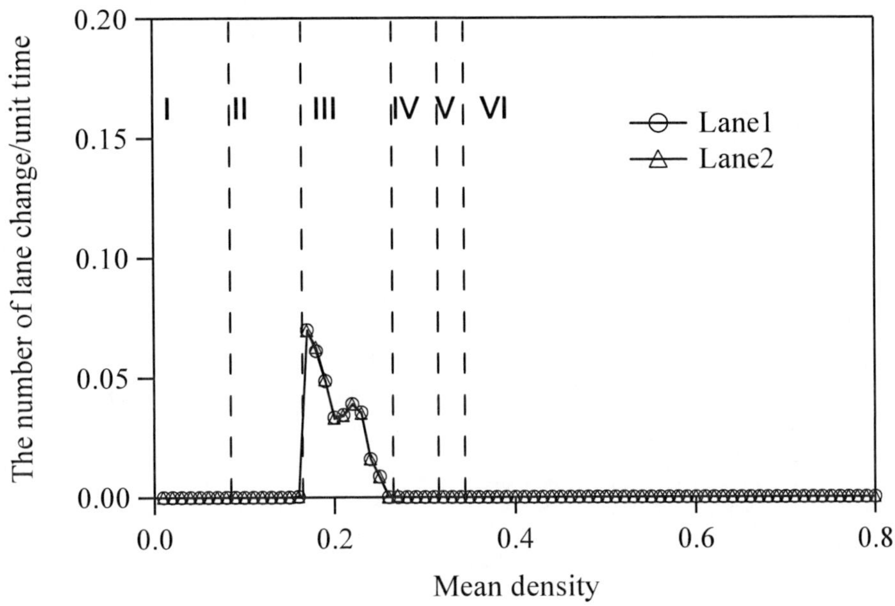

(b)

Figure 56. (a) Plot of vehicular occupancy (fraction) on each lane against density for the case in Figure 46(d). (b) Plot of lane-changing vehicular number (per unit time) against density for the case in Figure 46(d).

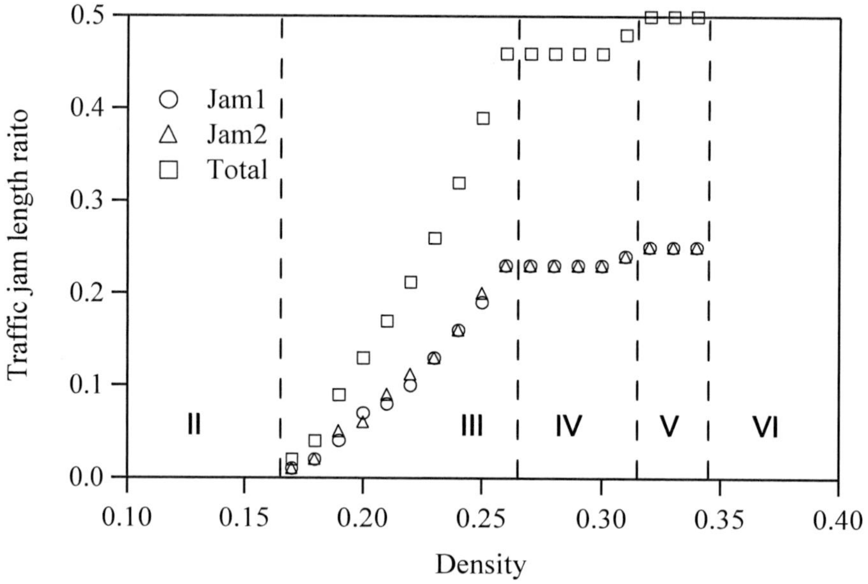

Figure 57. Plot of jam-length ratio (jam length/L) on the first and second lanes against density for the case in Figure 46(d). Open circles and triangles indicate, respectively, the jam-length ratios on the first and second lanes. Open squares represent the sum (total length ratio) of the jam-length ratio on the first and second lanes.

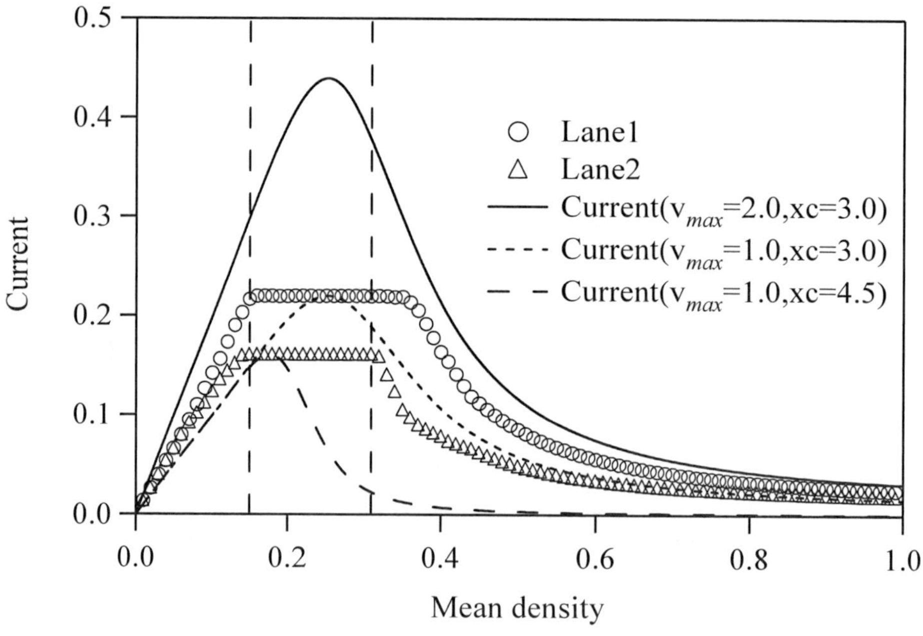

(a)

Figure 58. (Continued)

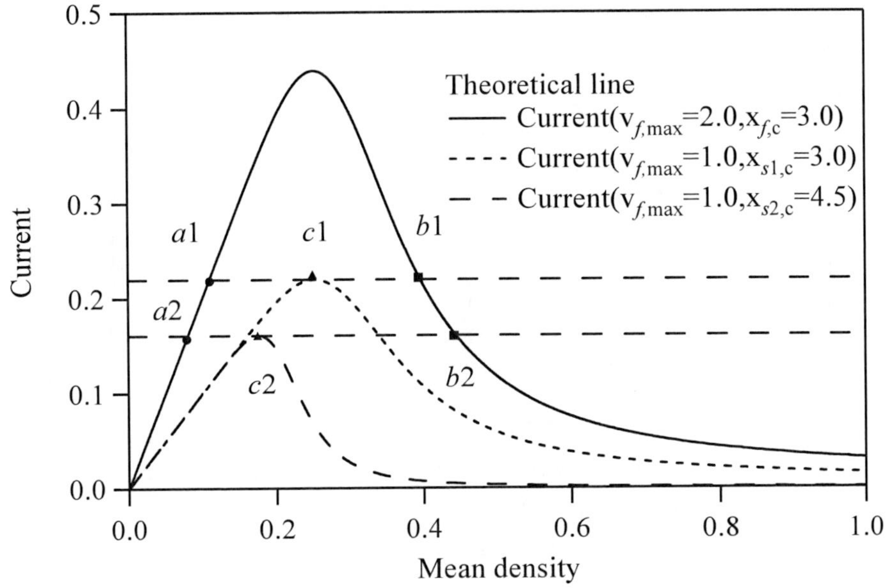

(b)

Figure 58. (a) Plots of currents on the first and second lane against density for the case in Figure 46(a) where the safety distances are the same ($x_{f,c} = x_{s,c,1} = 3.0$) except for that on the second lane at the slowdown section. Open circles indicate the current on the first lane and open triangles the current on the second lane. (b) Theoretical current curves. The solid curve represents the theoretical current for $v_{max} = 2.0$ and $x_c = 3.0$. The dotted line indicates the theoretical current for $v_{max} = 1.0$ and $x_c = 3.0$. The dashed line represents the theoretical current for $v_{max} = 1.0$ and $x_c = 4.5$.

Figure 57 shows the plot of jam-length ratio (jam length/L) on the first and second lanes against density. Open circles and triangles indicate, respectively, the jam-length ratios on the first and second lanes. Open squares represent the sum (total length ratio) of the jam-length ratio on the first and second lanes. The jam lengths increase with density in region Ⅲ. Two jams appear just before the slowdown sections on the first lane.

We study the traffic flow for such case that the safety distance on the second lane at the slowdown section is different from that on the first lane at the slowdown section for configuration (a) in Figure 46. We set the safety distance on the second lane at the slowdown section as $x_{s,c,2} = 4.5$. Figure 58(a) shows the plots of currents on the first and second lane against density where the safety distances are the same ($x_{f,c} = x_{s,c,1} = 3.0$) except for that on the second lane at the slowdown section. Open circles indicate the current on the first lane and open triangles the current on the second lane. The solid curve represents the theoretical current for $v_{max} = 2.0$ and $x_c = 3.0$. The dotted line indicates the theoretical current for $v_{max} = 1.0$ and $x_c = 3.0$. The broken line represents the theoretical current for $v_{max} = 1.0$ and $x_c = 4.5$. The current on the first lane saturates at the maximal value c1 of the theoretical curve in Figure 58(b). The current on the second lane saturates at the maximal value c2 of the theoretical curve in Figure 58(b).

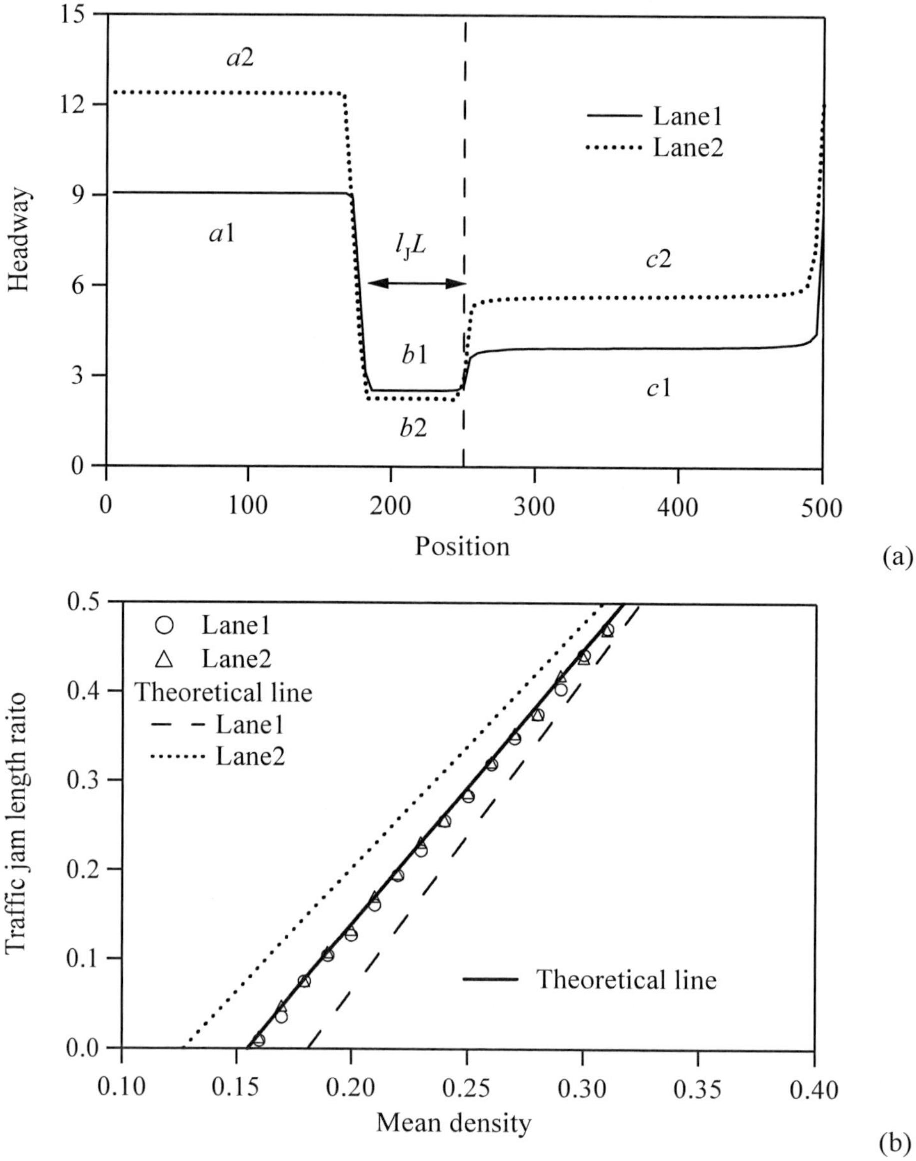

Figure 59. (a) Headway profiles on the first and second lanes at density $\rho = 0.2$ in Figure 58(a). (b) Plot of jam's length ratio against density. Open circles and triangles indicate, respectively, the jam's length ratio on the first and second lanes. The jam length on the first lane agrees with that on the second lane.

Figure 59(a) shows the headway profiles on the first and second lanes at density $\rho = 0.2$. The traffic jams are formed on both lanes before the slowdown section. The jam's length on the first lane agrees with that on the second lane. The densities before and after the jam on the first lane are given by values b1 and a1 in Figure 58(b). The densities before and after the jam on the second lane are given by values b2 and a2 in Figure 58(b). The densities on the first and second lanes at the slowdown section are given by values c1 and c2 in Figure 58(b).

Figure 59(b) shows the plot of jam's length ratio against density. Open circles and triangles indicate, respectively, the jam's length ratio on the first and second lanes. The jam length on the first lane agrees with that on the second lane.

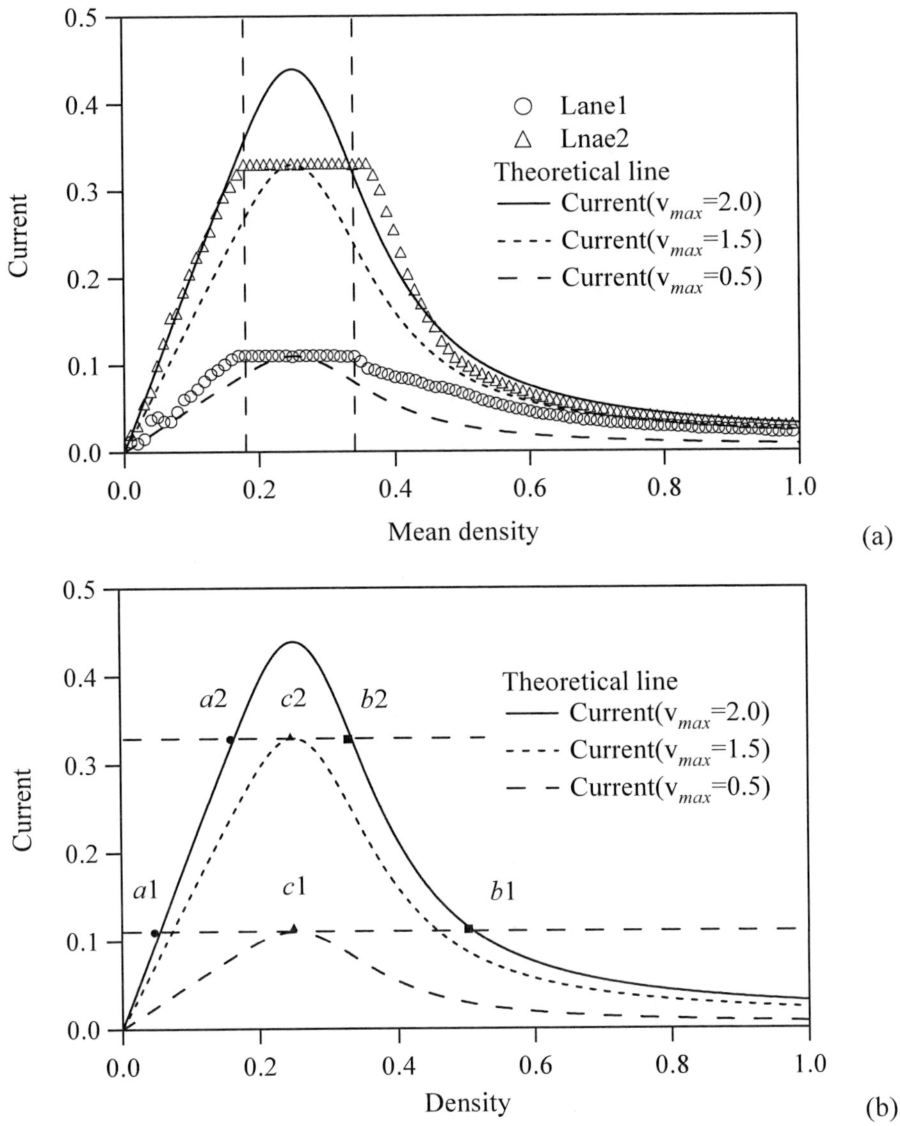

Figure 60. (a) Plots of currents on the first and second lane against density for the case in Figure 46(a) where $v_{s,\max,1} = 0.5$ and $v_{s,\max,2} = 1.5$. Open circles indicate the current on the first lane and open triangles the current on the second lane. (b) Theoretical current curves. The solid curve represents the theoretical current for $v_{\max} = 2.0$ and $x_c = 3.0$. The dotted line indicates the theoretical current for $v_{\max} = 1.5$ and $x_c = 3.0$. The dashed line represents the theoretical current for $v_{\max} = 0.5$ and $x_c = 3.0$.

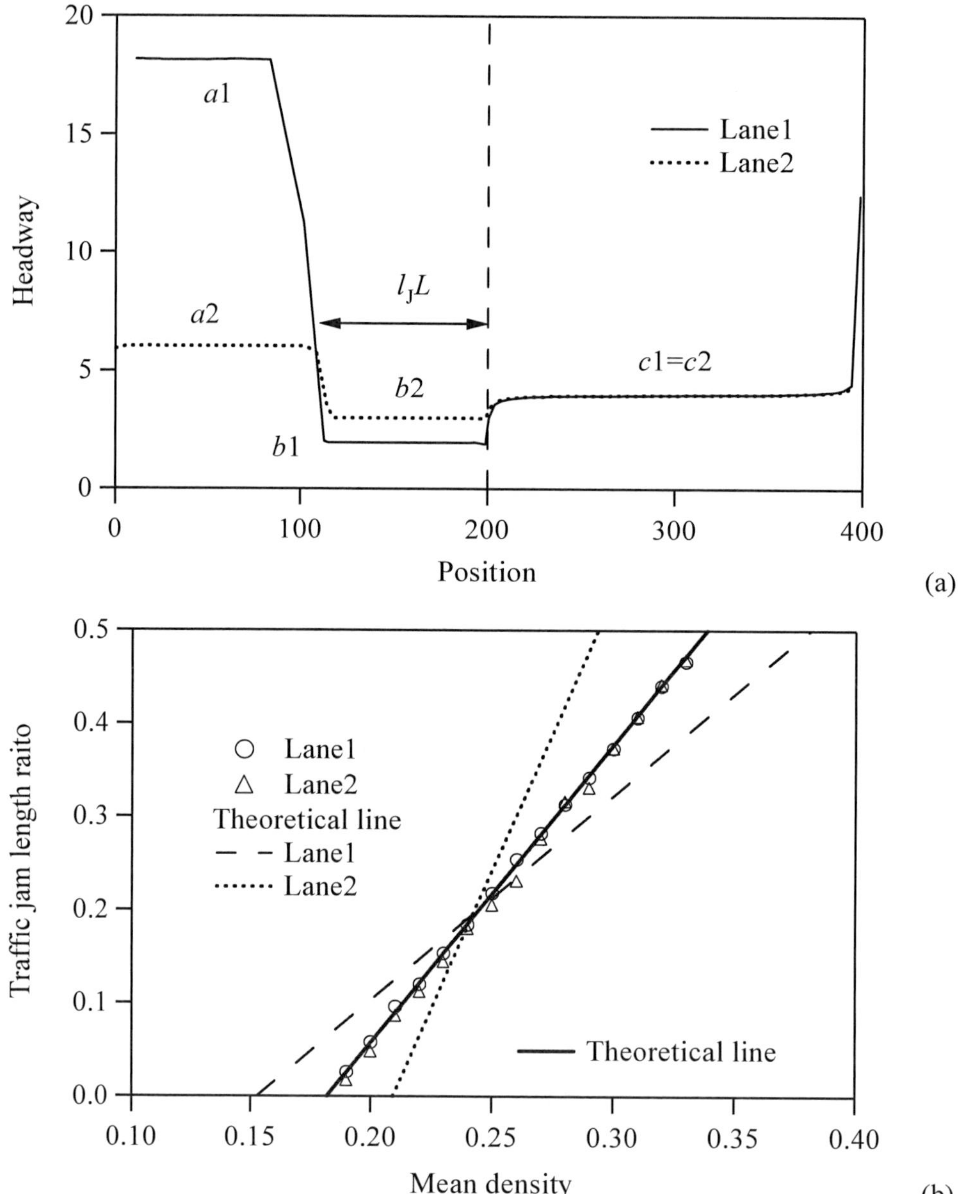

Figure 61. (a) Headway profiles on the first and second lanes at density $\rho = 0.25$ in Figure 60(a). (b) Plot of jam's length ratio against density. Open circles and triangles indicate, respectively, the jam's length ratio on the first and second lanes. The jam length on the first lane agrees with that on the second lane. The theoretical curve (67) is shown by the solid line in Figure 61(b).

We derive the relationship between the jam length and density analytically. The vehicular number is conserved on each lane because the vehicular number changing from lane 1 to lane 2 agrees with that from lane 2 to lane.

$$N_1 = \rho_{c1} L_S + \rho_{b1} l_J L + \rho_{a1} (L_N - l_J L), \tag{65}$$

$$N_2 = \rho_{c2} L_S + \rho_{b2} l_J L + \rho_{a2}(L_N - l_J L), \tag{66}$$

where $N_1(N_2)$ is the number of vehicles on the first (second) lane, $N = N_1 + N_2$, $\rho_{c1}(\rho_{c2})$ is the densities on the first (second) lane at the slowdown section, $\rho_{a1}(\rho_{a2})$ is the densities on the first (second) lane before the jam, $\rho_{b1}(\rho_{b2})$ is the densities on the first (second) lane within the jam, and $l_J L$ is the jam length. The jam length is obtained

$$l_J L = \frac{1}{(\rho_{b1} + \rho_{b2}) - (\rho_{a1} + \rho_{a2})} \{2L\rho - L_N(\rho_{a1} + \rho_{a2}) - L_S(\rho_{c1} + \rho_{c2})\}. \tag{67}$$

The theoretical curve (67) is shown by the solid line in Figure 59(b). The theoretical result agrees with the simulation result. The dashed line represents the jam length for the single-lane traffic flow with $v_{s,\max} = 1.0$ and $x_{s,c} = 3.0$. The dotted line represents the jam length for the single-lane traffic flow with $v_{s,\max} = 1.0$ and $x_{s,c} = 4.5$.

We study the traffic flow for such case that the maximal speed on the second lane at the slowdown section is different from that on the first lane at the slowdown section for configuration (a) in Figure 46. We set the maximal speeds on the first and second lane at the slowdown section as $v_{s,\max,1} = 0.5$ and $v_{s,\max,2} = 1.5$. Figure 60(a) shows the plots of currents on the first and second lane against density. Open circles indicate the current on the first lane and open triangles the current on the second lane. The solid curve represents the theoretical current for $v_{\max} = 2.0$ and $x_c = 3.0$. The dotted line indicates the theoretical current for $v_{\max} = 1.5$ and $x_c = 3.0$. The dashed line represents the theoretical current for $v_{\max} = 0.5$ and $x_c = 3.0$. The current on the first lane saturates at the maximal value c1 of the theoretical curve in Figure 60(b). The current on the second lane saturates at the maximal value c2 of the theoretical curve in Figure 60(b).

Figure 61(a) shows the headway profiles on the first and second lanes at density $\rho = 0.25$. The traffic jams are formed on both lanes before the slowdown section. The jam's length on the first lane agrees with that on the second lane. The densities before and after the jam on the first lane are given by values b1 and a1 in Figure 60(b). The densities before and after the jam on the second lane are given by values b2 and a2 in Figure 60(b). The densities on the first and second lanes at the slowdown section are given by values c1 and c2 in Figure 60(b) and they take the same value of the headway.

Figure 61(b) shows the plot of jam's length ratio against density. Open circles and triangles indicate, respectively, the jam's length ratio on the first and second lanes. The jam length on the first lane agrees with that on the second lane. The theoretical curve (67) is shown by the solid line in Figure 61(b). The theoretical result agrees with the simulation result. The dashed line represents the jam length for the single-lane traffic flow with $v_{s,\max} = 0.5$ and $x_{s,c} = 3.0$. The dotted line represents the jam length for the single-lane traffic flow with $v_{s,\max} = 1.5$ and $x_{s,c} = 3.0$.

5.2. Traffic Flow of Mixed Vehicles on Multi-lane Highway

In real traffic, slow and fast vehicles are mixed on a multi-lane highway. The traffic flow of mixed vehicles on the multi-lane highway has little studied until now from the point of view of the self-driven many-particle systems. Here, we study the traffic flow of mixed vehicles on the multi-lane highway where vehicles pass each other by lane-changing. Especially, we investigate the traffic flow at high sensitivity where the stop- and go-wave does not occur because one is interesting in the traffic states induced by both effects of slow vehicles and multi lanes. We study the jamming transition induced by the competition between the blocking of slow vehicles and passing of fast vehicles.

We consider such situation that many vehicles move ahead with changing lane on a multi-lane highway. Two kinds of vehicles with low and high velocities are introduced on the multi-lane highway. Traffic flow is under the periodic boundary condition. We assume that fast vehicles pass over the other vehicles by changing lane if the criteria of lane changing are satisfied.

We adopt symmetric lane changing rules. Lane changing is implemented as a pure sideways movement. We assume that the vehicular movement is divided into two parts: one is the forward movement and the other is the sideways movement. We apply the optimal velocity model to the forward movement. The optimal velocity model is described by Eq. (3). The optimal velocity function of fast vehicles is given by

$$V_f(\Delta x_i) = \frac{v_{f,\max}}{2}\left[\tanh(\Delta x_i - x_c) + \tanh(x_c)\right], \tag{68}$$

where $v_{f,\max}$ is the maximal velocity of fast vehicles and x_c the position of turning point.

The optimal velocity function of the slow vehicle is given by

$$V_s(\Delta x_i) = \frac{v_{s,\max}}{2}\left[\tanh(\Delta x_i - x_c) + \tanh(x_c)\right], \tag{69}$$

where $v_{s,\max}$ is the maximal velocity of the slow vehicle.

We adopt lane changing rule (64) for the two-lane highway. Also, we adopt the following lane changing rule except for the edge lanes for the multi-lane highway higher than two lanes:

$$\begin{aligned}
&\Delta x_i < 2x_c \quad \text{for the incentive criterion,}\\
&\Delta x_{fli} > \Delta x_i \text{ and } \Delta x_{bli} > x_c \text{ for the security criterion 1,}\\
&\Delta x_{fri} > \Delta x_i \text{ and } \Delta x_{bri} > x_c \text{ for the security criterion 2,}
\end{aligned} \tag{70}$$

where Δx_{fli} (Δx_{fri}) is the headway between vehicle *i* and the vehicle ahead on the left (right) target lane and Δx_{bli} (Δx_{bri}) is the headway between vehicle *i* and the vehicle behind on the left (right) target lane. Figure 62 shows the schematic illustration of lane changing on a lane except for the edge lanes for the multi-lane highway. A driver wants to change the lane when the headway is less than two times safety distance. In addition to the incentive criterion, when the headway between his vehicle and the front vehicle on the target lane is larger than his headway and the headway between his vehicle and the back vehicle on the target lane is larger than the safety distance, it is successful for his vehicle to change the lane. Furthermore, the driver chooses either left or right lanes by the following rules. If $\Delta x_{fli} > \Delta x_{fri}$ ($\Delta x_{fli} < \Delta x_{fri}$), the vehicle changes his lane to left (right) lane. If $\Delta x_{fli} = \Delta x_{fri}$ and $\Delta x_{bli} > \Delta x_{bri}$ ($\Delta x_{bli} < \Delta x_{bri}$), the vehicle changes his lane to left (right) lane. If $\Delta x_{fli} = \Delta x_{fri}$ and $\Delta x_{bli} = \Delta x_{bri}$, the driver changes his lane to either left or right lanes with probability 1/2.

For such case that there exit only fast vehicles, the current is given by

$$q = \rho_0 V_f(\rho_0) \text{ for } a > v_{f,\max}, \tag{71}$$

where mean density is $\rho_0 = 1/\Delta x_0$ (Δx_0: the mean headway). If $a > v_{f,\max}$, the stop- and go-waves do not occur.

For such case that there exit only slow vehicles, the current is given by

$$q = \rho_0 V_s(\rho_0) \text{ for } a > v_{s,\max}. \tag{72}$$

When $a > v_{s,\max}$, the stop- and go-waves do not occur.

We derive the theoretical current curves of mixed vehicular traffic under a simple assumption. We derive the current against density theoretically under such assumption that fast vehicles move at the optimal velocity with changing lanes freely and slow vehicles also move at the optimal velocity with being taken over. Then, the current is given by

$$q = c_f \rho_0 V_f(1/\rho_0) + c_s \rho_0 V_s(1/\rho_0), \tag{73}$$

where c_f (c_s) is the fraction of fast (slow) vehicles and ρ_0 is the mean density. Theoretical curve (73) is shown by the dashed line in Figure 64.

We derive the current against density under such condition that the fast vehicles move together with slow vehicles without changing lanes. Assume that all vehicles move together at velocity v. Then, the headway of slow vehicles is given by the inverse function

$$\Delta x_s = V_s^{-1}(v)$$

where $v = V_s(\Delta x_s) = \frac{v_{s,\max}}{2}(\tanh(\Delta x_s - x_c) + \tanh(x_c))$.

Also, the headway of fast vehicles is given by the inverse function

$$\Delta x_f = V_f^{-1}(v)$$

where $v = V_f(\Delta x_f) = \frac{v_{f,\max}}{2}(\tanh(\Delta x_f - x_c) + \tanh(x_c))$.

The mean density is given by

$$\rho_0 = 1/(c_f \Delta x_f + c_s \Delta x_s).$$

The current is obtained

$$q = \rho_0 v. \tag{74}$$

The theoretical curve (74) is shown by the solid curve on open circles in Figure 63. Thus, two theoretical currents (73) and (74) were derived for two limiting cases. Generally, the current curve can not be obtained analytically.

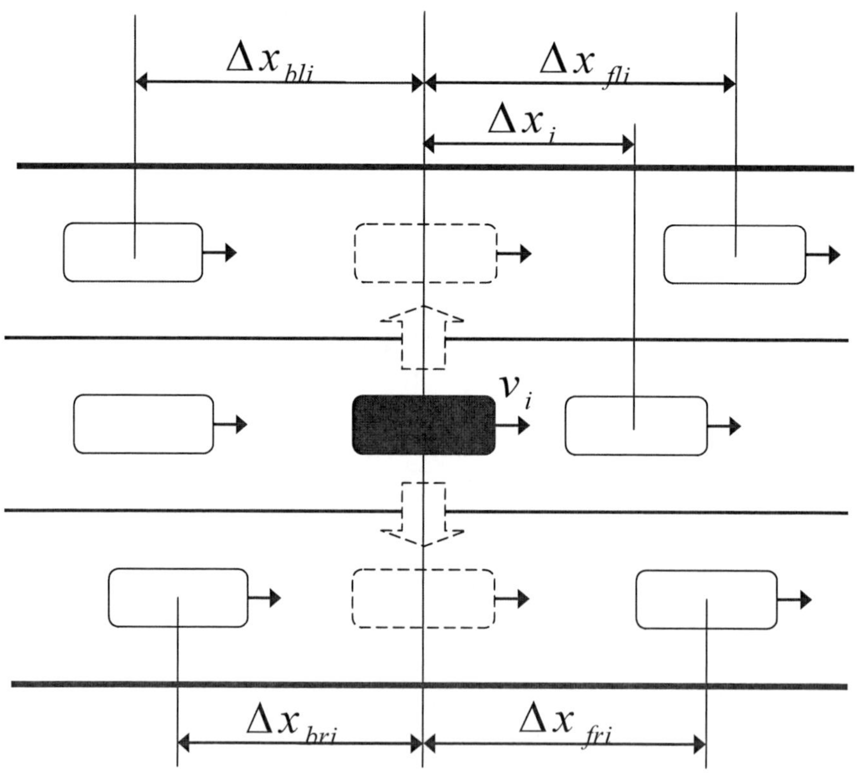

Figure 62. Schematic illustration of lane changing on a lane except for the edge lanes for the multi-lane highway. Δx_{fli} (Δx_{fri}) is the headway between vehicle *i* and the vehicle ahead on the left (right) target lane and Δx_{bli} (Δx_{bri}) is the headway between vehicle *i* and the vehicle behind on the left (right) target lane.

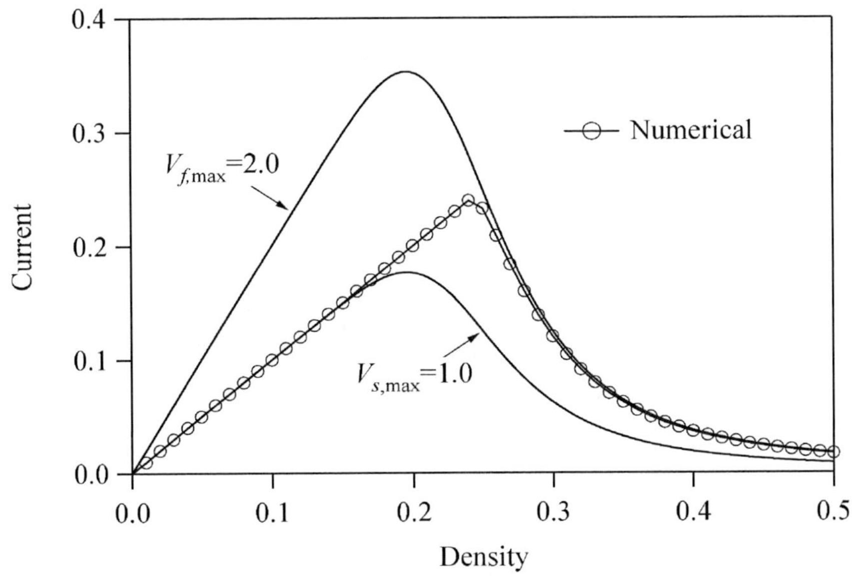

Figure 63. Fundamental diagram (current-density diagram) for the single-lane highway at $v_{s,\max} = 1.0$ where sensitivity $a = 3$, the total number of vehicles is N=100, and the number of slow vehicles is 5. Open circles indicates the simulation result. The upper curve represents theoretical current curve (71) for fast vehicles. The lower curve represents theoretical current curve (72) for slow vehicles.

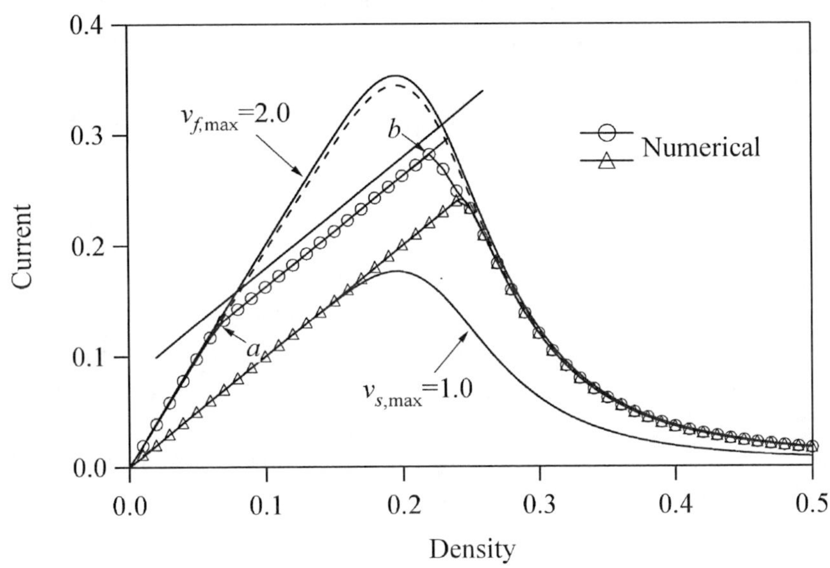

Figure 64. Plots of current against density for two-lane traffic flow for $v_{s,\max} = 1.0$ at sensitivity $a = 3.0$ where total number of all vehicles is N=200, the number of slow vehicles is 10, and $x_c = 4.0$. Open circles and open triangles indicate the simulation result. The upper solid curve indicates the theoretical curve (71) for the case of all fast vehicles. The lower curve represents the theoretical curve (72) for the case of all slow vehicles. The dashed line indicates the theoretical curve (73). The solid curve on open triangles represents the theoretical curve (74).

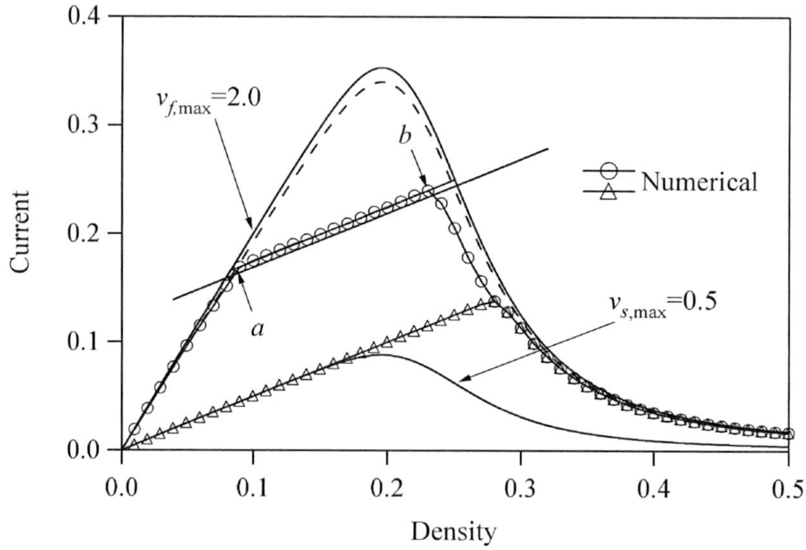

Figure 65. Plots of current against density for $v_{s,\max} = 0.5$. Open circles and open triangles indicate the simulation result. The dashed line indicates the theoretical curve (73). The solid curve on open triangles represents the theoretical curve (74).

We carry out simulation for 100 vehicles (*N*=100) on each lane and the maximal velocity $v_{f,\max} = 2.0$ of fast vehicles and the maximal velocity $v_{s,\max} = 0.5 - 1.5$ of slow vehicle. Initially, vehicles are positioned with the same headway on each lane. We solve numerically Eq. (3) with optimal velocity functions (68) and (69) by using fourth-order Runge-Kutta method where the time interval is $\Delta t = 1/128$. We apply Eqs. (64) and (70) to changing lanes.

First, we study the single-lane traffic flow for comparison with the multi-lane traffic flow. Figure 63 shows the fundamental diagram (current-density diagram) for the single-lane highway at $v_{s,\max} = 1.0$ where sensitivity $a = 3$, the total number of vehicles is *N*=100, and the number of slow vehicles is five. The slow vehicles are mixed randomly. Open circles indicate the simulation result. The upper curve represents the theoretical current curve (71) for fast vehicles. The lower curve represents the theoretical current curve (72) for slow vehicles. The fast vehicles are blocked by slow vehicles and move together with the slow vehicles because the fast vehicles can not pass the slow vehicles. The theoretical curve (74) agrees with the simulation result indicated by open circles.

We study the fundamental diagram for two-lane traffic flow. Figure 64 shows the plots of current against density for $v_{s,\max} = 1.0$ at sensitivity $a = 3.0$ where total number of all vehicles is *N*=200, the number of slow vehicles is ten, and $x_c = 4.0$. Open circles and open triangles indicate the simulation result. Two current curves are obtained. The current takes either high or low values for the same value of density. They depend on the initial configuration of mixed vehicles. The upper solid curve indicates the theoretical curve (71) for the case of all fast vehicles. The lower curve represents the theoretical curve (72) for the case of all slow vehicles. The dashed line indicates the theoretical curve (73). The solid curve on open triangles represents the theoretical curve (74). The simulation result represented by open triangles is consistent with the theoretical result (74). In the low-current curve, fast vehicles are blocked by slow vehicles and cannot pass slow vehicles. Therefore, the low-current curve is represented by theoretical curve (74). The simulation result indicated by open circles agrees with the theoretical curve (73) represented by the dashed curve until transition point *a*. In the high-current curve, the traffic state is in the free traffic for $\rho_0 < \rho_a$ (ρ_a: density at transition point *a*). At transition point *a*, the jamming transition occurs. The traffic state is in the inhomogeneous jammed state. However, fast vehicles change the lane sometimes. The jams are formed behind slow vehicles. The straight line indicates the line of the same slope as that of Eq. (74) starting from the point of $\rho_0 = 1/3x_c$. The value of $3x_c$ is the minimal value of the security criterion (64).

Figure 65 shows the plots of current against density for $v_{s,\max} = 0.5$. Open circles and open triangles indicate the simulation result. Similarly to Figure 64, two current curves are obtained. The dashed line indicates the theoretical curve (73). The solid curve on open triangles represents the theoretical curve (74). The simulation result represented by open triangles is consistent with the theoretical result (74). The simulation result indicated by open circles agrees with the theoretical curve (73) represented by the dashed curve until transition point *a*. For $\rho_0 < \rho_a$ (ρ_a: density at transition point *a*), the traffic state is in the free traffic. At transition point *a*, the jamming transition occurs. The traffic state is in the inhomogeneous jammed state. The jams are formed behind slow vehicles. The straight line indicates the line of the same slope as that of Eq. (74) starting from the point of $\rho_0 = 1/3x_c$. The straight line is very close to the simulation result indicated by open circles. Figure 66 shows the plot of transition point against maximal velocity $v_{s,\max}$ of slow vehicles. The open circles indicate the simulation result. The straight line represents $\rho_0 = 1/3x_c$. The transition points do not agree with $\rho_0 = 1/3x_c$ but are close to $\rho_0 = 1/3x_c$.

The transition point *a* increases a little with the number of slow vehicles. However, When slow vehicles increase, the current takes the lower value indicated by open triangles frequently. The high-current value indicated by open circles is obtained rare. Thus, the frequency of the high current increases with decreasing the number of slow vehicles.

We study the frequency at which the high current curve indicated by open circles appeared. We calculate the fraction of high current for 100 samples. Figure 67(a) shows the plots of fraction against density for concentrations 5% and 10% of slow vehicles when the vehicular position on the first lane is consistent with that on the second lane in the initial configuration but the configuration of fast and slow vehicles is random. The fraction decreases abruptly with increasing both density and concentration of slow vehicles. The low current state indicated by open triangles increases with both density and concentration of slow vehicles. Figure 67(b) shows the plots of fraction against density for concentrations 5% and 10% of slow vehicles when the vehicular position on the first lane shifts by half headway for that on the second lane in the initial configuration but the configuration of fast and slow vehicles is random. The fraction decreases abruptly at a lower density than that in Figure 67(a).

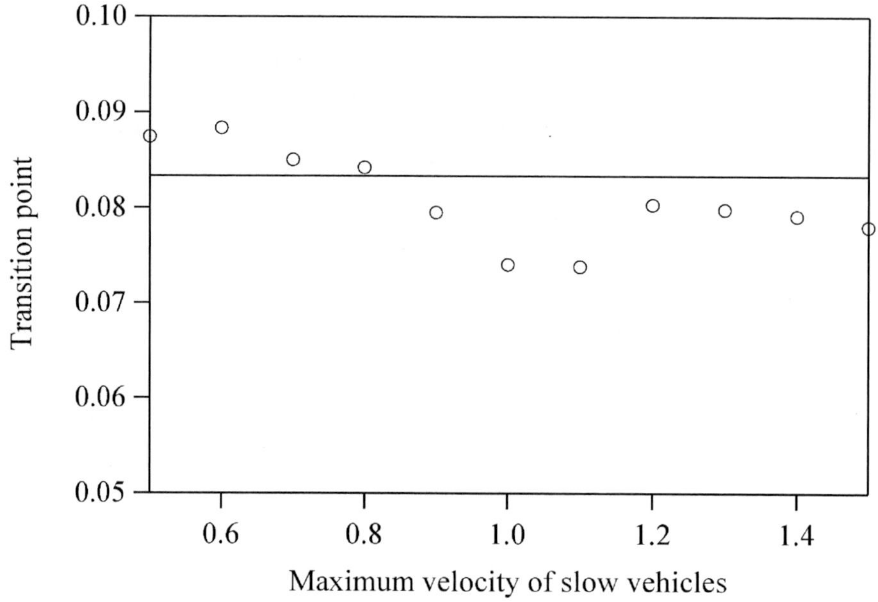

Figure 66. Plot of transition point a against maximal velocity $v_{s,\max}$ of slow vehicles. The open circles indicate the simulation result. The straight line represents $\rho_0 = 1/3x_c$.

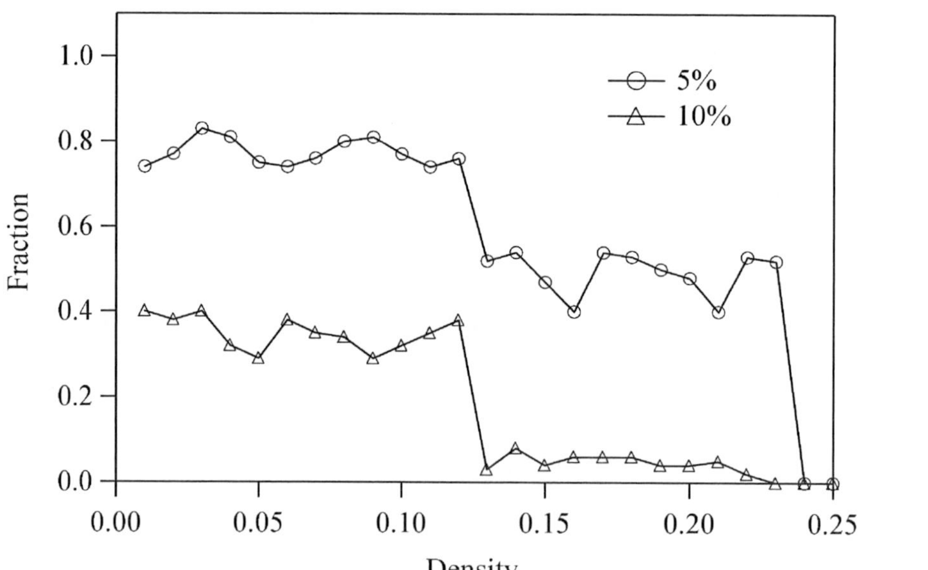

Figure 67. (Continued)

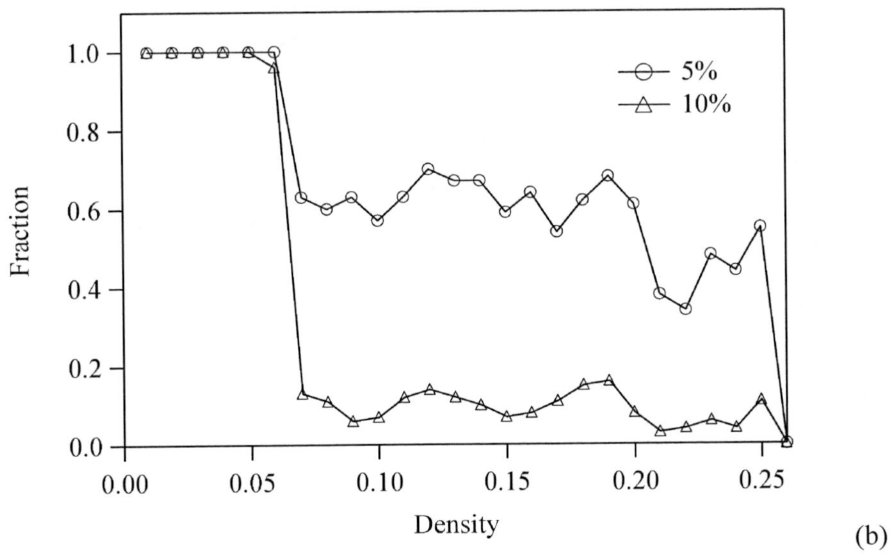

Figure 67. Relative frequency at which the high-current curve appeared. Fraction of high current for 100 samples. (a) Plots of fraction against density for concentrations 5% and 10% of slow vehicles when the vehicular position on the first lane is consistent with that on the second lane in the initial configuration. (b) Plots of fraction against density for concentrations 5% and 10% of slow vehicles when the vehicular position on the first lane shifts by half headway for that on the second lane in the initial configuration.

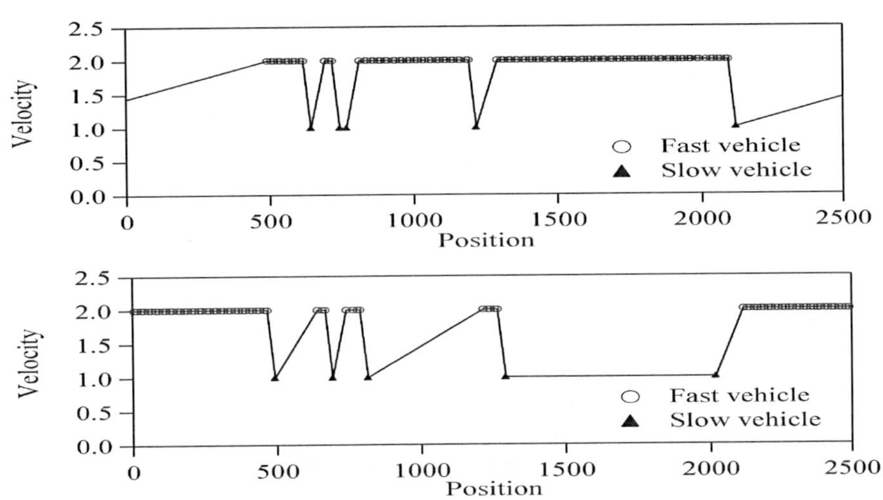

Figure 68. Velocity profiles corresponding to the high-current curve on open circles in Figure 64. Plots of velocity against position on the two-lane highway at low density $\rho = 0.04$ where $v_{s,\max} = 1.0$, sensitivity $a = 3.0$, total number of all vehicles is N=200, the number of slow vehicles is 10, and $x_c = 4.0$. Open circles indicate the velocity and position of fast vehicles. Full triangles indicate the velocity and position of slow vehicles. The upper and lower diagrams represent the velocity profiles on the first and second lanes respectively.

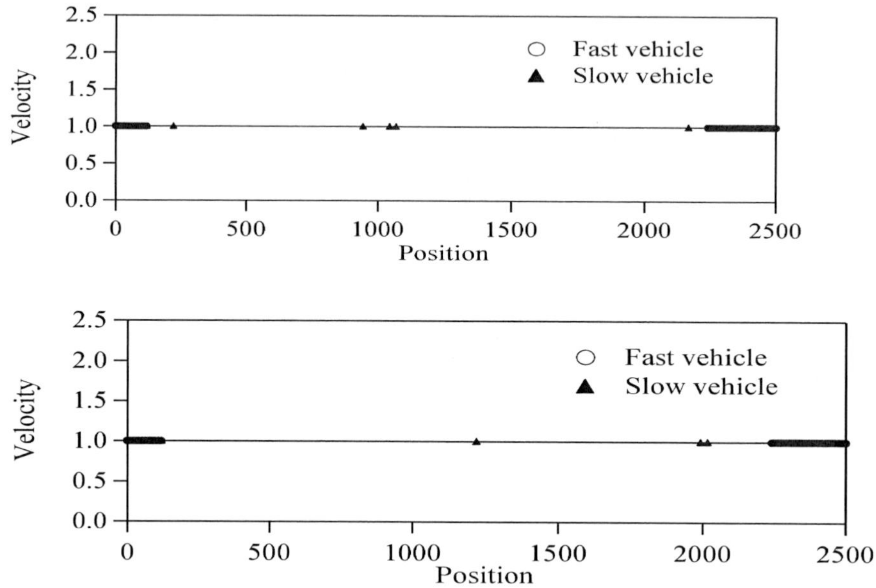

Figure 69. Velocity profiles corresponding to the low-current curve on open triangles in Figure 64. Plots of velocity against position on the two-lane highway at low density $\rho = 0.04$. The upper and lower diagrams represent the velocity profiles on the first and second lanes respectively.

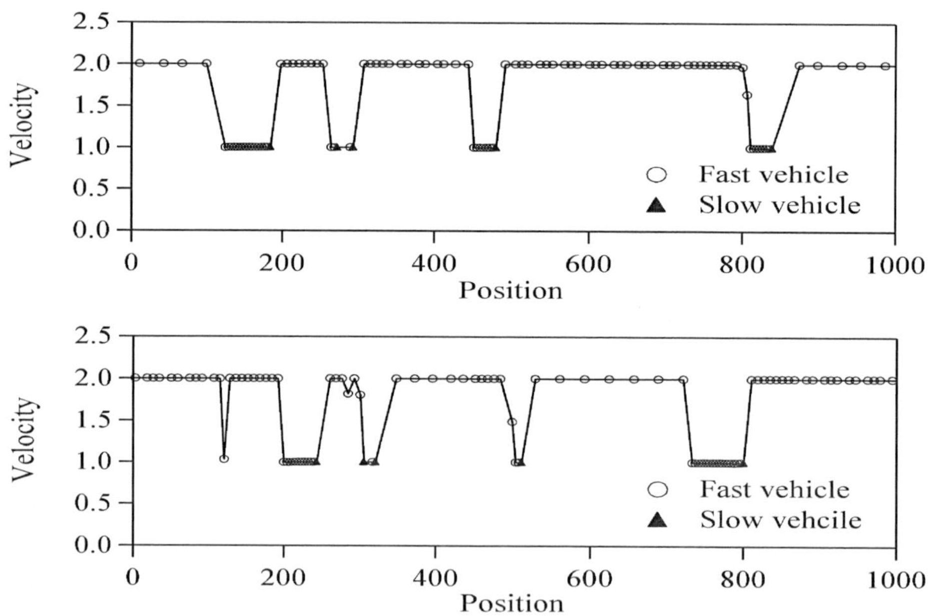

Figure 70. Velocity profiles corresponding to the high-current curve on open circles in Figure 64. Plots of velocity against position on the two-lane highway at low density $\rho = 0.10$. The upper and lower diagrams represent the velocity profiles on the first and second lanes respectively.

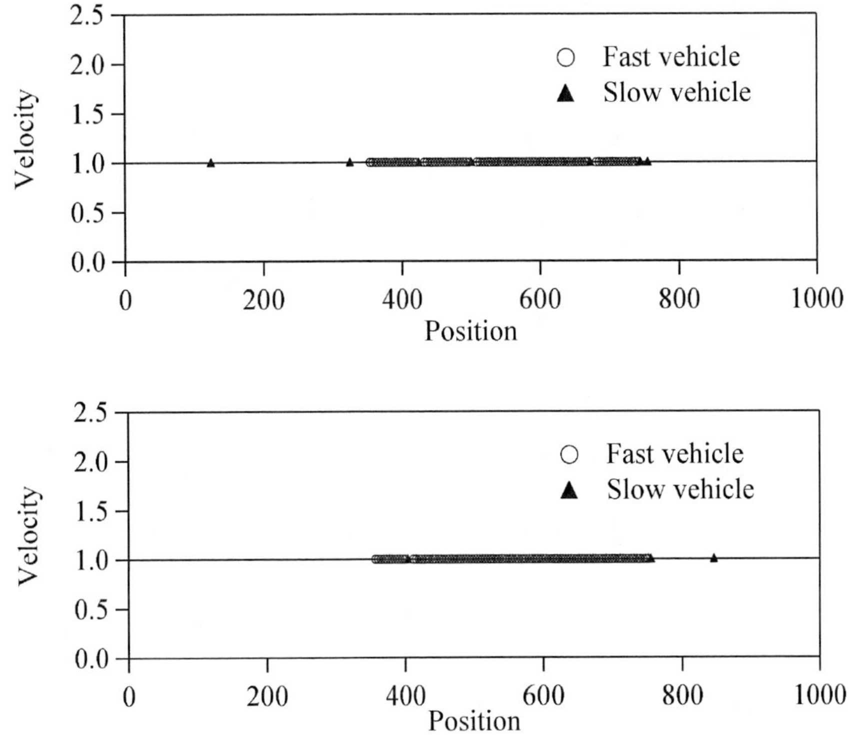

Figure 71. Velocity profiles corresponding to the low-current curve on open triangles in Figure 64. Plots of velocity against position on the two-lane highway at low density $\rho = 0.10$. The upper and lower diagrams represent the velocity profiles on the first and second lanes respectively.

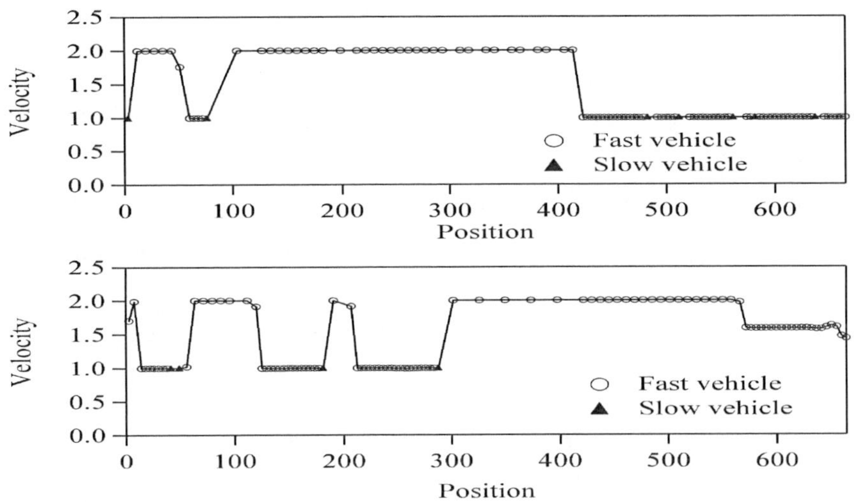

Figure 72. Velocity profiles at higher density $\rho = 0.15$, corresponding to the high-current curve on open circles in Figure 64. The upper and lower diagrams represent the velocity profiles on the first and second lanes respectively.

We study the velocity profile for various values of density. Figures 68 and 69 show the plots of velocity against position on the two-lane highway at low density $\rho = 0.04$ where $v_{s,\max} = 1.0$, sensitivity $a = 3.0$, total number of all vehicles is N=200, the number of slow vehicles is ten, and $x_c = 4.0$. Open circles indicate the velocity and position of fast vehicles. Full triangles indicate the velocity and position of slow vehicles. The upper and lower diagrams represent the velocity profiles on the first and second lanes respectively. The velocity profiles in Figure 68 correspond to the high-current curve on open circles in Figure 64. The fast vehicles move at maximal velocity with changing lanes. No jams appear because the density is less than transition point *a*. The velocity profiles in Figure 69 correspond to the low-current curve on open triangles in Figure 64. The fast vehicles are not able to change lanes and move together with slow vehicles at $v_{s,\max} = 1.0$.

Figures 70 and 71 show the plots of velocity against position at density $\rho = 0.1$ higher than transition point *a*. Open circles indicate the velocity and position of fast vehicles. Full triangles indicate the velocity and position of slow vehicles. The velocity profiles in Figure 70 correspond to the high-current curve on open circles in Figure 64. The part of fast vehicles is blocked by the slow vehicle and move together with the slow vehicle. The small jams are formed behind the slow vehicles. The remaining part of fast vehicles moves at maximal velocity with changing lanes. The velocity profiles in Figure 71 correspond to the low-current curve on open triangles in Figure 64. The fast vehicles are not able to change lanes and move together with slow vehicles at $v_{s,\max} = 1.0$. Figure 72 shows the velocity profiles at higher density $\rho = 0.15$, corresponding to the high-current curve on open circles in Figure 64. The large jams are formed behind the slow vehicles. By comparing with Figure 70, the jam grows. Thus, the jams are formed behind the slow vehicles but the lane changing occurs in the high-current curve on open circles in Figure 64. While the fast vehicles are blocked by slow vehicles without changing lane in the low-current curve on open triangles in Figure 64. Thus, when the density is higher than transition point *a*, the jamming transition occurs from the free traffic to the jammed state in the high-current curve.

We study the fundamental diagram (current-density diagram) of traffic flow on the three-lane highway. Figure 73 shows the plots of current against density for $v_{s,\max} = 1.0$ at sensitivity $a = 3.0$ where total number of all vehicles is N=300, the number of slow vehicles is 15, and $x_c = 4.0$. Open circles and open triangles indicate the simulation result. Two values of current are obtained. They depend on the initial configuration of mixed vehicles. Open circle represents the high current in such case that fast vehicles pass over slow vehicles with changing lanes. Open triangle represents the low current in such case that all fast vehicles move together with slow vehicles without changing lanes. The dashed line indicates the high-current curve of the two-lane traffic flow in Figure 64. The high-current curve is higher a little than that of two-lane traffic flow. The low currents represented by open triangles agree with those of the two-lane traffic flow in Figure 64. For $\rho_0 < \rho_a$ (ρ_a: density at transition point *a*) of the high-current curve, the traffic state is in the free traffic. At transition point *a*, the jamming transition occurs. The traffic state is in the inhomogeneous jammed state. The jams are formed behind slow vehicles but the part of fast vehicles can change the lane.

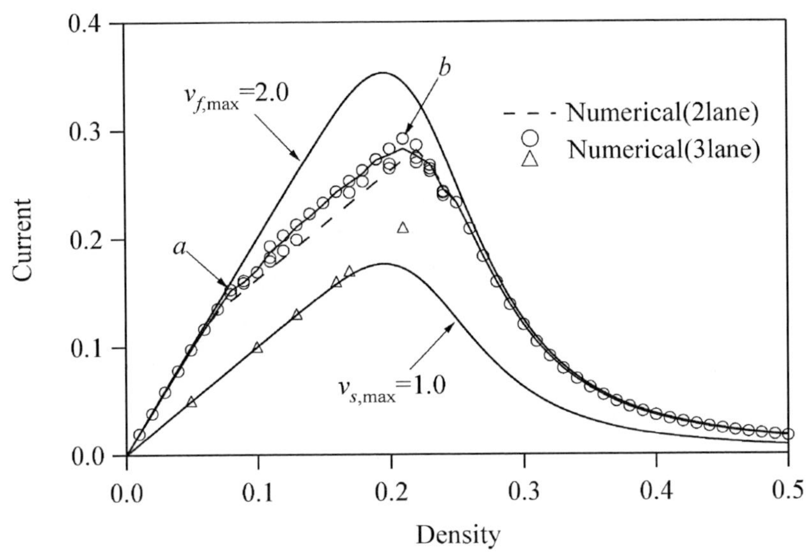

Figure 73. Plots of current against density for three-lane traffic flow for $v_{s,max} = 1.0$ at sensitivity $a = 3.0$ where total number of all vehicles is N=300, the number of slow vehicles is 15, and $x_c = 4.0$. Open circles and open triangles indicate the simulation result. The dashed line indicates the high-current curve of the two-lane traffic flow in Figure 64.

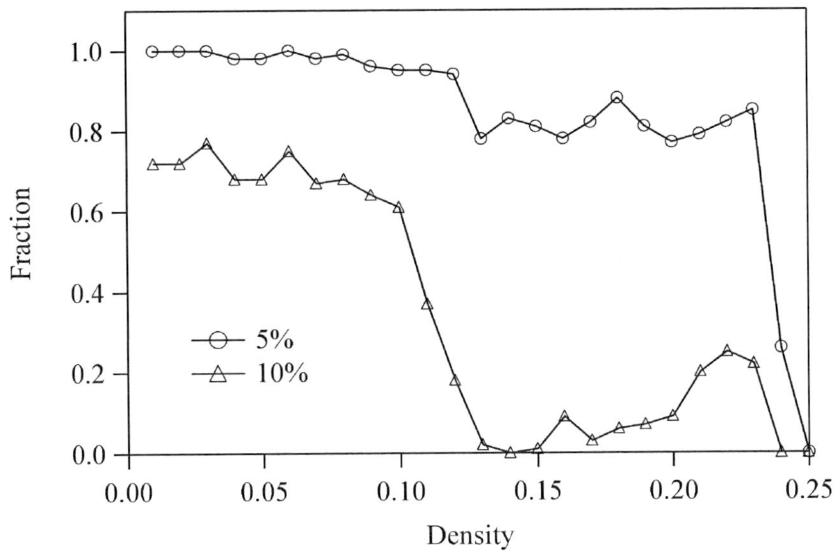

Figure 74. Relative frequency at which the high-current curve on open circles appeared. Fraction (relative frequency) of high current for 100 samples. Plot of fraction against density for concentrations 5% and 10% of slow vehicles.

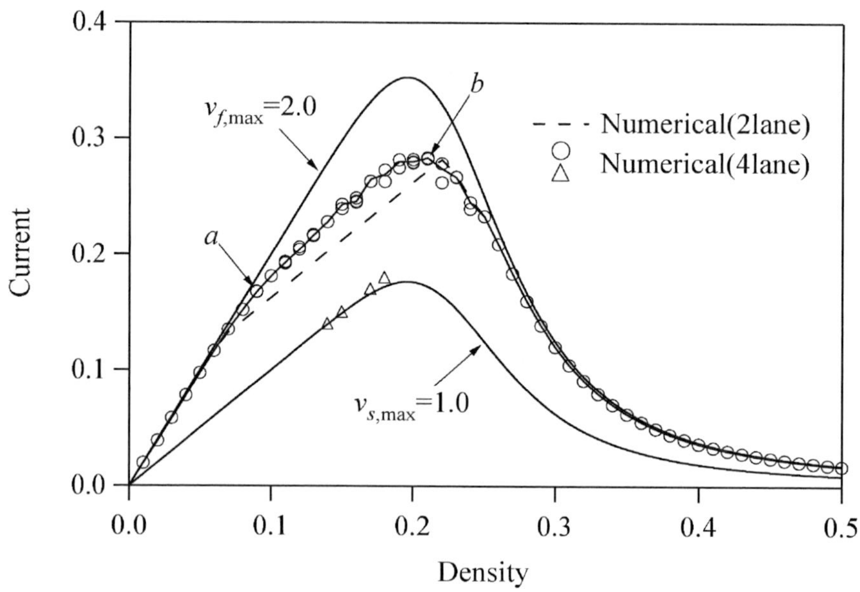

Figure 75. Fundamental diagram (current-density diagram) of traffic flow on the four-lane highway. Plots of current against density for $v_{s,\max} = 1.0$ at sensitivity $a = 3.0$ where total number of all vehicles is N=400, the number of slow vehicles is 20, and $x_c = 4.0$. Open circles and open triangles indicate the simulation result. The dashed line indicates the high- current curve of the two-lane traffic flow in Figure 64.

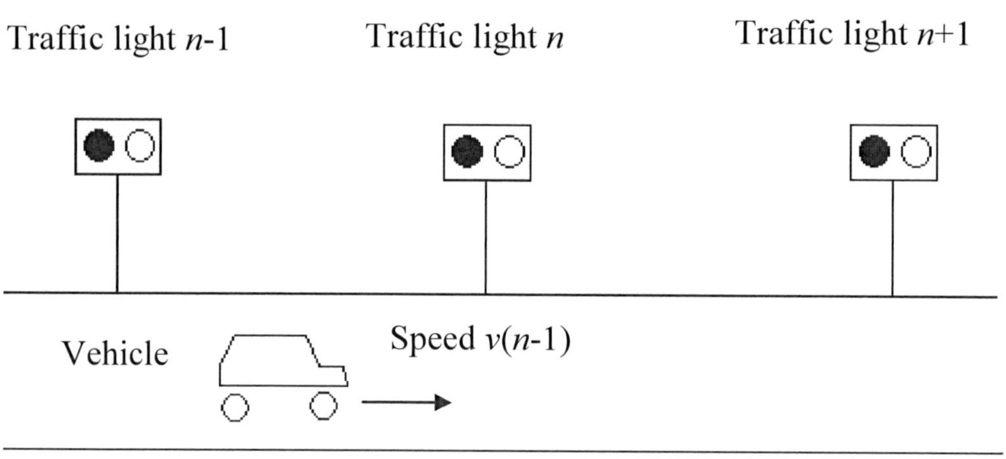

Figure 76. Schematic illustration of the single vehicle moving through a sequence of traffic lights. The traffic lights are numbered, from upstream to downstream, by 1, 2, 3, ---, *n*, *n*+1, ---. In the synchronized strategy, all the traffic lights change simultaneously from red (green) to green (red) with a fixed time period $t_s / 2$.

We study the frequency at which the high-current curve on open circles appeared. We calculate the fraction (relative frequency) of high current for 100 samples. Figure 74 shows the plots of fraction against density for concentrations 5% and 10% of slow vehicles when the vehicular position on the first lane is consistent with that on the second lane in the initial configuration but the configuration of fast and slow vehicles is random. The fractions of 5% and 10% are indicated by open circles and triangles respectively. The fraction takes about one until $\rho_0 < 0.12$ and then decreases a little with increasing density. The current of three-lane traffic flow takes high value almost over all density for concentration 5%. Fast vehicles move almost with changing lanes and passing over slow vehicles. Sometimes, the low current on open triangles appears. For concentration 10% of slow vehicles, the fraction decreases abruptly at $\rho_0 = 0.12$. For $\rho_0 > 0.13$, the low-current traffic occurs frequently.

We study the fundamental diagram (current-density diagram) of traffic flow on the four-lane highway. Figure 75 shows the plots of current against density for $v_{s,\max} = 1.0$ at sensitivity $a = 3.0$ where total number of all vehicles is *N*=400, the number of slow vehicles is 20, and $x_c = 4.0$. Open circles and open triangles indicate the simulation result. Two values of current are obtained. They depend on the initial configuration of mixed vehicles. Open circle represents the high current in such case that fast vehicles pass over slow vehicles with changing lanes. Open triangle represents the low current in such case that all fast vehicles move together with slow vehicles without changing lanes. The dashed line indicates the high-current curve of the two-lane traffic flow in Figure 64. The high-current curve is higher than that of two-lane traffic flow in Figure 64. Also, the high current-curve is higher a little bit than that of three-lane traffic flow in Figure 73. Furthermore, the frequency of low current is less than that in the three-lane traffic flow. Thus, the frequency of high current increases with number of lanes. The low current occurs rare in the four-lane traffic flow.

6. Vehicular Traffic Controlled by Traffic Signals

In urban traffic, the vehicles are controlled by traffic lights to give priority for a road because the city traffic networks often exceed the capacity. The flow throughout depends highly on both the cycle time and strategy. The dynamical state of traffic changes by varying the cycle time and strategy. Here, we study the traffic of a single vehicle moving through an infinite series of traffic lights [91-93].

6.1. Vehicular Behavior at the Synchronized and Green-wave Strategies

Until now, one has studied the periodic traffic controlled by a few traffic lights. It has been concluded that the periodic traffic does not depend on the number of traffic lights [94, 95]. Little works have been known for the vehicle traffic moving through an infinite series of traffic lights. Here, we study the traffic of a single vehicle moving through an infinite series of traffic lights, which are periodically positioned with a constant distance on a single-lane roadway and controlled by the synchronized and green-wave strategies. We present a nonlinear-map model to describe the dynamics of vehicle traffic controlled by traffic lights. We investigate the dynamical behavior of a single vehicle by iterating the nonlinear map.

We consider the motion of a single vehicle going through an infinite series of traffic lights. The traffic lights are periodically positioned with distance l. The vehicle moves with the mean speed v between a traffic light and its next light. Figure 76 shows the schematic illustration of the single vehicle moving through a sequence of traffic lights. The traffic lights are numbered, from upstream to downstream, by 1, 2, 3, ---, n, n+1, ---. In the synchronized strategy, all the traffic lights change simultaneously from red (green) to green (red) with a fixed time period $t_s/2$. The traffic lights flip periodically at regular time interval $t_s/2$. Time t_s is called the cycle time. When a vehicle arrives at a traffic light and if the traffic light is red, the vehicle stops at the position of the traffic light. Then, when the traffic light changes from red to green, the vehicle goes ahead. On the other hand, when a vehicle arrives at a traffic light and if the traffic light is green, the vehicle does not stop and goes ahead without changing speed.

We define the arrival time of the vehicle at traffic light n as $t(n)$. The arrival time at traffic light n+1 is given by

$$t(n+1) = t(n) + l/v + (r(n) - t(n))H(\sin(2\pi t(n)/t_s))$$

$$\text{with } r(n) = (\text{int}(t(n)/t_s) + 0.5)\cdot t_s\,, \tag{75}$$

where $H(t)$ is the Heaviside function: $H(t)=1$ for $t \geq 0$ and $H(t)=0$ for $t<0$. $H(t)=1$ if the traffic light is red, while $H(t)=0$ if the traffic light is green. $r(n)$ is such time that the traffic light just changed from red to green. The third term on the right hand side of Eq. (75) represents such time that the vehicle stops if traffic light n is red. The number n of iteration increases one by one accordingly the vehicle moves through the traffic light. The iteration corresponds to the going ahead on the highway.

In the green-wave strategy, the traffic light changes with a certain time delay τ between the traffic light phases of two successive intersections. The delay time is called the offset time. The change of traffic lights propagates backwards like a green wave. The arrival time at traffic light n+1 is given by

$$t(n+1) = t(n) + l/v + (r(n) - t(n) - n\tau)H(\sin(2\pi(t(n)+n\tau)/t_s))$$

$$\text{with } r(n) = (\text{int}((t(n)+n\tau)/t_s) + 0.5)\cdot t_s\,. \tag{76}$$

When $\tau = 0$, Eq. (76) reduces to Eq. (75).

Generally, a vehicle restarts after delay β when the traffic light changes from red to green. If one takes into account the restart's delay, the arrival time is given by

$$t(n+1) = t(n) + l/v + (r(n) - t(n) - n\tau + \beta)H(\sin(2\pi(t(n)+n\tau)/t_s))$$

with $r(n) = \left(\text{int}\left((t(n) + n\tau)/t_s\right) + 0.5\right) \cdot t_s$. (77)

By dividing time by the characteristic time l/v, one obtains the nonlinear equation of dimensionless arrival time:

$$T(n+1) = T(n) + 1 + \left(R(n) - T(n) - n\Pi + \Gamma\right) H\left(\sin(2\pi(T(n) + n\Pi)/T_s)\right)$$

with $R(n) = \left(\text{int}\left((T(n) + n\Pi)/T_s\right) + 0.5\right) \cdot T_s$, (78)

where $T(n) = t(n)v/l$, $R(n) = r(n)v/l$, $T_s = t_s v/l$, $\Pi = \tau \cdot v/l$, and $\Gamma = \beta \cdot v/l$. Thus, the dynamics of the vehicle is described by the nonlinear map (78). The motion of a vehicle depends on dimensionless cycle time $T_s = t_s v/l$, dimensionless delay (offset time) $\Pi = \tau \cdot v/l$, and dimensionless restart's delay $\Gamma = \beta \cdot v/l$. In the synchronized strategy, the vehicle depends only on dimensionless cycle time $T_s = t_s v/l$.

The length between two traffic lights or the mean speed of the vehicle vary from traffic light to traffic light. It is necessary and important to take into account the fluctuation of the length or speed. We consider the fluctuation as a noise. We extend the deterministic model to take into account the noise. The fluctuation is defined as $\xi(n) \equiv (l(n)/v(n) - \langle l \rangle / \langle v \rangle)/(\langle l \rangle / \langle v \rangle)$. The extended version of Eq. (78) is given by

$$T(n+1) = T(n) + 1 + \xi(n) + \left(R(n) - T(n) - n\Pi + \Gamma\right) H\left(\sin(2\pi(T(n) + n\Pi)/T_s)\right)$$

with $R(n) = \left(\text{int}\left((T(n) + n\Pi)/T_s\right) + 0.5\right) \cdot T_s$, (79)

where $\xi(n)$ is the white noise: $\langle \xi \rangle = 0$ and $\langle \xi(n)\xi(m) \rangle = \eta^2 \delta_{nm}$.

We study how the motion changes by varying the cycle time. We investigate the dynamical behavior of a single vehicle through an infinite series of traffic lights by iterating maps (78) or (79). We calculate the tour time between two traffic lights when the vehicle goes ahead on the highway. We study how the tour time varies with the cycle time and strategy. The tour time between a traffic light and its proceeding light is defined as *DT*.

First, we study the dynamical behavior of the vehicle at the synchronized strategy ($\Pi = 0$ and $\Gamma = 0$). Figure 77(a) shows the plot of the tour time *DT* against cycle time T_s for sufficiently large number $n = 1000 - 3000$ at the synchronized strategy. The tour time increases linearly from point (1,1) to point (2,2) accordingly the vehicle goes ahead. This means that the vehicle stops at all traffic lights and then goes. When the cycle time is higher than 2, the vehicle moves with two values of the tour time. The vehicle moves periodically. The vehicle stops at traffic lights or goes without stopping at traffic lights accordingly the traffic lights turn on and off. The horizontal line from point (2,1) to point $(\infty,1)$ represents the free movement due to the green of all traffic lights. For $T_s \geq 2$, the tour time *DT* is given by

$$DT = (T_s - 2i) + i \text{ or } DT = 1.0$$

with $i = \text{int}[T_s / 2]$. (80)

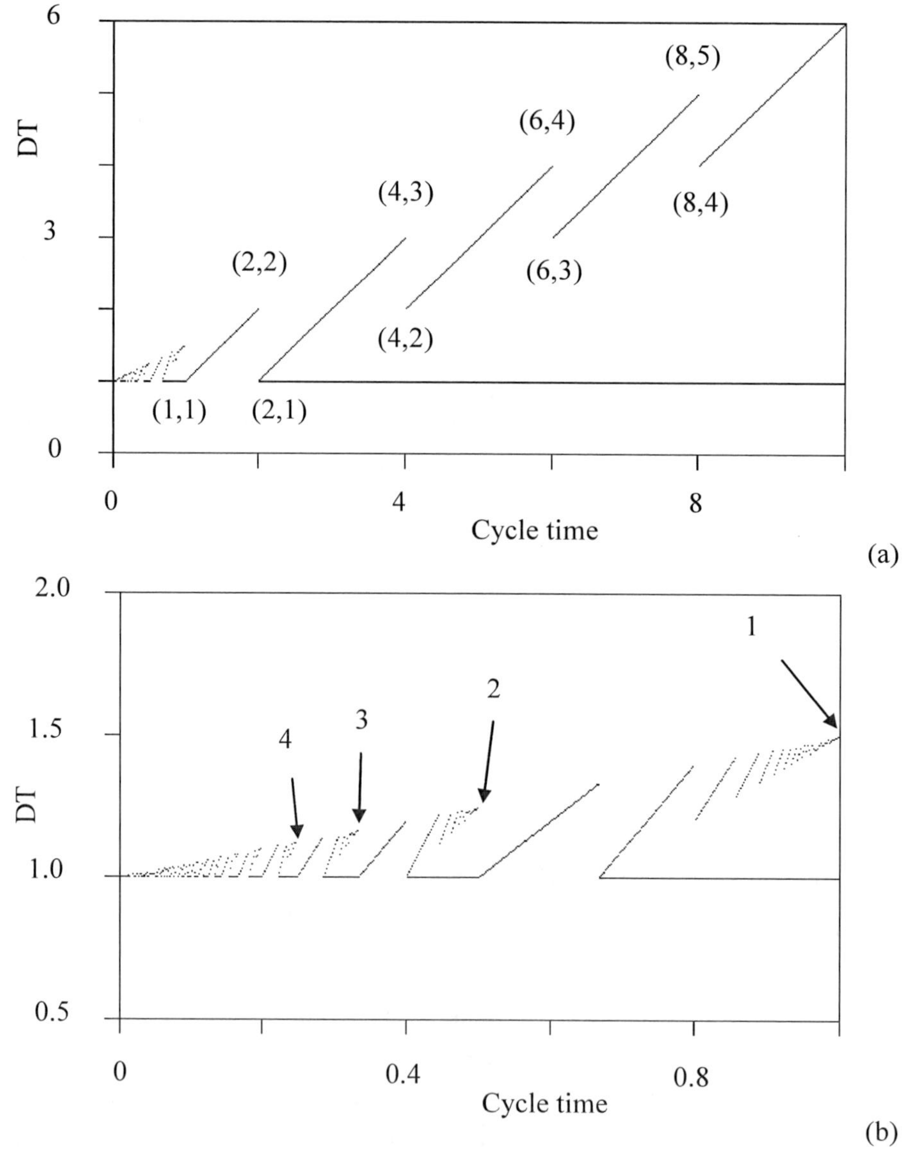

Figure 77. (a) Plot of the tour time *DT* against cycle time T_s for sufficiently large number $n = 1000 - 3000$ at the synchronized strategy. (b) Enlargement of diagram (a) for $0 \leq T_s \leq 1.0$. Points 1, 2, 3, --- represent the cumulative points.

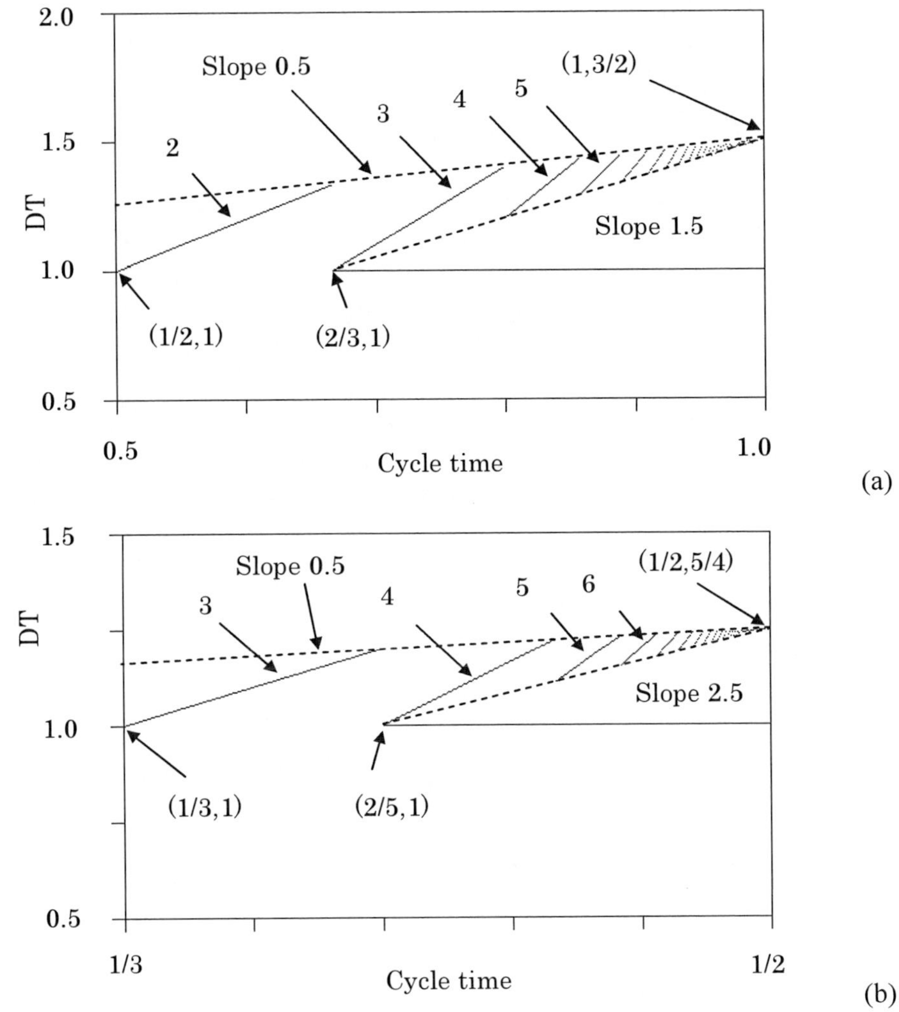

Figure 78. (a) Enlargement of Figure 77(b) for $1/2 \leq T_s \leq 1.0$. (b) Enlargement of Figure 77(b) for $1/3 \leq T_s \leq 1/2$.

The tour time exhibits the periodic structure for $T_s \geq 2$. The envelope is represented by the straight line which is given by $T_s/2+1$. The enlargement of diagram (a) is shown in Figure 77(b) for $0 \leq T_s \leq 1.0$. Points 1, 2, 3, --- represent the cumulative points. Their coordinates are given, respectively, by (1, 3/2), (1/2, 5/4), (1/3, 7/6), ---. Generally, the coordinate of cumulative point *k* is given by

$$\left(\frac{1}{k}, \frac{1}{2k}+1\right). \tag{81}$$

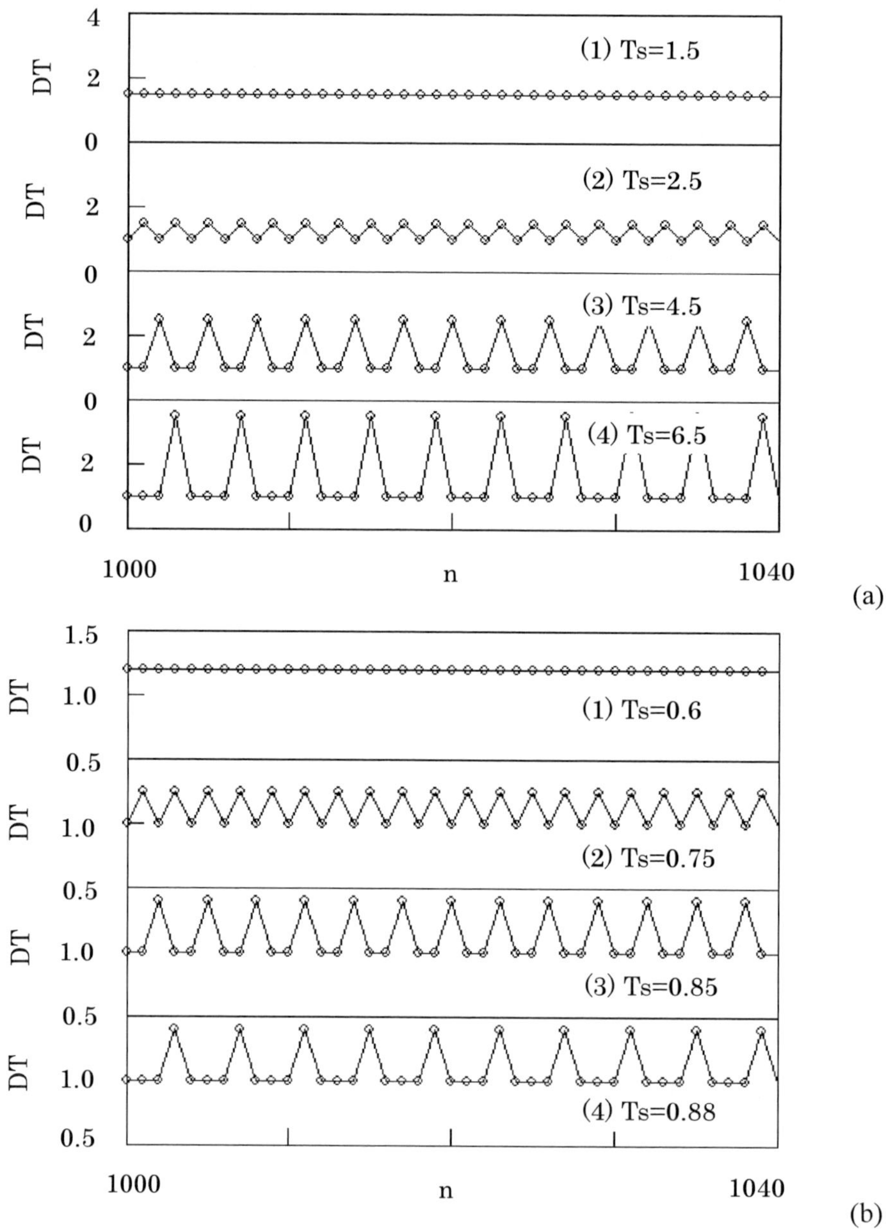

Figure 79. Plots of the tour time DT against traffic-light's number n for n=1000-1040. (a) Plots of the tour time DT are shown for (1) $T_s = 1.5$, (2) $T_s = 2.5$, (3) $T_s = 4.5$, and (4) $T_s = 6.5$. (b) Plots of the tour time DT for (1) $T_s = 0.6$, (2) $T_s = 0.75$, (3) $T_s = 0.85$, and (4) $T_s = 0.88$.

Their points are on the straight line of envelope: $T_s / 2 + 1$. The pattern from points 2 to 3 is self-similar to that from points 1 to 2. Furthermore, the pattern from points 3 to 4 is self-similar to that from points 2 to 3. The self-similar pattern is repeated. The enlargement of Figure 77(b) is shown in Figure 78(a) for $1/2 \le T_s \le 1.0$ and in Figure 78(b) for

$1/3 \le T_s \le 1/2$. The slopes of two envelops represented by dotted lines are given by 0.5 and 1.5 in Figure 78(a). The values of slopes of the segments are given, in order, by 2, 3, 4, 5, ---. The slopes of two envelops represented by dotted lines are given by 0.5 and 2.5 in Figure 78(b). The values of slopes of the segments are given, in order, by 3, 4, 5, 6, ---. The magnification factor of the self-similar pattern is not constant but varies with increasing *k*. The pattern of tour time for $0 \le T_s \le 1$ is not fractal but self-similar because the magnification factor is not constant. Thus, the vehicle exhibits the self-similar behavior for $T_s \le 1$ and the periodic behavior for $T_s > 1$. The dynamical behavior of the vehicle changes at $T_s = 1$.

We study the variation of tour time with increasing traffic-light's number *n*. Figure 79 shows the plots of the tour time *DT* against number *n* for *n*=1000-1040. The plots of the tour time *DT* are shown for (1) $T_s = 1.5$, (2) $T_s = 2.5$, (3) $T_s = 4.5$, and (4) $T_s = 6.5$ in Figure 79(a). In diagram (1), the vehicle stops at all the traffic lights and then goes ahead. Tour time *DT* is higher than one and constant. In diagram (2), the vehicle stops every two traffic lights because tour time *DT* takes two distinct values: one is 1 and the other the value higher than 1. In diagram (3), the vehicle stops every three traffic lights. In diagram (4), the vehicle stops every four traffic lights. These traffic states (1)-(4) correspond, respectively, to the lines of (1,1)-(2,2), (2,1)-(4,3), (4,2)-(6,4), and (6,3)-(8,5) in Figure 77(a). "Generally, the vehicle stops every (*i*+1)-th traffic light in the line of Eq. (80)." For a sufficiently long cycle time, the vehicle seldom stops at traffic lights. For $T_s < 1$, the plots of the tour time *DT* are shown for (1) $T_s = 0.6$, (2) $T_s = 0.75$, (3) $T_s = 0.85$, and (4) $T_s = 0.88$ in Figure 79(b). In diagram (1), the vehicle stops at all the traffic lights and then goes ahead. Tour time *DT* is higher than one and constant. In diagram (2), the vehicle stops every two traffic lights because tour time *DT* takes two distinct values: one is 1 and the other the value higher than 1. In diagram (3), the vehicle stops every three traffic lights. In diagram (4), the vehicle stops every four traffic lights. These traffic states (1)-(4) correspond, respectively, to the lines for $0.5 \le T_s \le 1$ in Figure 78(a). When cycle time T_s approaches one, the vehicle seldom stops at the traffic lights and the tour time approaches the cumulative point 1 in Figure 77(b).

We study the dynamical behavior of the vehicle at the green-wave strategy ($\Pi > 0$ and $\Gamma = 0$). Figure 80(a) shows the plot of the tour time *DT* against cycle time T_s for sufficiently large number $n = 1000 - 3000$ at the green-wave strategy of $\Pi = 0.2$. The tour time increases linearly from point $(1 + \Pi, 1)$ to point $(2(1 + \Pi), 2 + \Pi)$ accordingly the vehicle goes ahead. This means that the vehicle stops at all traffic lights and then goes. When the cycle time is higher than $2(1 + \Pi)$, the vehicle stops every two traffic lights and passes through green lights alternately. The horizontal line from point $(2(1 + \Pi), 1)$ to point $(\infty, 1)$ represents the free movement due to the green of all traffic lights. For $T_s \ge 2(1 + \Pi)$, the tour time *DT* is given by

$$DT = (T_s - 2\Pi - 2i) + i \text{ or } DT = 1.0$$

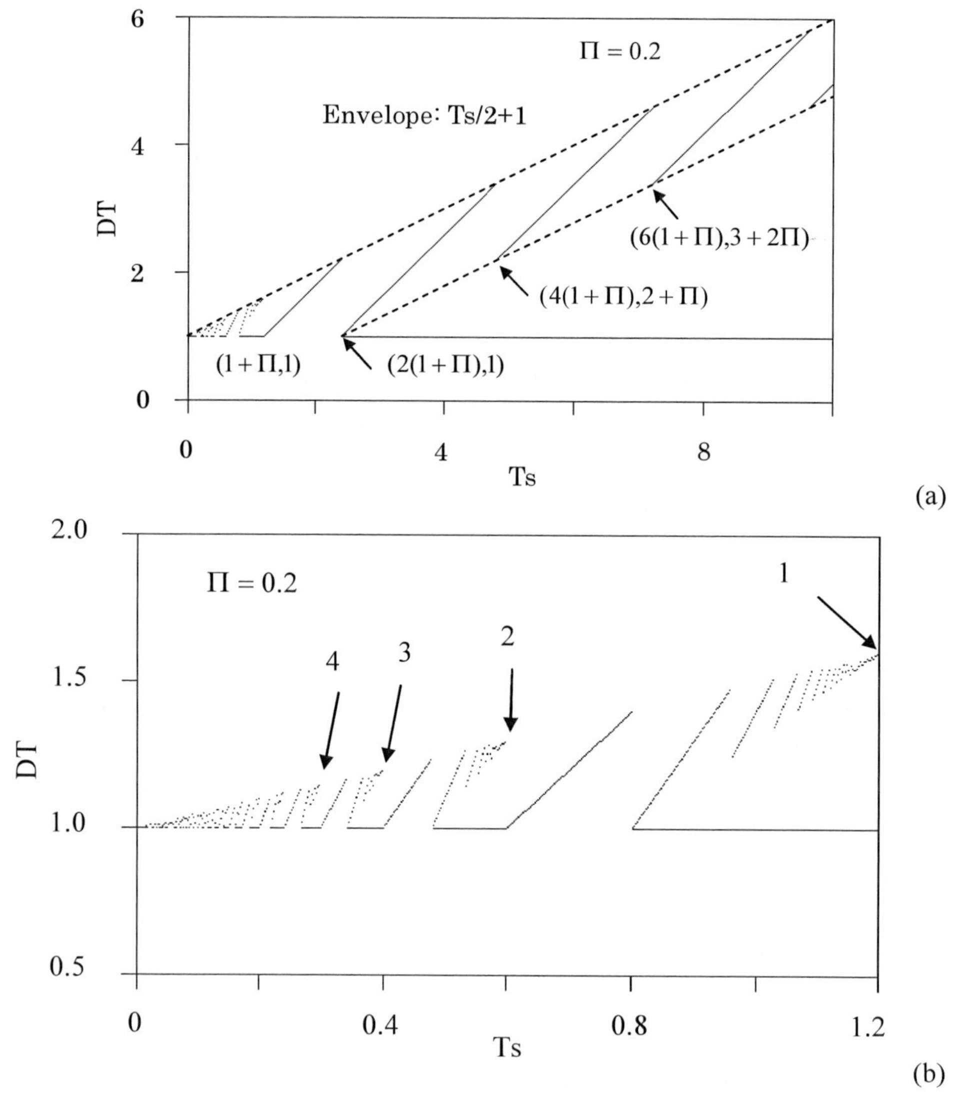

Figure 80. (a) Plot of the tour time DT against cycle time T_s for sufficiently large number $n = 1000 - 3000$ at the green-wave strategy of $\Pi = 0.2$. (b) Enlargement of diagram (a) for $0 \le T_s \le 1.2$.

with $i = \mathrm{int}[(T_s - \Pi)/2]$. (82)

The tour time exhibits the periodic structure for $T_s \ge 2(1+\Pi)$. The envelope is represented by the straight line which is given by $T_s/2+1$. The lines of tour time for the green-wave strategy is compared with Eq. (80) for the synchronized strategy. The enlargement of Figure 80(a) is shown in Figure 80(b) for $0 \le T_s \le 1.2$. Points 1, 2, 3, --- represents the cumulative points. The coordinate of cumulative point k is given by

$$\left(\frac{1+\Pi}{k}, \frac{1+\Pi}{2k}+1\right). \tag{83}$$

Their points are on the straight line of envelope: $T_s/2+1$. The pattern from points 2 to 3 is self-similar to that from points 1 to 2. Furthermore, the pattern from points 3 to 4 is self-similar to that from points 2 to 3. The self-similar pattern is repeated. The values of slopes of the segments are the same as those of the synchronized strategy. The pattern of tour time for $0 \le T_s \le 1+\Pi$ is not fractal but self-similar because the magnification factor is not constant. Thus, the vehicle exhibits the self-similar behavior for $T_s \le 1+\Pi$ and the periodic behavior for $T_s > 1+\Pi$. The dynamical behavior of the vehicle changes at $T_s = 1+\Pi$.

6.2. Split Effect on Vehicular Traffic

Traffic depends highly on both cycle time and split where the split of signal is the fraction of green time to the signal period. The effect of cycle time on vehicular traffic has been classified in the previous subsection. However, the split effect has not been done for the vehicular traffic through a series of signals.

We study the following case: the signals are periodically positioned with a constant distance on a single-lane roadway, controlled by the synchronized strategy, and turn on or off with a cycle time and a split. We present an extended model of nonlinear map to take into account both cycle time and split of traffic signals.

Each vehicle passes freely over other vehicles. Then, each vehicle does not depend on the other and is uncorrelated with the other vehicles. Therefore, we consider the dynamical behavior of a single vehicle. The traffic lights are periodically positioned with distance l. The vehicle moves with the mean speed v between a traffic light and its next light. In the synchronized strategy, all the traffic lights change simultaneously from red (green) to green (red) with a fixed time period $(1-s_p)t_s$ ($s_p t_s$). The period of green is $s_p t_s$ and the period of red is $(1-s_p)t_s$. Time t_s is called the cycle time and fraction s_p represents the split which indicates the ratio of green time to cycle time. When a vehicle arrives at a traffic light and if the traffic light is red, the vehicle stops at the position of the traffic light. Then, when the traffic light changes from red to green, the vehicle goes ahead. On the other hand, when a vehicle arrives at a traffic light and if the traffic light is green, the vehicle does not stop and goes ahead without changing speed.

The arrival time at traffic light $n+1$ is given by

$$t(n+1) = t(n) + l/v + \big(r(n) - t(n)\big)H\big(t(n) - (\mathrm{int}(t(n)/t_s)t_s) - s_p t_s\big)$$

with $r(n) = \big(\mathrm{int}\big(t(n)/t_s\big)+1\big)\cdot t_s$, (84) where $H(t)$ is the Heaviside function: $H(t)=1$ for $t \geq 0$ and $H(t)=0$ for $t<0$. $H(t)=1$ if the traffic light is red, while $H(t)=0$ if the traffic light is green. l/v is the time it takes for the vehicle to move between lights n and n+1. $r(n)$ is such time that the traffic light just changed from red to green. The third term on the right hand side of Eq. (84) represents such time that the vehicle stops if traffic light n is red. Eq. (84) is the extended version of nonlinear map model for split $s_p = 0.5$.

By dividing time by the characteristic time l/v, one obtains the nonlinear equation of dimensionless arrival time:

$$T(n+1) = T(n) + 1 + (R(n) - T(n))H(T(n) - (\text{int}(T(n)/T_s)T_s) - s_p T_s)$$

with $R(n) = (\text{int}(T(n)/T_s) + 1) \cdot T_s$, (85)

where $T(n) = t(n)v/l$, $R(n) = r(n)v/l$, and $T_s = t_s v/l$. Thus, the dynamics of the vehicle is described by the nonlinear map (85). The motion of a vehicle depends on both dimensionless cycle time $T_s = t_s v/l$ and split s_p.

Generally, the length between two traffic lights or the mean speed of the vehicle vary from traffic light to traffic light. It is necessary and important to take into account the fluctuation of the length or speed. One can take into account the fluctuation as a noise.

We investigate the effect of both cycle time and split on the motion of a single vehicle through an infinite series of traffic lights by iterating map (85). We calculate the tour time between two traffic lights when the vehicle goes ahead on the highway. We study how the tour time varies with both cycle time and split. The tour time between a traffic light and its proceeding light is defined as *DT*. We study the dynamical behavior of the signal traffic at the synchronized strategy. Figure 81(a)-(d) show the plots of the tour time *DT* against cycle time T_s for sufficiently large number $n = 400-500$ at (a) split $s_p = 0.25$, (b) $s_p = 0.5$, (c) $s_p = 0.6$, and (d) $s_p = 0.75$. In Figure 81(a) of split $s_p = 0.25$, the tour time increases linearly from point (1,1) to point (4,4) as the vehicle goes ahead. This means that the vehicle stops at all traffic lights and then goes. When the cycle time is higher than 4, the vehicle moves with two values of the tour time. The vehicle moves periodically. The vehicle stops at traffic lights or goes without stopping at traffic lights as the traffic lights turn on and off. The horizontal line from point (4,1) to point $(\infty,1)$ represents the free movement due to the green of all traffic lights. The straight line of slope 1 starting at point (4,3) represents the tour time at which the vehicle stops every two traffic lights. The envelope is represented by the dotted line. It is given by

$$DT = (1 - s_p)T_s + 1. \tag{86}$$

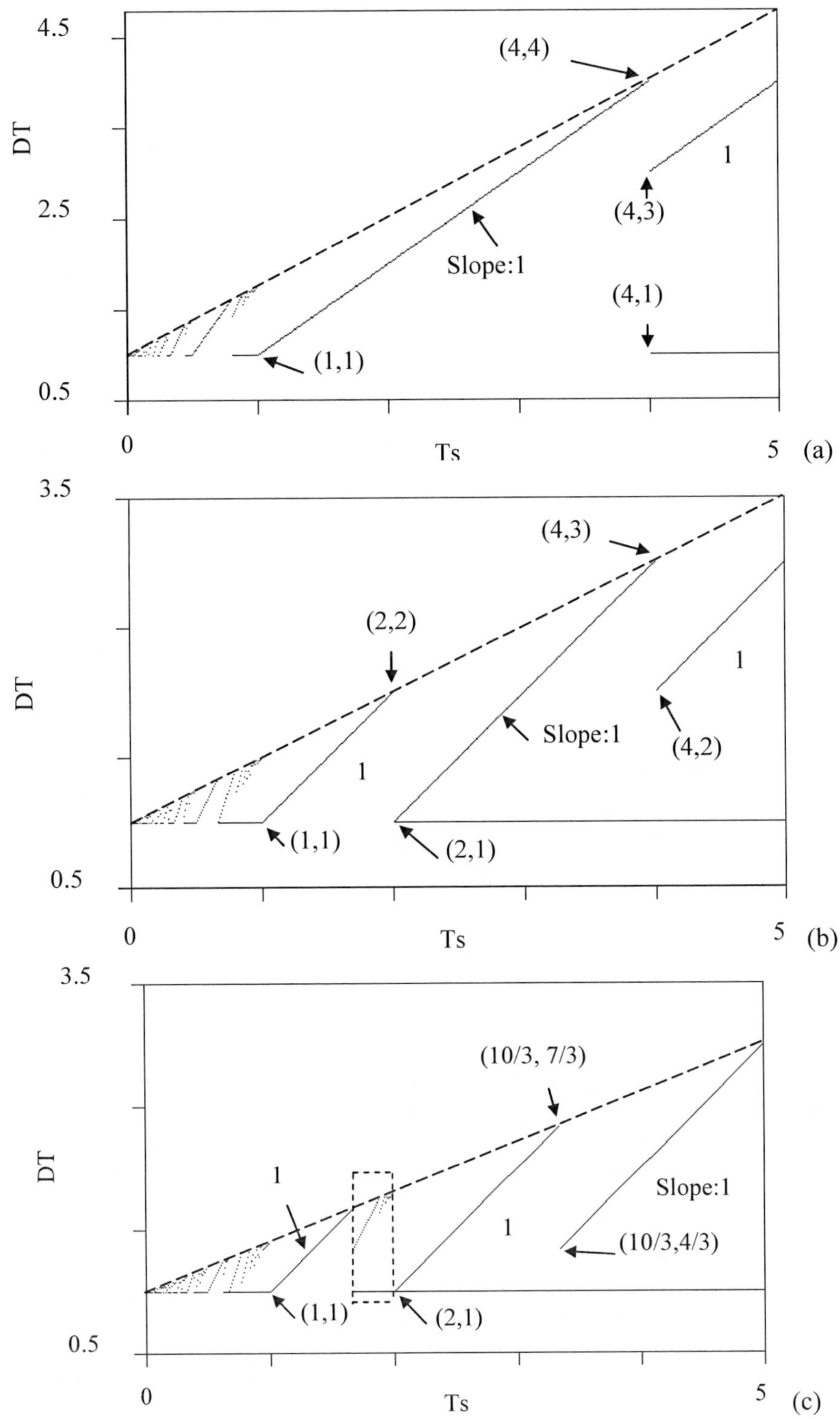

Figure 81. (Continued)

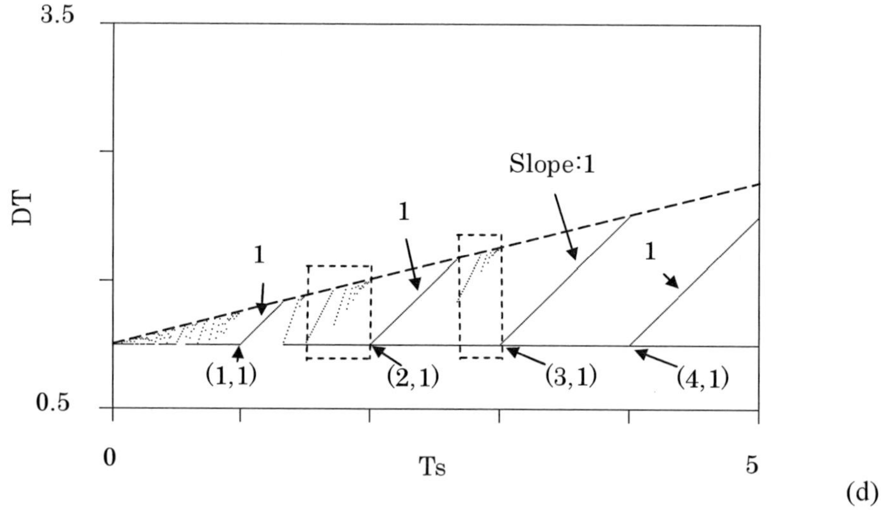

Figure 81. Plots of tour time DT against cycle time T_s for sufficiently large number $n = 400 - 500$ at (a) split $s_p = 0.25$, (b) $s_p = 0.5$, (c) $s_p = 0.6$, and (d) $s_p = 0.75$.

In Figure 81(b) of split $s_p = 0.5$, the tour time increases linearly from point (1,1) to point (2,2) as the vehicle goes ahead. This means that the vehicle stops at all traffic lights and then goes. When the cycle time is higher than 2, the vehicle moves with two values of the tour time. The vehicle moves periodically. The vehicle stops at traffic lights or goes without stopping at traffic lights as the traffic lights turn on and off. The horizontal line from point (2,1) to point $(\infty,1)$ represents the free movement due to the green of all traffic lights. The straight line of slope 1 starting at point (2,1) represents the tour time at which the vehicle stops every two traffic lights. The envelope is represented by the dotted line which is given by Eq. (86). For cycle time $T_s > 1$, the slope of all segments connecting the envelope have value 1.

In Figure 81(c) of split $s_p = 0.6$, the tour time increases linearly from point (1,1) to point (5/3,5/3) as the vehicle goes ahead. This means that the vehicle stops at all traffic lights and then goes. When the cycle time is higher than 5/3, the vehicle moves with two values of the tour time. The vehicle moves periodically. The vehicle stops at traffic lights or goes without stopping at traffic lights as the traffic lights turn on and off. The horizontal line from point (5/3,1) to point $(\infty,1)$ represents the free movement due to the green of all traffic lights. The straight line of slope 1 starting at point (2,1) represents the tour time at which the vehicle stops every two traffic lights. The envelope is represented by the dotted line which is given by Eq. (86). For cycle time $T_s > 2$, the slope of all segments connecting the envelope have value 1. In the region enclosed by the dotted square, a new branch appears in the tour time diagram.

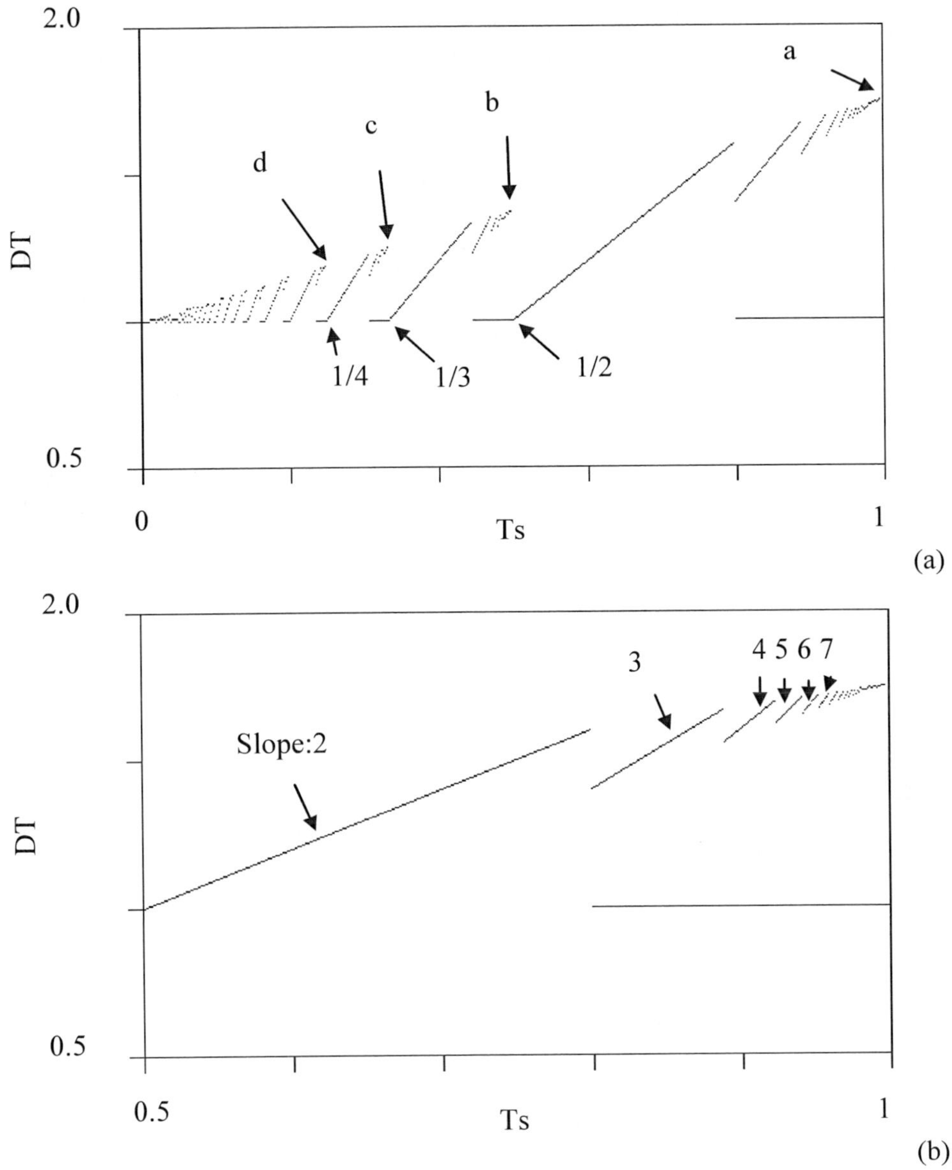

Figure 82. (a) Enlargement of Figure 81(a) (split $s_p = 0.25$) is shown for $0 \le T_s \le 1.0$. Points a, b, c, --- represent the cumulative points. (b) Enlargement of diagram (a) is shown for $0.5 \le T_s \le 1.0$.

In Figure 81(d) of split $s_p = 0.75$, the tour time increases linearly from point (1,1) to point (4/3,4/3) as the vehicle goes ahead. This means that the vehicle stops at all traffic lights and then goes. When the cycle time is higher than 4/3, the vehicle moves with two values of the tour time. The vehicle moves periodically. The vehicle stops at traffic lights or goes without stopping at traffic lights as the traffic lights turn on and off. The horizontal line from point (4/3, 1) to point $(\infty,1)$ represents the free movement due to the green of all traffic lights. The straight line of slope 1 starting at point (2, 1) represents the tour time at which the vehicle stops every two traffic lights. The envelope is represented by the dotted line which is given by Eq. (86). For cycle time $T_s > 3$, the slope of all segments connecting the envelope have value 1. In the regions enclosed by the dotted squares, two new branches appear in the tour time diagram.

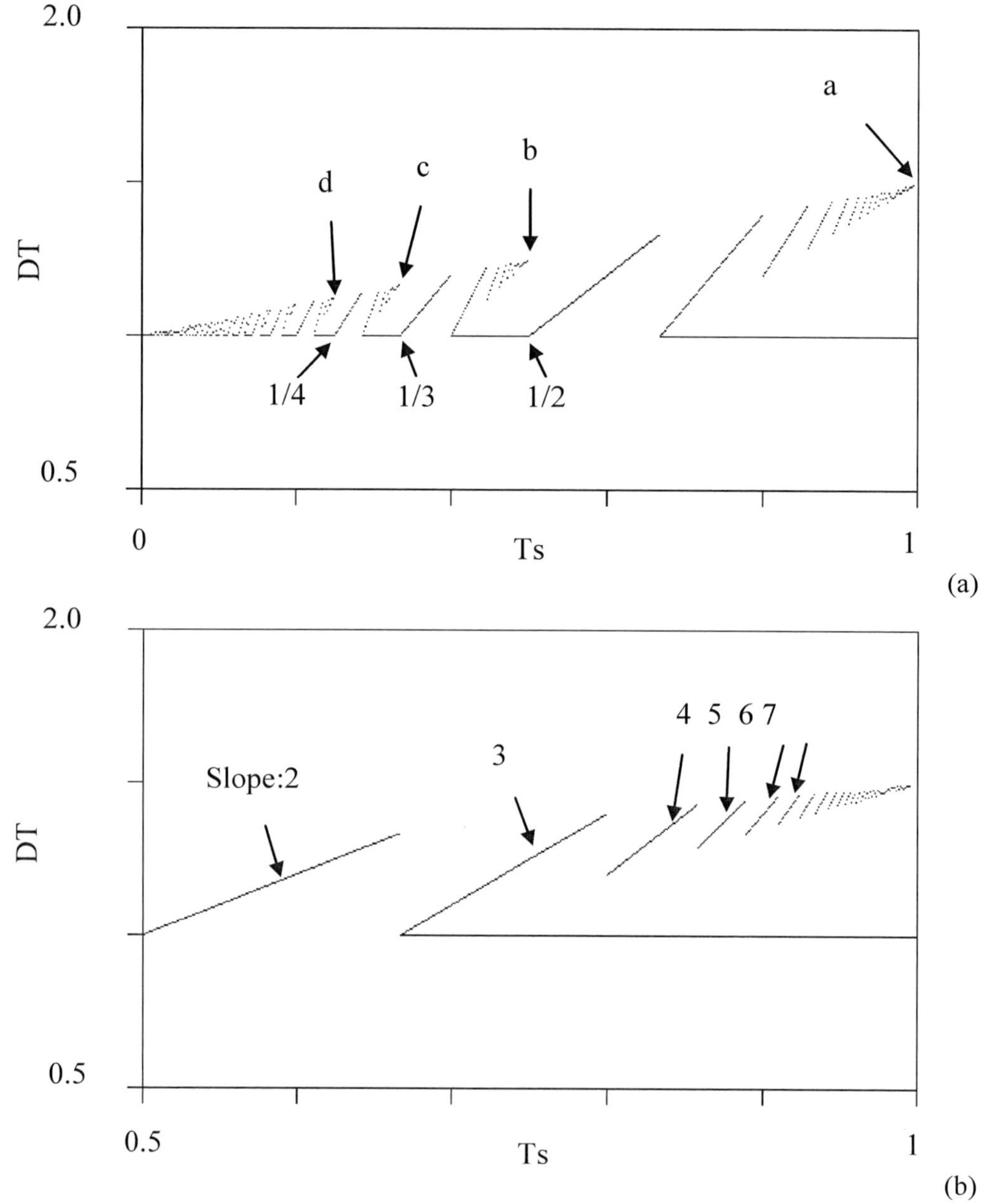

Figure 83. (a) Enlargement of Figure 81(b) ($s_p = 0.5$) is shown for $0 \le T_s \le 1.0$. Points a, b, c, --- represent the cumulative points. (b) Enlargement of diagram (a) is shown for $0.5 \le T_s \le 1.0$.

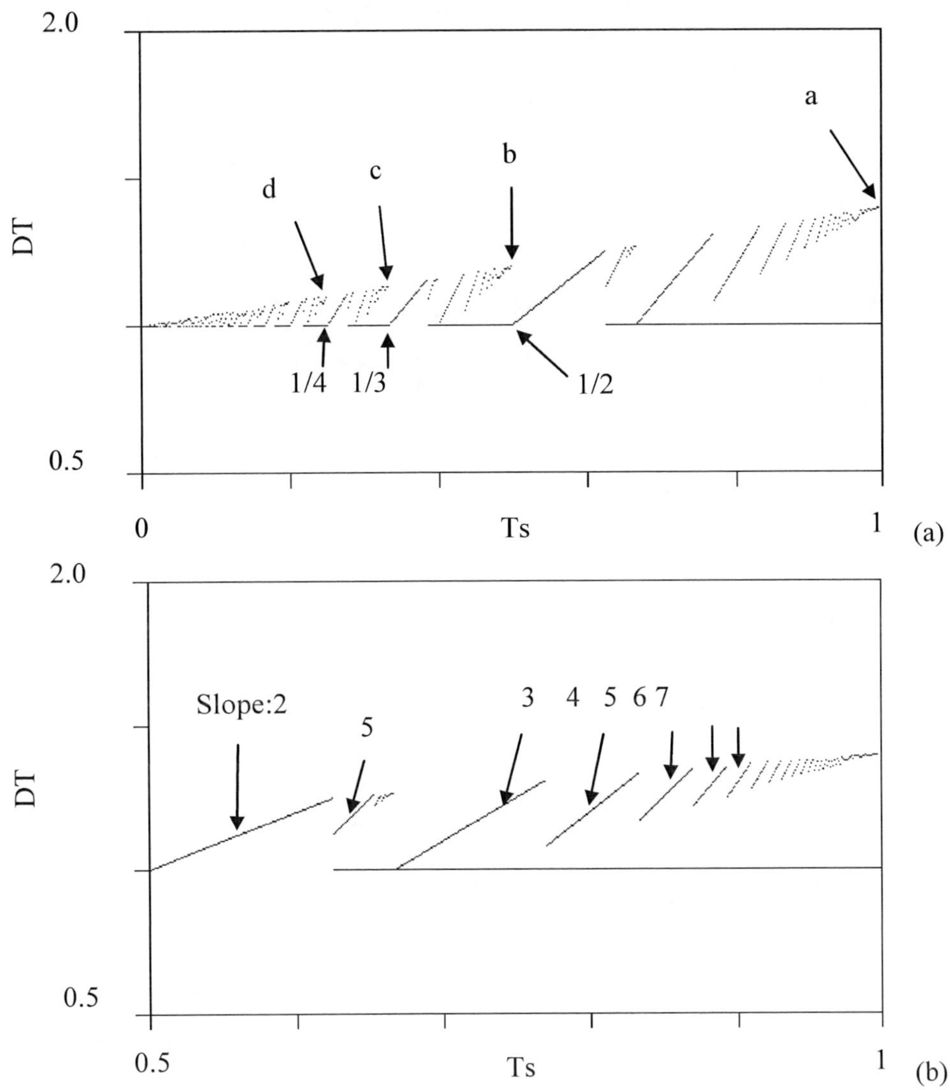

Figure 84. (a) Enlargement of Figure 81(c) ($s_p = 0.6$) is shown for $0 \le T_s \le 1.0$. Points a, b, c, --- represent the cumulative points. (b) Enlargement of diagram (a) is shown for $0.5 \le T_s \le 1.0$.

The enlargement of Figure 81(a) for split $s_p = 0.25$ is shown in Figure 82(a) for $0 \le T_s \le 1.0$. The enlargement of Figure 82(a) is shown in Figure 82(b) for $0.5 \le T_s \le 1.0$. Points *a*, *b*, *c*, --- represent the cumulative points. As a value of cycle time approaches a cumulative point from the below, a vehicle stops rare at signals and goes successfully through a sequence of green signals. The coordinate of a cumulative point is given by

$$\left(\frac{1}{k}, \frac{1-s}{k} + 1\right), \tag{87}$$

where k is an integer (k=1, 2, 3, ----).

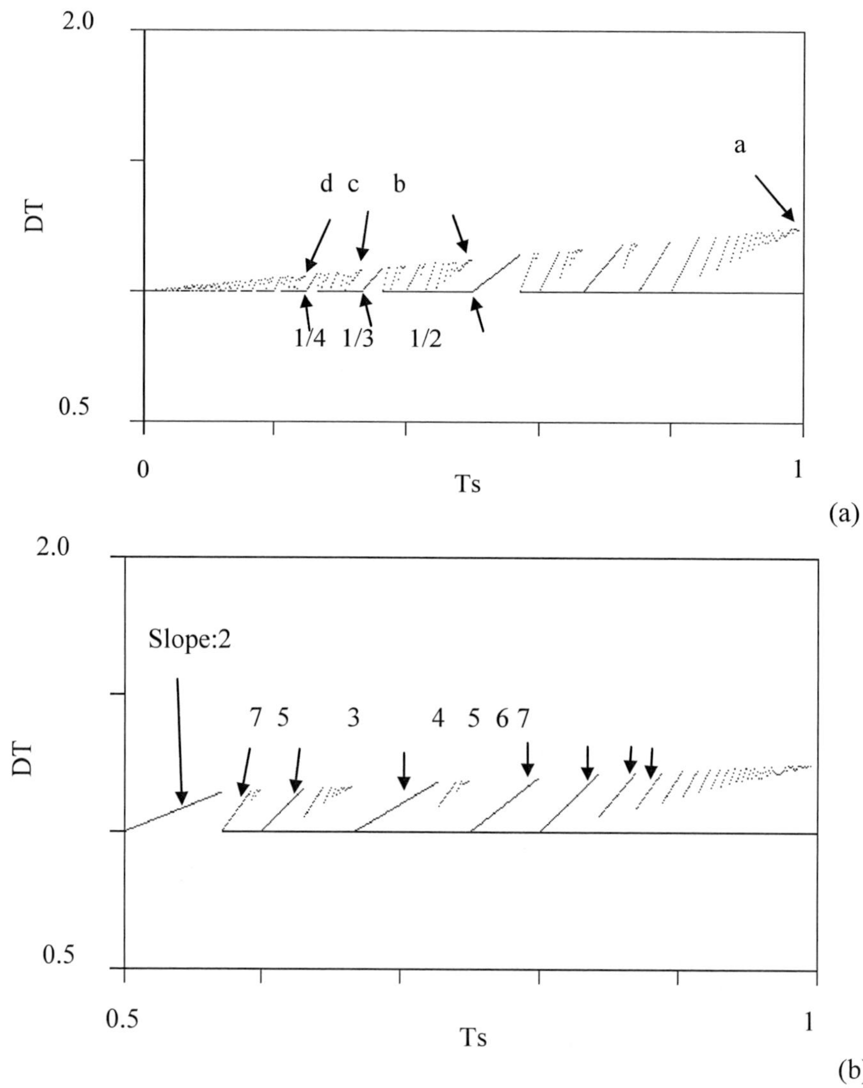

Figure 85. (a) Enlargement of Figure 81(d) ($s_p = 0.75$) is shown for $0 \le T_s \le 1.0$. Points a, b, c, --- represent the cumulative points. (b) Enlargement of diagram (a) is shown for $0.5 \le T_s \le 1.0$. Their points are on the straight line of envelope of Eq. (86). The pattern from points b to c is self-similar to that from points a to b. Furthermore, the pattern from points c to d is self-similar to that from points b to c. The self-similar pattern is repeated. However, the iterated pattern is not fractal because the scale factor is not constant. The numerical numbers 2, 3, 4, --- indicate, respectively, the slopes of segments of linear lines in Figure 82(b). The value of slope relates the period of vehicle's motion [28]. For example, the vehicle stops every two signals in the cycle time corresponding to slope 3. In the cycle time corresponding to slope 7, the vehicle stops every 6 signals.

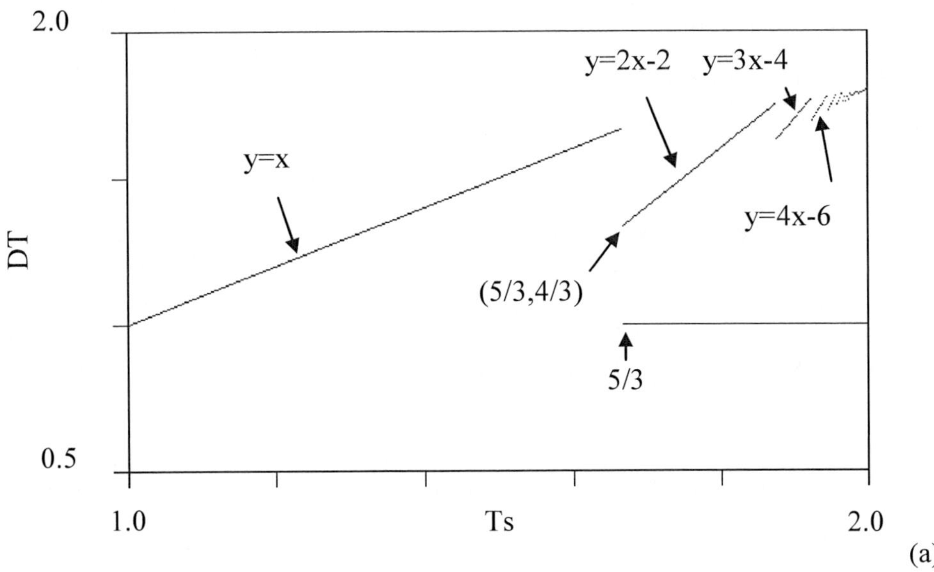

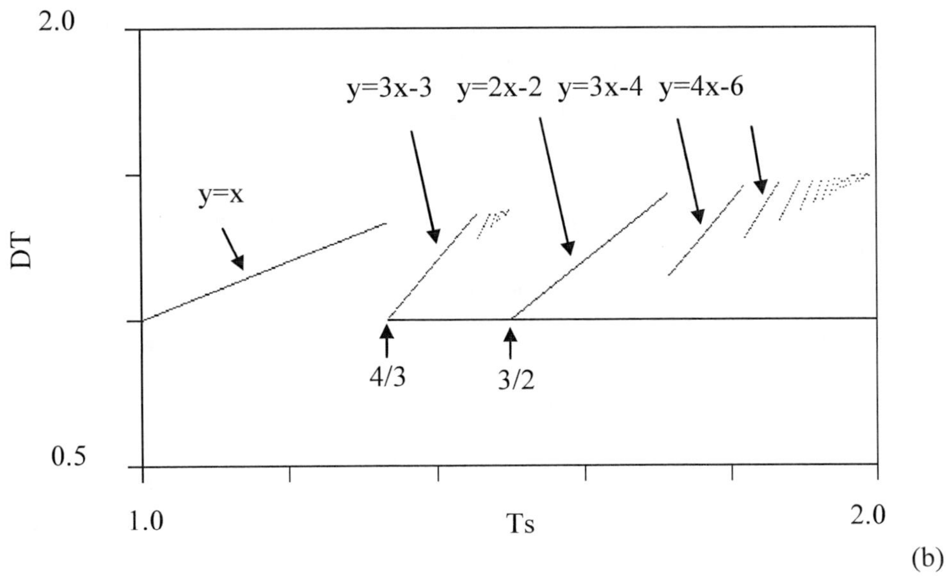

Figure 86. (a) Enlargement of Figure 81(c) ($s_p = 0.6$) for $1 \le T_s \le 2$. The segments of tour time are given, respectively, by $y = x$, $y = 2x - 2$, $y = 3x - 4$, $y = 4x - 6$, $y = 5x - 8$, ---. (b) Enlargement of Figure 81(d) ($s_p = 0.75$) for $1 \le T_s \le 2$. Similarly, the segments are given, respectively, by $y = x$, $y = 2x - 2$, $y = 3x - 4$, $y = 4x - 6$, $y = 5x - 8$, ---, except for region $4/3 \le T_s \le 3/2$. These segments are consistent with those in diagram (a). A new branch in region $4/3 \le T_s \le 3/2$ is added.

The enlargement of Figure 81(b) for $s_p = 0.5$ is shown in Figure 83(a) for $0 \le T_s \le 1.0$. The enlargement of Figure 83(a) is shown in Figure 83(b) for $0.5 \le T_s \le 1.0$.

The enlargement of Figure 81(c) for $s_p = 0.6$ is shown in Figure 84(a) for $0 \le T_s \le 1.0$. The enlargement of Figure 84(a) is shown in Figure 84(b) for $0.5 \le T_s \le 1.0$. The enlargement of Figure 81(d) for $s_p = 0.75$ is shown in Figure 85(a) for $0 \le T_s \le 1.0$. The enlargement of Figure 85(a) is shown in Figure 85(b) for $0.5 \le T_s \le 1.0$. Points *a*, *b*, *c*, --- represent the cumulative points. As a value of cycle time approaches a cumulative point from the below, a vehicle stops rare at signals and goes successfully through a sequence of green signals. The coordinate of a cumulative point is given by Eq. (87). Their points are on the straight line of envelope of Eq. (86). The value of *x* coordinate of a cumulative point does not depend on the value of split. Similarly, the iterated pattern occurs irrespective of split.

When the value of split is higher than 0.5, a vehicle displays a complex behavior. New branch appears in the diagram of tour time (see Figures 81(c) and 81 (d)). We study the fine structure of tour-time diagram in region between $1 \le T_s \le 2$ for $s_p > 0.5$. Figure 86(a) is the enlargement of Figure 81(c) of $s_p = 0.6$ for $1 \le T_s \le 2$. The segments of tour time are given, respectively, by $y = x$, $y = 2x - 2$, $y = 3x - 4$, $y = 4x - 6$, $y = 5x - 8$, ---. Figure 86(b) is the enlargement of Figure 81(d) of $s_p = 0.75$ for $1 \le T_s \le 2$. Similarly, the segments are given, respectively, by $y = x$, $y = 2x - 2$, $y = 3x - 4$, $y = 4x - 6$, $y = 5x - 8$, ---, except for region $4/3 \le T_s \le 3/2$. These segments are consistent with those in Figure 86(a). A new branch in region $4/3 \le T_s \le 3/2$ is added to the tour-time diagram in Figure 86(a).

We study the tour time by varying the split near a cumulative point. Figure 87 shows the plots of tour time *DT* against split s_p. Diagrams (a)-(d) display, respectively, the plots for (a) cycle time $T_s = 1.98$, (b) $T_s = 2.97$, (c) $T_s = 3.97$, and (d) $T_s = 4.96$. At cumulative point $T_s = 1.98$ in Figure 87(a), the tour time keeps a constant value until $s_p = 1/2$. When split is higher than 1/2, the tour time decreases tier upon tier with increasing split. At cumulative point $T_s = 2.97$ in Figure 87(b), the tour time keeps a constant value until $s_p = 1/3$, decrease discontinuously at $s_p = 1/3$, and again keeps a constant value until $s_p = 2/3$. Furthermore, when split is higher than 2/3, the tour time decreases tier upon tier with increasing split. At cumulative point $T_s = 3.97$ in Figure 87(c), the tour time keeps a constant value until $s_p = 1/4$, decrease discontinuously at $s_p = 1/4$, again keeps a constant value until $s_p = 2/4$, again decrease discontinuously at $s_p = 2/4$, and keeps a constant value until $s_p = 3/4$. Furthermore, when split is higher than 3/4, the tour time decreases tier upon tier with increasing split. At cumulative point $T_s = 4.96$ in Figure 87(d), the tour time keeps a constant value until $s_p = 1/5$, decrease discontinuously at $s_p = 1/5$, again keeps a constant value until $s_p = 2/5$, again decrease discontinuously at $s_p = 2/5$, keeps a constant value until $s_p = 3/5$, decrease discontinuously at $s_p = 3/5$, and keeps a constant value until $s_p = 4/5$. Furthermore, when split is higher than 4/5, the tour time decreases tier upon tier with increasing split.

Thus, we find the following relationships. When the split is higher than 0.5, new branches appear in the tour-time diagram for $T_s > 1$ and a vehicle exhibits more complex behavior than that of $s_p \leq 0.5$. The following relationships are obtained

(1) there is a branch in region $1.0 \leq T_s < 2.0$ of the tour-time diagram for $1/2 < s_p \leq 2/3$,

(2) there are two branches in regions $1.0 \leq T_s < 2.0$ and $2.0 \leq T_s < 3.0$ of the tour-time diagram for $2/3 < s_p \leq 3/4$,

(3) there are three branches in regions $1.0 \leq T_s < 2.0$, $2.0 \leq T_s < 3.0$, and $3.0 \leq T_s < 4.0$ of the tour-time diagram for $3/4 < s_p \leq 4/5$,

and (4) generally, for $\dfrac{k-2}{k-1} < s_p \leq \dfrac{k-1}{k}$ where $k \geq 3$, there are $k-2$ branches in regions $1.0 \leq T_s < 2.0$, $2.0 \leq T_s < 3.0$, ---, and $k-2 \leq T_s < k-1$ of the tour-time diagram.

We study the variation of tour time with increasing traffic-light's number *n*. Figure 88 shows the plots of the tour time *DT* against signal number *n* at split $s_p = 0.5$ for *n*=500-540. The plots of the tour time *DT* are shown for (1) $T_s = 1.5$, (2) $T_s = 2.5$, (3) $T_s = 4.5$, and (4) $T_s = 6.5$ in Figure 88(a). In diagram (1), the vehicle stops at all the traffic lights and then goes ahead. Tour time *DT* is higher than one and constant. In diagram (2), the vehicle stops every two traffic lights because tour time *DT* takes two distinct values: one is 1 and the other the value higher than 1. Value 1 means that the vehicle goes through green lights. In diagram (3), the vehicle stops every three traffic lights. In diagram (4), the vehicle stops every four traffic lights. These traffic states (1)-(4) correspond, respectively, to the lines of (1,1)-(2,2), (2,1)-(4,3), (4,2)-(6,4), and (6,3)-(8,5) in Figure 1(b). For a sufficiently large cycle time, the vehicle seldom stops at traffic lights. Slope *m* of segments in Figure 81(a) relates to period *h* of the vehicular motion. The following relationship holds

$$h = m \qquad \text{for } T_s > 1. \quad (88)$$

For $T_s < 1$, the plots of the tour time *DT* are shown for (1) $T_s = 0.6$, (2) $T_s = 0.75$, (3) $T_s = 0.85$, and (4) $T_s = 0.88$ in Figure 88(b). In diagram (1), the vehicle stops at all the traffic lights and then goes ahead. Tour time *DT* is higher than one and constant. In diagram (2), the vehicle stops every two traffic lights. In diagram (3), the vehicle stops every three traffic lights. In diagram (4), the vehicle stops every four traffic lights. These traffic states (1)-(4) correspond, respectively, to the lines for $0.5 \leq T_s \leq 1$ in Figure 83(b). When cycle time T_s approaches one, the vehicle seldom stops at the traffic lights and the tour time approaches the cumulative point a in Figure 83(b). Slope m of segments in Figure 83(b) relates to period h of the vehicular motion. The following relationship holds

$$h = m - 1 \text{ for } 0.5 < T_s < 1 . \tag{89}$$

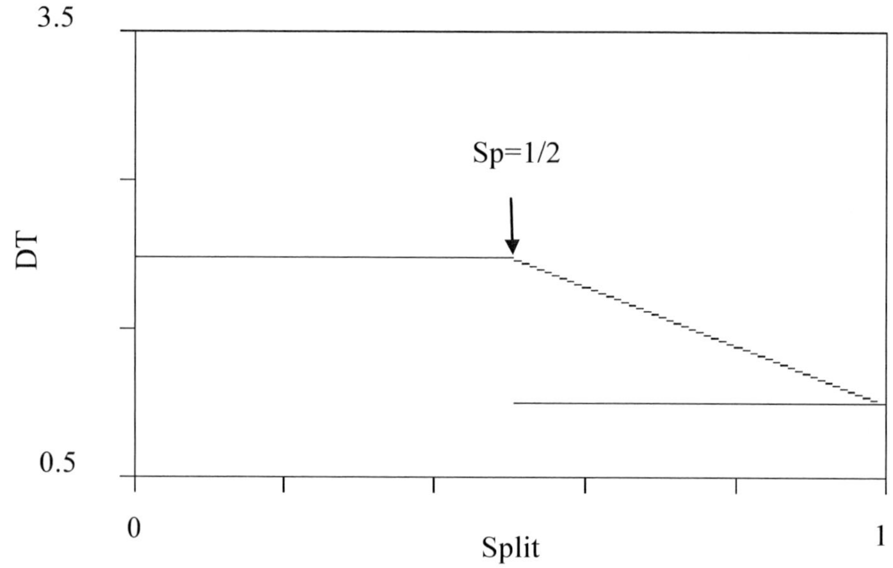

(a)

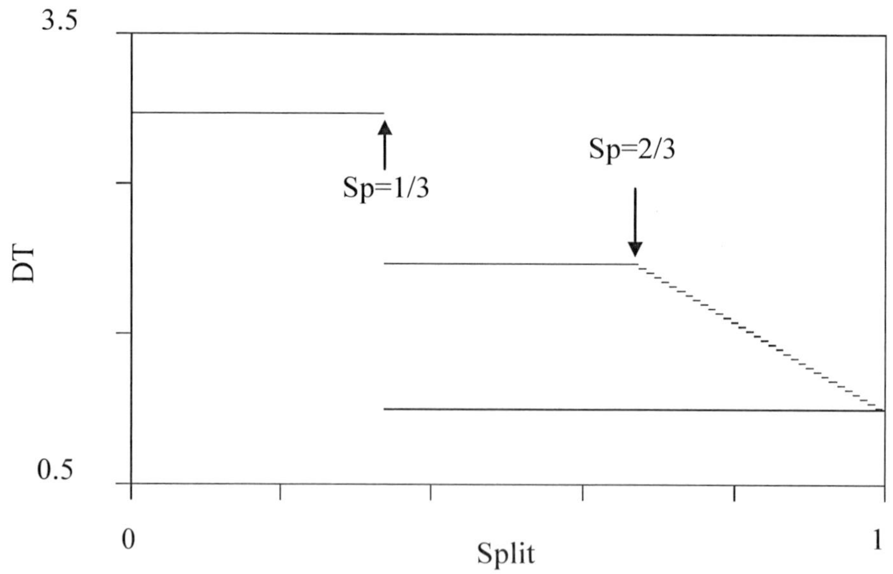

(b)

Figure 87. (Continued)

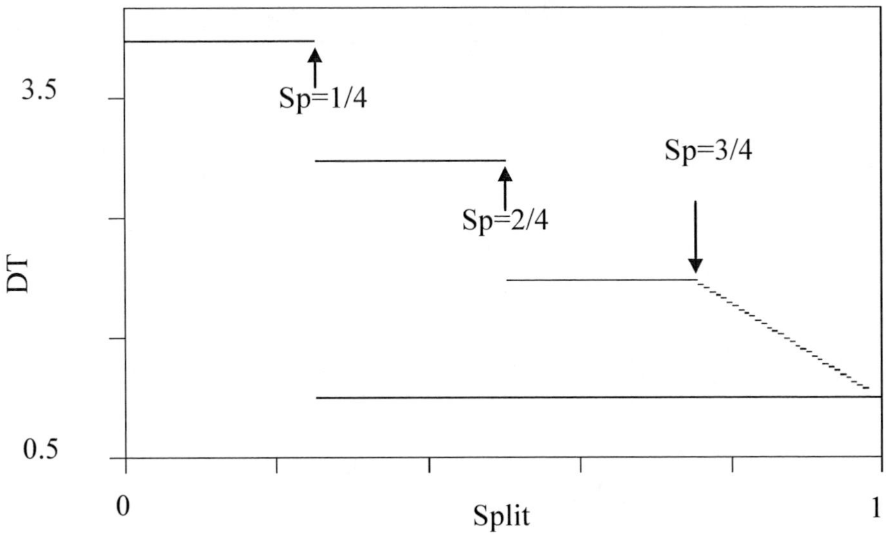

(c)

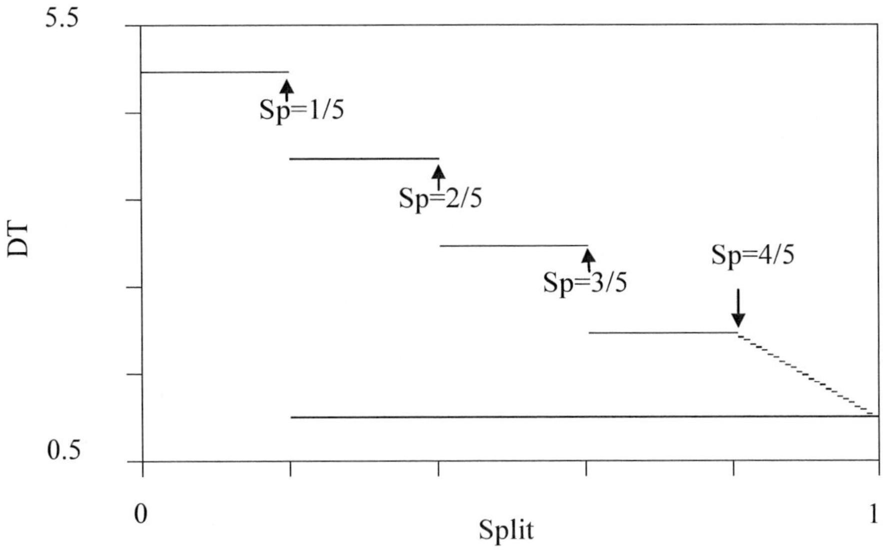

(d)

Figure 87. Plots of tour time DT against split s_p for (a) cycle time $T_s = 1.98$, (b) $T_s = 2.97$, (c) $T_s = 3.97$, and (d) $T_s = 4.96$.

Figure 89 shows the plots of the tour time DT against number n at split $s_p = 0.6$ for n=500-540. The plots of the tour time DT are shown for (1) $T_s = 1.5$, (2) $T_s = 2.5$, (3) $T_s = 4.5$, and (4) $T_s = 6.5$ in Figure 89(a). In diagram (1), the vehicle stops at all the traffic lights and then goes ahead. In diagram (2), the vehicle stops every two traffic lights. In diagram (3), the vehicle stops every three traffic lights. In diagram (4), the vehicle stops every four traffic lights. For a sufficiently large cycle time, the vehicle seldom stops at traffic lights.

These diagrams (1)-(4) at $s_p = 0.6$ in Figure 89(a) agree with those at $s_p = 0.5$ in Figure 88(a). The relationship (5) is also satisfied.

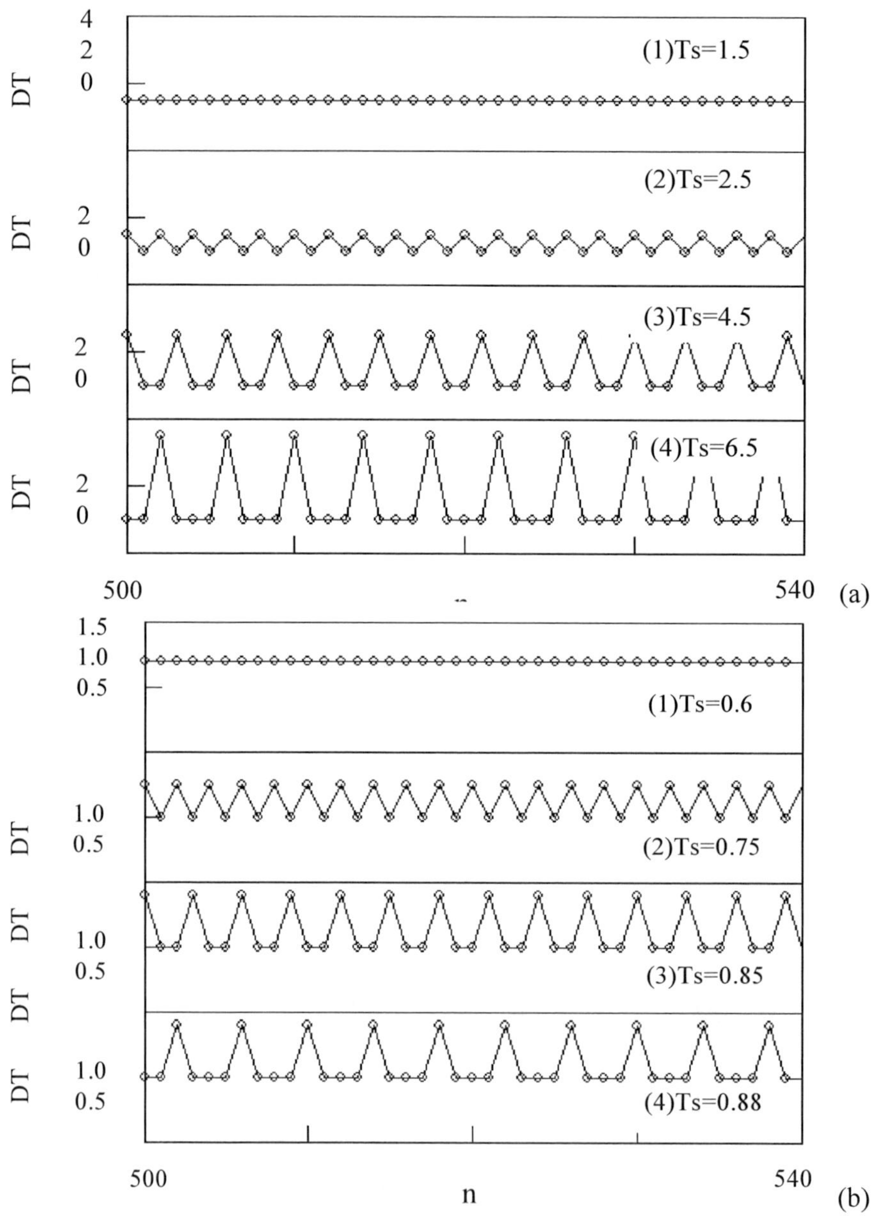

Figure 88. (a) When $T_s > 1$, plots of tour time DT against signal number n at split $s_p = 0.5$ for (1) cycle time $T_s = 1.5$, (2) $T_s = 2.5$, (3) $T_s = 4.5$, and (4) $T_s = 6.5$. (b) When $T_s < 1$, plots of tour time DT against signal number n at split $s_p = 0.5$ for (1) $T_s = 0.6$, (2) $T_s = 0.75$, (3) $T_s = 0.85$, and (4) $T_s = 0.88$.

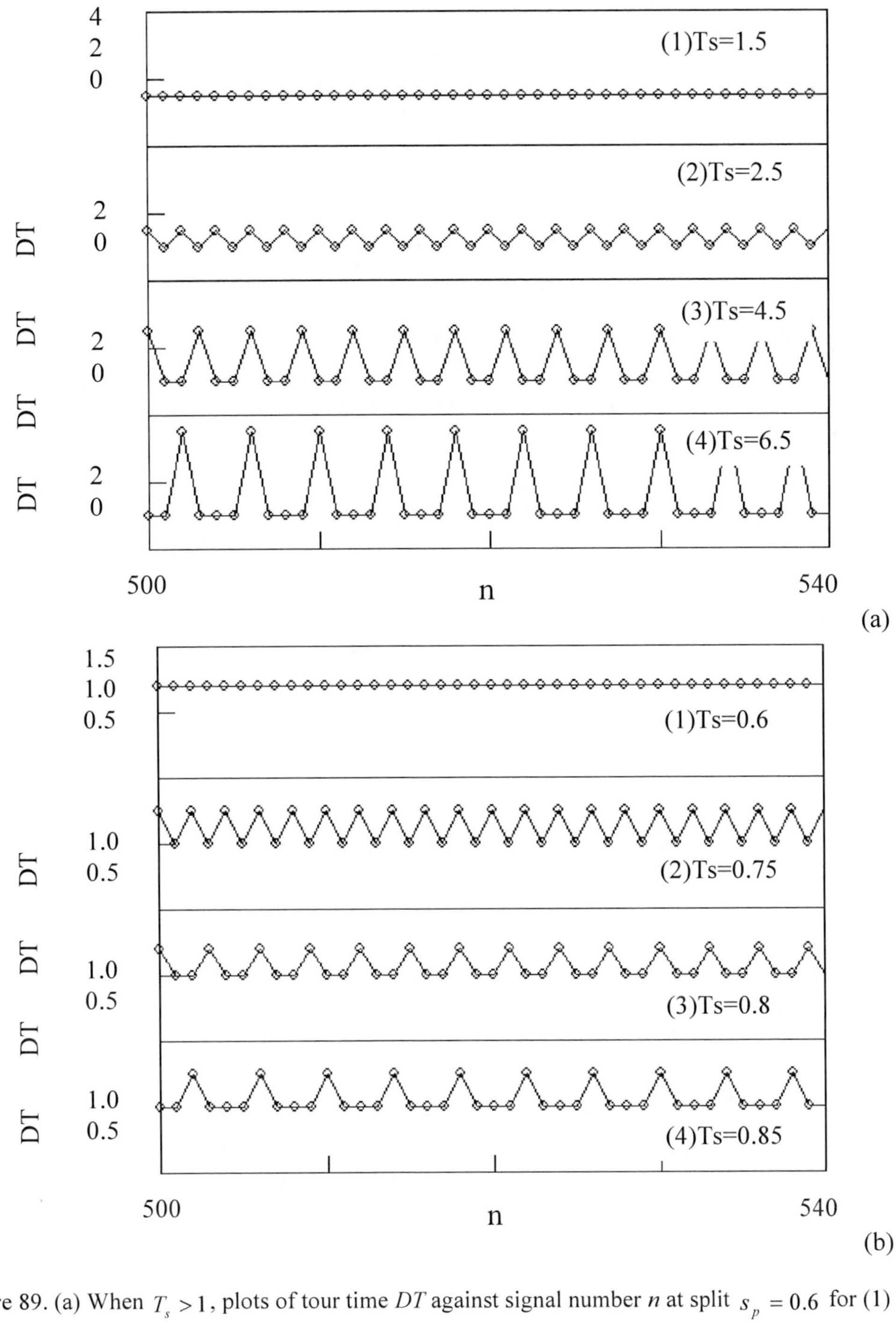

Figure 89. (a) When $T_s > 1$, plots of tour time DT against signal number n at split $s_p = 0.6$ for (1) cycle time $T_s = 1.5$, (2) $T_s = 2.5$, (3) $T_s = 4.5$, and (4) $T_s = 6.5$. (b) When $T_s < 1$, plots of tour time DT against signal number n at split $s_p = 0.6$ for (1) $T_s = 0.6$, (2) $T_s = 0.75$, (3) $T_s = 0.8$, and (4) $T_s = 0.85$.

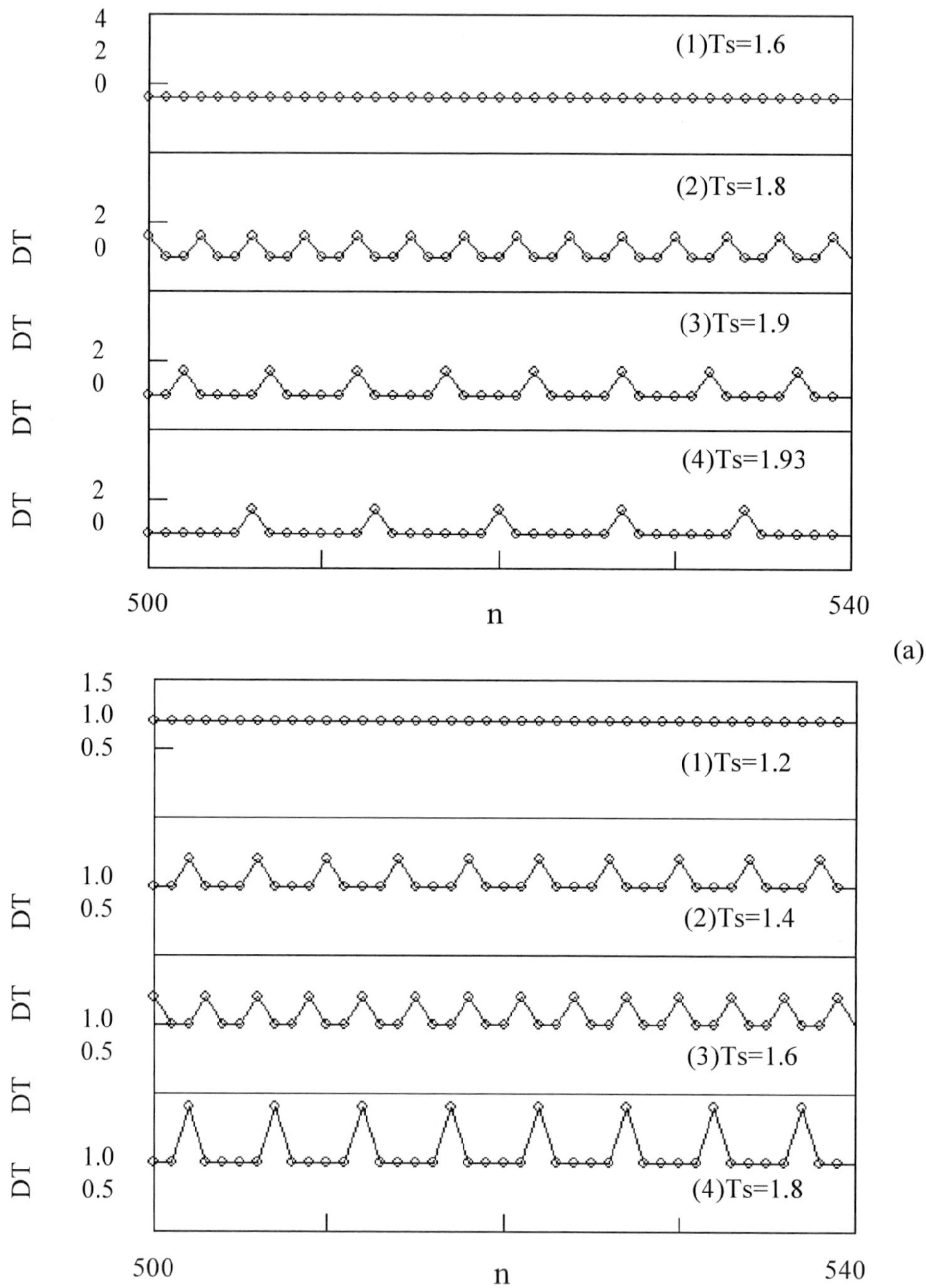

Figure 90. (a) Plots of tour time DT against signal number n at split $s_p = 0.6$ for (1) $T_s = 1.6$, (2) $T_s = 1.8$, (3) $T_s = 1.9$, and (4) $T_s = 1.93$ in new branch $1 < T_s < 2$. These plots correspond, respectively, to the tour times of the segments in Figure 86(a). (b) Plots of tour time DT against signal number n at split $s_p = 0.75$ for (1) $T_s = 1.2$, (2) $T_s = 1.4$, (3) $T_s = 1.6$, and (4) $T_s = 1.8$. These plots correspond, respectively, to the tour times of the segments in Figure 86(b).

For $T_s < 1$, the plots of the tour time *DT* are shown for (1) $T_s = 0.6$, (2) $T_s = 0.75$, (3) $T_s = 0.8$, and (4) $T_s = 0.85$ in Figure 89(b). In diagram (1), the vehicle stops at all the traffic lights and then goes ahead. In diagram (2), the vehicle stops every two traffic lights. In diagram (3), the vehicle stops every three traffic lights. In diagram (4), the vehicle stops every four traffic lights. These traffic states (1)-(4) correspond, respectively, to the lines for $0.5 \le T_s \le 1$ in Figure 84(b). When cycle time T_s approaches one, the vehicle seldom stops at the traffic lights and the tour time approaches the cumulative point *a* in Figure 84(b). Diagrams (1) and (2) at $s_p = 0.6$ in Figure 89(b) agree with those at $s_p = 0.5$ in Figure 88(b). The relationship (89) is also satisfied.

In new branch $1 < T_s < 2$ for $s_p > 0.5$, we study the variation of tour time with increasing traffic-light's number *n*. Figure 90(a) shows the plots of the tour time *DT* against number *n* at split $s_p = 0.6$ for *n*=500-540. The plots of the tour time *DT* are shown for (1) $T_s = 1.6$, (2) $T_s = 1.8$, (3) $T_s = 1.9$, and (4) $T_s = 1.93$ in Figure 90(a). These plots correspond, respectively, to the tour times of the segments in Figure 86(a). In diagram (1), the vehicle stops at all the traffic lights. In diagram (2), the vehicle stops every three traffic lights. In diagram (3), the vehicle stops every five traffic lights. In diagram (4), the vehicle stops every seven traffic lights. Slope *m* of segments in Figure 86(a) relates to period *h* of the vehicular motion. The following relationship holds

$$h = 2m - 1. \tag{90}$$

When cycle time T_s approaches the cumulative point of two, the vehicle seldom stops at the traffic lights.

Figure 90(b) shows the plots of the tour time *DT* against number *n* at split $s_p = 0.75$ for *n*=500-540. The plots of the tour time *DT* are shown for (1) $T_s = 1.2$, (2) $T_s = 1.4$, (3) $T_s = 1.6$, and (4) $T_s = 1.8$ in Figure 90(b). These plots correspond, respectively, to the tour times of the segments in Figure 86(b). In diagram (1), the vehicle stops at all the traffic lights. In diagram (2), the vehicle stops every four traffic lights. In diagram (3), the vehicle stops every three traffic lights. In diagram (4), the vehicle stops every five traffic lights. For $s_p = 0.75$, there are two cumulative points: 1.5 and 2. When cycle time T_s approaches the cumulative points of 1.5 and 2, the vehicle seldom stops at the traffic lights. Except for the region $4/3 \le T_s \le 3/2$, relationship (90) is satisfied for $s_p = 0.75$.

Thus, we find that the vehicular motion controlled by traffic lights displays an iterated (self-similar) periodic structure and the algebraic relation between tour time and period holds.

6.3. Effect of Irregularity on Vehicular Traffic

In real traffic, the vehicular traffic depends highly on the configuration of traffic lights and the priority of roadways. Until now, one has studied the vehicular traffic controlled by periodic traffic lights. The interval between signals was homogeneous and split was also the same for all signals. The more congested roadway has higher value of split in order to give a priority. The interval between signals is not homogeneous but changes from signal to signal. Also, the split changes irregularly from signal to signal. Thus, the inhomogeneity of signal's interval and irregular split have the important effects on vehicular traffic. However, the effects of inhomogeneity and irregularity have little been investigated. Vehicular traffic depends highly on signal configuration, cycle time, split, and offset time where the cycle time is the period of a traffic light and the offset time is the difference of signal phases between signals.

Here, we study the effect of irregularities on vehicular traffic through a series of traffic lights with inhomogeneous interval. We present a stochastic nonlinear-map model for traffic through the irregular series of signals. We investigate the dynamical behavior of a single vehicle by iterating the stochastic nonlinear map. We clarify the dynamical behavior of a single vehicle through a sequence of signals by varying cycle time.

The traffic lights are positioned irregularly (inhomogeneously) on a roadway. The interval between signals changes from signal to signal where the interval between signals n and $n+1$ is indicated by $l(n)$. The vehicle moves with the mean speed v between a traffic light and its next light.

In the synchronized strategy, all the traffic lights change simultaneously from red (green) to green (red) with a fixed time period $(1-s_p)t_s$ ($s_p t_s$). The period of green is $s_p t_s$ and the period of red is $(1-s_p)t_s$. The split also changes from signal to signal because the priority is different with each crossing. The split is represented by $s_p(n)$ in order to describe the dependence of split on signal n.

Furthermore, the signal timing is controlled by offset time t_{offset}. In the green wave strategy, it is given by $t_{offset} = n\tau$ where τ is the delay time. Then, the signal switches from red to green in green wave way. Here, we assume that the phase difference between two signals changes irregularly from signal to signal. The offset time is indicated by $t_{offset}(n)$.

When a vehicle arrives at a traffic light and the traffic light is red, the vehicle stops at the position of the traffic light. Then, when the traffic light changes from red to green, the vehicle goes ahead. On the other hand, when a vehicle arrives at a traffic light and the traffic light is green, the vehicle does not stop and goes ahead without changing speed.

We define the arrival time of the vehicle at traffic light n as $t(n)$. The arrival time at traffic light $n+1$ is given by

$$t(n+1)=t(n)+l(n)/v+\left(r(n)-t(n)\right)H\left(t(n)+t_{offset}(n)-\left(\mathrm{int}((t(n)+t_{offset}(n))/t_s)t_s\right)-s_p(n)t_s\right) \tag{91}$$

with $r(n) = \left(\text{int}\left((t(n) + t_{offset}(n))/t_s\right) + 1\right) \cdot t_s - t_{offset}(n)$, (92)

where $H(t)$ is the Heaviside function: $H(t) = 1$ for $t \geq 0$ and $H(t) = 0$ for $t < 0$. $H(t) = 1$ if the traffic light is red, while $H(t) = 0$ if the traffic light is green. $l(n)/v$ is the time it takes for the vehicle to move between lights n and n+1. $r(n)$ is such time that the traffic light just changed from red to green. The third term on the right hand side of Eq. (91) represents such time that the vehicle stops if traffic light n is red. The number n of iteration increases one by one when the vehicle moves through the traffic light. The iteration corresponds to the going ahead on the highway.

By dividing time by the characteristic time l_0/v, one obtains the nonlinear equation of dimensionless arrival time:

$$T(n+1) = T(n) + L(n) + \left(R(n) - T(n)\right)H\left(T(n) + T_{offset}(n) - \left(\text{int}((T(n) + T_{offset}(n))/T_s)T_s\right) - s_p(n)T_s\right)$$

with $R(n) = \left(\text{int}\left((T(n) + T_{offset}(n))/T_s\right) + 1\right) \cdot T_s - T_{offset}(n)$, (93)

where $T(n) = t(n)v/l_0$, $R(n) = r(n)v/l_0$, $L(n) = l(n)/l_0$, $T_{offset}(n) = t_{offset}v/l_0$ and $T_s = t_s v/l_0$. Parameters $L(n)$, $T_{offset}(n)$, and $s_p(n)$ are random variables. Thus, the dynamics of the vehicle is described by the stochastic nonlinear map (93).

We assume that the split is random and uncorrelated each other. Split $s_p(n)$ is given by

$$s_p(n) = 0.5 + a\xi_s(n), \quad (94)$$

where $<\xi_s(n)>= 0$ and $<\xi_s(n)\xi_s(m)>= \beta\delta_{nm}$ (β : a coefficient).

Similarly, we assume that the interval between signals is random and uncorrelated each other. Interval $L(n)$ is given by

$$L(n) = 1 + b\xi_L(n), \quad (95)$$

where $<\xi_L(n)>= 0$ and $<\xi_L(n)\xi_L(m)>= \beta\delta_{nm}$.

We also assume that the offset time is random and uncorrelated each other. Offset time $T_{offset}(n)$ is given by

$$T_{offset}(n) = c\xi_{offset}(n), \quad (96)$$

where $<\xi_{offset}(n)>= 0$ and $<\xi_{offset}(n)\xi_{offset}(m)>= \beta\delta_{nm}$.

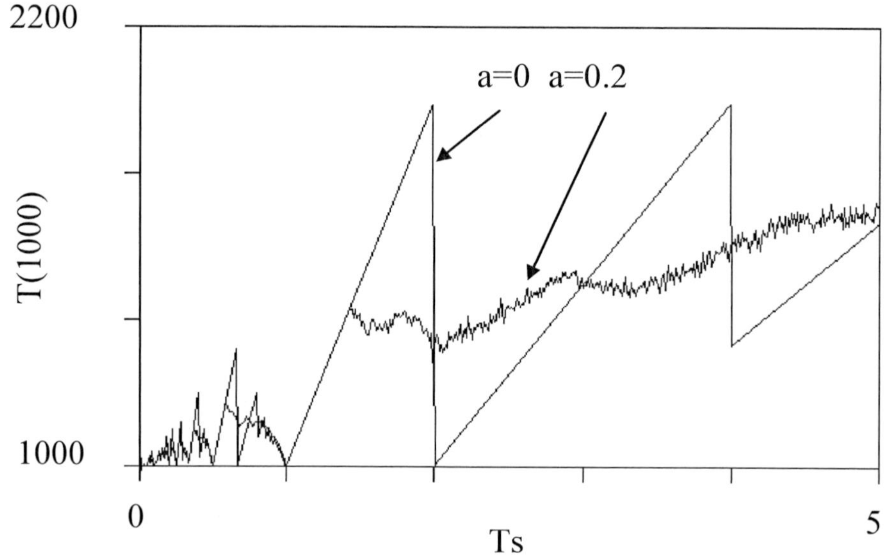

Figure 91. Plot of arrival time $T(n)$ against cycle time T_s at signal $n = 1000$ far from the origin for strength $a = 0.2$ of split's irregularity. For comparison, the arrival time of nonrandom split ($a = 0.0$) is also plotted.

We investigate the effect of irregularities on the motion of a single vehicle through the series of traffic lights by iterating stochastic map (93). We calculate the arrival time at traffic light n when the vehicle goes ahead on the roadway. We compare the traffic through the irregular series with that through the nonrandom series.

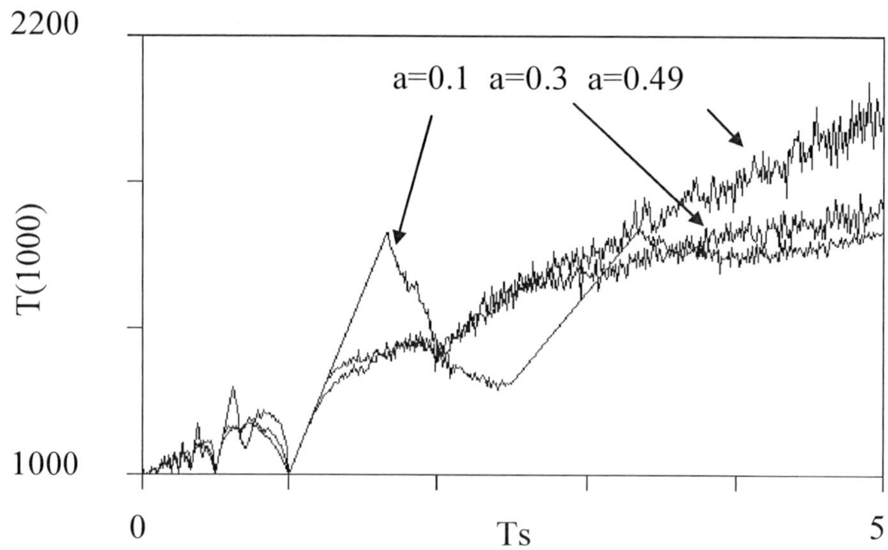

Figure 92. Plots of arrival time $T(n)$ against cycle time T_s at signal $n = 1000$ for strengths $a = 0.1, 0.3, 0.49$ of split's irregularity.

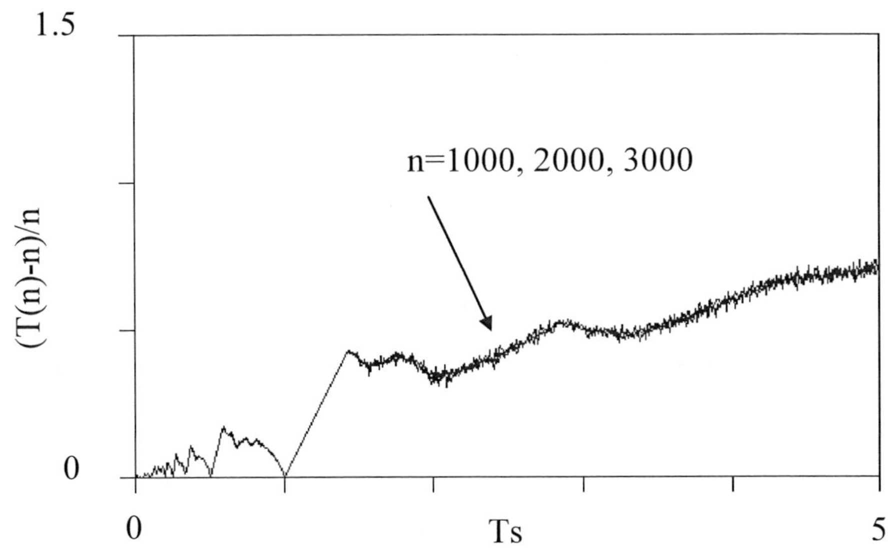

Figure 93. Plots of scaled arrival time $(T(n)-n)/n$ against cycle time for irregularity strength $a=0.2$ at n=1000, 2000, 3000. The arrival time collapse on a single curve.

First, we study the effect of irregularity on vehicular traffic when the split varies irregularly from signal to signal ($a \neq 0$), the interval between signals is the same ($b=0$), and the offset time is zero ($c=0$). The random variable $\xi_s(n)$ is given by $\xi_s(n)=2(rnd-1/2)$, where *rnd* is the random number extending uniformly from zero to unity, $<rnd>=0.5$, and $\beta=1/3$. At n=0, $T(0)=0$. Figure 91 shows the plot of arrival time $T(n)$ against cycle time T_s at signal $n=1000$ far from the origin for strength $a=0.2$ of split's irregularity. For comparison, the arrival time of nonrandom split ($a=0.0$) is also plotted. The arrival time changes highly by introducing the split's irregularity, while there exist the regions in which the arrival time does not change with split's irregularity. In the regions, the arrival time agrees with that of nonrandom split. Figure 92 shows the plots of arrival time $T(n)$ against cycle time T_s at signal $n=1000$ for strengths $a=0.1,0.3,0.49$ of split's irregularity. With increasing irregularity, the profile of arrival time becomes smooth. When irregularity strength a approaches to the maximal value 0.5, the arrival time exhibits the linear dependence for cycle time.

We study the dependence of arrival time on distance n of signals. Figure 93 shows the plots of scaled arrival time $(T(n)-n)/n$ against cycle time for irregularity strength $a=0.2$ at n=1000, 2000, 3000. The arrival time collapse on a single curve. The arrival time scales as

$$\frac{T(n)-n}{n}=f(T_s). \tag{97}$$

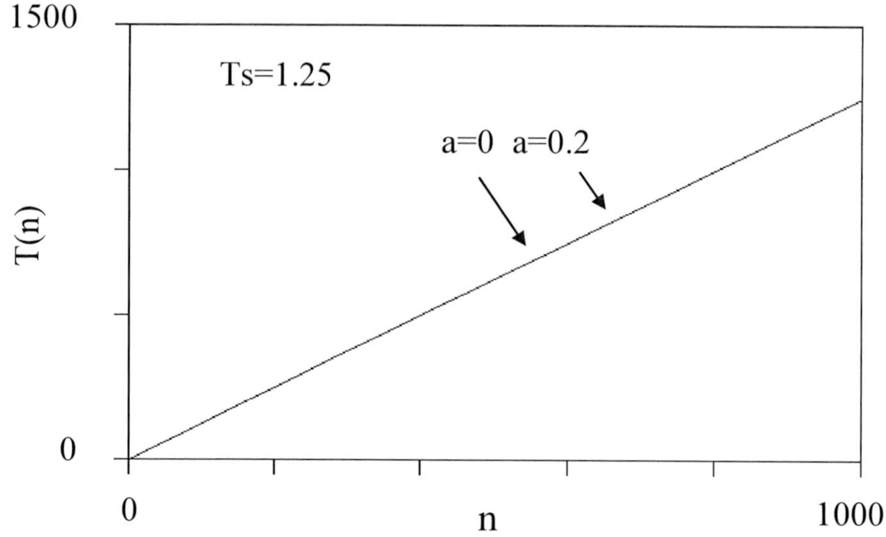

(a)

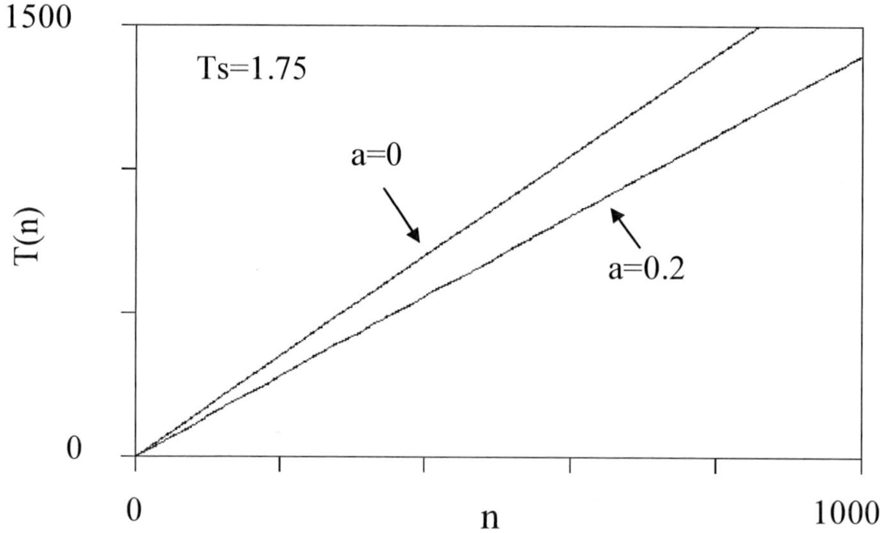

(b)

Figure 94. (a) Plots of arrival time against signal's position n at cycle time $T_s = 1.25$ for nonrandom split $a = 0$ and irregular split $a = 0.2$. The arrival time of irregular slit is consistent with that of nonrandom slit. (b) Plots of arrival time against signal's position n at cycle time $T_s = 1.75$ for nonrandom split $a = 0$ and irregular split.

Figure 94(a) shows the plots of arrival time against signal's position n at cycle time $T_s = 1.25$ for nonrandom split $a = 0$ and irregular split $a = 0.2$. The arrival time of irregular slit is consistent with that of nonrandom slit. Figure 94(b) shows the plots of arrival time against signal's position n at cycle time $T_s = 1.75$ for nonrandom split $a = 0$ and irregular split $a = 0.2$. The arrival time of irregular slit is lower than that of nonrandom slit. Thus, the difference between nonrandom and irregular splits occurs by varying cycle time.

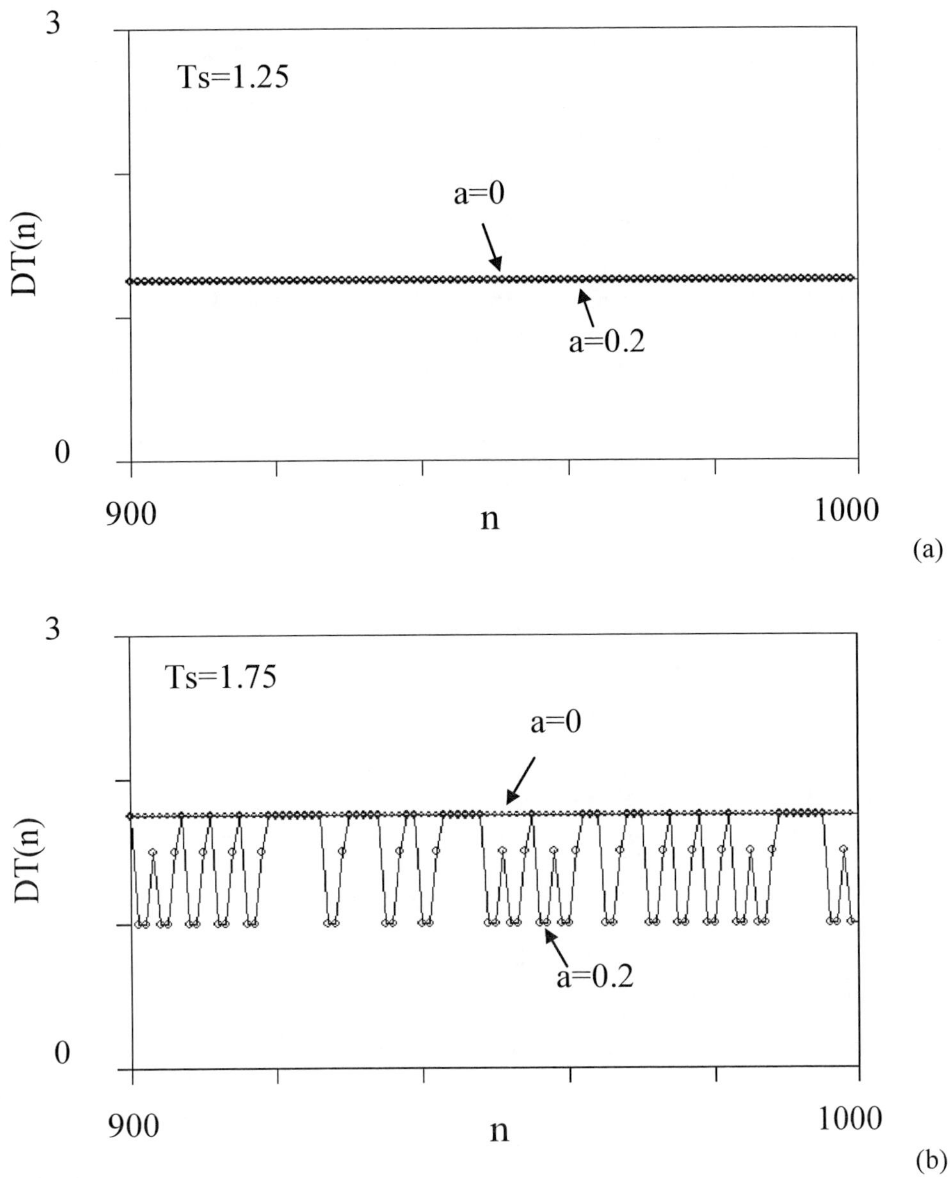

Figure 95. Plots of tour time against signal's position for nonrandom split $a = 0$ and irregular split $a = 0.2$ at (a) cycle time $T_s = 1.25$ and (b) $T_s = 1.75$.

We study the tour time $\Delta T(n) = T(n) - T(n-1)$ between two signals for various values of cycle time. Figure 95 shows the plots of tour time against signal's position for nonrandom split $a = 0$ and irregular split $a = 0.2$ at (a) cycle time $T_s = 1.25$ and (b) $T_s = 1.75$. At cycle time $T_s = 1.25$, the tour time of irregular split agrees with that of nonrandom split, while the tour time of irregular split is different from that of nonrandom split at cycle time $T_s = 1.75$. At $T_s = 1.75$, the tour time of irregular split takes three values but these values occur non-periodically. Figure 96 shows the plots of tour time against cycle time for (a) nonrandom split $a = 0$ and (b) irregular split $a = 0.2$. For the irregular split, some new branches are added to the tour time of nonrandom split.

We study such regions that the arrival time does not depend on the strength of split's irregularity. Figure 97 shows the region map (phase diagram) in (T_s, a)-space which represents the regions not depending on split's irregularity. The black region indicates that not depending on split's irregularity. Thus, the regions independent on split's irregularity decreases with increasing strength of split's irregularity.

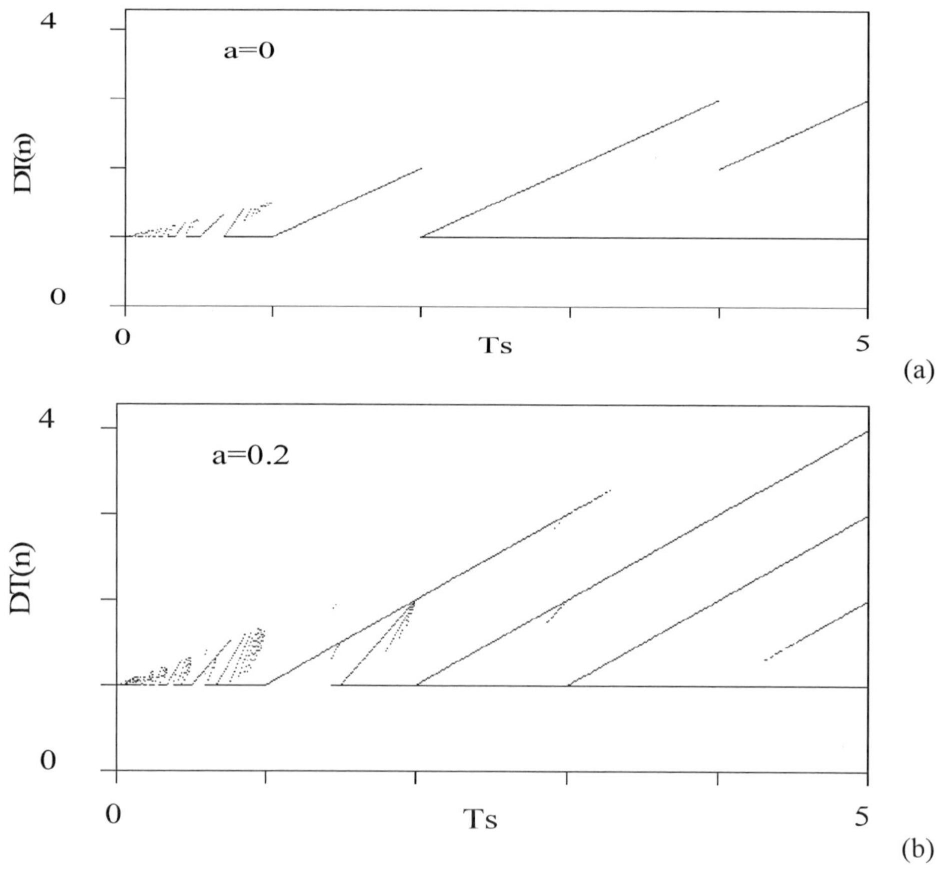

Figure 96. Plots of tour time against cycle time for (a) nonrandom split $a = 0$ and (b) irregular split $a = 0.2$. For the irregular split, some new branches are added to the tour time of nonrandom split.

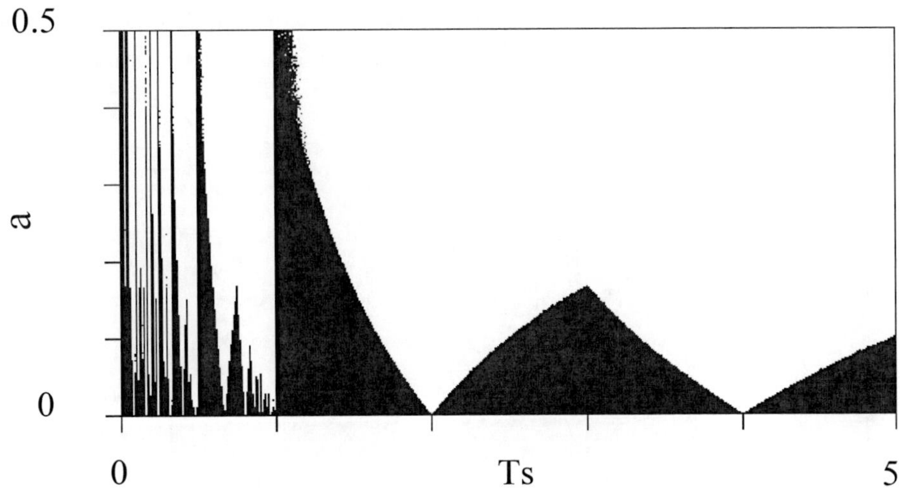

Figure 97. Region map (phase diagram) in (T_s, a)-space which represents the regions not depending on split's irregularity. The black region indicates that not depending on split's irregularity.

We study the effect of irregularity on vehicular traffic when the interval between signals varies irregularly from signal to signal ($b \neq 0$), the split is the same ($a = 0$), and the offset time is zero ($c = 0$). The random variable $\xi_L(n)$ is given by $\xi_L(n) = 2(rnd - 1/2)$, where *rnd* is the random number extending uniformly from zero to unity and $< rnd >= 0.5$. Figure 98(a) shows the plot of arrival time $T(n)$ against cycle time T_s at signal $n = 1000$ far from the origin for strength $b = 0.2$ of interval's irregularity. For comparison, the arrival time of the same interval ($b = 0.0$) is also plotted. The arrival time changes highly by introducing the interval's irregularity, while there exist the regions in which the arrival time does not change with interval's irregularity. In the regions, the arrival time agrees with that of the same interval. Figure 98(a) is compared with Figure 91. The effect of interval's irregularity is different from that of split's irregularity. Figure 98(b) shows the plots of arrival time $T(n)$ against cycle time T_s at signal $n = 1000$ for strengths $b = 0.1, 0.3, 0.49$ of interval's irregularity. With increasing irregularity, the profile of arrival time becomes smooth. When irregularity strength b becomes large, the arrival time exhibits the linear dependence for cycle time. Figure 98(b) is compared with Figure 92.

We study the dependence of arrival time on distance *n* of signals. Figure 99 shows the plots of scaled arrival time $(T(n) - n)/n$ against cycle time for irregularity strength $b = 0.2$ at *n*=1000, 2000, 3000. The arrival time collapse on a single curve. The arrival time scales as $\frac{T(n) - n}{n} = g(T_s)$. The scaling function $g(T_s)$ is different from $f(T_s)$.

Figure 100(a) shows the plots of arrival time against signal's position *n* at cycle time $T_s = 1.25$ for the same interval $b = 0$ and irregular interval $b = 0.2$. The arrival time of $b = 0.2$ is consistent with that of $b = 0$. Figure 100(b) shows the plots of arrival time against signal's position *n* at cycle time $T_s = 1.75$ for the same interval $b = 0$ and irregular interval $b = 0.2$. The arrival time of irregular interval is lower than that of the same interval. Thus, the difference between nonrandom and irregular intervals occurs by varying cycle time.

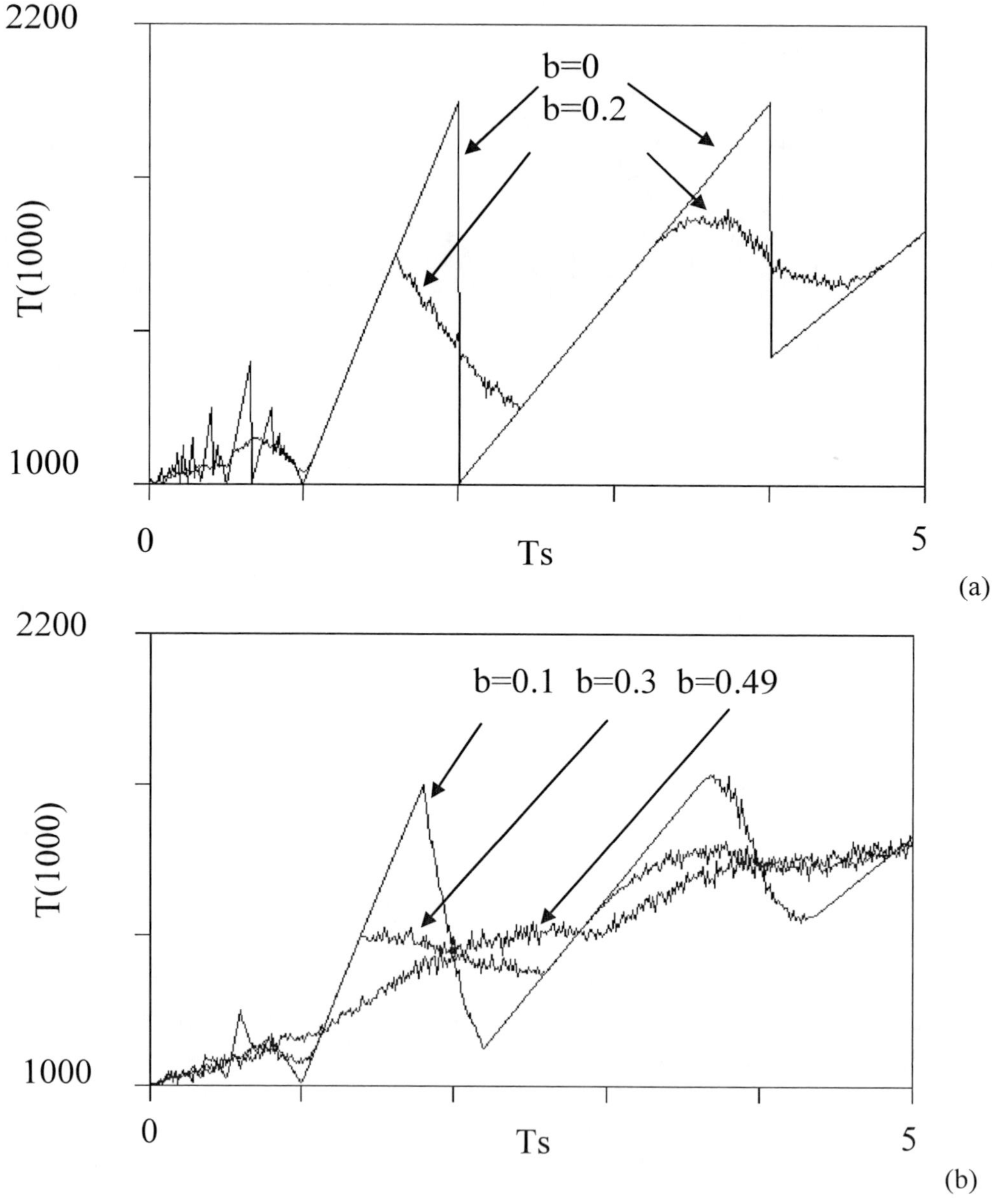

Figure 98. (a) Plot of arrival time $T(n)$ against cycle time T_s at signal $n = 1000$ far from the origin for strength $b = 0.2$ of interval's irregularity. For comparison, the arrival time of the same interval ($b = 0.0$) is also plotted. (b) Plots of arrival time $T(n)$ against cycle time T_s at signal $n = 1000$ for strengths $b = 0.1, 0.3, 0.49$ of interval's irregularity.

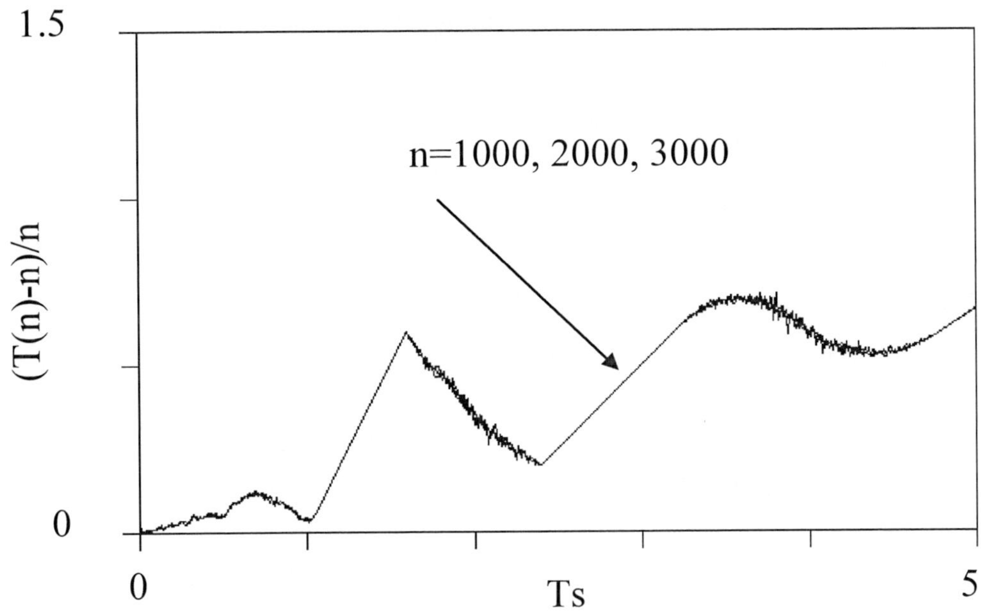

Figure 99. Plots of scaled arrival time $(T(n)-n)/n$ against cycle time for irregularity strength $b=0.2$ at n=1000, 2000, 3000. The arrival time collapses on a single curve.

We study the tour time $\Delta T(n)=T(n)-T(n-1)$ between two signals for various values of cycle time. Figure 101 shows the plots of tour time against signal's position for the same interval $b=0$ and irregular interval $b=0.2$ at (a) cycle time $T_s=1.25$ and (b) $T_s=1.75$. At cycle time $T_s=1.25$, the tour time of irregular interval varies from signal to signal but its mean value agrees with that of the same interval. While the tour time of irregular interval varies highly from signal to signal and its mean value is also different from that of the same interval at cycle time $T_s=1.75$. Figure 101 is compared with Figure 95. The behavior of irregular interval is definitely different from that of irregular split. The irregularity of interval induces the irregular motion, while the irregular split does not induce the irregular motion but induces periodic motions in Figure 95(b). Figure 102 shows the plots of tour time against cycle time for irregular interval $b=0.2$. For the irregular interval, the tour time takes various values and each branch is obscure when comparing with Figure 96(a) of the same interval.

We study such regions that the arrival time does not depend on the strength of interval's irregularity. Figure 103 shows the region map (phase diagram) in (T_s,b)-space which represents the regions not depending on interval's irregularity. The black region indicates that not depending on interval's irregularity. Thus, the regions independent on interval's irregularity decreases with increasing strength of interval's irregularity. Figure 103 is compared with Figure 97 of split's irregularity. The black region in Figure 103 is smaller than that in Figure 97 for $T_s < 1$, while the black region in Figure 103 is larger than that in Figure 97 for $T_s \geq 1$.

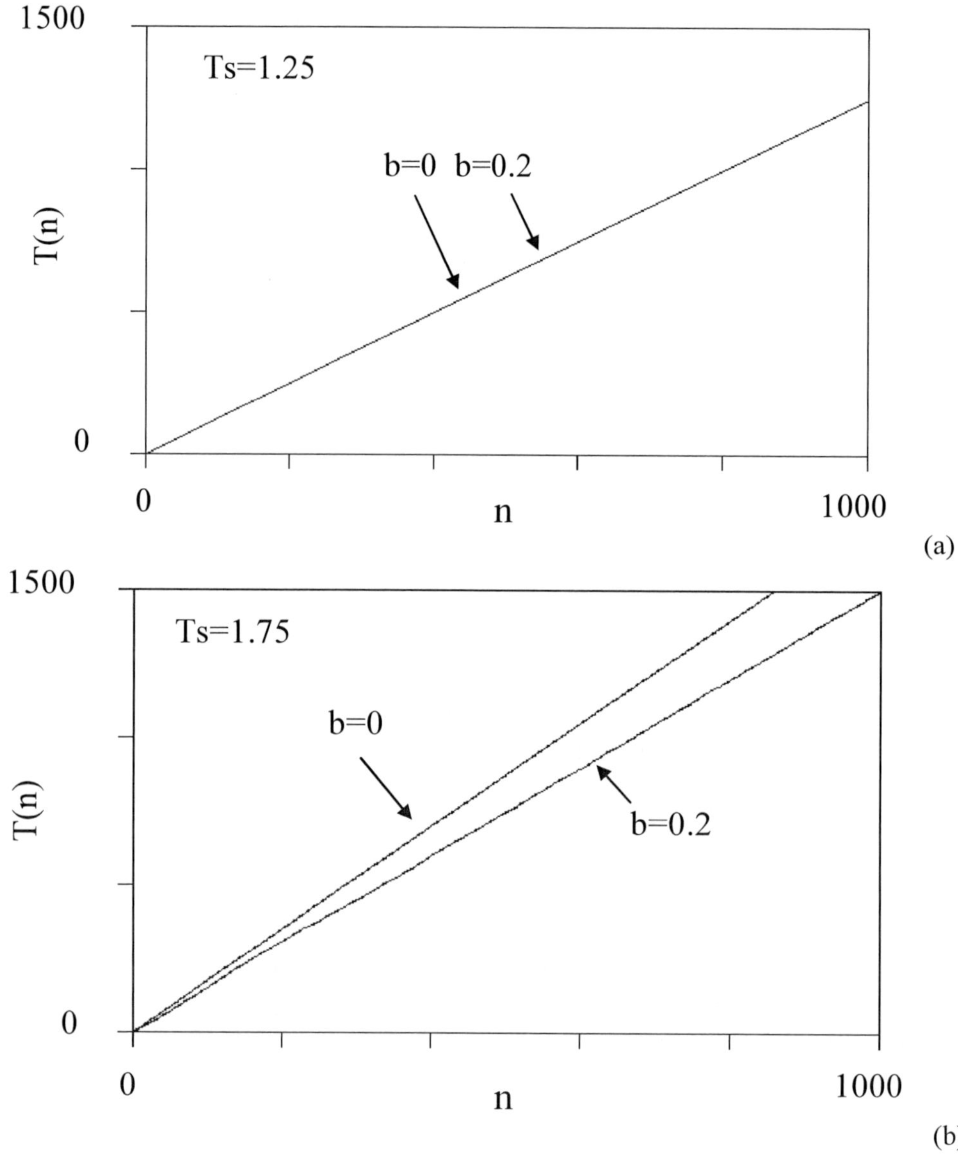

Figure 100. (a) Plots of arrival time against signal's position *n* at cycle time $T_s = 1.25$ for the same interval $b = 0$ and irregular interval $b = 0.2$. The arrival time of $b = 0.2$ is consistent with that of $b = 0$. (b) Plots of arrival time against signal's position *n* at cycle time $T_s = 1.75$ for the same interval $b = 0$ and irregular interval $b = 0.2$.

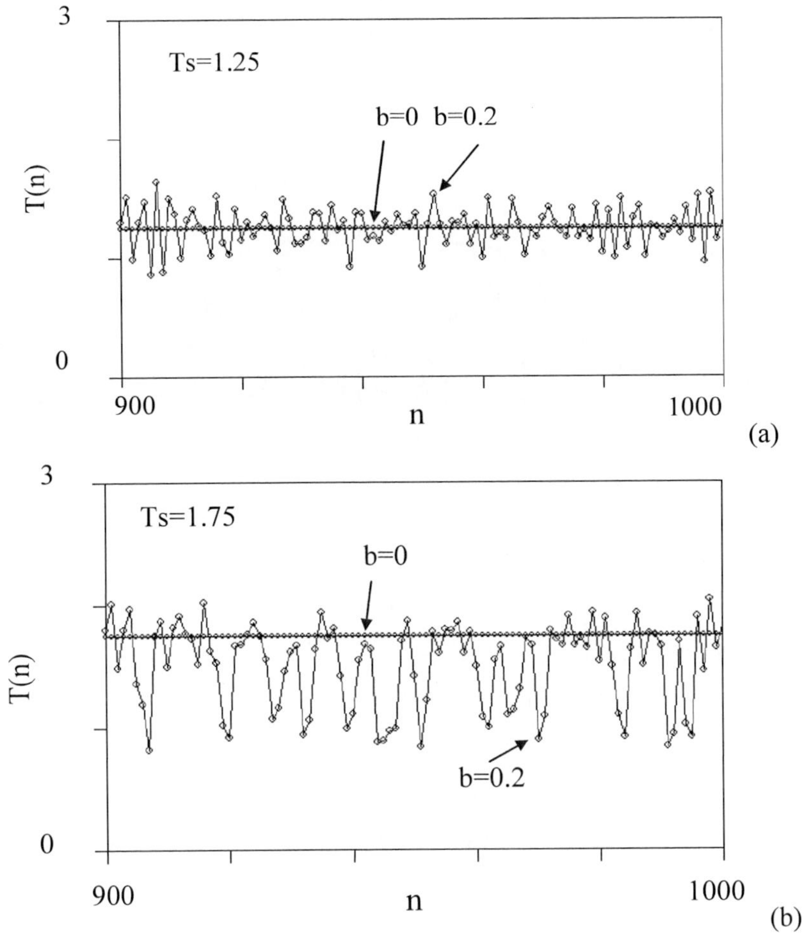

Figure 101. Plots of tour time against signal's position for the same interval $b = 0$ and irregular interval $b = 0.2$ at (a) cycle time $T_s = 1.25$ and (b) $T_s = 1.75$.

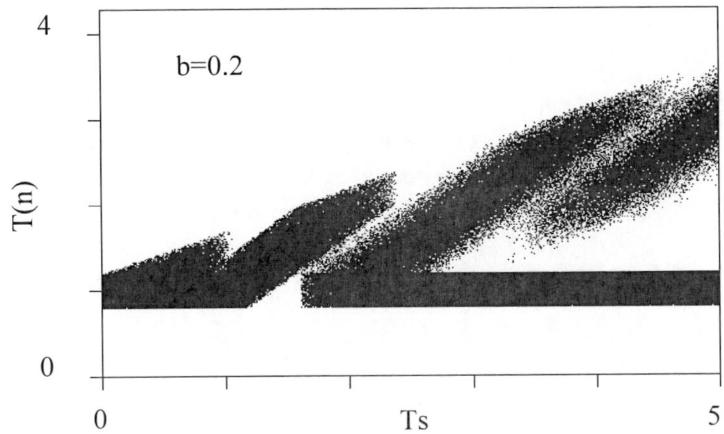

Figure 102. Plots of tour time against cycle time for irregular interval $b = 0.2$.

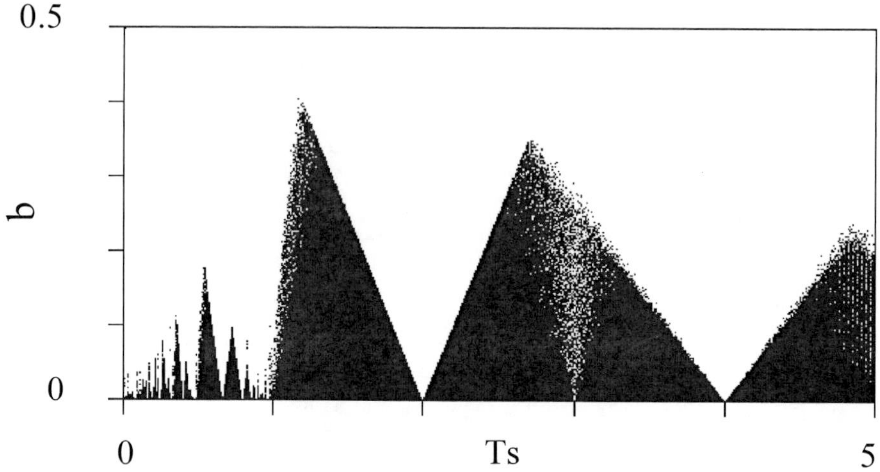

Figure 103. Region map (phase diagram) in (T_s, b) -space which represents the regions not depending on interval's irregularity. The black region indicates that not depending on interval's irregularity.

We study the effect of irregularity on vehicular traffic when the offset time between signals varies irregularly from signal to signal ($c \neq 0$), the split is the same ($a = 0$), and the interval is also the same ($b = 0$). The random variable $\xi_\tau(n)$ is given by $\xi_\tau(n) = 2(rnd - 1/2)$, where *rnd* is the random number extending uniformly from zero to unity and $< rnd >= 0.5$. Figure 104(a) shows the plot of arrival time $T(n)$ against cycle time T_s at signal $n = 1000$ far from the origin for strength $c = 0.2$ of offset-time's irregularity. For comparison, the arrival time of zero offset time ($c = 0.0$) is also plotted. The arrival time changes highly by introducing the offset-time's irregularity, while there exist the regions in which the arrival time does not change with offset-time's irregularity. In the regions, the arrival time agrees with that of zero offset time. Figure 104(a) is compared with Figures 91 and 98(a). The effect of offset-time's irregularity is similar to that of interval's irregularity. The irregularities of offset and interval induce the irregular motions, while the irregular split does not induce the irregular motion. Figure 104(b) shows the plots of arrival time $T(n)$ against cycle time T_s at signal $n = 1000$ for strengths $a = 0.1, 0.3, 0.49$ of offset-time's irregularity. With increasing irregularity, the profile of arrival time becomes smooth. When irregularity strength c becomes large, the arrival time exhibits the linear dependence for cycle time. Figure 104(b) is compared with Figures 92 and 98(b).

The arrival time scales as $\dfrac{T(n) - n}{n} = h(T_s)$. The scaling function $h(T_s)$ is a little different from $g(T_s)$.

Figure 105(a) shows the plots of tour time against signal's position n at cycle time $T_s = 1.25$ for zero offset time $c = 0$ and irregular offset time $c = 0.2$. The tour time of $c = 0.2$ varies irregularly with signal's position n but its mean value is consistent with that of $c = 0$. Figure 105(b) shows the plots of tour time against signal's position n at cycle time $T_s = 1.75$ for zero offset time $c = 0$ and irregular offset time $c = 0.2$. The mean value of tour time of irregular offset time is lower than that of zero offset time. Figure 105 is compared with Figures 95 and 101. Thus, the difference between nonrandom and irregular offset times occurs by varying cycle time.

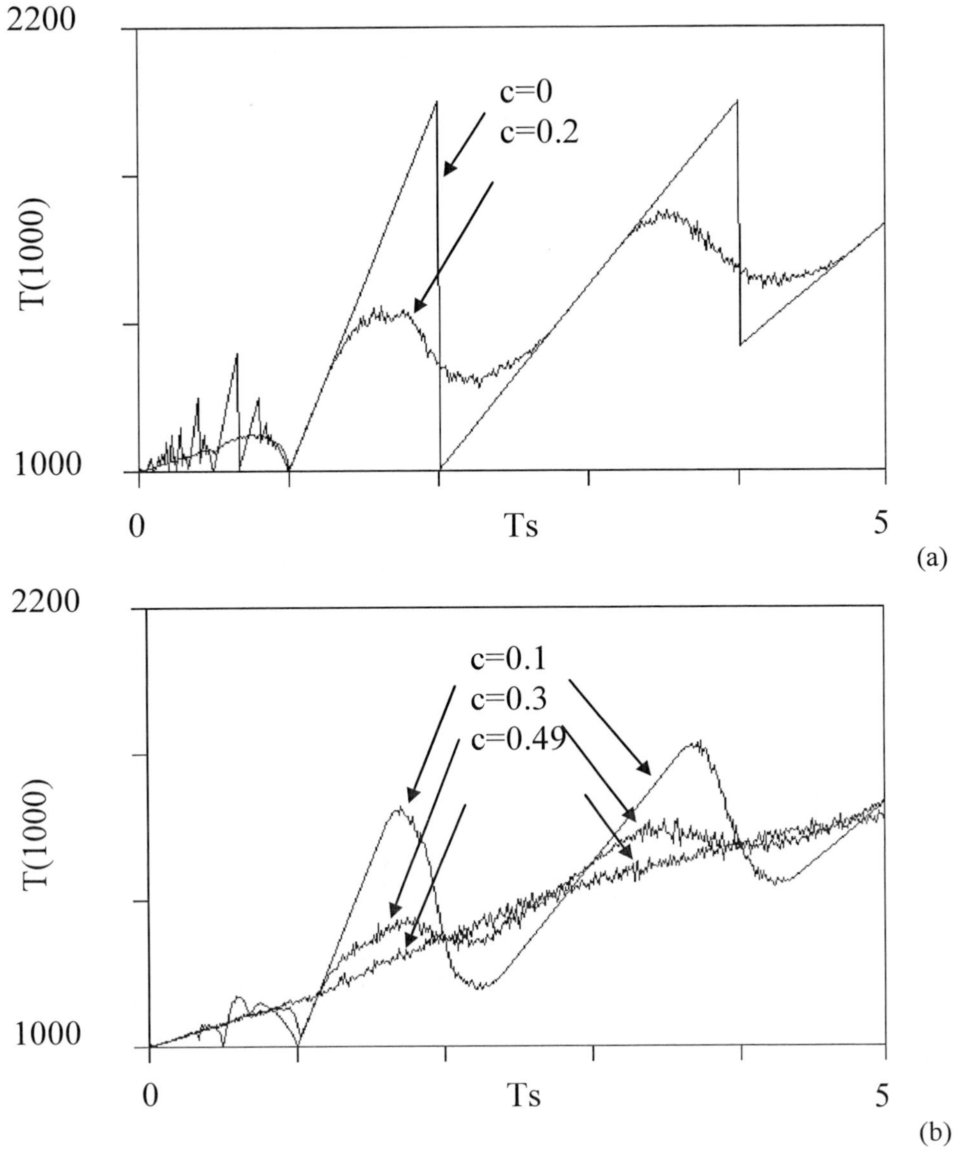

Figure 104. (a) Plot of arrival time $T(n)$ against cycle time T_s at signal $n = 1000$ far from the origin for strength $c = 0.2$ of offset-time's irregularity. (b) Plots of arrival time $T(n)$ against cycle time T_s at signal $n = 1000$ for strengths $c = 0.1, 0.3, 0.49$ of offset-time's irregularity.

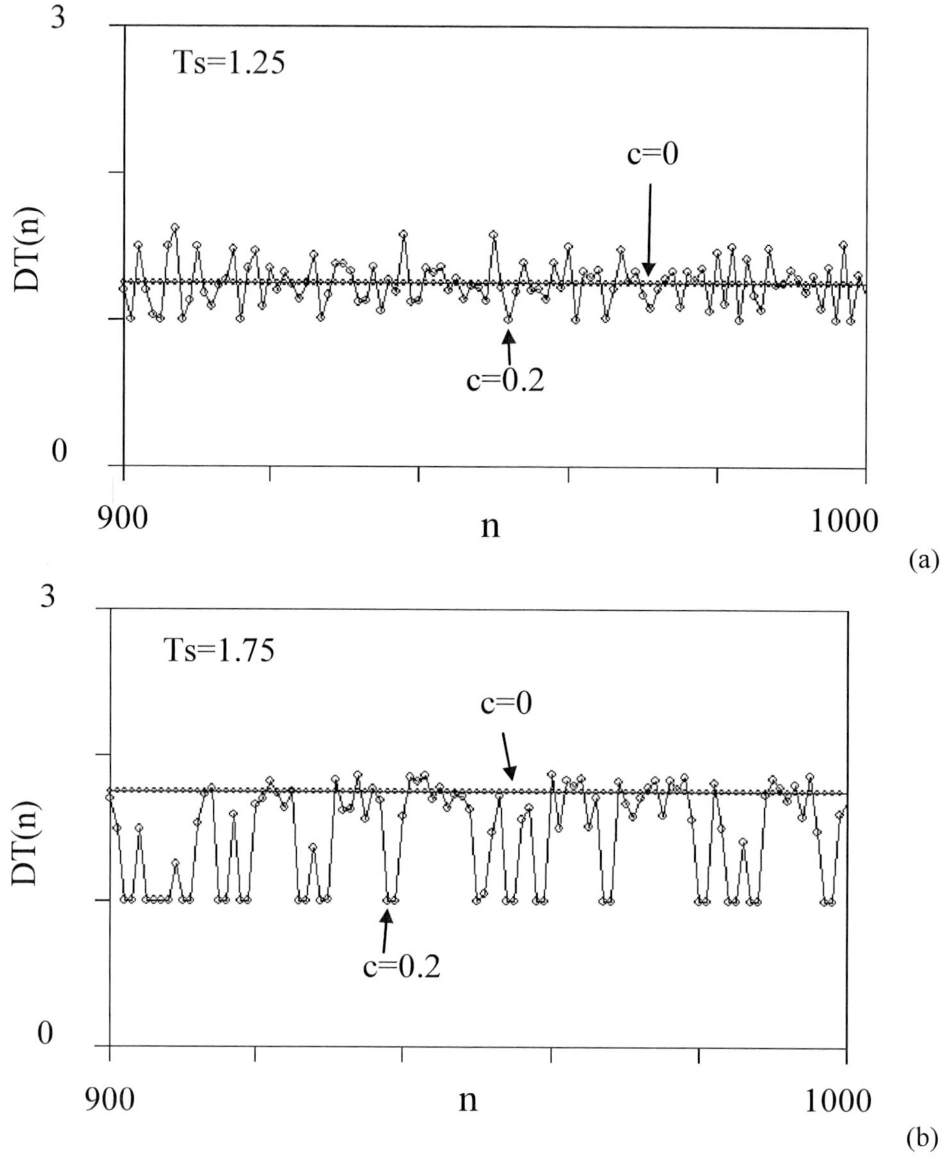

Figure 105. (a) Plots of tour time against signal's position *n* at cycle time $T_s = 1.25$ for zero offset time $c = 0$ and irregular offset time $c = 0.2$. (b) Plots of tour time against signal's position *n* at cycle time $T_s = 1.75$ for zero offset time $c = 0$ and irregular offset time $c = 0.2$.

We study the tour time $\Delta T(n) = T(n) - T(n-1)$ between two signals for various values of cycle time. Figure 106 shows the plots of tour time against cycle time for irregular offset times (a) $c = 0.05$ and (b) $c = 0.2$. For the irregular offset time, the tour time takes various values and each branch is obscure when comparing with Figure 96(a) of the same interval. However, branch of $T(n) = 0$ does not change with irregular offset time.

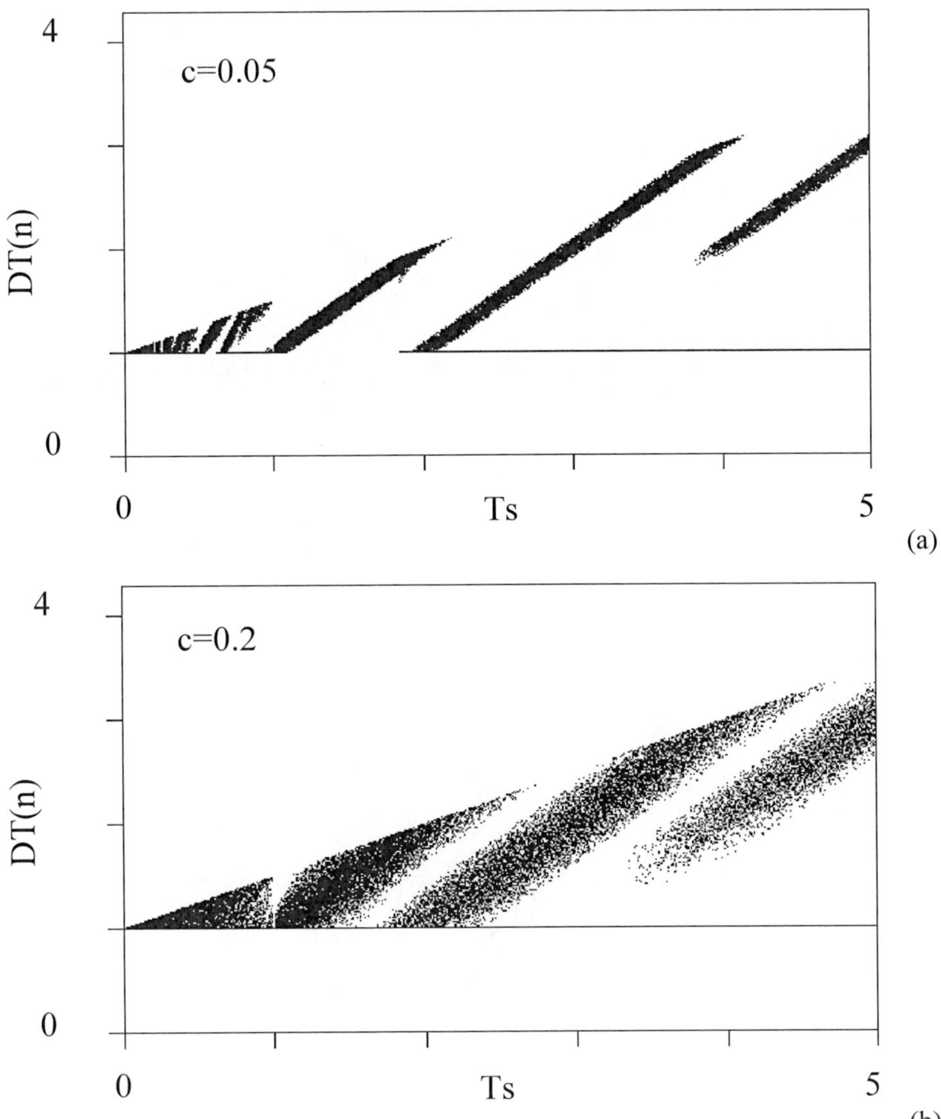

Figure 106. Plots of tour time against cycle time for irregular offset times (a) $c = 0.05$ and (b) $c = 0.2$.

We study such regions that the arrival time does not depend on the strength of offset-time's irregularity. Figure 107 shows the region map (phase diagram) in (T_s, c)-space which represents the regions not depending on offset-time's irregularity. The black region indicates that not depending on offset-time's irregularity. Thus, the regions independent on offset-time's irregularity decreases with increasing strength of irregularity. Figure 107 is compared with Figures 97 and 103 of split's and interval's irregularities.

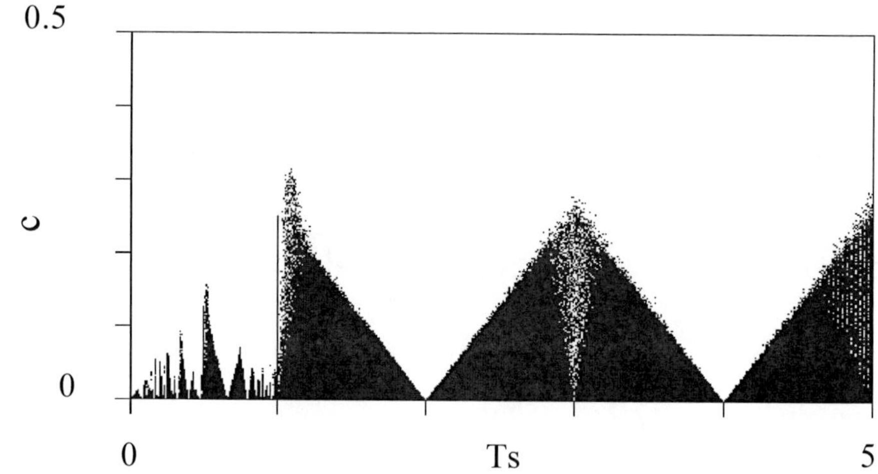

Figure 107. Region map (phase diagram) in (T_s, c)-space which represents the regions not depending on offset-time's irregularity. The black region indicates that not depending on offset-time's irregularity.

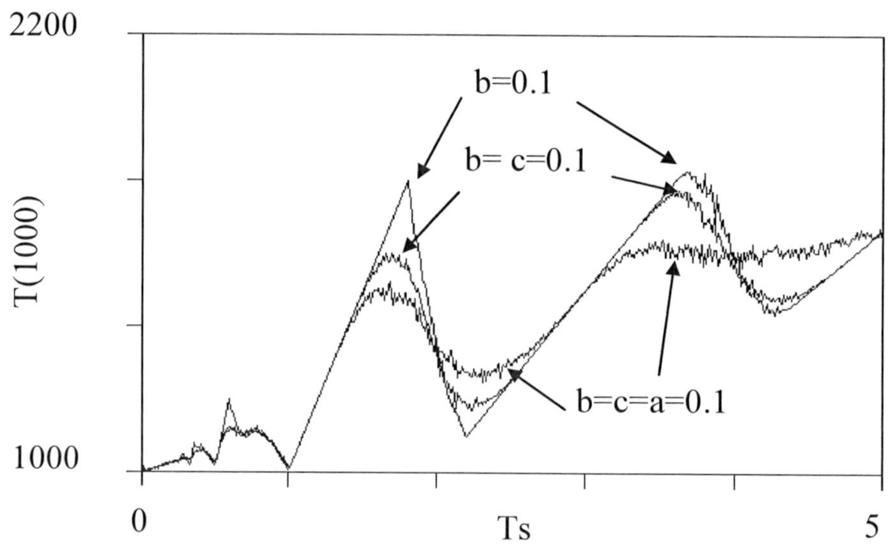

Figure 108. Plot of arrival time $T(1000)$ against cycle time T_s at n=1000 for the mixed case.

We study the mixed case that three irregularities are added simultaneously. Figure 108 shows the plot of arrival time $T(1000)$ against cycle time T_s at n=1000. The solid curve of b=0.1 represents the arrival time when there is only the irregular interval. The solid curve of b=c=0.1 represents the arrival time when there are the irregularities of both interval and offset. The solid curve of b=c=a=0.1 represents the arrival time when there are three irregularities of interval, offset, and split simultaneously. Each effect of irregularities is accumulated but the mixed effects are less than such sum that each effect is added independently. The mixed effect shows the similar behavior.

7. MAXIMAL FLOW AND CLUSTERING CONTROLLED BY SIGNALS

The traffic pattern changes by varying both split and cycle time of signals. The maximal current depends on the traffic pattern. The clustering of vehicles occurs by controlling the signals. The clustering is connected to the maximal current. However, the dependence of maximal current on the signal characteristics is known little. Here, we study the maximal current of vehicular traffic moving through an infinite series of traffic lights [96, 97].

7.1. Maximal Flow and Pattern at Synchronized Strategy

We consider the flow of vehicles going through an infinite series of traffic lights. Each vehicle is inhibited to pass other vehicles. Size of each vehicle is $l_{\min}$. A vehicle moves with the mean speed v if the way is not blocked by other vehicles and there are no signals. Then, the minimal time headway is $l_{\min}/v$. We consider the vehicular traffic on one-dimensional lattice. We set the lattice spacing as the minimal headway. The site is occupied by a single vehicle or empty. The overlapping of vehicles at a site is inhibited. Thus, the excluded-volume effect is taken into account. The traffic lights are periodically positioned with distance M. The lattice sites are numbered, from upstream to downstream, by 1, 2, 3, ---, n, n+1, ---. Also, vehicles are numbered, from the leader to the rear, by 1, 2, 3, --, i, i+1, --,N.

In the synchronized strategy, all the traffic lights change simultaneously from red (green) to green (red) with a fixed time period $(1-S_p)T_s$ (S_pT_s). The period of green is S_pT_s and the period of red is $(1-S_p)T_s$. Time T_s is called the cycle time and fraction S_p represents the split which indicates the ratio of green time to cycle time. When a vehicle arrives at a traffic light and if the traffic light is red, the vehicle stops at the position of the traffic light. Then, when the traffic light changes from red to green, the vehicle goes ahead. On the other hand, when a vehicle arrives at a traffic light and if the traffic light is green, the vehicle does not stop and goes ahead without changing speed.

We define the arrival time of vehicle i at the site nM+1 just after traffic light M as $T(i,nM+1)$. The arrival time of vehicle i is given by

$$T(i,nM+1) = T(i,nM) + 1 + \left(R(i,nM) - T(i,nM)\right) \times H\left(T(i,nM) - \left(\mathrm{int}(T(i,nM)/T_s)T_s\right) - S_pT_s\right)$$

with $R(i,nM) = \left(\mathrm{int}\left(T(i,nM)/T_s\right)+1\right)\cdot T_s$, (98)

$$T(i,nM+1) = \max\left[T(i,nM+1), T(i-1,nM+2)\right], \quad (99)$$

where $H(T)$ is the Heaviside function: $H(T)=1$ for $T \geq 0$ and $H(T)=0$ for $T<0$. $H(T)=1$ if the traffic light is red, while $H(T)=0$ if the traffic light is green. $R(i,nM)$

is such time that the traffic light just changed from red to green. The third term on the right hand side of eq. (98) represents such time that the vehicle stops if traffic light nM is red.

If vehicle i reaches the vehicle ahead i-1, it does not pass over the proceeding and it keeps the minimal headway. Equation (99) represents the condition of no passing.

In the region except for the signals, the arrival time of vehicle i at the site $nM + m$ ($1 < m \leq M$) is given by

$$T(i, nM + m) = T(i, nM + m - 1) + 1, \qquad (100)$$

$$T(i, nM + m) = \max[T(i, nM + m), T(i-1, nM + m + 1)]. \qquad (101)$$

Equation (100) means that a vehicle moves one site during unit time if the way is not blocked by other vehicles. Equation (101) represents the condition of no passing.

The number of iteration increases one by one as the vehicle moves ahead. The iteration corresponds to the going ahead on the highway. Thus, the arrival time can be calculated by iterating Eqs (98)-(101).

We calculate the traffic current at the open boundary when the input density of vehicles is very high. We study the maximal current being able to flow on the roadway with a series of traffic lights. We derive the dependence of maximal current on both cycle time and split. We study the connection between the maximal current and the tour time.

The traffic current is defined as the number of vehicles going through a position on the highway during the period of unit time. For a very high input density, maximal current Q is given by

$$Q = \lim_{N \to \infty} \frac{N}{T(N, n) - T(1, n)} \text{ for a sufficiently large } n, \qquad (102)$$

where N is the total number of vehicles. $T(N, n) - T(1, n)$ is the difference between the arrival times of rear vehicle N and leading vehicle 1 at position n far from the initial position. The maximal current is proportional to the inverse of the arrival-time difference.

The tour time between a traffic light and its proceeding light is defined as $DT_i(nM) \equiv T(i, nM) - T(i, (n-1)M)$.

We study the dynamical behavior of the signal traffic for the synchronized strategy. Initially, there exist all vehicles with minimal time headway 1. The initial density is the highest value, one. The initial condition of vehicle i is given by $T(i, 0) = i$.

Figure 109(a) shows the plot of the tour time $DT_1(n)$ against cycle time T_s for leading vehicle 1 and sufficiently large number $n = 1000 - 2000$ at split $S_p = 0.5$, where the total number of vehicles is $N = 200$ and the distance between two signals is $M = 20$. Figure 109(b) is the enlargement of Figure 1(a) for $0 \leq T_s \leq 20$.

Figure 110(a) shows the plot of the maximal current Q against cycle time T_s for sufficiently large number $n = 1000 - 2000$ at $S_p = 0.5$, where the total number of vehicles is $N = 200$ and the distance between two signals is $M = 20$. Figure 110(b) is the enlargement of Figure 110(a) for $0 \le T_s \le 20$.

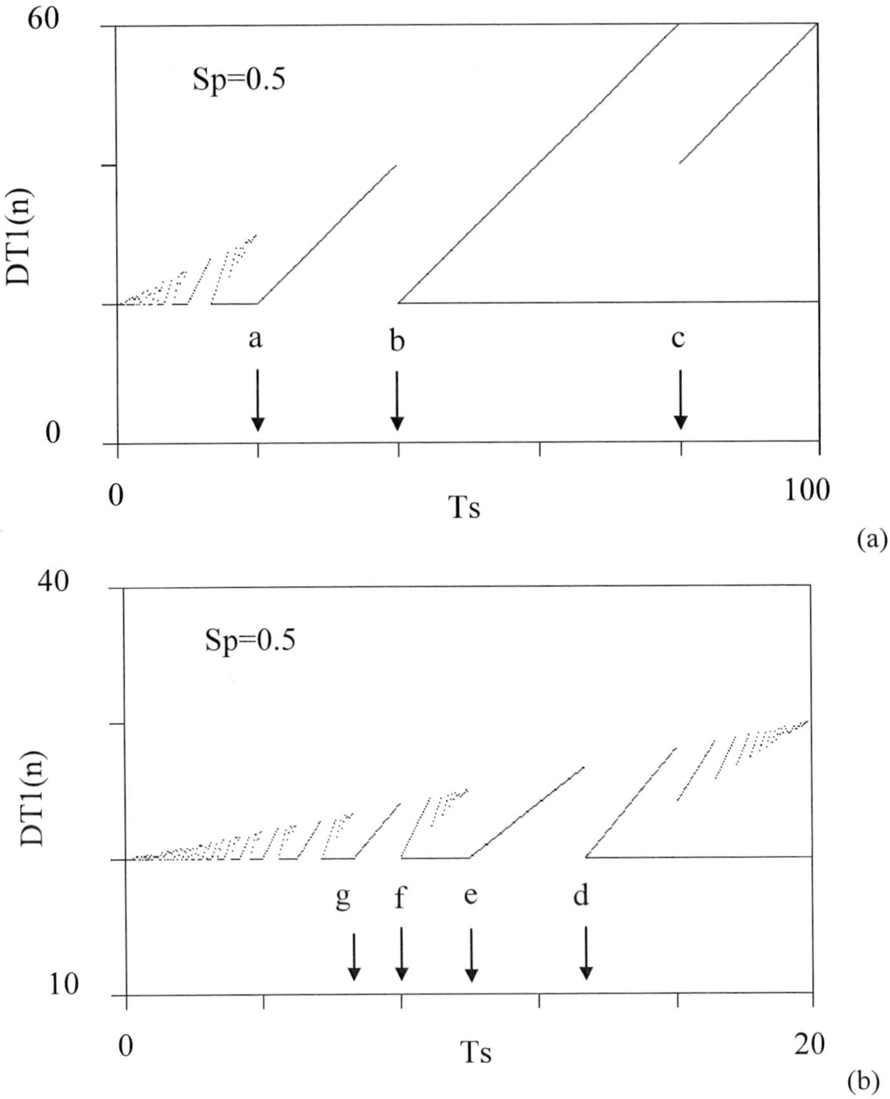

Figure 109. (a) Plot of the tour time $DT_1(n)$ against cycle time T_s for leading vehicle 1 and sufficiently large number $n = 1000 - 2000$ at $S_p = 0.5$, where the total number of vehicles is $N = 200$ and the distance between two signals is $M = 20$. (b) Enlargement of diagram (a) for $0 \le T_s \le 20$.

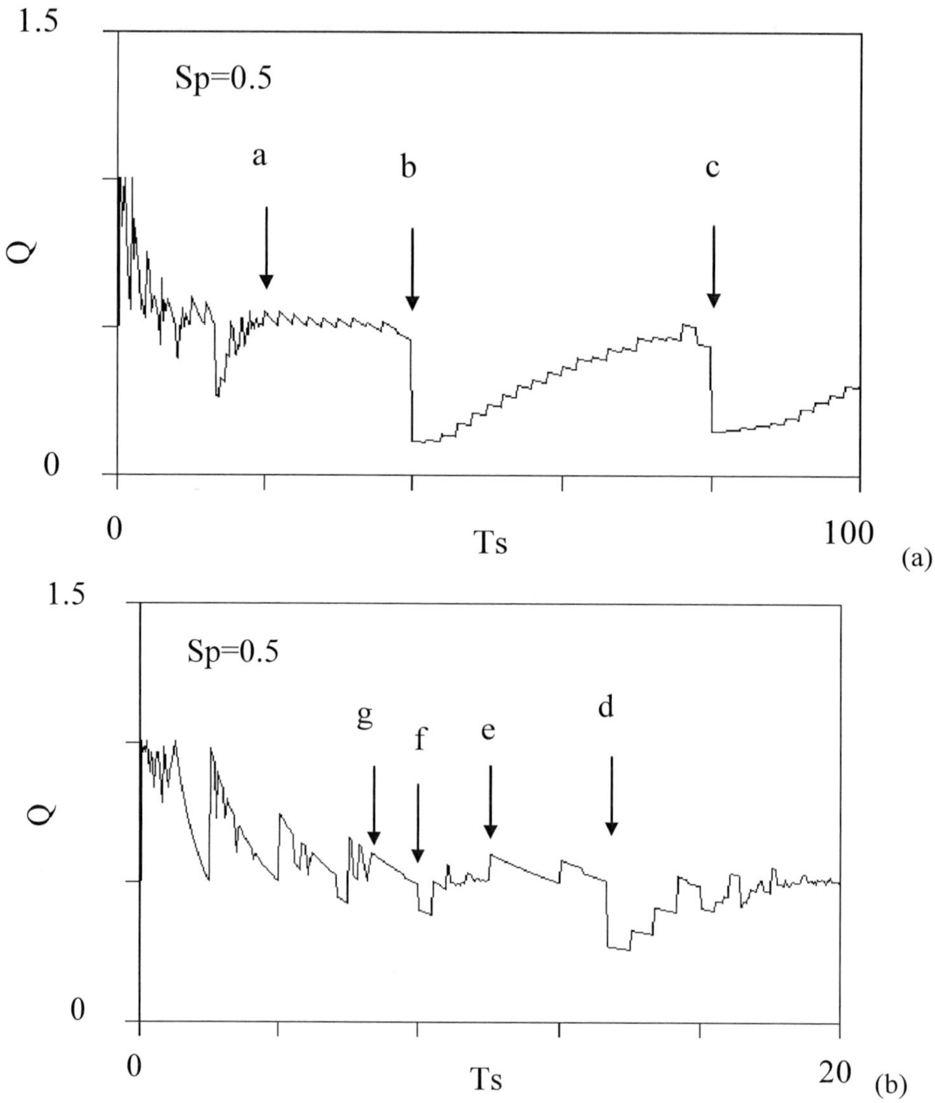

Figure 110. (a) Plots of the maximal current Q against cycle time T_s for sufficiently large number $n = 1000 - 2000$ at $S_p = 0.5$, where the total number of vehicles is $N = 200$ and the distance between two signals is $M = 20$. (b) Enlargement of diagram (a) for $0 \leq T_s \leq 20$.

The tour time increases linearly from point $a(T_s = 20)$ to point $b(T_s = 40)$ as the vehicle goes ahead. A vehicle stops at all traffic lights and then goes. Then, the current keeps almost constant value. When the cycle time is higher than 40, the tour time of vehicle takes two values. The vehicle displays a periodic motion. The vehicle stops at traffic lights or goes without stopping at traffic lights as the traffic lights turn on and off. The horizontal line from point (40,20) to point $(\infty,20)$ represents the free movement due to the green of all traffic lights. The straight line of slope 1 from point $b(T_s = 40)$ to point $c(T_s = 80)$ represents the tour time at which the vehicle stops every two traffic lights. The current decreases abruptly at point *b* and then increases with cycle time. When the cycle time is higher than 80, the vehicle stops every three signals and goes ahead. Again, the current decreases abruptly at point $c(T_s = 80)$ and increases again with cycle time.

In the region between points *e* and *d*, the vehicle stops at all the signals.

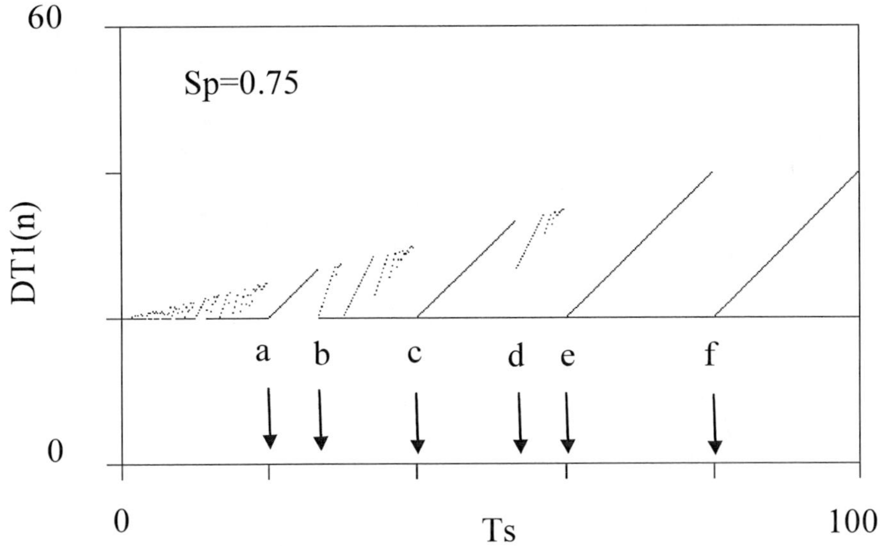

(a)

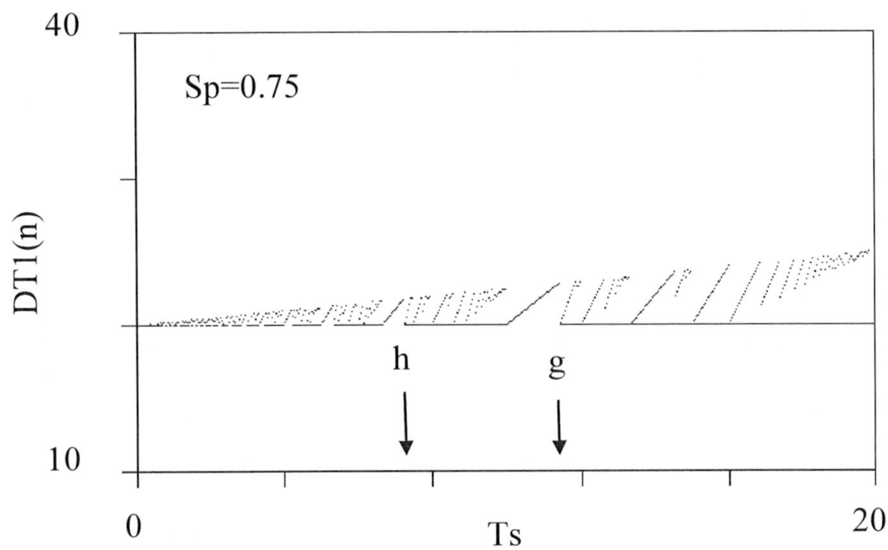

(b)

Figure 111. (a) Plot of the tour time $DT_1(n)$ against cycle time T_s for leading vehicle 1 and sufficiently large number $n = 1000 - 2000$ at split $S_p = 0.75$, where the total number of vehicles is $N = 200$ and the distance between two signals is $M = 20$. (b) Enlargement of diagram (a) for $0 \le T_s \le 20$.

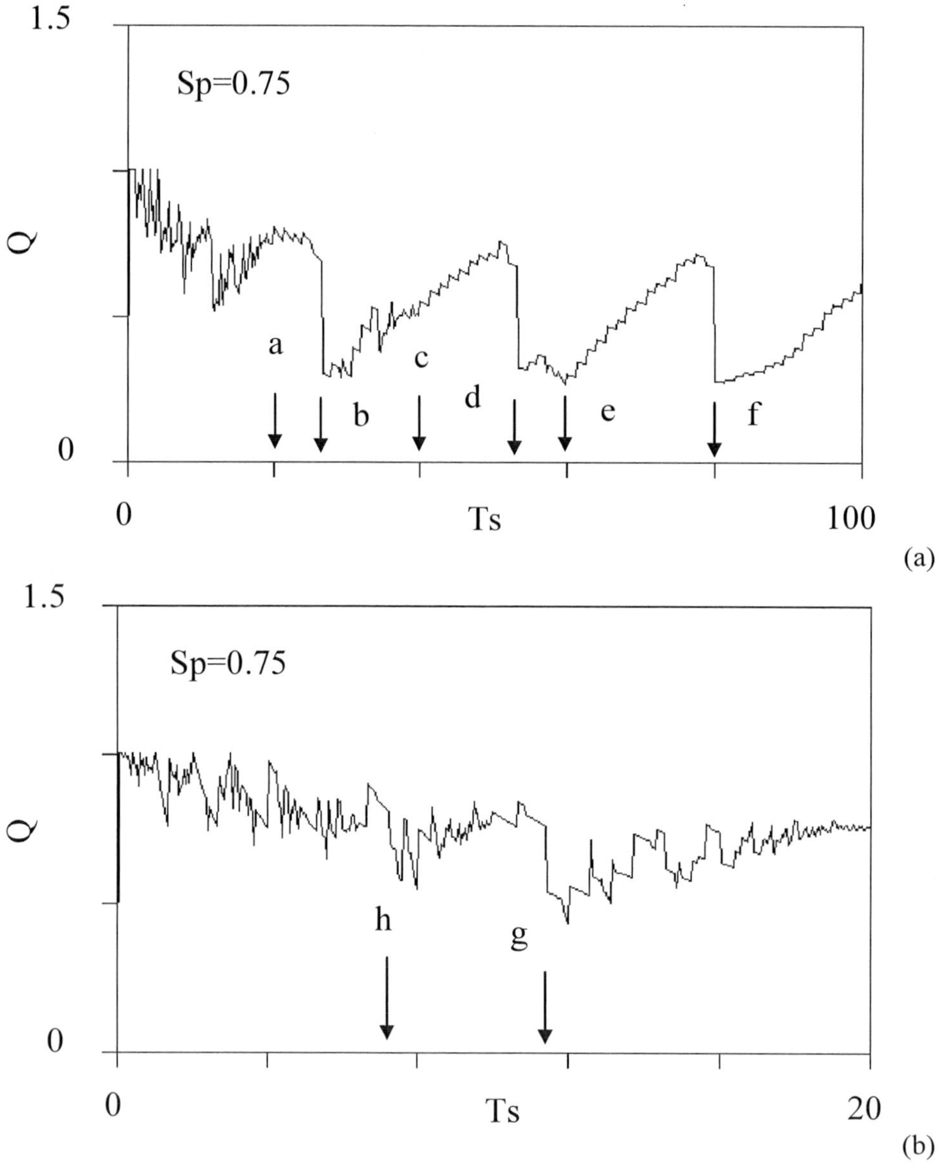

Figure 112. (a) Plot of the maximal current Q against cycle time T_s for sufficiently large number $n = 1000 - 2000$ at $S_p = 0.75$. (b) Enlargement of diagram (a) for $0 \le T_s \le 20$.

When the cycle time is higher than point *d*, the vehicle stops every two signals, every three signals, every four signals, --- corresponding to the segments of tour time. The current decreases abruptly at point *d*. When the cycle time increases through point *f*, the vehicular motion changes stopping at all signals to that every two signals. Again, at point *f*, the current decreases abruptly. Thus, the maximal current changes complexly with the tour time.

Figure 111(a) shows the plot of the tour time $DT_1(n)$ against cycle time T_s for leading vehicle 1 and sufficiently large number $n = 1000 - 2000$ at split $S_p = 0.75$, where the total number of vehicles is $N = 200$ and the distance between two signals is $M = 20$. Figure 111(b) is the enlargement of Figure 111(a) for $0 \le T_s \le 20$. Figure 112(a) shows the plot of the maximal current Q against cycle time T_s for sufficiently large number $n = 1000 - 2000$ at $S_p = 0.75$. Figure 112(b) is the enlargement of Figure 112(a) for $0 \le T_s \le 20$.

At split $S_p = 0.75$, the tour time increases linearly from point *a* (20,20) to point *b* (80/3,80/3) as the vehicle goes ahead. The vehicle stops at all traffic lights and then goes. When the cycle time is higher than 80/3 (point *b*), the vehicle stops every two signals and the tour time has two values. The vehicle moves with a period of two. The horizontal line from point (80/3,20) to point $(\infty,20)$ represents the free movement due to the green of all traffic lights. When the cycle time is higher than 40 (point *c*) the vehicle stops every three traffic lights and then goes ahead. When the cycle time is higher than 60 (point *e*), the vehicle stops every four signals and then goes ahead. Furthermore, when the cycle time is higher than 80 (point *f*), the vehicle stops every five signals and then goes ahead. At points *b*, *d*, and *f*, the current decreases abruptly with increasing cycle time. Thus, when the periodic motion of the vehicle changes, the current shows the abrupt decreases.

We study the traffic patterns of vehicles for split $S_p = 0.5$. We calculate the trajectories of all vehicles. Figure 113 shows the plots of vehicular position *n* against arrival time difference $T_i(n) - T_i(1000)$ for $n = 1000 - 1150$ where the total number *N* of vehicles is 50. Figure 113(a) shows the traffic patterns at cycle time $T_s = 20$ (point *a* in Figure 109(a)). All vehicles move without stopping at traffic signals. All vehicles go successfully through green signals. Vehicles make clusters. The maximal current displays the high value due to the clustering and free movement.

Figure 113(b) shows the traffic patterns at cycle time $T_s = 25$ in the region between points *a* and *b* in Figure 109(a). All vehicles stop at all signals. Vehicles make clusters. The size of a cluster increases with cycle time. The current decreases by stopping at signals but increases with cluster size. As the result, the maximal current keeps the high value.

Figure 113(c) shows the traffic patterns at cycle time $T_s = 40$ (point *b* in Figure 109(a)). All vehicles move without stopping at traffic signals. All vehicles go successfully through green signals. Vehicles do not make clusters. There exists only one vehicle between a signal and its proceeding signal. The vehicles extend widely on the roadway. Therefore, the maximal current shows the lowest value (point *b* in Figure 110(a)).

Figure 113(d) shows the traffic patterns at cycle time $T_s = 60$ in the region between points *b* and *c* in Figure 109(a). All vehicles stop every two signals. Vehicles make clusters. The size of a cluster increases with cycle time. The maximal current increases with cycle time because the cluster size increases with cycle time.

Figure 113(e) shows the traffic patterns at cycle time $T_s = 80$ (point *c* in Figure 109(a)). All vehicles stop every three traffic signals. Vehicles do not make clusters. There exists only one vehicle between a signal and its proceeding signal. The vehicles extend widely on the roadway. Therefore, the maximal current shows the lowest value (point *c* in Figure 110(a)).

Figure 113(f) shows the traffic patterns at cycle time $T_s = 100$. All vehicles stop every three signals. Vehicles make clusters. The size of a cluster increases with cycle time. The maximal current increases with cycle time because the cluster size increases with cycle time.

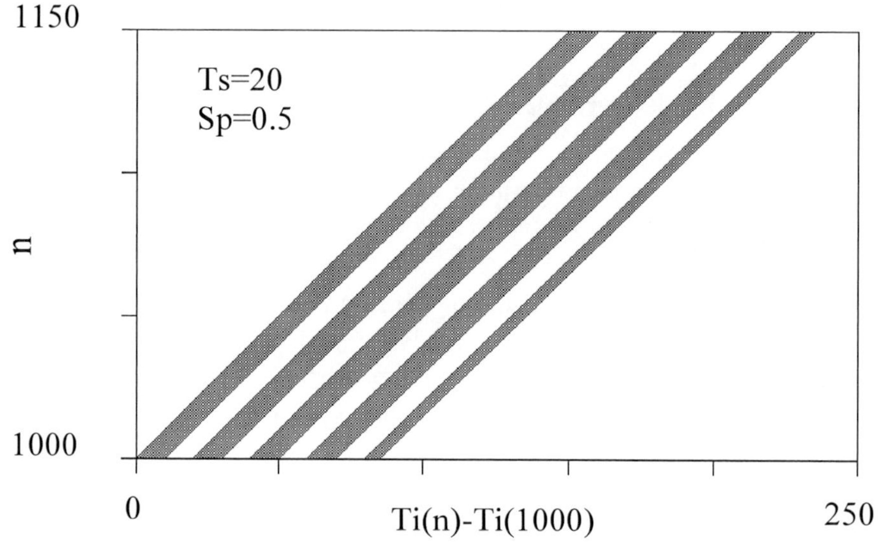

(a)

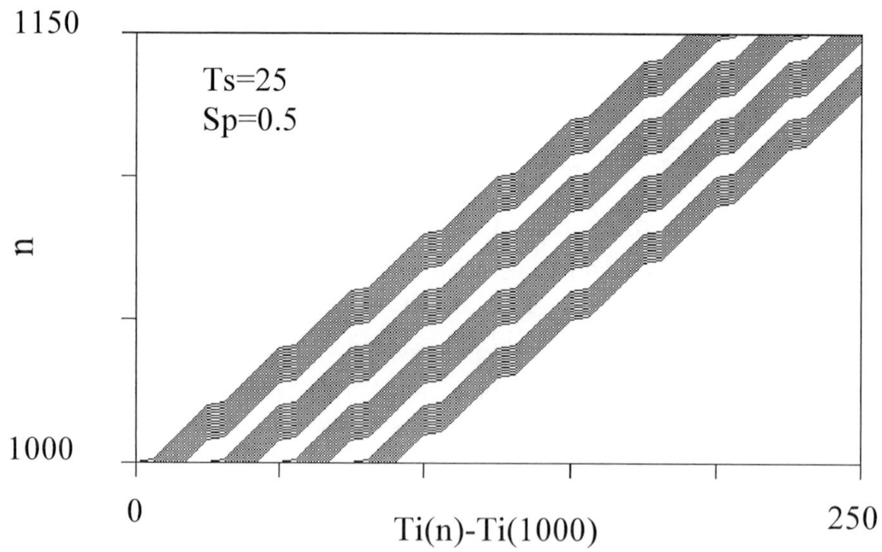

(b)

Figure 113. (Continued)

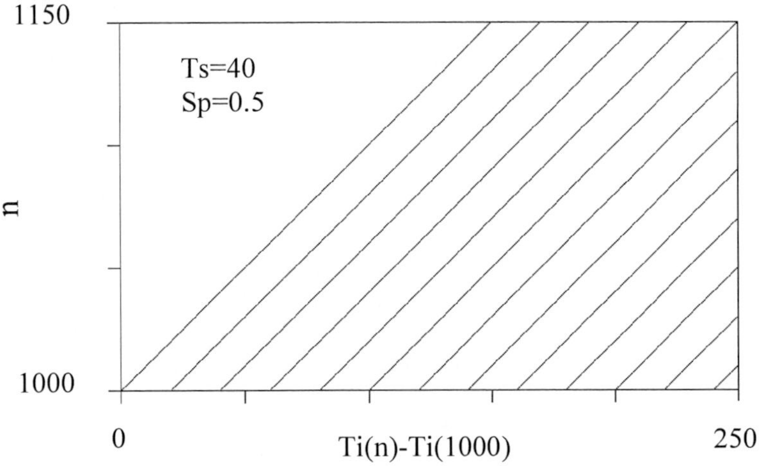

(c)

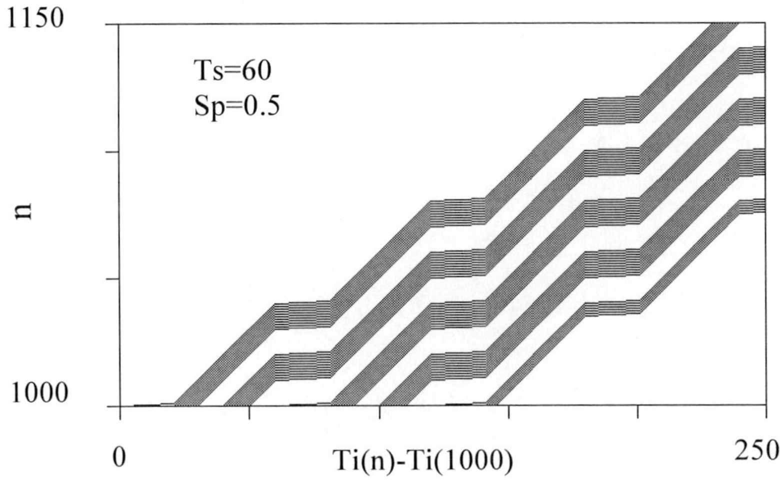

(d)

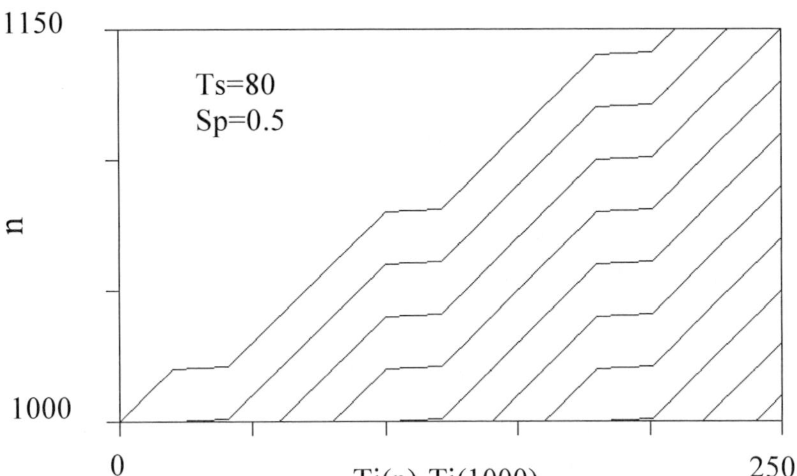

(e)

Figure 113. (Continued)

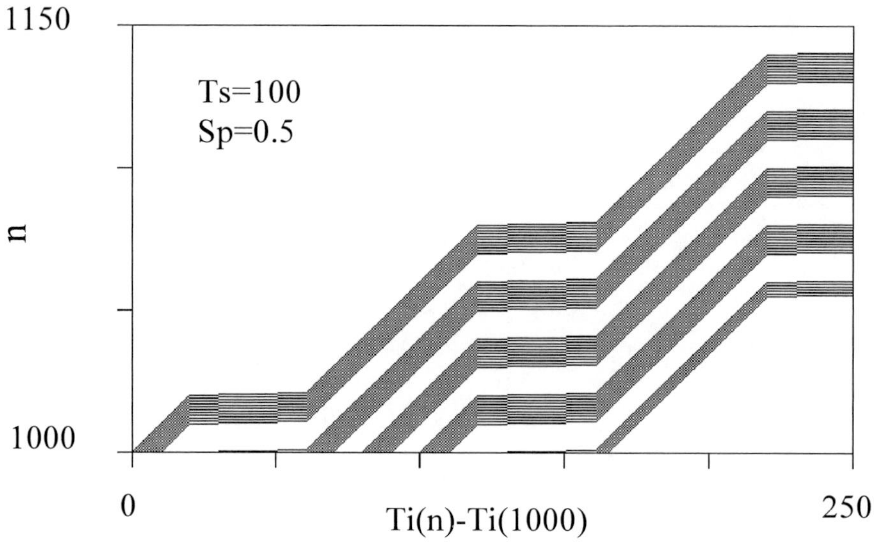

(f)

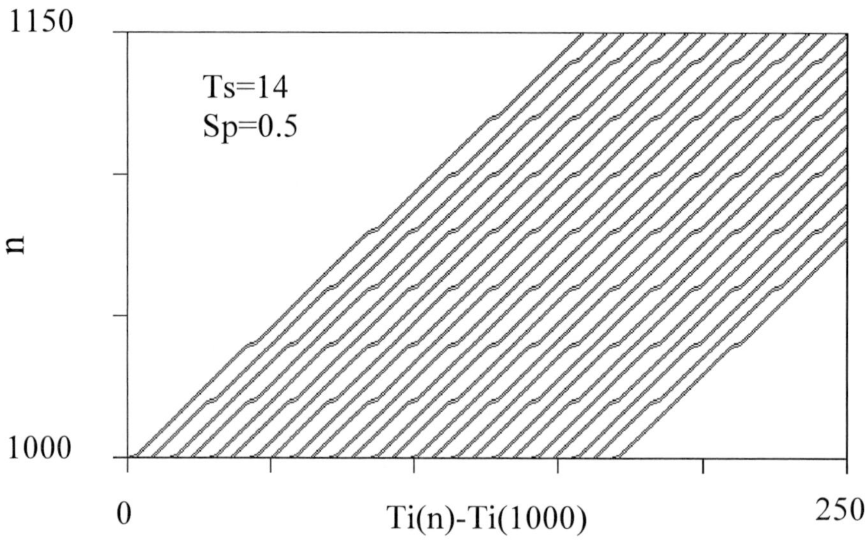

(g)

Figure 113. Plots of vehicular position *n* against arrival time difference $T_i(n)-T_i(1000)$ for $n=1000-1150$ where the total number *N* of vehicles is 50. (a) Traffic patterns at cycle time $T_s=20$ (point *a* in Figure 109(a)). (b) Traffic patterns at cycle time $T_s=25$ in the region between points *a* and *b* in Figure 109(a). (c) Traffic patterns at cycle time $T_s=40$ (point *b* in Figure 109(a)). (d) Traffic patterns at cycle time $T_s=60$ in the region between points *b* and *c* in Figure 109(a). (e) Traffic patterns at cycle time $T_s=80$ (point *c* in Figure 109(a)). (f) Traffic patterns at cycle time $T_s=100$. (g) Traffic patterns at cycle time $T_s=14$ in Figure 109(b).

Figure 113(g) shows the traffic patterns at cycle time $T_s = 14$ in Figure 109(b). All vehicles stop every two signals. Vehicles make clusters. When first, third, --- clusters stop at a signal, the second, fourth, --- clusters moves between two signals. The current decreases abruptly at point *d* and then increases with cycle time.

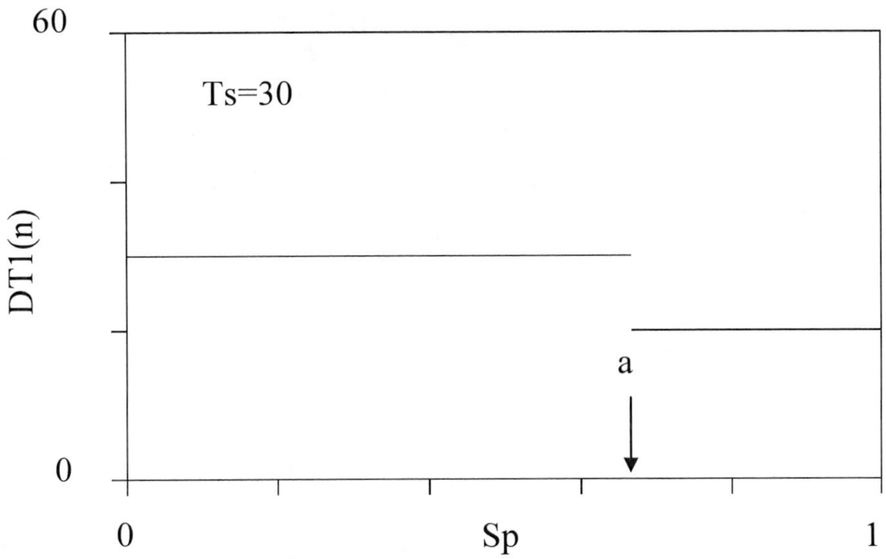

(a)

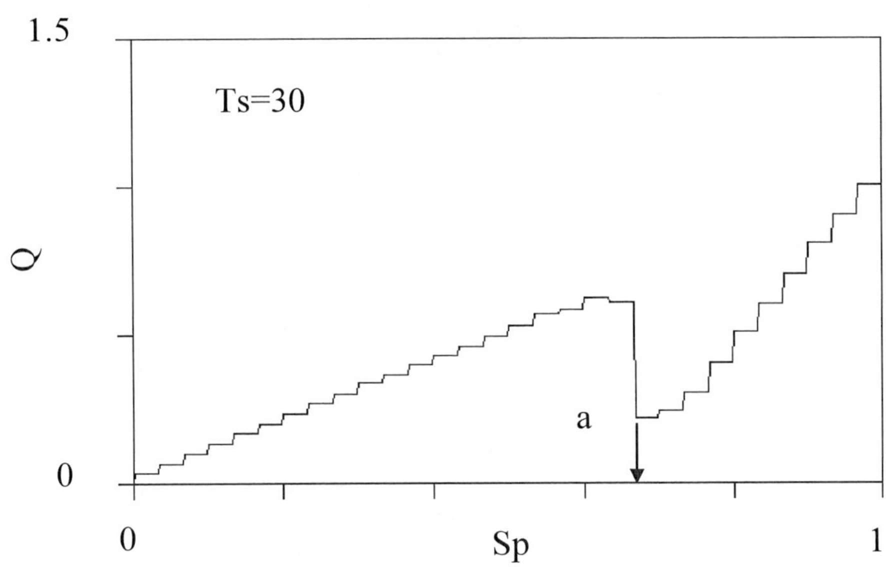

(b)

Figure 114. (a) Plot of the tour time $DT_1(n)$ against split S_p for leading vehicle 1 and sufficiently large number $n = 1000 - 2000$ at cycle time $T_s = 30$, where the total number of vehicles is $N = 200$ and the distance between two signals is $M = 20$. (b) Plot of the maximal current Q against split S_p for sufficiently large number $n = 1000 - 2000$ at $T_s = 30$.

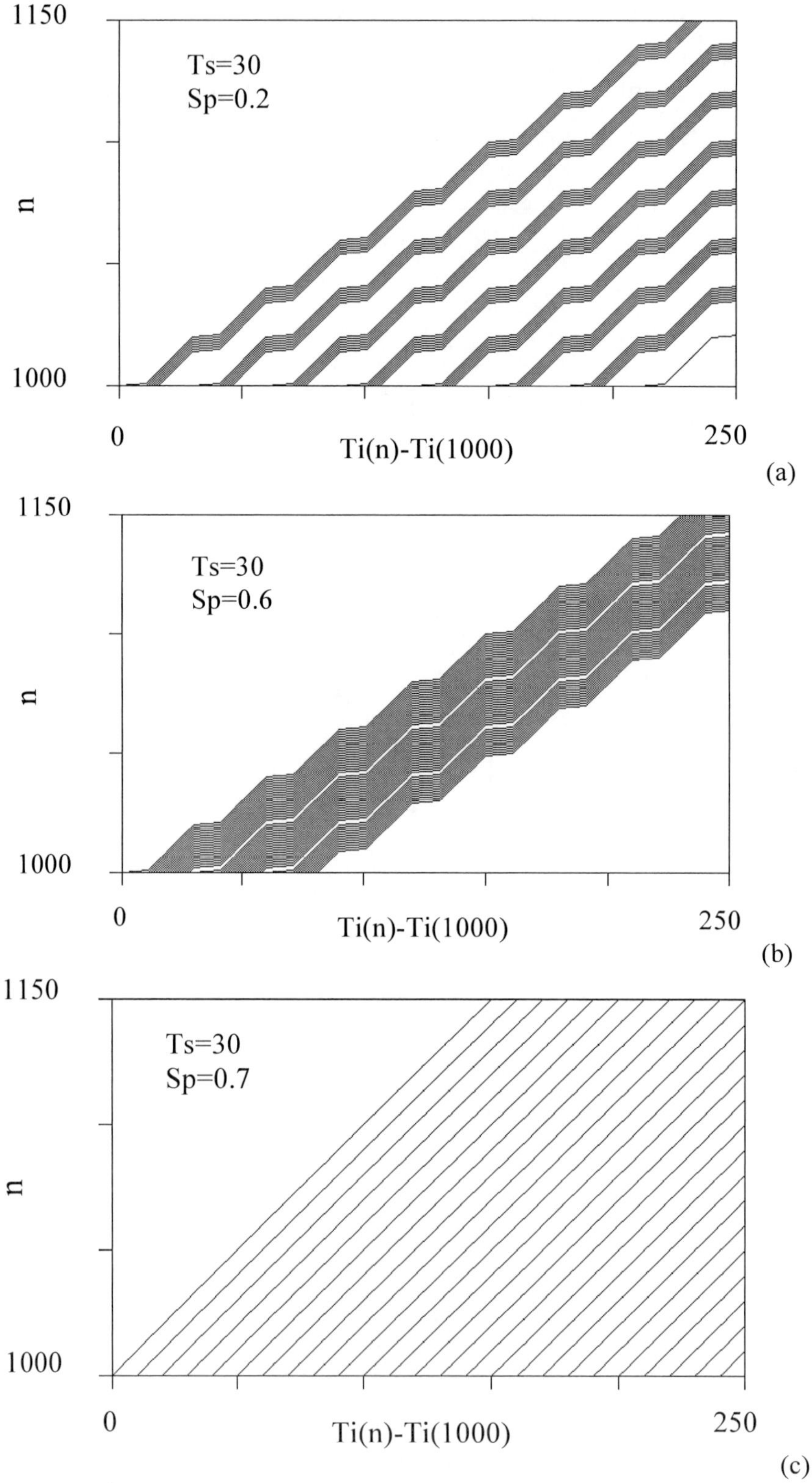

Figure 115. (Continued)

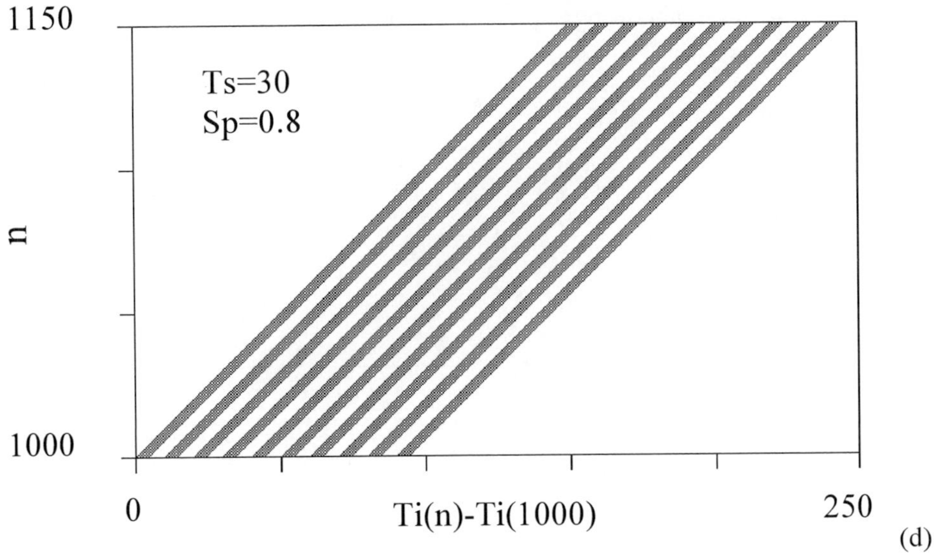

Figure 115. Traffic patterns at cycle time $T_s = 30$ for $1000 \le n \le 1150$. Patterns (a)-(d) indicate, respectively, the trajectories of all vehicles (*N*=50) at (a) $S_p = 0.2$, (b) $S_p = 0.6$, (c) $S_p = 0.7$, and (d) $S_p = 0.8$.

We study how maximal current Q varies with split S_p. Figure 114(a) shows the plot of the tour time $DT_1(n)$ against split S_p for leading vehicle 1 and sufficiently large number $n = 1000 - 2000$ at cycle time $T_s = 30$, where the total number of vehicles is $N = 200$ and the distance between two signals is $M = 20$. Figure 114(b) shows the plot of the maximal current Q against split S_p for sufficiently large number $n = 1000 - 2000$ at $T_s = 30$. The tour time keeps a constant value for $0 < S_p < 0.67$ (point *a*), while the current increases with split until $S_p = 0.67$. At $S_p = 0.67$, the tour time decreases abruptly and keeps a low value, while the current also decreases abruptly at point *a* and then increases with split.

Figure 115 shows the traffic patterns at cycle time $T_s = 30$ for $1000 \le n \le 1150$. Patterns (a)-(d) indicate, respectively, the trajectories of all vehicles (*N*=50) at (a) $S_p = 0.2$, (b) $S_p = 0.6$, (c) $S_p = 0.7$, and (d) $S_p = 0.8$. All vehicles stop always at all signals for $0 < S_p < 0.67$. Vehicles make clusters. The size of a cluster increases with split. The current increases with cluster size. While vehicles do not stop at signals for $S_p > 0.67$. At $S_p = 0.67$ (see Figure 115(c)), vehicles move one by one between a signal and its proceeding signal. The size of a cluster takes a minimum value, one. Therefore, the current takes the low value. With increasing split, the current increases for $S_p > 0.67$. This is due to increasing the cluster size.

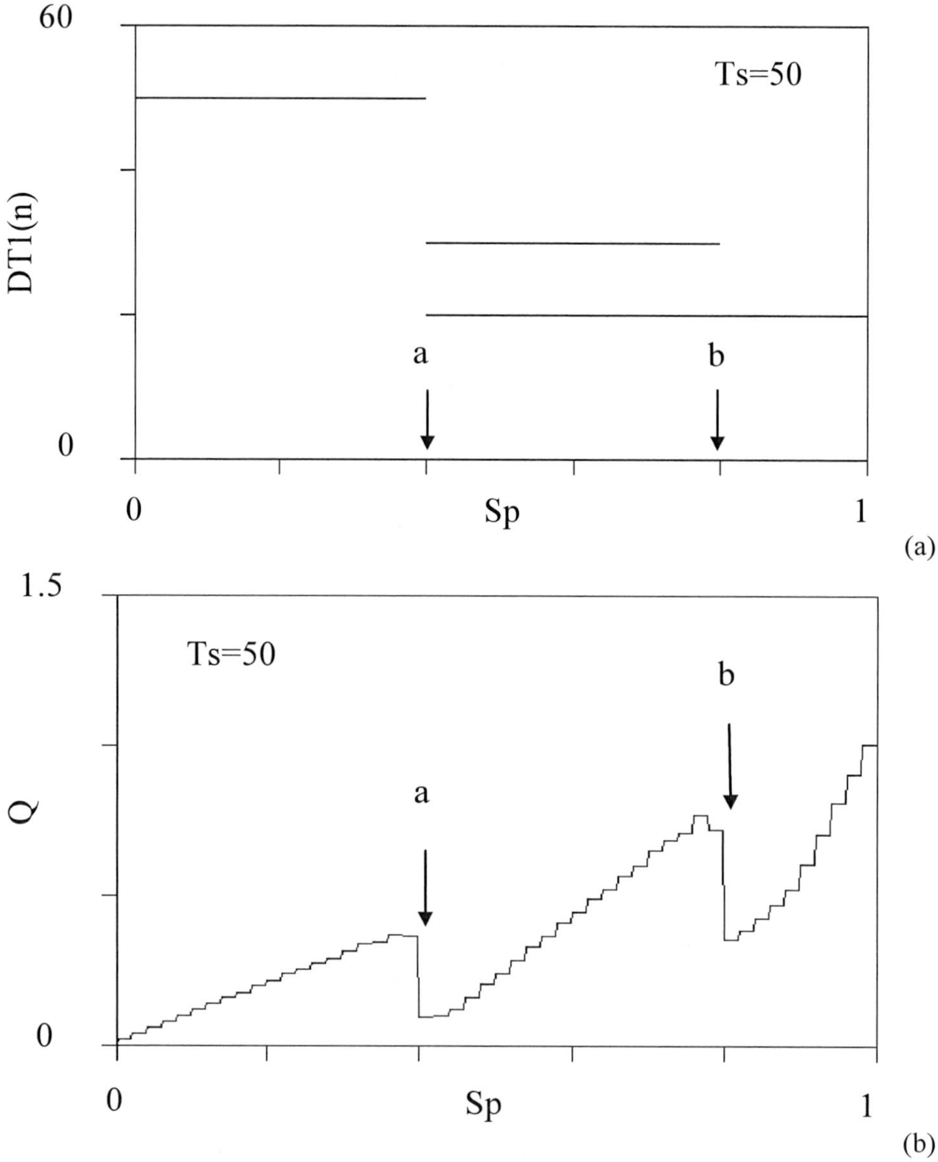

Figure 116. (a) Plot of the tour time $DT_1(n)$ against split S_p for leading vehicle 1 and sufficiently large number $n = 1000 - 2000$ at cycle time $T_s = 50$, where the total number of vehicles is $N = 200$ and the distance between two signals is $M = 20$. (b) Plot of the maximal current Q against split S_p for sufficiently large number $n = 1000 - 2000$ at $T_s = 50$.

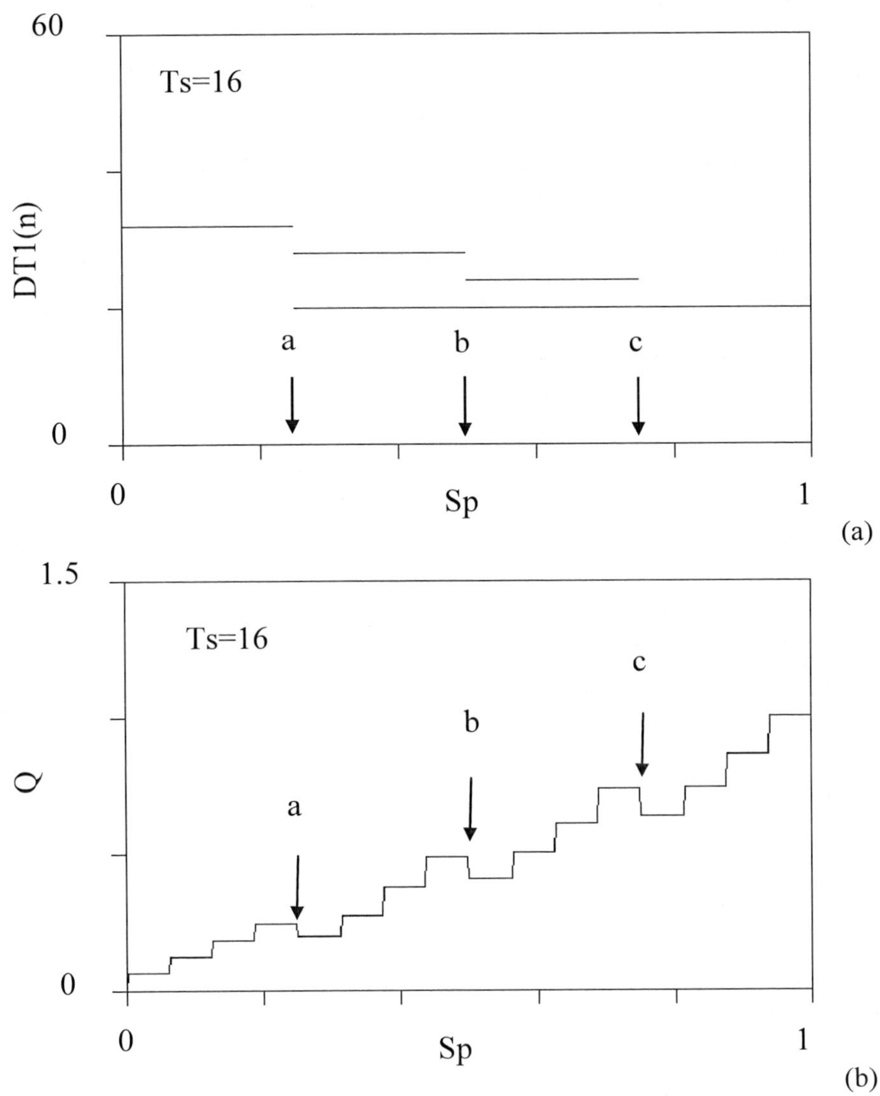

Figure 117. (a) Plot of the tour time $DT_1(n)$ against split S_p for leading vehicle 1 and sufficiently large number $n = 1000 - 2000$ at cycle time $T_s = 16$, where the total number of vehicles is $N = 200$ and the distance between two signals is $M = 20$. (b) Plot of the maximal current Q against split S_p for sufficiently large number $n = 1000 - 2000$ at $T_s = 16$.

Figure 116(a) shows the plot of the tour time $DT_1(n)$ against split S_p for leading vehicle 1 and sufficiently large number $n = 1000 - 2000$ at cycle time $T_s = 50$, where the total number of vehicles is $N = 200$ and the distance between two signals is $M = 20$. Figure 116(b) shows the plot of the maximal current Q against split S_p for sufficiently large number $n = 1000 - 2000$ at $T_s = 50$. The tour time keeps a constant value for

$0 < S_p < 0.4$, while the current increases with split until $S_p = 0.4$ (point *a*). At $S_p = 0.4$, the tour time decreases abruptly and takes two low values, while the current also decreases abruptly at point *a* and then increases with split until $S_p = 0.8$ (point *b*). When the split is higher than 0.8 (point *b*), the tour time takes the minimum value, 20 and all vehicles go through all signals without stopping. At $S_p = 0.8$ (point *b*), the current decreases abruptly and then increases with split. The increase of current is due to that of cluster size.

Figure 117(a) shows the plot of the tour time $DT_1(n)$ against split S_p for leading vehicle 1 and sufficiently large number $n = 1000 - 2000$ at cycle time $T_s = 16$, where the total number of vehicles is $N = 200$ and the distance between two signals is $M = 20$. Figure 117(b) shows the plot of the maximal current *Q* against split S_p for sufficiently large number $n = 1000 - 2000$ at $T_s = 16$. When the split is lower than the value at point *a*, all vehicles stops always at all signals and the tour time keeps a high value, while the current increases with split until point *a*. For the region of split between points *a* and *b*, vehicles stops every two signals. For the region of split between points *b* and *c*, vehicles stops every three signals. When the split is higher than that at point *c*, all vehicles go through green signals without stopping.

We study the effect of signal's distance *M* on the current and tour time. We change the signal's distance from 20 to 40. Figure 118(a) shows the plot of tour time $DT_1(n)$ against cycle time T_s for signal' distance *M*=40 at split $S_p = 0.5$, where the number of vehicles is *N*=200 and $n = 1000 - 2000$. Figure 10(a) is compared with Figure 1(a) of *M*=20. Figure 118(a) agrees exactly with the one obtained by enlarging both tour time and cycle time in Figure 109(a) two times. Generally, the tour time scales as

$$DT_1(n) = M^{-1} f(M^{-1}T_s), \tag{103}$$

where $f(x)$ is the scaling function of the tour time.

Figure 118(b) shows the plot of maximal current *Q* against cycle time T_s for signal' distance *M*=40 at split $S_p = 0.5$. Figure 118(b) is compared with Figure 110(a) of *M*=20. Figure 118(b) agrees nearly with the one obtained by enlarging cycle time in Figure 110(a) two times. Generally, the maximal current scales as

$$Q_{\max} = G(M^{-1}T_s), \tag{104}$$

where $Q_{\max}$ is the scaling function of the maximal current.

Thus, we find that the maximal current of vehicular traffic is connected highly to the clustering of vehicles induced by traffic signals.

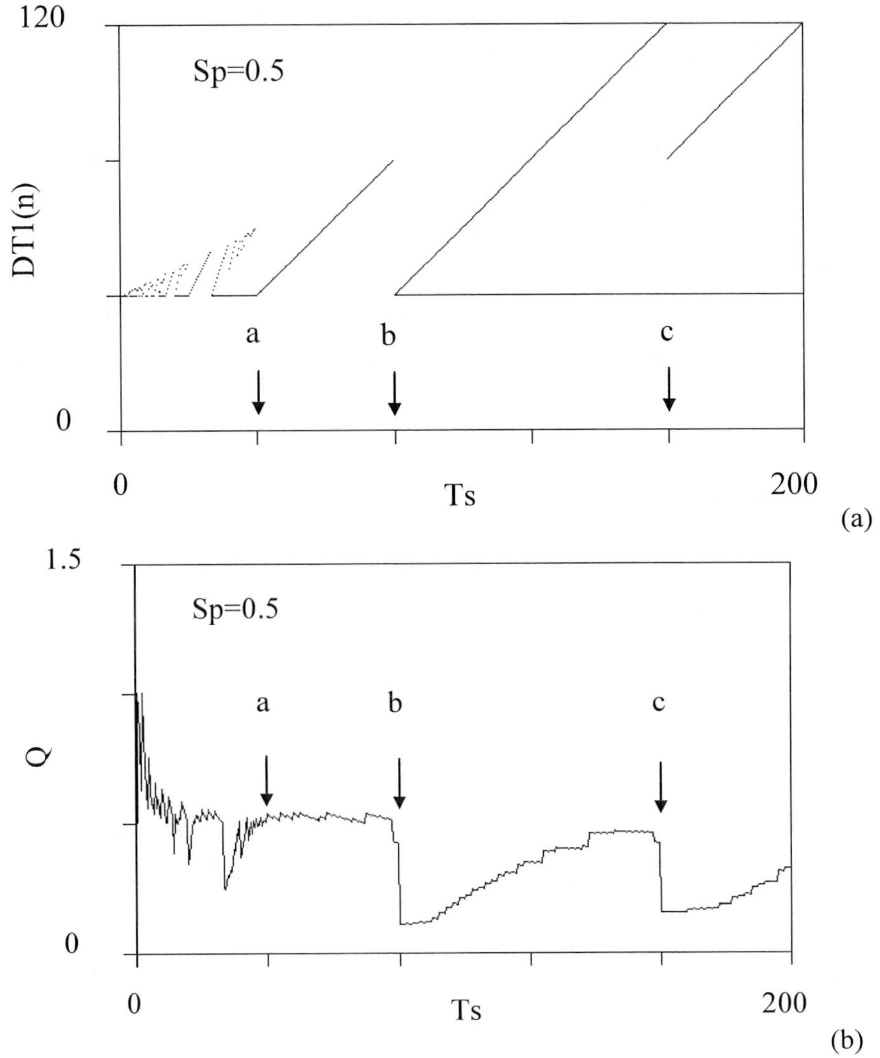

Figure 118. (a) Plot of tour time $DT_1(n)$ against cycle time T_s for signal' distance M=40 at split $S_p = 0.5$, where the number of vehicles is N=200 and $n = 1000 - 2000$. Figure 118(a) is compared with Figure 109(a) of M=20. (b) Plot of maximal current Q against cycle time T_s for signal' distance M=40 at split $S_p = 0.5$. Figure 118(b) is compared with Figure 110(a) of M=20.

7.2. Maximal Flow and Pattern at Green-wave Strategy

We study the vehicular traffic through the green-wave signals. We derive the dependence of tour time on offset time for various values of cycle time. We clarify the dependence of maximal current on offset time. We relate the traffic behavior at the green-wave strategy to that at the synchronized strategy.

We consider the flow of vehicles going through an infinite series of traffic lights. Each vehicle is inhibited to pass other vehicles. A minimal headway between a vehicle and that ahead is assumed to be $l_{\min}$. A vehicle moves with the mean speed v if the way is not blocked by other vehicles and there are no signals. Then, the minimal time headway is $l_{\min}/v$. The traffic lights are periodically positioned with distance M.

In the synchronized strategy, all the traffic lights change simultaneously from red (green) to green (red) with a fixed time period $(1-S_p)T_s$ (S_pT_s). In the green-wave strategy, the traffic light changes with a certain time delay t_{offset} between the traffic light phases of two successive intersections. The delay time is called the offset time. The change of traffic lights propagates backward like a green wave.

We define the arrival time of vehicle i at site nM+1 just after traffic light M as $t(i,nM+1)$. The arrival time of vehicle i is given by

$$\begin{aligned} t(i,nM+1) &= t(i,nM) + l_{\min}/v + \left(r(i,nM) - t(i,nM)\right) \\ &\times H\left(t(i,nM) + nt_{offset} - \left(\mathrm{int}((t(i,nM) + nt_{offset})/t_s\right)t_s) - S_p t_s\right) \end{aligned}$$

with $r(i,nM) = \left(\mathrm{int}\left((t(i,nM) + nt_{offset})/t_s\right) + 1\right)\cdot t_s - nt_{offset}$, (105)

$$t(i,nM+1) = \max\left[t(i,nM+1), t(i-1,nM+2)\right], \tag{106}$$

where $H(T)$ is the Heaviside function: $H(T)=1$ for $T \geq 0$ and $H(T)=0$ for $T<0$.

If vehicle i reaches the vehicle ahead i-1, it does not pass over the proceeding and it keeps the minimal headway. Equation (106) represents the condition of no passing.

At the position except for the signals, the arrival time of vehicle i at site $nM+m$ ($1 < m \leq M$) is given by

$$t(i,nM+m) = t(i,nM+m-1) + l_{\min}/v, \tag{107}$$

$$t(i,nM+m) = \max\left[t(i,nM+m), t(i-1,nM+m+1)\right]. \tag{108}$$

Equation (107) means that a vehicles moves one site during unit time $l_{\min}/v$ if the way is not blocked by other vehicles. Equation (108) represents the condition of no passing.

The number of iteration increases one by one as the vehicle moves ahead. The arrival time can be calculated by iterating Eqs (105)-(108).

By dividing time by characteristic time $l_{\min}/v_0$ (v_0: reference speed), one obtains the dimensionless arrival time:

$$\begin{aligned} T(i,nM+1) &= T(i,nM) + v_0/v + \left(R(i,nM) - T(i,nM)\right) \\ &\times H\left(T(i,nM) + nT_{offset} - \left(\mathrm{int}((T(i,nM) + nT_{offset})/T_s\right)T_s) - S_p T_s\right) \end{aligned}$$

with $R(i,nM) = \left(\mathrm{int}\left((T(i,nM) + nT_{offset})/T_s\right) + 1\right)\cdot T_s - nT_{offset}$, (109)

$$T(i,nM+1) = \max\left[T(i,nM+1), T(i-1,nM+2)\right], \quad (110)$$

$$T(i,nM+m) = T(i,nM+m-1) + v_0/v, \text{ for } 1 < m \le M \quad (111)$$

$$T(i,nM+m) = \max\left[T(i,nM+m), T(i-1,nM+m+1)\right], \quad (112)$$

where $T(i,n') = t(n')v_0/l_{\min}$, $R(n') = r(n')v_0/l_{\min}$, $T_s = t_s v_0/l_{\min}$, and $T_{offset} = t_{offset} v_0/l_{\min}$.

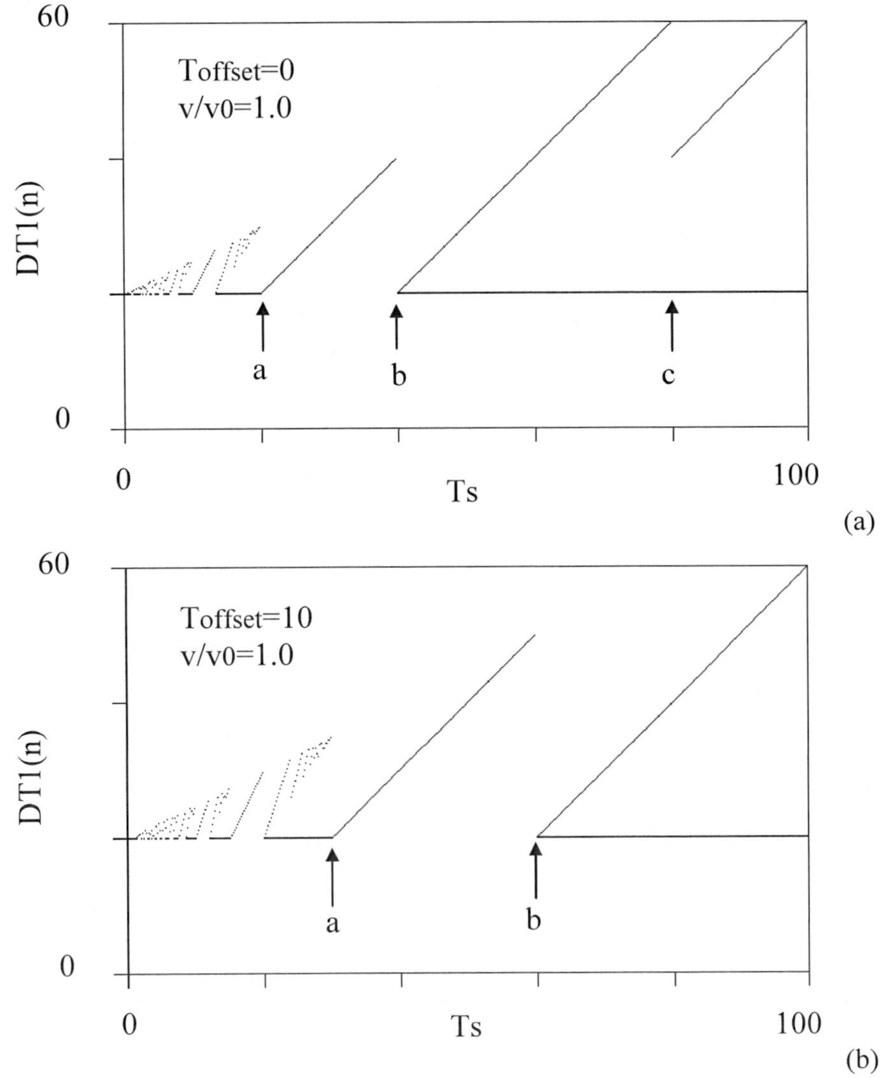

Figure 119. (Continued)

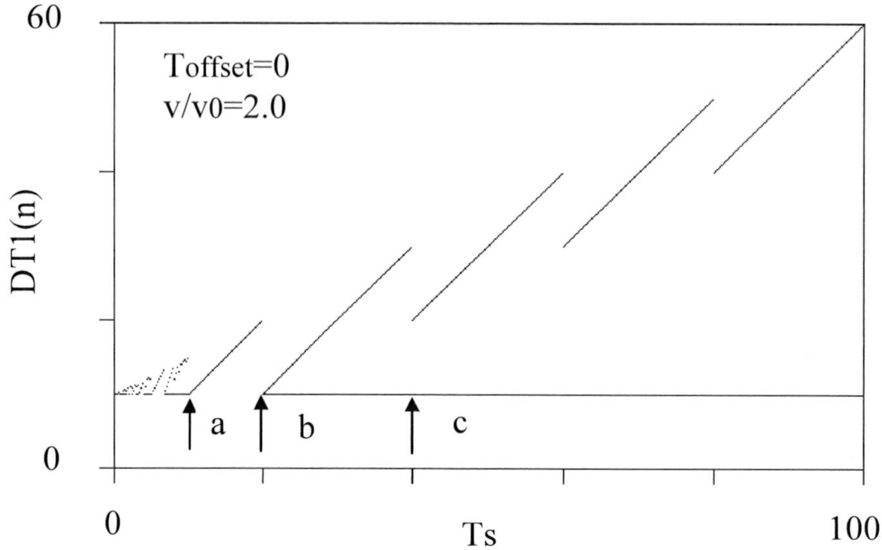

(c)

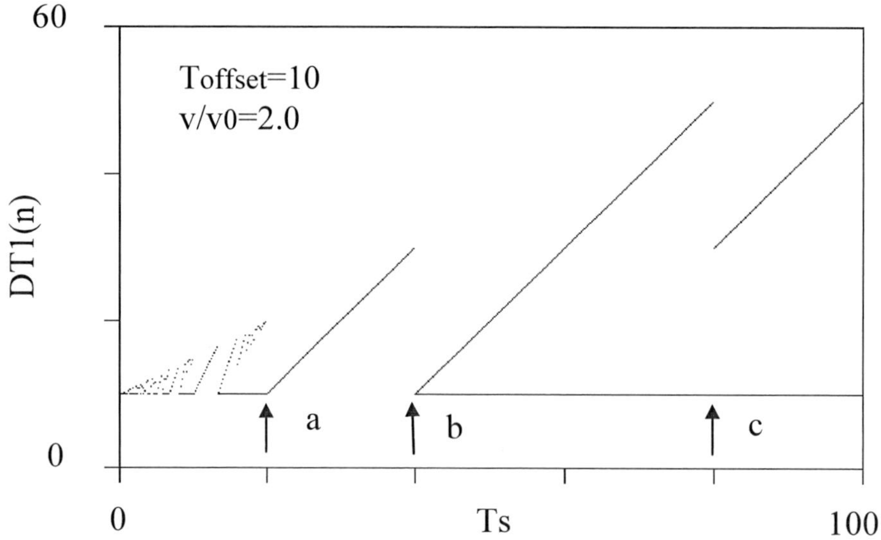

(d)

Figure 119. (a) Plot of the tour time $DT_1(n)$ against cycle time T_s for leading vehicle 1 at synchronized strategy $T_{offset} = 0$ and velocity $v/v_0 = 1.0$, where $n = 1000 - 2000$, split $S_p = 0.5$, the total number of vehicles is $N = 200$ and the distance between two signals is $M = 20$. (b) Plot of the tour time $DT_1(n)$ against cycle time T_s for leading vehicle 1 at green-wave strategy $T_{offset} = 10$ and velocity $v/v_0 = 1.0$. (c) Plot of the tour time $DT_1(n)$ against cycle time T_s for velocity $v/v_0 = 2.0$ at synchronized strategy $T_{offset} = 0$. (d) Plot of the tour time $DT_1(n)$ against cycle time T_s for velocity $v/v_0 = 2.0$ and offset time $T_{offset} = 10$ at the green-wave strategy.

We investigate the effect of both cycle time and offset time on the vehicular traffic through an infinite series of traffic lights by iterating map (109)-(112). We calculate the tour time, arrival time, and traffic current at the open boundary when the input density of vehicles is very high. We derive the dependence of the tour time, arrival time, and maximal current on both cycle time and offset time.

The tour time between a traffic light and its proceeding light is defined as follows

$$DT_i(nM) \equiv T(i,nM) - T(i,(n-1)M).$$

The traffic current is defined as the number of vehicles going through a position on the highway during the period of unit time $l_{\min}/v$. For a very high input density, maximal current Q is given by

$$Q = \lim_{N\to\infty} \frac{N}{T(N,n) - T(1,n)} \quad \text{for a sufficiently large } n,$$

where N is the total number of vehicles. $T(N,n) - T(1,n)$ is the difference between the arrival times of rear vehicle N and leading vehicle 1 at position n far from the initial position. The maximal current is proportional to the inverse of the arrival-time difference.

We study the dynamical behavior of the signal traffic for the green-wave strategy. Initially, there exist all vehicles with minimal time headway 1. The initial density is the highest value, one. The initial condition of vehicle i is given by $T(i,0) = i$. Figure 119(a) shows the plot of the tour time $DT_1(n)$ against cycle time T_s for leading vehicle 1 at synchronized strategy $T_{offset} = 0$ and velocity $v/v_0 = 1.0$, where $n = 1000 - 2000$, split $S_p = 0.5$, the total number of vehicles is $N = 200$ and the distance between two signals is $M = 20$. For comparison, we show the plot of tour time against cycle time for the synchronized strategy at which offset time T_{offset} is zero. For the value of cycle time at point *a*, the vehicle passes successfully through the green lights. For the value of cycle time between points *a* and *b*, the vehicle stops always at all signals. For the value of cycle time between points *b* and *c*, the vehicle stops every two signals. For the value of cycle time between 80 (points *c*) and 120, the vehicle stops every three signals. Figure 119(b) shows the plot of the tour time $DT_1(n)$ against cycle time T_s for leading vehicle 1 at green-wave strategy $T_{offset} = 10$, where the values of parameters are the same as those in Figure 119(a) except for the value of offset time. For the value of cycle time at point *a*, the vehicle passes successfully through the green lights. For the value of cycle time between points *a* and *b*, the vehicle stops always at all signals. For the value of cycle time between points *b* and 120, the vehicle stops every two signals. Thus, points *a* and *b* at the green-wave strategy shift horizontally by 10. By defining the tour time at the synchronized strategy as $\Delta T_1 = f(T_s)$ in Figure 119(a), the tour time at the green-wave strategy is given by

$$\Delta T_1 = \left(1 + \frac{T_{offset}}{M}\right)\left[f\left((1 + \frac{T_{offset}}{M})T_s\right) - M \right] + M, \tag{113}$$

where M is the distance between two successive signals.

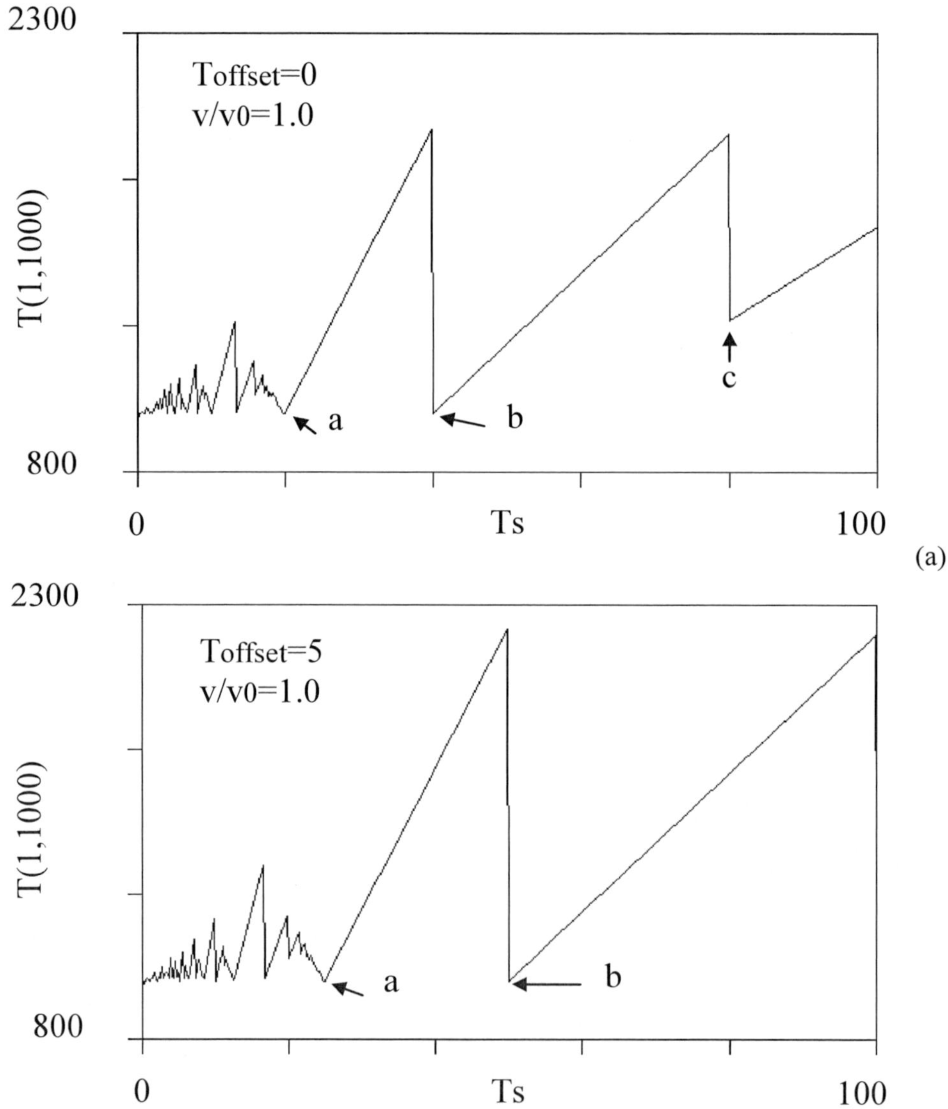

Figure 120. (a) Plot of arrival time of leading vehicle at position n=1000 against cycle time for the synchronized strategy where the velocity of vehicle is $v/v_0 = 1$, split $S_p = 0.5$, the total number of vehicles is $N = 200$ and the distance between two signals is $M = 20$. (b) Plot of arrival time of leading vehicle at position n=1000 against cycle time for the green-wave strategy with offset time $T_{offset} = 5$.

We study the dependence of tour time on the mean velocity. Figure 119(c) shows the plot of the tour time $DT_1(n)$ against cycle time T_s for mean velocity $v/v_0 = 2.0$ at synchronized strategy $T_{offset} = 0$, where the values of parameters are the same as those in Figure 119(a) except for the value of mean velocity. By changing the mean velocity, points *a*-*c* shift from Figure 119(a) to Figure 119(c). By defining the tour time at the synchronized strategy in Figure 119(a) as $\Delta T_1 = f(T_s)$, the tour time at v/v_0 is given by

$$\Delta T_1 = \frac{f\left(\frac{v_0 T_s}{v}\right)}{\frac{v}{v_0}}. \tag{114}$$

When both offset time and mean velocity change, the tour time changes as Figure 119(d). Figure 119(d) shows the plot of the tour time $DT_1(n)$ against cycle time T_s for mean velocity $v/v_0 = 2.0$ and offset time $T_{offset} = 10$ at the green-wave strategy, where the values of parameters are the same as those in Figure 119(a) except for the values of offset time and mean velocity. By changing the mean velocity, points *a*-*c* shift from Figure 119(a) to Figure 119(d). By defining the tour time at the synchronized strategy in Figure 119(a) as $\Delta T_1 = f(T_s)$, the tour time in Figure 119(d) is given by

$$\Delta T_1 = \left(1 + \frac{vT_{offset}}{v_0 M}\right)\left[\frac{f\left((1 + \frac{vT_{offset}}{v_0 M})\frac{v_0 T_s}{v}\right)}{\frac{v}{v_0}} - \frac{v_0 M}{v}\right] + \frac{v_0 M}{v}. \tag{115}$$

Thus, the tour time at the green-wave strategy is related to that at the synchronized strategy. When tour time $\Delta T_1 = f(T_s)$ at the synchronized strategy is given, the tour time of vehicles with velocity *v* at the green-wave strategy of offset time T_{offset} is obtained by Eq. (115).

We study the arrival time of leading vehicle for various values of cycle time at the green-wave strategy by comparing with that at the synchronized strategy. Figure 120(a) shows the plot of arrival time of leading vehicle at position *n*=1000 against cycle time for the synchronized strategy where the velocity of vehicle is $v/v_0 = 1$, split $S_p = 0.5$, the total number of vehicles is $N = 200$ and the distance between two signals is $M = 20$. Vehicles stops at all signals for cycle time between points *a* and *b*. Vehicles stops every two signals for cycle time between points *b* and *c*. Points a, b, and c correspond, respectively, to those in Figure 119(a). Figure 120(b) shows the plot of arrival time of leading vehicle at position

n=1000 against cycle time for the green-wave strategy with offset time $T_{offset} = 5$. Points a and b at the green-wave strategy shift horizontally by 5. By defining the arrival time at the synchronized strategy in Figure 120(a) as $T(1, L) = g(T_s)$, the arrival time at the green-wave strategy is given by

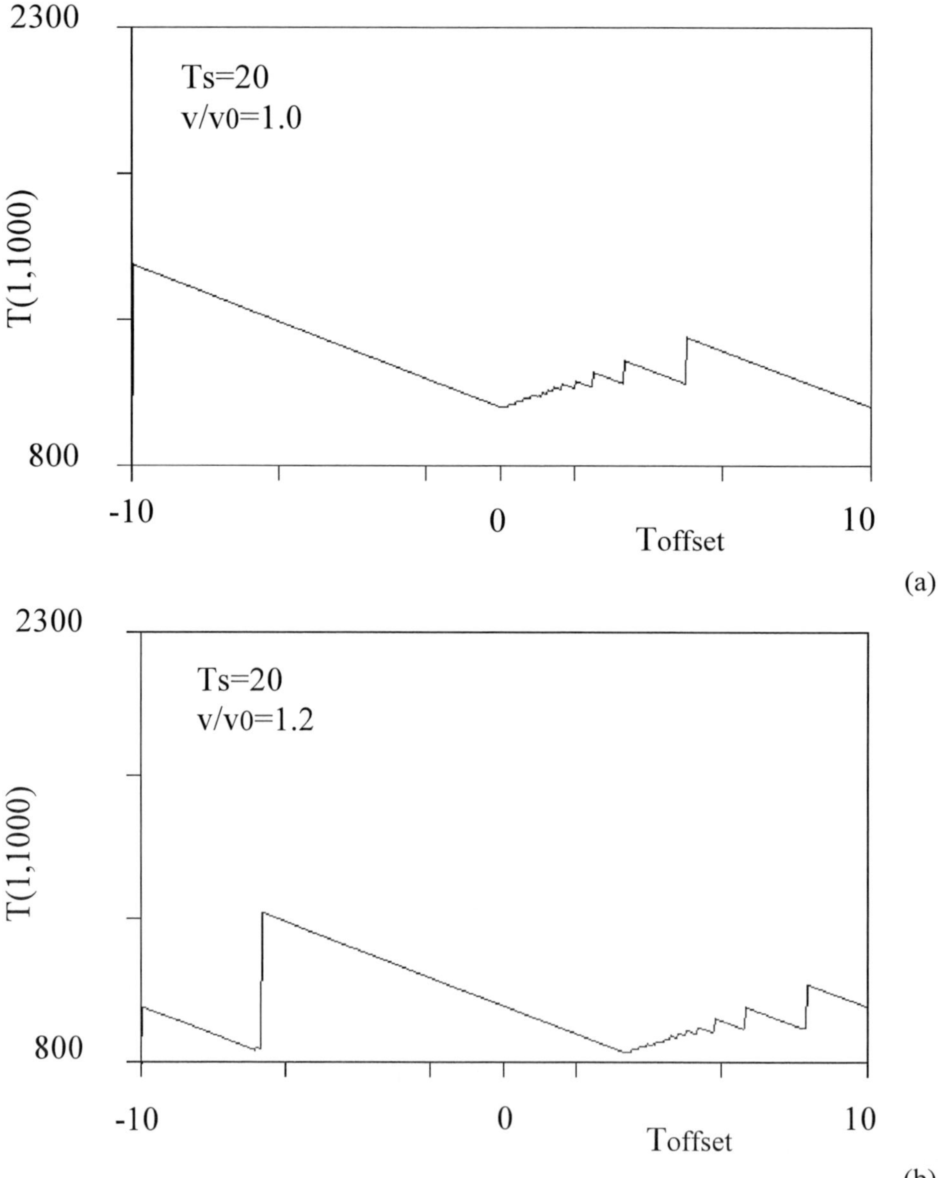

Figure 121. (Continued)

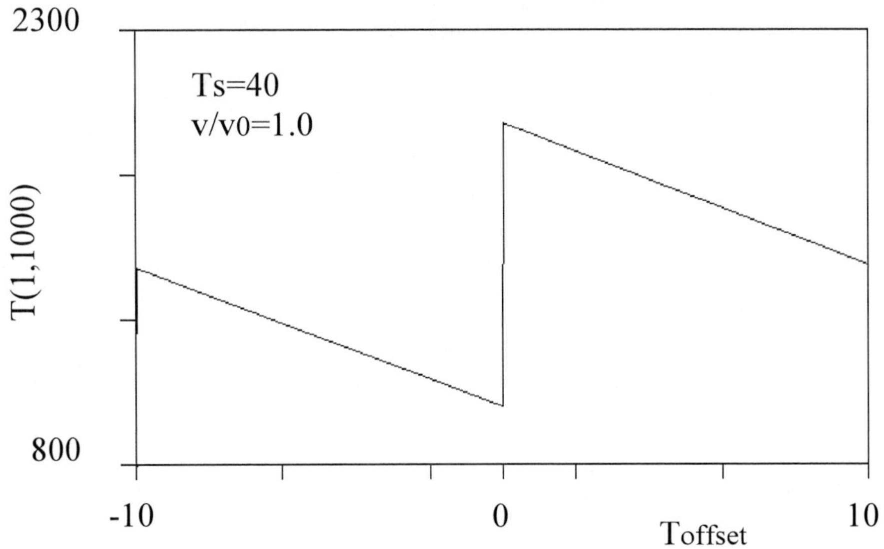

(c)

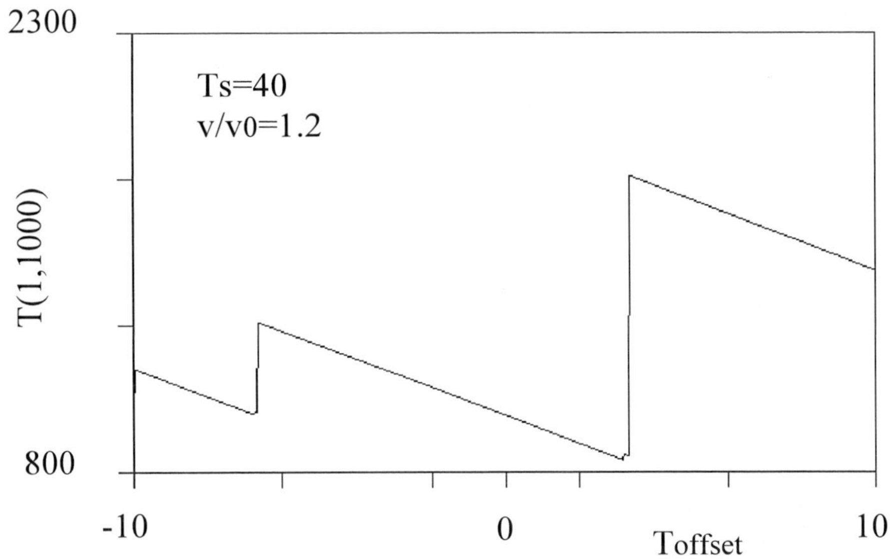

(d)

Figure 121. (a) Plot of arrival time against offset time for cycle time $T_s = 20$ and velocity $v/v_0 = 1.0$ where split $S_p = 0.5$, the total number of vehicles is $N = 200$ and the distance between two signals is $M = 20$. (b) Plot of arrival time against offset time for cycle time $T_s = 20$ and velocity $v/v_0 = 1.2$. (c) Plot of arrival time against offset time for cycle time $T_s = 40$ and velocity $v/v_0 = 1.0$. (d) Plot of arrival time against offset time for cycle time $T_s = 40$ and velocity $v/v_0 = 1.2$.

$$T(1,L)=\left(1+\frac{vT_{offset}}{v_0M}\right)\left[\frac{g\left((1+\frac{vT_{offset}}{v_0M})\frac{v_0T_s}{v}\right)}{\frac{v}{v_0}}-\frac{v_0M}{v}\right]+\frac{v_0M}{v}. \tag{116}$$

Thus, the relationship of arrival time at the green-wave strategy is similar to Eq. (115).

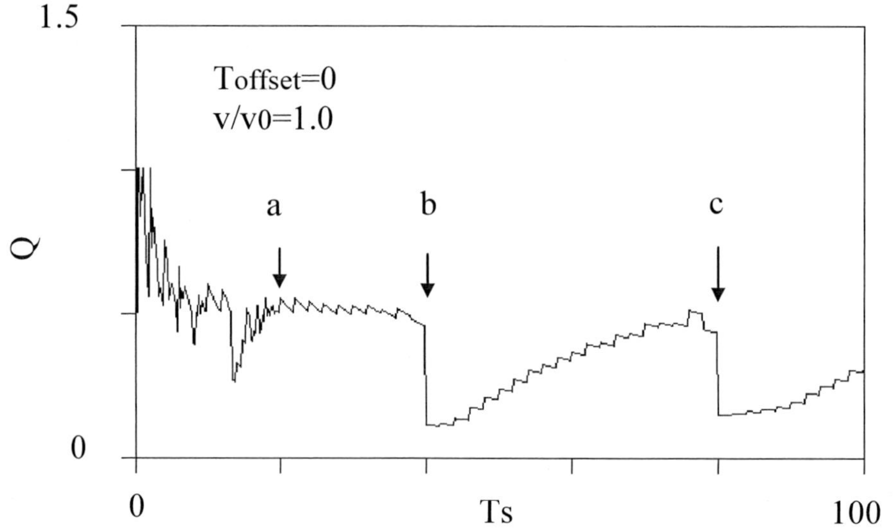

(a)

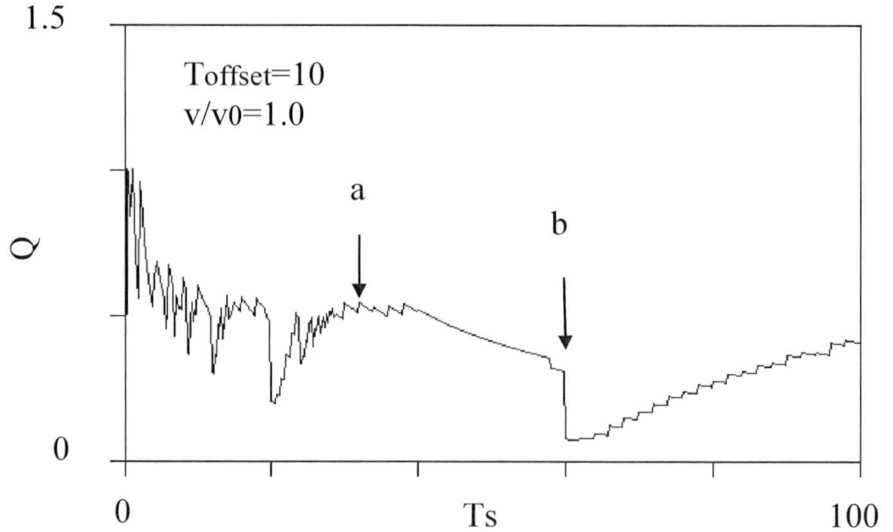

(b)

Figure 122. (Continued)

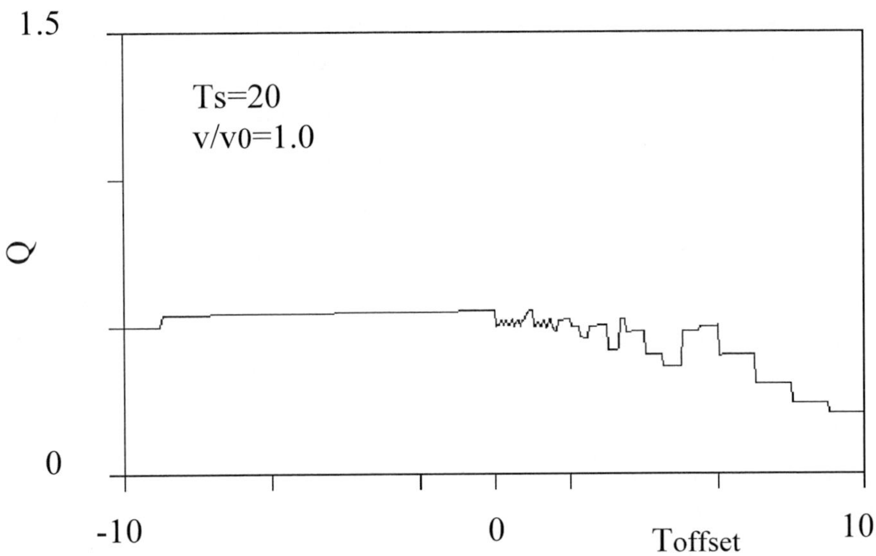

(c)

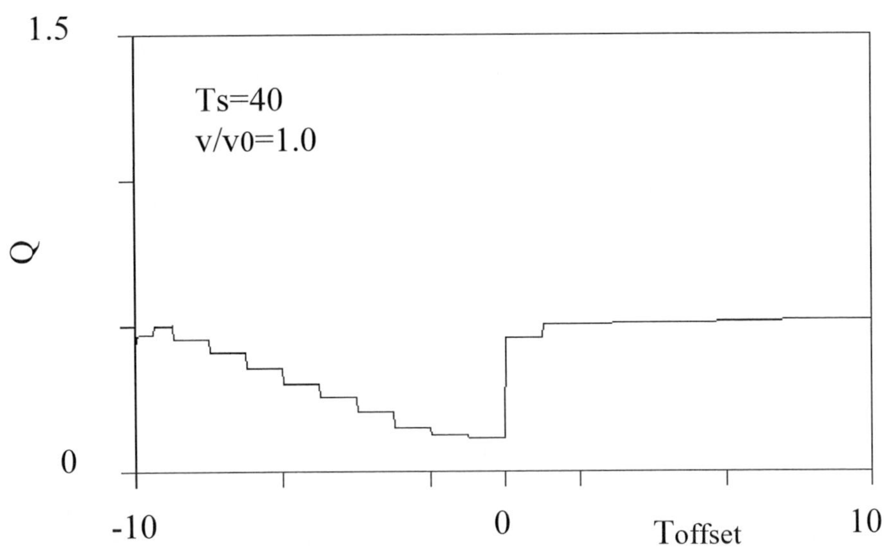

(d)

Figure 122. (a) Plot of the maximal current Q against cycle time T_s for $T_{offset} = 0$ and $v/v_0 = 1.0$, where $S_p = 0.5$, the total number of vehicles is $N = 200$ and the distance between two signals is $M = 20$. The maximal current profile corresponds to the tour time profile in Figure 119(a). (b) Plot of the maximal current Q against cycle time T_s for $T_{offset} = 10$ and $v/v_0 = 1.0$. The maximal current profile corresponds to the tour time profile in Figure 119(b). (c) Plot of the maximal current Q against offset time T_{offset} for cycle time $T_s = 20$ and velocity $v/v_0 = 1.0$. The maximal current profile corresponds to the arrival time profile in Figure 121(a). (d) Plot of the maximal current Q against offset time T_{offset} for cycle time $T_s = 40$ and velocity $v/v_0 = 1.0$. The maximal current profile corresponds to the arrival time profile in Figure 121(c).

We study the variation of arrival time by changing offset time for some values of cycle time. Figure 121(a) shows the plot of arrival time against offset time for cycle time $T_s = 20$ and velocity $v/v_0 = 1.0$ where split $S_p = 0.5$, the total number of vehicles is $N = 200$ and the distance between two signals is $M = 20$. Figure 121(b) shows the plot of arrival time against offset time for cycle time $T_s = 20$ and velocity $v/v_0 = 1.2$. The arrival time changes complexly with the offset. Figure 121(c) shows the plot of arrival time against offset time for cycle time $T_s = 40$ and velocity $v/v_0 = 1.0$. The arrival time changes abruptly when offset time changes from a negative value to a positive value. Figure 121(d) shows the plot of arrival time against offset time for cycle time $T_s = 40$ and velocity $v/v_0 = 1.2$. When velocity increases from $v/v_0 = 1.0$ to $v/v_0 = 1.2$, the arrival time profile shifts by about 3.3 horizontally. Thus, the arrival time changes complexly with the offset time.

We study the variation of maximal current by changing both cycle time and offset time. The current exhibits the maximal value under the open boundary when the input density is highest. Figure 122(a) shows the plot of the maximal current Q against cycle time T_s for $T_{offset} = 0$ and $v/v_0 = 1.0$, where $S_p = 0.5$, the total number of vehicles is $N = 200$ and the distance between two signals is $M = 20$. The maximal current profile corresponds to the tour time profile in Figure 119(a). The tour time increases linearly from point $a(T_s = 20)$ to point $b(T_s = 40)$ as the vehicle goes ahead. A vehicle stops at all traffic lights and then goes. Then, the current keeps almost constant value. When the cycle time is higher than 40, the tour time of vehicle takes two values. The straight line of slope 1 from point $b(T_s = 40)$ to point $c(T_s = 80)$ represents the tour time at which the vehicle stops every two traffic lights. The current decreases abruptly at point b and then increases with cycle time. When the cycle time is higher than 80, the vehicle stops every three signals and goes ahead. Again, the current decreases abruptly at point $c(T_s = 80)$ and increases again with cycle time.

Figure 122(b) shows the plot of the maximal current Q against cycle time T_s for $T_{offset} = 10$ and $v/v_0 = 1.0$. The maximal current profile corresponds to the tour time profile in Figure 119(b). Points a and b shifts horizontally by 10 in green-wave strategy $T_{offset} = 10$. The maximal current decreases abruptly at point b ($T_s = 60$).

Figure 122(c) shows the plot of the maximal current Q against offset time T_{offset} for cycle time $T_s = 20$ and velocity $v/v_0 = 1.0$. The maximal current profile corresponds to the arrival time profile in Figure 121(a). The maximal current keeps a constant value until $T_{offset} = 0$ and then decreases with increasing offset time. Figure 122(d) shows the plot of the maximal current Q against offset time T_{offset} for cycle time $T_s = 40$ and velocity $v/v_0 = 1.0$. The maximal current profile corresponds to the arrival time profile in Figure

121(c). The maximal current increases abruptly at $T_{offset} = 0$ and keeps a constant value with increasing offset time.

We study the traffic patterns of vehicles at offset time $T_{offset} = 10$ and split $S_p = 0.5$ for various values of cycle time. We calculate the trajectories of all vehicles. Figure 123 shows the plots of vehicular position n against arrival time difference $T_i(n) - T_i(1000)$ for $n = 1000 - 1150$ where the total number N of vehicles is 50. Figure 123(a) shows the traffic patterns at cycle time $T_s = 30$ (point a in Figure 119(b) and Figure 122(b)). All vehicles move without stopping at traffic signals. All vehicles go successfully through green signals. Vehicles make clusters. The maximal current displays the high value due to the clustering and free movement.

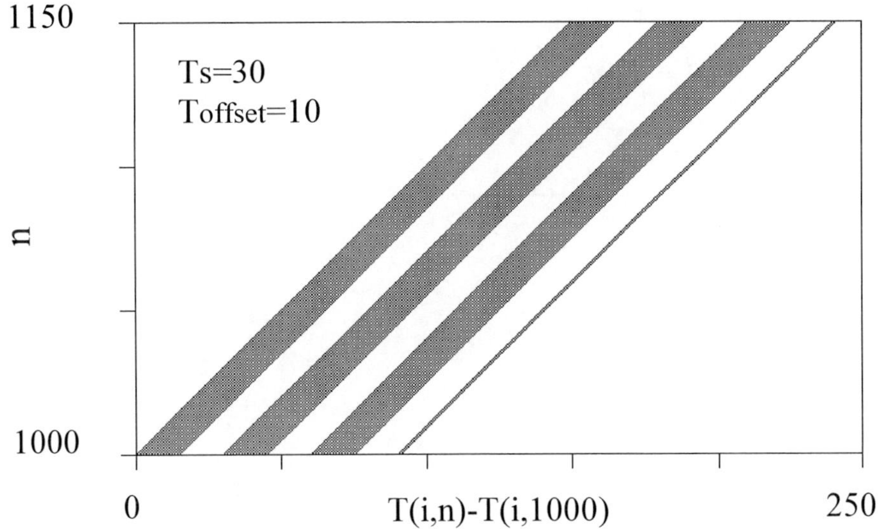

(a)

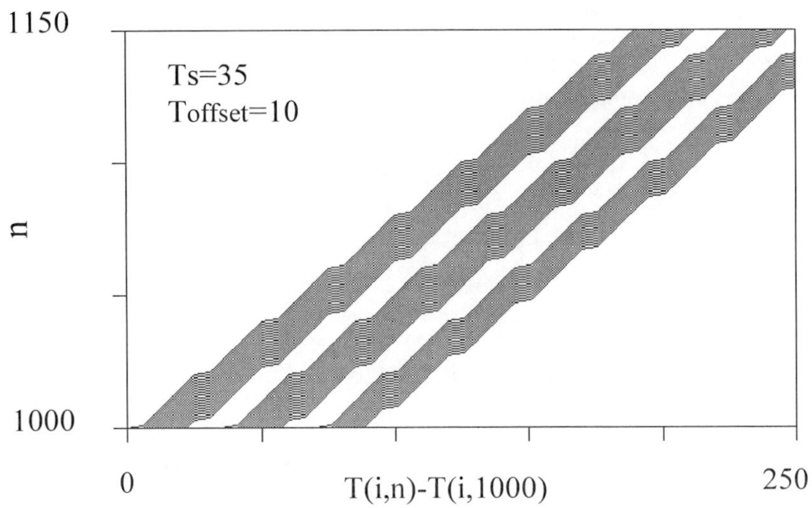

(b)

Figure 123. (Continued)

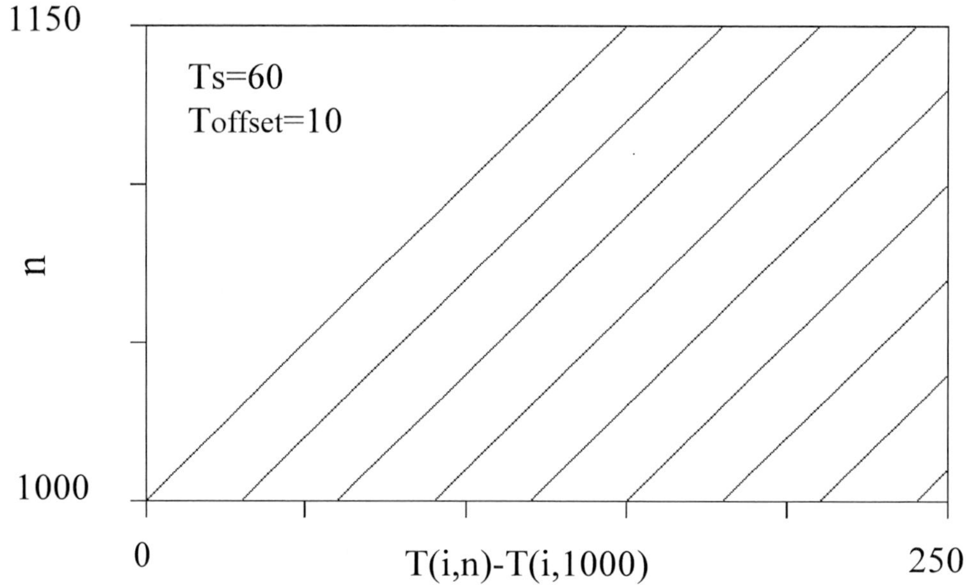

(c)

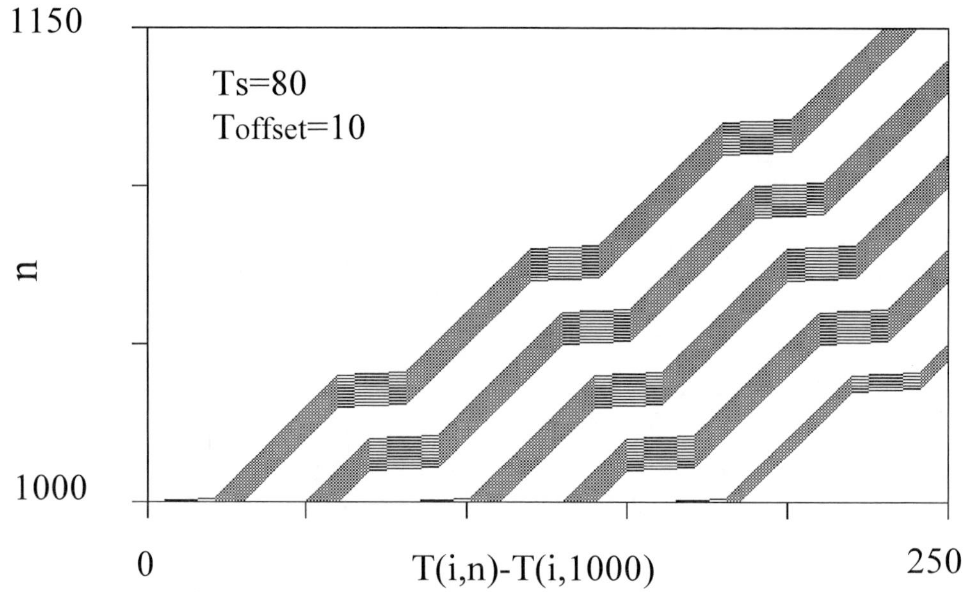

(d)

Figure 123. Plots of vehicular position *n* against arrival time difference $T_i(n) - T_i(1000)$ for $n = 1000 - 1150$ where the total number *N* of vehicles is 50. (a) Traffic patterns at cycle time $T_s = 30$ (point *a* in Figure 119(b) and Figure 122(b)). (b) Traffic patterns at cycle time $T_s = 35$ in the region between points *a* and *b* in Figure 119(b) and Figure 122(b). (c) Traffic patterns at cycle time $T_s = 60$ (point *b* in Figure 119(b) and Figure 122(b)). (d) Traffic patterns at cycle time $T_s = 80$ in the region on the right-hand of point *b* in Figure 119(b) and Figure 122(b).

Figure 123(b) shows the traffic patterns at cycle time $T_s = 35$ in the region between points *a* and *b* in Figure 119(b) and Figure 122(b). All vehicles stop at all signals. Vehicles make clusters. The size of a cluster increases with cycle time. The current decreases by stopping at signals but increases with cluster size. As the result, the maximal current keeps the high value.

Figure 123(c) shows the traffic patterns at cycle time $T_s = 60$ (point *b* in Figure 119(b) and Figure 122(b)). All vehicles move without stopping at traffic signals. All vehicles go successfully through green signals. However, vehicles do not make clusters. There exists only one vehicle between a signal and its proceeding signal. The vehicles extend widely on the roadway. Therefore, the maximal current shows the lowest value (point *b* in Figure 122(b)).

Figure 123(d) shows the traffic patterns at cycle time $T_s = 80$ in the region on the right-hand of point *b* in Figure 119(b) and Figure 122(b). All vehicles stop every two signals. Vehicles make clusters. The size of a cluster increases with cycle time. The maximal current increases with cycle time because the cluster size increases with cycle time.

We have derived the analytical expression of tour time (or arrival time) at the green-wave strategy in terms of that at the synchronized strategy. We have derived the dependence of maximal current on both cycle and offset times. We have shown how the tour time and maximal current are controlled by varying both cycle and offset times.

8. CHAOS INDUCED BY TRAFFIC SIGNAL AND SPEEDUP

We study the dynamical behavior of vehicles through a sequence of traffic lights when the vehicle speeds up to retrieve the delay induced by traffic lights. The vehicle moves chaotically even if the model is deterministic. By varying parameters, the vehicle displays the complex dynamical transitions among the regular, periodic, and chaotic motions [98-102].

8.1. Chaos of a Single Vehicle by Speedup

Generally, a vehicle speeds up in order to retrieve the delay induced by the stopping at traffic lights because a driver wants to arrive at his destination as soon as possible. It is unknown how the speedup affects the motion of a vehicle moving through the sequence of traffic lights. Does the vehicle move periodically or chaotically? How does the tour time of the vehicle vary with the speedup and cycle time?

Here, we study the traffic of a single vehicle moving through an infinite series of traffic lights when the vehicle speeds up to retrieve the delay of stopping at traffic lights.

We consider the motion of a single vehicle going through an infinite series of traffic lights. The distance between traffic lights n and $n+1$ is defined as l_n. The vehicle moves with the mean speed $v(n)$ between traffic lights n and n+1. We assume that the mean speed $v(n)$ increases proportionally to the stopping time at the traffic light n. Figure 76 shows the schematic illustration of the single vehicle moving through a sequence of traffic lights. In the synchronized strategy, all the traffic lights change simultaneously from red (green) to green (red) with a fixed time period $t_s/2$. We define the arrival time of the vehicle at traffic light *n* as $t(n)$. The arrival time at traffic light *n*+1 is given by

$$t(n+1)=t(n)+l_n/v(n)+\left(r(n)-t(n)\right)H\left(\sin(2\pi t(n)/t_s)\right)$$

$$\text{with } r(n)=\left(\text{int}\left(t(n)/t_s\right)+0.5\right)\cdot t_s\,, \tag{117}$$

where $H(t)$ is the Heaviside function. The vehicular speed increases proportionally to the stopping time at the traffic light. The stopping time at traffic light *n* is $t(n)-t(n-1)-l_{n-1}/v_0$. Then, the mean speed is given by

$$v(n)=v_0+s_p\left(t(n)-t(n-1)-l_{n-1}/v_0\right), \tag{118}$$

where v_0 is the mean speed without stopping at the traffic light and parameter s_p represents the speedup's rate.

In the green-wave strategy, the traffic light changes with a certain time delay τ between the traffic light phases of two successive intersections. The delay time is called the offset time. The change of traffic lights propagates backwards like a green wave. The arrival time at traffic light *n*+1 is given by

$$t(n+1)=t(n)+l_n/v(n)+\left(r(n)-t(n)-n\tau\right)H\left(\sin(2\pi(t(n)+n\tau)/t_s)\right)$$

$$\text{with } r(n)=\left(\text{int}\left((t(n)+n\tau)/t_s\right)+0.5\right)\cdot t_s\,. \tag{119}$$

We restrict ourselves to the case of the same distance for the traffic lights. By setting $l_n=l_0$ and dividing time by the characteristic time l/v_0, one obtains the nonlinear equation of dimensionless arrival time:

$$\begin{aligned}T(n+1)=T(n)&+1/\left(1+S_p\left(T(n)-T(n-1)-1\right)\right)\\&+\left(R(n)-T(n)-n\Pi\right)H\left(\sin(2\pi(T(n)+n\Pi)/T_s)\right)\end{aligned}$$

$$\text{with } R(n)=\left(\text{int}\left((T(n)+n\Pi)/T_s\right)+0.5\right)\cdot T_s\,, \tag{120}$$

where $T(n)=t(n)v_0/l_0$, $R(n)=r(n)v_0/l_0$, $T_s=t_sv_0/l_0$, $\Pi=\tau\cdot v_0/l_0$, and $S_p=s_pl_0/v_0^2$. Thus, the dynamics of the vehicle is described by the nonlinear map (120). The motion of a vehicle depends on dimensionless cycle time $T_s=t_sv/l$, dimensionless delay (offset time) $\Pi=\tau\cdot v/l$, and dimensionless speedup rate $S_p=s_pl_0/v_0^2$. In the synchronized strategy, the vehicle depends on both dimensionless cycle time $T_s = t_s v / l$ and dimensionless speedup rate $S_p = s_p l_0 / v_0^{\,2}$.

Generally, the distance between two traffic lights or the mean speed of the vehicle vary from traffic light to traffic light. We consider the fluctuation as a noise. We extend the deterministic model to take into account the noise. The fluctuation is defined as $\xi(n) \equiv (l_n / v_{0,n} - <l> / <v_0>)/(<l>/<v_0>)$. The extended version of eq. (120) is given by

$$T(n+1) = T(n) + 1/\left(1 + S_p\left(T(n) - T(n-1) - 1\right)\right) + \xi(n)$$
$$+ \left(R(n) - T(n) - n\Pi\right) H\left(\sin(2\pi(T(n) + n\Pi)/T_s)\right)$$

with $R(n) = \left(\mathrm{int}\left((T(n) + n\Pi)/T_s\right) + 0.5\right) \cdot T_s$, (121)

where $\xi(n)$ is the white noise: $<\xi> = 0$ and $<\xi(n)\xi(m)> = \eta^2 \delta_{nm}$.

We investigate the dynamical behavior of a single vehicle speeding up through an infinite series of traffic lights by iterating map (120). We study the dynamical behavior of the vehicle at the synchronized strategy ($\Pi = 0$). We calculate the tour time between two traffic lights when the vehicle goes ahead on the highway. We study how the tour time varies with the cycle time by speeding up to retrieve the delay induced at traffic lights. The tour time between a traffic light and its proceeding light is defined as *DT*.

We study the variation of tour time with going ahead through the traffic lights. Figure 124 shows the plots of the tour time *DT* against traffic light *n* for *n*=1000-1040. At cycle time $T_s = 0.6$, the plots of the tour time *DT* are shown for (1) $S_p = 0.4$, (2) $S_p = 0.6$, and (3) $S_p = 0.8$ in Figure 124(a). In diagram (1), the vehicle stops at all the traffic lights and then goes ahead. Tour time *DT* is higher than one and constant. In diagram (2), the vehicle moves periodically with period 7. The vehicle moves faster than non-stop every seven traffic lights since tour time *DT* is less than one. In diagram (3), the vehicle moves irregularly and the tour time varies randomly. In Figure 124(b), the plots of the tour time *DT* at cycle time $T_s = 0.9$ are shown for (1) $S_p = 0.0$, (2) $S_p = 0.5$, and (3) $S_p = 0.95$. In diagram (1), the vehicle stops every five traffic lights and then goes ahead. Tour time *DT* is higher than one when the vehicle stops at the traffic lights, while the tour time is one when the vehicle passes through the green lights. In diagram (2), the vehicle moves periodically with period 6. The vehicle moves faster than non-stop every six traffic lights since tour time *DT* is less than one. In diagram (3), the vehicle moves irregularly and the tour time varies randomly. Thus, the vehicle displays the irregular motion when the speedup rate is high.

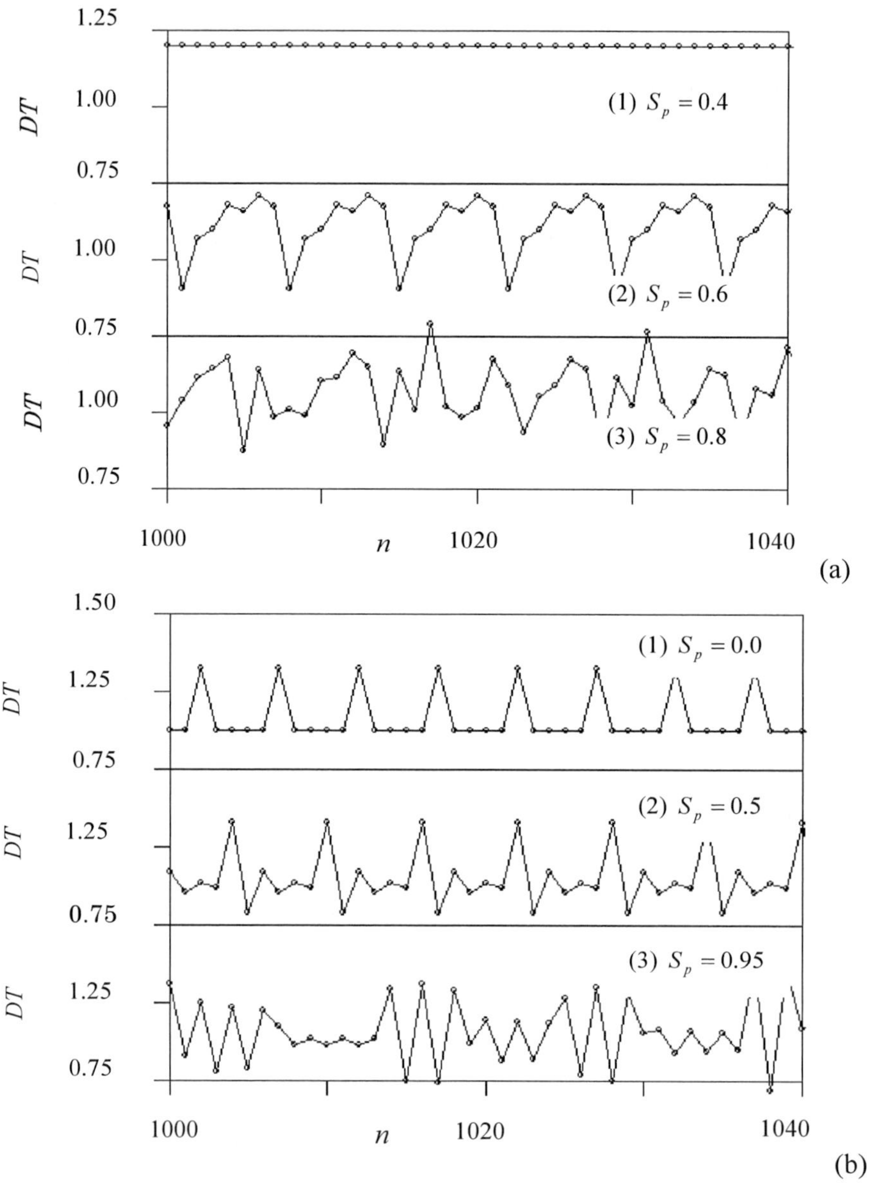

Figure 124. (a) Plots of the tour time DT against traffic light n at cycle time $T_s = 0.6$ for (1) $S_p = 0.4$, (2) $S_p = 0.6$, and (3) $S_p = 0.8$. (b) Plots of the tour time DT at cycle time $T_s = 0.9$ for (1) $S_p = 0.0$, (2) $S_p = 0.5$, and (3) $S_p = 0.95$.

Figure 125 shows the plot of the tour time DT against cycle time T_s at speedup parameter (a) $S_p = 0.0$, (b) $S_p = 0.4$, (c) $S_p = 0.6$, and (d) $S_p = 0.8$ for sufficiently large number $n = 1000 - 3000$. Figure 126(a)-(d) show, respectively, the enlargements of the diagrams (a)-(d) in Figure 125 for $0 \le T_s \le 1$. For case (a) of no speedup, the tour time displays the self-similar pattern. For case (b), the vehicle displays the periodic motions with different periods. The dynamic state of the vehicle varies with the cycle time. The dynamical transitions among the definite states occur at various cycle times. For cases (c) and (d), the black region indicates the irregular motion. In the other region, the vehicle moves periodically. When the value of cycle time equals $T_s = 1/k$, the value of tour time becomes one where k is a positive integer, because the vehicle can pass just through all green lights irrespective of value of speedup rate S_p. Figure 127(a) is the enlargement of Figure 126(d) for $1/2 \le T_s \le 1$. Figure 127(b) is the enlargement of Figure 126(d) for $1/3 \le T_s \le 1/2$. The pattern of Figure 127(b) is similar to that of Figure 127(a). Also, the pattern of tour time for $1/4 \le T_s \le 1/3$ is similar to that of Figure 127(b). The pattern of tour time for $1/2 \le T_s \le 1$ appears repeatedly for $1/(k+1) \le T_s \le 1/k$ where k is a positive integer larger than one. The pattern of tour time is not fractal but self-similar because the scaling factor is not constant.

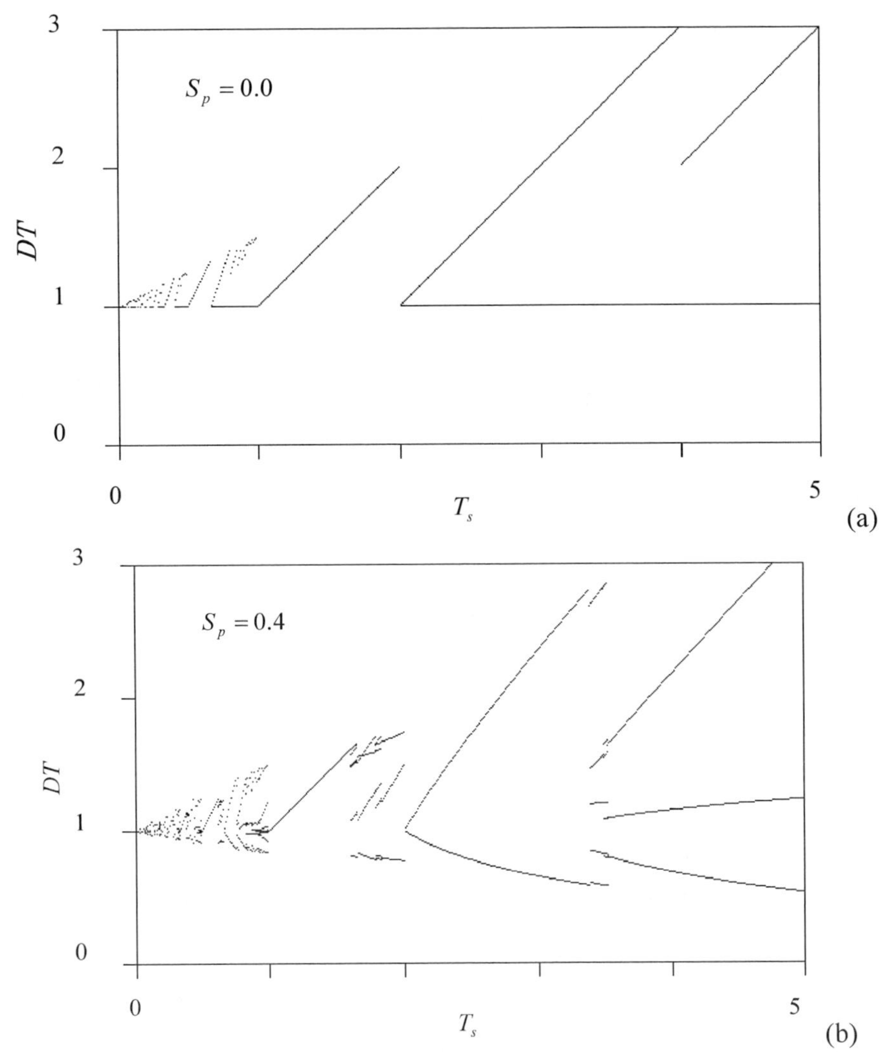

Figure 125. (Continued)

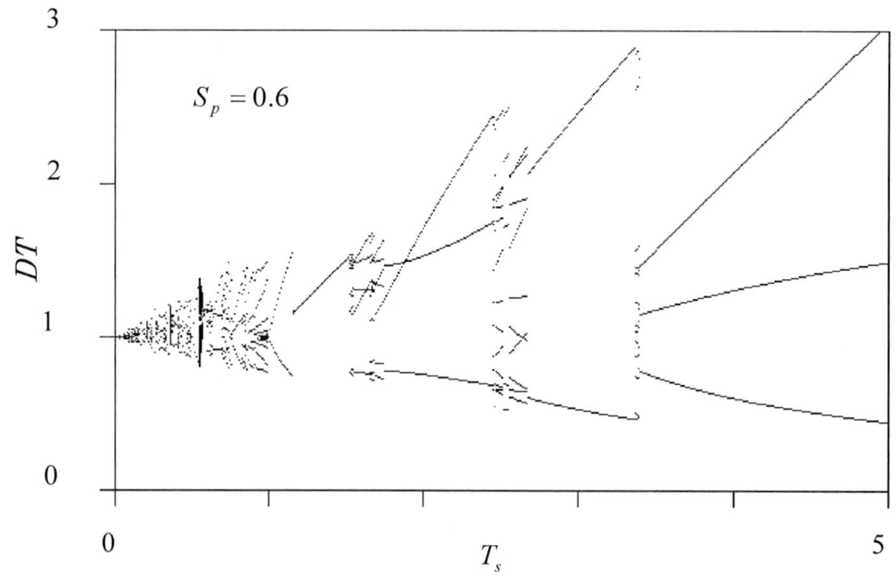

(c)

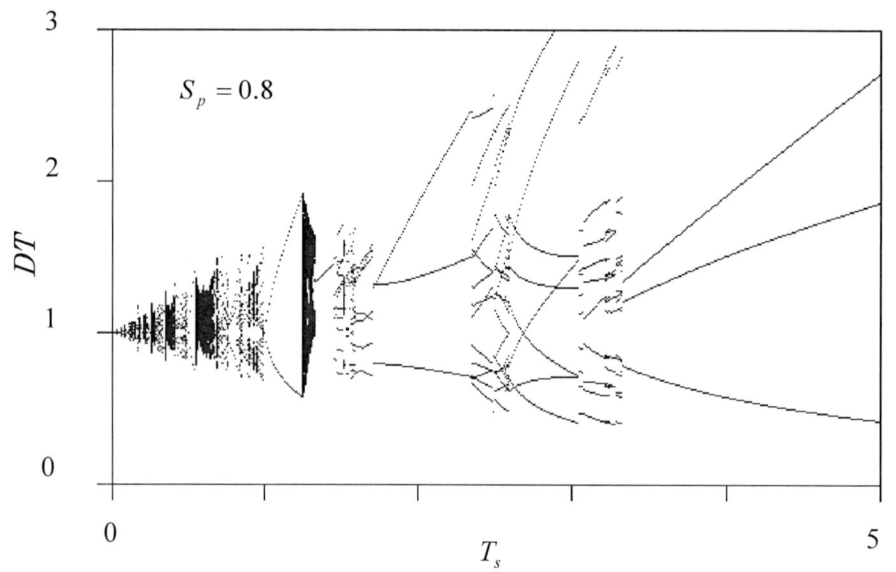

(d)

Figure 125. Plot of the tour time DT against cycle time T_s at speedup parameter (a) $S_p = 0.0$, (b) $S_p = 0.4$, (c) $S_p = 0.6$, and (d) $S_p = 0.8$ for sufficiently large number $n = 1000 - 3000$.

We calculate the Liapunov exponent of the tour time to study the irregular motions. Figure 128 shows the plots of Liapunov exponent λ against cycle time T_s for (a) $S_p = 0.6$ and (b) $S_p = 0.8$. The plots (a) and (b) correspond, respectively, to those of (c) and (d) in Figure 126. The values of Liapunov exponent are positive in the black regions in Figure 126 where the vehicle moves irregularly. Thus, the black regions in Figure 126 represent the chaotic motions of the vehicle.

We study the return map for the chaotic and periodic motions. Figure 129 shows the plots of tour time DT(n+1) against DT(n) for (a) $T_s = 0.52$, (b) $T_s = 0.55$, (c) $T_s = 0.65$, and (d) $T_s = 0.99$ at speedup rate $S_p = 0.8$. Return map (a) shows the periodic motion with period 2. Return maps (b) and (c) show the chaotic motions. Return map (d) shows the periodic motion with high period. The return maps are consistent with the behaviors of the plot of tour time against cycle time in Figure 127(a).

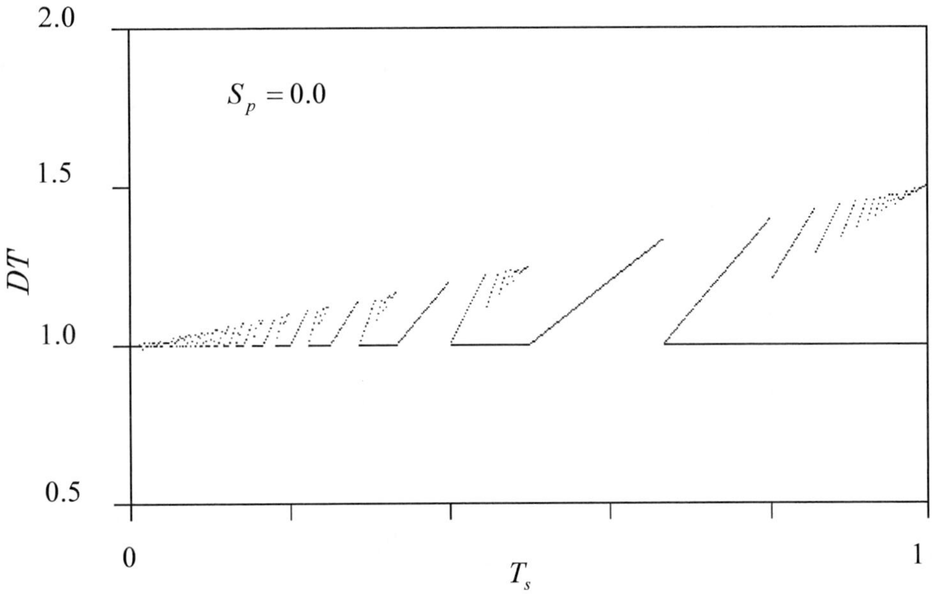

(a)

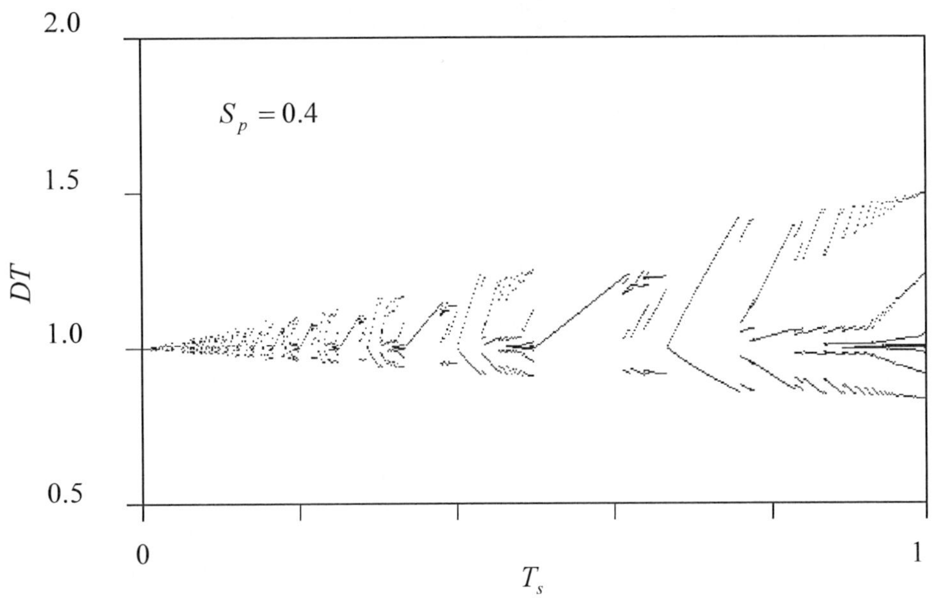

(b)

Figure 126. (Continued)

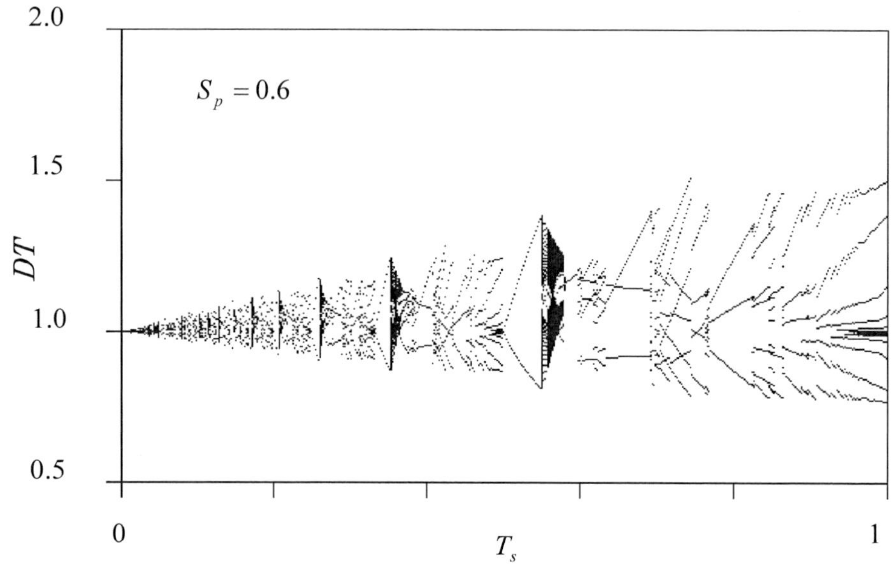

(c)

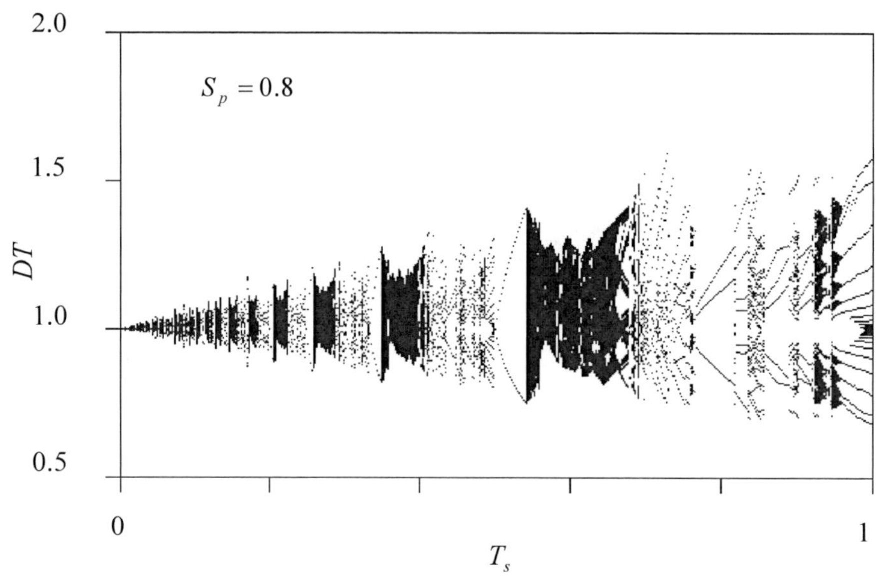

(d)

Figure 126. The diagrams (a)-(d) show, respectively, the enlargements of the diagrams (a)-(d) in Figure 125 for $0 \le T_s \le 1$.

We study the variation of tour time with speedup rate. Figure 130 shows the plots of tour time *DT* against speedup rate S_p at (a) $T_s = 0.52$, (b) $T_s = 0.6$, (c) $T_s = 0.8$, and (d) $T_s = 0.9$ for sufficiently large number $n = 1000 \sim 3000$. The black region in the plots represents a chaotic motion. In plot (a), the vehicle moves with a constant value of tour time for $S_p < 0.525$. When the speedup rate is higher than $S_p = 0.525$, the vehicle moves periodically with period 2. This motion corresponds to the return map in Figure 129(a). Furthermore, when the speedup rate is higher than $S_p = 0.88$, the dynamical transitions occur at some values of speedup rate. The motion bifurcates to chaos with increasing speedup rate. In plots (b)-(d), the similar dynamical transitions occur at some values of speedup rate. The behavior of the vehicle changes from the periodic to the chaotic motions.

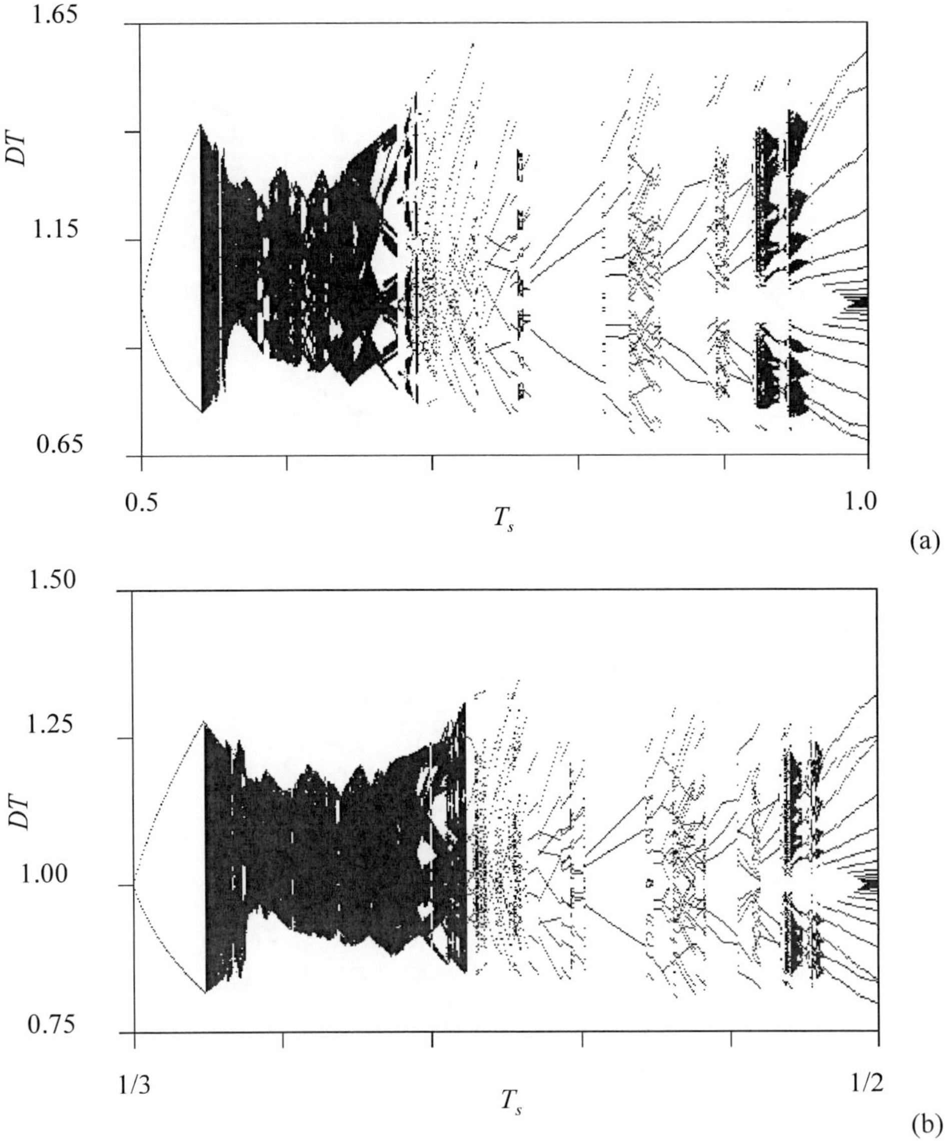

Figure 127. (a) Enlargement of Figure 126(d) for $1/2 \le T_s \le 1$. (b) Enlargement of Figure 126(d) for $1/3 \le T_s \le 1/2$.

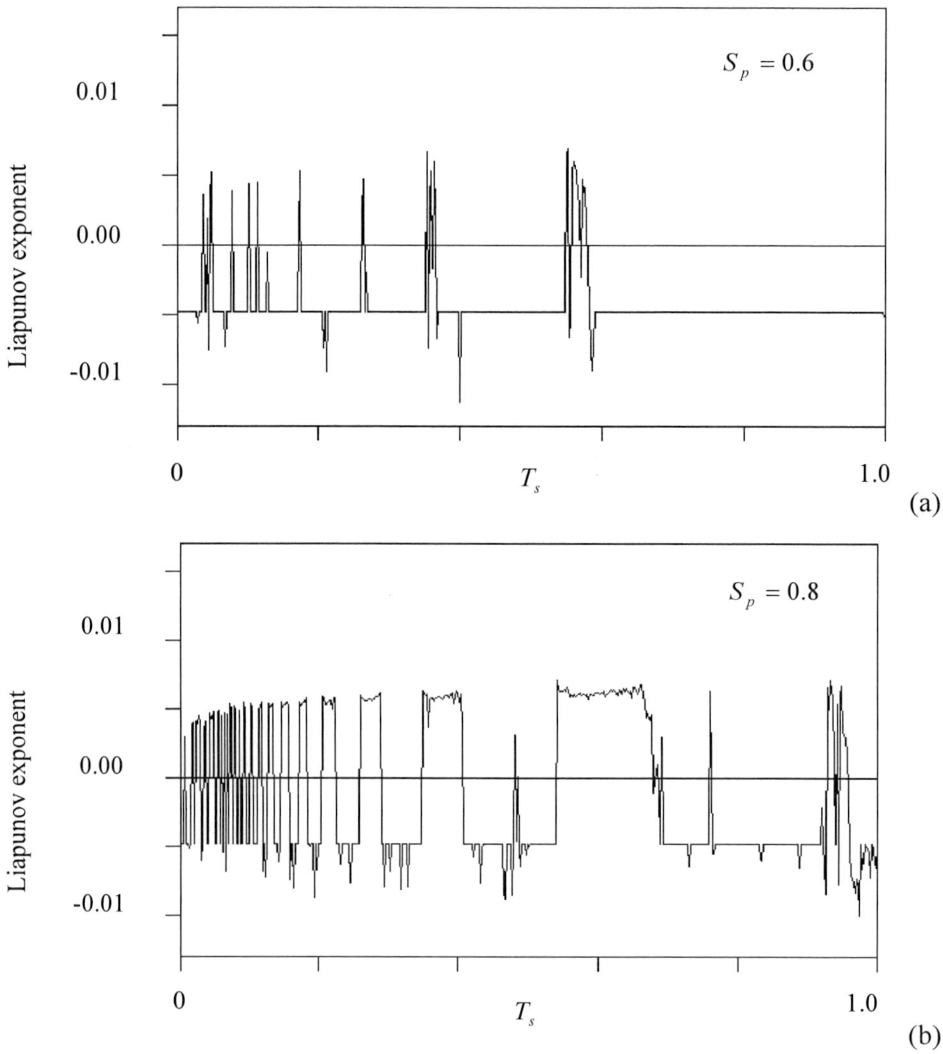

Figure 128. Plots of Liapunov exponent λ against cycle time T_s for (a) $S_p = 0.6$ and (b) $S_p = 0.8$. The plots (a) and (b) correspond, respectively, to those of (c) and (d) in Figure 126. The values of Liapunov exponent are positive in the black regions in Figure 126.

Figure 131 shows the plots of Liapunov exponent λ against speedup rate S_p for (a) $T_s = 0.6$ and (b) $T_s = 0.9$. The plots (a) and (b) correspond, respectively, to plots (b) and (d) in Figure 130. In plot (a), the Liapunov exponent changes from negative to positive values when the speedup rate is higher than $S_p = 0.715$. The region of positive Liapunov exponent corresponds to the black region in Figure 130(b). Similarly, the region of positive Liapunov exponent in plot (b) corresponds to the black region in Figure 130(d).

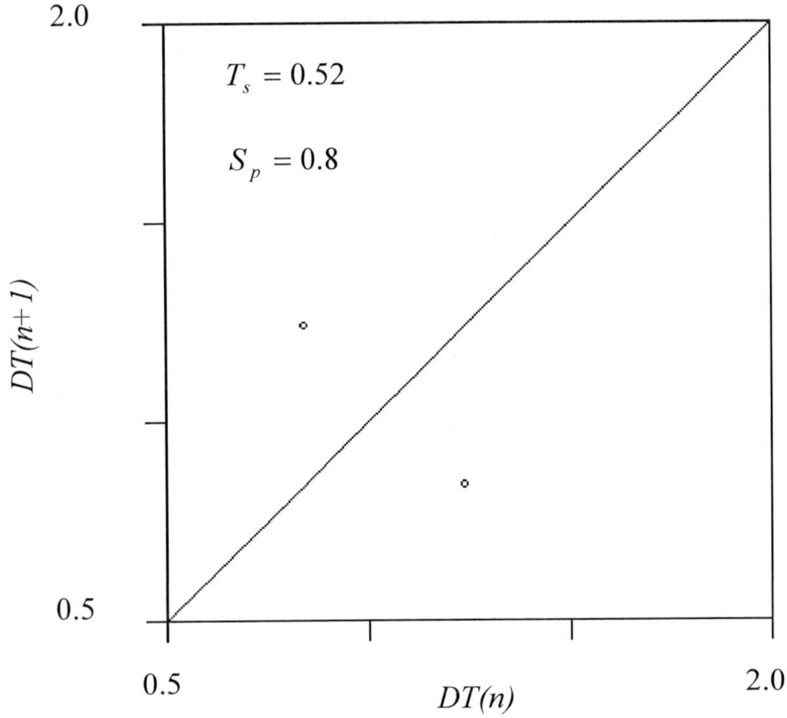

(a)

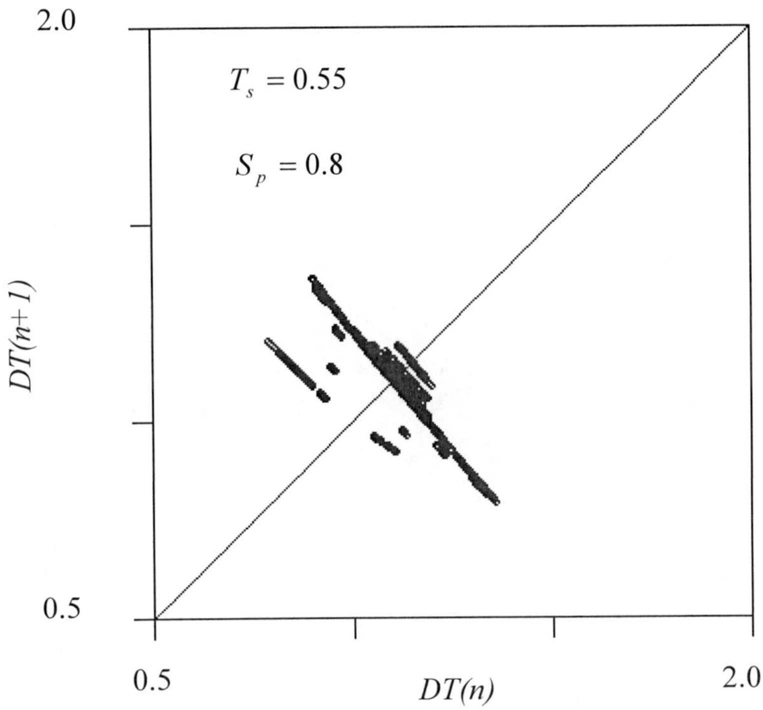

(b)

Figure 129. (Continued)

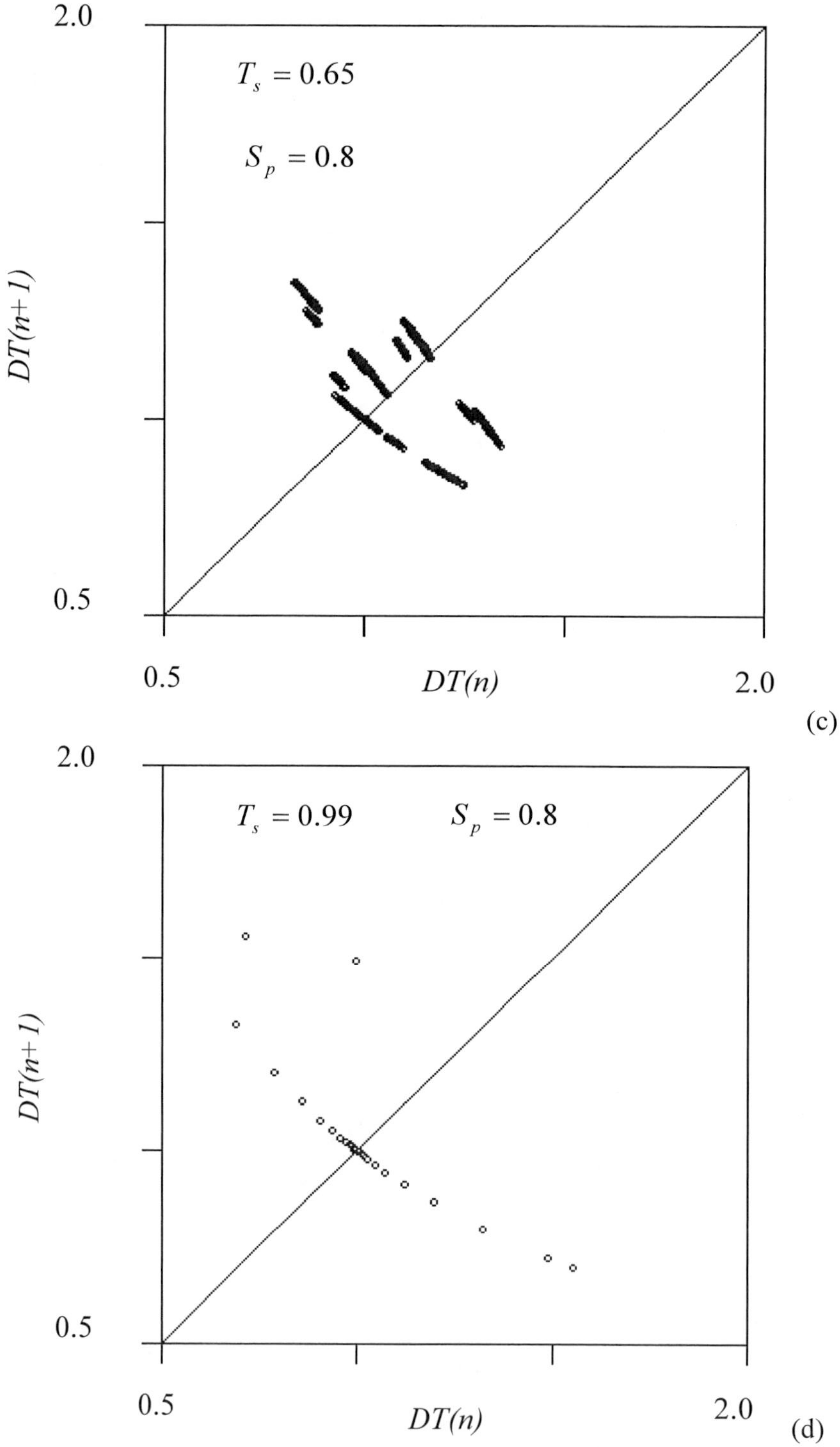

Figure 129. Plots of tour time DT(n+1) against DT(n) for (a) $T_s = 0.52$, (b) $T_s = 0.55$, (c) $T_s = 0.65$, and (d) $T_s = 0.99$ at speedup rate $S_p = 0.8$.

We have found that the vehicle exhibits the chaotic motion when the speedup rate is high, even if the model is deterministic. We have shown that the dynamical transitions between the chaotic and periodic motions occur at various values of cycle time for high speedup rate and the self-similar pattern of tour time appears for values of cycle time less than one. We have clarified the effect of the speedup for retrieving the delay on the tour time.

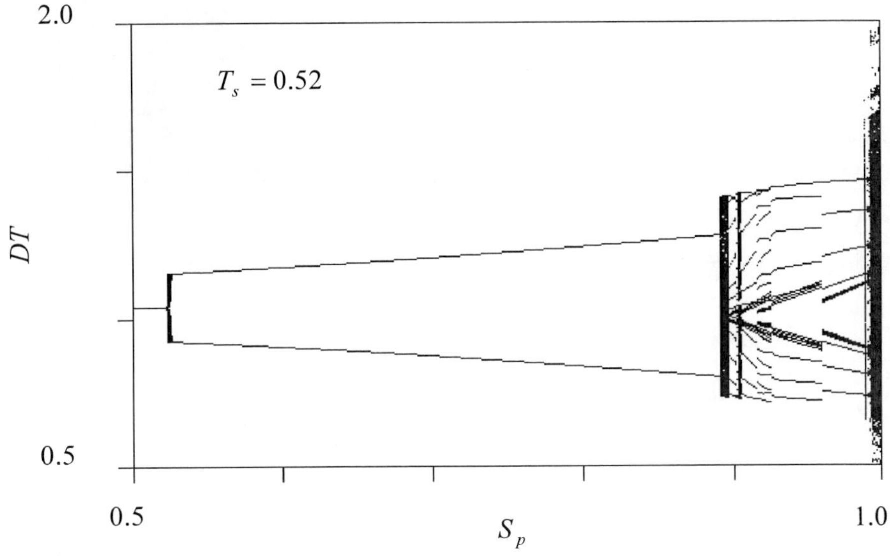

(a)

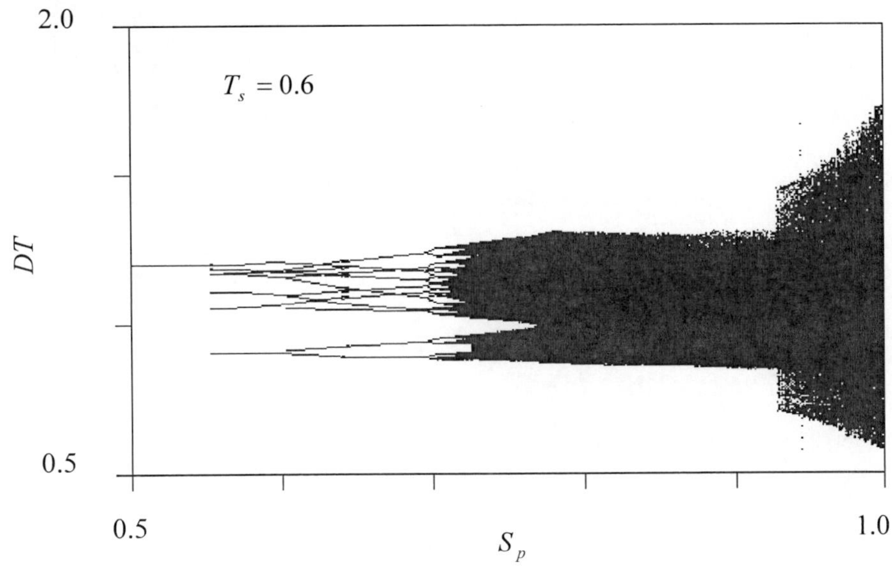

(b)

Figure 130. (Continued)

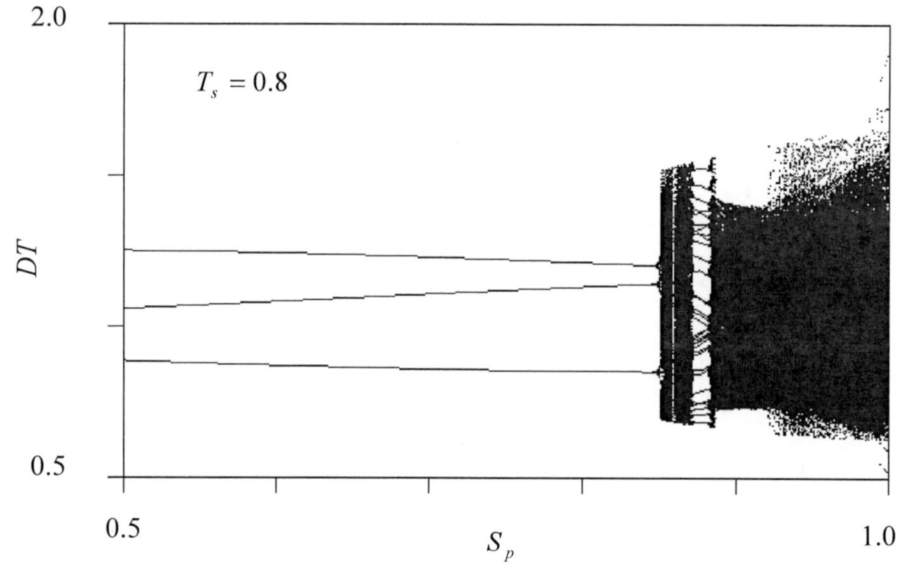

(c)

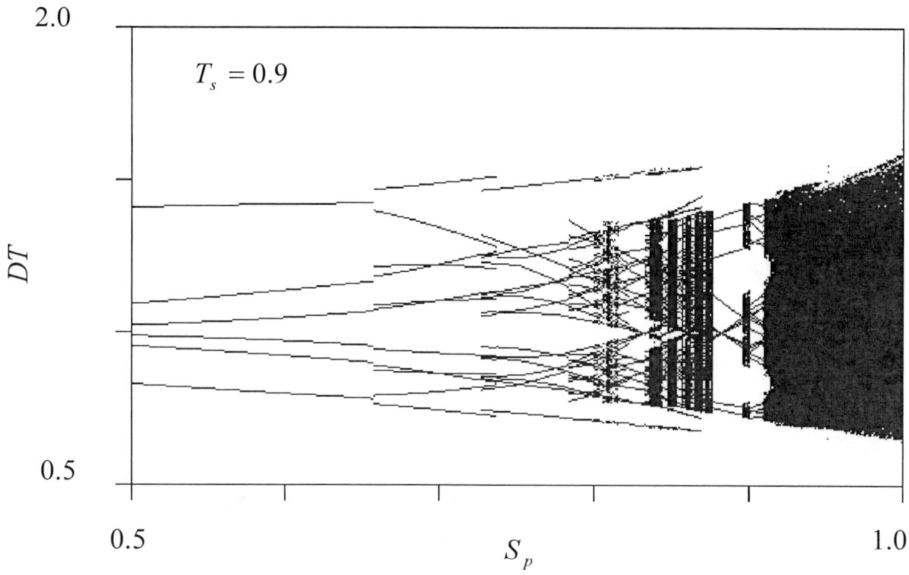

(d)

Figure 130. Plots of tour time DT against speedup rate S_p for (a) $T_s = 0.52$, (b) $T_s = 0.6$, (c) $T_s = 0.8$, and (d) $T_s = 0.9$.

8.2. Chaos of Two Competing Vehicles Controlled by Traffic Lights

When there are more vehicles than one in the neighborhood of the highway, they interact with the other vehicles. Then, the vehicles display the different behavior from the case of the single vehicle. Furthermore, the traffic lights affect the motion of vehicles. The traffic behavior of vehicles results in the complex motions. In the traffic flow controlled by traffic lights, vehicles interact and compete with the other vehicles. Do the vehicles move periodically or chaotically on the highway with traffic lights? How does the tour time of the vehicle vary with the speed variation and cycle time of traffic lights?

Here, we study the traffic of two vehicles moving through an infinite series of traffic lights when they speed up or down accordingly competing each other. We extend the dynamical model of a single vehicle to that of two competing vehicles. We clarify the dynamical states of two competing vehicles through a sequence of traffic lights.

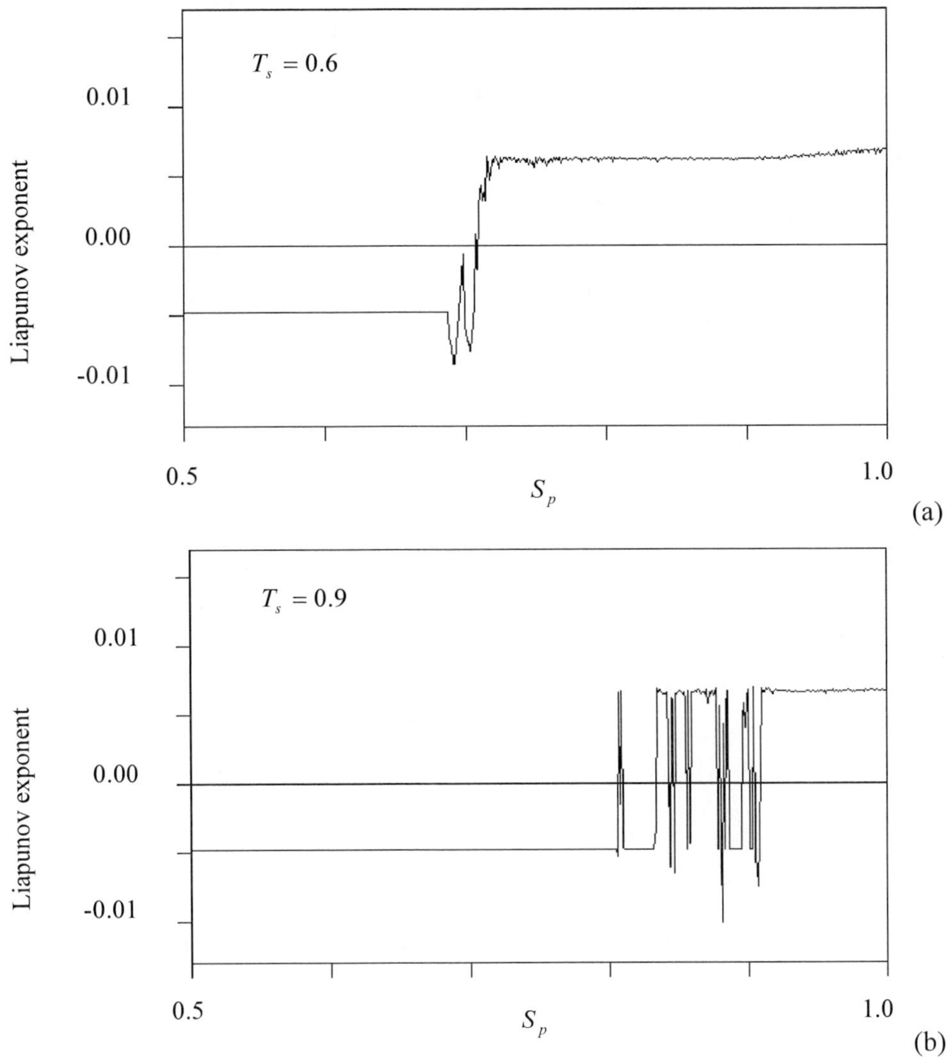

Figure 131. Plots of Liapunov exponent λ against speedup rate S_p for (a) $T_s = 0.6$ and (b) $T_s = 0.9$. Plots (a) and (b) correspond, respectively, to plots (b) and (d) in Figure 130.

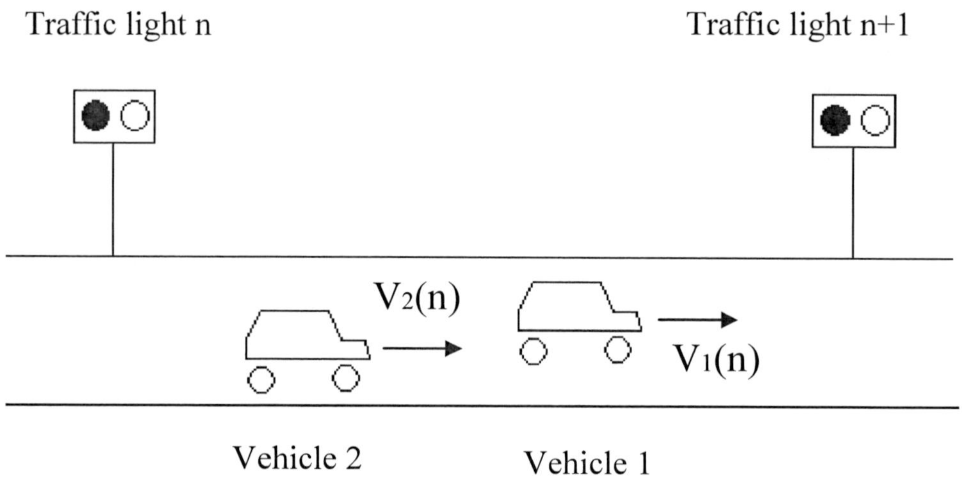

Figure 132. Schematic illustration of two competing vehicles moving through a sequence of traffic lights. The traffic lights are numbered, from upstream to downstream, by 1, 2, 3, ---, *n*, *n*+1, ---. In the synchronized strategy, all the traffic lights change simultaneously from red (green) to green (red) with a fixed time period $t_s/2$.

We consider the motion of two competing vehicles going through an infinite series of traffic lights. Vehicle 1(2) moves with the mean speed $v_{1(2)}(n)$ between traffic lights n and n+1. We assume that the mean speed $v_{1(2)}(n)$ increases proportionally to the difference between the arrival times at traffic light. Figure 132 shows the schematic illustration of two vehicles moving through a sequence of traffic lights. In the synchronized strategy of traffic light control, all the traffic lights change simultaneously from red (green) to green (red) with a fixed time period $t_s/2$. We define the arrival times of vehicles 1 and 2 at traffic light n as $t_1(n)$ and $t_2(n)$. The arrival times of vehicles 1 and 2 at traffic light n+1 are described by

$$t_1(n+1) = t_1(n) + l_n / v_1(n) + (r_1(n) - t_1(n))H(\sin(2\pi t_1(n)/t_s)) \tag{122}$$

and

$$t_2(n+1) = t_2(n) + l_n / v_2(n) + (r_2(n) - t_2(n))H(\sin(2\pi t_2(n)/t_s)) \tag{123}$$

with $r_1(n) = (\mathrm{int}(t_1(n)/t_s) + 0.5)\cdot t_s$ (124)

and $r_2(n) = (\mathrm{int}(t_2(n)/t_s) + 0.5)\cdot t_s$, (125)

where $H(t)$ is the Heaviside function. The vehicular speed increases proportionally to the difference between the arrival times of vehicles 1 and 2 at the traffic light. Then, the mean speeds of vehicles 1 and 2 are described by

$$v_1(n) = v_0 + s_p\left(t_1(n) - t_2(n)\right) \text{ if } v_0/2 < v_0 + s_p\left(t_1(n) - t_2(n)\right) < 2v_0,$$
$$v_1(n) = v_0/2 \text{ if } v_0 + s_p\left(t_1(n) - t_2(n)\right) \leq v_0/2,$$
$$v_1(n) = 2v_0 \text{ if } v_0 + s_p\left(t_1(n) - t_2(n)\right) \geq 2v_0, \tag{126}$$

and

$$v_2(n) = v_0 + s_p\left(t_2(n) - t_1(n)\right) \text{ if } v_0/2 < v_0 + s_p\left(t_2(n) - t_1(n)\right) < 2v_0,$$
$$v_2(n) = v_0/2 \text{ if } v_0 + s_p\left(t_2(n) - t_1(n)\right) \leq v_0/2,$$
$$v_2(n) = 2v_0 \text{ if } v_0 + s_p\left(t_2(n) - t_1(n)\right) \geq 2v_0, \tag{127}$$

where v_0 is the mean speed with no speed's variation, parameter s_p represents the rate of speed variation, and the minimal and maximal velocities are $v_0/2$ and $2v_0$.

In the green-wave strategy, the traffic light changes with a certain time delay τ between the traffic light phases of two successive intersections. The change of traffic lights propagates backwards like a green wave. The arrival time of vehicle 1 at traffic light n+1 is given by

$$t_1(n+1) = t_1(n) + l_n/v_1(n) + \left(r_1(n) - t_1(n) - n\tau\right)H\left(\sin(2\pi(t_1(n) + n\tau)/t_s)\right) \tag{128}$$

with $r_1(n) = \left(\text{int}\left((t_1(n) + n\tau)/t_s\right) + 0.5\right)\cdot t_s$. (129)

We restrict ourselves to the case of the same distance for the traffic lights. By setting $l_n = l_0$ and dividing time by the characteristic time l_0/v_0, one obtains the nonlinear equations of dimensionless arrival times:

$$T_1(n+1) = T_1(n) + 1/V_1(n) + \left(R_1(n) - T_1(n) - n\Pi\right)H\left(\sin(2\pi(T_1(n) + n\Pi)/T_s)\right) \tag{130}$$

and

$$T_2(n+1) = T_2(n) + 1/V_2(n) + \left(R_2(n) - T_2(n) - n\Pi\right)H\left(\sin(2\pi(T_2(n) + n\Pi)/T_s)\right) \tag{131}$$

with $V_1(n) = 1 + S_p\left(T_1(n) - T_2(n)\right)$ if $1/2 < 1 + S_p\left(T_1(n) - T_2(n)\right) < 1$,

$$V_1(n) = 1/2 \text{ if } 1 + S_p\left(T_1(n) - T_2(n)\right) \leq 1/2,$$
$$V_1(n) = 2 \text{ if } 1 + S_p\left(T_1(n) - T_2(n)\right) \geq 2,$$
$$V_2(n) = 1 + S_p\left(T_2(n) - T_1(n)\right) \text{ if } 1/2 < 1 + S_p\left(T_2(n) - T_1(n)\right) < 1,$$
$$V_2(n) = 1/2 \text{ if } 1 + S_p\left(T_2(n) - T_1(n)\right) \leq 1/2,$$
$$V_2(n) = 2 \text{ if } 1 + S_p\left(T_2(n) - T_1(n)\right) \geq 2,$$
$$R_1(n) = \left(\text{int}\left((T_1(n) + n\Pi)/T_s\right) + 0.5\right)\cdot T_s, \tag{132}$$

and $R_2(n) = \left(\text{int}\left((T_2(n) + n\Pi)/T_s\right) + 0.5\right) \cdot T_s$, (133)

where $T_{1(2)}(n) = t_{1(2)}(n) v_0 / l_0$, $R_{1(2)}(n) = r_{1(2)}(n) v_0 / l_0$, $T_s = t_s v_0 / l_0$, $\Pi = \tau \cdot v_0 / l_0$, and $S_p = s_p l_0 / v_0^2$. Thus, the dynamics of two vehicles is described by the nonlinear maps (130) and (131). The motion of vehicles depends on dimensionless cycle time $T_s = t_s v_0 / l_0$, dimensionless delay (offset time) $\Pi = \tau \cdot v_0 / l_0$, and dimensionless speed's variation rate $S_p = s_p l_0 / v_0^2$. In the synchronized strategy, the vehicle depends on both dimensionless cycle time $T_s = t_s v_0 / l_0$ and dimensionless speed's variation rate $S_p = s_p l_0 / v_0^2$.

We study how the motion changes by varying both cycle time and speed variation in the nonlinear-map model. We investigate numerically the dynamical behavior of two vehicles going ahead through an infinite series of traffic lights by iterating maps (130) and (131). Iteration of the maps corresponds to moving ahead of vehicles from traffic light to traffic light. We restrict ourselves to the dynamical behavior of vehicles at the synchronized strategy ($\Pi = 0$). We calculate the time headway between two vehicles and the tour time between traffic lights when the vehicles go ahead on the highway. We study how the tour time and headway vary with the cycle time and speed's variation rate. The tour time of vehicle 1(2) between a traffic light and its proceeding light is defined as $DT1(n) = T1(n) - T1(n-1)$ ($DT2(n) = T2(n) - T2(n-1)$). The tour time of vehicles is represented by $DT(n)$ when the dynamical behavior of vehicle 1 is not different from that of vehicle 2. The time headway of vehicle 1(2) between two vehicles is defined as $DT12(n) = T2(n) - T1(n)$ ($DT21(n) = T1(n) - T2(n)$).

We study the variation of tour time with going ahead through the traffic lights. Figure 133 shows the plots of the tour time *DT* of vehicle 2 against traffic light *n* for *n*=1000-1080. At speed's variation rate $S_p = 0.6$, the plots of the tour time *DT* are shown for (1) $T_s = 0.56$, (2) $T_s = 0.59$, and (3) $T_s = 0.63$ in Figure 133. In diagram (1), vehicles 1 and 2 go ahead by alternating their positions. Vehicle 1(2) passes vehicle 2(1) every two traffic lights. The sign of time headway of vehicle 1(2) changes alternately (every two traffic lights). In diagram (2), the vehicles do not move periodically but the motion displays the irregular behavior. Vehicle 2 passes vehicle 1 or is overtaken by vehicle 1 irregularly. In diagram (3), two vehicles move periodically with period 9.

Figure 134(a) shows the plot of the tour time *DT* against cycle time T_s at speed's variation rate $S_p = 0.6$ for sufficiently large number $n = 1000 - 3000$. Figure 134(b) shows the plot of the time headway against cycle time T_s at speed' variation rate $S_p = 0.6$ for sufficiently large number $n = 1000 - 3000$. For comparison, Figure 134(c) shows the plot of the tour time *DT* against cycle time T_s at speed's variation rate $S_p = 0.0$ for sufficiently large number $n = 1000 - 3000$. The black regions represent the irregular motion of vehicles in Figures 134(a) and (b). Except the black regions, the vehicles exhibit the periodic behavior with various periods. When the time headway is zero, the vehicles move together with the same speed and the tour time of vehicles is consistent with that of a single vehicle in Figure 134(c). In the regions of non-zero headway, the sign of time headway changes alternately. A vehicle passes another vehicle or is overtaken by another vehicle. At some specific values of the cycle time, the vehicles pass or are overtaken one another. Thus, the cycle time of traffic lights has the important effect on the motion of two vehicles.

When the time headway is zero, two vehicles move together accordingly the vehicles go through green traffic lights or stop at red lights. When the time headway is zero and the value of cycle time equals $T_s = 1/k$, the value of tour time becomes one where k is a positive integer, because the vehicle can pass just through all green lights irrespective of value of speed's variation rate S_p. When the cycle time becomes accordingly higher than $1/k$, the vehicles stop every traffic lights, stop every two lights, every three lights, ---, and stop every n lights. In the limit of $T_s \to 1/(k-1)$, the vehicles seldom stop at the traffic lights and go through almost traffic lights.

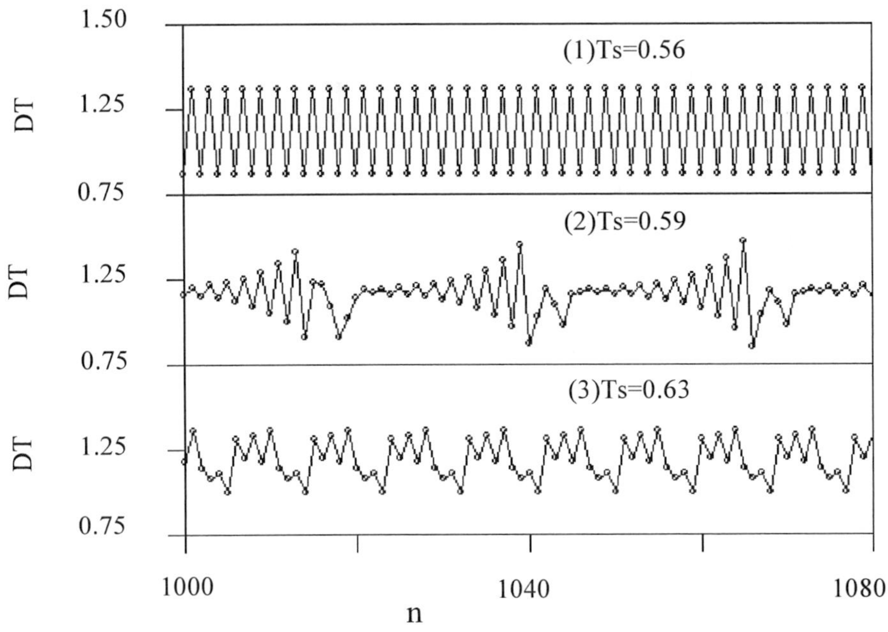

Figure 133. Plots of the tour time DT of vehicle 2 against traffic light n for n=1000-1080. At speed's variation rate $S_p = 0.6$, the plots of the tour time DT are shown for (1) $T_s = 0.56$, (2) $T_s = 0.59$, and (3) $T_s = 0.63$.

Figure 135(a) is the enlargement of Figure 134(a) for $1/2 \le T_s \le 1$. Figure 135(b) is the enlargement of Figure 134(a) for $1/3 \le T_s \le 1/2$. The pattern of Figure 135(b) is similar to that of Figure 135(a). Also, the pattern of tour time for $1/4 \le T_s \le 1/3$ is similar to that of Figure 135(b). The pattern of tour time for $1/2 \le T_s \le 1$ appears repeatedly for

$1/(k+1) \le T_s \le 1/k$ where k is a positive integer larger than one. The pattern of tour time is not fractal but self-similar because the scaling factor is not constant. Figure 136(a) is the enlargement of Figure 135(a) for $0.55 \le T_s \le 0.65$. Figure 136(b) shows the plot of the time headway corresponding to Figure 136(a). The region with a single value of tour time (headway) displays such regular state that the vehicles move with a constant value of speed. The region with two-valued headway (tour time) exhibits a periodic state with period 2. The irregular state appears in the black region. Multiple periodic states occur in the region with multi-valued headway (tour time).

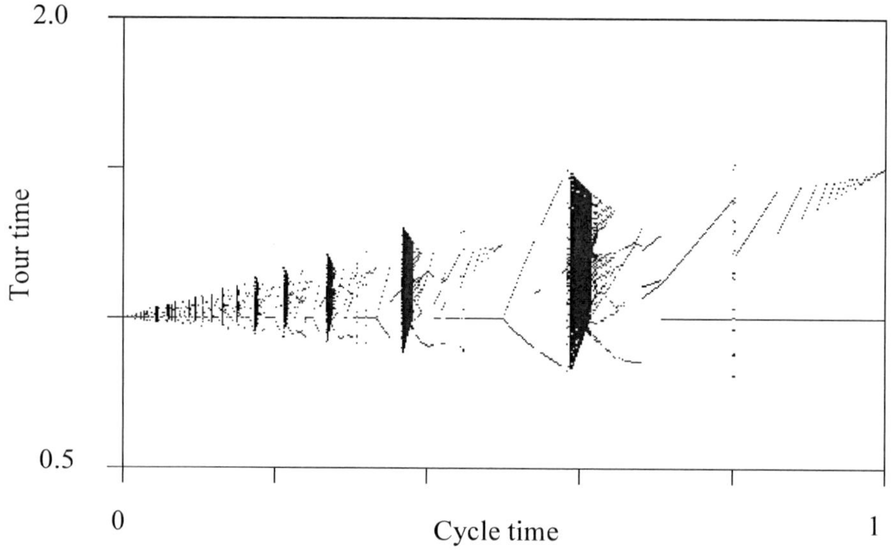

(a)

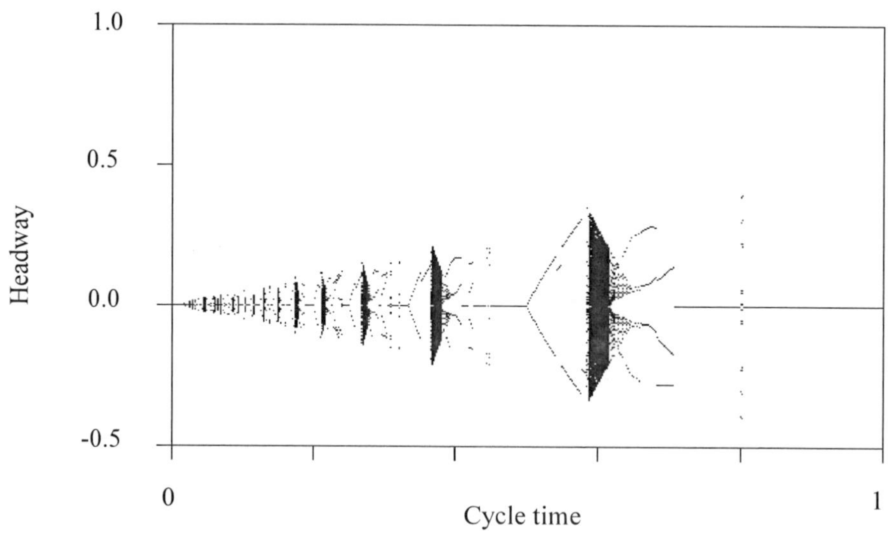

(b)

Figure 134. (Continued)

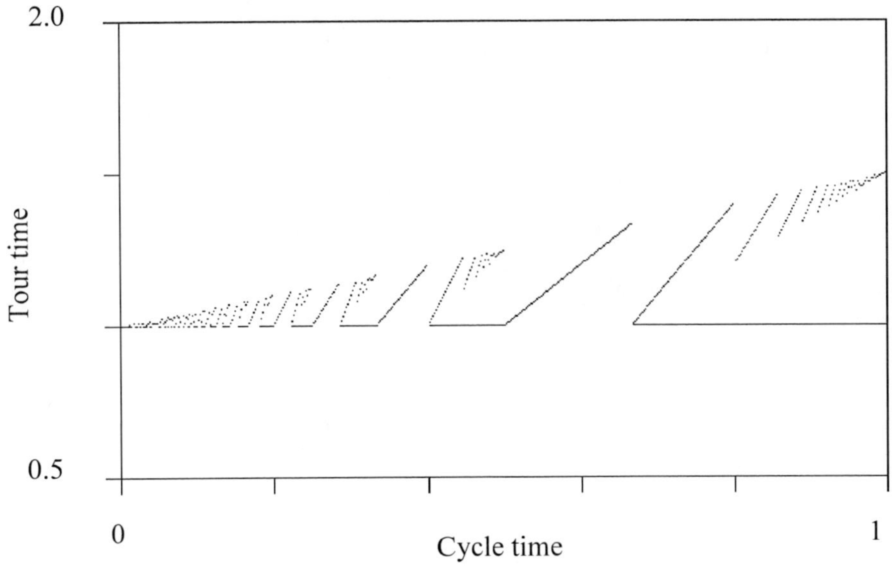

(c)

Figure 134. (a) Plot of the tour time *DT* against cycle time T_s at speed's variation rate $S_p = 0.6$ for sufficiently large number $n = 1000 - 3000$. (b) Plot of the time headway against cycle time T_s at speed's variation rate $S_p = 0.6$ for sufficiently large number $n = 1000 - 3000$. (c) Plot of the tour time *DT* against cycle time T_s at speed's variation rate $S_p = 0.0$ for sufficiently large number $n = 1000 - 3000$.

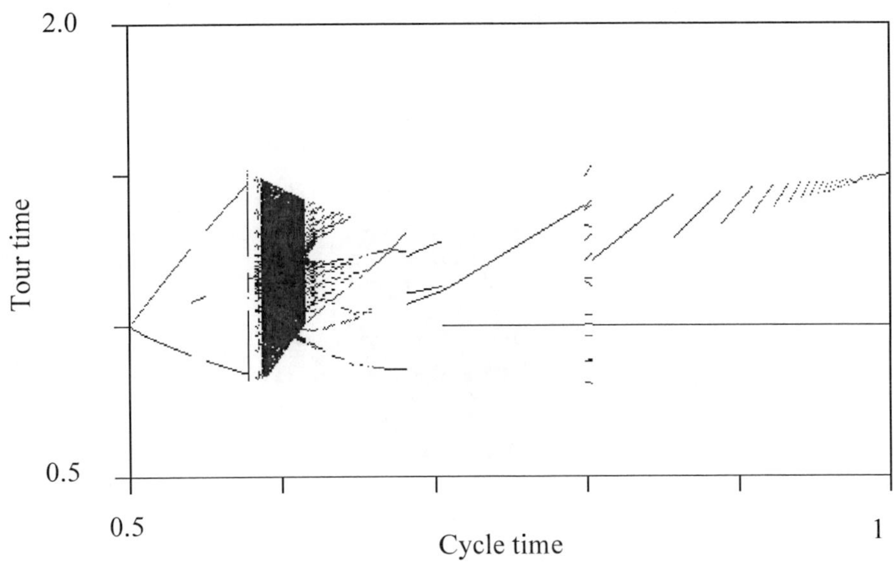

(a)

Figure 135. (Continued)

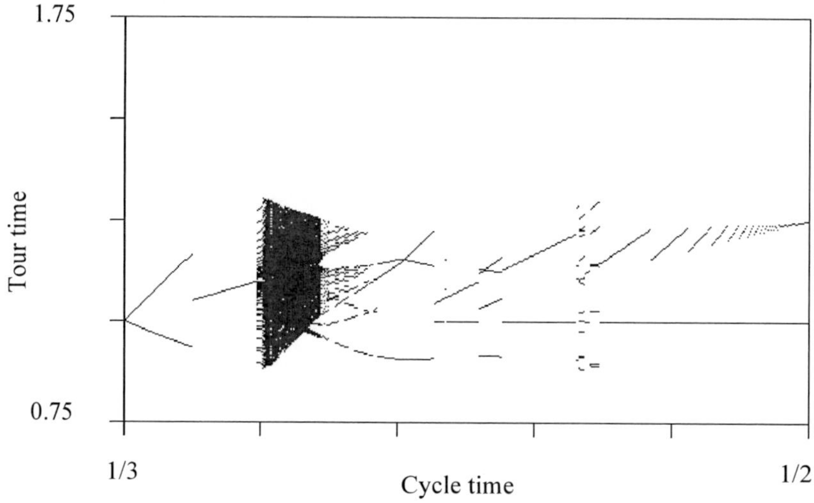

(b)

Figure 135. (a) Enlargement of Figure 134(a) for $1/2 \le T_s \le 1$. (b) Enlargement of Figure 134(a) for $1/3 \le T_s \le 1/2$.

We calculate the Liapunov exponent of the tour time to study the irregular motions. Figure 137(a) shows the plots of Liapunov exponent λ against cycle time T_s for $0 \le T_s \le 1$ at $S_p = 0.6$. Figure 137(b) is the enlargement of Figure 137(a) for $0.55 \le T_s \le 0.65$. The plots (a) and (b) correspond, respectively, to Figure 134(a) and Figure 136(a). The values of Liapunov exponent are positive in the black regions in Figure 134(a) and Figure 136(a) where the vehicles move irregularly. Thus, the black regions in Figures 134-136 represent the chaotic motions of vehicles.

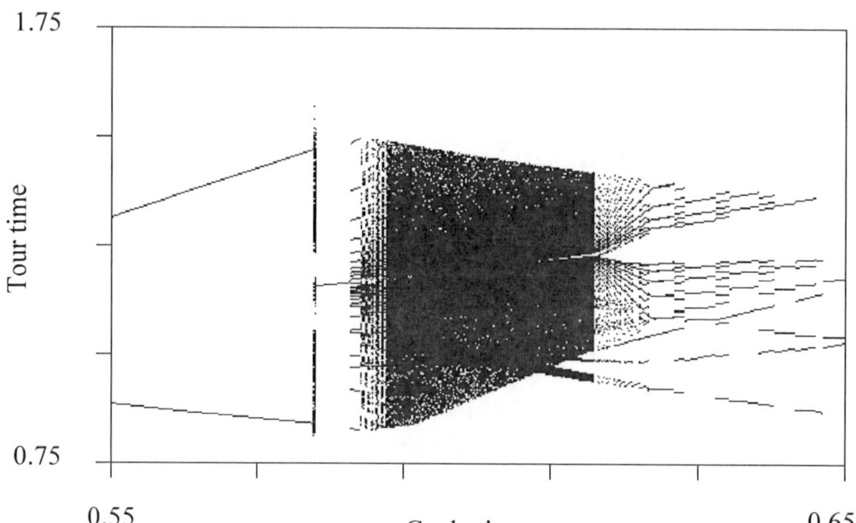

(a)

Figure 136. (Continued)

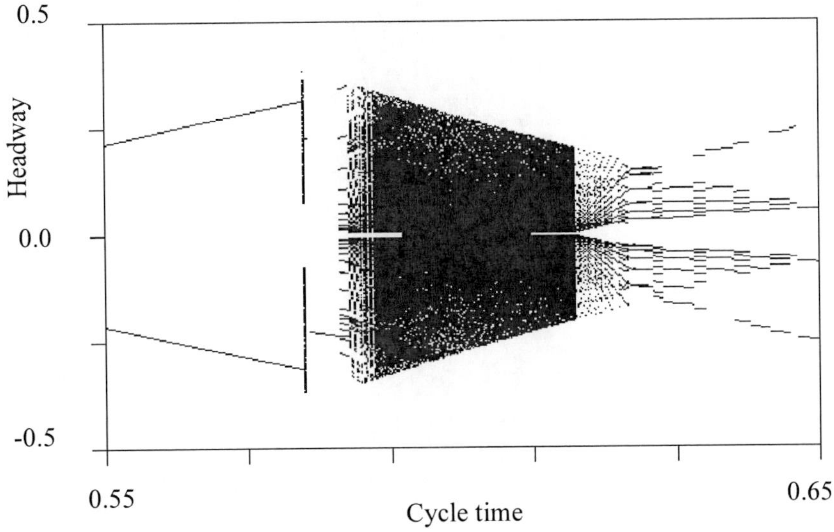

(b)

Figure 136. (a) Enlargement of Figure 135(a) for $0.55 \le T_s \le 0.65$. (b) Plot of the time headway corresponding to diagram (a).

We study the return map for the chaotic and periodic motions. Figure 138 shows the plots of tour time *DT*(n+1) against *DT*(n) for (a) $T_s = 0.6$ and (b) $T_s = 0.625$ at speed's variation rate $S_p = 0.6$ and sufficiently large number $n = 1000 \sim 3000$. Return map (a) shows the chaotic motion. The plotted points form some continuous curves. The return map displays the multi-valued function. Return map (b) shows the highly periodic motion. The plotted points do not form continuous curves. The return maps are consistent with the behaviors of the plot of tour time against cycle time in Figure 136(a).

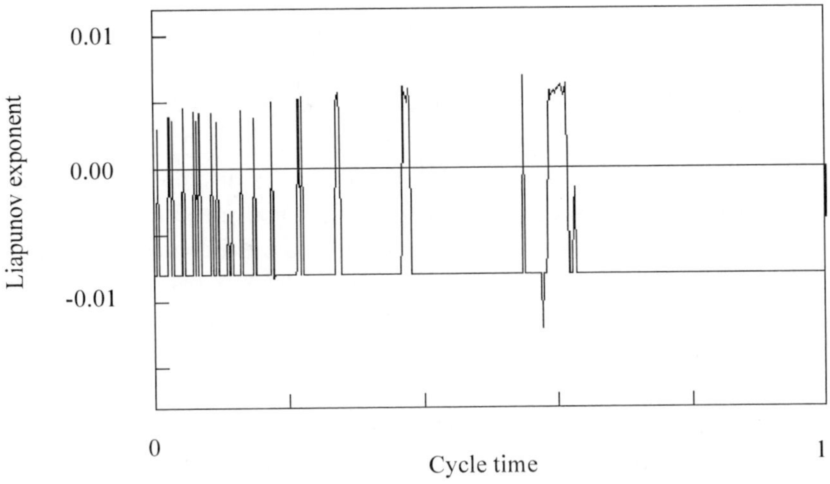

(a)

Figure 137. (Continued)

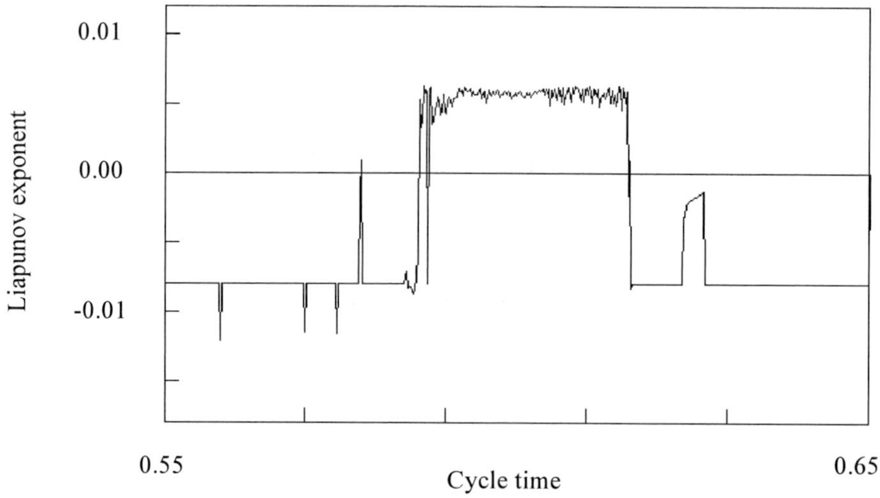

(b)

Figure 137. (a) Plot of Liapunov exponent λ against cycle time T_s for $0 \le T_s \le 1$ at $S_p = 0.6$. (b) Enlargement of diagram (a) for $0.55 \le T_s \le 0.65$. Plots (a) and (b) correspond, respectively, to Figure 134(a) and Figure 136(a).

We study the variation of tour time by speed's variation rate. Figure 139 shows the plots of tour time *DT* against speed's variation rate S_p at (a) $T_s = 0.6$ and (b) $T_s = 0.65$ for sufficiently large number $n = 1000 \sim 3000$. The black region in the plots represents a chaotic motion. In plot (a), the vehicles move together with a constant value of tour time for $S_p < 0.54$. The vehicles stop every traffic lights and go ahead when the traffic lights become green. When the speed's variation rate is higher than $S_p = 0.54$, the vehicles move chaotically. This motion corresponds to the return map in Figure 138(a). Furthermore, when the speedup rate is higher than $S_p = 0.97$, the dynamical transitions occur to the periodic motions. In plot (b), the different dynamical transitions occur at some values of speed's variation rate. The behavior of vehicles changes from the regular, through periodic, to chaotic, and again to periodic motions.

Thus, the motion of vehicles displays the regular, periodic, and chaotic states via the complex dynamical transitions by varying both cycle time and speed's variation rate. A single vehicle through a sequence of traffic lights does not exhibit the chaotic motion but moves regularly or periodically. By adding another vehicle to the highway, two vehicles compete one another and result in the chaotic motion. The dynamical transition to chaos is definitely different from that of the single vehicle.

We have shown that the vehicles exhibit the very complex behavior with increasing speed's variation rate for a short cycle time of traffic lights. We have clarified the effect of the speed's variation rate on the tour time when two vehicles move through traffic lights by competing one other.

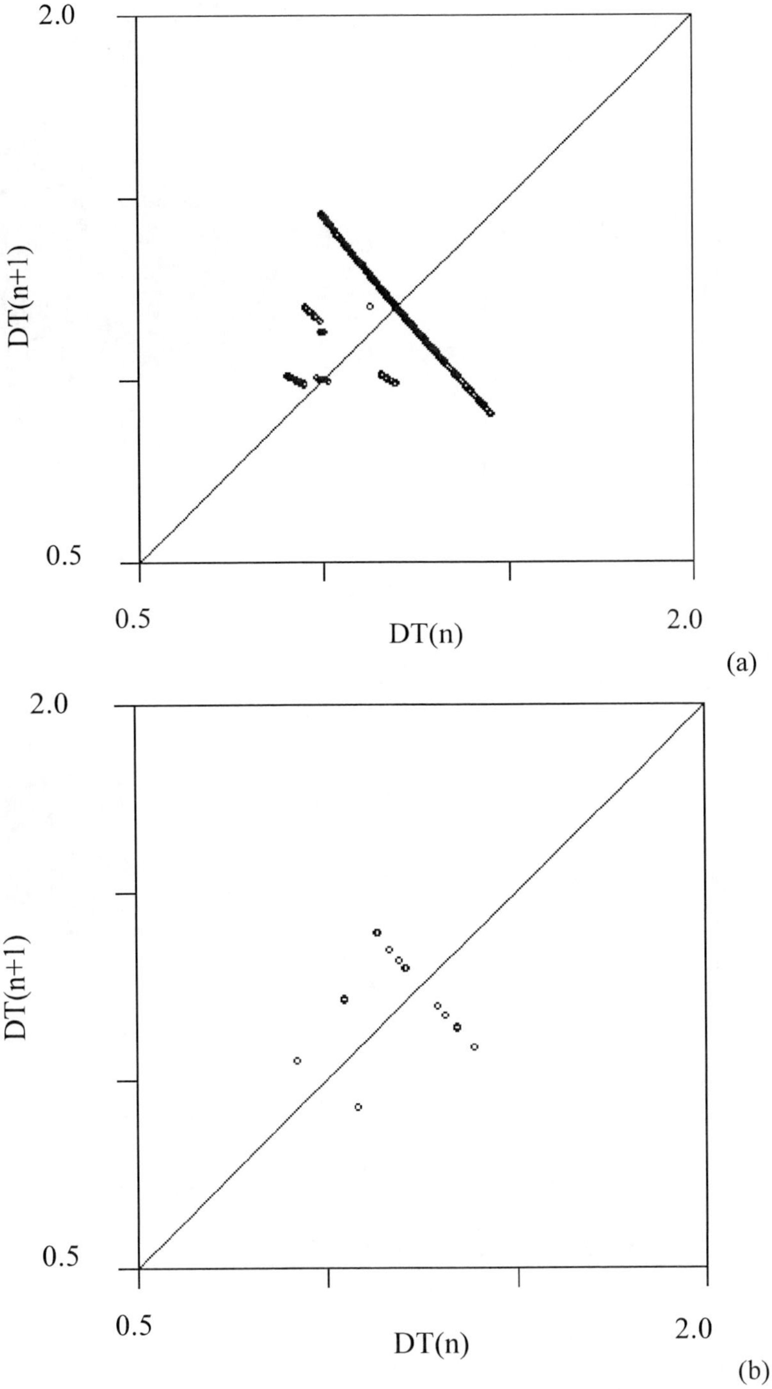

Figure 138. Plots of tour time DT(n+1) against DT(n) for (a) $T_s = 0.6$ and (b) $T_s = 0.625$ at speed's variation rate $S_p = 0.6$.

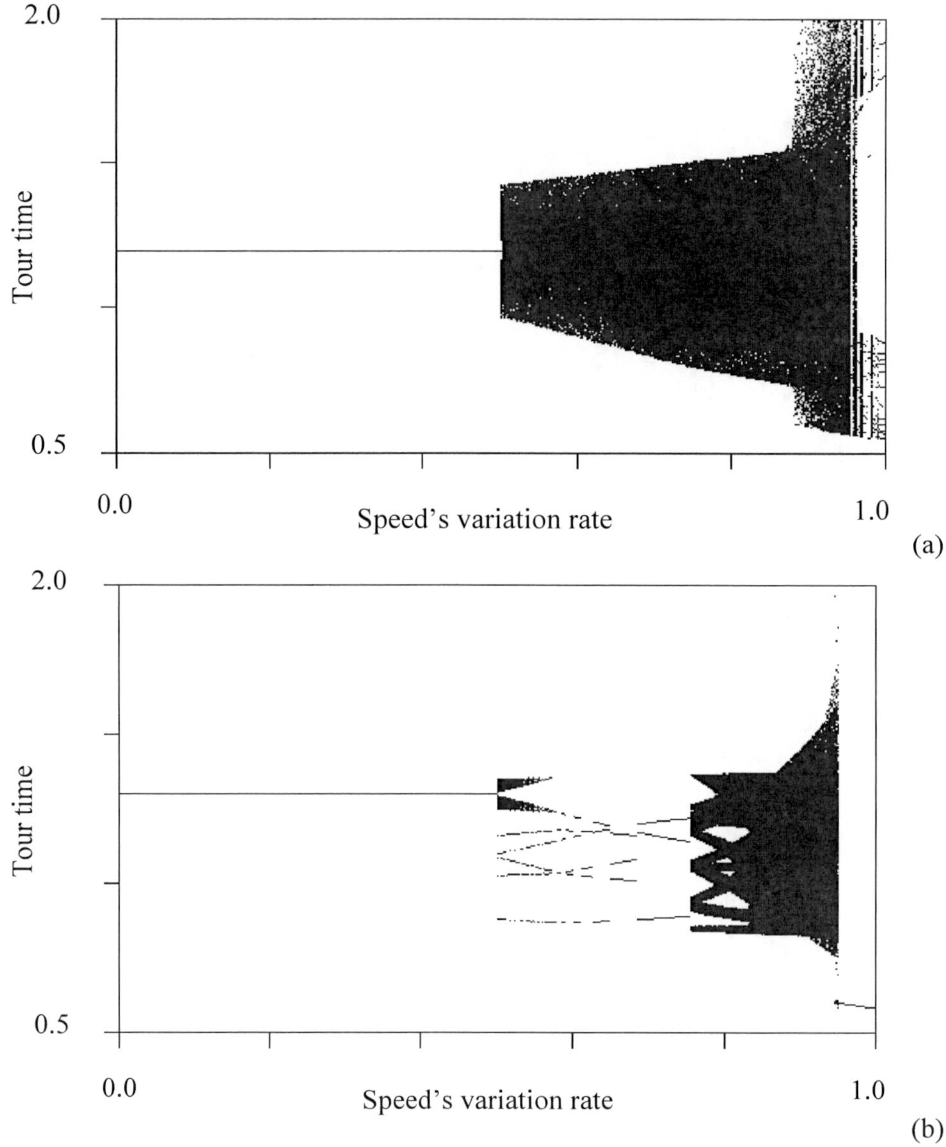

Figure 139. Plots of tour time DT against speed's variation rate S_p at (a) $T_s = 0.6$ and (b) $T_s = 0.65$ for sufficiently large number $n = 1000 \sim 3000$.

9. SUMMARY

In this article, we have focused attention mainly on the progress for traffic flow made in the recent years. We have discussed the main models of vehicular traffic including the car-following models, the cellular automaton models, the gas-kinetic models and the fluid-dynamic models. The relationships between different approaches of modeling have been explored in detail. The motion of vehicles has been described by Newton's equation of motion, complemented by driving forces and frictional dissipation. One has often found considerably different behavior compared with analogous systems from classical mechanics.

We have shown the fundamental diagrams (current-density diagrams) of traffic flow on two- and multi-lane highways. We have clarified the traffic jam and breakdown induced by speed limit at slowdown sections on a two-lane highway. One can predict the occurrence of a traffic jam at a specific place on a given highway at a particular instant of time by using the traffic modeling presented here.

We have presented the nonlinear-map model for city traffic controlled by traffic signal. We have clarified the dependence of traffic flow on signal characteristics. Furthermore, we have shown that vehicles exhibit the chaotic motion by traffic signals and speedup. One can evaluate the tour (travel) time of vehicles through a sequence of traffic lights by using the nonlinear-map modeling presented here.

To understanding of more complex traffic behavior, traffic researchers not only needs more realistic models but also needs more detailed and accurate empirical data from real traffic.

ACKNOWLEDGMENTS

I am grateful for fruitful collaborations with M. Muramatsu, S. Kurata, M. Sasaki, R. Nagai, K. Tanaka, H. Hanaura, S. Masukura

REFERENCES

[1] Nagel, K.; Esser, J.; Rickert, M., In *Ann. Rev. Comp. Phys.*; Stauffer, D. Ed.; World Science; Singapore ; 1999.

[2] Chowdhury, D.; Santen, L.; Schadschneider, A. *Phys. Rep.* 2000, 329, 199.

[3] Helbing, D. *Rev. Mod. Phys.* 2001, 73, 1067.

[4] Nagatani, T. *Rep. Prog. Phys.* 2002, 63, 1331.

[5] Kerner, B. S. *The Physics of Traffic; Springer*; Heidelberg, 2004.

[6] Greenshields, B. D. *Proc. Highway Research Board;* Washinton DC; Highway Research Board, 1935, 14, 448.

[7] Lighthill, M. J.; Whitham, G. B. *Proc. R. Soc. A* 1955, 229, 317.

[8] Prigogine, I.; Andrews, F. C. *Oper. Res.* 1960, 8, 789.

[9] Prigogine, I.; Herman, R. *Kinetic Theory of Vehicular Traffic;* Elsevier; New York, 1971.

[10] Newell, G. F. *Oper. Res.* 1961, 9, 209.

[11] Musha, T.; Higuchi, H. *Japan. J. Appl. Phys.* 1976, 15, 1271.

[12] Musha, T.; Higuchi, H. *Japan. J. Appl. Phys.* 1978, 17, 811.

[13] Biham, O.; Middleton, A. A.; Levine, D.; *Phys. Rev. A* 1992, 46, R6124.

[14] Nagel, K.; Schreckenberg, M. *J. Phys. I (France)* 1992, 2, 2221.

[15] Kerner, K; Konhauser, P. *Phys. Rev. E* 1993, 48, R2355.

[16] Hurdle, V. F.; Hauer, E.; Stewart, G. N.; Eds., *Proc. 8th International Symp. On Transportation and Traffic Flow Theory*; Univ. Tronto; Toronto; 1983.

[17] Daganzo, C. F.; Ed., *Proc. 12th International Symp. On Transportation and Traffic Flow Theory*; Elsevier, Amsterdom, 1993.
[18] Lesort, J. B.; Ed., *Proc. 13th International Symp. On Transportation and Traffic Flow Theory*; Pergamon, NY, 1996.
[19] Hall, R. W., *Handbook of Transportation Science*, Kluwer, Boston, 1999.
[20] Bando, M.; Hasebe, K.; Nakayama, A.; Shibata, A.; Sugiyama, Y. *Phys. Rev. E* 1995, 51, 1035.
[21] Nagatani, T.; Nakanishi, K; Emmerich, H. *J. Phys. A* 1998, 31, 5431.
[22] Nagatani, T. *Phys. Rev. E* 2000, 61, 3534.
[23] Treiber, M.; Hennecke, A.; Helbing, D. *Phys. Rev. E* 1999, 59, 239.
[24] Treiber, M.; Hennecke, A.; Helbing, D. *Phys. Rev. E* 2000, 62, 1805.
[25] Gipps, P. G. *Transpn. Res. B* 1981, 15, 105.
[26] Krauss, S.; Wagner, P.; Gawron, C. *Phys. Rev. E* 1996, 54, 3707.
[27] Krauss, S.; Wagner, P.; Gawron, C. *Phys. Rev. E* 1997, 55, 5597.
[28] Helbing, D. *Verkehrsdynamik*, Springer, Berlin, 1997.
[29] Helbing, D.; Tilch, B. *Phys. Rev. E* 1998, 58, 133.
[30] Wolf, D. E. *Physica A* 1999, 263, 438.
[31] Tomer, E.; Safonov, L.; Havlin, S. *Phys. Rev. Lett.* 2000, 84, 382.
[32] Nagatani, T. *Phys. Rev. E* 1999, 60, 6395.
[33] Mason, A. D.; Wood, A. W. *Phys. Rev. E* 1997, 55, 2203.
[34] Nakayama, A.; Sugiyama, Y.; Hasebe, K. *Phys. Rev. E* 2002, 65, 016112.
[35] Hayakawa, H.; Nakanishi, K. *Prog. Theor. Phys. Suppl.* 1998, 130, 57.
[36] Lubashevsky, I.; Kalenkov, S.; Mahnke, R. *Phys. Rev. E* 2002, 65 036140.
[37] Yukawa, S.; Kikuchi, M. *J. Phys. Soc. Japan* 1995, 64, 35.
[38] Tadaki, S.; Kikuchi, M.; Sugiyama, Y.; Yukawa, S. *J. Phys. Soc. Japan* 1998, 67, 2270.
[39] Wolfram, S. *Theory and Application of Cellular Automata*, World Scientific; Singapore, 1986.
[40] Wolfram, S. *Cellular Automata and Complexity*, Addison-Wesley; Reading, MA., 1994.
[41] Schmittmann, B.; Zia, R. K. P. *Phys. Rep.* 1998, 301, 45.
[42] Spohn, H. *Large Scale Dynamics of Interacting Particles*, Springer; Berlin, 1991.
[43] Fukui, M.; Ishibashi, Y. *J. Phys. Soc. Japan* 1996, 65, 1868.
[44] Nishinari, K. *J. Phys. A* 2001, 34, 10727.
[45] Nishinari, K.; Takahashi, D. *J. Phys. A* 2000, 33, 7709.
[46] Nagel, K.; Paczuski, M. *Phys. Rev. E* 1995, 51, 2909.
[47] Nagel, K.; Herrmann, H. J. *Physica A* 1993, 199, 254.
[48] Schadschneider, A.; Schreckenberg, M. *Ann. Physik*, 1997, 6, 541.
[49] Schreckenberg, M.; Schadschneider, A.; Nagel, K.; Itoh, N. *Phys. Rev. E* 1995, 51, 2939.
[50] Takayasu, M.; Takayasu, H. *Fractals* 1993, 1, 860.
[51] Schadscheider, A.; Schreckenberg, M. *J. Phys. A* 1997, 30, L69.
[52] Krauss, S.; Wagner, P.; Gawron, C. *Phys. Rev. E* 1996, 54, 3707.
[53] Krauss, S.; Wagner, P.; Gawron, C. *Phys. Rev. E* 1997, 55, 5597.
[54] Gipps, P. G. *Transpn. Res. B* 1981, 15, 105.
[55] Helbing, D.; Schreckenberg, M. *Phys. Rev. E* 1999, 59, R2505.

[56] Cheybami, S.; Kertetz, J.; Schreckenberg, M. *Phys. Rev. E* 2000, 63, 016107.
[57] Nagatani, T. *Phys. Rev. E* 1995, 51, 922.
[58] Krug, J.; Ferrari, P. A. *J. Phys. A* 1996, 29, L465.
[59] Ktitarev, D.; Chowdhury, D.; Wolf, D. E. *J. Phys. A* 1997, 30, L221.
[60] Paveri-Fontana, S. L. *Transpn. Res.* 1875, 9, 225.
[61] Ben-Naim, E.; Krapivsky, P. L.; Redner, S. *Phys. Rev. E* 1994, 50, 822.
[62] Ben-Naim, E.; Krapivsky, P. L. *Phys. Rev. E* 1997, 56, 6680.
[63] Helbing, D. *Phys. Rev. E* 1996, 53, 2366.
[64] Helbing, D.; Treiber, M. *Phys. Rev. Lett.* 1998, 81, 3042.
[65] Wagner, C.; Hoffmann, C.; Sollache, R.; Wagenhuber, J.; Schurmann, B. Phys. Rev. E 1996, 54, 5073.
[66] Wagner, C. *J. Stat. Phys.* 1998, 90, 1251.
[67] Nagatani, T. *J. Phys. Soc. Japan* 1996, 65, 3150.
[68] Nagatani, T. *Physica A* 1997, 237, 67.
[69] Lehmann, H. *Phys. Rev. E* 1996, 54, 6058.
[70] Helbing, D. *Physica A* 1995, 219, 391.
[71] Shvetsov, V.; Helbing, D. *Phys. Re. E* 1999, 59, 6328.
[72] Henneke, A.; Treiber, M.; Helbing, D. in *Traffic and Granular Flow '99* Helbing, D.; Herrmann, H. J.; Schreckenber, M.; Wolf, D.; Eds. 2000, Springer, Berlin, p. 383.
[73] Mahnke, R.; Pieret, N. *Phys. Rev. E* 1997, 56, 2666.
[74] Payne, H. J. in *Mathematical Models of Public Systems*, vol. 1, Bekey, G. A.;Ed. 1971, Simulation Council, La Jolla, CA, p. 51.
[75] Payne, H. J. *Transpn. Res. Rec.* 1979, 722, 68.
[76] Phillips, W. F. *Transpn. Plan. Technol.* 1979, 5, 131.
[77] Kerner, B. S.; Konhauser, P. *Phys. Rev. E* 1993, 48, R2355.
[78] Kerner, B. S.; Konhauser, P. *Phys. Rev. E* 1994, 50, 54.
[79] Lee, H. Y.; Lee, H-W.; Kim. D. *Phys. Rev. E* 1999, 59, 5101.
[80] Berg, P.; Mason, A.; Wood, A. *Phys. Rev. E* 2000, 61, 1056.
[81] Nelson, P. *Phys. Rev. E* 2000, 61, R6052.
[82] Nagatani, T. *Physica A* 1999, 264, 581.
[83] Nagatani, T. *Phys. Rev. E* 1999, 59, 4857.
[84] Kurtz, D. A.; Hong, D. C. *Phys. Rev. E* 1995, 52, 218.
[85] Lee, H. Y.; Lee, H-W.; Kim, D. *Phys. Rev. E* 2001, 64, 056126.
[86] Nagatani, T. *Physica A* 2001, 290, 501.
[87] Nagai, R.; Nagatani, T.; Yamada, A. *Physica A* 2005, 355, 530.
[88] Nagai, R.; Hanaura, H.; Tanaka, K.; Nagatani, T. *Physica A* 2006, 364, 464.
[89] Tanaka, K.; Nagai, R.; Nagatani, T. *Physica A* 2006, 370, 756.
[90] Hanaura, H.; Nagatani, T.; Tanaka, K. *Physica A* 2007, 374, 419.
[91] Nagatani, T. *Physica A* 2005, 347, 673.
[92] Nagatani, T. *Physica A* 2007, 374, 431.
[93] Nagatani, T. *Physica A* 2008, 387, 1637.
[94] Brockfeld, E.; Barlovic, R.; Schadschneider, A.; Schreckenberg, M. *Phys. Rev. E* 2001, 64, 056132.
[95] Sasaki, M.; Nagatani, T. *Physica A* 2002, 325, 531.
[96] Nagatani, T. *Physica A* 2007, 377, 651.
[97] Nagatani, T. *Physica A* 2007, 380, 503.

[98] Nagatani, T. *Physica A* 2005, 348, 561.
[99] Nagatani, T.; Nagai, R. *Chaos, Solitons & Fractals* 2005, 25, 245.
[100] Toledo, B. A.; Munoz, V.; Rogan, J.; Tenreiro, C. *Phys. Rev. E* 2004, 70, 016107.
[101] Toledo, B. A.; Cerda, E.; Rogan, J.; Munoz, V.; Tenreiro, C.; Zarama, R.; Valdivia, J. A. *Phys. Rev. E* 2007, 75, 026108.
[102] Wastavino, L. A.; Toledo, B. A.; Rogan, J.; Zarama, R.; Munoz, V.; Valdivia, J. A. *Physica A* 2007, 381, 411.

In: Road Traffic: Safety, Modeling and Impacts
Editors: S. E. Paterson and L. K. Allan, pp. 225-257
ISBN 978-1-60456-884-4

Chapter 2

FORENSIC INVESTIGATION OF TRAFFIC ACCIDENTS

Stavroula A. Papadodima, Emmanouil I. Sakelliadis, Sotirios A. Athanaselis and Chara A. Spiliopoyloy
Department of Forensic Medicine and Toxicology, Medical School, University of Athens, Greece

1. INTRODUCTION

In motor vehicle deaths, autopsies are performed to determine the cause and manner of death, detect any disease or factor that may precipitate or contribute to the death, document all findings for subsequent use in either criminal or civil actions and establish positive identification of the body, especially if it is burnt or severely mutilated (Saukko and Knight, 2004).

When victims of traffic accidents are autopsied, the standard autopsy procedure should be followed with detailed documentation of the injuries. It is very important that the body be seen clothed, if brought dead to the mortuary or hospital, so that injuries can be matched against soiling and damage to the garments. If this is not possible, as when survival allowed admission to a hospital or accident department, the clothing should be preserved and examined by the pathologist. In any event, the clothes should be retained by the police for submission to the forensic science laboratory, usually when criminal proceedings are likely. Toxicological analysis for alcohol is imperative, whether search for drugs depends on the information available. Blood samples should be retained for blood grouping and perhaps even 'DNA fingerprinting' (Saukko & Knight, 2004, DiMaio & DiMaio, 2001). All types of trace evidence may be found by a pathologist, from paint flakes and glass debris to parts of the vehicle structure and any of the above foreign bodies or particles, either in the clothing, hair, on the skin or in the wounds, must be carefully retained for forensic science examination (Smock et al., 1989, Curtin and Langlois, 2007).

2. Types of Motor Vehicle Accidents

2.1. Vehicle-vehicle Accident

2.1.1. Front Collision

When a *front collision* occurs, there is a deceleration of the vehicle as it hits another vehicle or a fixed object such as a tree. The forward motion of the unrestrained occupant (as he continues to move forward) is arrested as the occupant connects with the, by that time, stationary vehicle chassis. The initial impact point is often the lower extremities, resulting in fracture/dislocation of the ankles, knee or hip dislocations and femoral fractures. The driver will lean forward and the chest will be compressed against the steering wheel. Finally, the head will hit the windscreen. If the front impact is off-center, the driver (or passenger) might impact the A pillar with their head. The passenger in the front seat comes up against the windshield or the sun visor area. Unrestrained backseat passengers try to avoid injury during the crash by pushing their extended upper limbs against the front seat. Transmitted energy through their upper limbs classically causes upper limb fractures/dislocations. They considered as “flying bullets” within the vehicle as they usually hit front seat passengers (Saukko & Knight, 2004, DiMaio & DiMaio, 2001).

If the drivers and passengers have restraint devices and if the passenger compartment retains its integrity, then the occupants of the vehicle may survive without any significant injury. Seat belts are effective in preventing injuries and death in all types of motor vehicle crashes. They reduce head injuries, though increase abdominal injuries. Abdominal compression occurs when the restrained occupant is subjected to high speed deceleration (National Highway Traffic Safety Administration, 1996).

The intrusion of part of the vehicle or another object into the passenger compartment may be a serious cause of injury. The intrusion may be transitory, with the portion of car or the object springing back. Thus, it might not at first be obvious that there has been violation of the integrity of the passenger compartment (Saukko & Knight, 2004, DiMaio & DiMaio, 2001).

2.1.2. Side Collisions

Side collisions are vehicle crashes where the side of one or more vehicles is impacted. These crashes often occur at intersections, in parking lots, when two vehicles pass on a multi-lane roadway, or when a vehicle hits a fixed object. When a vehicle is hit on the side by another vehicle, the crumple zones of the striking vehicle will absorb some of the kinetic energy of the collision. The crumple zones of the struck vehicle may also absorb some of the collision's energy, particularly if the vehicle is not struck on its passenger compartment. Fatalities usually occur in the car impacted rather than the car impacting, because the engine protects the impacting driver and passengers. In side impact collisions with fixed objects, that is, when a car slides sideways into a fixed object, the driver or a passenger if not restrained may partially pop out the window, impact the fixed object, and then pop back into the vehicle. In such cases, dicing injuries can be found on either one or both sides of a driver, depending on whether the side glass is propelled into the driver; the driver into the glass, or both. The head can flex laterally through the side window, striking the impacting vehicle. It might also impact the A or B pillars.

If the impacting vehicle is a truck, the force is delivered from roof to floor level, and the intruding vehicle can make direct contact with the head. The region of the body which is close to the side of impact will be injured directly, while those away from the impact may hit the other side of the vehicle. Severe brain and thoracic injuries and mortality occur more frequently in side impact crashes compared with other types. The nearer the occupant is to side of the impact, the more serious his/her thoracic or abdominal injury will be (Eid & Abu-Zidan, 2007, Mikhail, 1995).

Both vehicles are frequently turned from their original directions of travel. If the collision is severe, the struck vehicle may be spun or rolled over potentially causing it to strike other vehicles, objects, or pedestrians. If the vehicles are traveling in the same direction and neither vehicle loses control, the collision energy is minimal and the vehicles may suffer only cosmetic damage. However, loss of control of either vehicle can have unpredictable effects and dramatically increases the expected crash severity (Saukko & Knight, 2004, DiMaio & DiMaio, 2001).

Seat belts tend to keep people in place rather than having them hurled from one side of the vehicle to the other, but they are not effective concerning the injuries resulting from direct impact. As already mentioned, side impact wrecks are more likely to involve multiple individual collisions or sudden speed changes before motion ceases. Side-impact airbags can provide protection, but only during the first collision; it may leave occupants unprotected during subsequent collisions in the crash. However, the first collision in a crash typically has the most severe forces, so an effective airbag provides maximum benefit during the most severe portion of a crash. Sideswipe collisions are where the sides of two parallel vehicles touch (Mikhail 1995).

2.1.3. Rear Collision

A *rear collision* is a traffic accident where a vehicle impacts another in front of it. Rear impact crashes are the least common form of fatal accident. This is because the occupants of the front seat of the impacted car are protected by the trunk and rear passenger portion of the vehicle. These usually decelerate the impacting vehicle sufficiently to protect individuals in the front seat. People in the impacting automobile are protected by their car's engine. Generally, injuries to the occupants are usually much worse for the impacted vehicle, because occupants of the following vehicle anticipate the imminent impact and take automatic measures. Although relatively uncommon, one of the potential dangers with the rear impact crash is rupture of the gas tank, with ignition of the fuel. Rupture of the tank is, of course, proportional to the speed of impact.

The front seat occupants of the impacted car are at high risk of getting whiplash injury especially in the absence of head restraints. If the head contacts with a head restraint, the hyperextension movement of the neck will be prevented and the severity of injury is reduced. Furthermore, presence of head restraint prevents collision between front and back seat occupants. In rear impact collisions, there may be seatback failure such that the back of the front seat goes horizontal. At the same time, the occupant of the seat can go backward, impacting the rear seat or the roof, or even be ejected out the rear window. This can result in serious, if not fatal, head and/or neck injuries. This can occur even if the individual is wearing a seatbelt (Saukko & Knight, 2004, DiMaio & DiMaio, 2001, Svensson et al., 2000).

2.1.4. Rollover

Rollover is a type of vehicle accident where a vehicle turns over on its side or roof. The main cause for rolling over is turning too sharply while moving too fast. Rollover crashes are generally less lethal than head-on and side impact collisions, provided the individual is not ejected or the vehicle rolls into an unyielding object such as a tree. Anything that prevents ejection of an occupant increases the probability of survival. As the vehicle rolls over, the roof can be compressed, and if the roof is not strong enough, the occupant can sustain head and spinal cord injuries. Unrestrained occupants inside the vehicle are susceptible to more serious injury as they are vulnerable to injuries caused by movements inside the car. They can also be ejected from the vehicle (Eid & Abu-Zidan, 2007).

2.1.5. Ejection

Ejection from a vehicle is associated with a significantly greater incidence of severe or critical injury (McCoy et al., 1988). As already mentioned, it is usually connected with rollover crashes. Ejection from the vehicle triples the injury severity and increases admissions to intensive care units and mortality. Ejected victims are unprotected and are hence susceptible to be run over by another running vehicle (Malliaris et al., 1996).

2.2. Motor Vehicle–Train Accidents

Collisions between trains and motor vehicles are virtually all of the side impact-type, with the train impacting the side of a vehicle that is either trying to beat the train through the intersection, or is stalled on the tracks. Less commonly, a speeding vehicle impacts the side of a train. The nature of the injuries varies from typical side- and front-impact automobile injuries to the more common nonspecific pattern of massive mutilating injuries (Saukko & Knight, 2004, DiMaio & DiMaio, 2001).

2.3. Motorcyclists

The commonest causes of injury are falling down, and striking against a vehicle or a fixed object. Head or neck injuries are the main cause of hospitalization and mortality. There are usually extensive skull fractures, predominantly basal. Extensive confluent scrape-like abrasions (as they slide across the pavement) may occur. An incision into this area typically reveals no underlying subcutaneous hemorrhage, because these injuries are very superficial and limited to the skin. Passengers falling off the backs of moving motorcycles typically have lacerations of the back of the head, fractures of the posterior fossa, contrecoup contusions of the frontal lobes of the brain, and abrasions of the back and elbows. If the person tumbles forward, there will be abrasions of the face. While motorcycle helmets reduce the incidence of head trauma in low-speed accidents (Johnson et al. 1995, Evans & Frick, 1988), at moderate and high speeds their sole function is to prevent brain matter from being spread over the highway.

Components of the bicycle itself can also injure the riders. Falling on the handlebars of the bicycle produces hepatic or pancreaticoduodenal injuries. Leg and foot injuries can be caused when the foot gets caught in the moving wheel of the bicycle. Fractures of the lower extremities are also common in motorcyclists. The spokes can also inflict injuries upon the rider. Genital and rectal injuries might be sustained in crashes involving the saddle and seat-post.

If a motor vehicle hits the bicyclist, the cyclist's head may sustain multiple impact injuries from hitting both the motor vehicle and the ground. Focal brain injuries as extradural haematomas are caused by direct blows to the head, while diffuse axonal injuries are produced by rotational movement, especially in the coronal plane. Occasionally, a motorcycle rider, seeing a car stop abruptly in front of him and knowing he will not be able to stop in time, will drop his motorcycle on its side and skid toward the vehicle in an attempt to prevent impacting it. Falling from the machine, especially at speed, can cause rib fractures and visceral damage, especially rupture of the liver and spleen. An injury common with motorcycles is the 'tail-gating' accident, where a rider drives into the back of a truck so that the machine passes underneath, but the head of the motorcyclist impacts upon the tail-board. Decapitation may occur in the most extreme cases, but severe head and neck injuries are almost inevitable.

Severe injury or even amputation may occur when the motorcycle operator does not see a cable or wire. Examination of the amputated heads and extremities shows the edges of the wounds to be sharp, almost as if they had been produced with a knife (Saukko & Knight, 2004, DiMaio & DiMaio, 2001).

2.4. Pedestrians

When a pedestrian is struck by a motor vehicle, the pattern and severity of the injuries, as well as their etiologic mechanisms, depend, for the most part, on four factors (Harruff et al., 1998, Atkins et al., 1988, Ashton, 1982)

1. The speed of the vehicle
2. The characteristics of the vehicle
3. Whether the vehicle was braking
4. Whether the victim was a child or an adult

Three impact phases are described during pedestrian injury:

1. Vehicular bumper impact: in the upright adult, initial impact is usually on the lower limbs (primary injuries)
2. Vehicular windscreen impact: torso and head injuries occur as the pedestrian impacts the body of the vehicle (primary injuries)
3. Ground impact: head, spinal and other collision injuries occur as the displaced pedestrian hits the ground (secondary injuries).

Primary injuries are caused by the first impact of the vehicle on the victim, additional contact with the vehicle, as when the pedestrian is hurled up against the windscreen. The height of the car bumper bar is well below the centre of gravity of the adult pedestrian, which lies in the abdominal region. Thus the first impact tends to knock the legs from under the victim and rotate them towards the oncoming vehicle. The tibia is often fractured in a wedge-shaped manner; the base of the wedge indicates the direction of the impact (often from behind), the front of the wedge pointing away from the side of contact. If the leg is weight-bearing at the time of the impact, the tibial fracture tends to be oblique, whereas if not stressed, as when being lifted during walking, the fracture line is often transverse. Not uncommonly, bumper fractures are at different levels on the two legs. This suggests that the individual was either walking or running at the time of impact, with the higher-placed injury indicating the leg that was in contact with the ground and supporting the body weight. If the individual was oriented sideways to the impacting vehicle, the "bumper fractures" might be confined to one leg. Sometimes the level of injury appears too low for the normal bumper height of most cars, but this may indicate that the vehicle was braking violently at the moment of impact, going down on its suspension as the front wheels decelerated or locked, unless dip compensators were fitted. In some instances, there are no fractures, just abrasions of the skin and hemorrhage into the calves. In other cases, there might be no visible injuries and it is not until an incision is made in the calf that one sees internal hemorrhage. Often, the body pivots on impact such that the buttocks and upper thigh region strike the front of the bonnet. The tangential force directed by the bonnet to the buttock and thigh may cause stripping of the skin and subcutaneous tissue from the muscles, creating a pocket in the upper thigh–buttock region. There can be extensive bleeding into these pockets, with collection of 1–2 L of blood. These pockets are often not visible externally. Depending on the profile of the front of the car, the struck pedestrian is either thrown forwards in the direction of travel if the bonnet-front is high and blunt — or scooped up onto the bonnet top, as with many slope-fronted modern vehicles. When a pedestrian is struck by a larger vehicle, such as a van, truck or bus, the initial point of impact is higher and may cause primary damage to pelvis, abdomen, shoulder-girdle, arm or head. Because of the profile of these vehicles, there is no scooping-up effect, and the victim is usually projected forwards to suffer secondary damage from road contact and sometimes to be run over (Matsui, 2005, Tanno et al., 2000, Ishikawa et al., 1994).

If scooped up, the victim will land on either the bonnet or against the windscreen or corner-supporting pillar (the 'A' frame). The flat bonnet usually does relatively little damage, though linear abrasions, brush grazes, or friction burns may be seen. Violent contact with the windscreen, especially the rim or side pillars, is the most frequent cause of severe head injury from primary impact. Parts of the vehicle may leave patterned imprints on the skin, such as headlamps, mirrors or other components. Many impacts are on the front corner of the car and the pedestrian may then be knocked diagonally out of the path of the car.

Scooping-up can occur at speeds as low as 23 km/hour (below 19 km/hour the body will usually be projected forwards). If the speed is high, the victim can be thrown up onto the car roof, sometimes somersaulting so that the head strikes the roof. He can then slide or be flung right over the back of the car, landing behind it in the roadway. This is more likely to happen if the car does not brake. In most cases, the scooped pedestrian falls or is flung off on one side of the car or the other, again to suffer secondary injuries in the road and perhaps be run over by another vehicle.

As the individual is picked up and thrown by the vehicle, he can literally be stripped of his clothing. Horizontal linear marks on the ankles may represent the edges of shoes that were torn off the feet (Saukko & Knight, 2004).

Secondary injuries are caused by subsequent contact with the ground. The usual pattern of events is that, at the instant of contact - or even slightly before — the driver will apply the brakes violently. The scooped-up victim will acquire the speed of the car by the time he lands on the bonnet, but then the vehicle decelerates. As the adhesion to the shiny surface is small, the newly acquired velocity of the body will cause it to slide off the front of the car as the latter brakes. Often the injuries are concentrated on one side, usually on the opposite side to the point of primary impact, because the body was thrown down onto the road. Sometimes he may be dragged by the under-belly of the car, and seriously soiled and injured, perhaps appearing at the rear if the vehicle does not stop quickly. If thrown into the centre of the roadway, the person can be run down by a different vehicle overtaking in another lane or by one coming in the opposite direction on a single carriageway. The secondary injuries may fracture the skull, ribs, pelvis, arm or thigh (Eid & Abu-Zidan, 2007, DiMaio & DiMaio, 2001).

If an individual is run over by a wheel, there are often tire tread marks on one surface of the body with scrape-like abrasions on the opposite side, i.e., the pavement side. The abrasions are caused by the body's scraping along the ground as the spinning tire pushes it backward ("flaying" injury). Tire tread marks are not invariably present, but if they are, may be on the clothing as well as, or instead of, the skin. If the wheel passes over a limb, the spinning movement of the tire may avulse skin and subcutaneous tissue from the fascia and muscle. When a wheel passes over the abdomen or pelvis, multiple parallel striae or shallow lacerations may occur near the contact area because of ripping tension in the skin. Great internal damage may occur even with little surface injury. The weight of a large vehicle can virtually flatten a head, crushing the cranial vault. Often the brain is extruded through scalp lacerations, as may be the intestine through an abdominal wound. The pelvis may flatten out when run over, the symphysis or superior rami breaking, and one or both sacroiliac joincs becoming detached. Any type of intraabdominal injury may occur from ruptured liver and spleen to perforated intestine, lacerated mesentery and fractured lumbar spine. In the chest, ribs, sternum and thoracic spine may fracture and heart and lung damage occur from crushing or laceration from jagged ribs. A "flail" chest is sometimes produced when a heavy wheel runs across the supine body, breaking all the ribs on each side in the anterior axillary line (Bloomer et al., 2004, DiMaio & DiMaio, 2001).

Occasionally, an individual may be found on a road or a parking lot with crushing injuries of the body, tire tread marks on one surface, brush abrasions on the opposite side and no evidence of an impact with the front of a motor vehicle. Toxicology virtually always shows acute alcohol intoxication. These are individuals who have gone to sleep or passed out on a road or in a parking lot, only to be subsequently run over by a vehicle whose driver did not see them. Actually, if a pedestrian is run over by a car, the question can always arise whether there was a preceding collision while the pedestrian was in an erect position or not. Karger et al. (2001) studied a total of 53 selected autopsy reports and concluded that wedge-shaped bone fractures, glass fragment injuries, traumatic amputations, traces of car paint on the lower extremities and abrasions of the shoe soles are specific, whereas sacroiliac dislocations and fractures of the thoracic spine are very indicative of primary impact in an erect position.

Trace evidence: Paint fragments and glass shards are trace evidence that must be carefully retained, as the forensic laboratory may be able to identify the make and model of vehicle involved, and match the fragments when a suspect car is examined. In hit-and-run deaths, photographs of the tread marks with a ruler in the field should be taken for subsequent comparison with a tire. In passing over a victim, grease or dirt from the undersurface of the car can be deposited on the body or clothing. Head hair, samples of grease from the body, and the clothing should be retained. Patterned injuries may be important, in that they can assist the police in identifying a vehicle in a 'hit-and-run' accident.

When a suspect vehicle is found, the undersurface should be examined for blood, hair, and clothing fibers. Any material recovered can then be compared with material removed from the body. The automobile will then present the classic picture of a pedestrian impact: cuffing or dents on the bumper, indentation of the front and top of the hood, and an impact site on the windshield. The weave pattern of clothing may be imprinted on the hood. In some instances, threads can have been caught by the deformed hood, which could be linked to clothing (Kuppuswamy & Ponnuswamy, 2000).

Impact velocity and injuries: In general, the severity of the injuries - both primary and secondary - will be the more severe the higher the speed. However, it is impossible to estimate the speed of impact from the nature of these injuries. They can be fatal even at slow speeds of the order of 10 km/h, yet occasionally high-speed impacts can produce only minor damage. If the car speed is appreciable (anything over 20 km/h is sufficient), the body can be thrown into the air or knocked down flat with a severe impact. In a high-speed impact, which may be anything over 50 km/hour, the body can be flung high in the air and for a considerable distance, either to the side or in the path of the car — or even backwards over the roof. This does not mean, however, that severe, even fatal, injuries cannot occur at lower speeds.

Karger (2000) et al. studied 47 pedestrian–passenger car fatalities in regard to impact velocity and injury. Primary and secondary injuries did not show a relationship to impact velocity. They found that four types of injury appeared to be correlated with impact velocity which is *fracture of the spine, rupture of the thoracic aorta, inguinal skin rupture and dismemberment.* A cautious interpretation of the data can be summarised in the following conclusions: If there is no spinal fracture, the velocity was below 70 km/h and probably below 50 km/h. Aortic and inguinal skin ruptures are always present if the velocity was above 100 km/h but never occurred below 50-60 km/h. If dismemberment occurs, the velocity was above 90 km/h. In regard to this last observation, Zivot and DiMaio (1993) reviewed 85 fatal motor vehicle–pedestrian deaths, and, in five cases, found amputation of a limb, and in two, transection of the torso. In all seven instances, the vehicles were going a minimum of 88.5 Km/h.

Concerning lower impact velocities, it has been supported that the main injury at an impact velocity of around 20-30km/h is to the knee ligament, whereas on the other hand, the main injury at an impact velocity of around 40km/h is a fracture of the lower extremities (Matsui, 2005).

If an automobile traveling at either high or moderate speed brakes hard prior to impact with an adult, there are two possibilities, in both of which the pedestrian is struck below the center of mass by a rapidly decelerating vehicle. In the first possibility, the pedestrian is thrown forward. In the second, the individual is struck by the vehicle, picked up, lands on the hood, and is then propelled forward, again coming to rest in front of the rapidly decelerating vehicle. In the latter case, there will be damage to the front and top of the hood (DiMaio & DiMaio, 2001).

Children are more prone to be run over by reversing vehicles, especially trucks, as they often play between parked vehicles and - being small — are less visible to the driver. In *child* victims, although the general pattern of injuries is similar, their shorter height and smaller weight affects the mechanics of impact. In children struck by non-braking or late-braking motor vehicles, impact with the front of the vehicle is above the body's center of gravity. The victim is impacted, slammed down, and run over. If the vehicle is braking hard prior to the impact, the front of the vehicle dips below the child's center of gravity and the child usually is thrown forward, though many do become scooped *up*. The primary contact is higher up their body, so that the femur may be fractured by the low bumper bar. Serious head injuries can also occur if the bumper hits their head (Brison et al., 1988).

3. Type and Location of Injury

3.1. Head Injury

Head injury is a major cause of mortality and serious morbidity in survivors; disability may occur whatever the initial severity of the head injury and surviving patients with brain injury are more impaired than patients with injuries to other parts of the body.

In moderate-speed motor-vehicle collisions (30 mph and above), the forward movement of the head is such that it might impact the steering wheel. The unrestrained driver rises and flexes forward so that the head crashes to the windshield, sun visor region above the windshield, or the frame (generally in this order). If the head of the driver or front seat passenger impacts the windshield, there will be abrasions and superficial cuts of the forehead, nose, and face, with the injuries having a vertical orientation. Thin slivers of windshield glass might be embedded in the wounds or be found loose on the clothing. The face frequently suffers multiple cuts from contact with the shattered safety glass. Damage to the eyes is common. In most present day vehicles, the glass is of the toughened, not laminate, variety and, when broken, it shatters into small cubes with relatively blunt edges. These still cause superficial lacerations, often in short V-shaped or sparrow-foot patterns. In themselves they are not a danger to life, but indicate an impact sufficient to hurl the driver on or through the glass. (Saukko & Knight, 2004). Blunt force impact on the windshield, even if not causing serious incised wounds, can however produce fairly severe soft tissue injuries. There can be partial avulsion of the skin with the avulsed skin anchored superiorly. These wounds, because of their location, often bleed very heavily, appearing very dramatic and life threatening (Turnage & Maull, 2000).

The laceration pattern may be indicative of the type of the accident. Two types of glass are used in automobiles: laminated, used in windshields; and tempered, used in the side and rear windows. Each of these types of glass will produce a pattern laceration unique to it. The windshield is composed of two layers of glass, laminated together, with a thin layer of clear plastic sandwiched between. This laminated glass will break into shards upon impact; wounds resulting from impact with laminated glass will be linear and incised. The tempered or 'safety glass' is a single layer of glass that breaks into small cubes when fractured. Wounds resulting from impact with tempered glass will appear 'diced', punctuate and rectangular in configuration.

In higher speed collisions, the impact against the windscreen rim or corner pillar, or after ejection, can cause any type or degree of head injury, including scalp laceration, fractured skull, intracranial hemorrhage or brain damage. The most common are fractures of the vault and basis of the skull (Menon et al. 2008). Basilar fractures tend to run along the length of the petrous ridges passing through the sella turcica ("hinge fractures"). Less common are ring fractures, which they are usually resulted from direct impact on the occiput, although anteroflexion has been described as a rare causal mechanism (Maeda et al., 1993).

Brain damage may also be an important cause of direct or delayed death and it includes two main categories: focal brain injuries and diffuse brain injuries. *Focal brain injuries* are usually caused by direct blows to the head, and comprise contusions, brain lacerations, and hemorrhage leading to the formation of haematoma in the extradural, subarachnoid, subdural, or intracerebral compartments within the head. *Diffuse brain injuries*, are usually caused by a sudden movement of the head, and cause diffuse axonal injury. Because secondary brain injury may also occur, as a result of hypotension and hypoxia, immediate treatment of a head-injured patient with early control of the airway, adequate ventilation and oxygenation and correction of hypovolaemia is fundamental (Adams, 1992).

Finally, *incomplete decapitation* has been described in the case of a motor-cyclist whose head was stationary at the moment of impact and the remainder of the body continued in a backward motion (Hitosugi et al., 2001).

3.2. Spinal Injury

Examination of the cervical spinal cord is a rare additional autopsy technique applied in forensic autopsies. A careful dissection of the neck, both anteriorly and posteriorly, in the upper cervical region must be performed. Death may occur by posterior fracture or dislocation of the upper cervical region, even in the absence of hemorrhage anteriorly or in the subarachnoid space of the brain stem. Injuries of the neck region are fatal in a considerable number of cases. However, such technique is neglected especially when a different cause of death is found. Woźniak & Rzepecka-Woźniak (2003), in their research on 316 autopsy cases, found about 40% of gross anatomy and about 46% of microscopic (blood suffusions) changes giving evidence of trauma of that region in groups of victims of traffic accidents. The authors concluded that techniques of examination of the cervical spinal cord and spinal column with the addition of microscopy in selected cases can prove the cause and mechanism of death.

The most common mechanism of spinal injuries is hyperflexion of deceleration followed by a rebound hyperextension when the head strikes an obstruction in front. Rear impacts are also cause the double 'whiplash" effect. Cervical spinal injuries are caused by the neck hyperextension and involve ligament disruption, dislocation and/or fracture of the cervical spine, and spinal cord injury (Fife & Kraus, 1986). Complete transection or crushing of the cord may also occur (Gautschi et al., 2007), whereas in rare instances, the cord may be violently pulled down, with partial or complete avulsion of the brain stem, ventrally, at the ponto-medullary junction (Kondo et al., 1995). The upper cervical spine may appear as the most vulnerable part, as atlanto-occipital dislocation or separation (Anderson et al., 2006), atlanto-axial dislocation (Carroll et al., 2001, Adams et al., 1992) have been reported.

In case of primary impact of a pedestrian, the resulting overextension/flexion, rotation and/or translation of the vertebral column can represent a forceful indirect injury mechanism sufficient to produce multiple fractures (Karger et al., 2000, Otte et al., 1990). In two cases reported in the literature, where the collision speed was approximately 140 km/h, this mechanism even caused severance of the trunk (Karger et al, 2000, Beier et al., 1974). Vertebral fractures occur frequently in frontal high speed collisions in an erect position (Karger et al., 2000). When the legs are knocked away violently by the front of the car, a rapid rotational acceleration is applied to the long axis of the body during the scooping-up motion. Fractures of the cervical and lumbar spine were found to be very indicative for impacts in an erect position (Karger et al., 2001).

On the other hand, fractures of the thoracic spine seem to be indicative of run over incidents, in which the spine is fixed and crushed between the road surface and the underside of the car (Karger et al., 2001). Of course, the above injuries are non-specific and conclusions about the circumstances of the accident may be derived only in combination with other findings (such extremities injuries, tire marks, sacroiliac dislocations).

3.3. Thoracic Injuries

Swan et al. (2001) concluded that thoracic injuries alone, with the exception of sternal fracture, were sufficiently life threatening to cause death. The chest of the driver can impact the steering wheel; the chest of the passenger, the dashboard. Evidence of injury from such an impact varies from imprinted abrasions/contusions of the wheel or instrument panel to complete absence of any evidence of external injuries. Rib fractures are common, though fatal visceral injuries can occur without rib fractures in young people because their ribs are more pliable. Transverse fracture of the sternum may also occur (usually at the third intercostal space). Serious chest injuries from impaction against the steering wheel became less frequent with the introduction of energy-absorbing compressible steering wheel, airbags and seatbelts (DiMaio & DiMaio, 2001).

Blunt tracheobronchial injuries constitute only a small fraction of admissions to trauma centres, as many patients die before they reach hospital. The presentation of thoracic tracheobronchial injury depends on whether the injury is confined to the mediastinum, or communicates with the pleural spaces. Injuries communicating with the pleural space usually present with subcutaneous emphysema and pneumothorax. If the injury is confined to the mediastinum then pneumomediastinum is usually present (Harrahill, 2002, Horinouchi et al., 1993).

The *lungs* often show areas of bleeding under the pleura, which may be from direct contusion, aspiration of blood from other damaged areas of lung or blood sucked down the air passages from injuries in the nose or mouth. Internal lacerations and rupture of the pulmonary parenchyma may occur, due to fractured ribs, direct impact or acceleration-decelleration forces. Blood blisters may also appear under the pleura overlying the bruised areas resulting in haemothorax. The interior of the lung may be pulped even in the presence of intact visceral pleura, from transmitted force or massive variations in intrathoracic pressure during the impact (Saukko & Knight, 2004).

One of the most common fatal thoracic injuries is *transsection of the aorta*. Typically, this occurs immediately distal to the origin of the left subclavian artery. It may be associated with a severe whiplash effect on the thoracic spine, as the aorta is tethered to the anterior surface of the vertebrae where the distal arch joins the straight descending segment. Probably the most common reason for aortic rupture, however, is the 'pendulum' effect of the heart within the relatively pliable thoracic contents. When the thorax is violently decelerated, the heavy cardiac mass attempts to keep moving ahead and may literally pull itself off its basal mountings, the most rigid part of which is the aorta. Separation takes place at the point where the aorta is attached to the spine at the termination of the arch.

The appearance of the aortic rupture is often of a clean-cut circular break, almost as sharp as if it had been transected with a scalpel. On opening the aorta, faint, horizontally oriented, linear scars on the intimal surface of the aorta distal to the left subclavian artery can be seen. These represent incomplete lacerations of the intima that subsequently healed. These may be present when no actual rupture has occurred and may be found as an incidental finding at autopsy. Sometimes they are deep enough to allow a local dissection of blood to seep into the intima, when death has not been virtually instantaneous. The frequency of such tears is common enough for a warning always to be offered to the autopsy prosector not to use undue force on the neck and thoracic structures when removing the organ pluck from the body. Rough handling during this stage can produce artefactual ladder tears in the aorta (Nikolic et al., 2006, Shkrum et al., 1999).

Injury to the *heart* is less common than aortic injuries. The heart may be damaged even in the absence of external marks or thoracic cage fractures. The most common injuries are myocardial contusion, laceration of the pericardial sac, rupture of the right atrium, rupture of the right ventricle anteriorly at the interventricular septum, rupture of the left atrium, and laceration of the interatrial septum. Penetrating injuries from sternum, ribs or external objects may lacerate the heart directly. The posterior surface may be damaged from impact against the spine. In high-speed impacts, the heart may be completely avulsed from its base and be found lying loose in the chest (Saukko & Knight, 2004).

Coronary artery thrombosis has been described following contusion over a coronary artery (Suhr et al., 2000, Unterberg et al., 1999). Rarely, traumatic dissecting aneurysms of the left anterior descending coronary artery may be caused by blunt trauma to the chest (Westaby et al., 1995). Myocardial necrosis, rupture of the ventricle, and pseudoaneurysm formation with subsequent rupture has also been reported (RuDusky, 2003).

Subendocardial haemorrhages on the left side of the interventricular septum and opposing papillary muscles are not a sign of impact, but an index of catastrophic hypotension and they are also seen in head injuries (Saukko & Knight, 2004).

Occasionally, there will be a motor vehicle accident in which the driver impacts the steering wheel and in which no anatomical cause of death presents after a complete autopsy and toxicological screen. There may be soft tissue trauma to the chest and a fractured sternum or ribs, but insufficient injuries to explain death. Such deaths are caused by fatal cardiac arrhythmia secondary to a cardiac contusion. Examination of the heart might fail to reveal any evidence of impact because of the suddenness of the death. Some individuals do not develop cardiac arrhythmias until hospitalized. In these individuals, the presence of a cardiac contusion is confirmed by enzymatic tests and EKGs. The patients usually recover, though occasionally, they will die from an arrhythmia. Death caused by cardiac function abnormalities without morphological or microscopically visible heart lesions after blunt impact to the chest is termed commotio cordis or cardiac concussion. Before death can be ascribed to cardiac contusion, positional or traumatic asphyxia must first be ruled out (Inoue et al., 2004, Michalodimitrakis & Tsatsakis, 1997).

3.4. Abdominal and Pelvic Injuries

The internal organs of the abdomen can also be injured by sudden acceleration, deceleration or strong compressive forces (Newgard et al., 2005). Liver and spleen are the most vulnerable organs. The injuries of the *liver* range from superficial capsular lacerations to bursting rupture. A common lesion is central tearing of the upper surface, which may extend deeply and even transect the organ. Less serious damage is often seen in the form of shallow, sometimes multiple, parallel tears on the upper surface of the right lobe. Subcapsular tears can occur with the formation of a subcapsular haematoma, which can rupture later. The *spleen* also shows shallow tears in some accidents, often around the hilum; in rare cases, it may be avulsed from the pedicle (Miltner et al., 1992). The *mesentery and omentum* may show bruising and, rarely, there is laceration and fenestration sufficient to cause a lethal haemorrhage (DiMaio & DiMaio, 2001).

Not uncommonly the pelvis is fractured, often at one or both sacroiliac joints. If the integrity of the passenger compartment is violated by the engine's being driven backward, the injuries produced can be the massive crushing injuries previously noted, plus injuries of the pelvis. Motorcyclists, also, often sustain pelvic injuries. Fractures of the pelvis are increasingly recognized as a marker of severe injury, as the force required to disrupt the pelvic ring is substantial (American College of Surgeons Committee on Trauma, 1997). Severe bleeding leading to hypovolaemic shock is often a feature of severe pelvic fractures. Unstable pelvic fractures can bleed torrentially, and uncertainty about coexistent intra-abdominal injury may cause significant problems in identifying the source of blood loss. Associated injury of the *perineum, rectum* (Jehle, 2001), *urethra* (Kommu et al., 2007), *vagina* (Okur et al., 1996), and *testes* (Wu et al., 2004) have been reported and their examination is especially important in order not to miss them.

3.5. Extremity Injuries

Impact against the fascia can cause abrasions, lacerations and fractures of the legs around knee or upper shin level. If the knees impact the dashboard, there may be fractures of the patella or the distal femur. There also can be dislocation at the hip joint or a fracture of the femur at its neck. Pressure of feet on the floor, especially when it is intruded by the engine, can cause fractures from foot to femur. Upper limb injuries are less common but may occur from transmitted force through gripping the steering wheel or from impact against the windscreen, pillars, intrusive roof, bonnet or ground when held up in a reflect protective position. In head-on crashes, the floorboards can be driven upward and inward, twisting the foot on the ankle and causing a fracture. In other instances, if the seat goes forward, the foot can be trapped beneath the seat, breaking the ankle (Huelke, 1970).

3.6. Psychological Impact of Motor Vehicle Accident

In trauma units around the world there is an increasing realization that trauma can have marked and sustained psychological effects. Up to 25% of severely injured patients experience significant early psychological reactions after trauma (Schnyder et al., 2000). In some patients, these reactions can be long lasting and have profoundly adverse effects on quality of life (Wang et al., 2005). Civil litigations on the basis of a psychological harm may arise from survivors of traffic accidents and forensic experts may be asked to give their expert opinion. Instruments for the detection and validation of such disturbances have been developed (Arce, 2006).

4. Natural Diseases

Natural deaths of drivers behind the wheel do happen (Büttner et al., 1999, Lau, 1996, Halinen & Jaussi, 1994). Drivers may have warning symptoms that allow them to take appropriate actions to prevent serious injuries to themselves and others (Kerwin 1984a, Kerwin 1984b, Ostrom & Eriksson, 1987). Some drivers lose consciousness before the vehicle stops (Ostrom & Eriksson, 1987), however, the trauma in this situation is insufficient to account for death.

Forensic pathologist should document any natural any natural disease, especially if it might have contributed to the accident, by affecting either the driving ability of the driver, or the behavior of the pedestrian in the roadway. Old and recent cardiac and cerebral lesions (such as myocardial infarcts, rupture of an aneurysm and intracerebral hemorrhage) are particular important, as cardiovascular disease is the predominant natural disease, implicated in traffic accidents (Büttner et al.1999). Epilepsy also accounts for a few number of traffic accidents, and any evidence of a fit, such as a bitten tongue, should be recorded. Old meningeal adhesions over cortical damage, mesial temporal sclerosis and cerebral atrophy as consequences of epilepsy and its treatment, and above all, a detailed medical history, are helpful in the postmortem diagnosis of epilepsy. Deterioration of visual and hearing acuity may have also influenced the ability of a driver or pedestrian, however, their assessment is almost impossible during autopsy, unless a gross abnormality is present (Saukko & Knight, 2004).

More rare natural diseases implicated in traffic accidents have also been reported. In an autopsy examination performed on a 75-year-old woman who had been in a minor road traffic accident an occipital lobar hematoma with subdural extension was revealed. Histological examination of brain tissue showed the presence of severe cerebral amyloid angiopathy (Dada & Rutherfoord, 1993).

It should be mentioned, however, that accidents caused by natural disease are rare and do not represent a serious danger to the public. Sudden natural deaths at the wheel generally result in the death of only the driver, with the cause of death usually the natural disease. The driver is often able to stop the vehicle and usually only minor injuries result (Büttner et al., 1999).

5. Complications after Traffic Injury/ Delayed Death

Although many deaths from accidents occur shortly after the incident, some deaths are delayed and they are due to various complications. The manner of death (pathological or violent) is the most important fact for the court and therefore, the most important part of the finding of autopsy reports. To recognize the manner of death in cases with long outliving period after injury could be difficult for forensic pathologists. In such cases, the forensic pathologist should be able to point out the direct relationship between initial injury and death by using his own experience and medical knowledge. Most specifically, the forensic pathologist must answer the following questions: whether the death was due to trauma; whether the precipitated cause of death was the consequence or complication of injury; what were the mechanism and mode of dying; whether the death was preventable; if there were possible malpractice and negligence, etc.

Nowadays, there are a few syndromes which could be the cause of death in relation with initial injuries. The most common are fat embolism syndrome (Nikolić & Micić, 2004), multiple organ failure (Lausevic et al., 2008), systemic inflammatory response syndrome (Previdi et al., 1996) and ventilator-associated pneumonia (Magnotti et al., 2004). The diagnosis of these syndromes is possible only clinically: the autopsy and histological findings are not specific. As long as a direct chain of events can be traced from the injury to the death, then the initial injury must be considered to be the basic cause of death, and this fact may have profound legal implications for both civil compensation and criminal responsibility. Some of the most difficult problems in forensic pathology concern deaths from which posttraumatic complications are disputed as being fatal causative factors. By autopsy, only the morphological consequences of these processes can be noted. The dynamics of dying, direct correlation between initial injury and death, as well as appearance and development of complications provoked by trauma could be established only by clinical medical data. Therefore, *medical clinical data* are crucial for forensic pathologists and for solving the problems about the mode and manner of death in cases with long outliving period. (Milić et al., 2002).

The regular use of trauma scores in forensic medicine may provide a standardized database of autopsy findings, which would be a tremendous contribution in the evaluation of the quality of trauma treatment and in the identification of preventable death (Sharma, 2005, Wyatt et al., 1999, Friedman et al., 1996).

The Abbreviated Injury Scale (AIS) was established in 1971 by the American Medical Association, the American Association for Automotive Medicine, and the Society of Automotive Engineers (Committee on Medical Aspects of Automotive Safety, 1971). It was revised five times, most recently in 1990 (American Association for Automotive Medicine, 1990). The early versions attempted to provide uniform grading for persons injured in motor vehicle accidents (MVAs). Later versions were refined to include penetrating injuries and blunt trauma in circumstances other then MVAs. In its present form, the AIS codes injuries are based on their anatomic site, nature, and severity. All injuries are assigned a six-digit score, in which the sixth digit represents the AIS severity. The minimal severity is 1 and the highest is 6 (1 Minor, 2 Moderate, 3 Serious, 4 Severe, 5 Critical and 6 Unsurvivable).

The Injury Severity Score (ISS) was developed from the AIS and was first published in 1974 (Baker et al., 1974). The system provides a summary severity score based on AIS coding. Each injury is assigned an AIS score and is allocated to one of six body regions (Head, Face, Chest, Abdomen, Extremities and Pelvis). The AIS-coded injuries are divided into six body regions. Only the highest AIS score in each body region is used. The three most severely injured body regions have their score squared and added together to produce the ISS score. Any injury coded as severity 6 (which is considered incompatible with life, such as a penetrating brain stem injury) automatically gives the ISS the maximum score of 75. This score can also result from three maximal severity scores that would each be compatible with life alone. The ISS score is virtually the only anatomical scoring system in use and correlates linearly with mortality, morbidity, hospital stay and other measures of severity.

Another, more recent approach to anatomic injury scoring is based on the International Classification of Disease, Ninth Edition (ICD-9) codes. This method is termed ICD-9 Injury Severity Score (ICISS) and uses survival risk ratios (SRRs) calculated for each ICD-9 discharge diagnosis. SRRs are derived by dividing the number of survivors in each ICD-9 code by the total number of patients with the same ICD-9 code. ICISS is calculated as the simple product of the SRRs for each of the patient's injuries (Riddick et al., 1998).

6. Injuries from Safety Systems

6.1. Seat-belts

There are three forms of automobile belt-type restraints: lap belts; shoulder (diagonal) belts and three-point belts (lap plus shoulder belt). Lap belts were the first form of restraint offered, becoming standard in automobiles in 1964. They are still found in older vehicles and in the back seats of some newer vehicles. All new vehicles use the three-point belt, moreover, most car belts are now of the 'inertia-reel' type, which allow slow movement but jam at a sudden tug. The advantage, apart from the comfort, is that they automatically tighten up around the body, as a slack belt is not only less effective but can actually constitute a danger.

The various forms of strap restraints act as follow (Saukko & Knight, 2004):

1. In frontal collisions, they prevent impact of the head of the driver or passenger with the windshield frame, the steering wheel and dashboard. The head, though still subject to hyperflexion, is prevented from smashing through the glass and the body cannot be projected through the screen onto the bonnet or roadway. The belt cannot cope with backward intrusion of the engine, floor, roof or corner pillar if those structures reach the occupant sitting in the original seat position. A belt is relatively ineffective in a side impact, except in that it reduces injuries from ejection (Kumaresan et al., 2006).

2. They extend the deceleration time and distance by substantial stretching of the belt fabric, which may lengthen by many centimeters during a violent arrest. To be effective the belt must be held tightly against the body to get the maximum restraint, either by adjusting the buckle or using an inertia reel.
3. They spread the area of application of deceleration forces.
4. Lap belts also prevent rear passengers from impacting or being propelled over the front seats.

Injuries from seatbelts can vary from the trivial to the fatal. A distinctive pattern of injury associated with lap belt restraints was first noted by Garrett and Braunstein in 1962 and termed the "seat belt syndrome." The constellation of injuries includes injuries to the abdominal wall, intestinal viscera and its mesentery, along with concomitant injuries to the lumbar spine. The mechanism of such an injury complex is the rapid deceleration caused by the lap belt resulting in compression of the lower abdomen and hyperflexion of the lumbar spine (Durbin et al., 2001). This hyperflexion focused anterior to the lumbar spine causes distraction of the posterior elements of the spine resulting in the classic Chance fracture, a horizontally oriented fracture through the spinous process, laminae, and vertebral body (Reid et al., 1990). At the same time, the flexion causes compression of the abdomen which may result in a multitude of abdominal visceral injuries. The bowel becomes trapped between the lap belt and the spine (Letts et al., 1999).

In the external examination, shoulder belt use may be reflected by a linear abrasion running downward and medially on the left side of the neck of the driver or the right side of the neck of the front passenger. A poorly defined area of abrasion and contusion indicating the distribution of the belt might be seen on the skin of the lower abdomen.

Soft tissue injuries produced by lap belts consist of contusions and lacerations of the duodenum, jejunum and ileum and lacerations of the spleen and pancreas. In intestinal injuries, the lacerations are on the anti-mesenteric side of the bowel. The abdominal aorta can also be crushed very rarely. While all of the aforementioned injuries occur from wearing the lap belt too high, such injury can still take place if it is worn properly through a phenomenon called "submarining," where, at impact, the pelvis sinks down into the seat and slides under the belt (Richards et al., 2006). The full bladder can be ruptured as can the caecum.

Compression fractures of the lumbar vertebrae; transverse fractures of the vertebral bodies; as well as fractures of the pedicles, transverse processes and lamina of the lumbar vertebrae may occur. Three-point restraint may also produce injuries such as rib fractures (single more likely than multiple), fractures of the clavicles, and sternum and cervical spine fractures.

Fractures of the sternum, rib cage and clavicles, with separation of the two halves of the rib cages and underlying trauma to the hearts and thoracic aorta have also been reported (Byard, 2002), although they are not frequently observed.

Children are more vulnerable to seat-belt injuries, because of factors such as higher center of gravity, smaller anteriorposterior dimension of the abdomen, and ill-fitting and improper use of seat belts. Unusual injuries, including abdominal wall disruption with herniation of bowel, blunt abdominal aortic injury, lumbar fracture with even complete avulsion of the spinal cord and cauda equine have been reported (Tubbs et al., 2006).

Some women find that the diagonal strap compresses the breast even in normal use, so that the greatly increased tension during deceleration is likely to injure the gland and breast trauma occasionally may occur (Majeski, 2007). Pregnant women also have problems with belts but, although uterine and fetal injuries have been recorded in accidents (Bunai et al., 2000), the incidence is relatively low and without the belt the consequences would have been as bad or probably worse (Pearlman et al., 2000, Pepperell et al., 1977).

6.2. Air Bags

In recent years, the use of airbags has spread from only the most expensive vehicles to many standard production cars. The air bag is composed of inflatable nylon fabric, and is activated when a sensor, generally located in the automobile bumper, registers a longitudinal acceleration of 20 km/h or more. The sensor then ignites the solid propellant sodium azide, the principal explosive device used to deploy air bags, and the ignition liberates nitrogen gas. This reaction yields byproducts including heat, sodium hydroxide, sodium carbonate and other metallic oxides, creating a highly corrosive alkaline aerosol. Nitrogen gas inflates the air bag to a volume of 50–60 l within <0.05 s in speeds of inflation exceeding 44 m/s. The air bag deployment can produce a concentrated force as high as 20 kN on the sternal area, which exceeds the established human tolerance of 3.3 kN against sternal impact (Sato et al., 2002).

Just like seat belts, air bags can cause injuries, but, unlike seat belts, the injuries can be immediately lethal. Deaths are usually associated with women of small stature and children below the age of 13 years, especially when the children are unrestrained or out of position. Rear-facing infant or child restraints should never be used in front seats, as they place the child's head and body very close to the air bag housing. Infants in the front seat, in rear-facing infant seats, have predominantly craniocerebral injuries. Drivers seated too close to the steering wheel (less than 10 inches) can be seriously injured or killed by deploying air bags. Short drivers are injured more frequently because they must sit closer to the steering wheel to reach the gas and break pedals (DiMaio & DiMaio, 2001).

Injuries may result from impact with the air bag, as well from impact with its module cover, and they can be inflected at any stage of the deployment process, depending on the posture and possible objects between the occupant and the deploying airbag or the module cover. In addition to above injuries, concern has been raised regarding the air bag's potential to either chemically or thermally burn an occupant. The rapid deployment of the airbag releases large amounts of heat energy, which can lead to thermal burns either directly or from melting clothes. The hot gases have been responsible for burns to the hand, particularly about the base of the thumb or the dorsum of the wrist or hand. Chemical materials released during airbag deployment may also cause alkali burns to the skin, upper chest and eyes.

Other *ocular* injuries include corneal abrasions from contact with the air bag, hyphema, vitreous hemorrhage, eye lens dislocation, retinal contusion or detachment and globe rupture from the blunt force trauma of the expanding bag. Orbital bone fractures can also be sustained from the deploying air bag (Kuhn et al., 1995).

Facial damage may also occur, and the most common injury associated with air bag deployment is that of *facial abrasion*. The abrasions result from a sliding contact between the bag and the face. The injuries are not chemical 'burns' but deep abrasions.

When acceleration forces are applied to the cranial vault, a variety of traumatic injuries to the *brain and surrounding structures* will result. These include sub-dural and subarachnoid hemorrhages, cortical contusions, atlanto-occi-pital dislocations, skull fractures and brainstem transections. Cranial injuries may result from contact with either the deploying of bag or module cover.

The blow from an air bag or module cover which produces a rapid and violent hyperextension of the cervical spine of the driver or passenger will have significant consequences for the *cervical spine*. Injuries commonly seen as a result of hyperextension include atlanto-occipital dislocation, fracture of cervical vertebrae, and rupture of the anterior and posterior longitudinal spinal ligaments, and cervical spine disarticulation with transection of the cervical cord. Skull ring fracture has also been reported. The majority of these injuries are associated with the upper cervical vertebrae, although lower cervical vertebrae injuries have been observed.

Multiple *ribs fractures*, usually bilateral and often with associated injuries to the underlying *thoracic and abdominal organs*, may also occur. Injuries, including lacerations, have been found mainly in the heart, spleen, liver, and aorta. Cardiac and pulmonary contusions and hemorrhages may be inflicted, even in the absence of rib fractures.

The u*pper extremities* are very vulnerable to traumatic injury from the deploying bag and its module cover. When an individual's hand or forearm is on or near the module cover at the moment of deployment, the occupant can expect to sustain multiple fractures, and/or tissue degloving or amputation of fingers, hand or forearm. The horn-button-within-the-module cover design significantly increases the risk of injury to the occupant's upper extremities at the moment of deployment. Many of these upper extremity injuries are associated with an occupant's attempt to blow the horn, the button of which is located within the module cover. Forces from air bag deployment may be transmitted to the hand, wrist or forearm and may even involve the humerus. It is not unusual to see significantly comminuted fractures involving the wrist, forearm, elbow and distal humerus (Sato et al., 2002).

Finally, the byproducts of combustion as well as other inert materials within the air bag may produce a white cloud within the vehicle. This whitish material is principally cornstarch, talc and the byproducts of sodium azide combustion. There may be a small percentage of unburned sodium azide present within this powder as well. Inhalation of these materials can result in a *chemical pneumonitis* and the induction of asthma-type symptoms (Hambrook & Fink, 2006). These byproducts may also cause a chemical irritation of open wounds and a basic burn to the eyes (Kuhn et al., 1995).

7. Motor Vehicle Fires

Fires secondary to a motor vehicle crash are uncommon. When they do occur, and involve fatalities or severe injuries, they almost invariably result in a major lawsuit against the automobile manufacturer. Because of this, in any motor vehicle accident where there is a fire and a burned body is removed from the vehicle, a complete and thorough autopsy should be performed. The fire ravaged body presents a formidable challenge to the pathologist who must conclude if the victim was dead or alive at the time of the fire and if he died as a result of *smoke inhalation and/or CO intoxication, thermal airway injury, or trauma.*

In such cases toxicological analysis for carbon monoxide analysis is fundamental as blood carboxyhemoglobin levels can be instrumental in the above differentiation. If no blood can be obtained, spleen, marrow, liver or muscle should be used in that order. Often the blood carboxyhemoglobin (COHb) level is sufficiently elevated above 50% to indicate that the person was clearly alive during the fire. In these cases, the carbon monoxide concentration is in and of itself sufficient to cause death. A carboxyhemoglobin level of > 30% also strongly suggests inhalation of combustion products as the cause of death (Saukko & Knight, 2004, DiMaio & DiMaio, 2001).

In contrast, a level of < 20% should prompt a search for other causes, as the absence or very low levels of COHb in the blood suggests that the subject was not breathing at the time of the fire. However, caution must be observed when interpreting similar cases with low COHb. Hirsch et al. (1977) described the phenomenon in which victims of an automobile fire might not have an elevated COHb if there was a flash fire or explosion. In an enclosed space, such as an automobile, a fire rapidly consumes O_2 and produces a variety of toxic combustion products, but most of them are not routinely measured (Stefanidou & Athanaselis, 2004, Alarie, 2002). Superheated air is often available for inhalation, and death can occur by asphyxia due to airway obstruction from edema, mucus, debris, and/or bronchospasm. Anatomic evidence of these mechanisms such as a burned larynx or a damaged pulmonary and mucosal lining can often be demonstrated at autopsy (Saukko & Knight, 2004, Eckert, 1981).

Meticulus description of injuries (including the above mentioned indicators of smoke inhalation) is most valuable in arriving at a cause of death. Careful examination of the neck, both anteriorly and posteriorly should be performed, especially, if there are no evident traumatic injuries to explain death. One must keep in mind that the severity of injury can be clouded by the effects of fire. The anatomic evidence of acute axonal injury, for example, can be subtle at best. In addition, due to extensive charring and incineration in most cases, the significance of thermal body injury is usually impossible to ascertain (Wirthwein & Pless, 1996).

Except from the cause of death, identification of the body may be another question in cases of motor vehicle fires. Postmortem procedures include a general external examination, routine photographs, dental examination, dental (intraoral and extraoral) and general radiographs (chest, ankle, etc.), and complementary biological methods for identification (DNA analysis).

8. Suicide or Homicide with Motor Vehicles

Although most cases of fatalities involving motor vehicles are caused by accidents, the pathologist must be aware that other manners of death do occur. True vehicular homicides are defined as those occurrences in which a motor vehicle is *intentionally* used as a weapon in the taking of a life (Copeland, 1986). Accidents occurring because of negligent or reckless driving, driving under the influence of alcohol or other drugs, failure to avoid an accident, or failure to render assistance to the injured are not considered in the above category. Because fatalities caused by genuine road accidents are so common, naturally the court of law will be cautious in arriving at a verdict of homicide (Nadesan, 2000).

Copeland (1986) reviewed nine cases of true vehicular homicide that occurred over a 5-year period in Metropolitan Dade Country. The conclusion about the homicidal manner of death was based mainly on the law enforcement investigation and scene circumstances in correlation with the forensic pathologist's findings. Indeed, in some instances, an engineering study on steering maneuvers was warranted, in others, intuition or a stray fingerprint pointed to the truth, whereas in others, the testimony of the witnesses led to a correct analysis.

Homicides disguised as road accidents happen even less often, but it is also a possibility that should be investigated. In June 1994, in the Rhône-Alpes region of France, a car was discovered at the bottom of a ravine. The driver, a married man with children, had been killed. The emergency physician noted craniocerebral trauma and signed the death certificate with no forensic objection to burial. In November 1995, the wife of the victim was arrested for a homicide. She then admitted that she and her lover had also assassinated her husband with a bullet in his head after having him drink a large quantity of whiskey mixed with dipotassium clorazepate. The sound of the gunshot had been deadened by placing a pillow between the barrel of the weapon and the victim's head. The two accomplices then disguised their crime as a road accident. The body was exhumed on an order from the prosecutor and, on November 9, 1995, the autopsy showed an entry wound on the right temple, associated with an exit wound on the left temple. The bullet was found in the soft tissue and exoskeleton in the area of the left temple. Toxicological examinations found the presence of dipotassium clorazepate in the larvae removed from the victim's body (Fanton et al., 1998).

In such circumstances, the pathologist's role is to match the injuries with a traffic accident and detect any which are atypical. For instance, focal depressed fractures of the skull of the type caused by a weapon are unusual in a car occupant unless there was a localized intrusion of the vehicle roof. The ante-mortem nature of the injuries should be demonstrated, though this is not always possible. When there is a fire, soot inhalation or carbon monoxide absorption should be sought, though — as mentioned previously - some flash petrol fires may kill before any monoxide is absorbed. Where the victim was unconscious though not dead, no such differentiation is possible (Saukko & Knight, 2004).

Attempts at deliberate self-destruction by the use of a motor vehicle are said to be not uncommon, though this is difficult to prove in most cases (Murray & DeLeo, 2007, Seilzer & Payne, 1962). A driver may deliberately drive either off the road (e.g., into a fixed object, another vehicle, water) or from a height. A vehicle may be deliberately stopped on railroad tracks. A pedestrian may step in front of a vehicle. Signs of possible suicidal impulses are suicide note, previously talked or attempted suicide, depression or other psychiatric disease. Typically, drivers who commit suicide by motor vehicle are alone and crash their cars head-on into a fixed object such as a concrete bridge, an embankment, or a utility pole, without using the brakes. It is usually obvious from a study of the tire tracks that such individuals had sufficient time to turn back onto the road or avoid the obstacle if they had accidentally gone off the road. In addition, if the death was witnessed, no brake lights would have been observed. The soles of the shoes of the driver should also be examined for transfer of the pedal pattern to the shoe sole. If the pattern is that of the gas pedal, then it can be concluded that, at the time of impact, the individual was still accelerating. It is obvious that in such cases, the evidence is more likely to be based on circumstantial rather than medical evidence - a matter for the investigating authorities rather than the pathologist.

The usual search for alcohol and other drugs must be made, as suicides often employ multiple methods to ensure self-destruction. Many such suicides go probably undetected because the cause of the "accident" is attributed to drinking or falling asleep at the wheel. A detailed medical history in such cases may be helpful, as usually, individuals committing suicide with a motor vehicle have a history of prior suicide attempts or treatment by a psychiatrist (Saukko & Knight, 2004, DiMaio & DiMaio, 2001). However, some psychiatric entities, such as borderline and antisocial personality disorders, have been linked with elevated risk for traffic accidents (Dumais et al., 2005), which renders the formulation of a conclusion concerning the manner of death in such cases, even more difficult.

9. Who Was the Driver?

Occasionally, accidents occur in which there are two or more occupants in a vehicle and it is not clear who the driver was. The occupants of the vehicle may be all dead and have been ejected or relocated within the vehicle, or the surviving driver may claim that a deceased individual was driving in order to avoid legal liability.

In such cases, the investigation that coordinates an examination of injury mechanisms, occupant kinematics, vehicle dynamics and the evaluation of trace evidence will facilitate the determination of each occupant's role. One of the most critical elements is the collection of trace evidence from the victims and the vehicle. Special efforts must be made to collect clothing and biological standards from all vehicle occupants. Examination of the soles of leather shoes may reveal the imprint of the accelerator or brake pedal, or the imprint of the leather floor mat. The preservation of clothing will permit a forensic examiner to compare clothing fibers to those fibers transferred to vehicle components during the collision. Fabric imprints may also be transferred to components within the vehicle. Contact with the front windshield will frequently result in the co-transference of hair and tissue to the glass and of glass to the tissue. Collection of this glass from a patient's wound can be matched with a particular window within the vehicle if a glass standard is collected from the vehicle involved (Smock et al., 1989, Curtin & Langlois, 2007).

10. Toxicology in Traffic Accidents

Where death occurs within 12 or even 24 hours of the time of the accident, blood analysis for alcohol is essential, whether in the driver, passenger or pedestrian. Where possible, screening for drugs of dependence and common medicinal substances that might have caused drowsiness should be carried out. In combination with alcohol, even low levels of sedative, hypnotic and antihistamine drugs may be relevant in the causation of an accident. Whether a drug is a cause of an accident, either wholly or in part, can be decided only by individual analysis of a case. Drug testing on passengers is recommended for two reasons — first, a "passenger" occasionally turns out to have been the driver; second, the presence of a drug or alcohol in a passenger often reflects the toxicological status of the driver.

10.1. Collection of Specimens

The tissue of most importance for analysis is blood. This is logical, given that the blood level of the drug that has the effect on the individual. Collection of blood through venepuncture of a cubital vein in an arm is the standard technique used clinically for extracting venous blood from a live person. In deceased subjects, venous blood extracted from a vein in the arm or leg provides the optimum sample for interpretive purposes. Other sites for collection are also available, however, present several interpretive difficulties. On occasion, in instances of massive trauma to a body, no blood can be collected from the vasculature, though there is free blood in the body cavities. If this blood is collected, tested for alcohol and drugs and found to be negative then one is safe in assuming the individual was not under the affects of alcohol or drugs at the time of death. A positive test, on the other hand, must take into account the possibility of contamination. In such a case, another material such as vitreous or muscle must be analyzed to evaluate the accuracy of the test results on the blood.

Blood must always be drawn under medically accepted conditions by a qualified individual. It is important to apply a nonalcoholic disinfectant before the suspect's skin is penetrated with a sterile needle or lancet in order to negate any argument that an alcoholic disinfectant may have inadvertently contributed to a falsely high blood-alcohol result. Anticoagulant and preservative is added in the sample.

Drugs are excreted in the urine. Analysis of urine for drugs is easy because there is no protein binding to hinder extraction and many drugs are concentrated in the urine. Their collection is of major importance because they are necessary for the screening. It should be realized, however, that the level of a drug in the urine is usually of no significance in the interpretation of any effect on the driving ability. It is the level in the blood that determines whether an individual is affected or not (Saukko & Knight, 2004, DiMaio & DiMaio, 2001).

10.2. Analysis of Specimens

Alcohol analysis is performed in whole blood samples utilizing a Gas Chromatography head-space method (Loffe & Vittenberg, 1984).

Analysis of biological tissues for toxicological purposes involves three fundamental steps applicable to any specimen: separation of the drug from the biological tissue, purification of the drug, analytical detection and quantification. With some drugs, specimens and methodologies, the first and second step can be eliminated and direct analytical analysis may be performed. Thus, analysis for drugs of abuse in urine using immunoassay techniques does not require the above steps. Separation of a drug from the biological specimen, e.g., blood, is usually accomplished using a solvent. Purification is carried out by additional extraction procedures using alkaline and acid solutions. Analysis is then conducted by gas chromatography (GC), gas chromatography-mass spectrometry (GC-MS), high performance liquid chromatography, immunoassay or UV spectrophotometry. Except for GC-MS, none of the methods is totally specific and another test must be performed for positive identification. The confirmatory test must involve a totally different method of analysis from the one originally used. If a method of analysis other than GC-MS is used for initial identification, it is often easier to make positive identification and even quantification using the GC-MS. If initial analysis is made with the GC-MS, there is no necessity to redo the identification, because this method is specific.

One of the more developed tools of the forensic toxicologist is the immunoassays. There are four types: radioimmunoassay, enzyme immunoassay, fluorescent immunoassay and kinetic interaction of microparticles. The major advantage to these systems is that large numbers of tests can be performed in a small amount of time using an extremely small volume of specimen, with semi-automated and automated systems to speed the rate of analysis. While radio-immuoassays can be used on blood, the other types of immunoassay should be confined to analysis of urine. These systems were never intended for the analysis of blood. There are two disadvantages to immunoassay techniques. First, the analysis is usually very narrow in scope; i.e., one analyzes for either a specific drug or a specific family of drugs, rather than for the several hundred drugs that can be analyzed for in one test with a GC or GC-MS. In addition, this method of analysis is not absolutely specific, although, with some of the newer kits, the specificity is extremely good. A positive test result must be confirmed by another analytical method, usually GC-MS. In a high-volume forensic lab, immunoassay methods can be used to screen for opiates, cocaine, amphetamines and methamphetamines, barbiturates, and cannabinoids in the urine. Negative results indicate that the compounds are not present; positive results indicate that the compound may be present. *A positive identification with any of the immunoassay tests should never be reported unless it has been confirmed by another method of analysis (*Recommended Methods for the Detection and Assay of Barbiturates and Benzodiazepines in Biological Specimens, 1997, Recommended Methods for the Detection and Assay of Heroin, Cannabinoids, Cocaine, Amphetamine, Methamphetamine, and Ring-Substituted Amphetamine Derivatives in Biological Specimens, 1995).

10.3. Driving under the Influence of Alcohol

Driving under the influence of alcohol is the act of operating a motor vehicle (and even a bicycle, boat or horse in some jurisdictions) after having consumed alcohol, to the degree that mental and motor skills are impaired. Correlation of blood alcohol concentration with behavioural and Central Nervous System effects has been well established (table 1). The unfortunate paradox is that alcohol gives a feeling of wellbeing but actually depresses brain function, lessening muscular control and co-ordination, lengthening reaction time. Vision is blurred and awareness decreased, especially in the dark. The ability to judge speed and distance is impaired, as is the capacity to deal with the unexpected. Not only do all of these factors adversely affect driving performance, by alcohol impairs judgment to the extent that drivers under the influence genuinely believe that they are driving better that they are (Ferrara et al., 1994).

Alcohol seems to affect in a predictable and uniform manner on the brain, thus permitting relatively safe interpretations from the experts. Legal limits have been established almost worldwide and the range from zero limit to 0.08 g/dl. In some countries special provisions exist for professional, new and young drivers. In most international jurisdictions, anyone who is convicted of injuring or killing someone while under the influence of alcohol can be heavily fined, in addition to being given a lengthy prison sentence.

In case a period of time has elapsed between the accident and the blood sampling, the issue of the estimation of blood alcohol concentration at the precise time of the accident is posed. The quantity of alcohol consumed and the relevant time frame are also useful information to obtain when investigating cases of traffic accidents. In forensic science practice, Widmark's method is widely used to translate a measured blood alcohol into the amount of alcohol consumed. Widmark developed an equation for estimating the amount of alcohol in the body from values of specific input variables. The Widmark equation states that the amount of alcohol absorbed and distributed in the body (A) is a function of several independent variables: A = *(*w,p,C,,& t), where: A = amount of alcohol ingested (g), w = body weight (kg), ρ = volume of distribution (Widmark's rho, here expressed in terms of l/kg), C_1 = BAC at time t (g/l), β= zero-order elimination rate of alcohol from the blood (g/l/h), and t = time in hours after the start of drinking. Accordingly, the value A is a simple linear function of each variable expressed as: A mean = $w\rho(C_1 + \beta t)$ (Gullberg & Jones, 1994).

BAC (g/dl)	Signs/ Symptoms
0.01-0.05	Slight physiological impairment.
0.05-0.07	Slight physiological impairment detectable on careful testing by 0.05 g/100 mL. Euphoria; increased self-confidence.
0.07-0.10	Increasing impairment of reaction responses, attention, visual acuity, sensory-motor coordination, and judgment. Individual may still appear sober.
0.10-0.20	Increasing impairment of sensory-motor activities, reaction times, attention, visual acuity, and judgment. Progressive increase in drowsiness, disorientation, emotional instability.
0.20-0.30	Loss of coordination, staggering gait, slurred speech, grossly impaired, drunk; may be lethargic and sleepy or hostile and aggressive.
0.30-0.40	Impaired consciousness, stupor, unconsciousness.
0.40+	Unconsciousness, coma. Possible death.

10.4. Driving under the Influence of Drugs

In addition to driving under the influence of alcohol, driving under the influence of drugs is generally covered in existing legislation in most nations around the globe. The definition of a drug is very broad, and typically includes any substance that can affect a person's mental or physical capacities to drive. The drugs causing or contributing to the impairment need not be illegal, but can consist of lawfully prescribed or over-the-counter medication.

In order to determine whether a driver, involved in an accident or stopped at a roadside checkpoint, is impaired or under the influence of a certain substance there are three basic approaches used by the current DUID laws (Jones, 2005). One is the *effect-based* or *impairment* approach where the fitness of the drivers is observed and assessed. This requires that each suspect should be examined by a physician, who will look for signs and symptoms of drug influence and will also conduct various clinical tests of impairment. When the above examination has not taken place timely, or the driver has fatally been injured, the afterwards evaluation of driving ability based on the drug blood concentration presents several problems.

Indeed, in contrast to alcohol, the interpretation of drug concentrations in biological fluids requires some knowledge about the dose, the route of administration, the pattern or frequency of drug use, and the dispositional kinetics (distribution, metabolism, and excretion) of the drug. Interpreting the meaning of either drug/metabolite concentration in a single biological specimen with reference to impaired driver performance is therefore an extremely difficult task for a scientist. Experts must be able to show that an impairing substance appreciably adversely affected the driver's physical and/or mental faculties and the above opinion must be supported by scientific documentation, experience and by other evidence. However, there are thousands of drugs available and millions of combinations of these drugs. The wide ranges of drug concentrations resulting from therapeutic doses and the lack of studies of therapeutic and higher doses for most drugs and combinations make expert opinions of drug effects on driving performance questionable (Jones, 2005).

The above mentioned difficulties have lead many jurisdictions to adopt a *per se* legislation. In one version of this approach, as in the case of alcohol, a science-based finite limit is used for the "tolerable" concentration of a drug or its metabolites in driver's blood. The other version is the *zero tolerance* case where the detection of any detectable amount of the drug in driver's blood is penalized and any amount of prohibited drug found in the blood or urine of drivers while operating a motor vehicle is a per se violation of DUI statutes. *De facto* zero limits is practically the limit of detection (LOD) of the analytical method used and as the different forensic laboratories worldwide do not use the same analytical methodology so, obviously, this *de facto* limit varies (Walsh & DeGier, 2004, Moeller & Kraemer, 2002).

10.5. Carbon Monoxide

Carbon monoxide levels should be performed in most motor vehicle deaths, since occasionally death might be caused by, or the accident precipitated by, acute carbon monoxide poisoning. Carbon monoxide has been reported to affect the reaction time and cause headaches, irritability, vertigo and palpitation to drivers (Jovanović et al., 1999). Its combination with heat may further deteriorate driving performance (Walker et al., 2001).

The source of the carbon monoxide is usually a defective exhaust system, which allows gas to percolate through the floor or engine bulkhead into the interior. Rarely, a strong following wind blows the external exhaust gas into the open doors of a van or truck. Another motoring cause is a leak in the heat exchanger in those vehicles that use a direct air supply from around the exhaust manifold to provide passenger heating (Saukko & Knight, 2004).

11. Conclusions

Traffic accidents are a major cause of mortality and morbidity, mainly among young people. When a traffic accident occurs, a matter of penal and/or civil liability almost always arises. Insurance issues may also be posed. Moreover, although most cases of traffic fatalities are caused by accidents, other manners of death (suicidal or homicidal) may occur. A complete and careful forensic investigation in such cases may reveal findings which when combined with circumstantial, trace and other evidence may lead to the formulation of a final conclusion.

REFERENCES

Adams, V.I. (1992). Neck injuries: II. Atlantoaxial dislocation--a pathologic study of 14 traffic fatalities. *J Forensic Sci,* 37(2), 565-573.

Alarie, Y. (2002). Toxicity of fire smoke. *Crit Rev Toxicol*, 32(4), 259-289.

American Association for Automotive Medicine (1990). *The Abbreviated Injury Scale-1990.* De Plains, IL: American Association for Automotive Medicine.

American College of Surgeons Committee on Trauma (1997). *Advanced Trauma Life Support Course.* Chicago: American College of Surgeons.

Anderson, A.J., Towns, G.M. & Chiverton, N. (2006). Traumatic occipitocervical disruption: a new technique for stabilisation. Case report and literature review. *J Bone Joint Surg Br,* 88(11): 1464-1468.

Arce, R., Fariña, F., Carballal, A. & Novo M. (2006). [Evaluating psychological injury in motor vehicle accidents (MVA): development and validation of a protocol for detecting pretense] *Psicothema,* 18(2), 278-283.

Ashton, S.J. (1982) Vehicle design and pedestrian injuries. In: Chapman AJ, Wide FM, Foot HC (eds) *Pedestrian accidents* (pp 169–202).

Atkins, R.M., Turner, W.H., Duthie, R.B. & Wilde, B.R. (1988) Injuries to pedestrians in road traffic accidents. *BMJ,* 297, 1431–1434.

Baker, S.P., O'Neill, B., Haddon, W. Jr., Long, W.B. (1974). The injury severity score: a method for describing patients with multiple injuries and evaluating emergency care. *J Trauma,* 14(3):187-96.

Beier, G. & Pfriem, D. (1974). [Impact-geometry induced pecularities in fatally injured pedestrians]. *Beitr Gerichtl Med,* 32, 73-77.

Bloomer, R., Willett, K. & Pallister, I. (2004). The stove-in chest: a complex flail chest injury. *Injury,* 35(5), 490-493.

Brison, R.J., Wicklund, K. & Mueller, B.A. (1988) Fatal pedestrian injuries to young children: a different pattern of injury. *Am J Public Health*, 78, 793–795.

Büttner, A., Heimpel, M. & Eisenmenger, W. (1999). Sudden natural death 'at the wheel': a retrospective study over a 15-year time period (1982-1996). *Forensic Sci Int,* 103(2), 101-112.

Bunai, Y., Nagai, A., Nakamura, I. & Ohya, I. (2000). Fetal death from abruptio placentae associated with incorrect use of a seatbelt. *Am J Forensic Med Pathol,* 21(3), 207-209.

Byard, R.W. (2002). Shoulder-lap seat belts and thoracic transection. (2002). *J Clin Forensic Med,* 9(2), 92-95.

Carroll, E.A., Gordon, B., Sweeney, C.A., Joy, S. & Connolly, P.J. (2001). Traumatic atlantoaxial distraction injury: a case report. *Spine,* 26(4), 454-457.

Committee on Medical Aspects of Automotive Safety. (1971). Rating the severity of tissue damage. I. The Abbreviated Scale, *JAMA,* 215, 277-280.

Copeland, A.R. (1986). True vehicular homicide. *Am J Forensic Med Pathol*, 7, 305-307.

Curtin, E. & Langlois, N.E. (2007). Predicting driver from front passenger using only the postmortem pattern of injury following a motor vehicle collision. *Med Sci Law,* 47(4), 299-310.

Dada MA, Rutherfoord GS. Medicolegal aspects of cerebral amyloid angiopathy. A case report. *Am J Forensic Med Pathol.* 1993 Dec;14(4):319-22.

Dada, M.A. & Rutherfoord, G.S. (1993). Medicolegal aspects of cerebral amyloid angiopathy. A case report. *Am J Forensic Med Pathol,* 14(4), 319-322.

DiMaio, V.J. & DiMaio D. (2001). Deaths caused by Motor Vehicle Accidents. In: *Forensic Pathology* (pp 279-317) CRC Press.

Dumais, A., Lesage, A.D., Boyer, R., Lalovic, A., Chawky, N., Ménard-Buteau, C., Kim, C. & Turecki, G. (2005). Psychiatric risk factors for motor vehicle fatalities in young men. *Can J Psychiatry,* 50(13), 838-844.

Durbin, D.R., Arbogast, K.B. & Moll, E.K. (2001). Seat belt syndrome in children: a case report and review of the literature. *Pediatr Emerg Care,* 17, 474–477.

Eckert, W.G. (1981). The medicolegal and forensic aspects of fires. *Am J Forensic Med Pathol,* 2(4):347-357.

Eid, H.O. & Abu-Zidan, F.M. (2007). Biomechanics of road traffic collision injuries: a clinician's perspective. *Singapore Med J,* 48(7), 693-700.

Evans, L. & Frick, M.C. (1988). Helmet effectiveness in preventing motorcycle driver and passenger fatalities. *Accid Anal Prev,* 20(6), 447-458.

Fanton, L., Miras, A., Tilhet-Coartet, S., Achache, P. & Malicier, D. (1998). The perfect crime: myth or reality? *Am J Forensic Med Pathol,* 19(3), 290-293.

Ferrara, S.D., Zancaner, S. & Giorgetti, R. (1994). Low blood alcohol concentrations and driving impairment. A review of experimental studies and international legislation. *Int J Legal Med,* 106(4), 169-177.

Fife, D. & Kraus, J. (1986). Anatomic location of spinal cord injury. *Relationship to the cause of injury. Spine,* 11(1), 2-5.

Friedman, Z., Kugel, C., Hiss, J., Marganit, B., Stein, M. & Shapira, S.C. (1996). The Abbreviated Injury Scale. A valuable tool for forensic documentation of trauma. *Am J Forensic Med Pathol*, 17(3), 233-238.

Garrett, J.W. & Braunstein, P.W. (1962). The seat belt syndrome, *J Trauma,* 1962, 2, 220–238.

Gautschi, O.P., Woodland, P.R. & Zellweger, R. (2007). Complete medulla/cervical spinal cord transection after atlanto-occipital dislocation: an extraordinary case. *Spinal Cord, 45(5), 387-393.*

Gullberg R.G. & Jones, A.W. (1994). Guidelines for estimating the amount of alcohol consumed from a single measurement of blood alcohol concentration: re-evaluation of Widmark's equation. *Forensic Sci Int,* 69(2), 119-130.

Halinen, M.O. & Jaussi, A. (1994). Fatal road accidents caused by sudden death of the driver in Finland and Vaud, Switzerland. *Eur Heart J,* 15(7), 888-894.

Hambrook, D.W., Fink, J.N. (2006). Airbag asthma: a case report and review of the literature. *Ann Allergy Asthma Immunol*, 96(2), 369-372.

Harrahill, M. (2002). Tracheobronchial injuries. *J Emerg Nurs,* 28(3), 265-266.

Harruff, R.C., Avery, A. & Alter-Pandya, A.S. (1998) Analysis of circumstances and injuries in 217 pedestrian traffic fatalities. *Accid Anal Prev,* 30, 11–20.

Hirsch, C.S., Bost, R.O., Gerber, S.R., Cowan, M.E., Adelson, L., Sunshine, I. (1977). Carboxyhemoglobin concentrations in flash fire victims: report of six simultaneous fire fatalities without elevated carboxyhemoglobin. *Am J Clin Pathol,* 68(3), 317-320.

Hitosugi, M., Fukui, K. & Takatsu, A. (2001). Incomplete decapitation of a motorcyclist from hyperextension by inertia: a case report. *Med Sci Law,* 41(2), 174-177.

Horinouchi, H., Kato, R., Kaseda, S. & Maenaka, Y. (1993). [Tracheobronchial injury: clinical analysis of 17 consecutive cases]. *Kyobu Geka,* 46(9), 756-761.

Huelke, D.F. (1970). Extremity injuries produced in motor vehicle collisions. *J Trauma, 10(3), 189-200.*

Inoue, H., Ikeda, N., Tsuji, A., Kudo, K. & Nishida, N. (2004). Case reports. 1. An autopsy case of fatal arrhythmia induced by injuries of the atrioventricular conduction system: a case report. *Med Sci Law,*44(4), 353-358.

Ishikawa, H., Kajzer, J., Ono, K. & Sakurai, M. (1994). Simulation of car impact to pedestrian lower extremity: influence of different car-front shapes and dummy parameters on test results. *Accid Anal Prev,* 26, 231–242.

Jehle, E.C. (2001). [Penetrating injuries of the rectum and anus: primary management, reconstruction, rehabilitation]. *Kongressbd Dtsch Ges Chir Kongr*, 118, 417-420.

Johnson, R.M., McCarthy, M.C., Miller, S.F. & Peoples, J.B. (1995). Craniofacial trauma in injured motorcyclists: the impact of helmet usage. J Trauma, 38(6), 876-878.

Jones, A.W. (2005). Driving under the influence of drugs in Sweden with zero concentration limits in blood for controlled substances. *Traffic Inj Prev,* 6, 317-322.

Jovanović, J., Jovanović, M. & Dordević, D. (1999). Professional exposure of drivers to carbon monoxide as a possible risk factor for the occurrence of traffic accidents in the road traffic. *Vojnosanit Pregl,* 56(6), 587-592.

Karger, B., Teige, K., Bühren, W. & DuChesne, A. (2000). Relationship between impact velocity and resulting injuries in fatal pedestrian car collisions. I*nt J Legal Med,* 113, 84–88.

Karger, B., Teige, K., Fuchs, M. & Brinkmann, B. (2001). Was the pedestrian hit in an erect position before being run over*? Forensic Sci Int,* 15, 119(2), 217-220.

Kerwin, A. J. (1984)a. Sudden death while driving. *Can Med Assoc J,* 131, 312–314.

Kerwin, A. J. (1984)b. The electrophysiologic features of sudden death. *Can Med Assoc J* 131, 315–317.

Kommu, S.S., Illahi, I, Mumtaz, F. (2007). Patterns of urethral injury and immediate management. *Curr Opin Urol,* 17(6), 383-389.

Kondo, T., Saito, K., Nishigami, J. & Ohshima, T. (1995). Fatal injuries of the brain stem and/or upper cervical spinal cord in traffic accidents: nine autopsy cases. *Sci Justice,* 35(3), 197-201.

Kuhn, F., Morris, R., Witherspoon, C.D. (1995). Eye injury and the air bag. *Curr Opin Ophthalmol.* 6(3), 38-44.

Kumaresan, S., Sances, A. Jr., Carlin, F., Frieder, R., Friedman, K. & Renfroe, D (2006). Biomechanics of side impact injuries: evaluation of seat belt restraint system, occupant kinematics and injury potential. *Conf Proc IEEE Eng Med Biol Soc,* 1, 87-90.

Kuppuswamy, R. & Ponnuswamy, P.K. (2000). Note on fabric marks in motor vehicle collisions. *Sci Justice,* 40(1), 45-47.

Lau, G. (1996). Natural disease and alcohol intoxication amongst drivers of motor vehicles in Singapore from 1989 to 1993: a study of 140 necropsies. *Ann Acad Med Singapore,* 25(4), 516-521.

Lausevic, Z., Lausevic, M., Trbojevic-Stankovic, J., Krstic, S. & Stojimirovic, B. (2008). Predicting multiple organ failure in patients with severe trauma. *Can J Surg,* 51(2), 97-102.

Letts, M., Davidson, D. Fleuriau-Chateau, P. & Chou, S. (1999). Seat belt fracture with late development of an enterocolic fistula in a child. *Spine,* 24, 1151–1155.

Loffe, B. & Vittenberg, A.G. (1984). *Headspace Analysis and Related Methods in Gas Chromatography,* John Wiley, New York.

Maeda, H., Higuchi, T., Imura, M., Noguchi, K. & Yokota, M. (1993). Ring fracture of the base of the skull and atlanto-occipital avulsion due to anteroflexion on motorcycle riders in a head-on collision accident. *Med Sci Law,* 33(3), 266-269.

Magnotti, L.J., Croce, M.A. & Fabian, T.C. (2004). Is ventilator-associated pneumonia in trauma patients an epiphenomenon or a cause of death? *Surg Infect* (Larchmt), 5(3), 237-242.

Majeski, J. Shoulder restraint injury of the female breast. (2007). *Int Surg,* 92(2), 99-102.

Malliaris, A.C., DeBlois, J.H. & Digges, K.H. (1996). Light vehicle occupant ejections--a comprehensive investigation. *Accid Anal Prev,* 28(1), 1-14.

Martin-de las Heras, S., Valenzuela, A., Villanueva, E., Marques, T., Exposito, N. & Bohoyo, J.M. (1999) Methods for identification of 28 burn victims following a 1996 bus accident in Spain. *J Forensic Sci,* 44(2), 428-431.

Matsui, Y. (2005). Effects of vehicle bumper height and impact velocity on type of lower extremity injury in vehicle-pedestrian accidents. *Accid Anal Prev,* 37(4), 633-640.

McCoy, G.F., Johnstone, R.A., Nelson, I.W., Kenwright, J. & Duthie, R.B. (1988). Incidence and consequences of ejection in motor vehicle accidents. *BMJ,* 297(6658), 1244-1245.

Menon, A., Pai, V.K. & Rajeev A. (2008). Pattern of fatal head injuries due to vehicular accidents in Mangalore. *J Forensic Leg Med,* 15(2), 75-77.

Michalodimitrakis, E.N. & Tsatsakis, A.M. (1997). Vehicular accidents and cardiac concussion. A traumatic connection. *Am J Forensic Med Pathol,* 18(3), 282-284.

Mikhail, J.N. (1995). Side impact motor vehicular crashes: patterns of injury. *Int J Trauma Nurs,* 1(3):64-69.

Milić, J., Nikolić, S. & Mihailović, Z. (2002). [Analysis of causes of death in long-term survivors of injuries sustained in traffic accidents] *Srp Arh Celok Lek,* 130(5-6), 149-153.

Miltner, E., Wiedmann, H.P., Leutwein, B., Hepp, H.P., Fischer, R., Salwender, H.J., Frobenius, H. & Kallieris, D. (1992). Liver and spleen ruptures in authentic car-to-car side collisions with main impact at front door or B-pillar. *Am J Forensic Med Pathol* 13(1), 2-6.

Moeller, M.R. & Kraemer, T. (2002). Drugs of abuse monitoring in blood for control of driving under the influence of drugs. *Ther Drug Monit,* 24(2), 210-221.

Murray, D. & de Leo, D. (2007). Suicidal behavior by motor vehicle collision. *Traffic Inj Prev,* 8(3), 244-247.

Nadesan, K. (2000). Murder and robbery by vehicular impact: true vehicular homicide. *Am J Forensic Med Pathol,* 21(2), 107-113.

National Highway Traffic Safety Administration. *Third report to Congress: effectiveness of occupant protection systems and their use.* Washington, DC: NHTSA; 1996. New York, Wiley & Sons.

Newgard, C.D., Lewis, R.J. & Kraus, J.F. (2005). Steering wheel deformity and serious thoracic or abdominal injury among drivers and passengers involved in motor vehicle crashes. *Ann Emerg Med*, 45(1), 43-50.

Nikolić, S. & Micić, J. (2004). [Correlation between the survival duration and trauma severity in persons who died from posttraumatic fat embolism]. *Srp Arh Celok Lek,* 132(5-6), 167-170.

Nikolic, S., Atanasijevic, T., Mihailovic, Z., Babic, D. & Popovic-Loncar, T. (2006). Mechanisms of aortic blunt rupture in fatally injured front-seat passengers in frontal car collisions: an autopsy study. *Am J Forensic Med Pathol,* 27(4), 292-295.

Okur, H., Küçïkaydin, M., Kazez, A., Turan, C. & Bozkurt, A. (1996). Genitourinary tract injuries in girls. *Br J Urol,* 78(3), 446-449.

Ostrom, M. & Eriksson, A. (1987). Natural death while driving. *J Forensic Sci,* 32, 988–998.

Otte, D., Sandor, L. & Zwipp, H. (1990). Significance and mechanism of thoracic and lumbar spine injuries in traffic accidents]. *Unfallchirurg,* 93(9), 418-425.

Pearlman, M.D., Klinich, K.D., Schneider, L.W., Rupp, J., Moss, S. & Ashton-Miller, J. (2000). A comprehensive program to improve safety for pregnant women and fetuses in motor vehicle crashes: a preliminary report. *Am J Obstet Gynecol,* 182(6), 1554-1564.

Pepperell, R.J., Rubinstein, E., MacIsaac, I.A. (1977). Motor-car accidents during pregnancy. *Med J Aust*, 12;1(7), 203-205.

Previdi, J.K., Cayten, C.G. & Byrne, D.W (1996). Early predictors of sepsis in the motor-vehicle crash trauma victim. *Prehosp Disaster Med,* 11(1), 27-36. *Erratum in: Prehospital Disaster Med,* 11(2), 122.

Recommended Methods for the Detection and Assay of Barbiturates and Benzodiazepines in Biological Specimens, *Manual for Use by National Laboratories,* (1997) ST/NAR/28, United Stations.

Recommended Methods for the Detection and Assay of Heroin, Cannabinoids, Cocaine, Amphetamine, Methamphetamine, and Ring-Substituted Amphetamine Derivatives in Biological Specimens (1995). *Manual for Use by National Laboratories,* ST/NAR/27, United Stations.

Reid, A.B., Letts, R.M. & Black, G.B. (1990). *Pediatric* Chance fractures: association with intraabdominal injuries and seatbelt use. *J Trauma,* 30, 384–391.

Richards, D.M., Carhart, M., Raasch, C., Pierce, J., Steffey, D. & Ostarello, A. (2006). Incidence of thoracic and lumbar spine injuries for restrained occupants in frontal collisions. *Annu Proc Assoc Adv Automot Med,* 50, 119-133.

Riddick, L., Long, W.B., Copes, W.S., Dove, D.M. & Sacco, W.J. (1998). Automated coding of injuries from autopsy reports. *Am J Forensic Med Pathol,* 19(3), 269-274.

RuDusky, B.M. (2003). Myocardial contusion culminating in a ruptured pseudoaneurysm of the left ventricle--a case report. *Angiology,* 54(3), 359-362.

Sato, Y., Ohshima, T., Kondo, T. (2002). Air bag injuries--a literature review in consideration of demands in forensic autopsies. *Forensic Sci Int,* 28, 128(3), 162-167.

Saukko, P. Knight B. (2004). Transportation injuries. In: *Knight's Forensic Pathology* (pp.281-300). London, Edward Arnold Publishers.

Schnyder, U., Morgeli, H., Nigg, C., Klaghofer, R., Renner, N., Trentz, O. & Buddeberg, C. Early psychological reactions to life-threatening injuries. *Crit Care Med,* 28(1), 86-92.

Seilzer, M.L. & Payne, C.E. (1962). Automobile accidents, suicide and unconscious motivation. *Am J Psychiatry,* 119, 237-240.

Sharma, B.R. (2005). The injury scale--a valuable tool for forensic documentation of trauma. *J Clin Forensic Med,* 12(1), 21-28.

Shkrum, M.J., McClafferty, K.J., Green, R.N., Nowak, E.S. & Young, J.G. (1999). Mechanisms of aortic injury in fatalities occurring in motor vehicle collisions. *J Forensic Sci,* 44(1), 44-56.

Smock, W.S., Nichols, G.R. 2nd, Fuller P.M. & Weakley-Jones B. (1989). The forensic pathologist and the determination of driver versus passenger in motor vehicle collisions. The need to examine injury mechanisms, occupant kinematics, vehicle dynamics, and trace evidence. *Am J Forensic Med Pathol,* 10(2), 105-114.

Stefanidou M & Athanaselis S. (2004). Toxicological aspects of fire. V*et Hum Toxicol, 46(4), 196-199.*

Suhr, H., Hambrecht, S., Mauser, M., Fleischmann, D. & Foesel, T. (2000). [Blunt chest trauma with severe pulmonary contusion and traumatic myocardial infarction] *Anasthesiol Intensivmed Notfallmed Schmerzther,* 35(11), 717-720.

Svensson, M.Y., Boström, O., Davidsson, J., Hansson, H.A., Håland, Y., Lövsund, P., Suneson, A. & Säljö, A. (2000). Neck injuries in car collisions--a review covering a possible injury mechanism and the development of a new rear-impact dummy. *Accid Anal Prev,* 32(2), 167-175.

Swan Jn, K.G. Swan, B.C. & Swan, K.G. (2001). Decelerational thoracic injury. *J Trauma* 51, 970–974.

Tanno, K., Kohno, M., Ohashi, N., Ono, K., Aita, K., Oikawa, H., Myo-Thaik-Oo, Honda, K. & Misawa, S. (2000). Patterns and mechanisms of pedestrian injuries induced by vehicles with flat-front shape. *Leg Med* (Tokyo), 2(2), 68-74.

Tubbs, R.S., Golden, B., Doyle, S., Grabb, P.A. & Oakes, W.J. (2006). Lap-belt injury with complete avulsion of the spinal cord and cauda equina. *Clin Anat,* 19(7), 665-668.

Turnage, B. & Maull, K.I. (2000). Scalp laceration: an obvious 'occult' cause of shock. *South Med J,* 93(3), 265-266.

Unterberg, C., Buchwald, A. & Wiegand, V. (1989). Traumatic thrombosis of the left main coronary artery and myocardial infarction caused by blunt chest trauma. *Clin Cardiol,* 12(11), 672-674.

Walker, S.M., Ackland, T.R. & Dawson, B (2001). The combined effect of heat and carbon monoxide on the performance of motorsport athletes. *Comp Biochem Physiol A Mol Integr Physiol,* 128(4), 709-718.

Walsh, J.M., de Gier, J.J., Christopherson, A.S. & Verstraete, A.G.(2004). *Drugs and driving. Traffic Inj Prev,* 5(3), 241-253.

Wang, C.H., Tsay, S.L. & Bond, A.E. (2005). Post-traumatic stress disorder, depression, anxiety and quality of life in patients with traffic-related injuries. *J Adv Nurs,* 52(1), 22-30.

Westaby, S., Drossos, G., Giannopoulos, N. (1995). Posttraumatic coronary artery aneurysm. *Ann Thorac Surg,* 60(3):712-713.

Wirthwein, D.P. & Pless, J.E. (1996). Carboxyhemoglobin levels in a series of automobile fires. Death due to crash or fire? *Am J Forensic Med Pathol,* 17(2), 117-123.

Woźniak, K. & Rzepecka-Woźniak, E. (2003). [Routine examination of cervical spinal cord and spinal column during forensic autopsies]. *Arch Med Sadowej Kryminol,* 53(2), 91-107.

Wu, C.J., Tsai, W.F., Tsai, J.L., Chue, Y.C. & Chang, H.C. (2004). Bilateral traumatic dislocation of testes. *J Chin Med Assoc,* 67(6), 311-313.

Wyatt, J.P., Beard, D. & Busuttil, A. (1999). Use of the Abbreviated Injury Scale to identify preventable deaths. *Am J Forensic Med Pathol,* 20(1), 102.

Zivot, U. & Di Maio, V.J. (1993). Motor vehicle-pedestrian accidents in adults. Relationship between impact speed, injuries, and distance thrown. *Am J Forensic Med Pathol,* 14(3), 185-186.

In: Road Traffic: Safety, Modeling and Impacts
Editors: S. E. Paterson and L. K. Allan, pp. 259-278
ISBN 978-1-60456-884-4

Chapter 3

MANAGEMENT OF DEPRESSED SKULL FRACTURE: EXPERIENCE OF GENERAL SURGEONS IN NORTHERN NIGERIA

A. Ahmed* and M. A. Jimoh
Department of Surgery
Ahmadu Bello University Teaching Hospital
Zaria, Nigeria

ABSTRACT

Background

Head injury is a common cause of accidental deaths and of severe disabilities. It usually results from road traffic accident especially in developing countries. The presence of skull fracture is an important indicator of the nature and severity of the impact and risk of an operable intracranial lesion. Whereas in developed countries patients with skull fractures are managed by neurosurgeons, such patients are usually managed by general surgeons in developing countries. We present the experience of general surgeons on the management of depressed skull fractures in Nigeria.

Patients and Method

This study was conducted in the department of surgery Ahmadu Bello university teaching hospital Zaria, Nigeria. Adult patients seen between 1995 and 2005 with clinical and radiological diagnosis of depressed skull fracture were retrospectively reviewed. Patient evaluation included assessment of level of consciousness using the Glasgow coma scale, and other neurologic findings. All patients had skull x-rays. CT scan was done in

* Correspondence: Dr Adamu Ahmed, Department of Surgery, Ahmadu Bello University teaching Hospital, Zaria, Nigeria. Email: mrahmed1010@yahoo.com; Phone: +2348037200894

some patients. Clinical and radiological features were used to select patients that required operative intervention. The records of these patients were analysed in respect of age and sex distribution and mechanism of injury. The type and location of skull fracture, clinical course, operative findings and neurologic outcome were also reviewed.

Results

There were 235 patients with depressed skull fractures which represent 3.6% of head injured patients. Their ages ranged 15 to 68 years, mean of 30 ± 5.7SD. Male to female ratio was 3.8:1. Road traffic accident caused fractures in 155 (66.0%) patients. Blows and missiles accounted for 15.7% and 11.5% respectively. Road traffic accident caused the most severe skull fractures. Of the 235 depressed skull fractures 152 (64.7%) were compound. The frontal bones were fractured in 115 (48.9%) patients while the parietal bone was involved in 61 (26.0%) patients. In 13 (5.5%) patients the fractures were located on cranial venous sinuses. The admission Glasgow coma score was ≤ 8 in 28 (11.9%) patients and 9-12 in 67 (28.5%). Elevation of depressed skull fracture was performed in 128 (54.5%) patients of which 80 (62.5%) had additional treatment of intracranial pathology. At discharge from hospital, 185 (78.7%) patients had complete recovery while additional 32(13.6%) had residual neurologic deficit but live an independent life. Overall, mortality was 10.2%.

Conclusion

Skull fracture is common among head injured patients and is usually caused by road traffic accident. The complications and sequelae of depressed skull fracture can be minimised by early diagnosis and treatment. With careful selection many of these patients can be safely manage non-operatively.

Key words: depressed skull fracture, management, outcome, general surgeon

INTRODUCTION

Traumatic Brain Injury is a leading cause of death and severe disability among young people worldwide. Most of the burden is in developing countries because of lack of resources and organized and integrated trauma care system. In United States, more than 2 million people sustain traumatic brain injury annually of which 15% have prolonged physical and psychological impairment [1]. In most trauma series, road traffic accident (RTA) is the leading cause of traumatic brain injury [2, 3]. In Nigeria, a fivefold increase in traffic-related injuries was observed in the last 30 years [4, 5]. Given the high prevalence of motor vehicle accident in Nigeria, the care of head injured patients is of great concern. Head injury (HI) can be defined as the application and consequences of an external mechanical insult to the scalp, skull and intracranial contents. Skull fracture therefore, comprises a significant component of the surgical pathology of HI. The discovery of an isolated skull fracture rarely warrants intervention. However, the presence of a skull fracture has consistently been shown to be associated with increased incidence of intracranial lesions, neurological deficit and poorer outcome [6,7,8,9]. A study of skull fracture sustained can give information about the nature and severity of the impact and the type of brain injury to expect. In addition, the location of the fracture has a bearing on the development of certain complications of HI such as extradural haematoma, CSF leakage and cranial nerve involvement. Only about 3.0% of patients with mild head injury have skull fractures most of which are not associated with intracranial abnormality requiring surgery [10]. However in the conscious patient, the presence of skull fracture increases the risk of intracranial haematoma about 400 times [11,12]. Chan et al. found skull fracture to be the only independent significant risk factor in predicting intracranial haematoma in a cohort of 1178 patients [13]. Other reports have shown that the presence of both loss of consciousness and skull fracture significantly increase the risk of surgically significant intracranial haematoma compared to when one or neither condition exist [12,14]. Depressed skull fractures (DSF) are more frequently associated with intracranial sequelae. They are often associated with dural tear which significantly increases the risk of intracranial infection, neurological deficit and seizures [15,16]. Several reports have indicated that DSF is the most frequent indication for operative intervention in head injured patients [17, 18]. Depressed skull fractures associated with intracranial haematoma is frequently found in fatal head injuries [19]. It was found in 73.2- 80.0% of fatal head injuries in Nigeria [20, 21].

The skull is more prone to fracture at the squamous temporal and parietal bones which are thin [22]. The fracture occurs when local deformation of the vault exceeds regional bony tolerance. The nature of the fracture depends on the magnitude of the force applied, the site impacted, and the area over which the force is applied. Linear fractures are a result of the out bending of bone at a distance from impact site [22]. The fracture line takes the path of least resistance usually running towards the point of contact. Depressed skull fracture results from a high energy direct blow to a small surface area of the skull with a blunt object [22]. Such fracture becomes clinically significant when the fragment is depressed below the inner table of the surrounding intact skull. A DSF can be open or closed. Open fractures have either a scalp laceration over the fracture or the fracture runs through paranasal sinuses or the middle ear structures, resulting in a communication between the external environment and intracranial cavity [23, 24].

Neurosurgery was one of the first disciplines to emerge as a distinct sub-specialty within modern surgery. DSF is one of the common conditions needing urgent operation in neurosurgical practice. In developed countries, management of HI patients includes a computerized tomography (CT) scan to determine the presence and extent of intracranial pathology. Cases requiring surgical intervention are then transferred to the care of neurosurgeons in appropriate centres [11,12]. Several studies support the safe and competence performance of emergency burr hole or craniotomy by general surgeons [2,9,25]. A recent report indicates that in United States, the availability of neurosurgeons to care for injured patients is precipitously diminishing because of limited number and distribution particularly in rural areas [26]. In addition many practicing neurosurgeons have abandoned trauma care in their hospitals because of liability insurance crisis [26]. Indeed, the scarcity of neurosurgeons is more severe in developing countries. The neurosurgeon to population ratio is 1: 1,000,000 in India compared to 1: 1,300,000 in Pakistan [27,28]. In the West African subregion, management of neurotrauma patients is very difficult because of factors that complicate evaluation and care. Firstly, most hospitals do not have CT scanners. Transfer to neurosurgical units is not practical because of distance and lack of ambulance services. Most importantly, there is limited neurosurgical specialty care. The ratio of neurosurgeon to population ranges between 1: 4,000,000 and 1:12,500, 000 [29]. Several countries in the region do not have a resident neurosurgeon. In Nigeria, with a population of more than 140 million, there are only 20 neurosurgeons on whom lies the onus of providing neurosurgical expertise as well as salvaging the lives and shaping the fates of these severely head injured patients [29]. Therefore, a general surgeon who is more readily available is usually called upon to provide acute management of these patients thus gaining time and saving these patients before neurosurgical help is available. This was the case in our institution before the establishment of a dedicated neurosurgical unit in 2006.

The pattern of head injury in Zaria, Northern Nigeria has been previously reported [2]. Zaria is located at the confluence of the major highways linking the Northern to the Southern parts of Nigeria. Thus our institution receives for care many of the patients injured in RTA in this region. Many of these patients have associated injuries to the chest and abdomen. In the absence of a neurosurgical unit, head injured patients were managed by general surgeons. The use of CT in the evaluation of these patients was limited by cost and frequent malfunction of equipments. Due to lack of immediate rescue many of these patients present to the hospital several hours or days after injury. The absence of pre-hospital care also means that there would be little skilled airway management to prevent secondary brain damage from hypoxia and hypotension, and many patients with devastating injuries would have died before arrival at hospital [30]. In this paper we present the pattern, management and outcome of depressed skull fracture as managed by general surgeons in Zaria, Northern Nigeria.

Clinical Patients and Method

This study was carried out in the department of surgery Ahmadu Bello University Teaching Hospital Zaria Nigeria, a tertiary medical care facility. All patients above 14 years that presented at the accident and emergency (AE) department between January 1995 and December 2005 with clinical and radiological diagnosis of DSF were retrospectively reviewed. Patients whose fractures resulted from firearm injury or penetrating objects were not included. Patients that died during admission before they were clinically evaluated were also excluded. All patients were admitted regardless of whether they were for surgical intervention or observation. Patients admitted for observation that remained well [Glasgow Coma Scale (GCS) score 15/15] were discharged after 24 to 48 hours on head injury advice. Information was obtained from patient's case notes, operation records and discharge summaries. A standard proforma was used to collect information on patient demographics, extent and aetiology of injuries, their acute management and neurological outcome.

Treatment Protocol

The management protocol of patients with head injury in our institution has been previously described [2]. On admission at AE the patients had airway management, control of external haemorrhage, appropriate fluid support and cervical spine stabilization. Attempts were always made to maintain blood pressure between 90 and 120mmHg to maintain adequate cerebral perfusion pressure. Glasgow coma scale score at the time of admission was used to assess the degree of head injury. Head injury was defined as severe when the GCS score was ≤ 8, moderate as 9-12 and mild as GCS score of 13-15 [31]. Other neurological findings noted include cranial nerves abnormalities, extremity paresis or paralysis, dysphasia and seizures. Physical examination, X-ray and CT findings were used to select patients that would require operative treatment of their depressed skull fractures. Lateral and frontal view skull X-rays were performed in all cases. Angiographic examination was not performed because it was not available on emergency basis. The patient's neurological status, temperature, respiratory rate, pulse rate and blood pressure were monitored regularly. Serum glucose, blood gasses and urea and electrolytes were regularly determined for patients with severe head injury. There were no facilities for intracranial pressure monitoring. Intravenous infusion of mannitol (0.5-1g/Kg) was given in appropriate patients to reduce intracranial pressure. All patients with compound DSF were given intravenous antibiotics for 5-7 days. Tetanus prophylaxis was also given. Patients with simple DSF that had operative treatment were similarly given antibiotics.

Operative treatment consisted of debridement of scalp wound, craniotomy and elevation of depressed bone fragments, evacuation of underlying haematoma and repair of dural tear in appropriate cases. As per protocol, attention was always given to haemostasis to prevent postoperative epidural collection and bone fragments were not replaced. Non-operative treatment included local wound care consisting of wound irrigation with copious normal saline and hydrogen peroxide, debridement and closure under local anaesthetic in the emergency room.

Assessment of Outcome

Each patient's neurological outcome was assessed according to Glasgow outcome scale (GOS) [32]. Outcome was scored as 5, if there was good recovery (resumption of normal life despite minor deficit); 4, moderate disability (disabled but independent); 3, severe disability (conscious but disabled and dependent for daily support); 2, persistent vegetative (minimal responsiveness) or 1, death. Assessment of outcome was performed at the time of discharge from hospital and at three to six months after discharge.

Follow-Up

Following discharge from hospital, patients were seen 2 weekly for 2 months and 1-6 monthly thereafter. Clinical assessment of scalp wound for sepsis, CSF leak and neurological examination of the patient ware done at every visit. Particular attention was paid to the presence of epilepsy, hemiparesis, paraparesis, dysphasia, cranial nerves paralysis and cosmetic effects. Check X-rays and CT scan were also done in some patients.

Statistical Analysis

Data was entered into SPSS (version 14.0, SPSS, Chicago, IL) statistical software. Frequencies, means and standard deviations were determined. Associations between depressed skull fracture and intracranial squealae were assessed by using Mann-Whitney test, chi-squire test and Fisher's exact test when indicated. Logistic regression modelling was performed to identify independent factors significant for the prediction of outcome of treatment. Factors included in the model were age, sex, admission GCS score, intracranial haematoma and operative treatment. A p-value of < 0.05 was taken as significant.

RESULTS

During the period of study 235 patients with depressed skull fractures were managed. Overall, depressed skull fracture was seen in 3.6% of head injured patients. Of the patients with depressed skull fractures, there were 186 males and 49 females, male to female ratio of 3.8:1. Their ages ranged between 15 and 68 years with a mean of 30± 5.7 SD years. The highest incidence was in the third decade (figure 1). Interval between trauma and presentation at our hospital ranged between 17 minutes and 22 days, median of 5 hours. One hundred fifty seven (67.0%) patients presented within 6 hours of injury. Forty- five (19.1%) patients were admitted and resuscitated at other medical facilities before they were transferred to our hospital. In all, 152 (64.7%) patients had open fractures while 83 (35.3%) had closed fractures. Twenty five (30.0%) patients with closed fractures had surgical elevation of their fractures compared to 103 (67.8%) patients with open fractures. The indications for surgical intervention were evidence of significant intracranial pathology in 80 (62.5%) patients including 60 (46.8%) with intracranial heamatoma, sharp in-driven fragment with neurological deficit in 30 (23.4%), depression causing disfigurement in 10(7.8%) and gross contamination of wound in 8 (6.3%). The interval between trauma and operative intervention ranged between 2 hours and 31 days, median 4 days. In 45 (35.1%) patients operation was performed within 6 hours of injury. The timing of operation was often determined by severity of head injury and presence of open fracture. CT scan was performed in only 34 (26.6%) of patients. It was not performed in others mainly because of financial constraints.

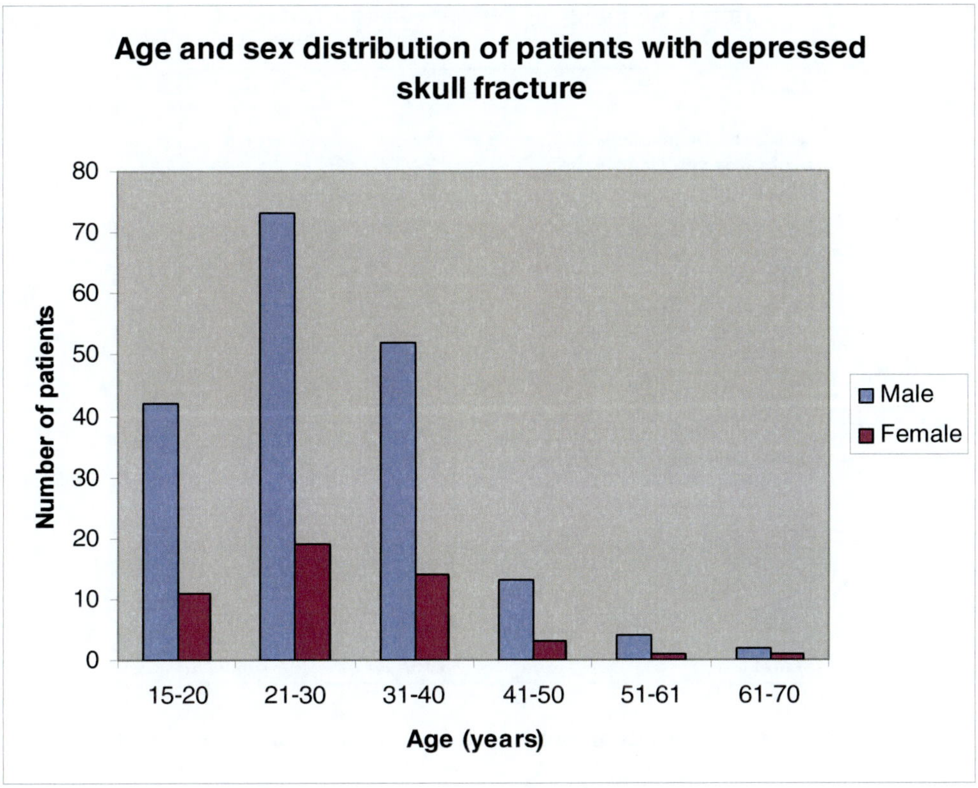

Figure 1. Age and sex distribution of patients with depressed skull fracture.

Aetiology of Trauma

The most common cause of DSF was RTA which accounted for fractures in 155 (66.0%) patients (table 1). Road traffic accident caused the most severe injuries, particularly in patients involved in motor vehicle accidents. None of the motor cyclists wore protective helmet. Falls were mainly from mango trees or trees being cut for firewood. Civil strife affected males and was usually a result of fights between supporters of rival political parties. Other causes include horse riding accidents and sport injuries.

Table 1. Aetiology of depressed skull fractures

Cause	Number	%
Road traffic accident	155	66.0
Motor vehicle injuries	77	
Motor cycle injuries	68	
Pedestrian	10	
Assault	37	15.7
Civil strife	22	
Armed robbery attack	11	
Other	4	
Fall	27	11.5
From tree	18	
At construction site	5	
Other	4	
Other	16	6.8
Falling object	9	
Sport injury	3	
Horse racing	4	
Total	235	100.0

Table 2. Location of depressed skull fractures on the cranial vault

Location	Side				Total	
	Right		Left			
	No	%	No	%	No	%
Frontal	69	29.4	46	19.5	115	48.9
Parietal	36	15.3	25	10.7	61	26.0
Occipital	14	6.0	10	4.2	24	10.2
Temporal	15	6.4	5	2.1	20	8.5
Other	9	3.8	6	2.6	15	6.4
Total	143	60.9	92	39.1	235	100.0

Location of Fractures

The frontal bone was the commonest site of fracture (figure 2) which together with the parietal bone accounted for fractures in 176 (74.9%) patients (table 2). In 15 (6.4%) patients the fractures involved more than 1 bone. In 13 patients the fractures were located on cranial venous sinuses. The commonest site was the anterior third of superior sagittal sinus (figures 3 and 4).

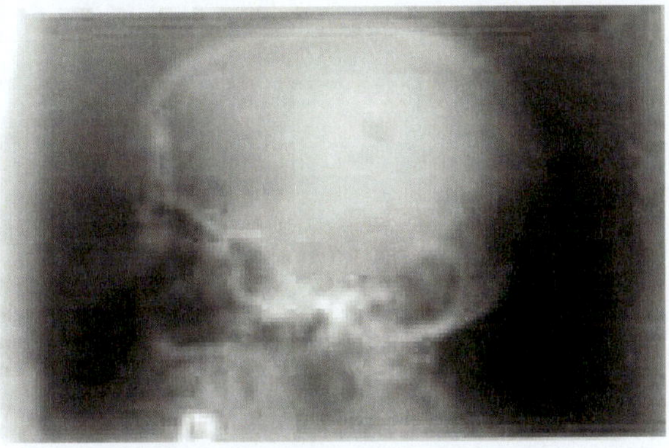

Figure 2. Depressed skull fracture on the frontal bone.

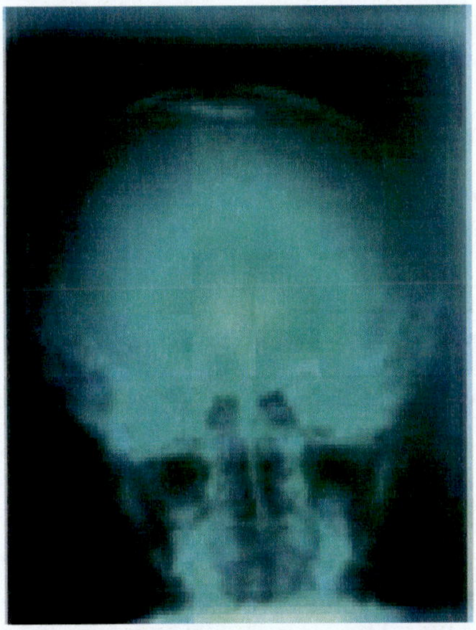

Figure 3. Anterior- posterior view of depressed skull fracture on superior sagittal sinus.

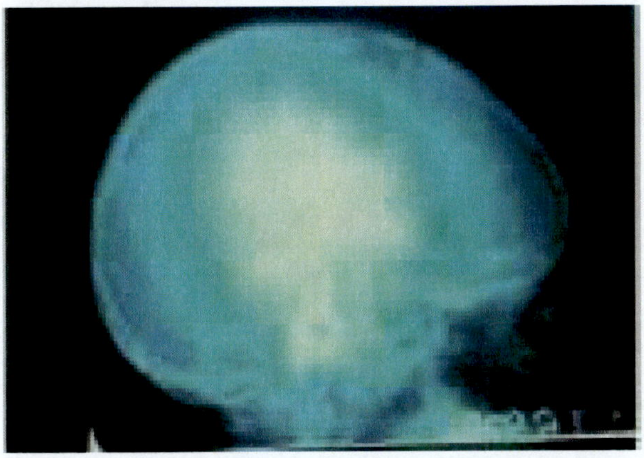

Figure 4. Lateral view of the depressed fracture on superior sagittal sinus.

Three of these patients had GCS score of less than 9 on admission. Four patients had neurological deficit three of whom were evaluated with a CT scan. Two of the patients that had elevation of their fractures required intraoperative blood transfusion of three units each. The management of these patients is shown in table 3.

Table 3. Management of patients with fractures over cranial venous sinuses

Age	Sex	Aetiology	Location	Type	Treatment	Outcome
19	M	RTA	Ant 1/3 SSS	Closed	Nonoperative	Good recovery
27	M	Assault	Transverse sinus	Open	Operative	Good recovery
39	M	RTA	Ant 1/3 SSS	Open	Nonoperative	Died of associated chest injuries
25	F	Assault	Ant 1/3 SSS	Closed	Nonoperative	Good recovery
22	M	Falling object	Mid 1/3 SSS	Open	Operative	Good recovery
47	M	Assault	Ant 1/3 SSS	Open	Nonoperative	Moderate disability
36	M	Assault	Ant 1/3 SSS	Open	Operative	Good recovery
18	M	Falling object	Ant 1/3 SSS	Open	Nonoperative	Good recovery
24	M	Assault	Mid 1/3 SSS	Open	Nonoperative	Died. Had severe head injury
35	F	RTA	Ant 1/3 SSS	Closed	Nonoperative	Good recovery
16	M	Falling object	Mid 1/3 SSS	Open	Nonoperative	Good recovery
42	M	Assault	Ant 1/3 SSS	Open	Operative	Good recovery
23	M	RTA	Transverse sinus	Open	Nonoperative	Good recovery

Ant 1/3 = Anterior one-third; Mid 1/3 = Middle one-third; SSS= Superior sagittal sinus.

Severity of Head Injury and Outcome of Treatment

At the time of admission the GCS score ranged between 3 and 15, mean 12 ± 1.7SD. Majority of the patients (59.6%) had mild head injury. In 19 patients the admission GCS score was 15 but deteriorated to between 7 and 13 within 6 hours of admission. Associated intracranial pathology found is shown in table 4. Associated injuries in areas other than the head were found in 26 (11.1%) patients. The outcome of treatment is shown in table 5. No patient surviving in persistent vegetative state. The admission GCS score was a good predictor of outcome of treatment ($p<0.001$). There was a strong correlation between severity of head injury and GOS score ($r = 0.55$; $p= 0.001$). Similarly, the presence of intracranial pathology significantly affected the outcome of treatment ($p= 0.001$). Open fractures were more frequently associated with intracranial pathology (49.4%) than closed fractures (27.7%) and had poorer outcome of treatment. Patients whose fractures resulted from RTA had a worse outcome compared to those whose injuries resulted from other causes ($p= 0.001$).

Wound infection occurred in 36 (15.3%) patients consisting of 4(4.8%) with closed DSF and 32(21.0%) of those with open fractures.

Table 4. Head injury pathology and mortality in 235 patients

Pathology	Patients No.	Patients %	Mortality No.	Mortality %
DSF only				
Closed	60	25.5	0	0.0
Open	77	32.8	1	1.3
DSF and:				
Dural laceration only	34	14.5	0	0.0
Extra dural Haematoma	18	7.7	1	5.6
Sub dural Haematoma	32	13.6	11	34.4
Intra cerebral Haematoma	10	4.3	4	40.0
Other	5	2.1	0	0.0

DSF =depressed skull fracture.

Table 5. Severity of head injury and outcome of treatment

GCS score	Good recovery No	Good recovery %	Moderate disability No	Moderate disability %	Severe disability No	Severe disability %	Persistent vegetative No	Persistent vegetative %	Death No	Death %
3-8 (n=28; 11.9%)	7	25.0	6	21.4	1	3.6	0	0	14	50.0
9-12 (n=67; 28.5%)	50	74.6	15	22.4	0	0	0	0	2	3.0
13-15 (n=140; 59.6%)	128	91.4	11	7.9	0	0	0	0	1	0.7

Among the patients with open fractures that were operated 22(21.3%) had infection (including two meningitis) compared to 10(20.4%) of the observation group. At the time of discharge from hospital 185 (78.7%) patients had complete recovery while 33 (14.1%) had residual focal neurological deficit including hemiparesis 6, paraparesis 1, dysphasia 3, and seizures 8. Thirty two of these patients had significant improvement and were living an independent life when seen 1 to 6 months after discharge. Of the 17 (7.2%) patients that died in this study, 10 died within 8 hours of injury including the 4 patients with GCS 12-15 that died of associated blunt chest and abdominal injuries. The highest mortality was in patients with GCS score of 3 to 5 where mortality rate was 78.0%. One patient that presented with an open fracture seven days after trauma died of intracranial infection before operation.

Follow-up ranged between 1 and 42 months with a mean of 8.5 months. In 40.0% of patients follow –up was for more than 12 months.

Discussion

In this study, DSF occurred in 3.6% of head injured patients which agrees with 2.0-6.0% reported by others [23,33]. The sociodemographic characteristics were directly related to the incidence of HI. Highest risk of DSF was found in the third decade, accounting for 39.2% of our patients. Similar results were reported from other studies [2,17,20]. The third decade is the most active and adventurous age group and therefore more involved in RTA, the major cause of DSF in this study. Annegers et al and Mackenzie et al described 15 to 24 years as the most adventurous age group carrying higher risk of head injury in the United States [34,35]. The male to female ratio of 3.8 to 1 agrees with other reports [19,20,25]. This reflects male preponderance for exposure to head injury compared to females.

Road traffic injury was the most common cause of DSF in this study accounting for 66.0%. This agrees with 50-83% reported from developing countries [3,23,30]. However a report from Pakistan showed that RTA caused 31.3% of skull fractures [18]. Most injuries were from motor vehicle crashes (32.8%). Motor cyclists also constituted a significant portion of our patients (29.0%). This indicates an inappropriately managed road traffic system. In Nigeria, there is poor knowledge and practice of road safety measures by the general population. While use of seat belt and crash helmet is not mandatory, high speed driving is very common. Licensing authorities and the mechanisms of checking road worthiness of vehicles are ineffective. These in part, accounted for high incidence of RTA which frequently results in head injuries. Recent trends in developed countries indicate a reducing incidence of vehicular-related severe head injury [36]. This is attributed to improved traffic control, compliance with road safety measures and public prevention campaign [36, 37]. In addition, there have been improvements in vehicle design to make them safer for occupants in the event of a collision. These include automatic air bags, head restrain and automatic roll bars for vehicles that overturn in a collision [36]. Violence is the second most frequent cause of DSF in our patients. Majority of these patients were young males presenting after social or political events. Violence-related HI is also an important health issue in developed countries. In England, Gilthorpe et al reported 3756 patients with head injury caused by assault over a 2 -year period [38]. Most of these patients were young males aged 15 to 44 years generally presenting after social events [38]. In China, the incidence of violent HI increased from 2.72% between 1991 and 1998 to 9.46% a decade later [39].

In our patients, 74.9% of DSF were located on the frontal and parietal bones. Similar results have been documented in other reports [19,23,33]. This is because the location is prone to an assailant attack and is the most common site of impact following acceleration-deceleration injury in RTA. While severe neurological signs were observed with temporal and parietal depressed fractures, occipital fractures (figure 5) were associated with high mortality.

Imaging Studies

About three decades ago, Desmet et al and Mastres shifted the emphasis in diagnostic imaging of head injured patient away from identification of skull fracture as an end point to the diagnosis of intracranial squealae [40,41]. This resulted in a shift from skull radiograph to CT scan of the head when clinical suspicion warranted imaging.

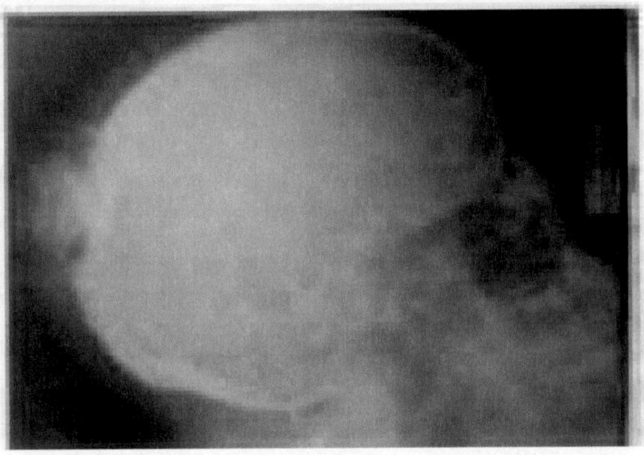

Figure 5. Depressed skull fracture on the occipital region.

Therefore, CT scan is the imaging modality of choice in patients with DSF and is invaluable for rational surgical intervention [42]. Skull fracture on the calvarium can be well demonstrated. Extent of depression of the displaced fragment can be accurately assessed and lesions that may indicate dural tear may be seen [42,43]. More importantly, CT is both sensitive and specific in detecting the associated intracranial lesions that are frequently seen in patients with DSF [43,44,45]. Location of these lesions in relation to the skull base and lobes of the brain can be precisely determined [46,47]. Fractures that involved air sinuses or mastoid air cells may show fluid levels. If an emergency CT is planned then there is no need for skull radiography. Following skull X-ray, DSF are seen as bone condensation, linear or curvilinear when seen enfaced and as depression tangentially. Aerocoeles may also be seen [46]. However, intracranial haematoma are not seen on plane skull radiographs. In this study, 26.4 % of patients had CT scan as part of the evaluation of DSF. This is lower than 70.4% reported by others [23]. In our setting, the use of CT was limited by cost and frequent failure of equipments. National Institute for Clinical Excellence (NICE) guidelines suggests that skull radiographs have important role when combined with careful inpatient observations where CT scan is not available [43]. Therefore, skull X-ray remains an invaluable and efficacious tool in the management of DSF in our patients [45,46,48].

Treatment of Depressed Skull Fracture

In this study, 67.8% of patients with open DFS were treated operatively compared to 30.0% of those with closed fractures. The most common indication of operative treatment was associated intracranial haematoma in 60 (46.8%) patients. Cosmetic defect was the indication for operation in 7.8% of our patients compared to 14.7% and 67.8% reported by Mumtaz et al and Mlay and Sayi respectively [18,48].

In the past, surgical elevation of closed DSF was usually recommended if the fractured fragment is depressed beyond the full thickness of the adjacent bone with the theoretical benefit of better cosmesis, decreased late onset epilepsy and reduction in incidence of persistent neurological deficit [15,24,49]. However, several reports have shown that such fractures can be safely managed by observation because DSF alone rarely causes cerebral compression and that epilepsy remains uninfluenced whether the depressed bone is elevated or not [16,50]. In this study, surgical elevation was considered when there was evidence of underlying brain injury, intracranial haematoma or signs of cerebral compression such as hemiplegia or aphasia. Of our 83 patients with closed fractures, 25 were operated on while 58(70.0%) were safely observed. Our patients were observed even when the depression was greater than full thickness of adjacent skull provided the patient was conscious and did not have neurological deficit.

Open DSF is a neurosurgical emergency that should be treated promptly and properly. Traditionally the presence of open DSF is a mandatory indication for elevation of the depressed fragment [15,23,48]. This is because of high risk of infection which is associated with persistent neurological deficit, late epilepsy and mortality. Several reports have argued against automatic operative treatment of all open fractures [16,24,33]. In this study, 49 (32.2%) patients with open DSF were managed nonoperatively. Forty five of these patients had no clinical or radiological evidence of significant intracranial haematoma or underlying brain injury. In these patients, surgery was limited to scalp wound debridement, irrigation and closure. The incidence of infection in the observed group (20.4%) was similar to the operative group (21.3%). Others have also reported observational treatment of open DSF with minimal risk of infection [24,33]. In addition, in our patients there was no significant difference in outcome between the operated and observed patients with respect to seizures, persistent neurological deficit and cosmesis. This and other studies demonstrated that a select group of patients with open DSF can be safely managed without surgery [15,33,50]. A recently published guidelines for surgical management of traumatic brain injury recommends conservative management of open DSF in patients that do not have evidence of dural penetration, significant intracranial haematoma, pneumocephalus or gross wound contamination [24]. This has been our practice for the past decade. The median interval between injury and operative intervention was 4 days in our patients similar to 4.8 days reported from Pakistan but lower than 12.5 days reported from Ibadan, Nigeria [18,23]. This is more than the 72 hours post-injury time lapse for successful replacement of bone fragment [51]. This and the severe wound contamination preclude replacement of fractured bone fragments in our patients. Bone fragment removal often necessitates a second operation to repair the resultant calvarial defect. This is usually not performed in our setting [23,48]. Replacement of bone fragments have been advocated since1888 by Macewen [52]. Kriss et al reported 2 instances of infection after a long-term follow-up of 79 patients with open DSF treated by immediate replacement of bone fragments [53]. Both Braakman and Jannett have also demonstrated the safety of immediate bone fragment replacement [15,16]. In contrast, other reports have shown a significant association between bone fragment replacement and infection [18, 54]. Curry and Frim demonstrated that optimizing the cerebral perfusion pressure allows for as long as 12 days of preoperative delay under antibiotic cover without neurological sequelae [55]. Whether this allows for safe replacement of bone fragments remain to be established.

Depressed Skull Fractures Overlying Venous Sinuses

The importance of DSF overlying a dural venous sinus lies in the probability of laceration of the vessel wall with resulting profuse bleeding, which may occur at the time of injury or at the time of elevation in the operating room. In our patients as in others, blow on top of the head with an implement by an assailant was the most common cause of these fractures [56,57]. Because assailants often face each other, injury to the anterior cranium is more common. This explains the high frequency (61.5%) of fractures on the anterior one-third of the superior saggital sinus. Neurological deficit and prolonged decreased level of consciousness were seen in 30.7% and 23.0% of patients. Raised intracranial pressure resulting from DSF occluding the saggital sinus has been described but was not seen in our patients [58,59]. A preoperative angiography with venous flow phase has been recommended by some authors [56,57]. This provides information regarding the position and extent of occlusion and transverse sinus dominance. Ten (77.0%) of our patients were treated based on clinical and skull x-ray (figures 3and 4) findings alone. In these patients, the decision to operate is based on the neurological status of the patient, the location of the sinus involved and the extent of venous flow compromise. Fractures with dural violation that have transgressed venous sinuses ware treated conservatively if the patient was neurologically stable. In fractures associated with injury to the brain surgical management is deemed appropriate. However, surgery is often associated with high incidence of severe intraoperative haemorrhage [56,59]. Of the 13 patients with DSF overlying venous sinuses, 9 (69.3%) were treated nonoperatively. This is similar to reports from other centres [56,57]. The obvious bias in favour of conservative treatment is admitted. However, conservative treatment is not only acceptable but also a safer course of treatment.

Neurological Outcome

The outcome of management of patients with DSF is largely dependent upon the severity of any associated primary brain injury. In the present series GCS score was a reliable predictor of outcome, there being a strong correlation between head injury severity, as determined by GCS scores, and the Glasgow Outcome Scale. Of the 28 patients that presented with severe HI, 7 were discharged with persistent neurological deficit while 14(50.0%) died, whereas 99.3% of patients who presented with mild head injury ended up with a moderate disability or good recovery (Table 5). Fractures resulting from road traffic injury had poor outcome because of the severity of HI and frequently associated abdominal or chest injuries. Our infection rate of 15.3% compares favourably to 31.5-34% reported from Nigeria [23,60]. However, it is higher than 1.9-10.6% reported by others [16,50,54,61]. To reduce the risk of infection all patients with open DSF were given antibiotics. However, Al-Haddad and Kirollis reported no statistically significant association between use of antibiotics and reduction of infection [61]. We noted prolonged injury-to-arrival time at hospital, due to delayed rescue from the scene of accident. The time element and absence of pre-hospital care impact adversely on the outcome of treatment. The absence of patients surviving in a persistent vegetative state may indicate inadequate intensive and surgical ward care that does not enable such patients to survive. Of all our patients, 78.7% had good recovery with complete resumption of normal life while additional 13.6.0% had moderate disability though live an independent life. Of the 17 (7.2%) patients that died in this study, 4 died of injuries in areas other than the head. Multitrauma patients may require further investigations and surgery for injuries to other systems, and this may influence the course and outcome of management. Emejulu et al reported mortality of 5.6% following operative treatment of open DSF. However, 92.6% of their patients were referred from other medical facilities several days after injury [23]. Our mortality of 7.2% agrees with 1.4-19.0% reported by others [15,18,33,61]. Our results of management of DSF compare favorably to other reports [8,18,23,61]. Therefore, management of DSF by general surgeons remains an acceptable method of health provision in the West African Subregion as it is unlikely that neurosurgeons will become more readily available in the near feature [29]. Similar observations have been made from other parts of the world [25,26,27,62].

Implications for Surgical Training

Against the growing concern that specialist procedures should not be carried out by infrequent operators, HI patients would continue to be treated by general surgeons in our hospitals. This has important implications for training. The need for basic competence in procedures for neurotrauma has been recognized to be part of general surgical training by several training programs [26,63,64]. The Royal Collage of Surgeons of England recommends that surgeons appointed to remote areas should be trained to carry out burr hole and craniotomy safely and effectively in selected cases [63]. The Royal Australian Collage of Surgeons also requires a trainee surgeon to acquire adequate knowledge of the management of the patient with severe HI and craniotomy for acute intracranial collection [64]. In Nigeria, Fellowship trainees in general surgery are required to have broad-based training necessary for service in rural areas including skills in surgical management of head injury [65]. Such competency is acquired by spending several months training in a neurosurgical registrar level. However, it is unfortunate to note that neurosurgery has become an unpopular elective for surgery trainees who prefer rotations in areas constituting greater portion of general surgical practice. Thus in many instances, the level of neurotrauma surgical experience gained during training is often inadequate [27,62]. Therefore, it is necessary to train our residents in the basics of medical treatment of HI and improve their operative skills so that they can handle any emergency surgical procedure in head injury. The scope of training should include clearly defined protocol for assessment of the patient, decision making, scope of surgery and the follow-up [27,65]. Ajayi et al discussing surgery in Nigeria said "training programs must ensure that the specialist is adequately equipped to deal with conditions that may not be considered general surgery" [65]. He added that "Superspecialization and increasing fragmentation of surgical disciplines are not the pressing issues of surgical development in Nigeria at the moment" [65]. Perhaps, this applies to many other developing countries too.

Conclusion

The results of our study show that DSF are not uncommon in patients with head injury. Because of economic, social and geographical factors the urgent surgical management of the head injured patient for optimal outcome must necessarily be undertaken by general surgeons. We recommend that patients with closed DSF without neurological deficit should be managed nonoperatively unless there is significant cosmetic deformity. The complications and sequelae of open DSF are minimized by early diagnosis and treatment. Visual examination of the skull through the scalp laceration may not reveal a fracture. Careful digital examination of the scalp wound with a globed finger under local anaesthetic and aseptic conditions may reveal a bone edge, depression or a mobile bony fragment. Skull radiograph is an adjunct to effective and safe treatment but should not be performed if a CT can be performed immediately. These patients can be safely managed nonoperatively if there is no clinical or radiological evidence of dural laceration, significant intracranial haematoma, gross cosmetic deformity or gross wound contamination. However, it is essential to initially obtain an accurate assessment as reference point for gauging improvement or deterioration in the patient to enable prompt changes in management to optimize neurological recovery. The wound should be thoroughly irrigated with copious amount of normal saline and hydrogen peroxide to remove hair and debris. Debridement of apparently non-viable tissue should be carried out but wide removal of healthy tissue should be avoided to prevent creating a defect that may require more extensive surgery to repair. The scalp can then be sutured in one layer using a monofilament suture material. All management strategies for open DSF should include antibiotics. Using these principles, the majority of our patients had operative management, although a select group was managed nonoperatively based on clinical and radiological findings. In over 92% of our patients the outcome of management was good with regard to functional and cosmetic effects. Head injury prevention is the primary goal in the management of these patients. RTA, the leading cause of DSF in this study can be prevented by enactment and enforcement of road safety regulations, public enlightenment and proper road networks. The morbidity and mortality in these patients can be minimized by appropriate prehospital care and prompt and safe transportation of patients to a hospital where hypoxia and hypotension can be treated. Hospital-based improvements would significantly reduce mortality and prolong disability.

REFERENCES

[1] Sosin D, Sniezek J, Thuman D. Incidence of mild and moderate brain injury in the United States. *Brain Inj.* 1996; 10: 47-54.

[2] Muhammad I. Management of head injuries at the ABU hospital, Zaria. *East Afr. Med. J.* 1990;67: 447-451.

[3] Perel PA. Predicting outcome after traumatic brain injury: practical prognostic models based on large cohort of international patients. *Br. Med. J.* 2008; 336: 425-429.

[4] Ezenwa AD. Trends and characteristics of road traffic accidents in Nigeria. *J. R. Soc. Health* 1986; 106: 27-29.

[5] Oluwasanmi AJ. Road accident trends in Nigeria. *Accid. Prev.* 1993; 25:485-487.

[6] Borczuk P Predictors of intracranial injury in patients with mild head trauma. Ann *Emerg. Med.* 1995; 25: 731- 736.

[7] Macpherson BC, Macpherson P, Jennett B. Evidence of intracranial contusion and haematoma in relation to the presence, site and type of skull fracture. *Clin. Radiol.* 1990; 42:321-326.

[8] Servadei F, Ciucci G, Pagano F, Rebucci GG, Ariano M, Piazza G, Gaist G. Skull fracture as a risk factor of intracranial complications in minor head injuries: A prospective CT study in a series of 98 adult patients. *J. Neurol. Neurosurg. Psychiatry* 1988; 51: 526-528.

[9] Liko O, Chalau P, Rosenfield JV, Watters DAK. Head injuries in Papua New Guinea *PNG Med. J.* 1996;39:100-104.

[10] Ono K, Wada K, Takahara T, Shirotani T. Indications for computed tomography in patients with mild head injury. *Neurol. Med. Chir* (Tokyo) 2007;47: 291-298.

[11] Stein SC. Management of minor closed head injury. *Neurosurgery* Q 1996; 6:108-115.

[12] Kay A, Teasdale G. Head injury in United Kingdom. *Word J. Surg*. 2001;25:1210-1220.

[13] Chan KH, Mann KS, Yue CP, Fan YW, Cheung M. The significance of skull fracture in acute traumatic haematoma in adolescents: A prospective study. *J. Neurosurg.* 1990; 72: 189- 194.

[14] Hung CC, Chiu WT, Lee LS, Lin LS, Shih CJ. Risk factors predicting surgically significant intracranial haematoma in patients with head injuries. *J. Formos Med. Assoc.* 1996; 95: 294-297.

[15] Braakman R. Depressed skull fracture: Data, treatment and follow-up in 225 consecutive cases. *J. Neurol. Neurosurg. Psychiatry* 1972; 35: 395-402.

[16] Jennett B, Miller JD. Infection after depressed fracture of the skull: Implications fro management of non-missiles injuries. *J. Neurosurg.* 1972; 32:333-339.

[17] Adeloye A, Olumide AA Pattern of skull fractures in Nigerians. *Medicine d'Afrique Noire* 1976; 23:109-113.

[18] Mumtaz A, Liaqat A, Inayat SR Surgical management of depressed skull fracture. *J. P. M. I.* 2007; 17: 116-123.

[19] Sunay YM, Mahmut A, Gurcel C, Yasemin GB, Muzaffer A. The correlation between skull fractures and intracranial lesions due to traffic accidents. *Am. J. Forensic Med. Pathol.* 2003;24 :339- 345.

[20] Flesha SO, Daramola AO Fatal head injuries: The Lagos University Teaching Hospital experience (1993-1997). *Nig. Postgrad. Med. J.* 2002; 9: 38-42.

[21] Akang EE, Kuti MA, Osunkoya AO, Komolafe EO, Malomo AO, Shokunbi MT, Amutta SB. Pattern of fatal head injuries in Ibadan; a 10 year review. *Med. Sci. Law* 2002;42: 160-166.

[22] Oh S Clinical and experimental morphological study of depressed skull fracture. *Acta Neurochirur* 1983; 68: 111-121.

[23] Emejulu JKC, Shokunbi MT, Malomo AO, Adeloye A Causes of delay in the definitive treatment of compound depressed skull fracture: A five year study from Nigeria. *East* Cent. Afr. J. Surg. 2006;12 :116-122.

[24] Bullock RM, Chesnut R, Ghajar J, Gordon D, Hartl R, Newel DW, Servadei F, Walters BC, Wilberger J. Surgical management of depressed cranial fractures. *Neurosurgery* 2006; 58 (3 Suppl): S56-60.

[25] Treacy JP, Reilly P, Brophy B. Emergency neurosurgery by general surgeons at a remote major hospital. *ANZ J. Surg*. 2005; 75;852-857.

[26] Esposito TJ, Luchette FA, Gamelli RL. Do we need neurosurgical coverage in trauma centre? *Advances in Surgery* 2006;40: 213-221.

[27] Dubey AK Head injury: Should a trauma or general surgeon operate? *I J N T 2006*; 3:69-71.

[28] Raja IA, Vohra AH, Ahmed M. Neurotrauma in Pakistan. *World J. Surg*. 2001;25 : 1230-1237.

[29] Yeboah ED Middle level man power development. *Paper presented at the 46th conference of West African Collage of Surgeons*. February 4th-11th 2006. Accra, Ghana.

[30] Adesunkanmi ARK, Akinkuolie AA, Badru OS. A five year analysis of death in accident and emergency room of a semi-urban hospital. *West Afr. J. Med.* 2002;2:99-104.

[31] Teasdale G, Jennett B. Assessment of coma and impaired consciousness: a practical scale. *Lancet* 1974;2 : 81-84.

[32] Jennett B, Bond M. Assessment of outcome after severe brain damage: A practical guide. *Lancet* 1975; i :480-484.

[33] Heary RF, Hunt CD, Krieger AJ, Schulder M, Vaid C. Nonsurgical treatment of compound depressed skull fractures. *J. Trauma* 1993; 35:441-447.

[34] Annegers JF, Grabow HD, Kurland LT, Laws ER. The incidence causes and secular trends in head injury in Olmsted County, Minnesota, 1935-1974. *Neurology* 1980;30:912-919.

[35] Mackenzie EJ, Edelstein SL, Flynn JP. Hospitalized head injured patients in Maryland: incidence and severity of injuries. *Md Med. J.* 1989;38: 725-732.

[36] Ross A, Bagunley C, Hills B, McDonald M, Silcock D. Towards safer roads in developing countries. A guide for planners and engineers. *Transport Research Laboratory*, 1991 Crowthorne UK.

[37] Odero W, Garner P, Zwi A. Road traffic injuries in developing countries-a comprehensive review of epidemiological studies. *Trop. Med. Int. Health* 1997;2:445-460.

[38] Gilthorpe M, Wilson R, Moles D. Variations in admission to hospital for head injury and assault to the head. Part 1: Age and gender. *Br. J. Oral Maxillofac. Surg.* 1999; 37:294-300.

[39] Jiyang JY, Feng H, Fu Z, Guo G, Wei-ping L, Wei-guo L et al Violent head trauma in China: Report of 2254 cases. *Surgical Neurology* 2007; 68 (S2): S2-S5.

[40] Desmet AA, Fryback DG, Thombury JR. A second look at the utility of radiographic skull examination for trauma. *Am. J. Radiol.* 1979;132: 95-97.

[41] Masters SJ. Evaluation of head trauma: efficacy of skull films. *Am. J. Radiol.* 1980; 135: 539-547.

[42] Harrison M, Jones S. Tangential views or computed tomography in suspected depressed skull fracture. *Emerg. Med. J.* 2002; 19: 329-330.

[43] National Institute foe Clinical Excellence (NICE). Head Injury. Triage, assessment, investigation and early management of head injuries in infants, children and adults. *Clinical guideline* number 4, London: NICE, 2003.

[44] Kristen IM, Thomas LS, Alexa IC, David JA, Cynthia LA. Appropriate imaging in children with skull fractures and suspicion of abuse. *Radiology* 1998;208:521524.

[45] Ogunseyinde AO, Obajinmi MO, Ogundare SM. Radiological evaluation of head trauma by CT in Ibadan, Nigeria. *West Afr. J. Med.* 1999; 18: 33-38.

[46] Njeze NR, Umerah BG, Mgbor SO Skull radiograph in head injuries. *West Afr. J. Radiol.* 2001;8 : 1-11.

[47] Obajimi MO, Ogunseyinde R, Sa'ad ST. Computed tomography (CT) - Evidence of skull fracture in relation to the presence and site of intracranial contusion and haematoma. *Nig. Postgrad. Med. J.* 1999;6 :53-55.

[48] Mlay SM, Sayi EN. The management of depressed skull fracture in children at Muhimbili medical centre Dar es Salam, Tanzania. *East Afr. Med. J.* 1993;70:291-293.

[49] Shokunbi MT, Bamgbade AO, Malomo AO, Ogunseyinde OA, Olumide AA, Adeloye A. Focal neurological deficits are predictive of laceration of the dura in children with compound depressed skull fracture. *Nig. Q. J. Hosp. Med.* 1997; 7:95-98.

[50] Van der Heever CM, Van der Merwee DJ. Management of depressed skull fracture. Selective conservative management of nonmissile injuries. *J. Neurosurg.* 1989;71:186-190.

[51] Wylen EL, Willis BK, Nanda A. Infection rate with replacement of bone fragment in compound depressed skull fracture. *Surg. Neurol.* 1999; 51:452-457.

[52] Macewen W. Address on the surgery of the brain and spinal cord. *Lancet* 1888; 2: 254-261.

[53] Kriss FC, Taren JA, Kahn AE. Primary repair of compound skull fracture by replacement of bone fragments. *J. Neurosurg.* 1969; 30: 698-702.

[54] Rehman L, Ghani E, Hussain A, Shah A, Noman MA, Khaleeq-Uz-Zaman. Infection in compound depressed fracture of the skull. *J. Coll. Physicians Surg. Pak.* 2007; 17: 140-143.

[55] Curry DJ, Frim DM. Delayed repair of open depressed skull fracture. *Paediatr. Neurosurg.* 1999;39:294-297.

[56] LeFeuvre D, Taylor A, Peter JC. Compound depressed skull fractures involving a venous sinus. *Surg. Neurol.* 2004; 62: 121-126.

[57] Ozer FD, Yurt A, Sucu HA, Tektas S. Depressed fracture over cranial venous sinus. *J. Emerg. Med.* 2005; 29: 137-139.

[58] Dalgic A, Secer M, Ergungor F, Okay O, Akdag R, Ciliz D. Dural sinus thrombosis following head injury: Report of two cases and review of the literature. *Turkish Neurosurg.* 2008; 18: 70-77.

[59] Fuentes S, Metellus P, Levrier O, Adetchessi T, Dufour H, Grisoli F. Depressed skull fracture overlying the superior saggital sinus causing benign intracranial hypertension. Description of two cases and review of the literature. *Br. J. Neurosurg.* 2005; 438-442.

[60] Shokunbi MT, Komolafe EO, Malomo AO, Amanor-Boadu DS, Sanusi A, Olumide AA, Adeloye A. Scalp closure without fracture elevation does not reduce the risk of infection in patients with compound depressed skull fracture. *Afr. J. Med. Sci.* 2000; 29: 293-296.

[61] Al-Haddad SA, Kirollos R. A 5-year study of the outcome of surgically treated depressed skull fractures. *J. R. Coll. Surg. Engl.* 2002; 84: 196-200.

[62] Bishop CV, Drummond KJ. Rural neurotrauma in Australia: implications for surgical training. *ANZ J. Surg.* 2006;76: 53-59.

[63] The Royal Collage of Surgeons of England. *Report of the working party on the management of patients with head injury.* 1999.

[64] The Royal Australian Collage of Surgeons. *Surgical Education and Training Handbook*. 2004; Sec 11, pp 67-82.

[65] Ajayi OO, Adebamawo CA. Surgery in Nigeria. *Arch. Surg.* 1999; 134: 206-211.

In: Road Traffic: Safety, Modeling and Impacts
Editors: S. E. Paterson and L. K. Allan, pp. 279-300
ISBN 978-1-60456-884-4

Chapter 4

CONSTRUCT AND CRITERION VALIDITY OF AN OBJECTIVE MEASURE OF RESPONDENTS' SUBJECTIVELY ACCEPTED LEVEL OF RISK IN ROAD TRAFFIC

Andreas Hergovich[1,*], Martin E. Arendasy[2], Markus Sommer[2] and Bettina Bognar[3]
[1]Social Psychology Research Group
[2]Dept. of Psychology, Differential & Personality Research Group
[3]Fa. Dr. G. Schuhfried, Mödling
University of Vienna, Austria

ABSTRACT

This chapter outlines a theory-based approach to the construction and psychometric evaluation of more behavioural based measures of willingness to take risks in road traffic. Based on the risk homeostasis theory (Wilde, 1994) several traffic situations varying on the degree of objective danger where filmed. Respondents were asked to indicate at which point the action that is contingent to the described situation will become too dangerous for them to carry out. Latencies at the item level were obtained. The first study deals with the dimensionality of this newly developed measure which is referred to as the Vienna Risk Taking Test Traffic. The second study investigates the generalizability of these results to a sample of professional driver applicants. The third study locates the newly developed measure within the context of the Giant Three model of personality. Confirmatory factor analyses indicated that the Vienna Risk Taking Test Traffic loads on the Giant Three factor 'Adventurousness', thereby confirming the construct validity of the behavioural based measures of willingness to take risks. Two additional studies investigate the criterion validity of the newly constructed measure using accident rates as well as respondents' performance in a standardized driving test as criterion measure. Despite the wealth of literature supporting the predictive validity of personality-related and ability-related determents of fitness to drive these two kinds of predictor variables have previously been studied in isolation. In order to overcome this methodological shortcoming the incremental validity of the Vienna Risk Taking Test over and above a rather comprehensive set of ability and personality tests was investigated. In line with theoretical considerations the newly developed objective personality test contributed incrementally to the prediction of respondents' performance in a standardized driving test as well as to the prediction of their accident rates. The theoretical implications of the results presented in this chapter are discussed in the light of current theoretical model on fitness to drive.

INTRODUCTION

Driving a motor vehicle is one of the most dangerous things we do in everyday life (Evans, 1991). Austrian accident statistics (Statistik Austria, 2007) show that in 2005 alone there were 33,711 recorded accidents in which cars were involved. In that year there were 432 deaths as a result of accidents involving a motor car. Even though the number of road traffic accidents has fallen over the past ten years, incidence remains high and such accidents cause serious financial and personal damage to society as a whole. In Austria the direct financial damage due to road traffic accidents amounts to 5 billion euros per year. Furthermore, the annual follow-up costs of road traffic accidents have been estimated at 3.6 billion euros. Due to the fact that human failure had been identified as the principle cause in a significant proportion of traffic accidents (cf. Evans, 1991), it is hardly surprising, that the determinating factors of risky accident proneness and fitness to drive have been the focus of attention in traffic psychological research for a long time.

One problem with the current available research on the personality and ability determants of accident proneness and fitness to drive is that driving-related ability traits and personality traits have frequently been studied in isolation. While some studies focus entirely on the predictive validity of ability tests (e.g. Bukasa, Wenninger & Brandstätter, 1990; Bukasa, Christ, Ponocny-Seliger, Smuc, & Wenninger 2003; Karner & Neuwirth, 2000; McKnight and McKnight, 1999; Myers, Ball, Kalina, Roth & Goode, 2000), others have investigated the predictive validity of personality measures without including ability tests in their test batteries (e.g. Burns & Wilde, 1995; Cellar, Nelson, & York, 2000; Dahlen & White, 2006; Dahlen, Martin, Ragan & Kuhlman, 2005; Iversen & Rundmo, 2002; Jonah, 1997; Jonah, Thiessen & Au-Yeung, 2001; Oltedal & Rundmo, 2006). This separation of the research on personality and ability determinants of safe driving behaviour is also apparent in current theoretical models of driving. According to Ranney (1994) ability-based models, on the one hand, focus entirely on perceptual, cognitive and senso-motor abilities, while motivational models on the other hand concentrate almost entirely on personality traits such as lack of acceptance of norms, sensation-seeking and subjective accepted degree of risk. There is at present no coherent and unified theoretical model that takes both determinants into account and clearly

* Correspondence address: Prof. Dr. Andreas Hergovich, Universität Wien, Fakultät für Psychologie, Liebiggasse 5, 1010 Wien, Austria. e-mail: andreas.hergovich@univie.ac.at

specifies the structural relations between the various predictor variables and criterion measures of fitness to drive and safe driving behaviour (Risser, 2000).

Furthermore, research on the personality determants of accident proneness and fitness to drive resorted predominately to conventional personality questionnaires. The main problem associated with these kind of measures is their vulnerability to impression management (Paulhus, 1984) which clearly limits the generalizability of research findings to applied traffic psychological assessment.

OBJECTIVE PERSONALITY TESTS IN TRAFFIC PSYCHOLOGICAL ASSESSMENT

In the past several attempts have been made to develop more behavioural based measures of driving-related personality traits. For instance, Horswill and McKenna (1999) presented video recordings of seven driving situations to their respondents and asked them to indicate the extent to which they would drive faster or slower than the vehicle they saw. The deviation coefficient of the driving speed was used as the measure of willingness to take risks. Although the authors were able to demonstrate that their measure explain a significant proportion of the variance of the number of accidents which could not yet be explained with a paper-and-pencil questionnaire about speed selection, the lack of any reference to current models of risky driving behavior and the absence of studies of its convergent and discriminant validity limits the applicability of this test in applied traffic psychological assessments.

Santacreu, Rubio, & Hernandéz (2006) administered a "Roulette test" as well as a "Betting dies test" and "Crossing the streets test" to their respondents. The task of the respondents was to choose one out of a pre-defined number of alternatives, which differed with regard to their probability of a success and the reward assosiated with a success for each of the alternative choices. The values for these two parameters were set in a manner which results in the same expected reward values (=probability of a success * reward) for each alternative. The risk value for each alternative was defined as the inverse of the likelihood of a success while the 'willingness to take risks'. Although the authors were able to demonstrate satisfactory internal consistencies and a satisfactory convergent validity for their measures, they failed to demonstrate the predictive validity with regard toi traffic safety.

On the other hand, Hergovich, Arendasy, Sommer, Bognar, & Olbrich (2004) used the Risk Homeostasis Theory of Wilde (1994, Burns & Wilde, 1995; Simonet & Wilde, 1997) to construct an objective personality test measuring willingness to take risks. The model assumes that a respondent's risk-taking behaviour in a driving situation is derived homeostatically from the interplay between a subjectively accepted level of risk (target value) and the actual level of risk (actual value) inherent in the traffic situation. The subjectively accepted level of risk is assumed to be a one-dimensional latent trait. Hergovich et al. (2004) measured the subjectively accepted level of risk by showing subjects a series of videos of traffic situations involving road junctions, overtaking, and speed selection situations in good weather conditions. The videos were constructed so that the objective risk level of the situation continually increases; the increase in risk level is brought about by varying features of the situation that affect "time to collision" (Vogel, 2003). Subjects were required to press a button to indicate the point in time at which they would no longer execute a pre-defined driving manoeuvre. The latent personality trait of subjectively accepted level of risk corresponds on a manifest level to the latency until the button is pressed. This latency unambiguously defines the position of the vehicles involved in the particular driving situation and thus constitutes a valid measure of the respondents' subjectively accepted degree of risk. The studies carried out to date have been able to demonstrate that the Vienna Risk-taking Test Traffic indeed measures a one-dimensional latent personality trait (cf. Arendasy, Sommer, Bognar & Hergovich, in revision; Hergovich et al., 2004). Furthermore, initial research on the construct validity of this measure indicated that the Vienna Risk-taking Test Traffic is significantly related to conventional questionairs measuring 'willingness to take risks' while being virtually uncorrelated with measures of reaction speed (Arendasy et al., in revision). In addition, research conducted by Sommer, Herle, Häusler, Risser, Schützhofer & Chaloupka (in press) revealed that the Vienna Risk-taking Test Traffic contributes incrementally to the identification of risky drivers over and above other driving related ability and personality traits.

Formulation of the Problem

Even though the results obtained with the Vienna Risk-taking Test Traffic (WRBTV: Hergovich, Bognar, Arendasy & Sommer, 2005) thus far were encouraging, a more detailed examination of the available empirical research indicated some omissions with regard to the empirical evidence on the construct and criterion valitiy of this measure. Although Arendasy et al. (in revision) were able to demonstrate that the Vienna Risk-taking Test Traffic is significantly related to conventional questionnaires measuring 'willingness to take risks' while being virtually uncorrelated with measures of reaction speed the authors failed to locate their measure within a more conprehensive structure of personality. With regard to the criterion validity current available research indicates that the Vienna Risk-taking Test Traffic is significantly related to measures of fitness to drive such as standardized driving tests but failed to provide empirical evidence on whether this measure is able to simultaniously predict fitness to drive as well as accident proneness. However, such evidence is explicitly required both under Section 18 para. 1 of the health regulations which form part of Austrian driving licence law (FSG-GV) and under the Driving Licence Ordinance (FeV) Appendix 5 in Germany. There is also doubt as to whether the item parameters and norms obtained under laboratory conditions can be generalised to the real-life situation of a traffic psychological assessment. It is these omissions which the present study aims to rectify by examining the dimensionality and validity of an extended item pool for the Vienna Risk-taking Test Traffic.

Overview of the Studies

The aim of the present article is to investigate the psychometric quality of the Vienna Risk-taking Test Traffic (WRBTV) in a series of studies. The first study seeks to reoplicate the findings on the dimensionality and reliability of the Vienna Risk-taking Test Traffic outlined above, while the second study investigated the generalisability of the item parameters and the possibility of intentional faking in the context of a traffic psychological assessment of professional driver applicants. The third study deals with the construct validity of the Vienna Risk-taking Test Traffic by examining the factorial structure of a test battery comprising several scales taken from the Eysenck Personality Profiler (EPP: Eysenck, Wilson, & Jackson, 2000) in addition to the Vienna Risk Taking Test Traffic. The final two studies aim to investigate the criterion validity of the Vienna Risk Taking Test Traffic using retrospective accident frequencies and global evaluations of respondents' performance in a standardized driving test. In all studies the Vienna Risk-taking Test Traffic was administered by trained test administrators in small groups of not more than five individuals, using the Vienna Test System.

Description of Extended Vienna Risk-taking Test Traffic

Each traffic situation video (see figure 1) is shown to the subject twice. Before the first presentation the forthcoming traffic situation is briefly described in words, as is an action contingent on that situation (here: overtaking the vehicle in front when there is snow on the road).

Subjects are shown 23 different traffic situations, which can be divided into two categories based on their content. One category (k=10) comprises videos of traffic situations involving speed selection and overtaking, while the remaining traffic situations (k=13) depict scenarios involving decisions at road junctions. The traffic situations also differ with regard to visibility and weather conditions. In some of the video clips (k=17) visibility and weather conditions are good, while the remaining traffic situations (k=6) depict scenarios in which, as a result of rain, snow, or darkness, visibility and weather conditions are poor. In extending the item pool emphasis was placed on more systematically varying features of the driving situations that affect "time to collision" measues. The mean latency measured and used to assess respondents willingness to take risks, or more precisely, their subjectively accepted level of risk.

Study I

This study investigated the dimensionality and reliability of the Vienna Risk-taking Test Traffic. The data were collected between 2004 and 2007 as part of a nomalization study at the research laboratory of the SCHUHFRIED GmbH. The data was obtained with the aid of trained test administrators; the subjects were given the Vienna Risk-taking Test Traffic, which they completed on computers using the Vienna Test System as administration software.

Sample

The sample consists of 440 (49.2 %) men and 455 (50.8 %) women. aged between 16 and 91. The median age is 44 with a mean age of 45 years 11 months and a standard deviation of 17 years 8 months. The sample can be considered to be representative for Austria and Germany with regard to gender (χ^2 [1] = .02; p = .890) and age ($\square^2$ [14] = 20.84; p = 0.106). A total of 2 (0.2%) individuals have not completed compulsory schooling (EU educational level 1), 90 (10.1%) have completed compulsory schooling or intermediate secondary school but without obtaining a career qualification (EU educational level 2), 331 (37.0%) individuals have completed a vocational qualification or attended technical school (EU educational level 3), 377 individuals (42.1%) have a school-leaving qualification at university entrance level or a higher education diploma (EU educational level 4) and 95 individuals (10.6%) have a university level qualification (EU educational level 5).

Statistical Methods

Rasch (1980) developed an IRT-based model for speed tests in order to investigate the one-dimensionality of all kinds of assessments where the response latency is the variable of interest. In such tests a sample of N subjects is presented with k items and the latency is measured for each item. These item-by-item latencies follow a distribution characterised by the intensity parameter λ. Rasch (1980) specified the model in its general form as follows:

$P(t \mid 1) = \lambda\, exp\text{-}\lambda t$

The probability that an item i will be solved within a time period t is modelled by the intensity parameter λ. According to Rasch (1980) and Jansen (1997), the intensity parameter λ can be decomposed in the multiplicative combined parameters θ_v and ε_i. The parameter θ_v represents the speed of a person v, while parameter ε_i stands for the easiness of item i. This means the higher θ_v and ε_i, the shorter the expected reaction times.

Fischer and Kisser (1983) demonstrated that this multiplicative decomposition using Marginal Maximum Likelihood (MML) estimatttes has the disadvantage that sufficient statistics for θ_v and ε_i cannot be derived. The authors therefore propose an additive decomposition of the intensity parameter λ resorting to a Conditional Maximum Likelihood (CML) estimation method.

Figure 1. Example of a picture taken from the video clip showing an overtaking situation in the Vienna Risk-Taking Test Traffic (WRBTV).

In order to examine the fit of the model to empirical data sets, Scheiblechner (1985) recommends the use of a Likelihood Quotient Test (LQT) according to Andersen (1973) on the basis of a CML estimation of the item parameter. In this LQT the likelihoods of model estimates of varying restrictiveness are related to each other, and this estimate is transferred into a χ^2 statistic for inferential statistical purposes. According to Rost (2004) According to Rost (2004), the calculation of this fit statistic represents an examination of the assumption of person homogeneity, that is, the statistical equivalence of item parameter estimations in different subgroups of people compared to the overall sample. Analogous to this, the assumption of item homogeneity (Rost, 2004) of the latency model, i.e. the one-dimensionality of a scale's hypothetically selected item subsets, can be examined via the LRT developed by Martin-Löf (1973). In case the Martin-Löf (1973) statistic fails to reach the significance level, the null hypothesis that the theoretically formed item subsets measure essentially the same one-dimensional latent trait can be retained. This essentially constitutes a check of the generalizeability over the investigated item or situation sets. In case aaall of these model tests fail to reach the significance level the assumption that the test measures a one-dimensional latent trait can be retained. In the present study the Latency Model is applied to the item-by-item reaction times obtained for the enlarged item pool of the Vienna Risk-taking Test Traffic.

Results

The mean latencies of the 23 items varied between 2.32 seconds and 19.08 seconds with a mean of 7.58 seconds and a standard deviation of 3.62 seconds. Internal consistency as measured by Cronbach's α is .916.

The assumption of person homogeneity (Rost, 2004) was tested by means of the Latency Model using the classic partitioning criteria raw score, gender, age and educational level. Due to the number of model tests carried out α was set a priori at .01. The results of the Likelihood Ratio Test are shown in Table 1.

Table 1. Study I: Goodness-of-Fit statistics and effect size of the Latency Model (LRTs)

Partitioning criterion	n	χ^2	p	ω
Raw score[a]	895	28.14	.171	.125
≤ 7.09 sec.	449			
> 7.09 sec.	446			
Gender	895	23.98	.348	.116
female	455			
male	440			
Age	895	27. 72	.185	.124
≤ 44 years	452			
> 44 years	443			
Educational level	895	18.50	.676	.102
EU 1-3	423			
EU 4-5	472			

Note: All dfs = 22. [at]the raw score is calculated as the mean latency.

As Table 1 shows, all of the model tests failed to reach statistical significance at □ = .05. It can thus be assumed that the parameter estimates are generalisable over the sub-samples studied. The assumption of person homogenity can thus be retained.

In addition, the assumption of item homogeneity (Rost, 2004) was tested by dividing the driving situations into two groups: (1) situations involving speed choices and overtaking or decision-making at junctions and (2) situations in good or bad weather situations. For either situations involving speed choices and overtaking, or decision-making at junctions this the assumption of item homogeneity can be retained ($\chi^2[129]= 42.3$; p=.999). The same held true for the The test of item homogeneity based on the division into driving situations in good vs. poor visibility and weather conditions ($\chi^2[101]= 48.5$; p=.999). Thus, the latencies of the three types of driving situation are therefore suitable for measuring a one-dimensional latent trait.

Study II

This study investigated whether the item parameters obtained in the laboratory situation can be transferred to the real-life situation of a traffic psychological assessment of professional driver applicants. This involved comparing the test performance of a sample of individuals, parallelised for age, gender and educational level, who completed the Vienna Risk-taking Test Traffic (WRBTV) in the context of a norming study with the test performance of professional driver applicants, who were tested in the course of a traffic psychological assessment. In contrast to individuals who undergo a traffic psychological assessment as a consequence of having committed driving offences professional driver applicants cannot be assumed to differ significantly from the norm with regard to their subjectively accepted level of risk. One would thus expect no significant differences in the mean latencies between the norm sample and professional driver applicants. The design of this study is therefore also suitable for investigating the effect of any potential impression management (Paulhus, 1984) of the test results in the real-life situation of a traffic psychological assessment.

In order to clarify the question of whether the item parameters can be generalised to the real-life situation of a traffic psychological assessment, the data of the professional drivers and the individuals tested under laboratory conditions were first analysed together with the Latency Model. To investigate whether individuals in the real-life situation are systematically advantaged or disadvantaged by individual video sequences, a model test for the partitioning criterion "setting" (laboratory vs. traffic psychological assessment) was carried out. A significant model test for the splitting criterion "setting" would be expected if intentional faking is more readily facilitated by some video sequences rather than others, enabling individuals in the real-life situation to put themselves at an advantage. The next evaluation step involved using a T-test for independent samples to check whether the two sub-samples differ significantly in their mean latency time. Because the two sub-samples have been parallelised accoring to relevant socio-demographic characteristics, a significant difference in favour of the sub-sample from the real-life situation can be interpreted as an indicator of successful deliberate faking of the objective personality test.

Sample

The sample consists of 180 (75.0%) men and 60 (25.0%) women aged 18 – 64 with an average age of 33.05 and a standard deviation of 12.58 years. The median age is 33. Thirty-eight people (15.80%) in the norm sample had completed compulsory schooling or basic secondary school but without completing vocational training (EU educational level 2), 122 people (50.83%) had completed vocational training or a course at a technical college (EU educational level 3), 60 people (25.00%) had a school-leaving qualification at university entrance level or a qualification from a technical university (EU educational level 4) and 20 people (8.33%) had a university degree (EU educational level 5). Of these subjects, 124 (51.67%) completed the Vienna Risk-taking Test Traffic as part of a norming study in the research laboratory of the SCHUHFRIED GmbH, while 116 (48.33%) individuals took the Vienna Risk-taking Test Traffic as part of a medical/psychological assessment which is mandatory for professional driver applicants in Austria. The two sub-samples did not differ from each other with regard to gender (χ^2 [1] = 1.44; p = .231), educational level (χ^2 [4] = 2.71; p = .438) and age (Z = -.087; p = .930).

Results

The mean latency times of the 23 items varied between 2.35 seconds and 19.29 seconds with a median of 7.82 seconds and an interquartile range of 2.98 seconds. Internal consistency as measured by Cronbach's α is .911. The assumption of person homogeneity (Rost, 2004) was tested by means of the Latency Model using the classic partioning criteria of raw score, gender, age and educational level in addition to the partitioning criterion 'setting'. Due to the number of model tests carried out α was set a priori at .01. The results indicated that neither the partitioning criteria raw score (χ^2[22]= 18.00, p=0.706, ω=.194), gender (χ^2[22]= 19.80, p=.596, ω=.203), age (χ^2[22]= 14.30, p=.890, ω=.173) and educational level (χ^2[22]= 12.00, p=.957, ω=.158), nor the partitioning criteria 'setting' (χ^2[22]= 12.00, p=.957, ω=.158) reach the significance level, supporting the assumption that the parameter estimates can be generalised to the real-life situation of a traffic psychological examination of professional driver applicants. The application of a t-test for independent samples to test for differences between the means also produced a non-significant result (Levene Test: F=.335; p=.563; t-test: t=.618; df=285; p=.537; Cohen's d=.073). The result thus argues against the assumption, that the Vienna Risk-taking Test Traffic can be intentionally faked in real-life situations.

Study III

Previous studies already indicated that the Vienna Risk Taking Test traffic is significantly correlated with conventional measures of risk taking (cf. Hergovich et al., 2004; Arendasy et al., in revision; Vogelsinger, 2005) , while being simultaneously statistically uncorrelated with measures of reaction speed (Arendasy et al., in revision). However, the currently available studies failed to integration of the Vienna Risk-taking Test Traffic into current global models of personality (e.g. Andersen, 1995). This study was thus conducted to fill this gap by investigating the convergent and discriminant validity of the Vienna Risk-taking Test Traffic, utilising other construct-related and non-construct-related subscales taken from the short version of the Eysenck Personality Profiler (EPP: Eysenck, Wilson, & Jackson, 2000) by means of confirmatory factor analysis.

Test Battery

The test battery consists of Vienna Risk-taking Test Traffic (WRBTV), and the subscales 'activity', 'sociability', 'expressiveness', 'inferiority', unhappiness', 'anxiety', 'risk taking', 'impulsiveness' and 'irresponsibility' from the short version of the Eysenck Personality Profiler (EPP: Eysenck, Wilson, & Jackson, 2000). All the tests are computerised and were administered using the Vienna Test System in a single session lasting around 45 minutes.

Sample

The sample consists of 72 (42.6%) men and 97 (57.4%) women aged 18 to 73 years (mean=44.80; SD=14.38). The median age is 44 years. Seventeen (10.06%) respondents had completed compulsory schooling or basic secondary school but without completing vocational training (EU educational level 2), 65 (38.46%) respondents had completed vocational training or a course at a technical college (EU educational level 3), 68 (40.24%) had a school-leaving qualification at university entrance level (EU educational level 4) and 19 (11.24%) had a university degree (EU educational level 5).

Models Tested

On the basis of theoretical considerations and previous studies on the construct validity of the Vienna Risk-taking Test Traffic (for a summary: Hergovich et al., 2005) and the short version of the Eysenck Personality Profiler (EPP) it was assumed that the intercorrelations of the nine EPP subscales and the main variable 'subjectively accepted level of risk' of the Vienna Risk-taking Test Traffic can be explained by three latent factors, which represent the Giant Three. In the tested model it was thus assumed that the EPP subscales 'activity', 'sociability' and 'expressiveness' form a factor referred to as 'Extraversion', while the three EPP subscales 'inferiority', 'unhappiness' and 'anxiety' loading on the factor 'Emotionality'. It was further assumed that the EPP subscales 'risk taking', 'impulsiveness' and 'irresponsibility' form a thrird factor referred to as 'Adventurousness' and that the main variable 'subjectively accepted level of risk' of the Vienna Risk-taking Test Traffic (WRBTV) would load on this factor. Furthermore, it was assumed that the Giant Three factors would be correlated.

Results

The calculations were carried out using AMOS 5.0 (Arbuckle, 2003). Maximum Likelihood was used as the estimation algorithm. Table 2 (above) gives the means, standard deviations and reliabilities as well as the skew and kurtosis of the nine EPP subscales and the main variable 'subjectively accepted level of risk' taken from the Vienna Risk Taking Test Traffic (WRBTV). As can be seen in Table 2 (above), the numerical value for the skewness and kurtosis of all the variables is < 2, so that the standard criteria for a univariate normal distribution are met (cf. Kline, 1998). The fit of postulated model to the data was investigated using the following cut-off values for the global fit indices: non-significant χ^2 test (Marsh, Hau & Wen, 2004), $\chi^2/df < 2$ (Byrne, 2001), RSMEA $\leq .06$ (Browne & Cudeck, 1993; Hu & Bentler, 1999; Marsh et al., 2004), SRMR $\leq .06$ (Byrne, 2001; Hu & Bentler, 1999; Marsh et al., 2004) and CFI $\geq .95$ (Byrne, 2001; Hu & Bentler, 1999; MacCallum, Browne, & Sugawara, 1996; Marsh et al., 2004). The fit statistics are given in Table 2 (below).

Table 2. Study III: Means, standard deviation (SD), scewness, curtosis and internal consistency (Cronbach α) of the individual scales (above), and Goodness-of-Fit statistics of the confirmatory factor analysis (below)

Scale	Mean	SD	Scewness	Curtosis	Cronbach α
EPP-AC	22.360	6.702	.124	.354	.80
EPP-SO	21.030	7.674	-.413	-.276	.75
EPP-EX	16.370	5.691	.184	-.115	.73
EPP-LO	11.490	9.220	.947	-.035	.83
EPP-UN	11.170	10.028	.887	-.345	.85
EPP-AN	13.800	9.657	.607	-.620	.86
EPP-RI	18.600	6.813	.197	-.344	.70
EPP-IM	16.510	7.240	-.288	-.654	.76
EPP-IR	15.870	7.189	-.065	-.878	.73
WRBTV	7.601	1.620	.542	.321	.91

Model	χ^2	df	p	χ^2/df	CFI	RSMEA	SRMR
Tested Model	42.912	32	.094	1.341	.986	.050	.056

Note: EPP-AC: EPP subscale 'activity'; EPP-SO: EPP subscale 'sociability'; EPP-EX: EPP subscale 'expressiveness'; EPP-LO: EPP subscale 'inferiority'; EPP-UN: EPP subscale 'unhappiness'; EPP-AN: EPP subscale 'anxiety'; EPP-RI: EPP subscale 'risk taking'; EPP-IM: EPP subscale 'impulsiveness'; EPP-IR: EPP subscale 'irresponsibility'; WRBTV: Vienna Risk-taking Test Traffic main variable 'subjectively accepted level of risk' (in sec).

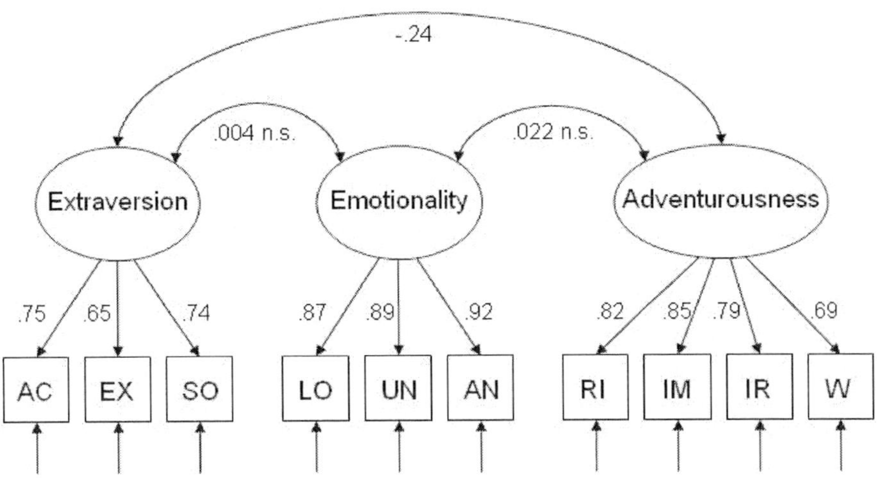

Note: AC: EPP subscale 'activity'; SO: EPP subscale 'sociability'; EX: EPP subscale 'expressiveness'; LO: EPP subscale 'inferiority'; UN: EPP subscale 'unhappiness'; AN: EPP subscale 'anxiety'; RI: EPP subscale 'risk taking'; IM: EPP subscale 'impulsiveness'; IR: EPP subscale 'irresponsibility'; W: Vienna Risk-taking Test Traffic main variable 'subjectively accepted level of risk' (in sec).

Figure 2. Standardized factor loadings of the theoretically postulated model.

As can be seen in Table 2, the theoretically postulated model shows a good fit to the data. Therefore the statistical significance of the path coefficients was tested in a subsequent step. The loadings of the individual scales were medium to high and statistically significant at α=.05. The standardized factor loadings are presented in Figure 2.

In line with previous research on the factorial validity of the short version of the Eysenck Personality Profiler the subscales 'activity', 'sociability' and 'expressiveness' form a factor that can be described as the Giant Three factor 'Extraversion', while the subscales 'inferiority', 'unhappiness' and 'anxiety' load on the Giant Three factor 'Emotionality'. The three EPP subscales 'risk taking', 'impulsiveness' and 'irresponsibility'as well as the main variable 'subjectively accepted level of risk' taken from the Vienna Risk Taking Test Traffic (WRBTV) load on the Giant Three factor 'Adventurousness'. Furthermore, the two Giant Three factors 'Extraversion' and 'Adventurousness' turned out to be significantly correlated ($r=-.24$; $p<.05$), while being statistically unrelated to the Giant Three factor 'Emotionality'.

In summary it can therefore be stated that the present results argue for the construct validity of the Vienna Risk Taking Test Traffic (WRBTV) as a measure of respondents' willingness to take risks. The results of this study are in line with previous studies (cf. Hergovich et al., 2004; Arendasy et al., in revision; Vogelsinger, 2005), which indicated that the Vienna Risk Taking Test Traffic is correlated with conventional measures of risk taking. Furthermore, the present results extend previous studies by locating the Vienna Risk Taking Test within the framework of the Giant Three model.

Study IV

This study deals with the incremental validity of the Vienna Risk Taking Test Traffic over and above other driving related ability and personality traits in predicting the number of driving accidents in which the respondents had been at fault within the past five years.

Measurement of the Predictor Variables

In addition to the Vienna Risk taking Test Traffic the following tests taken from the Expert System Traffic (Schuhfried, 2005) have been administered:

Adaptive Matrices Test (AMT/S11)

This test was administered in order to assess 'fluid intelligence (G_f)'. The items resemble classical matrices, but they are based on explicit construction rules (Hornke, Etzel & Rettig, 2003). The respondent's task is to identify from eight answer alternatives the figural pattern that completes the given matrix item. Internal consistency can be assumed due to the fit of the Rasch model (Rasch, 1980). The test was administered as a computerised adaptive test. The target reliability was set to $\alpha=.70$.

Determination Test (DT/S1)

This test was used to measure 'resilience of attention and reaction speed under conditions of sensory stress'. The respondent's task is to identify various stimuli and to react to them by pressing the corresponding response buttons. In this study the main variable 'correct reactions' was used (Schuhfried, 1998). The reliability (Cronbach α) of this variable amounts to $\alpha=.98$.

Reaction Test (RT/S3)

This test measures 'decision speed' using a simple choice reaction time paradigm (Schuhfried & Prieler, 1997). Various auditory and visual stimuli are presented to the respondent. During presentation the respondent places his finger on a rest button. His task is to leave the rest button and to press a defined button whenever a yellow circle and a certain auditory signal appear in combination. All other types of signal and their combinations must be ignored. The mean reaction time from the onset of the relevant stimulus to the lifting of the finger from the rest button serves as a measure of 'decision speed'. 'Physical motor speed' is measured by latency from the onset of the lifting of the finger to the moment when the response key is pressed. The reliability coefficients (Cronbach α) amount to α=.94 and α=.98 respectively.

Tachistoscopic Traffic Perception Test (TAVTMB)

In this test pictures of traffic scenes are presented to the respondent for one second. The respondent must then select the objects that he/she remembers seeing in the picture from a list of five different objects. The main variable 'overview' serves as a measure of 'perceptual speed' (Biehl, 1996). Internal consistency can be assumed due to the fit of the Rasch model (Rasch, 1980). The reliability coefficient (Cronbach α) amounts to α=.82.

Cognitrone (COG/S11)

In order to measure 'selective attention' the Cognitrone test (form S11) was administered. The respondent's task is to compare an abstract figure with the model to determine whether it matches or not. The items are administered without any time restriction. The mean time of correct rejections serves as a measure of 'selective attention' (Wagner & Karner, 2001). The reliability coefficient (Cronbach α) amounts to α=.95.

Inventory of Driving-Related Personality Traits (IVPE)

This standardised questionnaire consists of 39 items measuring four driving-related personality traits: sensation-seeking, social responsibility, self-control and emotional stability (Herle, Sommer, Wenzl & Litzenberger, 2004). Respondents assess the extent to which 39 different statements apply to them. The answer is entered on an answer bar with a sliding marker. This response is recalculated dichotomously by the evaluation program. The internal consistency of each scale can be assumed due to the fit of the Rasch model (Rasch, 1980). The reliability coefficients (Cronbach α) range from α=.69 (self-control) to α=.76 (social responsibility and sensation-seeking).

Measurement of the Criterion Variable

Several prior studies have focused on accident rates as a criterion measure since this measure accounts for the strategic, the tactical and the operational level (Michon 1985) alike. However, the fact that accidents are multi-causally determined (cf. Evans, 1991; Klebelsberg, 1982) and the finding that various measurement methods of accident proneness such as data gathered from official sources or insurance company records and self-reports, exhibited correlation coefficients that vary considerably (r=.11 to r=.67; e.g. Owsley, Ball, Sloane, Roenker & Bruni, 1991; Szlyk, Seiple & Viana, 1995) constitutes a major problem. The various measurement methods have different benefits and advantages. While accident rates gathered from official sources or insurance company records are less prone to impression management and self deception (Paulhus, 1984) than self reported accidents they do not necessarily cover the whole range of traffic accidents committed. In order to cirumvent this measurement problem of accident frequencies a standardized interview on the type and frequency of road accidents he which s/he had been involved in in the past five years was carried out (Litzenberger, 2003). The self-report data on respondents' accident involvment was cross-checked with their insurrance company records. Respondents for whome the self-reported accident frequencies did not match the insurrance company records were excluded from the analysis to ensure a reliable and valid measure of the criterion variable.

Sample

The sample consists of 86 (56.2%) men and 67 (43.8%) women aged 22 to 45 years (mean=31.73; SD=6.25). The median age is 31 years. The members of this sample are volunteers who were recruited in cooperation with an Austrian insurance company and a traffic psychological assessment centre. Seven (4.60%) respondents had completed compulsory schooling or basic secondary school but without completing vocational training (EU educational level 2), 42 (27.5%) respondents had completed vocational training (EU educational level 3), 83 (54.2%) respondents had a school-leaving qualification at university entrance level (EU educational level 4) and 21 (13.70%) respondents had a university degree (EU educational level 5). All the participants had held a driving licence for at least five years and had driven at least 50,000 kilometres.

Results

A stepwise multiple regression analysis was used to evaluate the incremental validity of Vienna Risk Taking Test Traffic over and above the other driving-related ability and personality traits measured in the test battery described above. The model summary of the stepwise multiple regression, as well as the regression weights of the predictor variables entered into the regression, are presented in Table 3.

Table 3. Study IV: Model summary and regression weigths for the significant predictors of accident proneness

model	R	R^2	adj. R^2	ΔR^2	ΔF	df1	df2	p
1	.265	.070	.064	.070	11.361	1	151	.001
2	.338	.114	.102	.044	7.441	1	150	.007
predictor variable				B	S.E.	β	T	p
IVPE: emotional stability				.221	.062	.274	3.562	<.001
WRBTV: subjectively accepted level of risk				.423	.155	.210	2.728	.007

As can be seen in Table 3 the variable 'emotional stability' from the IVPE was entered into the regression model first, accounting for 7% of the explained variance in respondents' self reported accident frequencies. The inclusion of the variable 'subjectively accepted level of risk' obtained from the Vienna Risk Taking Test Traffic contributed significantly to the validity of the overall model and accounted for another 4.4% of the variance in predicting respondents' accident frequencies. The standardized ß weights indicate that a lower subjective accepted level of risk (lower raw scores) and a higher emotional stability (lower raw scores) is associated with less accident proneness. None of the other driving-related personality or ability traits contributed incrementally to the predictive model over and above these two personality traits. Taken together these two personality traits account for 11.4% of the variance in respondents' self reported accident frequencies. The result of this study is in line with a supposition of Risser (2000) that young or middle-aged drivers who have had multiple accidents are distinguished from those who have had none primarily by their personality traits. Furthermore, the present results are also in accordance with previous findings on the personality differences between young and middle-aged drivers who have had no accidents and those who have had several (e.g. Moe, 1994; Iversen & Rundmo, 2002, Sommer & Häusler, 2005; Sommer et al., in press).

Study V

This study deals with the incremental validity of the Vienna Risk Taking Test Traffic over and above other driving related ability and personality traits in predicting respondents' performance in a standardized driving test. In contrast to accident rates standardized driving tests are capable to take account of precursors of traffic accidents which might remain unnoticed when fitness to drive is solely measured by means of self-reported accidents (cf. Evans, 1991; Klebelsberg, 1982). In line with this theoretical consideration Chapman and Underwood (2000) recently demonstrated that about 80% of less severe accidents and a major proportion of near-accidents had been forgotten within a period of two weeks. Furthermore, currently available research seems to indicate that standardised driving tests are also less prone to impression management than self-reported accident rates (for an overview: Hjälmdahl & Várhelyi, 2004). However, standardized driving tests are unable to account for the strategic aspects of driving proposed by Michon (1985) since vehicle and route choices are already pre-determined in advance. From a theoretical point of view standardized driving tests are thus limited to measure the tactical and operational level of driving (Michon, 1985) which predominately reflects the error component while accident rates are more linked to the violation component of safe driving (cf. Rim & Berg, 1999).

Measurement of the Predictor Variables

The test battery used in this study consists of the following tests taken from the Expert System Traffic (Schuhfried, 2005): Adaptive Matrices Test (AMT/S11: Hornke et al., 2003), Reaction Test (RT/S3: Schuhfried & Prieler, 1997), Determination Test (DT/S1: Schuhfried, 1998), Tachistoscopic Traffic Perception Test (TAVT: Biehl, 1996), Cognitione (COG/S11: Wagner & Karner, 2001), Vienna Risk-taking Test Traffic (WRBT: Hergovich et al., 2005) and the Inventory of Driving-related Personality Traits (IVPE: Herle, et al., 2004). The test battery is identical to the one used in the fourth study reported in the article.

Measurement of the Criterion Variable

Each respondent took a standardised driving test referred to as the Vienna Driving Test (Bukasa & Risser, 1985; Chaloupka & Risser, 1995). The Vienna Driving Test was chosen as a criterion measure based on prior research on the criterion validity of this measure. Chaloupka and Risser (1995) investigated the correlation between police-registered accidents for 51 road sections along a standardised route in Vienna and behavioural data obtained with the Vienna Driving Test. The authors reported significant correlation coefficients from r=.38 to r=.48 between observational variables obtained in the Vienna Driving Test and police-recorded accidents. Similar results have been reported by Bukasa and Risser (1985), who investigated the correlation between self-reported accidents and the observational variables obtained in the Vienna Driving Test. Taken together these results argue for the validity of Vienna Driving Test as a measure of respondents' fitness to drive.

The Vienna Driving Test was carried out immediately after completion of the test battery. Each standardized driving test took about 45 minutes and was carried out in driving-school cars in the presence of driving instructors. The observation took place along a route that involved a wide range of driving tasks, such as lane keeping, lane choice before obstacles and intersections, speed selection, maintaining distance from the cars ahead and from persons and objects on the roadside, overtaking manoeuvres etc. Respondents' driving behaviour was evaluated by two independent observers with the aid of a standardized assessment form. The oberservers scored each driving task along the route as well handled (2 points), handled (1 point), task failed (0 points) or not observeable (missing value). The observers had previously received intensive training in conducting behavioural observations using the Vienna Driving Test and were asked not to discuss their ratings with each other. Furthermore, the observers were blind to the respondents' test results. A global score was calculated from the ratings provided by both observers on each task along the route. The inter-rater reliability of this measure amounts to .88. In addition, both observers made a global evaluation of the respondent's driving behaviour on a five-point rating scale. This scale is normally used for government driving exams in Austria. A score of 4 or 5 corresponds to driving behaviours that would fail to obtain the candidate a driving license, while 1 is very good, 2 is good and 3 is sufficient. The inter-rater reliability of the global evaluation provided by both observers amounted to .89. Based on the considerable high inter-rater reliabilities for the global score as well as the global evaluation the scores provided by both raters were averaged. The correlation between the more mean objective global score and the mean global evaluation of respondents' driving behaviour on the five-point scale amounted to .90. In the following analysis we will thus resort to the mean global score as our main criterion measure.

Sample

The sample consisted of 127 (79.9 %) male and 32 (20.1 %) female respondents aged 19 to 73 years (mean = 39.41; SD = 11.92). The median age is 39 years. The members of this sample were recruited in cooperation with an Austrian traffic psychological assessment centre. All respondents completed the test battery and the standardised driving test as part of a traffic-psychological assessment which is compulsory in Austria in cases where a driver's license has been suspended. A total of 29 (18.2 %) respondents had completed compulsory schooling or basic secondary school but without completing vocational training (EU educational level 2), 87 (54.7 %) respondents had also completed vocational training (EU educational level 3), 30 (18.9 %) respondents had a school-leaving qualification at university entrance level (EU educational level 4) and 13 (8.2 %) respondents had a university degree (EU educational level 5).

Table 4. Study V: Model summary and regression weigths for the significant predictors of respondents' global evaluation of their driving performance in a standardized driving test

Model	R	R^2	adj. R^2	ΔR^2	ΔF	df1	df2	p
1	.423	.179	.173	.179	34.145	1	157	<.001
2	.475	.226	.216	.047	9.541	1	156	.002
3	.505	.255	.241	.029	6.139	1	155	.014
4	.524	.275	.256	.019	4.046	1	154	.046
Predictor variable				B	S.E.	β	T	p
WRBTV: subjectively accepted level of risk				-.089	.024	-.315	-3.757	<.001
DT: resilience of reaction speed				.007	.002	.280	3.340	.001
IVPE: emotional stability				-.083	.033	-.175	-2.516	.013
AMT: fluid intelligence				.144	.072	.155	2.012	.046

Results

A stepwise multiple regression analysis was used to evaluate the incremental validity of Vienna Risk Taking Test Traffic over and above the other driving-related ability and personality traits measured in the test battery. The model summary of the stepwise multiple regression, as well as the regression weights of the predictor variables entered into the regression, are presented in Table 4.

As can be seen in Table 4 the variable 'subjectively accepted level of risk' obtained from the Vienna Risk Taking Test Traffic was entered into the regression model first, accounting for 17.9% of the variance in respondents' global scores in the standardized driving test. In a subsequent step the main variable 'correct' from the Determination Test (DT/S1) measuring resilience of reaction speed contributed incrementally to the predictive validity of the regression model and explained another 4.7% of the variance in respondents' global score in the standardized driving test. In a third step the main variable 'emotional stability' from the IVPE was included and the variable 'fluid intelligence' measured by means oif the Adaptive Matrices Test (AMT/S11) was entered in a final step. These two variables accounted for 2.9% and 1.9% of the variance in respondents' global scores in the standardized driving test, respectively. No other variable from the present test battery contributed incrementally to the validity of the regression model in predicting respondents' performance in the standardized driving test over and above these four main variables. The regression model obtained in this study accounts for 25.6% of the variance respondents' global scores in the standardized driving test. An inspection of the standardized ß weights indicates that a lower subjectively accepted level of risk (lower raw scores) and a higher resilience of reaction speed (higher raw scores) as well as a higher emotional stability (lower raw scores) and fluid intelligence (higher raw scores) are associated with a better performance in the Vienna Driving Test. In line with previous studies (cf. Bukasa et al., 1990; Bukasa et al., 2003; Karner & Neuwirth, 2000; Risser et al., 2008; Sommer & Häusler, 2005; Sommer et al., in press), the mean variable 'correct' obtained in the Determination Test (DT) measuring resilience of reaction speed contributes significantly to the predictive validity of the overall model. Furthermore, the finding that the personality taits 'subjectively accepted level of risk (WRBTV)' and 'emotional stability (IVPE-PS)' contribute incrementally to the predictive validity of the overall model is in line with previous empirical studies (e.g. Arendasy et al., in revision;

Cellar et al., 2000; Dahlen & White, 2006; Dahlen et al., 2005; Iversen & Rundmo, 2002; Sommer et al., in press; Ulleberg, 2001). Contrary to previous studies using both ability and personality predictors, fluid intelligence contribute incrementally to the predictive validity of the overall model (Sommer et al., in press). Possible explanation for this initially surprising finding are differences level of fluid intelligence between the sample reported in this article and the one investigated by Sommer et al. (in press) and differences in the calculation of the criterion measure. While the present study used a mean global score derived from the individual ratings of both observers Sommer et al. (in press) dichotomized their criterion measure by resorting to the observers global evaluations on a five-point rating scale in order to classify respondents as fit or unfit to drive based on their performance in the standardized driving test.

Discussion

The aim of studies reported in this article was to investigate the dimensionality, reliability and validity of an objective personality test measuring respondents' 'subjectively accepted level of risk' in road traffic. Building on the results of Hergovich et al. (2004; Arendasy et al., in revision) the one-dimensionality of the traffic videos shown in WRBTV was tested using the Latency Model. The results support the assumption of the fit of the Latency Model. Reliability measured by Cronbach's α is also satisfactory high. In addition, Study II demonstrated that the item parameters estimated in a laboratory setting can be generalized to the real-life situation of a traffic psychological assessment of professional driver applicants. It was also shown that professional driver applicants examined in the context of a traffic psychological assessment did not differ significantly from individuals parallelised for age, education and gender who were tested in a laboratory situation. This finding can be interpreted as an indicator of the test's resistance to impression management and is entirely in line with the findings of Großmann (2004), who used an experimental design (instructed 'honest' and 'faking good' condition) to investigate the resistance to impression management of a short form of the WRBTV. A further point of criticism was the lack of evidence of convergent and discriminant validity within current global models of personality (e.g. Andersen, 1995). In order to fill this gap the third study investigated the factorial validity of the Vienna Risk taking Test Traffic within the context of the Giant Three. The results confirm the initial hypothesis that the main variable 'subjectively accepted level of risk' can be subsumed under the broad personality factor 'Adventurousness'. The final two studies investigated the criterion validity of this new objective personality test using (1) self-reported accident rates and (2) global scores obtained in a standardized driving test as criterion measures to cover both the strategic and tactical and operational level of driving proposed by Michon (1985) alike. In line with previous studies the results seem to indicate that respondents' subjectively accepted level of risk is associated both with the strategic and tactical level of driving; indicating the relevance of this personality traits and setting the goals of driving and adjusting them to the current traffic conditions. Despite these encouraging results the authors acknowledge, that the results of the present studies need to be cross-validated using an independent data set. Furthermore, additional research should be conducted on the construct validity of the Vienna Risk Taking Test traffic to gain a more comprehensive picture of the location of this measure within current models of personality.

REFERENCES

Andersen, E. B. (1973). A goodness of fit test for the Rasch model. *Psychometrica, 8*, 123-140.

Andersen, B. (1995). Risikobereitschaft (R)- der sechste Basisfaktor der Persönlichkeit: Konvergenz multivariater Studien und Konstruktexplikation [Willingness to take risks – the sixth personality factor: Konvergence of multivariate studies and construct explication]. *Zeitschrift für Differentielle und Diagnostische Psychologie, 16*, 210-236.

Arbuckle, J. L. (2003). *Amos 5.0 Update to the Amos User's Guide*. Smallwaters Corporation, Chicago IL.

Arendasy, M. E., Sommer, M., Bognar, B., & Hergovich, A. (in revision). Subjectively accepted level of risk: Dimensionality, fakeability and criterion validity of a video-based objective personality test. *Personality and Individual Differences*.

Biehl, B. (1996). *Manual Tachistoscopic Traffic Perception Test* (*TAVTMB*). Mödling: SCHUHFRIED GmbH.

Browne, M. W. & Cudeck, R. (1993). *Alternative ways of assessing model fit*. In K. A. Bollen & J. S. Long (Eds.), Testing structural equation models (pp. 445-455). Newbury Park: Sage.

Bukasa, B. & Risser, R. (1985). *Fahreignungsdiagnostik am Verkehrspsychologischen Institut* [Traffic psychological assessment at the institut of traffic psychology]. Wien: Literas.

Bukasa, B., Wenninger, U. & Brandstätter, C. (1990). *Validierung verkehrspsychologischer Testverfahren* [Validation of traffic psychological test batteries]. Wien: Literas.

Bukasa, B., Christ, R., Ponocny-Seliger, E., Smuc, M. & Wenninger, U. (2003). Validitätsüberprüfung verkehrspsychologischer Leistungstests für die Fahreignungs-begutachtung [Validation of traffic psychological ability tests]. *Zeitschrift für Verkehrssicherheit, 49*, 191-197.

Burns, C.P. & Wilde, G.J.S. (1995). Risk taking in male taxi drivers: relationships among personality, observational data and driver records. *Personality and Individual Differences, 18* (2), 267-278.

Byrne, B. M. (2001) *Structural equation modeling with AMOS: Basic concepts, application and programming*. London: Lawrence Erlbaum.

Cellar, D. F., Nelson, Z. C., & Yorke, C. M. (2000). The five-factor model and driving behaviour: Personality and involvement in vehicular accidents. *Psychological Report, 86*, 454-456.

Chaloupka, Ch., & Risser, R. (1995). Don't wait for accidents – possibilities to assess risk in traffic by applying the „Wiener Fahrprobe". *Safety Science, 19*, 137-147.

Chapman, P., & Underwood, G. (2000). Forgetting near-accidents: The roles of severity, culpability and experience in the poor recall of dangerous driving situations. *Applied Cognitive Psychology, 14*, 31-44.

Dahlen, E. R., & White, R. P. (2006). The Big Five factors, sensation seeking, and driving anger in the prediction of unsafe driving. *Personality and Individual Differences, 41*, 903-915.

Dahlen, E. R., Martin, R. C., Ragan K., & Kuhlman, M. (2005). Driving anger, sensation seeking, impulsiveness, and boredom proneness in the prediction of unsafe driving. *Accident Analysis and Prevention, 37*, 341-348.

Evans, L. (1991). *Traffic safety and the driver*. New York: Fan Nostrand Reinhold.

Eysenck, H., Wilson, G. & Jackson, C. (2000). *Manual of the Eysenck Personality Profiler*. Guilford: Psi-Press

Fischer, G. H. & Kisser, R. (1983). Notes on the Exponential Latency Model and an empirical Application. In. H. Wainer & S. Messick (Eds), *Principles of modern psychological measurement* (pp. 139-157). Hillsdale, NJ: Lawrence Erlbaum.

Großmann, S. (2004). *Wer ist gefährlich? Eine empirische Untersuchung zu Möglichkeiten und Grenzen der Diagnose der Eignung zum Umgang mit Schusswaffen*. Unveröff. Dipl.Arbeit, Justus-Liebig-Universität Gießen.

Hergovich, A., Bognar, B., Arendasy, M. & Sommer, M. (2005). *Manual Vienna Risk-taking Test Traffic* (*WRBTV*). Mödling: Schuhfried GmbH.

Hergovich, A., Arendasy, M. E., Sommer M., Bognar, B. & Olbrich, A. (2004). Zur Dimensionalität und Konstruktvalidität eines videobasierten, objektiven Persönlichkeitstests zur Messung der Riskobereitschaft im Straßenverkehr [On the dimensionality and construct validity of a video-based objective personality test to assess „willingness to take risks“]. *Zeitschrift für Differentielle und Diagnostische Psychologie, 25* (2), S. 79-86.

Herle, M., Sommer, M., Wenzl, M. & Litzenberger, M. (2004) *Manual Inventory of Driving-related Personality Traits* (*IVPE*). Mödling: SCHUHFRIED GmbH.

Hjälmdahl, M., & Várhelyi, A. (2004). Validation of in-car observations, a method for driver assessment. *Transportation Research Part A, 38*, 127-142.

Hornke, L. F., Etzel, S. & Rettig, K. (2003). *Manual Adaptive Matrices Test* (*AMT*). Mödling: SCHUHFRIED GmbH.

Horswill, M. S., & McKenna, F. P. (1999). The development, validation, and application of a video-based technique for measuring an everyday risk-taking behaviour: drivers speed choice. *Journal of Applied Psychology, 84*, 977-985.

Hu, L. T. & Bentler, P. M. (1999). Cutoff criteria for fit indices in covariance structure analysis: Conventional criteria versus new alternaties. *Structural Equaion Modeling: A Multidisciplinary Journal, 6*, 1-55.

Iversen H. & Rundmo, T. (2002). Personality, risky driving and accident involvement among Norwegian drivers. *Personality and Individual Differences, 33* (8), 1251-1263.

Jansen M.G.H. (1997). Rasch's Model for reading speed with manifest explanatory variables. *Psychometrika, 62*(3), 393-409.

Jonah, B. A. (1997). Sensation seeking and risky driving: A review and synthesis of the literature. *Accident Analysis and Prevention, 29*, 651-665.

Jonah, B. A., Thiessen, R., & Au-Yeung, E. (2001). Sensation seeking, risky driving, and behavioural adaptation. *Accident Analysis and Prevention, 33*, 679-684.

Karner, T., & Neuwirth, W. (2000). Validation of traffic psychology tests by comparing with actual driving. International Conference on Traffic and Transport Psychology, 4-7 September, Berne Switzerland.

Klebelsberg, D. (1982). *Verkehrspsychologie* [Traffic psychology]. Berlin: Springer.

Kline, R. B. (1998). *Principles and practice of structural equation modeling*. New York: Guilford Press.

Litzenberger, M. (2003). *Structured interview to assess previous traffic accident involvement.* Unpublished manuscript, Institut for Developmental Psychology and Psychological Assessment, Faculty of Psychology, University of Wien.

MacCallum, Browne, & Sugawara, (1996). Power analysis and determination of sample size for covariance structure modeling. *Psychological Methods, 1*, 130-149.

Marsh, H. W., Hau, K. T., & Wen, Z. (2004). In search of golden rules: Comments on hypothesis testing approaches to setting cutoff values for fit indices and dangers of overgeneralising Hu & Bentler's (1999) findings. *Structural Equation Modeling, 11*, 320-341.

Martin-Löf, P. (1973). *Statistiska Modeller. Anteckningar fran seminarier lasaret 1969-1970 utarbetade av Rolf Sundberg. Obetydligt ändrat nytryck, october 1973* (photocopied manuscript). Institutet för säkringsmatematik och matematisk statistik vid Stockholms universitet.

McKnight, A. J., & McKnight, A. S. (1999). Multivariate analysis of age-related driver ability and performance deficits. *Accident Analysis and Prevention, 31*, 445-454.

Michon, J. A. (1985). A critical review of driver behaviour models. What do we know, what should we do? In L. Evans & R. Schwing (Eds.), *Human behaviour and traffic safety* (pp. 485-520). New York: Plenum Press.

Moe, D. (1994). Junge Fahrerinnen und Fahrer, Risikoverhalten und Unfälle [Young drivers, risk taking behaviour and accident involvement]. *Berichte der Bundesanstalt für Straßenwesen. Referate der Ersten Interdisziplinären Fachkonferenz 12.-14. Dezember in Köln,* 127-131.

Myers, R. S., Ball, K. K., Kalina, T. D., Roth, D. L., & Goode, K. T. (2000). Relation of useful field of view and other screening tests to on-road driving performance. *Perceptual and Motor Skills, 91*, 279-290.

Oltedal, S., & Rundmo, T. (2006). The effect of personality and gender on risky driving behaviour and accident involvement. *Safety Science, 44*, 621-628.

Owsley, C., Ball, K., Sloane, M. E., Roenker, D. L., & Bruni, J. R. (1991). Visual/cognitive correlates of vehicle accidents in older drivers. *Psychology and Ageing, 6*, 403-415.

Paulhus, D. L. (1984). Two-component models of socially desirable responding. *Journal of Personality and social Psychology, 15*, 383-388.

Ranney, T. A. (1994). Models of driving behaviour. A review of their evolution. *Accident Analysis and Prevention, 26*, 733-750.

Rasch, G. (1980). *Probabilistic models for some intelligence and attainment tests*. Chicago: The University of Chicago Press. (Original published in 1960).

Rim, P.-A., & Berg, L. (1999). On the distinction between violations and errors: sensation seeking associations. *Transportation Research Part F, 2*, 151-166.

Risser, R. (2000). *Assessing the Driver*. Braunschweig: Rot-Gelb-Grün.

Risser, R., Chaloupka, Ch., Grundler, W., Sommer, M., Häusler, J. & Kaufmann, C. (2008). Using non-linear methods to investigate the criterion validity of traffic psychological test batteries. *Accident Analysis and Prevention, 40*, 149-157.

Rost, J. (2004). *Lehrbuch Testtheorie, Testkonstruktion* [Textbook test-theory test-construction]. Bern: Huber.

Santacreu, J., Rubio, V. J., & Hernandéz, J. M. (2006) The objective assessment of personality: Cattells's T-data revisited and more. *Psychology Science*, *48*, 53-68.

Scheiblechner, H. (1985). Psychometric models for speed-test construction: The linear exponential model. In S. E. Embretson (Ed.), *Test design: Developments in psychology and psychometrics* (pp.219-244). New York: Academic Press.

Schuhfried, G.. (1998). *Manual Determination Test* (*DT*). Mödling: SCHUHFRIED GmbH.

Schuhfried, G.. (2005). *Manual Expert System traffic* (*XPSV*). Mödling: SCHUHFRIED GmbH.

Schuhfried, G. & Prieler, J. (1997). *Manual Reaction Test* (*RT*). Mödling: SCHUHFRIED GmbH.

Simonet, S. & Wilde, G. J. (1997). Risk: Perception, acceptance and homeostasis. *Applied Psychology: An International Review*, *46* (3), 235-252.

Sommer, M. & Häusler, J. (2005). Non-linear methods for the identification of drivers at risk to cause accidents. In L. Dorn (Ed.), *Driver behaviour and training. Volume II* (pp. 425-436). Burlington: Ashgate.

Sommer, M., Herle, M., Häusler, J., Risser, R., Schützhofer, B. & Chaloupka, Ch. (in press). Cognitive and personality determinants of fitness to drive. *Transportation Research Part F*.

Statistik Austria (2007). *Statistisches Jahrbuch Österreich 2007* [Statistical year-book Austria 2007]. [Online] URL http://www.statistik.at/jahrbuch_2007/pdf/K25.pdf [06.29.2007].

Szlyk, J. P., Seiple, W., & Viana, M. (1995). Relative effects of age and compromised vision on driving performance. *Human Factors*, *37*, 430-436.

Ulleberg, P. (2001). Personality subtypes of young drivers. Relationship to risk-taking preferences, accident involvement, and response to a traffic safety campaign. *Transportation research Part F*, *4*, 279-297.

Vogel, K. (2003). A comparison of headway and time to collision as safety indicators. *Accident Analysis & Prevention*, *35*, 427-433.

Voglsinger. J. (2005). *Prüfung der Dimensionalität und Konstruktvaliditdt des erweiterten Subtests "Risikobereitschaft im Straßenverkehr" des WRBT* [Assessment of dimensionality and construct validity of the Vienna Risk-Taking Test - Traffic]. Unpublished diploma thesis, Universitat Wien.

Wagner, M. & Karner, T. (2001). *Manual Cognitrone* (*COG*). Mödling: SCHUHFRIED GmbH.

Wilde, G. (1994). *Target Risk. Dealing with danger of death, disease and damage in everyday decisions*. Toronto: PDE Publications.

In: Road Traffic: Safety, Modeling and Impacts
Editors: S. E. Paterson and L. K. Allan, pp. 301-320
ISBN 978-1-60456-884-4

Chapter 5

SIMULATION OF TRAVELLERS' DYNAMIC RESPONSE TO REAL TIME TRAFFIC INFORMATION

Kun Zhang, Branko Stazic and Michael A. P. Taylor
Transport Systems Centre, Institute for Sustainable Systems and Technologies,
University of South Australia, Adelaide, Australia

ABSTRACT

The increased availability of traveller information and onboard navigation systems is dramatically changing the characteristics of urban traffic. Traveller information (e.g. real time traffic information) helps individual travellers in minimising their travel costs through pre-trip scheduling and en-route dynamic route choice. On the other hand, traveller information systems provide traffic managers with an alternative and powerful means to implement transport policy and handle congestion by influencing travellers' dynamic route choice. This research investigates travellers' dynamic responses to real time traffic information and proposes a cost-effective way to support traveller information system development. In this research, we have developed a portable application program to implement the variable message sign (VMS) based route guidance system in a simulated urban city environment. Expected travel delay on the targeted road links is the primary information distributed by the VMS. Individual drivers' inherent characteristics (i.e. awareness and aggression), their minimum acceptable travel delay and perceived travel cost variations determine their dynamic route choice decisions. The program has been tested using the Adelaide CBD microscopic traffic simulation model which was developed in our previous research. The preliminary results demonstrate the capability of the program in mimicking travellers' dynamic response to VMS (and traffic information in general). Meanwhile, the tests reveal that the impact of a traveller information system on urban traffic should not be overestimated, and empirical results from travel behaviour studies are essential for any traveller information system development.

1. INTRODUCTION

The increased availability of traveller information and onboard navigation systems is dramatically changing characteristics of urban traffic [Iguchi 2002; VICS 2008]. Traffic information broadcasting, pre-trip electronic route planning, on-board navigation systems, electronic route guidance systems and strategically located variable message signs (VMS) are among the methods used for providing travellers with real time traffic information. The principal aim of these advanced traveller information systems (ATIS) is to influence travellers' behaviour on route choice and departure time decisions in order to improve mobility and reduce traffic congestion.

The installation and operation of such systems is not cheap and there is a widespread belief that overuse, or inappropriate use, of the system (e.g. VMS messages) may lead to them losing their credibility with the motorists and thus ceasing to be effective [Wardman et al. 1997]. Current advances in transport and traffic modelling and simulation offer us a cost-effective and very efficient way to evaluate the alternatives of a proposed ATIS system before conducting operational tests, thus resulting in more effective operational tests and implementation. The research presented in this chapter focuses on how to accurately simulate individual travellers' response to traffic information at real time, the major issue concerning ATIS system evaluation.

The Revealed Preference (RP) and Stated Preference (SP) approaches are two basic methods to examine drivers' route choice behaviour. The RP approach is used to examine existing roads or conditions and to investigate how travellers make their choice in real situations. An SP experiment offers travellers a series of hypothetical scenarios to be evaluated, usually in the form of discrete choices between travel alternatives, and the alternatives are characterized by variables whose effects on travel behaviour we wish to examine. The attractions of the SP approach largely stem from its ability to control the choice context and the independent variables that will enter the demand model [Kroes and Sheldon 1988]. Meanwhile, the SP surveys are relatively cheap to run. SP techniques have a significant number of applications to travellers' route choices [Abdel-Aty et al. 1997; Dia 2002; Hensher et al. 1990; Wardman 1991; Louviere et al. 2000].

The analysis of travel behaviour is typically disaggregated, and discrete choice models are constructed to analyse and predict individual traveller's route choice decisions using RP or SP survey data. The decision rule used in the discrete choice model is the process used by the decision-maker to evaluate the attributes of the alternatives in the choice set and determine a choice [Ben-Akiva and Bierlaire 1999]. The complexity of human behaviour suggests that the decision rule should include a probabilistic dimension. Discrete choice models based on random utility can be used to estimate the probability that an individual will choose one alternative from a choice set.

Assessing the impact of traveller information systems on network performance is a complex task, due to the intricacy of modelling the time-dependent movements of vehicles in a network and the uncertainty of modelling drivers' route choice behaviour in response to traffic information. Traffic assignment is a macro-representation of route choice behaviour, and a dynamic, combined macroscopic assignment-simulation approach [Luk and Yang 2003; Taylor 1990] offers an advantage in computation efficiency. Dynamic traffic assignment is in fact a natural extension of the static assignment such as the user-equilibrium in which the drivers can be assumed 'selfish' and would select routes that minimize their own travel times until all alternative routes used have equal travel times [Wardrop 1952]. The speed-flow relationships are used to calculate link delay and queues as traffic movements evolve over time in network. The mesoscopic approach is to combine a micro-simulation of individual vehicles or packets of vehicles with macroscopic relationships (e.g. travel time-flow functions) to move the vehicles in a network [Ben-Akiva et al. 1997; Mahmassani and Jayakrishnan 1991] . This approach uses a 'macro-particle' concept to simulate traffic movements and queue formation/dissipation procedures to determine delay at intersections.

Microscopic traffic models are capable of simulating individual vehicle's route choice and movement at second-by-second level. Meanwhile, travel delay can be determined reasonably accurate from cumulative arrivals and departures at intersections. The micro-simulation approach provides us a better chance to integrate the empirical travel behaviour study results into traffic simulation, because it can reproduce the levels of variability inherent in real traffic systems, and thus produce more accurate results of travellers' dynamic route choice behaviour. To carry out this research, we need 1) to identify the key factors which influence travellers' route choice decisions from the empirical results of travel behaviour studies, and 2) to simulate traffic operations at the microscopic level and to incorporate those factors into each individual driver-vehicle-unit in order to produce more realistic simulation results of traveller information systems. Hence, the integration of empirical results of travel behaviour studies into an appropriate modelling and simulation framework becomes the key to our simulation problem.

The remainder of this chapter is organized in five sections. Section 2 focuses on the discrete choice modelling and stated preference survey which aim to identify key factors influencing travellers' route choice decision. The typical Adelaide travellers' behaviour which was obtained from our previous study is presented here. Section 3 deals with the microscopic traffic simulation, the platform used in this research to simulate travellers' dynamic route choice decisions. The key features of the Adelaide CBD micro-simulation model and issues concerning the reliability of simulating real-time travellers' route choice are discussed here. In section 4, the detailed VMS bases route guidance program is presented. This program is used to exemplify how individual travellers' dynamic route choice is simulated using our proposed micro-simulation approach. The simulation results of the VMS system are presented in section 5. Conclusion and future research directions are highlighted in section 6.

2. Traveller Route Choice Behaviour

2.1. Discrete Choice Modelling

The analysis of travel behaviour is typically disaggregated, and discrete choice analysis is the general methodology used to analyse and predict travel decisions. In the framework for a discrete choice model, a decision-maker is provided alternatives to choose from. Each alternative contains attributes measuring the benefits and costs of an alternative to the decision-maker; and decision rules are used by the decision-maker to choose an alternative. The *utility theory* assumes that the decision-maker's preference for an alternative is captured by a value, called utility, and the decision-maker selects the alternative in the choice set with the highest utility.

The utility can be modeled as a random variable in order to reflect the uncertainty: unobserved alternative attributes, unobserved individual characteristics, measurement errors, and proxy or instrumental variables [Ben-Akiva and Bierlaire 1999]. The utility that individual n associates with alternative i in the choice set C_n is given by

$$U_{in} = V_{in} + \varepsilon_{in} \tag{1}$$

where V_{in} is the deterministic (or systematic) part of the utility, and ε_{in} is the random term, capturing the uncertainty.

The deterministic term V_{in} of each alternative is a function of the attributes of the alternative itself z_{in} and the characteristics of the decision-maker S_n. The simplified linear formation of this term is

$$V_{in} = \sum_k \beta_k x_{ink} \tag{2}$$

where x_{ink} is the kth element of a new vector of attributes from both z_{in} and S_n. The deterministic term of the utility is therefore fully specified by the vector of parameters β_k. Logistic probability based models (Logit models) have been widely used for travel demand analysis including travel behaviour studies. These models are derived from the assumption that the error terms of the utility functions are independent and identically Gumbel distributed. That is, ε_{in} for all i, n is distributed as:

$$F(\varepsilon) = \exp[-e^{-\mu(\varepsilon-\eta)}], \mu > 0 \tag{3}$$

$$f(\varepsilon) = \mu e^{-\mu(\varepsilon-\eta)} \exp[-e^{-\mu(\varepsilon-\eta)}] \tag{4}$$

where η is a location parameter and μ is a strictly positive scale parameter.

The probability that a given individual n chooses alternative i within the choice set C_n is given by

$$P(i \mid C_n) = \frac{e^{\mu V_{in}}}{\sum_{j \in C_n} e^{\mu V_{jn}}} \tag{5}$$

This means if we can identify the attributes x_{ink} (which are associated with both the alternatives and decision-makers) and work out their associated parameters β_k (see Equation (2)) of the user utility function, we would be able to predict individual traveller's route choice decision. To this end, both the RP and SP approaches are two basic methods which can help us to obtain travel behaviour data and make the Equation (5) workable.

2.2. Traveller's Route Choice Decision Responding to VMS Message

As mentioned in the previous section, the attractions of the SP approach largely stem from its ability to control the choice context and the independent variables that will enter the demand model. The major shortcomings of this approach are related to the fact that individuals are not committed to behave in accordance with their SP responses. However, the empirical evidence is quite encouraging in that stated preferences have generally been found to correspond reasonably well with revealed preferences [Hensher 1992; Louviere et al. 1980; Louviere et al. 2000; Wardman 1988; Wardman 1991]. Typical techniques of the SP approach use travel simulators and questionnaires. Questionnaire surveys are capable of representing the real world situation in relation to the drivers' actual trip. Researchers have used the questionnaire survey approach and reported many useful results for fundamental characteristics of route choice behaviour [Chatterjee et al. 2002; Hidas and Awadalla 2001; Wardman et al. 1997].

Wardman et al. [1997] indicated that additional delays mentioned on the VMS panel were found to be valued more highly than normally expected travel time. They found that the value of delays increased with the size of the delay, and the variation in the value of delay time between 5 and 30 min is considerable. This finding indicates that minimising travel time motivates the individual travellers to switch to an alternative route from their usual paths. Meanwhile, different stated causes of delay had different impacts. Delays attributed to accidents have the biggest impact on route choice whilst if no cause was quoted the messages have a relatively low effect. Visible queues were also found to have a significant effect on route choice, particularly for more experienced drivers. This finding implies that travellers do have their own means to estimate expected travel cost as alternatives to the VMS message.

The London experience of VMS reported by Chatterjee et al. [2002] indicated that the low proportion of drivers noticing VMS information has implications for the future placement of VMS so that the very best opportunities for drivers to take information on board are exploited. This finding points to another key factor, the awareness of individual traveller, which has a significant influence on travellers' route choice decision. Another important finding from this literature was that the only display advance warnings may be detracting from its effectiveness as a means of disseminating current real time information. Therefore, the real-time traffic information which is directly related to the individual traveller's potential travel time reduction would be the most appropriate contents of the VMS.

2.3. Travellers' Route Choice Behaviour in Adelaide

The Transport Systems Centre (TSC) at the University of South Australia conducted an SP survey for one of the major arterial road corridors serving the Adelaide CBD to investigate travellers' route choice behaviour response to VMS [Furusawa et al. 2004]. This study sought to confirm the earlier findings from the literature and to develop a model capable of substantiating travellers' route choice behaviour in the Adelaide CBD environment. The route choice model focused on the relationship between the content of VMS and the degree of diversion resulting from information provided to travellers. The research results could feed in the detailed traffic micro-simulation study to evaluate the impact of traveller information systems on traffic operation.

The questions used for the SP questionnaire survey were classified into the following category: VMS contents, Familiarity with the targeted roads, Experience with VMS, Socio-economic factors of the respondents and Verification of qualitative messages. Both qualitative information (cause of incident, status of congestion, location of incident, route recommendation) and quantitative information (delay time information and queue length of congestion) were used as the VMS contents. A total of 266 of the 1000 questionnaires were returned with 21 of the returned questionnaires were incomplete. This study was only interested with localised traffic diversions along a target route on the Adelaide CBD network. The route was carefully chosen to represent two real alternatives out of the city, namely using the Britannia Roundabout or not using the roundabout.

The descriptive analysis of the SP data indicated that the respondents think Minor Delay is the shortest time of delay followed by Expect Delay, Long Delay and Major Delay with approximately 7 minutes, 12 minutes, 30 minutes and 43 minutes respectively. The VMS message of Expect Delay and Minor Delay was perceived with less variance time amongst the respondents, whereas, the message of Long Delay and Major Delay had large variance. It was very encouraging that NO gap in relation to the delay time information between drivers' perceptions and that of the transport authority (Transport SA) was found. The above findings are the valuable information for detailed microscopic traffic simulation, especially when we need to set travel delay thresholds to simulate travellers' dynamic route choice decisions.

Since only two alternative route *i* and *j* were used in this study, the binary logit model (see Equation (6)) was developed from Equation (5) when μ is set to 1.

$$P_{in} = \frac{1}{1+e^{-(V_{in}-V_{jn})}} \tag{6}$$

The difference of the utility between the alternatives *i* and *j* of individual *n* can be represented as the following

$$V = V_{in} - V_{jn} = ASC + \sum_k \beta_k X_{ink} + \sum_m \alpha_m Y_{inm} \tag{7}$$

where ASC is Alternative Specific Constant (the mean of the error term in the utility functions), β_k is coefficient of attribute *k* for alternative *i*, X_{ink} is explanatory variable of attribute *k* for alternative *i* except VMS message, α_m is coefficient of VMS type *m* for alternative *i*, and Y_{inm} is dummy explanatory variable of VMS type *m* for alternative *i*. As shown in Table 1, quantitative Model A applies to the full parameter set and Model B applies to the reduced parameter set [Furusawa et al. 2004].

The models reveal that the probability of diversion is increased if the delay time is increased. The coefficient value of "5 min Delay" is small, but the coefficient value of "15 min Delay", "20 min Delay" and "Long Delay" is very large indicating that the possibility of taking alternatives is very high. This coincides with the finding from Wardman et al. [1997] that the *travel delay* is the key motivate for travellers to change route. The model result of a negative sign in terms of familiarity with the CBD indicates that travellers who know road network and alternatives in the CBD well are likely to choose alternatives. This indicates different groups of travellers (e.g. Familiar and Unfamiliar) react to VMS message differently. Experience of VMS is also negative implying that travellers who saw VMS messages before react more sensitively than inexperienced travellers, which highlights the importance of integrating the *awareness* factor of individual travellers into their route choice decisions. The model also indicated that socio-economic factors do influence route choice decision. However, it would be very difficult to incorporate such detailed information into traffic modelling even at the microscopic level.

Table 1. Quantitative model for travellers' route choice in Adelaide

	Model A		**Model B**	
	Coefficient	t-statics	Coefficient	t-statics
Constant	1.341	4.594	1.548	6.262
Purpose of trip	0.161	1.317 **		
Passengers	0.216	1.810 **		
Familiarity with the CBD	-0.029	-0.173 **		
Experience of VMS	-0.584	-3.486	-0.54	-3.287
Frequency of VMS	-0.404	-2.296 *	-0.409	-2.345 *
Gender	0.193	1.705 **		
Age	0.512	4.154	0.490	4.167
Education	0.111	0.896 **		
Income	0.511	4.436	0.599	5.452
Area	0.380	3.407	0.381	3.460
VMS (Condition of V2-V10: Congestion at BR)				
V1: Congestion Ahead	-2.967	-12.261	-2.948	-12.229
V2: 5 min Delay	-2.016	-8.906	-2.002	-8.879
V3: 10 min Delay	-3.762	-13.725	-3.741	-13.691
V4: 15 min Delay	-5.429	-11.959	-5.405	-11.922
V5: 20 min Delay	-6.550	-8.918	-6.525	-8.889
V6: Expect Delay	-4.146	-13.866	-4.124	-13.830
V7: Long Delay	-6.139	-10.055	-6.114	-10.022
V8: Queue 50 m Long	-1.684	-7.483	-1.672	-7.458
V9: Queue 100 m Long	-2.245	-9.829	-2.229	-9.800
V10: Queue 150 m Long	-3.183	-12.803	-3.163	-12.771
V11: Queue 200 m Long	-3.894	-13.816	-3.872	-13.781
V12: Queue 400 m Long	-5.130	-12.653	-5.107	-12.615
Number of observations	3016		3016	
L(0)	-1613.4		-1613.4	
L(β)	-1095.7		-1100.4	
Adjusted R square	0.3157		0.3139	

BR: Britannia Roundabout.

* Insignificant at 1 %. ** Insignificant at 5 %.

3. MICROSCOPIC TRAFFIC SIMULATION

The microscopic traffic simulation method has been widely used to develop and evaluate advanced urban traffic signal control systems [Hansen et al. 2000; Hicks and Carter 2000; Lucas et al. 2000], freeway traffic management systems [Ozbay et al. 2006], incident detection and management systems [Dia and Cottman 2005; Zhang and Taylor 2006ab]. It contributes significantly to the evaluation of the impacts of potential ITS deployment [Taylor et al. 2000]. Taking advantages of current advanced microscopic traffic simulation techniques, we are able to simulate different traffic scenarios caused by both traffic demand variation and traffic signal adjustment. The traffic simulation approach used in this research aims to simulate individual travellers' dynamic response to real time traffic information distributed by VMS, which would minimise costs and risks associated with the investigation in driver responses to such ATIS systems.

3.1. Microscopic Traffic Simulation

Micro-simulation models try to model the movement and behaviour of individual vehicles on urban and highway road networks. Three core mathematical models, namely car-following model, lane-changing model and route choice model, form the base for simulating detailed interaction between individual vehicles and their traffic environment. Meanwhile, the characteristics of an individual driver-vehicle-unit DVU (i.e. vehicle kinematics and physical attributes, driver awareness and aggression factors) influence the real time operation of each of the three core models, which results in different reactions from different DVUs given similar traffic conditions. In this way, micro-simulation reproduces daily traffic flow on a second-by-second basis. Paramics [Quadstone 2000] is a suite of high performance software tools which was used to develop the Adelaide CBD micro-simulation model.

In the context of real time driving behaviour, it is the target headway that affects an individual DVU's car-following and lane-changing manoeuvre. In fact, the two key attributes of the DVU (driver awareness and aggression) influence significantly on its target headway settings. The table below shows how the target headway used by the individual DVU i is modified in Paramics based on the two attributes [Quadstone 2006]

Table 2. Headway adjusting factor for individual DVU in Paramics

Headway adjusting factor f_i	Aggression (every where)		Awareness (Near lane-drop)	
	$Ag_i \le 4$	$Ag_i \ge 4$	$Aw_i \le 4$	$Aw_i \ge 4$
	$\frac{36-5Ag_i}{16}$	$\frac{12-Ag_i}{8}$	$\frac{Aw_i+4}{8}$	$Aw_i - 3$

where $Ag_i \in [0,8]$ and $Aw_i \in [0,8]$.

Generally, an individual DVU's route choice decisions are based on travel costs (distance, time, and toll) and network restrictions. The progression of vehicles through the network is dynamic since DVUs have only partial information on their preferred route to the destination and the information can change depending on road conditions ahead. In this study, we extend the usage of the above two attributes (awareness and aggression) of the individual DVUs to their dynamic route choice process, which is one of the major elements of our VMS route guidance program and will be detailed in the next section.

3.2. Adelaide City Council Area Micro-simulation Model

Recently, the TSC in partnership with the Adelaide City Council (ACC) has developed a micro-simulation model of the entire city of Adelaide road network including the inner ring route. It includes all arterial roads and local streets in the municipality. The ACC area which includes the Adelaide central business district (CBD) consists of a regular grid network with many alternative routes (see Figure 1). The area exists as a separate entity from the rest of the metropolitan road network due to surrounding parkland with a manageable number of roads entering/exiting the Adelaide CBD. It is also an area of high traffic attraction and subject to many special events that require the implementation of special traffic management plans. Especially, VMS has not been implemented in the Adelaide CBD, whereas the potential need for traffic information is high. Therefore, it presents a good case study for analysing travellers' route choice and simulating traveller information systems such as VMS.

The ACC area road network has been modelled in three stages (the CBD South, the CBD North, and the CBD Ring Route) to eventually become the one combined model: Adelaide CBD micro-simulation model or ACC model [Stazic et al. 2005]. The CBD South network constitutes the most feature intensive of the three networks. It is a grid type road network with a large number of alternative routes. The CBD Ring Route system is an important part of the model, which allows route guidance and the potential effect of traveller information systems to be investigated, as well as possible traffic impacts on the surrounding metropolitan area.

The construction of the ACC model involves the following procedures: road network construction, travel demand estimation, and model calibration and validation. The road network data is represented by detailed network geometry and the comprehensive sets of network attribute specifications. These include signalised and un-signalised junction data, public transport data, road attributes, etc. Another important task is to develop a vehicle fleet for South Australian urban conditions based on Australian Bureau of Statistics' Motor Vehicle Census, as vehicle physical and kinematics characteristics have a potential to impact on road capacity, traffic flow rate and level of service, especially in urban areas. Simulated vehicles are then sampled from this vehicle fleet. In addition to network and vehicle attribute specifications, an essential task is the development of an origin-destination (OD) matrix that describes where vehicles are travelling from and to in the road network and when they are travelling. This required the OD matrix to be synthesised using data from several sources and validated against traffic counts collected at different locations throughout the CBD area. To insure that the model accurately mimics the real life traffic behaviour, model parameters have been calibrated and validated against the real life traffic data.

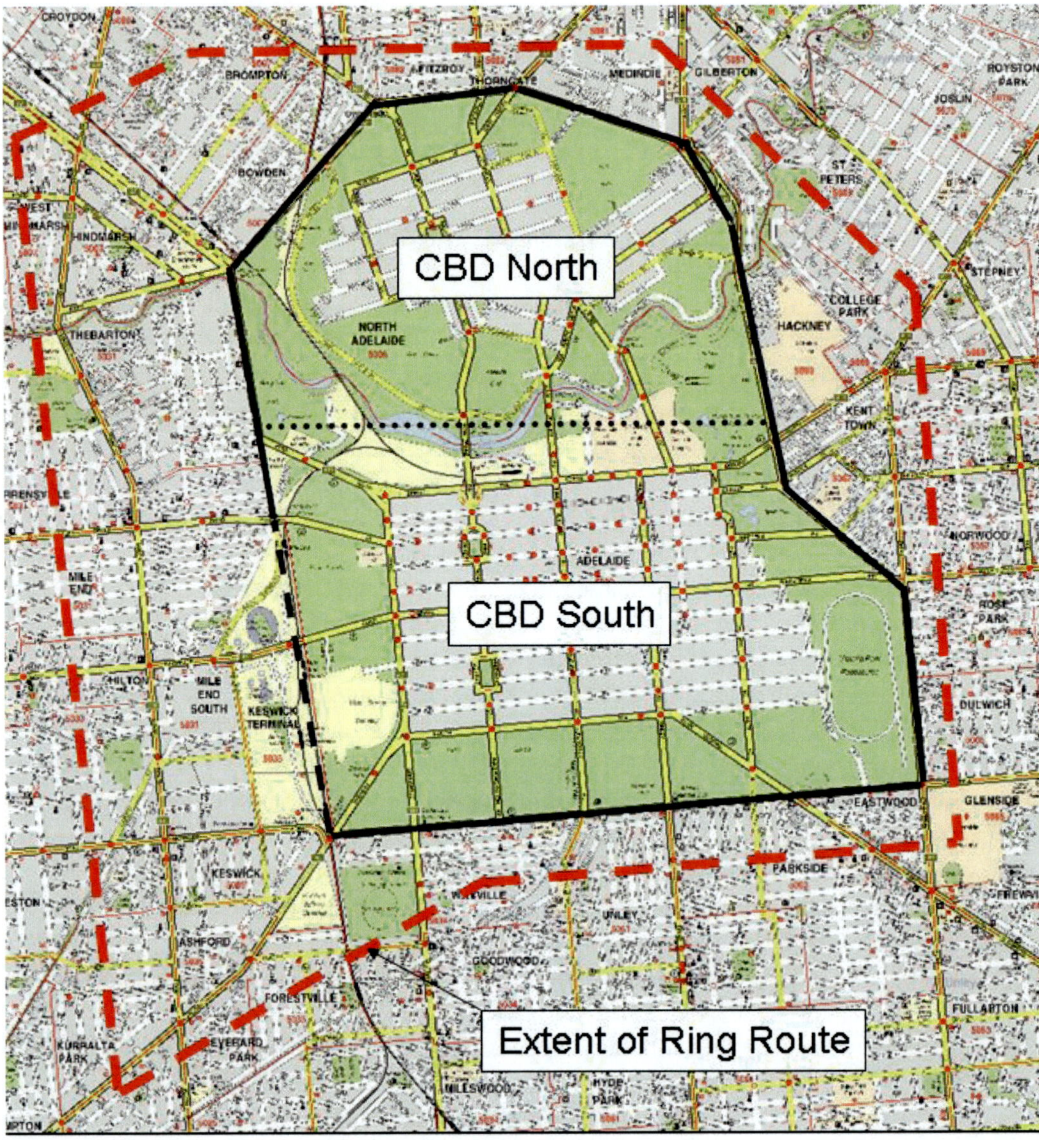

Figure 1. Adelaide City Council area road network.

The Paramics is among the most powerful software packages for micro-simulation studies. The standard functions provided by Paramics are suitable for time-responsive traffic modelling. Vehicles are released from their origins into the road network according to time-dependent (not elastic) OD matrices. With in each specific time period which is defined in the vehicle release profile and is associated to the OD matrices, traffic management strategies of road infrastructure (e.g. lane closure, restrictions, traffic signals, etc.) are relatively fixed. Paramics does provide dynamic traffic assignment function taking current traffic conditions into account. However, this is performed during a fixed feedback period (e.g. 5 min). The ATIS strategies such as VMS route guidance are more event-responsive. The VMS responds to unpredictable traffic events (e.g. incidents) instead of specific time periods, which in turn requires individual travellers respond to the VMS message dynamically when they reach the

VMS sign. To model such VMS applications, the Application Programming Interface (API) of Paramics was utilised in this study and specific codes were developed to extend and override (if needed) the default standard functions of Paramics. The resultant application program is portable, which works with the ACC model as a 'plug-in' to implement the VMS functionalities.

4. VMS Route Guidance Application

The two major objectives of developing the VMS route guidance program include: to enhance the capability of simulating individual DVUs' dynamic response to VMS messages based on the Adelaide travel behaviour studies, and to make the program portable so that it can work with large scale ACC model to assist ATIS strategy development and evaluation.

4.1. Dynamic Micro-simulation Framework for VMS Application

The nature of VMS route guidance involves traffic information gathering, traveller information distribution, and most importantly, individual travellers' route choice decisions (to stay on the original route or select an alternative route). In the micro-simulation context, we prefer the term DVU to represent the moving object (i.e. the vehicle and its driver) in the traffic network. The primary stimulus for a DVU to change its normal route is expected travel delay, and the final route choice decision relies on the inherent characteristics of each individual DVU. These are supported by earlier travel behaviour studies which have been discussed in the previous section. In addition, the way in which traveller information is distributed also influences the effectiveness of the VMS route guidance application.

The proposed dynamic micro-simulation framework for evaluating the VMS route guidance system consists of the following three elements:

1. Traffic network. This element prepares the detailed traffic network and its initial network condition for the micro-simulation. A time-dependent OD matrix is an input to the module. The OD matrix is treated as inelastic during entire simulation while traffic control strategies may change.
2. Traffic information generation. The delay on the targeted route is the primary information distributed by the VMS. Real time traffic data are collected from the network at each simulation time step (e.g 0.5 sec). The travel time estimation is performed at our predefined time interval (i.e. 2 min) to reflect the real world traveller information system operations.
3. Travellers' (DVUs') route choice. This element plays the key role in our VMS application. Individual DVUs' characteristics, their minimum acceptable delay and the expected travel delay interact with each other to influence the DVUs' route choice second-by-second when they are passing the VMS sign.

The effects of travellers' route choice can be monitored and assessed through interactive micro-simulation interface during traffic simulation.

4.2. VMS Route Guidance Program

4.2.1. Travel Time Estimation

Travel time represents spatial traffic information related to a specific link (or set of consecutive links) of the traffic network. Different methods have been presented in literature on arterial road travel time estimation [Anderson and Bell 1997; Ishak and Al-Deek 2002; Wu et al. 2004]. We also proposed a different approach to estimate detailed travel time distribution which take into account both prevalent traffic conditions and traffic signal settings at the same time [Lin et al. 2006; Zhang et al. 2008]. In this study, we propose a sampling method to obtain average travel time of a specific link to simplify the travel time estimation process.

Taking advantage of the Paramics API, we can obtain travel speed v_i of each individual DVU at each simulation time step (e.g. 0.5s) on our targeted link. If we set the travel time estimation interval to 2 min, then we can have 240 sets of speed samples with each of these sets containing speed samples of all vehicles on the link at a specific time step. Hence, the average travel speed v_a of the link for the last two minutes would be:

$$v_a = \frac{1}{n}\sum_{i=1}^{n} v_i \tag{8}$$

where n is the total number of samples collected from the link during entire two minutes periods, which represents all moving vehicles.

Note that we are dealing with a CBD environment and intersection travel delay (both geometric delay and stop-line delay) imposed on each individual vehicle needs to be taken into account. We deliberately separate the moving vehicles from the stationary ones in the samples using a cut off threshold of 1.0 m/s. This threshold is chosen to replicate GPS systems that also use this value to identify stationary objects. We now have two groups of speed samples: one group contains n speed samples for moving vehicles (see n in equation (8)), and the other group has m samples for stationary vehicles during the same travel time estimation period.

The length of the targeted link l is known. By introducing a parameter r_m

$$r_m = \frac{n}{m+n} \tag{9}$$

the proportion of link travel time used for actual travel, we may have an estimate of the average link travel time t_a as follow [Zhang et al. 2007]

$$t_a = \frac{l / v_a}{r_m} \tag{10}$$

This method is very efficient except when gridlock happens. In that case, we set up a cap as the maximum delay time for a specific link based on our experience.

4.2.2. Information Distribution

In the traffic simulation environment, traveller information distribution via VMS is quite different from the real world VMS application. In the real world, VMS messages are exposed to every traveller who passes the VMS sign. It is up to the individual travellers to determine whether they notice and then want to read the message and whether they accept the information and change their normal routes. In contrast to this process, we have to fulfil the following three tasks in order to make our VMS route guidance program work during micro-simulation.

1. *Distributing traffic information at the right place*. It is quite clear that traveller information should be delivered to DVUs on the link where the VMS is located (on the VMS link).Since the VMS in Paramics is a passive display tool instead of a control device that communicates with each DVU, we have to make sure that the information become available to a DVU when it reaches the VMS link. We only want the real time traffic information be solely used for en-route guidance on specific links, which is distinguished from dynamic traffic assignment provided by Paramics. In fact, we control how a DVU is going to receive the VMS message rather than where to distribute it.
2. *Distributing the information at the right time*. As we know, the primary stimulus for a DVU to change its normal route is increased individual delay. This implies that if the estimated link travel time is close to the normal free flow travel time, there is no incentive for the drivers to make route changes. On the other hand, it will distract drivers from their driving tasks. The VMS experience from London [Chatterjee et al. 2002] also suggested that the real time, important and pertinent traffic information would enhance the effectiveness of VMS application. In our VMS program, we start to distribute travel time information when the estimated travel delay on the targeted link exceeds a certain threshold (e.g. 5 min). Our former SP study on travellers' route choice [Furusawa et al. 2004] is used as a guide to determine the appropriate thresholds used in the program.
3. *Choosing the right DVUs to absorb the VMS message*. Not every DVU that reaches the VMS link either has a strong tendency to use or has feasible alternatives to shun the targeted link (e.g. in the case of incidents) where the travel delay is estimated. The potential users of the targeted link who do have alternatives need to be selected as the candidate DVUs. We focus on the destination zone of each DVU appearing on the VMS link. After analysing the traffic network, we are able to construct a list of destination zones so that the targeted link is more attractive to DVUs whose destination zone is included in the list. These DVUs become our VMS control targets.

The first and third tasks indicate the distribution of traffic information is based on user groups in our program. This fact opens an opportunity for us to test the different ways of information distribution (e.g. user prescribed real time traffic information) using the similar program architecture.

4.2.3. Driver's Route Choice Decision

Individual DVU's dynamic route choice forms the core of our VMS route guidance program. As discussed in the previous section, individual DVUs' characteristics (i.e. awareness, aggression) greatly influence their car-following and lane-changing behaviour. In our program, these characteristics play a key role in route choice decision making.

We know that aware drivers tend to have plenty of time to judge traffic conditions (e.g. they will change lanes much earlier in order to make a turning manoeuvre) and obtain information from VMS to make more informed route choices. If the awareness of drivers is low, they may not have time to read VMS messages and the likelihood for them to follow VMS guidance is low. Aggressive drivers keep seeking opportunities to reduce their travel cost by frequent lane changing and selection of faster routes. Conservative drivers tend to adhere to their usual lane or familiar route as much as possible. Hence, both awareness and aggression factors of an individual DVU will influence its patience when expected travel delay is encountered.

The primary stimulus for a driver to change route is expected increase in delay. The route choice decision is the result of balancing between the individual DVU's patience and its expected travel delay. Since the minimum acceptable delay for certain traveller groups (e.g. familiar travellers) may be obtained from RP or SP studies, the patience p_i of individual DVU i can be estimated as follow

$$p_i = (2 - \frac{Ag_i + Aw_i}{8}) p_a \tag{11}$$

where p_a is the average minimum acceptable travel delay for a certain user group, and Aw_i and Ag_i represent the awareness and aggression factors of the DVU i respectively. Here, both the awareness and aggression of DVUs follow a normal distribution $N(4,1)$ which is truncated to the range [0,8]. Paramics allows users to customise the distribution of these factors. If the estimated travel delay exceeds the patience of the DVU, then the chance for the DVU to accept VMS guidance increases.

In addition, two other factors, trust put on the VMS and individual DVU's perceived link travel cost, are also used for route choice decisions. The linear combination of the awareness and aggression of each candidate DVU forms its trust level of the VMS message. The DVU's perceived travel cost on the targeted link is decided using both the perturbation rate of the DVU and a random number generated at the time when the DVU reaches the VMS link.

The VMS route guidance program is developed using C language. The Paramics API allows our program to augment the core Paramics simulation with new functions, driver behaviours and practical features. Through the API function calls, travel time estimation, information distribution and individual DVU route choice decisions can be performed within Paramics micro-simulation environment at each simulation time step. The VMS guidance program is implemented via Dynamic Linked Libraries (DLL files). Each time when the Paramics Modeller starts to load the ACC model, it automatically searches for user developed DLL files, such as the VMS route guidance program, and loads them on request. The Paramics treats the user-defined program as its 'plug-in' function.

5. Preliminary Simulation Results

5.1. Study Site

The test site of the VMS route guidance program is the north-eastern quarter of the Adelaide CBD road network (see Figure 2), which also includes a section of the Ring Route system (Hackney Road). Our targeted road is North Terrace, one of the major arterial road linking city east and the west. Along this road, there are two university campuses, a major hospital, the central railway station, a number of government buildings, together with a major shopping mall just next to it. All these make the North TCE a very busy road, especially during morning peak period. The average daily traffic entering and leaving the Adelaide CBD using the east section of North TCE are 19418 veh/day and 17207 veh/day respectively (2006 data).

During the AM peak period (8:00 – 9:00), the average volume of the city bound traffic using this route is 1864 veh/h, which includes 31 buses and 21 heavy commercial vehicles. If any congestion happens along the North TCE inside the Adelaide CBD during this period, it would greatly affect both the city bound traffic and the normal operation of the Ring Route system. The alternative routes to North TCE could be Rundle Road, Bartels Road and Wakefield Street, which are parallel to North TCE down the south. The average volumes of city bound traffic using these roads are 997 veh/h, 1314 veh/h and 1682 veh/h respectively.

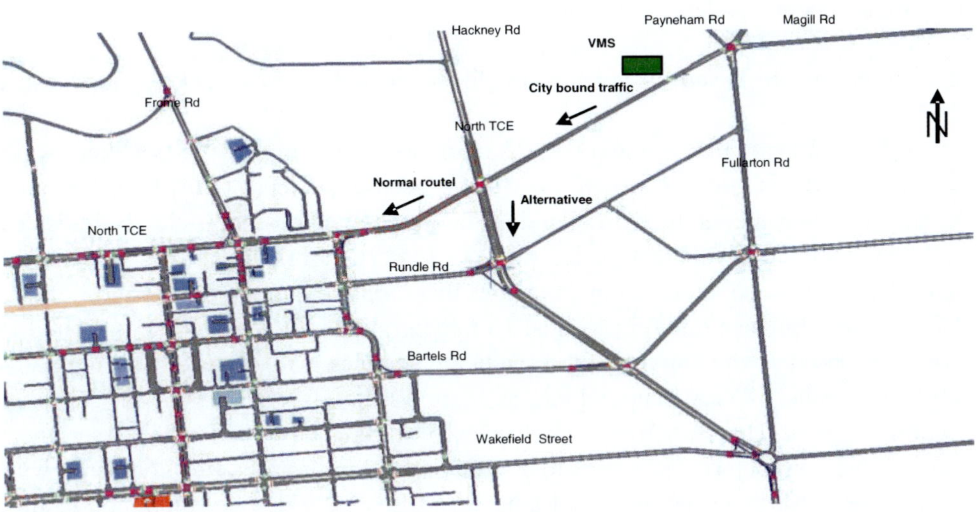

Figure 2. Test site for VMS-route guidance program.

In our VMS route guidance program, the link travel time of the east section of North TCE (between King William Road to the east and Hackney Road to the west) to influence the route choice of city-bound traffic heading to North TCE from Magill Road, Payneham Road and Fullarton Road. The VMS panel is located next to the intersection between Magill Road and Payneham Road further downstream. 'Drive Safely' is the normal display of the VMS panel. Once the delay on our targeted link exceeds a predefined threshold, the VMS panel starts to display the estimated travel delay to travellers.

Table 3. Number of vehicle diverting (every 2 mins) vs Expected travel delay

Delay (min)	5	5	6	13	14	20	19	16
Number of vehicle diverting	0	0	2	6	8	16	18	11
Number of candidate vehicles	36	44	32	34	50	30	41	42
Total vehicle passed VMS	61	74	53	56	84	51	86	70

5.2. Results

The VMS route guidance program is implemented in the ACC model during the morning peak period (8:00 – 9:00). We create a temporary incident which blocks one lane of North TCE inside the Adelaide CBD. This incident starts at 8:10 to allow the ACC model sufficient time to load vehicles into the network. The simulation results are shown in Table 3, which contains a typical 16-minute VMS operation. Table 3 is obtained when the parameters are set as follows

- Simulation time step: 0.5 sec
- Distribution of awareness and aggression: Normal
- Minimum delay required to activate VMS route guidance system: 5 min
- Mean acceptable delay for familiar vehicles: 10 min
- Mean acceptable delay for unfamiliar vehicles: 15 min.

These settings are based on general travellers' behaviour in Adelaide [Furusawa et al. 2004].

The figures indicate that as the travel delay increases, the number of travellers willing to take alternative routes goes up which could have been expected. Meanwhile, the nonlinear relationship between the estimated delay and the number of diversion clearly indicates that the individual travellers' characteristics play an important role in his/her real time route choice decision. This is one of the key outcomes this research has achieved.

The VMS compliance rate for a typical 16 min period of VMS operation is around 19 per cent, which is based on the number of the candidate vehicles who actually have alternatives to choose. This number is very encouraging, as it shows the effectiveness of the VMS and its application potential. However, if the total volume of vehicles which have passed the VMS link is used, the compliance rate drops to 11 per cent. This figure is low compared with that found in earlier Australian research by Ramsay and Luk [1997]. The simulation results also indicate that a very small traffic diversion occurs when the expected travel delay is less than 10 min. When the expected travel delay is greater than 15 min, the number of vehicles that follow the VMS route guidance increases. The above findings imply that we should not overestimate the influence of the VMS on travellers' route choice decision, especially when our focus is on the entire traffic rather than the ones who do have alternatives to choose and the delay is not big enough.

Micro-simulation approach provides us the flexibility to access and adjust characteristics of individual DVU using empirical results from traveller route choice behaviour studies. In contrast to setting a VMS compliance rate in advance and randomly picking up vehicles to obey the VMS route guidance, we use individual DVU's characteristics and its surrounding traffic conditions to decide whether the DVU should follow the VMS route guidance or not. In this way, we can improve the simulation of driver's route choice in response to VMS (and to traffic conditions in general) using empirical findings and our travel experiences. Meanwhile, we can explain the traffic simulation results with increased confidence. These critical features of our proposed micro-simulation approach are essential for any ATIS strategy development and testing.

6. Conclusion

In this research, we present a micro-simulation approach to model traveller's real time route choice decisions in response to VMS. We have developed a portable application program which implements the VMS based traveller information system in a simulated urban traffic environment. The expected travel delay on the targeted route is the primary information distributed by the VMS. Individual travellers' inherent characteristics (i.e. awareness and aggression), their minimum acceptable travel delay and perceived travel cost variations determine their dynamic route choice. The empirical results from the previous study on travellers' route choice behaviour in Adelaide are used as a guide to set up a number of thresholds used by the program.

The program has been tested using the Adelaide CBD microscopic traffic simulation models which were developed in our previous study. The preliminary results demonstrate the capability of the micro-simulation approach in mimicking travellers' dynamic response to VMS (and traffic information in general). Meanwhile, the test reveals that the impact of the traveller information systems on urban traffic should not be overestimated, and empirical results from travel behaviour studies are essential for any traveller information system development. Currently, we are planning to expand the VMS application to the key intersections along the Adelaide ring route system. This would provide us a better chance to carry out detailed investigation into the impact of the ATIS systems on urban traffic networks.

Acknowledgements

The authors would like to thank the Australian Research Council and Adelaide City Council for their financial support to this research project, and Transport SA for providing us up-to-date traffic data essential for the project.

References

Abdel-Aty, M. A., Kitamura, R., and Jovanis, P. P. (1997). Using stated preference data for studying the effect of advanced traffic information on drivers' route choice. *Transportation Research Part C*, 5, 39 - 50.

Anderson, J., and Bell, M. G. H. (1997). Travel time estimation in urban road networks. IEEE Conference, Boston, USA.

Ben-Akiva, M., and Bierlaire, M. (1999). Discrete choice methods and their applications to short-term travel decisions. *Handbook of Transportation Science*, R. W. Hall, ed., Kluwer.

Ben-Akiva, M., Bierlaire, M., Bottom, J., Koutsopoulos, H., and Mishalani, R. (1997). Development of a route guidance generation system for real-time application. 8th IFAC Symposium on Transportation Systems, Chania, Greece.

Chatterjee, K., Hounsell, N. B., Firmin, P. E., and Bonsall, P. W. (2002). Driver response to variable message sign information in London. *Transportation Research Part C*, 10, 149 - 169.

Dia, H. (2002). An agent-based approach to modelling driver route choice behaviour under influence of real-time information. *Transportation Research Part C*, 10, 331 - 349.

Dia, H., and Cottman, N. (2005). Evaluation of arterial incident management impacts using traffic simulation. IEE Proceedings - Intelligent Transport Systems, 153, 242 - 252.

Furusawa, H., Woolley, J., and Yue, W. L. (2004). A stated preference survey for investigating route choice behaviour in Adelaide. 27th Australasian Transport Research Forum, Adelaide.

Hansen, B. G., Martin, P. T., and Perrin, H. J. (2000). SCOOT Real-time Adaptive Control in a CORSIM Simulation Environment. *Transportation Research Record*, 1727, 27 - 31.

Hensher, D. A. (1992). Investigating revealed preference and stated response data into a jointly estimated hierarchical mode choice model. Working Paper ITS-WP-92-9, Institute of Transport Studies, Graduate School of Business, University of Sydney.

Hensher, D. A., Milthorpe, F. W., Smith, N. C., and Barnard, P. (1990). Urban toll roads and the value of travel time savings. *Economic Record*, 66, 146 - 156.

Hicks, B., and Carter, M. (2000). What Have We Learned About Intelligent Transportation Systems? -- Arterial Management. US Department of Transportation / Federal Highway Administration.

Hidas, P., and Awadalla, E. (2001). Investigation of route choice in response to variable message signs. *Journal of the Eastern Asia Society for Transportation Studies*, 4, 39 - 54.

Iguchi, M. (2002). A perspective on ITS deployment. *JSAE Review*, 23, 173 - 176.

Ishak, S., and Al-Deek, H. (2002). Performance evaluation of short-term time-series traffic prediction model. *Journal of Transportation Engineering*, 128(6), 490-498.

Kroes, E. P., and Sheldon, R. J. (1988). Stated preference methods: an introduction. *Journal of Transport Economics and Policy*, 22, 11 - 25.

Lin, H.-E., Zito, R., and Taylor, M. A. P. (2006). The Application of Micro-simulation in a Travel Time Study on a SCATS Arterial Road. 2nd International Symposium of Transport Simulation, Lausanne.

Louviere, J. J., Henley, D. H., Woodworth, G., Meyer, R. J., Levin, I. P., Stoner, J. W., Curry, D., and Anderson, D. A. (1980). Laboratory simulation versus revealed preference methods for estimating travel demand models. *Transportation Research Record*, 794, 42 - 51.

Louviere, J. J., Hensher, D. A. and Swait, J. D. (2000). *Stated Choice Methods: Analysis and Application*. Cambridge, Cambridge University Press.

Lucas, D. E., Mirchandani, P. B., and Head, L. K. (2000). Remote Simulation to Evaluate Real-Time Traffic Control Strategies. *Transportation Research Record*, 1727, 95 - 100.

Luk, J. Y. K., and Yang, C. (2003). Comparing driver information systems in a dynamic modeling framework. *Journal of Transportation Engineering*, 129(1), 42 - 50.

Mahmassani, H. S., and Jayakrishnan, R. (1991). System performance and user response under real-time information in a congested traffic corridor. *Transportation Research Part A*, 25, 293 - 307.

Ozbay, K., Doshi, H., Mudigonda, S., Bongiovanni, T., Chen, J., Boile, M., Shah, J., and Desai, J. (2006). Evaluation of Adaptive Control Strategies for NJ Highways. FHWA-NJ-2006-001, Federal Highway Administration, US Department of Transportation, Washington, D.C.

Quadstone. (2000). www.paramics-online.com.

Quadstone. (2006). Quadstone Paramics V5.2 Modeller Reference Manual.

Ramsay, E. D., and Luk, J. Y. K. (1997). Route choice under two Australian travel information systems. *ARR312*, ARRB Transport Research Ltd, Melbourne.

Stazic, B., Taylor, M. A. P., and Woolley, J. E. (2005). Traffic microsimulation of the Adelaide CBD road network. Intelligent Vehicles and Road Infrastructure Conference, SAE-Aust, Melbourne.

Taylor, M. A. P., Woolley, J. E., and Zito, R. (2000). Integration of the Global Positioning System and Geographical Information systems for Traffic Congestion Studies. *Transportation Research Part C*, 8, 257 - 285.

Taylor, N. B. (1990). CONTRAM 5: an enhanced traffic assignment model. *TRRL Research Report RR249*, Transport and Road Research Laboratory, Crowthorne, UK.

VICS. (2008). www.vics.or.jp/navi/shipment.html.

Wardman, M. (1988). A comparison of revealed preference and stated preference models of travel behaviour. *Journal of Transport Economics and Policy*, 22, 71 - 91.

Wardman, M. (1991). Stated preference methods and travel demand forecasting: an examination of the scale factor problem. *Transportation Research Part A*, 25, 79 - 89.

Wardman, M., Bonsall, P. W., and Shires, J. D. (1997). Driver response to variable message signs: a stated preference investigation. *Transportation Research Part C*, 5, 389 - 405.

Wardrop, J. G. (1952). Some theoretical aspects of road traffic research. Proceedings of Institute of Civil Engineers, 1, 325 - 378.

Wu, C.-H., Ho, J.-M., and Lee, D. T. (2004). Travel-Time Prediction with Support Vector Regression. *IEEE Transactions on Intelligent Transportation Systems*, 5, 276 - 281.

Zhang, K., Lin, H.-E., and Taylor, M. A. P. (2008). Travel time estimation in urban traffic environments. *New Transportation Research Progress*, F. N. Gustavsson, ed., Nova Science Publishers, New York, 181 - 210.

Zhang, K., and Taylor, M. A. P. (2006a). Effective arterial road incident detection: a Bayesian network based algorithm. *Transportation Research Part C*, 14(6), 403 - 417.

Zhang, K., and Taylor, M. A. P. (2006b). Towards Universal Freeway Incident Detection Algorithms. *Transportation Research Part C*, 14(2), 68 - 80.

Zhang, K., Woolley, J., and Stazic, B. (2007). An effective tool for advanced traveller's information systems development. State of Australian Cities Conference, Adelaide.

In: Road Traffic: Safety, Modeling and Impacts
Editors: S. E. Paterson and L. K. Allan, pp. 321-345
ISBN 978-1-60456-884-4

Chapter 6

NEW TRENDS IN ROAD TRAFFIC SAFETY RESEARCH ON SENIOR DRIVERS

Joceline Rogé[1], Laurence Paire-Ficout, Catherine Gabaude, Ladislav Mota and Claude Marin-Lamellet
Laboratory of Ergonomics and Cognitive Sciences applied to Transport (LESCOT)
The French National Institute for Transport and Safety Research (INRETS), France

ABSTRACT

People aged over 65 represent the most rapidly growing segment of the population in most industrialized countries. Compared with previous generations, these people will, in future, demand greater transport mobility. As the general population ages, the percentage of elderly drivers on the road will increase. It is well established that age per se is not the best predictor of driving performance. However, it is recognized that aging has an effect on visuo-attentional and cognitive abilities, which can have an impact on driving ability and could be potentialy compensated by metacognitive abilities. The purpose of this chapter is to highlight the main trends in traffic safety research concerning senior drivers with health conditions, as a cross-disciplinary issue.

The first part of the chapter is concerned with advances in research related to neurocognitive abilities and driving. In past decades, research in this area has evolved from a classical medical approach to a more neuropsychological one. Numerous research protocols including neuropsychological tests (paper-and-pencil or computer-based), specific scenarios implemented in driving simulators and field observations have been used to study the impact of neurocognitive impairments on the driving performance of older drivers. Most of these studies have been conducted to explore the value of these tests in predicting future driving performance. A review of the various research protocols developed in these studies provides an opportunity to discuss and comment on some neuropsychological models that are capable of explaining some of the underlying mechanisms.

---

[1] Contact: joceline.roge@inrets.fr.

This chapter also addresses visuo-attentional issues in aging, focusing mainly on the useful visual field concept. Several experimental approaches have been used to study this concept, which has a number of clinical and practical applications. Different tests developed in these studies are presented, allowing a better understanding of numerous scientific approaches that highlight two main points: how is the useful visual field affected by aging?; and what are the consequences of this deterioration for driving activity?

In the final section of the chapter, we suggest an explanation on the potential adaptation of the driving behavior of some elderly drivers to their visual or cognitive deficiencies. The increasing body of research on metacognition is presented. Here, the objective is to emphasize the potential benefit for the development of more efficient training procedures for elderly drivers.

Simultaneous consideration of these three aspects provides an overarching framework for the conduct of future research on older drivers' fitness to drive.

Introduction: Evolution of Research Work in Aging and Driving Context

The objectives and the purpose of research work based on the effects of aging on driving have changed considerably from 1940 to the present day. DaSilva, in 1938 is something of a trail-blazer in this field (DaSilva, 1938). He spoke about the steadily increasing number of drivers over 40 years of age. It is interesting to notice that the notion of accompanying older persons in order to continue to drive as long as they possibly can was already being discussed. In 2008, almost sixty years later, such words can raise a smile because we follow the same objective but today the main concern is rather to focus on sexagenarians rather than on forty-year-old people.

In the late 1960s and early 1970s, the older driver problem was recognized as a scientific and social issue. Much research has been conducted on older drivers mainly in the U.S (Finesilver, 1969; Grow, 1972; McFarland, 1964; Planek, 1972). In this first wave of research, the focus was on visual and medical disabilities. In particular, the research focused on physical and/or sensory decline in order to determine their impact on driving behavior. Muscular, locomotor and articular difficulties were identified as factors that reduce a person's ability to drive (Planek, 1972). There was also a focus on visual problems that occur with age, such as the reduction in visual acuity and sensitivity to contrasts (Hill and Burg, 1977). As Hakamies-Blomqvist wrote in her review of 1996, the driver's "deficiencies" were seen exclusively as the cause of all problems, and the characteristics of the traffic system were taken largely for granted (Hakamies-Blomqvist, 1996). Naturally, this point of view dates from a period before the "cognitive revolution" (see (Miller, 2003) for a review).

In the 1980s and 1990s, many studies were carried out in the area of accident epidemiology in order to gain a better understanding of the causes of accidents. These works have shown that older drivers have more accidents per distance driven (Evans, 1987; Stutts and Martell, 1992) and are overrepresented in accidents occurring at intersections and other complex traffic situations (Fontaine and Gourlet, 1992; Hakamies-Blomqvist, 1994; Hakamies-Blomqvist, 1993; OECD, 1985). Alcohol is involved less often for older drivers, they have less single-vehicle accidents (Campbell, 1966; Hakamies-Blomqvist, 1993), and accidents mostly occur during the day time (Campbell, 1966; Fontaine and Gourlet, 1992).

There has been no research to demonstrate that regular medical check-ups have a positive, direct impact on road safety. In fact, researchers recognize that mandatory assessment programs targeting older drivers are unlikely to produce safety benefits and may produce counter-productive results (Hakamies-Blomqvist, 1996). A comparison between Finland, which requires regular medical check-ups with license renewal starting at age 70, and Sweden where no such age-related control was required, indicates that Finland has a higher rate of fatalities among unprotected older road users. It is arguably the result of an increase in the number of older pedestrians who had lost driving licenses. Moreover, some drivers prefer to stop driving (maybe too prematurely) rather than undergo a medical examination. At this time, the doubt surrounding the effectiveness of medical check-ups remains.

In Europe, increased interest in aging and automobile research was encouraged by a report by the OECD in 1985. This report reviewed accident studies and research needs were specified. This is also the beginning of the repercussions of the cognitive revolution: the information-processing framework viewpoint dominated (Baddeley, 1992; Broadbent, 1962; Norman and Shallice, 1980; Norman and Shallice, 1986). There was a shift towards more cognitive conceptualizations and high-level functions as visuo-attentional rather than sensory phenomena and attentional processes rather than reaction times or motor speed (Gabaude, 2003; Hakamies-Blomqvist, 1996). During the same period, there was much research on aging and cognitive processes (Salthouse, 1994; Salthouse and Meinz, 1995). Selective attention and divided attention were central to the research's focus. It was shown that older drivers always have a significantly decreased ability to divide attention between a main task (driving for example) and a secondary, visual or hearing task (Brouwer, Waterink, Van Wolffelaar, and Rothengatter, 1991; Crook, West, and Larrabee, 1993). The speed of processing was also examined and major results show that older persons were generally slower than others (Salthouse, 1994; Salthouse and Meinz, 1995). The outcome of these studies on the relation between aging and cognitive process is that attentional components are particularly age-related and disease-related (Hasher, Stoltzfus, Zacks, and Rypma, 1991; Tipper, 1991). Their respective relevance and the relationship with driving performance were also demonstrated (Daigneault, Joly, and Frigon, 2002; Whelihan, DiCarlo, and Paul, 2005).

During the 1990s and 2000s, demographic projections predicted an increase in the number of older people suffering from pathologies, particularly mental pathologies such as Alzheimer's disease or Parkinson's disease, raising new questions about road safety. In general, mental pathologies affect the brain, causing cognitive difficulties. With some diseases such as Alzheimer's (the most common) and Parkinson's, attentional problems appear at a relatively early stage. Various studies have shown that there is an increased risk of accidents and poorer driving performance among populations suffering from Alzheimer's (Carr, 1997; Carr, Duchek, and Morris, 2000; Dubinsky, Stein, and Lyons, 2000; Fox and Bashford, 1997; Friedland, et al., 1988; Johansson, Bogdanovic, Kalimo, Winblad, and Viitanen, 1997; Uc, Rizzo, Anderson, Shi, and Dawson, 2006) and Parkinson's (Dubinsky, et al., 1991; Homann, et al., 2003). Research in the 1990s therefore further concentrated on moving from a general approach towards a differentiated approach, focusing on high-risk sub-groups. The evaluation of cognitive abilities is becoming increasingly precise and the number of cognitive evaluation tools, borrowed from neuropsychology, is constantly growing

(Brown, et al., 2005; Duchek, Hunt, Ball, Buckles, and Morris, 1998; Duchek, et al., 2003; Marin-Lamellet, et al., 2003; Uc, Rizzo, Anderson, Shi, and Dawson, 2006; Whelihan, DiCarlo, and Paul, 2005). The best correlations between the risk of accident and perceptive-cognitive ability have been observed through measuring the useful visual field. This concept has been widely used in the field of driving behavior both in terms of the relationship between vision and attention, and to evaluate the abilities of senior drivers under normal or pathological aging conditions. This topic will be detailed in a forthcoming section.

Progress in the field of neuro-cognitive imagery means that it is now possible to propose new hypotheses on the relationship between cognitive functions, particularly those used when driving. These imagery techniques, combined with a driving simulator, are promising for the future and should allow further advances in our understanding of the cognitive functions involved in driving.

At the same time that research focused on high-risk sub-groups in order to be sure that they do not increase the accident risk for others, some researches focused on the question of monitoring some older drivers in order to optimize their driving competency. A British survey showed that older drivers are in favor of monitoring for improving their ability to drive (Parker, McDonald, Rabbitt, and Sutcliffe, 2003). Eby et al. (2003) have proposed a driving self-assessment called *The Driving Decisions Workbook* which involves completing a questionnaire about driving habits and driving difficulties (Eby, Molnar, Shope, Vivoda, and Fordyce, 2003). Staplin et al. (2003) propose a complete model program (Staplin, Lococo, Gish, and Decina, 2003). This model includes procedures for detecting functionally-impaired drivers who pose an elevated risk to themselves and others, to support the remediation of functional limitations and to provide mobility counseling to inform and connect individuals with local alternative transportation options. The goal of these training programs is to allow older drivers to continue to drive safely for longer in safe conditions and to increase general knowledge about driving and self-awareness.

These training programs are part of an emerging field of research, that of metacognition, which studies how individuals understand their own thought processes. . The idea is that the development of metacognitive abilities improves performance. The greater an individual's ability to self-evaluate, the easier it will be for them to adjust their behavior. In the context of driving, Brouwer suggests that drivers with the ability to evaluate their own performance and functional limitations will be able to regulate their own activity accordingly (avoiding night driving or complex areas). Where a deficit exists (attentional deficit or slight visual deficit, for example), good driving ability can be maintained through adaptation or compensation strategies (Brouwer, 2002; Brouwer and Ponds, 1994). This self-regulation is conditioned partly by the recognition and understanding of the individual's own difficulties, and partly by their ability to regulate their own activities. This is why authors have tried to bring together the fields of metacognition and executive functions (Fernandez-Duque, Baird, and Posner, 2000). Considering the metacognitive approach in order to develop a coherent and unified theoretical model of driving behavior is promising. This is the object of the third part of this chapter.

USEFUL VISUAL FIELD AND DRIVING: STATUS REPORT AND EVOLUTION

Driving is mainly based on visual information processing. When the driver searches for relevant information, visual and cognitive processes are called upon, especially attentional abilities. As a result of aging, these capacities decrease and, in some cases, this can have an impact on daily-life activities such as driving.

Several authors have sought to link these changes with driving performance. First, the link between sight deterioration and accident risk has been studied (Owsley, et al., 1998; Sims, McGwin, Allman, Ball, and Owsley, 2000) but apart from in Marottolli's study (Marottoli, et al., 1998), the results were not significant and the conclusion of these studies was that driving safety does not depend so much on what is seen, but rather on how quickly and adequately drivers respond (Brinig, Wilkinson, Daly, Jogerst, and Stone, 2007). Then, the focus of research moved from sight to vision, and cognitive functions start being studied. Work has shown that decision-making under duress was more impaired among drivers aged 65 and over than among younger drivers (Caird, Edwards, Creaser, and Horrey, 2005). Different kinds of difficulties are encountered by elderly drivers: perceptive errors (Caird and Hancock, 2002; Staplin, 1995), visual attention failures (Owsley, 2004), memory loss (Delorme and Marin-Lamellet, 1998) and cognitive overload (Janke, 2001; Van Elslande, 2003). More detailed studies indicate that the difficulties faced by the elderly when crossing an intersection are due to a shared attention deficit (Caird and Chugh, 1997) and/or a lack of visual information searching (McDowd and Shaw, 2000).

Concerning this last point, the assessment of the useful visual field (using the UFOV test that we will present below) has a number of clinical and practical applications that seem promising for elderly drivers (Edwards, et al., 2006). It would be sensitive to the cognitive decline experienced by older adults with dementias such as Alzheimer's (Duchek, Hunt, Ball, Buckles, and Morris, 1997; Duchek, Hunt, Ball, Buckles, and Morris, 1998). The UFOV test is used to evaluate driving capacity in stroke patients (Fisk, Novack, Mennemeier, and Roenker, 2002) and in patients with traumatic brain injury (Fisk, Owsley, and Mennemeier, 2002; Novack, et al., 2006). The score in the UFOV test could also be useful for ophthalmologists and optometrists treating older patients who complain of functional visual difficulties in the absence of abnormalities detected by a standard ophthalmological exam (Ball, Owsley, Sloane, Roenker, and Bruni, 1993).

CONCEPT AND EXPERIMENTAL DEVICES

Peripheral vision plays a decisive role in all human activities. The detection of an object in the peripheral field enables the eyes to move so that this object is focused in the fovea and thus more precisely analyzed. Peripheral vision therefore plays a major role in the analysis of the environment. The spatial area of the visual field can be measured by asking the subject to stare at a central spot while detecting or identifying the display of a peripheral signal at the same time. The furthest eccentricity at which the subject can see the signal determines their visual field (Verriest, Barca, and Dubois-Poulsen, 1983).

The term "useful field of view" was first used by Mackworth in a task in which the subject had to detect similarities between uppercase letters presented simultaneously in periphery and in the fovea (Mackworth, 1965). The useful field of view is the area around the fixation point from which information is briefly stored and read out during a visual task. In other words, this field corresponds to the part of the peripheral visual field around the fixation point inside which the sources of information can be processed in a single glance, i.e. without any movement of the eyes or the head (Ball, Beard, Roenker, Miller, and Griggs, 1988; Sanders, 1970; Scialfa, Kline, and Lyman, 1987). This field is also called functional field, working visual field or useful visual field (Ball and Owsley, 1993; Rogé, Kielbasa, and Muzet, 2002). Due to attentional workload, this field is smaller than the visual field (Ball, Owsley, Sloane, Roenker, and Bruni, 1993; Scialfa, Thomas, and Joffe, 1994). It is generally measured by instructing the subject to carry out two simultaneous tasks: a central task and a peripheral task. Some researchers assess the useful visual field by instructing the subjects to detect the presence of a peripheral signal and to identify it (Leibowitz, Abernethy, Buskirk, Bar-Or, and Hennessy, 1972; Mackworth, 1965; Rogé, Kielbasa, and Muzet, 2002; Williams and Lefton, 1981), whereas others demand the participant to localize it (Ball, Beard, Roenker, Miller, and Griggs, 1988; Ball and Owsley, 1993; Chan and Courtney, 1993; Chan and Courtney, 1994; Chan and Courtney, 1998; Ikeda and Takeuchi, 1975; Owsley, Burton-Danner, and Jackson, 2000; Rogé, Pébayle, Campagne, and Muzet, 2005; Seiple, Szlyk, Yang, and Holopigian, 1996; Sekuler, Bennett, and Mamelak, 2000). Various scientific approaches with a number of different goals have been used to study the useful visual field and consequently different tests for its assessment have been developed. Two of these are presented in this chapter.

The *UFOV* (useful field of view) *Visual Attention Analyzer test* measures the ability to process rapidly presented information within a single glance through three visual subtests involving the measurement of visual processing speed (subtest1), selective (subtest2) and divided (subtest3) visual attention (Ball, Owsley, Sloane, Roenker, and Bruni, 1993). The first subtest requires the subject to identify a target (a pictogram of either a car or truck) briefly flashed at varying durations in the centre of the screen. The second subtest measures divided attention. The subject performs the same central task and also has to localize a peripheral target. This peripheral target (same pictogram as above) is presented unpredictably at one of eight different meridians at 10, 20 or 30 degrees of eccentricity at the same time that the central stimulus is presented. The duration of the display is varied to measure the speed of visual processing for this divided attention task. The third subtest consists of these same two tasks, but the peripheral stimuli are embedded in a series of visual distracters (same size and luminance as the target) that fill the rest of the visual display. For subtest 1, the minimum duration at which the subject could perform the task with 75% accuracy is measured. For subtests 2 and 3, the line of best fit reflecting the relationship between eccentricity and localization errors is first calculated for each test duration (presentation times varies from 50 to 125 milliseconds), and the size of the UFOV is defined for that stimulus duration as that eccentricity at which the participant could localize the peripheral target correctly 50% of the time. Performance in each of the three subtests is then scaled, to arrive at three scores representing the extent of difficulty with respect to speed of processing (subtest 1), divided attention (subtest 2) and selective attention (subtest 3). The three scaled scores are added together to yield a composite score between 0 and 90, which represents the total percentage reduction of the UFOV for Ball and her colleagues.

The PECVU (Poste d'Etude du Champ Visuel Utile - Useful Visual Field Study) device has been designed to assess the area inside which the information is processed in a single glance while the participant carries out a dual task (Muzet, 1998; Muzet, Rogé, Jouny, and Muzet, 1999; Rogé, Kielbasa, and Muzet, 2002; Rogé, Pébayle, Campagne, and Muzet, 2005). The subject was seated in the centre of a hemi-cylindrical panel (covering 140° horizontally) with a monitor screen in the middle. The central task involves detecting a more luminous spot among eight spots displayed in a circle on the monitor screen. The peripheral signals (green light emitting diodes) appear along arcs of circles (at 20°, 30°, 40°, 50°, 60° and 70°) on the hemi-cylindrical panel. The instruction is to detect the signal as quickly and accurately as possible. The furthest eccentricity from which the subject perceives at least 50% of the peripheral signals is considered to be the limit of the useful visual field. The percentage of peripheral signals perceived or/and localized by eccentricity in the peripheral task is also analyzed to define the shape of the deterioration of the useful visual field (Muzet, Rogé, Jouny, and Muzet, 1999; Rogé, Kielbasa, and Muzet, 2002; Rogé, Pébayle, Campagne, and Muzet, 2005).

The size of the useful visual field is not constant even if an individual has no ophthalmological problem or disease. Its size depends mainly upon the nature of the perceptive task and upon the situation (Sanders, 1970; Verriest, Bailey, Calabria, and Campos, 1985; Verriest, Barca, and Dubois-Poulsen, 1983). It can also depend upon individual characteristics such as age.

Several Forms of Deterioration

The ability to detect peripheral signals in a double task deteriorates with age. But the extent of this deterioration varies from one study to the next (figure 1). Some studies indicate a general interference (Rogé, Pébayle, Hannachi, and Muzet, 2003; Rogé, et al., 2004; Seiple, Szlyk, Yang, and Holopigian, 1996; Sekuler, Bennett, and Mamelak, 2000). Performance of older people is worse overall than that of younger persons, whatever the eccentricity of the peripheral signal to be detected (figure 1 a). Deterioration similar to the phenomenon of tunnel vision is noticed in other studies (Ball, Beard, Roenker, Miller, and Griggs, 1988; Ball, Roenker, and Bruni, 1990; Rogé, Pébayle, Hannachi, and Muzet, 2003; Rogé, et al., 2004; Scialfa, Kline, and Lyman, 1987; Sekuler and Ball, 1986). Older people always perform less well than younger people, and the difference in performance becomes more pronounced when the eccentricity of the peripheral signal to be detected increases (figure 1 b).

Tunnel vision is sometimes considered as a 'contraction' of the attentional capacity, similar to that of the pupil when there is too much light. The purpose of the contraction could be to prevent an overload of the visual system when the amount of information becomes too great as is probably the case for older people.

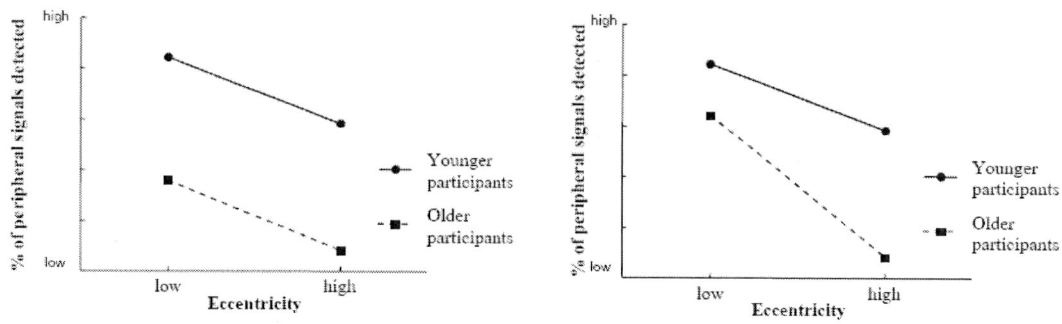

a: General Interference
(Age did not interact with the eccentricity of the peripheral signal)

b: Tunnel Vision
(Age interacted with the eccentricity of the peripheral signal)

Figure 1. General Interference and Tunnel Vision.

General interference could also be a variation of attentional capacity, similar to a lens of variable power. This lens would not be powerful for older people and the visual field would be available but with very poor definition. For younger people, this lens would be more powerful, allowing sharp resolution across the entire visual field. This metaphor implies that visual perceptive and cognitive ability should be stable for an individual according to their personal characteristics and the features of the tasks. But two remarks question this premise of stability of the useful visual field. First, the performance in the Useful Field of View test of elderly adults can be improved with a speed-of-processing training program (Roenker, Cissell, Ball, Wadley, and Edwards, 2003). Secondly, we suggest that some adaptation strategies developed by older people (Marquié and Isingrini, 2001) may be used to compensate for their limited useful visual field during driving (see future perspectives, part 2.4).

The diversity of the tasks employed for the assessment of the useful visual field do not allow for formulation of clear rules regarding the deterioration of the useful visual field with aging according to the characteristics of the task (e.g. instructions, duration of signals, length of the experiment…). One can wonder whether the size of the useful visual field could vary according to a continuous process which goes from an extended useful visual field (that is observed when the cognitive abilities of the participant are sufficient to carry out a task prescribing a weak attentional load) to a deteriorated one (when the cognitive abilities of the participant are too limited to obtain good performance in a complex task). The deterioration could begin from the external part of the field resulting in a tunnel vision phenomenon, progressively corrupting the entire field, showing general interference. In this perspective, the different models of deterioration observed by researchers may only reflect several levels of deterioration of the useful visual field on this continuum. The level of deterioration could be determined by the interaction of some characteristics related to the task and some characteristics of the participant.

Whatever form of deterioration the useful visual field takes (tunnel vision or general interference) with aging and whatever the stability of the size of the useful visual field, the modification of this field has major consequences for many everyday activities, particularly car driving.

Relationship between Indices Related to Visual Field and Road Safety

Several elements of visual information must be analyzed during driving, in relation to vehicle trajectory, the behavior of other road users, and indications displayed on the road and inside the vehicle (dashboard, driving assistance systems). The processing of this information involves both central and peripheral vision and can be carried out under impaired visual conditions, as is the case whilst driving at night, in the rain, snow or fog. If the useful visual field is reduced, the driver should be slower to detect this information (on the road or inside the car), or should not perceive them at all, exposing himself (or herself) and the other road users to a potential accident. Some researchers have therefore studied the relationship between indices in visual tests and data related to road safety according to the age of the drivers.

Road Accidents and the Risk of Having an Accident

Several visual indices (such as visual acuity, contrast sensitivity, color discrimination, peripheral vision, dynamic visual acuity, motion perception and visual attention) have been studied. Most of these indices are weakly associated with older drivers' involvement in road accidents (Ball and Owsley, 1993; Owsley, 1994; Sims, McGwin, Allman, Ball, and Owsley, 2000). Among them, the measurement of visual attention seems to be the most significant factor related to road accidents in older drivers (Ball, Roenker, and Bruni, 1990; Myers, Ball, Kalina, Roth, and Goode, 2000; Owsley, 1994; Owsley, et al., 1998; Owsley and McGwin, 1999).

Several researchers observe a relationship between the reduction of useful visual field (as measured with the UFOV test) and the number of accidents on the road, counted over a period of several years either before the UFOV test is carried out (Ball and Owsley, 1993; Ball, Owsley, Sloane, Roenker, and Bruni, 1993; Owsley, 1994; Owsley, et al., 1998; Sims, McGwin, Allman, Ball, and Owsley, 2000) or just after this test (Owsley, 1994; Owsley, et al., 1998). These results mean that the performance in this test could be predictive of driving performance. Myers uses a different method: older drivers carry out a driving test on the road and are evaluated by two assessors (Myers, Ball, Kalina, Roth, and Goode, 2000). Poor UFOV test performance is correlated to a high number of driving errors (failure to stop at a stop sign, failure to see important road signs, errors of judgment or bad road position). Finally, a meta-analysis highlights the useful field of view assessed with the UFOV test as a valid and reliable index of driving performance and safety (Clay, et al., 2005).

Rather than studying actual accidents, some researchers examine the ability of drivers to manage a potential high-risk situation on a simulator: a truck leaves a rest-area, refuses to yield right of way and cuts sharply in front of the participant's vehicle (Rogé, Pébayle, Campagne, and Muzet, 2005). Analyzing data obtained in this critical situation in a simulated driving test indicates that there are correlations between driver age, the time taken to swerve and hence avoid a collision with the truck and the reduction of useful visual field. The older the drivers are, the more their useful visual field (as measured by asking the driver to detect and locate peripheral signals) decreases, and the longer they take to react, thus increasing their risk of being involved in an accident.

These studies lead to the conclusion that the deterioration of useful visual field with age has important consequences for road safety. One can question whether the assessment of the useful visual field with UFOV test and/or PECVU test is genuinely significant to study the visuo-attentional ability of the elderly driver. Indeed, tasks carried out in these both tests are far from driving activity (no tracking, no complex visual background as a road environment, no temporal pressure due to a continuous activity…). In order to estimate the useful visual field of the driver in a more ecological situation, some researchers suggest assessing it while driving.

Useful Visual Field Assessment during Driving

The initial protocol for assessing the useful visual field during driving was carried out with a driving simulator (Rogé, Pébayle, Kiehn, and Muzet, 2002). The road image covered 50° of the horizontal visual field and 25° of the vertical visual field. The participants drove on a motorway. The instruction was to follow a vehicle going at an average speed of 110 km per hour, while staying at a constant distance. The subject had to stare at a colored circle that regularly appeared on the rear window of the lead vehicle. The central task consisted in detecting as quickly as possible the changes in color of this signal. Simultaneously, they had to detect the brief displays of a red point, which appeared randomly in time on the road scene. This peripheral signal could be located at different eccentricities (5°, 10°, 15° and 20°) in the visual field. The 0° eccentricity was located at the same level as the orange circle the subject had to stare at. Another version of this test was designed to assess the size of the driver's useful visual field in a simulated road traffic situation (Rogé, et al., 2004). The central task was the same and simultaneously the driver had to detect the brief display of a yellow luminous signal that was located in one of the rear lights of the visible vehicles making up the traffic. This peripheral signal appeared on the backlight of the vehicle when it was located in different areas (between 1° and 4°, between 6° and 10°, between 12° and 16° and between 18° and 22°). The useful visual field of the driver was assessed via the percentage of peripheral signals detected.

The same impact of age on the ability to detect visual signals observed in tests carried out with a computer has been noted in experiments using a driving task. The performance of two age groups (young drivers and middle-aged drivers) has been compared in order to study the effect of age on the ability of the driver to detect visual information on the road scene whilst driving (Rogé, Pébayle, Hannachi, and Muzet, 2003; Rogé, et al., 2004). The older drivers detected fewer peripheral signals (in the road scene or in the traffic) than the younger drivers and this difference increased when the signal was further from the fixation point, thus revealing a tunnel vision phenomenon induced by age.

When the peripheral signal appeared at a place where the information relative to other vehicles braking or cutting-in was usually located (as is the case when the driver has to detect a signal located on the rear-lights of the vehicles in traffic), rapid processing of this information is necessary to ensure the driver's safety.

FUTURE PERSPECTIVES

Two critical remarks can be made about the protocol of the useful visual field assessment during driving. Older drivers could adopt a lower speed to compensate for their cognitive and visual deficiencies at the behavioral level (Szlyk, Seiple, and Viana, 1995). If driving slowly is an adaptation strategy adopted to compensate for the reduction of the useful visual field, then the instruction to follow a vehicle which drives at high speed (110 km per hour) is probably more prejudicial to older than to younger drivers. So, it could be interesting to modify this protocol in order to let drivers regulate their speed freely. In order to come near to actual driving, it would also be necessary to assess the useful visual field during driving with a protocol that allows eyes movements in order to allow the driver free visual search in the road environment.

Various populations are also concerned with attentional failures (due to Attention-Deficit Hyperactivity Disorder -ADHD, brain injury, dementia of various types: Alzheimer's and Parkinson's diseases). It may be of interest to study the correlations between the useful visual field and specific attentional component (selective attention and attentional sharing performance) of these populations to better understand the underlying mechanisms of attentional failures. However, it is important to have a flexible protocol in order to assess the useful visual field of these specific populations because the classical protocols are certainly too complex for them (signal duration too short, complex instructions, test that is too long, etc).

In future research, it could be of significant interest to see if some adaptive strategies that older drivers use (like adopting a lower speed, avoiding complex divided attention conditions, not driving in busy hours) genuinely compensate for the deterioration of their useful visual field. If these strategies are effective, they would have a significant effect on their ability to detect and analyze visual information during driving. Moreover, we can question whether specific training in the use of some adaptive strategies could also lead to a long term improvement of the useful field of view.

In relation to the effectiveness of a visual attention retraining, it would be interesting to explore ways of improving processing speed, divided attention and selective attention. Mazer develops such a program in order to increase the driving success rate of stroke patients (Mazer, et al., 2003). Unfortunately, rehabilitation that targets visual attention skills using the UFOV test was not significantly more beneficial than traditional perceptual training in improving the outcome of an on-road driving evaluation. Nevertheless some patients with right hemisphere stroke may benefit from this type of training. With a training procedure based on the UFOV test, processing speed training seems to be efficient but the same cannot be said for divided attention and selective attention training. These results are unexpected and it is perhaps necessary to train people with more ecological materials to predict their potential contributions.

Various forms of retraining procedure exist. Among them, the knowledge acquired on the development of more strategic abilities through to the use of metacognitive abilities is very promising, especially for elderly drivers with broad driving experience.

Metacognition: A Key Factor in the Development of Drivers' Adaptive Strategies

What Kind of Regulation for Elderly Drivers?

If an elderly driver does not compensate spontaneously, or retraining on some specific cognitive function is not sufficient enough, there may still be another way to compensate for individuals' mildly reduced cognitive capacities: becoming aware of one's limitations and adjusting one's behavior consciously by optimizing the use of the cognitive resources that are still available. For example, an elderly driver aware of his shared attention difficulties will probably tend to plan in advance his passage through an unknown town, instead of trying to find the desired direction when already approaching a complex crossroad.

In past decades, the emphasis on the benefits of an active and adapted conscious regulation of one's behavior was analyzed under the auspices of metacognitive theory (Brown, 1987; Flavell, 1979; Metcalfe and Kornell, 2005; Nelson and Narens, 1990; Thiede and Dunlosky, 1999). In the remainder of this study, we will briefly introduce some of its basic notions and explain the usefulness of this approach for drivers and particularly for senior driver retraining programs.

Some Basic Notions of Metacognition

Designing cognition about cognition (Flavell, 1979) or an understanding of cognition (Brown, 1987): the concept of metacognition can be explained in two different, even if closely related ways. First is the concept of metacognition as knowledge of one's own cognitive processes. As we can see below, this knowledge is not only relatively stable and identifiable, but also fallible. Second is the concept of metacognition as regulation of one's (cognitive) activities including planning activities (predicting outcomes, choosing appropriate strategies), monitoring activities (monitoring, scheduling strategies), and checking outcomes (evaluation of efficacy and effectiveness).

Overall, although metacognitive theory is still under development, metacognitive theorists emphasize that the maturation of metacognition contributes to one's intellectual development, and allows individuals to achieve set goals through monitoring (bottom-up) and control (top-down) processes. It can be also viewed as a tool to optimize performance through the clever management of one's cognitive resources. For example, the above-mentioned well-planned passage through an unknown town might help you to avoid an attentional overload which could prove dangerous when driving. How can metacognitive research help us, then, to understand the ways in which elderly people drive?

METACOGNITION AND DRIVING

There is very little literature on the role of metacognition in driving (Gregersen, 1996; Hatakka, Keskinen, Gregersen, Glad, and Hernetkoski, 2002; Liourta and Empelen, 2008; Owsley, Stalvey, and Phillips, 2003; Schlag, 1993). Note that some of these studies mentioned do not explicitly use the terms of metacognition but their issues fit in with this point of view. Additionally, no detailed metacognitive model including cognitive components linked to driving (*i.e.*, attention, working memory, decision making, etc.) is available currently. The existing theory on metamemory and learning allows us, however, to draw some preliminary implications, either for driving in general or in particular for elderly driver retraining procedures (Cavanaugh and Borkowski, 1979; Koriat, 2007; Pressley, Borkowski, and O'Sullivan, 1985).

Focused mainly on the relationship between self-evaluation, self-regulation and the identification of favorable factors permitting good transfer, classical metacognitive research concludes that accurate self-evaluation (monitoring) allows for optimal self-regulation (control) if well established from a motivation point of view. In the following section, this statement will be investigated in order to assess its implementation in the context of elderly drivers.

Accurate Self-Evaluation (Monitoring) ...

Accurate self-evaluation and on-line monitoring in general play a decisive role in every human activity and permit the efficient adjustment of cognitive strategies. A driver needs to monitor his cognitive processing effectively every time anticipation is required or some adaptation has to take place. He will probably slow down if he recognizes that when maintaining his speed, he can not sufficiently analyze the approaching intersection. Maybe he will slow down if he is a young person, but he will undoubtedly slow down further if he is an elderly person aware of his visuo-attentional deficit or, simply, of his slower reaction time. In fact, elderly people seem to monitor their driving more, although this may lengthen their reaction time: "such an increased control may improve the possibility of rectifying errors, provided that enough time is available," (Schlag, 1993).

The importance of self-evaluation is also mentioned in the driving context especially concerning driver education. Previous research conducted by Keskinen (Keskinen, 1996), Laapotti, Keskinen, Hatakka, and Katila (Laapotti, Keskinen, Hatakka, and Katila, 2001) and especially Hatakka *et al.* (Hatakka, Keskinen, Gregersen, Glad, and Hernetkoski, 2002) presents a descriptive model of driver behavior as a hierarchy in which the goals and motives of the driver play a crucial role. This hierarchy is accompanied by three essential areas of content for driver education: knowledge and skills, risk-increasing factors and self-evaluation (see Figure 2). Together, this matrix is known as the Goals for Driver Education framework (GDE matrix), especially in European countries trying to implement it in their educational curricula (Hatakka, Keskinen, Gregersen, Glad, and Hernetkoski, 2002). According to the authors, self-evaluation is important in driver training and also in development of driving skill after training. Because these self-evaluative skills do not develop automatically, they should be trained (Hatakka, Keskinen, Gregersen, Glad, and Hernetkoski, 2002).

hierarchical levels of behavior	essential content		
	knowledge and skills	*risk-increasing factors*	*self-evaluation*
goals for life and skills for living (general)	lifestyle, group norms, personal values	sensation seeking, social pressure, acceptance of risks	awareness of personal skills for risky habits
goals and context of driving (trip related)	planning routes, evaluation of necessity of trip	purpose of driving, extra motives (competing)	typical goals of driving, typical risky driving motives
mastery of traffic situations	traffic rules, anticipation of course of situation	wrong expectations, unsuitable speed adjustment	personal safety margins, personal driving style
vehicle maneuvering	control of direction, vehicle properties, physical phenomena	insufficient skills, physical conditions	strong and weak point of basic maneuvering skills

Figure 2. GDE framework (adapted from Hatakka, Keskinen, Gregersen, Glad, and Hernetkoski, 2002).

For the last 30 years, there has been a wide range of debate on self-enhancement bias (Delhomme, 1991; Näätänen and Summala, 1975; Svenson, 1981; Waylen, Horswill, Alexander, and McKenna, 2004), unreal optimism (Causse, Kouabenan, and Delhomme, 2004; DeJoy, 1989), or illusion of control (Deery, 1999; Holland, 1993; McKenna, 1993), all in reference to an overestimation of drivers' own skills, of drivers' safety driving, or of the likelihood of avoiding an accident. Results from these studies are not unequivocal and we find this overestimation even in populations of elderly drivers (Freund, Colgrove, Burke, and McLeod, 2005; Holland, 1993), not only in younger males as was assumed for a long time (Finn and Bragg, 1986; Matthews and Moran, 1986).

Nevertheless, we must consider two aspects of these studies. First, in majority of these studies only questionnaires were used and it is possible that people answered in such a way as to give the best possible image of themselves. The self-enhancement bias was not explored in relationship with real driving situations and drivers compared themselves to unknown others (the so called "average driver"). Moreover, this kind of "mean" evaluation seems not to be equal with how a driver would evaluate his or her own performance (Koriat, 2007).

Secondly, these studies do not always clearly identify what cognitive capacities are accounted for, be it from a driver's point of view or in general (Hakamies-Blomqvist, 1998). Therefore, precise conceptualizations, more developed methodologies, and cautious interpretation of results are indispensable in determining the real nature of drivers' self-evaluation.

An accurate self-evaluation becomes more and more important as some cognitive deficiencies may appear in elderly drivers. Typically, however, self-evaluative skills are not included in recent elderly driver retraining programs. Hatakka et al. (2002) stress that self-evaluation tend to be ignored and miss effective methods and that driving methods improving drivers' metacognitive skills are equally scarce (Hatakka, Keskinen, Gregersen, Glad, and Hernetkoski, 2002). Therefore, even though the contributions of other domains may be required to improve self-evaluation, the GDE framework offers not only advices to assess drivers' competencies, but also delineates some essential themes of road safety research.

...Allows Optimal Self-Regulation (Control)...

Following Amalberti, the manner in which a person judges his own (cognitive) competencies with regard to task requirements will determine an initial level of accepted risk (so-called "safety margins" considered as sufficient to avoid accident) and behavior in risky situations (Amalberti, 1996). This relationship, however, is not always as straightforward as it may seem.

In line with Amalberti's assumption, the self-enhancement bias researchers conclude that a driver's overconfidence in his or her skilled or safety driving will certainly lead to underestimation of risks (Causse, Kouabenan, and Delhomme, 2004; McKenna, Stanier, and Lewis, 1991; Svenson, 1981). However, this is not the result of all studies: for example, Katila et al. (2004) found that high confidence in one's personal skills does not necessarily imply negative safety, the crucial factor is how these learned skills are used and for what purpose (Katila, Keskinen, Hatakka, and Laapotti, 2004). In the same vein, Gregersen (Gregersen, 1996) considered the effects of two different kinds of skid training, the insight skid training (focusing on attitudinal and motivational skills) and the traditional one (focusing on physical skills). As predicted, the "insight" group learned more about anticipation and about their own limits, and, as a result, proved less overconfident in such situations. On the other hand, the real driving performance of both groups was similar, regardless of the degree of overconfidence.

Metacognitive theorists assume that accurate self-evaluation increases the probability of optimal self-regulation (Koriat, 2007; Thiede, Anderson, and Therriault, 2003). Although, in a learning context, some cases of inefficient self-regulation can be found in literature: *labor-in-vain effect* (Nelson and Leonesio, 1988), and moreover, self-regulation does not always occur spontaneously. In general, the more detailed your knowledge of how to achieve your goals, the more often you will behave in this way (Lefebvre-Pinard and Pinard, 1985).

Notice that all of above discussed terms (knowledge of one's goals, possible behaviors, or one's own strengths or weaknesses) fit well into a metacognitive model - they are what we call metacognitive knowledge used when monitoring and controlling one's behavior. However, as well as accurate metacognitive knowledge, other factors have to be mentioned in order to train the skill of high quality self-regulation in elderly drivers.

...If Well Established!

While the accuracy of monitoring appears to be important, other factors are indeed essential in order not only to introduce a positive change (safer driving), but also to maintain it over time and to promote it in more or less similar situations. To illustrate the benefit metacognitive theory can bring to research on elderly drivers, and, in fine, to designing new retraining curricula, consider the following experimental study and the answers that metacognitive theory provides. Molesworth et al. (2006) raised the question of consciousness of safe behavior learning: they argued that fighter pilots could change their behavior in a positive manner without being aware of it, so the role of experienced risk in a flight simulator seemed to be decisive (Molesworth, Wiggins, and O'Hare, 2006). Nevertheless, we do not know if this behavioral change would subsist over time.

Although some learning may prove implicit and non-conscious, if we want somebody to maintain the use of some new strategy for longer time, explicit learning should take place and feedback on its usefulness must be provided (a driver must understand why he or she succeeded and recognize the benefit of learned behavior) (Cavanaugh and Borkowski, 1979). As emphasized by Dunlosky, Kubat-Silman, and Hertzog, the learning of self-regulatory strategies alone may prove inefficient if it is not linked to explicit learning of possible ways of monitoring one's performance (Dunlosky, Kubat-Silman, and Hertzog, 2003). Importantly, the use of monitoring may be easy for older adults to learn because monitoring accuracy appears to be relatively spared by aging (Dunlosky, Kubat-Silman, and Hertzog, 2003). This is well-documented in existing literature on educational interventions in older drivers. More precisely, these interventions prove successful because drivers are not only informed about the risk that older drivers face in general, but also how to find their own concrete solutions and the ways in which to apply them in real life (Owsley, Stalvey, and Phillips, 2003). Little work has been done, however, on motivation and its role in driving.

Indeed, when looking for transfer to new situations, a more essential change in motivational factors has to occur. The need for a positive motivational change in driving is well documented in the GDE framework (Figure 2). The upper levels of the hierarchy predetermine the motivation and the use of skills on the lower levels. It is therefore beneficial to consider one's "superior" motivations when trying to introduce some behavioral change, even on the lower levels.

Implication for Future Research: New Retraining Curricula for Elderly People?

For more than 30 years, metacognitive theorists have tried to highlight some basic principles of human beings' active self-regulation and explore the ways in which to enhance it. The metacognitive approach has only recently been used in contexts outside the educational domain, for example in social psychology (Jost, Kruglanski, and Nelson, 1998) or in problem-solving and creativity (Metcalfe, 1998). Extreme caution is needed when implementing metacognitive concepts in driving. Although operationalization may prove difficult in this domain, it seems promising in terms of retraining for the elderly and we will focus on this issue shortly.

In our opinion it is useful to begin research into metacognitive processes related to driving in the elderly driver population. Firstly, according to some authors, novices are not able to use their metacognition because of elevated cognitive demands at the beginning of comprehension (Lefebvre-Pinard and Pinard, 1985). In this respect, elderly drivers are generally experts with a lifetime of experience. Secondly, with aging, elderly people may face many involutive processes and declines, be it impaired vision, attention, or others. If metacognitive regulation offers the advantage of optimal resource management, the quality of such resource supervision will be decisive in elderly people's willingness to maintain their mobility. If safe driving does not occur spontaneously with aging, well-planned metacognitive training may facilitate the maintenance of mobility. Finally, research into elderly populations, frequently considered in literature as the age of wisdom, may shed some light on the characteristics of metacognitive processing from a developmental point of view.

CONCLUSION

The aim of this chapter was to highlight the main trends in traffic safety research concerning elderly drivers and to propose an overarching framework to lead to future research works. The chapter detailed the evolution of research on this topic in the past decades, from a classical medical approach to a more comprehensive one including neuropsychological and behavioral components. We focused on two promising directions for driver rehabilitation. The first one set up an overview of useful visual field studies and their implications on elderly driver issues. The useful visual field concept was first developed to explore fitness to drive and training opportunities were subsequently revealed. The second direction, less in relation to skills development but more in terms of conscious regulation of behavior, raises new questions for research on driving license renewal and/or rehabilitation and falls within the scope of metacognition.

We argue that it is important to conduct simultaneous research in these two directions in order to propose a relevant model for the assessment and rehabilitation of elderly drivers. In all the industrialized countries facing an aging driver population, relevant tools and methods should be developed in the near future to find the right balance between strict control of driving abilities and an accompanying program to preserve safe mobility for the elderly. This kind of program should not be based solely on skills training but also on knowledge acquisition when driving. Of course, the success of the remediation would vary according to the pathology and the ability to learn, assuming that an accurate self-evaluation increases the probability of optimal self-regulation.

REFERENCES

Amalberti, R. (1996). *Conduite des systèmes à risques*. Paris: PUF.

Baddeley, A.D. (1992). Attention et contrôle en mémoire. In A.D. Baddeley (Ed.), *La Mémoire Humaine, Théorie et Pratique* (pp. 133-158). Grenoble: Presses Universitaires de Grenoble.

Ball, K., Beard, B., Roenker, D., Miller, R., and Griggs, D. (1988). Age and visual search: expanding the useful field of view. *Journal of the Optical Society of America*, 5, 2210-2219.

Ball, K., and Owsley, C. (1993). The useful field of view test: a new technique for evaluating age related declines in visual function. *Journal of the American Optometric Association*, 64, 71-79.

Ball, K., Owsley, C., Sloane, M.E., Roenker, D.L., and Bruni, J.R. (1993). Visual attention problems as a predictor of vehicle crashes in older drivers. *Investigative Ophthalmology and Visual Science*, 34, 3110-3123.

Ball, K., Roenker, D., and Bruni, J. (1990). Developmental changes in attention and visual search throughout adulthood. In J.T. Enns (Ed.), *The development of attention: research and theory* (pp. 489-508). North Holland: Elsevier Science Publishers B.V.

Brinig, M.F., Wilkinson, M.E., Daly, J.M., Jogerst, G.J., and Stone, E.M. (2007). Vision standards for licensing and driving. *Optometry - Journal of the American Optometric Association*, 78, 439-445.

Broadbent, D.E. (1962). Attention and the perception of speech. *Sci Am*, 206, 143-151.

Brouwer, W.H. (2002). Attention and driving: A cognitive neuropsychological approach. In P. Zimmermann and M. Leclercq (Eds.), *Applied Neuropsychology of Attention* (pp. 223-248). Hove: Psychology Press.

Brouwer, W.H., and Ponds, R.W. (1994). Driving competence in older persons. *Disability and Rehabilitation*, 16, 149-161.

Brouwer, W.H., Waterink, W., Van Wolffelaar, P.C., and Rothengatter, T. (1991). Divided attention in experienced young and older drivers: lane tracking and visual analysis in a dynamic driving simulator. *Human Factors*, 33, 573-582.

Brown, A. (1987). Metacognition, executive control, self-regulation, and other more mysterious mechanisms. In F.E.W.R.H. Kluwe (Ed.), *Metacognition, motivation, and understanding* (pp. 65-116). Hillsdale NJ: Erlbaum.

Brown, L.B., Ott, B.R., Papandonatos, G.D., Sui, Y., Ready, R.E., and Morris, J.C. (2005). Prediction of on-road driving performance in patients with early Alzheimer's disease. *Journal of the American Geriatrics Society*, 53, 94-98.

Caird, J.K., Edwards, C.J., Creaser, J.I., and Horrey, W.J. (2005). Older drivers failures of attention at intersections : Using change blindness methods to assess turn decision accuracy. *Human Factors*, 47, 235-249.

Caird, J.K., and Hancock, H.E. (2002). Left-turn and gap acceptance crashes. *Human Factors in Traffic Safety* (pp. 613-652): Lawyers and Judges publishing company.

Campbell, B.J. (1966). Driver age and sex related to accident time and type. *Traffic Safety*, 10, 36-42.

Carr, D.B. (1997). Motor vehicle crashes and drivers with DAT. *Alzheimer Disease and Associated Disorders*, 11 Suppl 1, 38-41.

Carr, D.B., Duchek, J., and Morris, J.C. (2000). Characteristics of motor vehicle crashes of drivers with dementia of the Alzheimer type. *Journal of the American Geriatrics Society*, 48, 18-22.

Causse, P., Kouabenan, D.R., and Delhomme, P. (2004). Perception du risque d'accident lié à l'alcool chez des jeunes automobilistes : quelques déterminants de l'optimisme comparatif. *Travail Humain*, 67, 235-256.

Cavanaugh, J.C., and Borkowski, J.G. (1979). The metamemory-memory "connection": effects of strategy training and maintenance. *The Journal of General Psychology*, 101, 161-174.

Chan, H.S., and Courtney, A.J. (1993). Effects of cognitive foveal load on a peripheral single-target detection task. *Perceptual and Motor Skills*, 77, 515-533.

Chan, H.S., and Courtney, A.J. (1994). Effects of priority assignment of attentional resources, order of testing and response sequence on tunnel vision. *Perceptual and Motor Skills*, 78, 899-914.

Chan, H.S., and Courtney, A.J. (1998). Stimulus size scaling and foveal load as determinants of peripheral target detection. *Ergonomics*, 41, 1433-1452.

Clay, O., Wadley, V., Edwards, J., Roth, D., Roenker, D., and Ball, K. (2005). Cumulative Meta-analysis of the Relationship between Useful Field of View and driving performance in older adults: current and future Implications. *Optometry and vision science*, 82, 724-731.

Crook, T.H., West, R.L., and Larrabee, G.J. (1993). The driving-reaction time test: assessing age declines in dual-task performance. *Developmental Neuropsychology*, 9, 31-39.

Daigneault, G., Joly, P., and Frigon, J.Y. (2002). Previous convictions or accidents and the risk of subsequent accidents of older drivers. *Accident Analysis and Prevention*, 34, 257-261.

DaSilva, H.R. (1938). Age and Hoghway accidents. *Scientific Monthly*, 47, 536-546.

Deery, H.A. (1999). Hazard and Risk Perception among Young Novice Drivers. *Journal of Safety Research*, 30, 225-236.

DeJoy, D.M. (1989). The optimism bias and traffic accident risk perception. *Accident Analysis and Prevention*, 21, 333-340.

Delhomme, P. (1991). Comparing one's driving with others': assessment of abilities and frequency of offences. Evidence for a superior conformity or self-bias? *Accident Analysis and Prevention*, 23, 493-508.

Delorme, D., and Marin-Lamellet, C. (1998). Age-related effects on cognitive processes: Application to decision majing under uncertainty and time pressure. In J. Graafmans, V. Taiple, and N. Charness (Eds.), *Gerontechnology* (pp. 124-127). Burke: VA: IOS Press.

Dubinsky, R.M., Gray, C., Husted, D., Busenbark, K., Vetere-Overfield, B., Wiltfong, D., Parrish, D., and Koller, W.C. (1991). Driving in Parkinson's disease. *Neurology*, 41, 517-520.

Dubinsky, R.M., Stein, A.C., and Lyons, K. (2000). Practice parameter: risk of driving and Alzheimer's disease (an evidence-based review): report of the quality standards subcommittee of the American Academy of Neurology. *Neurology*, 54, 2205-2211.

Duchek, J., Hunt, L., Ball, K., Buckles, V., and Morris, J. (1997). The role of selective attention in driving and dementia of the Alzheimer type. *Alzheimer disease and associated disorders*, 11, 48-56.

Duchek, J., Hunt, L., Ball, K., Buckles, V., and Morris, J. (1998). Attention and driving performance in Alzheimer's disease. *The journals of gerontology. Series B, Psychological sciences and social sciences*, 53B, 130-141.

Duchek, J.M., Carr, D.B., Hunt, L., Roe, C.M., Xiong, C., Shah, K., and Morris, J.C. (2003). Longitudinal driving performance in early-stage dementia of the Alzheimer type. *Journal of American Geriatrics Society*, 51, 1342-1347.

Dunlosky, J., Kubat-Silman, A.K., and Hertzog, C. (2003). Training Monitoring Skills Improves Older Adults' Self-Paced Associative Learning. *Psychology and Aging*, 18, 340-345.

Eby, D.W., Molnar, L.J., Shope, J.T., Vivoda, J.M., and Fordyce, T.A. (2003). Improving older driver knowledge and self-awareness through self-assessment: The driving decisions workbook. *Journal of Safety Research*, 34, 371-381.

Edwards, J., Ross, L., Wadley, V., OJ. Clay, O., Crowe, M., Roenker, D., and Ball, K. (2006). The useful field of view test: Normative data for older adults. *Archives of Clinical Neuropsychology*, 21, 275-286.

Evans, L. (1987). Fatal and severe crash involvement versus driver age and sex. *31st annual proceedings-American association for automotive medicine* (pp. 59-76). New Orleans, Louisiana.

Fernandez-Duque, D., Baird, J.A., and Posner, M.I. (2000). Executive Attention and Metacognitive Regulation. *Consciousness and Cognition*, 9, 288-307.

Finesilver (1969). The older driver: a statistical evalustion of licensing and accident involvement in 30 states. Denver: University of Denver College of Law.

Finn, P., and Bragg, B.W.E. (1986). Perception of the risk of an accident by young and older drivers. *Accident Analysis and Prevention*, 18, 289-298.

Fisk, G., Novack, T., Mennemeier, M., and Roenker, D. (2002). Useful field of view after traumatic brain injury. *The Journal of head trauma rehabilitation*, 17, 16-25.

Fisk, G., Owsley, C., and Mennemeier, M. (2002). Vision, attention, and self-reported driving behaviors in community-dwelling stroke survivors. *Archives of Physical Medicine and Rehabilitation*, 83, 469-477.

Flavell, J. (1979). Metacognition and cognitive monitoring: A new area of cognitive-developmental inquiry. *American Psychologist*, 34, 906-911.

Fontaine, H., and Gourlet, Y. (1992). Elderly people, mobility and safety., *European Road Safety Conference* Berlin: FERCI and VTI.

Fox, G.K., and Bashford, G.M. (1997). Driving and dementia: balancing personal independence and public safety. *The Medical journal of Australia*, 167, 406-407.

Freund, B., Colgrove, L.A., Burke, B.L., and McLeod, R. (2005). Self-rated driving performance among elderly drivers referred for driving evaluation. *Accident Analysis and Prevention*, 37, 613-618.

Friedland, R.P., Koss, E., Kumar, A., Gaine, S., Metzler, D., Haxby, J.V., and Moore, A. (1988). Motor vehicle crashes in dementia of the Alzheimer type. *Annals of Neurology*, 24, 782-786.

Gabaude, C. (2003). Exploration of the visual and attentional abilities of elderly drivers: Benefits and techniques. *Recherche Transport et Sécurité*, 81, 165-176.

Gregersen, N.P. (1996). Young drivers' overestimation of their own skill—an experiment on the relation between training strategy and skill. *Accident Analysis and Prevention*, 28, 243-250.

Grow, N.L. (1972). The literature of the pedestrians and drivers: a bibliography 1962-1972. In E.W. T.W. Planek (Ed.), *Aging and Highway Safety: the elderly in a mobile society Highway Safety.* (pp. 1-38). Chaptel Hill, N.C.: North Carolina Symposium on Highway Safety.

Hakamies-Blomqvist, L. (1994). Compensation in older drivers as reflected in their fatal accidents. *Accident Analysis and Prevention*, 26, 107-112.

Hakamies-Blomqvist, L. (1996). Research on older drivers : a review. *IATSS Research*, 20, 91-101.

Hakamies-Blomqvist, L. (1998). Older drivers' accident risk: Conceptual and methodological issues. *Accident Analysis and Prevention*, 30, 293-297.

Hakamies-Blomqvist, L.E. (1993). Fatal accidents of older drivers. *Accident Analysis and Prevention*, 25, 19-27.

Hasher, L., Stoltzfus, E.R., Zacks, R.T., and Rypma, B. (1991). Age and inhibition. *Journal of Experimental Psychology: Learning, Memory, and Cognition*, 17, 163-169.

Hatakka, M., Keskinen, E., Gregersen, N.P., Glad, A., and Hernetkoski, K. (2002). From control of the vehicle to personal self-control; broadening the perspectives to driver education. *Transportation Research Part F*, 5, 201-215.

Hill, B.L., and Burg, A. (1977). A reanalysis of California driver vision data: general findings. In T.L.R. No.768 (Ed.): Transport and road Research Laboratory.

Holland, C.A. (1993). Self-bias in older drivers' judgments of accident likelihood. . *Accident Analysis and Prevention*, 25, 431-441.

Homann, C.N., Suppan, K., Homann, B., Crevenna, R., Ivanic, G., and Ruzicka, E. (2003). Driving in Parkinson's disease - a health hazard? *Journal of Neurology*, 250, 1439-1446.

Ikeda, M., and Takeuchi, T. (1975). Influence of foveal load on the functional visual field. *Perception and Psychophysics*, 18, 255-260.

Janke, M.K. (2001). Assessing older drivers: Two studies. *Journal of Safety Research*, 32, 43-74.

Johansson, K., Bogdanovic, N., Kalimo, H., Winblad, B., and Viitanen, M. (1997). Alzheimer's disease and apolipoprotein E epsilon-4 allele in older drivers who died in automobile accidents. *The Lancet*, 349, 1143-1144.

Jost, J.T., Kruglanski, A.W., and Nelson, T.O. (1998). Social metacognition: An expansionist review. *Personality and Social Psychology Review*, 2, 137-154.

Katila, A., Keskinen, E., Hatakka, M., and Laapotti, S. (2004). Does increased confidence among novice drivers imply a decrease in safety? The effects of skid training on slippery road accidents. *Accident Analysis and Prevention*, 36, 543-550.

Keskinen, E. (1996). *Why do young drivers have more accidents?* Bergisch Gladbach: BASt.

Koriat, A. (2007). Metacognition and Consciousness. In P.D. Zelazo, M. Moscovitch, and E. Thompson (Eds.), *The Cambridge handbook of consciousness* (pp. 289-325). Cambridge, UK: Cambridge University Press.

Laapotti, S., Keskinen, E., Hatakka, M., and Katila, A. (2001). Novice drivers' accidents and violations—a failure on higher or lower hierarchical levels of driving behaviour. *Accident Analysis and Prevention*, 33, 759-769.

Lefebvre-Pinard, M., and Pinard, A. (1985). Taking care of one's cognitive activity: A moderator of competence. In D. Neimark and R.D. Lisi (Eds.), *Moderators of competence* (pp. 191-211). Hillsdale, NJ: Lawrence Erlbaum Associates.

Leibowitz, H.W., Abernethy, C.N., Buskirk, E.R., Bar-Or, O., and Hennessy, R.T. (1972). The effect of heat stress on reaction time to centrally and peripherally presented stimuli. *Human Factors*, 14, 93-97.

Liourta, E., and Empelen, P.v. (2008). The importance of self-regulatory and goal-conflicting processes in the avoidance of drunk driving among Greek young drivers. *Accident Analysis and Prevention*, 40, 1191-1199.

Mackworth, N.H. (1965). Visual noise causes vision tunnel. *Psychonomic Science*, 3, 67-68.

Marin-Lamellet, C., Paire-Ficout, L., Lafont, S., Amieva, H., Laurent, B., Thomas-Anterion, C., Dirson, S., and Fabrigoule, C. (2003). The creation of a tool for evaluating attentional deficits that affect driving skills during normal and pathological ageing: The SEROVIE study. *Recherche - Transports - Securite*, 81, 177-189.

Marottoli, R.A., Richardson, E.D., Stowe, M.H., Miller, E.G., Brass, L.M., Cooney, L.M., and Tinetti, M.E. (1998). Development of a test battery to identify older drivers at risk for self-reported adverse driving events. *Journal of the American Geriatrics Society*, 46, 562-568.

Marquié, J.-C., and Isingrini, M. (2001). Aspects cognitifs du vieillissement normal. In E. Aubert and J.-M. Albaret (Eds.), *Vieillissement et Psychomotricité* (pp. 77-113). Marseille: Solal Editeur.

Matthews, M.L., and Moran, A.R. (1986). Age differences in male drivers' perception of accident risk: the role of perceived driving ability. *Accident Analysis and Prevention*, 18, 299-313.

Mazer, B.L., Sofer, S., Korner-Bitensky, N., Gelinas, I., Hanley, J., and Wood-Dauphinee, S. (2003). Effectiveness of a visual attention retraining program on the driving performance of clients with stroke. *Archives of Physical Medicine and Rehabilitation*, 84, 541-550.

McDowd, J.M., and Shaw, R.J. (2000). Attention and aging: A functional perspective. In F.I.M. Craik and T.A. Salthouse (Eds.), *The handbook of aging and cognition* (pp. 221-292). Mahwah, NJ: Lawrence Erlbaum.

McFarland, R.A. (1964). On the driving of automobiles by older people. *Journal of Gerontology*, 19, 190-197.

McKenna, F.P. (1993). It won't happen to me: Unrealistic optimism or illusion of control? *British Journal of Psychology*, 84, 39-50.

McKenna, F.P., Stanier, R.A., and Lewis, C. (1991). Factors underlying illusory self-assessment of driving skill in males and females. *Accident Analysis and Prevention*, 23, 45-52.

Metcalfe, J. (1998). Cognitive optimism: Self-deception or memory-based processing heuristics? *Personality and Social Psychology Review*, 2, 100-110.

Metcalfe, J., and Kornell, N. (2005). A Region of Proximal Learning model of study time allocation. *Journal of Memory and Language*, 52, 463-477.

Miller, G.A. (2003). The cognitive revolution: a historical perspective. *Trends in Cognitive Sciences*, 7, 141-144.

Molesworth, B., Wiggins, M.W., and O'Hare, D. (2006). Improving pilots' risk assessment skills in low-flying operations: The role of feedback and experience. *Accident Analysis and Prevention*, 38, 954-960.

Muzet, V. (1998). *Conception, réalisation et validation d'un système de mesure du champ visuel utile chez l'homme*. Thesis in Sciences, Psychophysique et Génie Biomédical, Université Louis Pasteur.

Muzet, V., Rogé, J., Jouny, C., and Muzet, A. (1999). Changes in central and peripheral vision before and after a simulated driving task. In A.G. Gale, I.D. Brown, C.M. Haslegrave, and S.P. Taylor (Eds.), *Vision in vehicles VII* (pp. 139-147): Elsevier.

Myers, R., Ball, K., Kalina, T., Roth, D., and Goode, K. (2000). Relation of useful field of view and other screening tests to on-road driving performance. *Perceptual and Motor Skills*, 91, 279-290.

Näätänen, R., and Summala, H. (1975). *Road-user behavior and traffic accidents*. Amsterdam: North-Holland.

Nelson, T.O., and Leonesio, R.J. (1988). Allocation of self-paced study time and the "labor-in-vain effect". *Journal of Experimental Psychology: Learning, Memory, and Cognition*, 14, 676-686.

Nelson, T.O., and Narens, L. (1990). Metamemory: a theoretical framework and new findings. *The Psychology of Learning and Motivation*, 26, 125-173.

Norman, D., and Shallice, T. (1980). Attention to action: Willed and automatic control of behavior. In U.o. California (Ed.), *CHIP Report #99*. San Diego.

Norman, D., and Shallice, T. (1986). *Attention to action: Willed and automatic control of behavior*. New York,: Plenum.

Novack, T., Banos, J., Alderson, A., Schneider, J., Weed, W., Blankenship, J., and Salisbury, D. (2006). UFOV performance and driving ability following traumatic brain injury. *Brain Injury*, 20, 455-461.

OECD (1985). La sécurité des personnes âgées dans la circulation routière. Paris: OCDE et OMS.

Owsley, C. (1994). Vision and driving in the elderly. *Optometry and Vision Science*, 71, 727-735.

Owsley, C. (2004). Drivers Capabilities. In T.R. Board (Ed.), *Transportation in an Ageing Society* (pp. 44-65). Maryland.

Owsley, C., Ball, K., McGwin, G., Sloane, M.E., Roenker, D.L., White, M.F., and Overley, E.T. (1998). Visual processing impairment and risk of motor vehicle crash among older adults. *Journal of the American Medical Association*, 279, 1083-1088.

Owsley, C., Burton-Danner, K., and Jackson, G. (2000). Aging and spatial localization during feature search. *Gerontology*, 46, 300-305.

Owsley, C., and McGwin, G. (1999). Vision impairment and driving. *Survey of Ophthalmology*, 43, 535-550.

Owsley, C., Stalvey, B.T., and Phillips, J.M. (2003). The efficacy of an educational intervention in promoting self-regulation among high-risk older drivers. *Accident Analysis and Prevention*, 35, 393-400.

Parker, D., McDonald, L., Rabbitt, P., and Sutcliffe, P. (2003). Older drivers and road safety: the acceptability of a range of intervention measures. *Accident Analysis and Prevention*, 35, 805-810.

Planek, T.W. (1972). The aging driver in today's traffic: a critical review. In W.A.M. T.W. Planek, EL. Wiener (Ed.), *Aging and Highway Safety: the elderly in a mobile society* (pp. 1-38). Chaptel Hill, N.C.: North Carolina Symposium on Highway Safety.

Pressley, M., Borkowski, J.G., and O'Sullivan, J.T. (1985). Children's Metamemory and the Teaching of Memory Strategies. In D.L. Forrest-Pressley, G.E. Mackinnon, and T.G. Waller (Eds.), *Metacognition, Cognition, and Human Performance* (Vol. 1, pp. 111-153). Orlando, FL: Academic Press.

Roenker, D.L., Cissell, G.M., Ball, K.K., Wadley, V.G., and Edwards, J.D. (2003). Speed-of-processing and driving simulator training result in improved driving performance. *Human Factors*, 45, 218-233.

Rogé, J., Kielbasa, L., and Muzet, A. (2002). Deformation of the useful visual field with state of vigilance, task priority, and central task complexity. *Perceptual and Motor Skills*, 95, 118-130.

Rogé, J., Pébayle, T., Campagne, A., and Muzet, A. (2005). Useful visual field reduction as a function of age and risk of accident in simulated car driving. *Investigative Ophthalmology and Visual Science*, 46, 1774-1779.

Rogé, J., Pébayle, T., Hannachi, S.E., and Muzet, A. (2003). Effect of sleep deprivation and driving duration on the useful visual field in younger and older subjects during simulator driving. *Vision Research*, 43, 1465-1472.

Rogé, J., Pébayle, T., Kiehn, L., and Muzet, A. (2002). Alteration of the useful visual field as a function of state of vigilance in simulated car driving. *Transportation Research, Part F: Traffic Psychology and Behaviour*, 5, 189-200.

Rogé, J., Pébayle, T., Lambilliotte, E., Spitzenstetter, F., Giselbrecht, D., and Muzet, A. (2004). Influence of age, speed and duration of monotonous driving task in traffic on the driver's useful visual field. *Vision Research*, 44, 2737-2744.

Salthouse, T.A. (1994). Age-related differences in basic cognitive processes: implications for work. *Experimental Aging Research*, 20, 249-255.

Salthouse, T.A., and Meinz, E.J. (1995). Aging, inhibition, working memory, and speed. *Journal of Gerontology*, 50, P297-306.

Sanders, A.F. (1970). Some aspects of the selective process in the functional visual field. *Ergonomics*, 13, 101-117.

Schlag, B. (1993). Elderly drivers in Germany—Fitness and driving behavior. *Accident Analysis and Prevention*, 25, 47-55.

Scialfa, C., Kline, D., and Lyman, B. (1987). Age differences in target identification as a function of retinal location and noise level: examination of the useful field of view. *Psychology and Aging*, 2, 14-19.

Scialfa, C.T., Thomas, D.M., and Joffe, K.M. (1994). Age differences in the useful field of view: an eye movement analysis. *Optometry and Vision Science*, 71, 736-742.

Seiple, W., Szlyk, J.P., Yang, S., and Holopigian, K. (1996). Age-related functional field losses are not eccentricity dependent. *Vision Research*, 36, 1859-1866.

Sekuler, A.B., Bennett, P.J., and Mamelak, M. (2000). Effects of aging on the useful field of view. *Experimental Aging Research*, 26, 103-120.

Sekuler, R., and Ball, K. (1986). Visual localization: age and practice. *Journal of the Optical Society of America*, 3, 864-867.

Sims, R.V., McGwin, G.J., Allman, R.M., Ball, K., and Owsley, C. (2000). Exploratory study of incident vehicle crashes among older drivers. *Journal of Gerontology: Biological Sciences and Medical Sciences*, 55, M22-M27.

Staplin, L.J. (1995). Simulator and field measures of driver age differences in left-turn gap judgments. *Transportation Research Record*, 1325, 23-33.

Staplin, L.J., Lococo, K., Gish, K., and Decina, L. (2003). Model driver screening and evaluation program.

Stutts, J.C., and Martell, C. (1992). Older driver population and crash involvement trends, 1974-1988. *Accident Analysis and Prevention*, 24, 317-327.

Svenson, O. (1981). Are we all less risky and more skillful than our fellow drivers? *Acta Psychologica*, 47, 143-148.

Szlyk, J.P., Seiple, W., and Viana, M. (1995). Relative effects of age and compromised vision on driving performance. *Human Factors*, 37, 430-436.

Thiede, K.W., Anderson, C.M., and Therriault, D. (2003). Accuracy of Metacognitive Monitoring Affects Learning of Texts. *Journal of Educational Psychology*, 95, 66-73.

Thiede, K.W., and Dunlosky, J. (1999). Toward a General Model of Self-Regulated Study: An Analysis of Selection of Items for Study and Self-Paced Study Time. *Journal of Experimental Psychology: Learning, Memory, and Cognition*, 25, 1024-1037.

Tipper, S.P. (1991). Less attentional selectivity as a result of declining inhibition in older adults. *Bulletin of the Psychonomic Society*, 29, 45-47.

Uc, E.Y., Rizzo, M., Anderson, S.N., Shi, Q., and Dawson, J.D. (2006). Unsafe rear-end collision avoidance in Alzheimer's disease. *Journal of the Neurological Sciences*, 251, 35-43.

Van Elslande, P. (2003). Les erreurs des conducteurs ages: Elderly drivers errors. *Recherche - Transports - Securite*, 81, 190-202.

Verriest, G., Bailey, I.L., Calabria, G., and Campos, E. (1985). The occupational visual field: II. Practical aspects: the functional visual field in abnormal conditions and its relationship to visual ergonomics, visual impairment and job fitness. In H. A. and G. E.L. (Eds.), *6th International Visual Field Symposium* (pp. 281-327). The Hague: Boston Lancaster: Dr W. Junk Publishers, Dordrecht, The netherlands.

Verriest, G., Barca, L., and Dubois-Poulsen, A. (1983). The occupational visual field: I. Theoretical aspects: the normal functional visual field. In E.L. Greve and A. Heijl (Eds.), *6th International Visual Field Symposium* (pp. 165-185). The Hague: Boston Lancaster: Dr W. Junk Publishers, Dordrecht, The netherlands.

Waylen, E.A., Horswill, M.S., Alexander, J.L., and McKenna, F.P. (2004). Do expert drivers have a reduced illusion of superiority? *Transportation Research Part F*, 323-331.

Whelihan, W.M., DiCarlo, M.A., and Paul, R.H. (2005). The relationship of neuropsychological functioning to driving competence in older persons with early cognitive decline. *Archives of Clinical Neuropsychology*, 20, 217-228.

Williams, L.J., and Lefton, L.A. (1981). Processing of alphabetic information presented in the fovea or the periphery: functional visual field and cognitive load. *Perception*, 10, 645-650.

In: Road Traffic: Safety, Modeling and Impacts
Editors: S. E. Paterson and L. K. Allan, pp. 347-365
ISBN 978-1-60456-884-4

Chapter 7

DEVELOPMENT AND EVALUATION OF AN EFFECTIVE SEQUENTIAL APPROACH FOR DYNAMIC ACCIDENT DURATION FORECASTING

Ying Lee*

Department of Hospitality Management, Ming Dao University
No.369, Wen-Hua Road, Pee-Tow Township, Chang-Hua County, Taiwan

Chien-Hung Wei†

Department of Transportation and Communication Management Science,
National Cheng Kung University, No.1, Ta-Hsueh Road, Tainan City, Taiwan

ABSTRACT

This study creates a sequential approach to represent the dynamic update forecast of accident duration. This procedure includes two Artificial Neural Network-based models. Model A is used to forecast the duration time at the instant of accident notification while Model B provides multi-period updates of duration time after the moment of accident notification. These two models together provide a sequential forecast of accident duration from the accident notification to the accident site clearance. With these two models, the estimated duration time can be provided by plugging in relevant traffic data as soon as an incident is reported. This study shows very promising practical applicability of the proposed models in the Intelligent Transportation Systems context.

Key words: sequential forecast, freeway accident duration analysis, data fusion techniques

* Fax: +886-4-8879035; E-mail: yinglee1017@gmail.com
† Fax: +886-6-2753882; E-mail: louiswei@mail.ncku.edu.tw

1. INTRODUCTION

Incident is one of the reasons causing traffic congestion. The total amount of extra travel time and fuel consumption caused by congestion, both recurring and occasional, has been valued at an estimated $35 billion for 50 large U.S. urban areas in 1988 (Small, 1997). Freeway traffic in Taiwan is often congested due to the increasing number of vehicles and in particular due to incidents. The uncertainty of travel time increases when the travel conditions are unknown, although with advanced traffic data collection techniques, traffic data can be collected much more easily. Traffic patterns could be adequately characterized if the data could be processed and analyzed effectively, thereby providing useful information such as travel time information for those planning to travel (D' angelo et al., 1999; Chen & Chien, 2001; Krikke, 2002). Many studies have discussed travel time forecast issues, although few of them have considered the influence of incidents in their models (Wei & Lee, 2003; Zhang & Rice, 2003).

The traffic condition varies from minute to minute. Through the on-board unit or other mobile devices, the driver can receive the latest traffic information. The effect is quite likely that travelers might change their routes or rearrange the schedule. Regardless of the duration of an accident, the traffic condition changes continuously with time. The accident duration forecasting information needs to be updated frequently with the traffic change. The accident duration with sequentially forecasted information can really fulfill the road users' need. Literature seems relatively rare proposing continuous update of accident duration till it is cleared.

The innovative sequential approach proposed in this study is composed of two types of accident duration models –Model A and Model B. Figure 1 illustrates the time points when these two models are employed to perform forecasts in accident duration. When an accident is being notified for the first time, Model A provides a preliminary forecast by using the traffic data which represent the traffic situation right before the accident. After the accident notified, Model B takes over Model A to perform forecasts with updated data. Model A and Model B together provide a sequential forecast for accident duration.

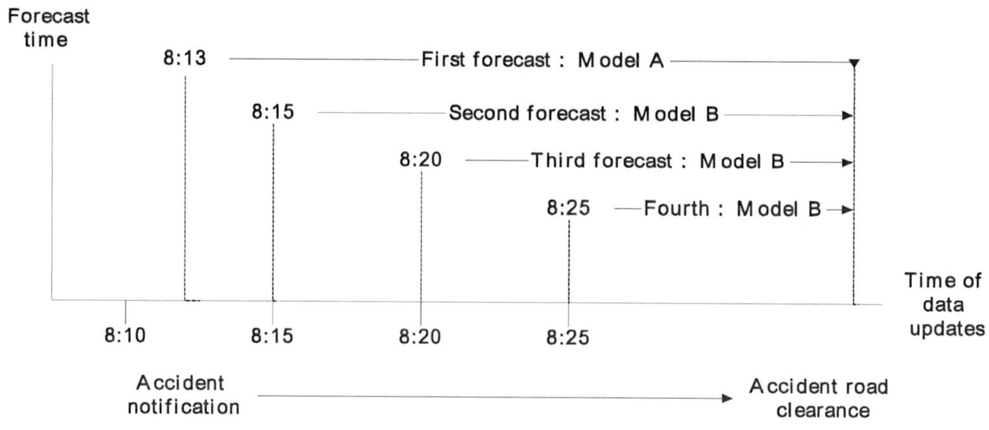

Figure 1. Multi-period forecast of accident duration.

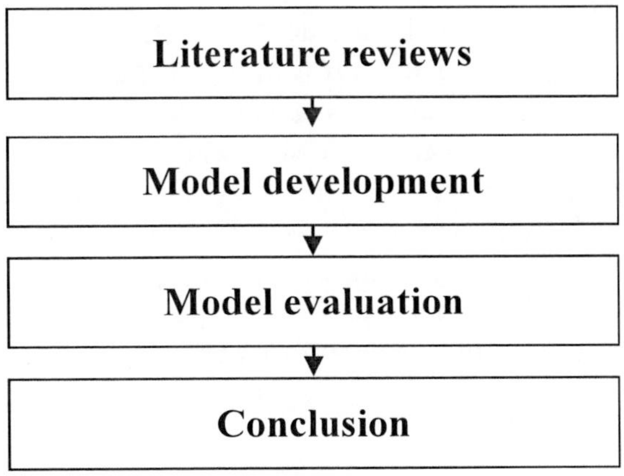

Figure 2. The study flowchart of the proposed model.

The study flowchart, Figure 2, is shown as follow. The relevant literatures about incident duration prediction will be reviewed. This study will analyze the traffic environment, collect the traffic data, build the data process procedure and conduct the sequential accident duration forecasting models. After model training, the appropriate mapping between variables and accident duration is formed. The mean absolute error (MAE), R^2 and mean absolute percentage error (MAPE) have been chosen for model evaluation. The content of this study is generalized from the previous researches (Wei & Lee, 2007a; Lee & Wei, 2007; Wei & Lee, 2005).

2. Literature Review

The accident duration is usually defined as the time between accident occurrence and roadway clearance. This duration can be divided into three parts, namely, reporting time (the time between accident occurrence and accident notification), response time (the time between accident notification and rescuer arrival) and clearance time (the time between rescuer arrival and accident road clearance). As illustrated in Figure 3, the accident duration can be forecasted at one of the three moments—accident occurrence, accident notification or rescuer arrival.

Garib et al. (1997) employs a multiple regression model to forecast the clearance time of an accident. The model shows that 81% of variations in accident duration can be forecasted by number of lanes affected, number of vehicles involved, truck involvement, time of the day, police response time, and weather condition. Nam and

Nam and Mannering (2000) moreover applies a hazard-based analysis to build three accident duration models from the three moments, i.e. accident occurrence, accident notification and rescuer arrival. The input factors of three models are similar, such as temporal characteristics, environmental characteristics, geographic information and accident characteristics. Only the third model which forecasts the clearance time of accident adds the lead agency as the model input. The model estimation results show that a wide variety of factors significantly affect accident duration. Factors such as month, raining condition, geographic information, and accident characteristics (e.g., fatality and injury) are important variables in forecasting duration at the accident occurrence moment. On the other hand, temporal characteristics (e.g., timing and traffic volume) and geographic information are the two major variables for the accident notification moment.

Wei and Lee (2005) developed the accident duration model by data fusion techniques. This model provides a forecasted duration at the accident notification moment. The mean absolute percentage error (MAPE) of forecasted results of the accident duration are mostly under 20%. By the factor sensitivity analysis, the significant factors can be identified. The factors, time of day, number of vehicles & vehicle type involved in accident and geometry, have high significance in conducting the duration model.

3. Model Building

Previous studies found several highly significant factors for the present development of an accident duration model. Different from the hazard-based model and the regression approach, the study employs data fusion techniques in modeling and builds two models to forecast the accident duration from the moment of accident notification to the accident clearance.

These two forecast models develop the functional relationship between real-time traffic data as the input variables and accident duration as the output variable. Real-time traffic data are collected from loop vehicle detectors (VD) and accident advisory databases. In the model development, the VD data from neighboring sections are considered to present the time-space relation for accident existence. In order to recognize the change of traffic situation in the interval when traffic data updated and accident first notified, the time gap and the space gap associated with these data are considered as the model inputs. The model concept is shown in Figure 1. In order to represent the variation of traffic situation, two models will be built. Model A provides a preliminary forecast at accident notification. Model B updates the forecasted duration with the latest traffic data till accident road has been cleared.

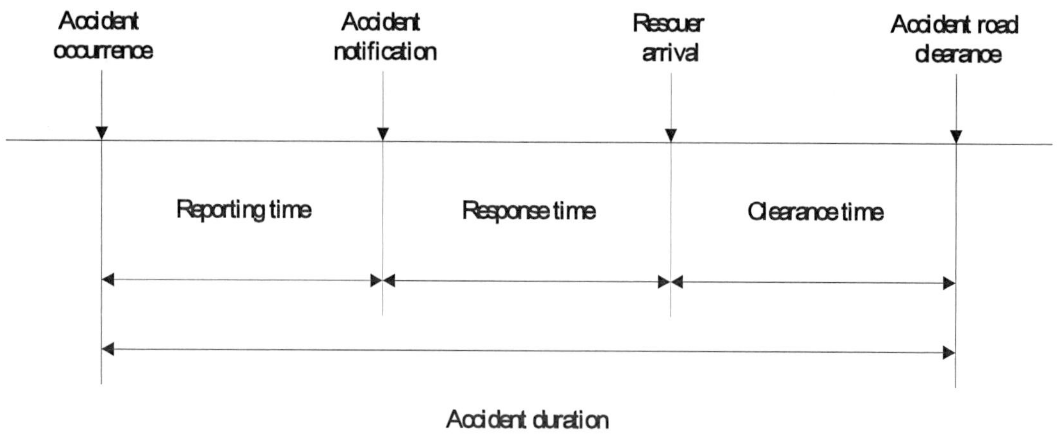

Figure 3. Components of accident duration.

3.1. Data Analysis

This study selected the southern part of the No.1 national freeway in Taiwan, which is from An-ding interchange to Lu-jhu interchange, as the experiment site. This double lane site has a length of 27.2 km, 1 toll stations, 3 interchanges and 2 system interchange, as shown in Figure 4. The southbound also has 7 VD's, and a speed limit of 100 km/hr. The raw data are collected for 24 hours a day over a 6-month period from December 2004 to September 2005 (not include February, March, April, May 2005). The time period which covers weekdays/weekends and peak/non-peak hours adequately reflects real traffic conditions. In this period, the road lane between An-ding and Lu-jhu were not broadened as 3 lanes. Therefore, this study did not consider the data and traffic scenario with 3 lanes.

3.1.1. VD Data

The VD data and incident data are recorded with a preset time interval. From Taiwan Area National Freeway Bureau (TANFB), the VD data are recorded and accumulated every 300sec for each lane. The VD records the time, the vehicle running speed, the vehicle volume and the vehicle occupancy. The VD data in original form are shown in Table 1. Although the VD can recode the traffic data every minute, the quantity of traffic data is too huge to storage. In regular, TANFB storages the aggregated traffic data every 5-minute.

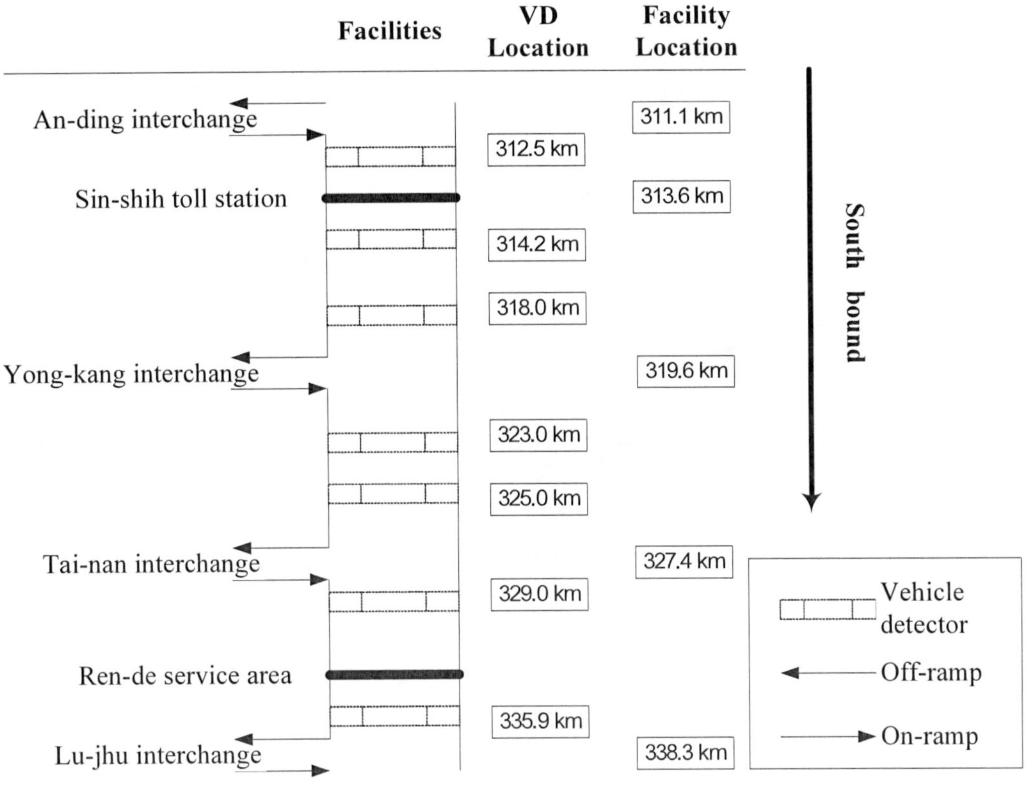

Figure 4. Freeway layout from An-ding to Lu-jhu (South bound).

Table 1. The VD data in original form

Date (Year, Month, Day)	20050201	20050201	20050201	20050201	...
Time (Hour, Minute)	1945	1950	1950	1955	...
VD ID	29	29	29	29	...
Lane number	2	1	2	1	...
Controller Situation (0 is true)	0	0	0	0	...
Car volume (vehicle / 5 min)	40	71	40	83	...
Bus & truck volume (vehicle / 5 min)	12	13	13	9	...
Trailer volume (vehicle/ 5 min)	16	4	11	8	...
Car speed (km/ hour)	94	97	96	95	...
Bus & truck speed (km/ hour)	86	96	91	97	...
Trailer speed (km/ hour)	91	99	87	100	...
Average speed (km/ hour)	92	97	93	96	...
Average vehicle length (cm)	77	55	73	55	...
Average occupancy rate (%)	7	6	6	7	...
Average usage rate (lane / volume)	41	57	42	63	...
Average headway (1/10 sec)	43	33	46	29	...
Congestion situation	1	1	1	1	...
System situation (N is normal)	N	N	N	N	...

3.1.2. Incident Data

The incident data are supplied by the Police Radio Station and updated every 180sec. Each incident is shown repeatedly in the database until it is cleared away from the traffic lane occupied. Table 2 shows the incident data in original form. The time, the incident type and incident location are recorded. The duration time of each incident can be calculated from its first and last record. Through data filtering process, this study sieves out the accident data from incident database for model development.

Table 2. The incident data in original form

Date	2005/1/31	2005/1/31	2005/1/31	...
Time	23:23:59	23:28:53	23:29:48	...
Location	Taipei city	No. 1 national freeway	No. 1 national freeway	...
Direction		Southern bound	Northern bound	...
Origin (KM)		51.8	52.9	...
Destination (KM)				...
Incident type	Accident	Maintenance	Maintenance	...
Incident situation	Collision with two cars at cheng-de bridge	Road maintenance at shoulder and outside lane	Road maintenance at shoulder and outside lane	...
Provider	Traveler	Maintenance staff	Maintenance staff	...

Table 3. Relative frequency of duration

Accident duration (sec)	Samples	Frequency	Cumulative frequency
<2000	3	8%	8%
2001~3000	9	23%	31%
3001~4000	7	18%	49%
4001~5000	4	10%	59%
5001~6000	8	20%	79%
6001~7000	2	5%	84%
7001~8000	5	13%	97%
>8001	1	3%	100%

3.1.3. Accident Duration Data

The relative frequency of duration for 39 accidents is constructed in Table 3. About 49% of the accidents, the duration is less than 4000 seconds. The proportion of duration between 4001 seconds and 6000 seconds is 30%. The relative frequency of accident duration with more than 6001 seconds is around 21%. The 39 accidents are divided into two parts, 24 accidents for model training and 15 accidents for model testing. The 15 accidents for model testing are sampled randomly according to the relative frequency of duration.

3.2. Model Structure

The flowchart of accident duration model is shown in Figure 5. The data sources are mostly from VD and accident database. The VD data and the accident data are updated regularly for a fixed period of time. In modeling, Model A is used to forecast duration period at the moment of accident notification while Model B provides dynamic updates of accident duration after an accident is being notified. The input variables for both models are different. Relevant data are aligned according to available inputs and accident locations.

3.3. Model Inputs and Output

Accident data are updated every three minutes in the accident database while the VD data are recorded and accumulated every five minutes for each lane. The time-space relationship between the accident and the VD data is shown in Figure 6. In the figure, the accident occurs in Section 2 at 8:13 A.M. The figure further indicates that several gaps exist between the accident and the VD data. Therefore, the gaps must be considered when using the VD data to describe the traffic situation during the accident.

In Model A, input selection is based on the reporting time at accident notification moment. The input variables for Model B include all the input variables in Model A as well as the updated VD data after accident notification. In summary, as illustrated in Table 4, input variables include accident characteristics, traffic data, time relationship, space relationship, and geometry characteristics.

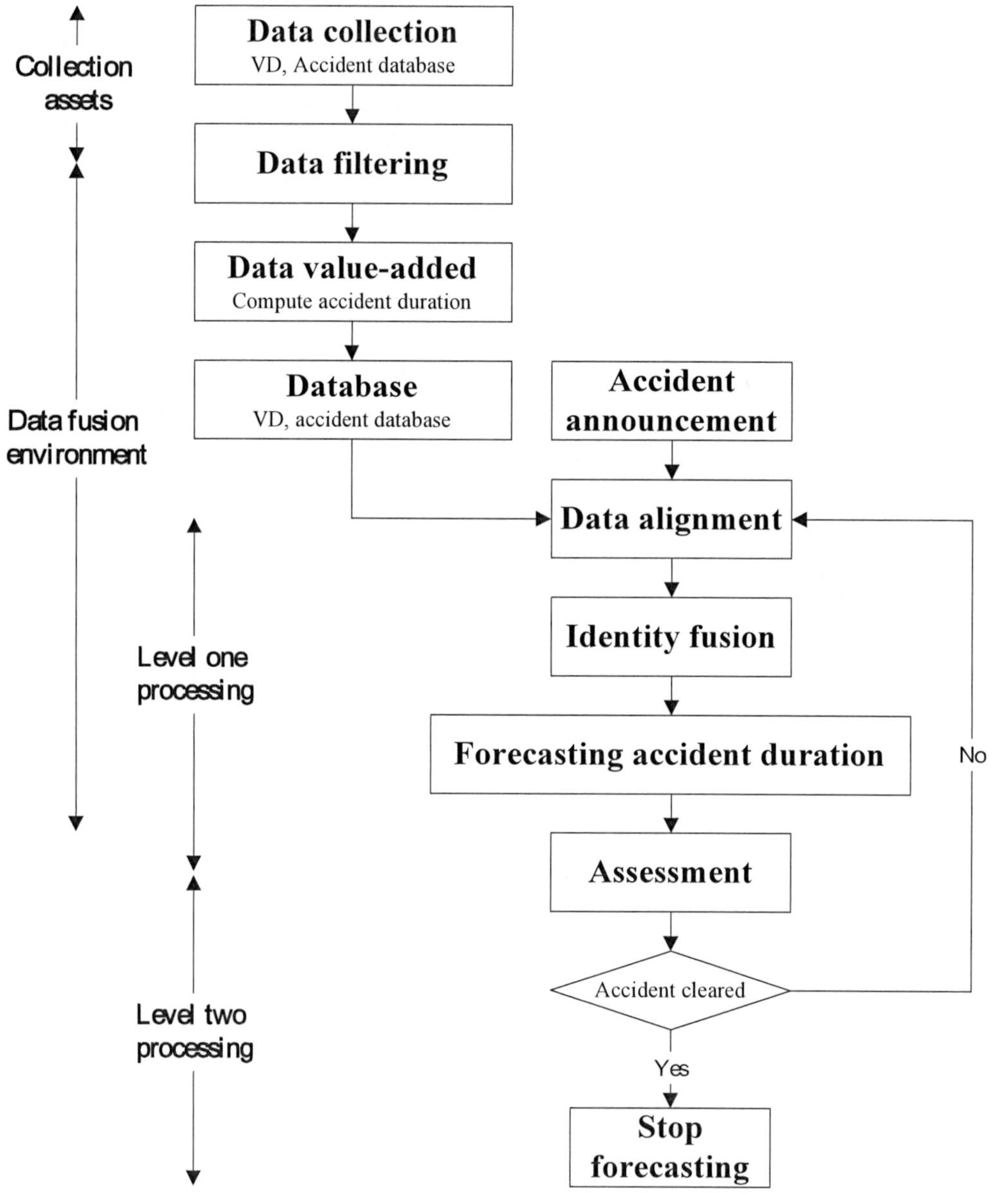

Figure 5. Accident duration forecast flowchart.

Accident Characteristics

In our previous study (Wei & Lee, 2005), the factors significant to the accident duration are discussed by the factor sensitivity analysis. The factors, time of day, number of vehicles and vehicle type involved in accident and geometry, have high significance in conducting the duration model. In this way, the variables adopted in this study are based on our previous research. Accident characteristics include occupied lane, number of vehicles, and vehicle type involved in an accident.

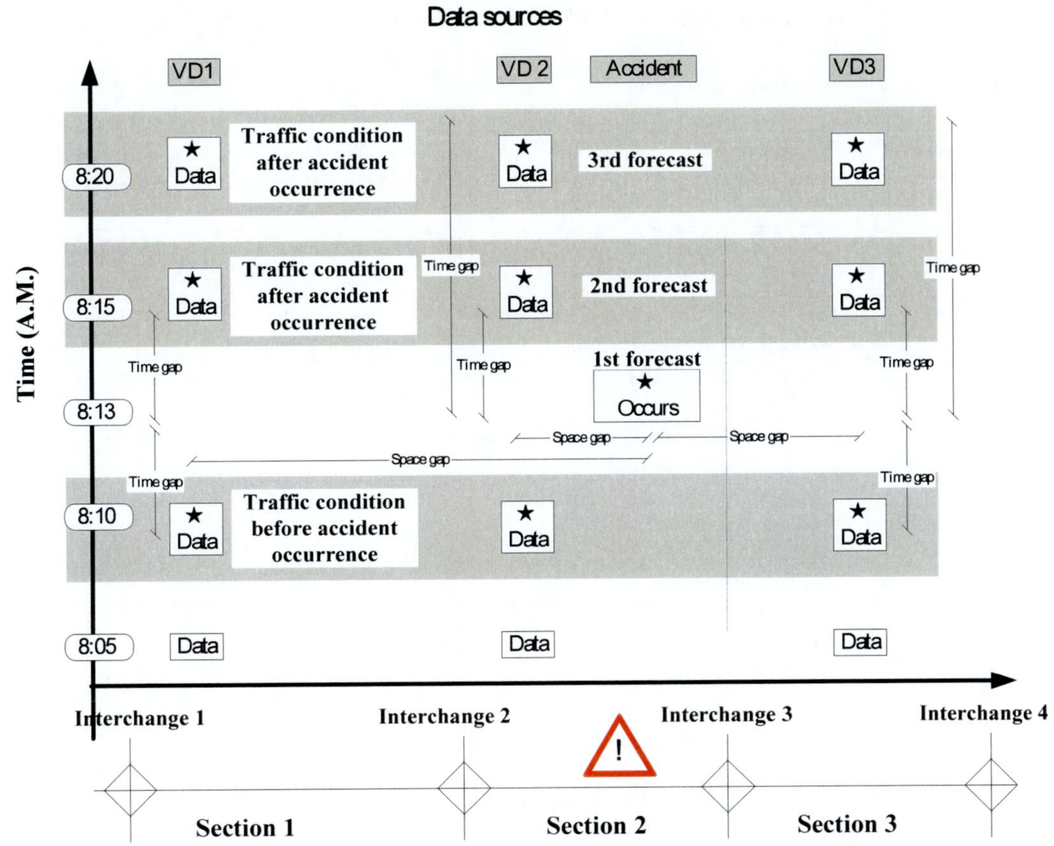

Figure 6. Time-space relationship between an accident and the VD data.

Traffic Data

Based on accident location, traffic data collected by the first and the second VD at upstream and the first VD at downstream are used to present the traffic situation in neighbor of an accident. According to the allocation spacing of VDs in Taiwan, considering two VDs at the upstream can cover the affected area by an accident. Model B is formulated to provide an update of forecasted duration. Therefore, at every forecast, Model B considers real-time traffic data which presents the latest traffic situation during an accident. The input variables in Model B reflect a contrast between the traffic data before the occurrence of an accident and the real-time traffic data after the occurrence of an accident.

Time Relationship

The time of accident notification is daytime and nighttime. The daytime is from 7:00 a.m. to 6:30 p.m. and the nighttime is from 6:30 p.m. to 7:00 a.m. The VD data are accumulated and recorded every five minutes for each lane. There probably exists a time gap between the time of an accident notification and the recording time of VD data. The time gap is considered as an input to represent the possible variation of traffic situation between the time of accident notification and the recording time of VD data.

Table 4. Inputs of accident duration forecast models

	Factors	Variables	Descriptions	For the Model A or B
Inputs	Accident	Turn over	1:Yes, 0:No	Model A, B
		Occupied lane	1:Shoulder, 2:One lane, 3:Two lane	Model A, B
		# vehicles and types of vehicle involved in accident	# passenger car	Model A, B
			# non-passenger car	Model A, B
	Traffic data	Data from the first VD upstream before accident occurrence	Volume, Speed	Model A, B
		Data from the first VD upstream after accident occurrence	Volume, Speed	Model B
		Data from the second VD upstream before accident occurrence	Volume, Speed	Model A, B
		Data from the second VD upstream after accident occurrence	Volume, Speed	Model B
		Data from the first VD downstream before accident occurrence	Volume, Speed	Model A, B
		Data from the first VD downstream after accident occurrence	Volume, Speed	Model B
	Time	The occurrence time	1:Day, 2:Night	Model A, B
		The time gap between the accident occurrence and the recording time of VD data before accident occurrence	Seconds	Model A, B
		The time gap between the accident occurrence and the recording time of VD data after accident occurrence	Seconds	Model B
	Space	Distance between the accident and the first VD upstream	Kilometers	Model A, B
		Distance between the accident and the second VD upstream	Kilometers	Model A, B
		Distance between the accident and the first VD downstream	Kilometers	Model A, B
		Distance between the accident and the interchange upstream	Kilometers	Model A, B
		Distance between the accident and the interchange downstream	Kilometers	Model A, B
	Geometry	Location of accident	1:Interchange area, 2:Service facility area, 3:Other areas	Model A, B
		An interchange exists between the accident and the first VD upstream	1: Yes, 0: No	Model A, B
		An interchange exists between the accident and the first VD downstream	1: Yes, 0: No	Model A, B
		A toll plaza or service area exists between the accident and the first VD upstream	1: Yes, 0: No	Model A, B
		A toll plaza or service area exists between the accident and the first VD downstream	1: Yes, 0: No	Model A, B
Output	Accident Duration		Seconds	Model A, B

is the number of.

Space Relationship

The Distance between an accident and VD site presents likely variation of traffic situation. Generally speaking, the greater the distance, the longer the time is required to detect the influence of an accident in VD data. When an accident occurs, rescuers arrive at the accident location from neighboring interchanges. The distance between the accident and the neighboring interchanges is used to represent the time that rescuers take to arrive at an accident location.

Geometry Characteristics

Highway possesses several kinds of geometry characteristics. In different geometry, similar accidents may result in different durations. Accident locations can be classified as interchange area, service area, and other types of area. If an interchange is close to an accident and VD sites, using the VD data to present the traffic situation after the occurrence of an accident will be affected by that interchange. This is because the existence of interchange between the accident and the VD in neighboring is to explain the influence of interchange to the VD data. A similar reason is for considering the existence of service facility between the accident and the neighboring VD.

When an accident occurs, the accident duration can be computed in two ways. First way is calculating from the recording time of accident notification and accident road clearance in the daily record of the National Freeway Bureau. Second, the accident duration can be observed by the change of volume in VD data. In these two ways, accident durations obtained are sometimes different. In this study, the resulting accident duration is integrated from these two ways. Comparing the trend of volumes and records, the earlier occurrence notifications and the latter clearance notifications are preserved. When the actual duration of an accident is not available, this integration is a plausible way to obtain an accident duration period as close as possible to a real one.

3.4. Identity Fusion

In identity fusion, ANNs are the key technique. The ANNs approach is a data-driven, self-adaptive, and nonlinear methodology. After model training, the mapping between variables and accident duration is formed. Then by inputting the data from tested examples into this trained model, the forecasted accident duration is produced. Figure 7 is the schematic representation of a general ANN network. This study only estimates one output y, accident duration.

When inputting variables into the network, the weights from input layer to hidden layer are calculated. Through the transfer function in the hidden layer, the input data are rescaled as inputs to the output layer. This process is shown in Equation 1, where the output "y" is the estimated accident duration. Since a discrepancy might occur between the estimated output and the actual accident duration, the weights are adjusted repeatedly by a suitable training method until the resulting error is stabilized and negligible. The accident duration function inside the ANN model is formed after this training procedure is completed (Zurada, 1992).

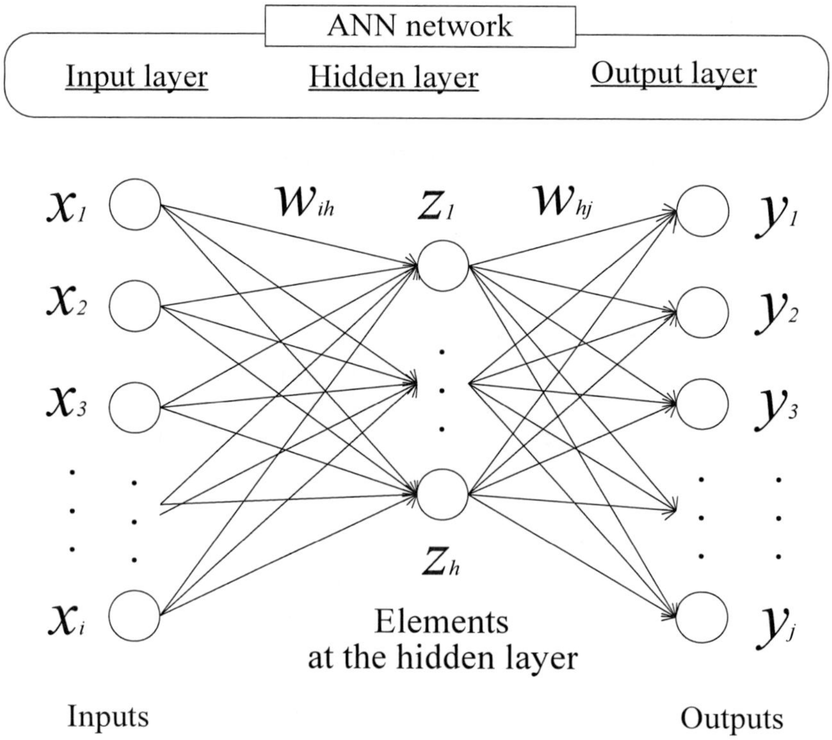

Figure 7. Scheme of an ANN model.

$$y = g\left(\sum_j w_{hj} \times f\left(\sum_i w_{ih} \times x_i - \theta_h\right) - \theta_j\right) \quad . (1)$$

Where

y: Output variable (Accident duration)
i: Elements at input layer
x_i: Input variables (Shown in Table 1)
h: Elements at hidden layer
θ_h: Threshold values at hidden layer
w_{ih}: Weights between input layer and hidden layer
f: Transfer function at hidden layer
j: Elements at output layer
θ_j: Threshold values at output layer
w_{hj}: Weights between hidden layer and output layer
g: Transfer function at output layer

After model training, the mapping between variables and accident duration is formed. Inputting the data from the tested examples into this trained model, the forecasted accident duration is produced. The plots, mean absolute error (MAE), R2 and mean absolute percentage error (MAPE) have been chosen for model evaluation.

The plots are used to show the relationship between the forecasted duration and the actual duration. When the plot configuration shows linear relationship on the diagonal vector, this result indicates that most differences between the forecasted duration and the actual duration are small.

The MAE has been chosen for each test example to present the difference between actual and forecasted duration, as shown in equation 2. MAE value is smaller, and the forecasted accident duration is closer to the actual accident duration.

$$MAE = \frac{1}{M}\sum_{k=1}^{M}\left|\hat{x}(k) - x(k)\right| \qquad \ldots (2)$$

Where

M : Total number of examples

k : The k_{th} example

$\hat{x}$: Forecasted accident duration

x : Actual accident duration

The R^2 is shown in equation 3. The R^2 means the explanation capability from inputs to output. R^2 is between 0 and 1. As the R^2 is closer to 1, and the information supplied by inputs is more helpful to output.

Table 5. Criteria of MAPE for model evaluation

MAPE (%)	Assessment
<10	Highly accurate forecasting
10-20	Good forecasting
20-50	Reasonable forecasting
>50	Inaccurate forecasting

$$R^2 = \frac{\sum_{k=1}^{M}\left(\hat{x}(k) - \bar{x}(k)\right)^2}{\sum_{k=1}^{M}\left(x(k) - \bar{x}(k)\right)^2} \qquad (3)$$

$\bar{x}$: Mean of actual accident duration

The MAPE has been chosen as the primary criterion for model evaluation, as shown in equation 4.

$$MAPE = \frac{1}{M}\sum_{k=1}^{M}\left|\frac{\hat{x}(k) - x(k)}{x(k)}\right| \times 100\% \qquad (4)$$

Typical MAPE values for data and their assessment are shown in Table 5 (Lewis, 1982). As the MAPE becomes closer to 0, the forecasted value becomes more accurate.

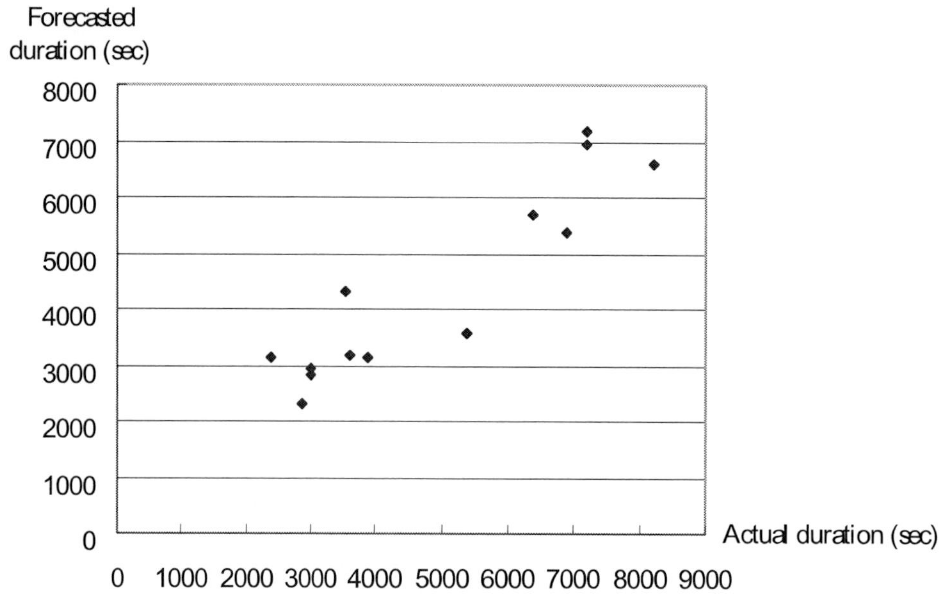

Figure 8. Plot for both actual and forecasted durations (static forecast).

Table 6. Performance of static forecast

MAE(sec)	585.838
R^2	0.896
MAPE(%)	12.371
Computing time (sec)	10.235

4. Model Evaluation Results

4.1. Static Forecast

Static forecast adopts Model A to forecast the accident duration when an accident is being notified for the first time. Figure 8, the plot, shows the relationship between the forecasted duration and the actual duration. The plot configuration shows a linear relationship on the diagonal vector, indicating that the forecasted accident duration may reasonably represent the actual accident duration. Near the two extremities of the diagonal vector, the plots are bilateral and balance the error.

Table 6 shows the performance of static forecast by MAE, R^2, MAPE and computing time. In terms of MAE, the difference between the actual accident duration and forecasted accident duration is lower than 10 minutes. R^2 almost approach 0.9 which shows good forecast. The MAPE value is well below 13%.

4.2. Semi-dynamic Updated Forecast and Dynamic Updated Forecast

In this study, Model B provides a dynamic updated forecast for accident duration. In contrast to dynamic updated forecasting, semi-dynamic updated forecasting provides the forecasted duration of each time point by subtracting the past time from the forecasted duration of Model A. The definition of past time is the time from accident notification to the time point of forecast. Figure 9 shows the plot for actual duration and forecasted duration of one accident at each time point of forecast from the accident notification to the accident road clearance and traffic recovery. With time past, the forecasted accident duration decreased gradually and the forecasted accident duration with dynamic update get closer to the actual duration. This result shows the dynamic update of accident duration model can revise the prediction according to the real-time traffic data.

The plot for both actual and forecasted duration with semi-dynamic update forecast is shown in Figure 10. The plots of Figure 10 are semi-dynamic updated forecast which subtracting the past time from the forecasted duration of Model A in the Figure 8. The plot configurations roughly show a linear relationship on the diagonal vector. Most predictions plot on the right side of diagonal vector. Through this semi-dynamic update method, the forecasted accident durations are lower than the actual accident duration.

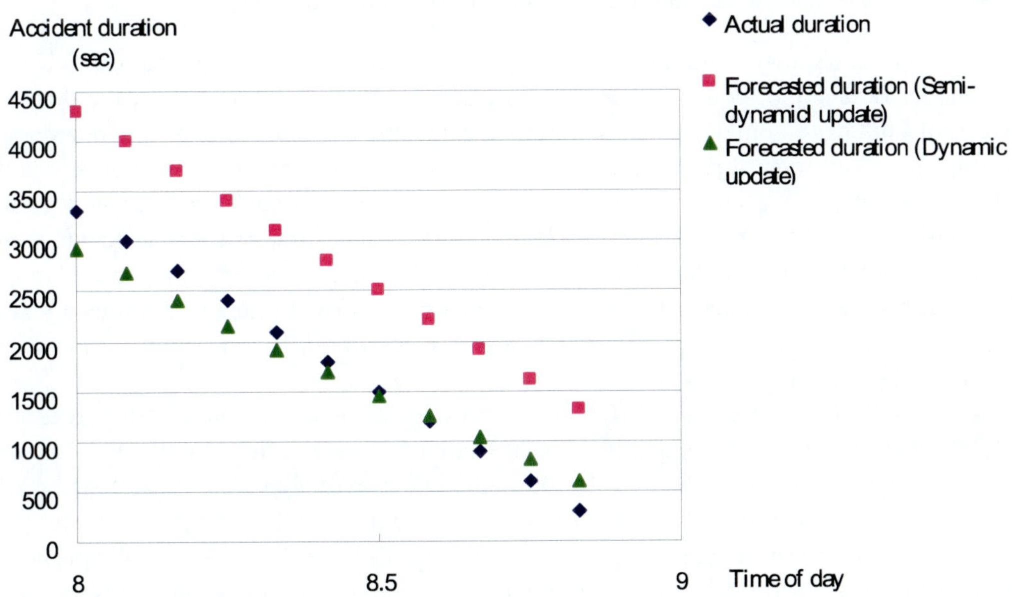

Figure 9. Plots of actual durations and forecasted durations at each time point.

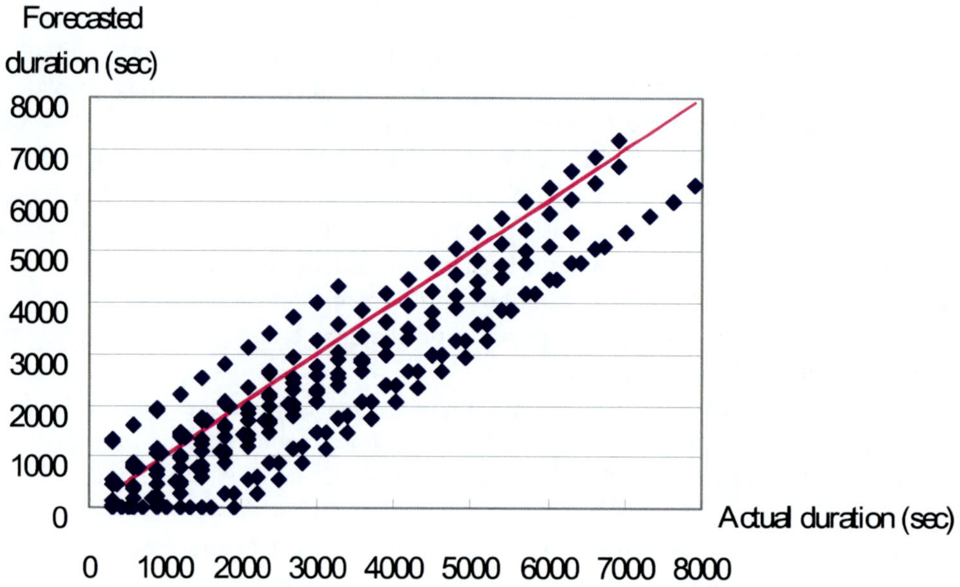

Figure 10. Plot for both actual and forecasted durations (Semi-dynamic update).

Figure 11 shows the plots both for actual and forecasted duration with dynamic update by using Model B. The plot configurations roughly show a linear relationship on the diagonal vector, but still have some unsuitable forecasted results. Several outliers are not on the diagonal vector and some forecasted durations are much smaller than the actual durations. In the end stage of some accidents, the traffic condition recovers vary fast, but it does not completely back to the normal traffic condition. Traffic congestion of these accidents is not very serious, but the accident durations are long in the database. Therefore, the forecast model provides a short prediction according to the un-congested traffic. Generally, Figure 11 shows that dynamic update method could consider the changes of traffic condition to adjust the prediction value at each time point.

Table 7 shows the performance of semi-dynamic update and dynamic update forecast by MAE, R^2, MAPE and computing time. The R^2 value of model effects with semi-dynamic updated method is higher than the R^2 value of model effects with dynamic updated method. In terms of the MAE and MAPE, the model effects of dynamic updated method are better than the model effects of semi-dynamic updated method. Semi-dynamic update prediction which uses only Model A to provide forecasted duration is simple and convenient. However, the updated prediction does not consider the real time traffic conditions after accident notification. Dynamic update forecasting is more costly for modeling and data collection, but it can provide more accurate prediction for accident duration. Thus, trade-offs exists between these two approaches.

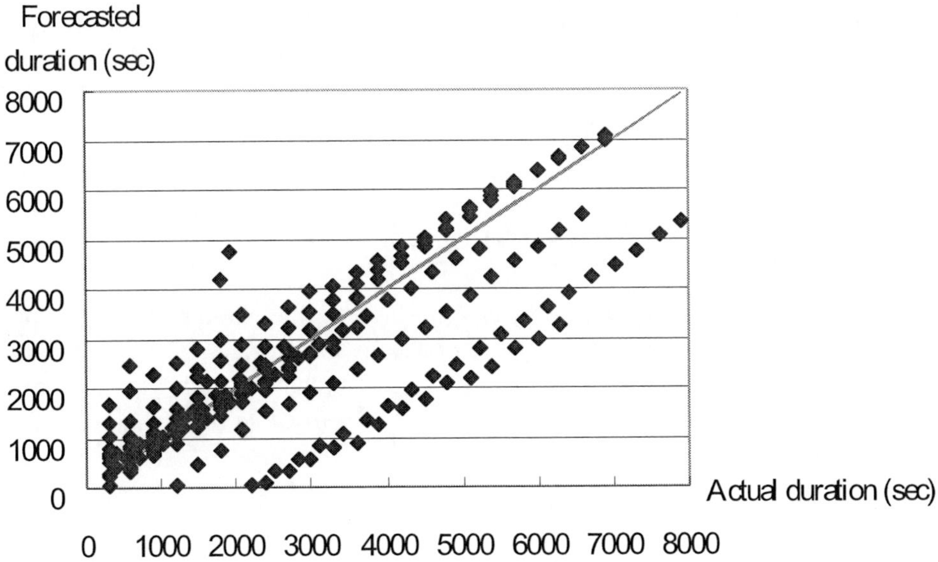

Figure 11. Plot for both actual and forecasted durations (Dynamic update).

Table 7. Performance of semi-dynamic update and dynamic update forecast

	Semi-dynamic updated	Dynamic updated
MAE(sec)	784.204	**751.625**
R^2	**0.824**	0.695
MAPE(%)	42.638	**34.995**
Computing time- fusion(sec)	**0.000**	16.678

SUMMARY

This study presents two accident duration models for multi-period forecast applying Artificial Neural Networks. Using sequential approach to conduct this dynamic update accident duration forecasting model is an innovation approach. In terms of the plot, MAE, R^2 and MAPE, the proposed models have shown reasonable model performance. The plot represents the relationship between the forecasted duration and actual duration and the plot configuration shows linear relationship on the diagonal. The MAPE percentages are mostly under 43%. This shows that the developed model fits the actual accident duration well during the accident, and that Artificial Neural Networks effectively smooth data noise of the model.

In terms of the model effect, the accident characteristics, VD data, time relationship, space relationship, and geometry characteristics are feasible as the inputs of accident duration forecasting model.

The proposed two models can provide the forecasted accident duration with the updated traffic data from the accident notification to the accident road clearance sequentially and periodically. This is one of the characteristics in this study.

With these two models, the estimated duration can be provided by plugging in relevant traffic data as soon as an incident is noted. The travelers and traffic management units can generally realize the impact by the forecasted incident duration. From the assessment of model effects, this study shows that the proposed models are feasible in the Intelligent Transportation Systems (ITS) context.

ACKNOWLEDGMENTS

This study is based on a research project sponsored by the National Science Council of Taiwan R.O.C. [NSC93-2218-E-006-094]. We gratefully acknowledge the data providers for this research, including Taiwan Area National Freeway Bureau and Police Radio Station. The helpful comments of Dr. Steven Chien at the New Jersey Institute of Technology, USA, are sincerely appreciated.

REFERENCES

Chen, M., Chien, I. J., 2001. Dynamic freeway travel time forecast using probe vehicle data: link-based vs. path-base. *Transportation Research Record* 1768. 157-161.

D'Angelo, M. P., Al-Deek, H. M., Wang, M. C., 1999. Travel time prediction for freeway corridors. *Transportation Research Record* 1676. 184-191.

Garib, A., Radwan, A. E., Al-Deek, H., 1997. Estimating magnitude and duration of incident delays. *Journal of Transportation Engineering*. 123(6), 459-466.

Krikke, R. 2002. Short-range travel time prediction using an artificial neural network. In *Proceedings of the 9th ITS World Congress* (CD-ROM), Chicago, USA.

Lewis, C. D., 1982. *Industrial and Business Forecasting Method.* Butterworth Scientific, London.

Nam, D., Mannering, F., 2000. An exploratory hazard-based analysis of highway incident duration. Transportation Research Part A. 34(2), 85-102.

Small, K. A., 1997. Economic and urban transportation policy in the United States. *Regional Science and Urban Economics*. 27(6), 671-691.

Wei, C. H., Lee, Y., 2003. Development of freeway travel time forecasting models using artificial neural networks. In *Proceedings of the 10th ITS World Congress* (CD-ROM), Madrid, Spain.

Wei, C. H., Lee, Y., 2005. Appling data fusion techniques to traveler information services in highway network. *Journal of the Eastern Asia Society for Transportation Studies.* 5, 2457-2475.

Wei, C. H., Lee, Y., 2007. Data fusion and feature composition approach to sequential accident duration forecasting. *Journal of the Eastern Asia Society for Transportation Studies* 7, 2546-2560.

Wei, C. H., Lee, Y., 2007. Sequential forecast of incident duration using artificial neural network models. *Accident Analysis and Prevention* 39(5), 944-954.

Zhang, X., Rice, J. A., 2003. Short-term travel time prediction. *Transportation Research Part* C. 11(3-4), 187-210.

Zurada, J. M., 1992. *Introduction to Artificial Neural Systems.* West Publishing company, St. Paul.

In: Road Traffic: Safety, Modeling and Impacts
Editors: S. E. Paterson and L. K. Allan, pp. 367-383
ISBN 978-1-60456-884-4

Chapter 8

OPTIMIZATION ALGORITHMS FOR SIGNALIZED ROAD NETWORK DESIGN PROBLEM

Suh-Wen Chiou*

Department of Information Management, National Dong Hwa University
1, Sec. 2, Da Hsueh Rd., Shou-Feng, Hualien 97401 Taiwan

ABSTRACT

A signalized road network design problem is to find the optimal signal settings and link capacity expansions while taking into account the route choice of users. This problem can be formulated as a bi-level optimization problem by taking the user equilibrium traffic assignment as a constraint. At the upper level an optimization problem of signal timings and link capacity expansions for road networks is determined, among which the signal timing plan for coordinated fixed time control is defined by the common cycle time, the start and duration of greens. At the lower level the user equilibrium traffic assignment obeying Wardrop's first principle can be formulated as a minimization problem where the link travel time function is defined as the sum of the undelayed travel time on the link and the average delay incurred by vehicles at the downstream end of the link. In this chapter, the methods solving the signalized road network design problem are investigated via numerical calculations on real data road network. Improvement on a locally optimal search by combining the technique of parallel tangents with conjugate gradient projections is particularly investigated. A Hybrid Search heuRistic (HSR) is therefore proposed by combining the technique of parallel tangents with conjugate gradient projections. As it shows, the proposed HSR method achieved substantially better performance than did conventional approaches when solving the signalized road network design problem with equilibrium flows.

Key words: bi-level optimization, signalized road network, equilibrium constraints, conjugate gradients

* E-mail:chiou@mail.ndhu.edu.tw

1. INTRODUCTION

A signalized road network design problem with expansion of link capacity is considered where the sum of total travel times and investment cost of link capacity expansion is minimized. The decision variables are the set of link capacity expansions and signal settings. In past decades, many researchers via the techniques of optimization have investigated the network design problem with equilibrium flows [1, 3, 5, 13, 21, 26]. In order to determine the signal settings, a good performance value can be achieved by optimizing a chosen objective function, in which the users' behavior of choice of routes is supposed to follow Wardrop's first principle. That is, a road user will choose his or her route between a specified origin-destination pair with the minimal travel cost, which is in turn dependent on the choice of the network signal settings and link capacity expansions. Therefore, in the optimization process for network flows, not only the signal settings and link capacity expansions need to be considered but also the consequential effects on the equilibrium flows caused by the signal settings and link capacity expansions need to be taken into account as well. Abdulaal and LeBlanc [1] were the first ones who proposed the *Hooke-Jeeves (HJ)* method to solve the equilibrium network design problem directly via continuous decision variables and obtained feasible solutions, which only confined to a small-scaled test network in practice due to the computation limit. Suwansirikul, Friesz and Tobin [22] proposed a simple heuristic called the *Equilibrium Decomposed Optimization (EDO)* and performed this heuristic on several example networks. The computing efficiency of *EDO* results from the decomposition of the network design problem into a set of interacting optimization problems and at iteration only one user equilibrium flow needs to be calculated when updating the improvements of all links of the network. Using the results from sensitivity analysis for equilibrium network flows (Tobin and Friesz [23]), Suh and Kim [21], Clegg et al. [11] and Meng, Yang and Bell [17] employed the bilevel programming technique to tackle a road network design problem. For example, in Meng et al. [17] an A*ugmented Lagrangian Algorithm (ALA)* is presented to solve a continuous network design problem and simple calculations are made in moderate networks with demanding computational efforts. Recently, Ceylan and Bell [5] presented a genetic algorithm to solve a signal timing optimization problem taking drivers' route choice into account. Chiou [7-9] also presented new algorithms to solve a continuous transportation network design problem where users route choice was modeled as a mathematical optimization problem.

Consider the optimization problem of signalized road networks, *TRANSYT model* [24] has been widely recognized as one of the most useful tools in studying the optimization of signalized road networks. As it has been widely indicated, the equilibrium flows and travel times are strongly influenced by the operation of signals. Ways of using mathematical programming to solve the problem of signal settings and network flow have been well researched. As far as the ways in optimizing signal settings for road networks are concerned, Allsop and Charlesworth [2] proposed a mutually consistent calculation, which was also known as the *Iterative Optimization Assignment (IOA)* for the signal setting minimization problem and equilibrium network flows. The resulting mutually consistent signal settings and link flows will, however, in general be a non-optimal solution as has been discussed by Gershwin and Tan [14] and Dickson [12]. Smith [20] presented a capacity-maximizing policy where a conventional signal-setting problem can be formulated as Wardrop equilibrium.

Heydecker and Khoo [15], on the other hand, proposed a *Linear Constraint Approximation (LCA)* to the equilibrium flows and solved a signal settings problem as a constrained optimization problem in a sequence of linear approximation. Yang and Yagar [26] proposed a *sensitivity analysis based (SAB)* algorithm to solve the signalized road network design problem where a linearized sub-problem for the objective function of the signalized road network design problem is formulated via the results of sensitivity analysis from Tobin and Friesz [23] and is solved by the simplex method. Cipriani and Fusco [10] presented a new algorithm which applies the Armijo step length rule to general projected gradient algorithm for a global signal setting problem with user equilibrium network flows. Investigations on the shape of the objective function together with algorithm properties are conducted via numerical calculations on a small test network. Furthermore, Cascetta, Gallo and Montella [4] discussed models and algorithms for the optimization of urban signal settings with stochastic traffic assignments where numerical tests are reported on a small-scale real network. Because of the non-convexity of the signalized road network design problem, solution algorithms mentioned above can solve the signalized road network design problem only locally.

A network design problem is a problem of finding the optimal design parameters while taking into account the route choice of users. This problem can be formulated as a special case of the Optimization Program with Equilibrium Constraints (OPEC) problem, which has been extensively discussed by Luo, Pang and Ralph [16], Shimizu, Ishizuka and Bard [19] and Outrata, Kocvara and Zowe [18]. In this paper, an OPEC program for a signalized road network with equilibrium flows subject to link capacity expansion is investigated, among which the coordinated signal timing plan is defined by the common cycle time, the start and duration of greens. The performance index is defined as the sum of a weighted linear combination of rate of delay and number of stops per unit time for all traffic streams and the investment for link capacity expansions. A new *Hybrid Search heuRistic (HSR)* is proposed to efficiently determine the optimal signal settings and link capacity expansions for signalized road network design problem. A hybrid search heuristic combines the technique of parallel tangents with conjugate gradient projections. By doing so, a good descent direction for locally optimal search can be determined along which the value of objective function is greatly improved. A benchmark real data road network is illustrated for investigating the effectiveness of *HSR* in solving the signalized road network design problem with expansions of link capacity. As it shows, the proposed *HSR* outperforms conventional approaches with consistent effectiveness in computations and stability when various initial data are taken into account.

The rest of the chapter is organized as follows. In next section, notation and problem formulation are given for the signalized road network design problem subject to link capacity expansions. In Section 3, a new hybrid search is proposed by combining the technique of parallel tangents with conjugate gradient projections. The Karush- Kuhn-Tucker (KKT) points are identified where a near optimum can be found effectively. In Section 4, numerical computations are conducted on a real data Sioux Falls network where promising results demonstrate the performance and robustness of the proposed method. Conclusions and discussions for this paper are made in Section 5.

2. Problem Formulation

The problem for solving a signalized road network design problem with link capacity expansions subject to travelers' route choice can be formulated as an OPEC problem in the following way. First, a generalized traffic assignment with decision variables of link capacity expansions and signal settings is presented. Next, an OPEC program is presented where the KKT system of equations is considered. Notation used throughout this paper is summarized below.

2.1. Notation

$G(N, L)$	a directed road network, where N is the set of nodes and L is the set of links.
y	the vector of link capacity expansions.
u	the vector of upper bound for link capacity expansions.
η	the conversion factor from investment cost to travel times.
$G(y)$	the vector of investment costs.
$\Psi = (\zeta, \theta, \phi)$	the set of signal setting variables, respectively for the reciprocal of cycle time, start and duration of greens.
Λ_a	the duration of effective green for link a.
g_{jm}	the minimum green for signal group j at junction m.
γ	the clearance time between the end of green for group j and the start of green for incompatible group l at junction m.
$\Omega_m(j,l)$	a collection of numbers 0 and 1 for each pair of incompatible signal groups at junction m; where $\Omega_m(j,l) = 0$ if the start of green for signal group j proceeds that of l, and $\Omega_m(j,l) = 1$, otherwise.
D_a	the rate of delay on link a.
S_a	the number of stops per unit time on link a.
T	the travel demands for origin-destination (OD) pairs.
f	the vector of path flows.
q	the vector of link flows.
ν	saturation flow.
δ	the link-path incidence matrix.
Δ	the OD-path incidence matrix.
c	the vector of link flow travel time.
C	the vector of path flow travel time.

2.2. User Equilibrium Traffic Assignment

A general formulation for a user equilibrium traffic assignment problem can be expressed in terms of a variational inequality. To find values q such that

$$c^t(q)(z-q) \geq 0 \tag{1}$$

for all $z \in K = \{q : q = \delta f, T = \Delta f, f \geq 0\}$ where superscript t is matrix transpose operator. The equivalent variational inequality in terms of path flow can be given as follows. To find values f such that

$$C^t(f)(z-f) \geq 0 \tag{2}$$

for all $z \in K' = \{f : T = \Delta f, f \geq 0\}$

2.3. Parametric Variational Inequality

For user equilibrium traffic assignment problem with link capacity expansions y and signal settings $\Psi = (\zeta, \theta, \phi)$, the parametric variational inequality for (1) can be expressed as follows.

$$c^t(q, y, \Psi)(z - q(y, \Psi)) \geq 0 \tag{3}$$

for all $z \in K(y, \Psi) = \{q(y, \Psi) : q(y, \Psi) = \delta f(y, \Psi), T = \Delta f(y, \Psi), f(y, \Psi) \geq 0\}$

Similarly, the parametric variational inequality in terms of path flow for (2) can be expressed as follows.

$$C^t(f, y, \Psi)(z - f(y, \Psi)) \geq 0 \tag{4}$$

for all $z \in K'(y, \Psi) = \{f(y, \Psi) : T = \Delta f(y, \Psi), f(y, \Psi) \geq 0\}$

2.4. Sensitivity Analysis

Suppose (f, y, Ψ) solves problem (4), the KKT system, H, of equations of parametric variational inequality (4) can be expressed. Let $w = (f, \pi, \mu)$ and the associated Lagrange multipliers π and μ, we have

$$H(w, y, \Psi) = 0 \tag{5}$$

$$\boldsymbol{\pi} \geq 0 \tag{6}$$

i.e.

$$C(f, y, \Psi) - \boldsymbol{\pi} - \Delta^t \boldsymbol{\mu} = 0 \tag{7}$$

$$diag(\boldsymbol{\pi}) f(y, \Psi) = 0 \tag{8}$$

$$\Delta f(y, \Psi) - T = 0 \tag{9}$$

$$\boldsymbol{\pi} \geq 0 \tag{10}$$

where $diag(\cdot)$ denotes the diagonal matrix. Suppose we work on path with positive flow, i.e. $f > 0, \boldsymbol{\pi} = 0$, the KKT system of equations (7)-(10) can be re-expressed as

$$C(f, y, \Psi) - \Delta^t \boldsymbol{\mu} = 0 \tag{11}$$

$$\Delta f(y, \Psi) - T = 0 \tag{12}$$

The first order sensitivity results for (11)-(12) with respect to link capacity expansions and signal settings are as follows. Let $\boldsymbol{\varepsilon} = \begin{pmatrix} y \\ \Psi \end{pmatrix}$ and $w = (f, \boldsymbol{\mu})$, it implies

$$\nabla_{\boldsymbol{\varepsilon}} H(w, y, \Psi) + \nabla_w H \nabla_{\boldsymbol{\varepsilon}} w = 0 \tag{13}$$

where $\nabla_w H = \begin{pmatrix} \nabla C & -\Delta^t \\ \Delta & 0 \end{pmatrix}$ and $\nabla_{\boldsymbol{\varepsilon}} H = \begin{pmatrix} \nabla_{\boldsymbol{\varepsilon}} C \\ 0 \end{pmatrix}$. Therefore the first order partial derivatives of equilibrium flow and associated Lagrange multiplier with respect to the decision variables are of the following form.

$$\nabla_{\boldsymbol{\varepsilon}} w = -\nabla_w H^{-1}(w, y, \Psi) \nabla_{\boldsymbol{\varepsilon}} H(w, y, \Psi) \tag{14}$$

2.5. Delay-minimizing Problem with Link Capacity Expansions

A delay-minimizing signalized road network design with link capacity expansions can be formulated as follows. Let P be the performance index for the signalized road network design problem, which is expressed via function P_0 in terms of link capacity expansions $\mathbf{y}$, signal timings $\boldsymbol{\Psi}$ and network flow $\mathbf{f}$. The delay-minimizing problem can be expressed as to

$$\underset{\mathbf{y},\boldsymbol{\Psi},\mathbf{f}}{Min} \quad P = P_0(\mathbf{y},\boldsymbol{\Psi},\mathbf{f}) \tag{15}$$

subject to $\boldsymbol{\Psi} \in \boldsymbol{\Pi}$, $\mathbf{0} \leq \mathbf{y} \leq u$

and $\mathbf{f} \in K'(\mathbf{y},\boldsymbol{\Psi})$

where the set $\boldsymbol{\Pi}$ defines the constraints of signal settings $\boldsymbol{\Psi}$, which can be expressed as a system of linear inequalities for minimum green, maximum cycle time and capacity constraints as in (16-19) below and u denotes the maximum expansions of link capacity. For a signalized road network, let θ_{jm} and ϕ_{jm} respectively denote the start and duration of green at junction m with signal group j and let $\zeta_{\min}$ and $\zeta_{\max}$ constrain the feasible bound for the common cycle time, the set $\boldsymbol{\Pi}$ can be detailed as follows.

$$\zeta_{\min} \leq \zeta \leq \zeta_{\max} \tag{16}$$

$$g_{jm}\zeta \leq \phi_{jm} \leq 1, \qquad j = 1,2,... \tag{17}$$

where g_{jm} denotes the minimum green time. The feasible demand on a link is bounded by the capacity

$$q_a \leq v_a \boldsymbol{\Lambda}_a \tag{18}$$

and the clearance time for incompatible signal groups j and l can be expressed as

$$\theta_{jm} + \phi_{jm} + \gamma\zeta \leq \theta_{lm} + \boldsymbol{\Omega}_m(j,l) \tag{19}$$

where γ and $\boldsymbol{\Omega}_m(j,l)$ respectively denotes the clearance time and binary variable set of 0 and 1, in which 0 if start of green for signal group j precedes that of l, and 1 otherwise. In problem (15), $K'(\mathbf{y},\boldsymbol{\Psi})$ defines the solution set of equilibrium flows when $\boldsymbol{\Psi}$ and $\mathbf{y}$ are specified. Since the equilibrium traffic assignment can be expressed as a KKT system, the problem (15) can be regarded as a standard non-linear problem and therefore is solved by normal non-linear programming techniques. Also, the performance index P is taken to be a linear combination of the rate of delay and the number of stops per unit time together with the investment cost, which can be defined as follows. Let W_D and W_S be the weighting factors for the rate of delay and the number of stops per unit time and M_D, M_S the corresponding monetary factors common to all links, thus

$$P_0(\mathbf{y},\boldsymbol{\Psi},\mathbf{f}) = \sum_a D_a W_D M_D + S_a W_S M_S + \eta G_a(y_a) \tag{20}$$

In constraints (17)-(19), let $\mathbf{A}_i$ and $\mathbf{a}_i$ be the junction-specific matrix and vector with junction i respectively related to the coefficients of $\boldsymbol{\theta}$, $\boldsymbol{\phi}$ and $\boldsymbol{\zeta}$. Also in constraint (16) let $\mathbf{a}_0$ denote the coefficient vector of $\boldsymbol{\zeta}$, thus the constraints (16)-(19) can be re-expressed as the following form.

$$\mathbf{A\Psi}^t \leq \mathbf{B}$$

where $\mathbf{A} = \begin{bmatrix} \mathbf{a}_0 & \mathbf{0} \\ \mathbf{a} & \mathbf{A}' \end{bmatrix}$ and $\mathbf{A}'$ is the diagonal supermatrix with $\mathbf{A}_i$. Let $\mathbf{B} = [\mathbf{b}_0, \mathbf{b}_\mathbf{i}]^t$ where $\mathbf{b}_i$ denotes a junction-specific vector with junction i related to constraints in (17)-(19) and $\mathbf{b}_0$ the constant vector for the common cycle time constraint as in (16). Following the results in sensitivity analysis, the problem (15) now can be rewritten as a single-level problem.

$$\underset{\mathbf{y},\mathbf{\Psi},\mathbf{f}}{Min} \quad P = P_0(\mathbf{y}, \mathbf{\Psi}, \mathbf{f}) \tag{21}$$

subject to $\mathbf{C}(\mathbf{f}, \mathbf{y}, \mathbf{\Psi}) - \mathbf{\Delta}^t \boldsymbol{\mu} = \mathbf{0}$ (22)

$$\mathbf{\Delta f}(\mathbf{y}, \mathbf{\Psi}) - \mathbf{T} = \mathbf{0} \tag{23}$$

$$\mathbf{0} \leq \mathbf{y} \leq \mathbf{y}_{\max} \tag{24}$$

$$\mathbf{A\Psi}^t \leq \mathbf{B} \tag{25}$$

3. Solution Method

In this section, a Hybrid Search heuRistic (HSR) combines the technique of parallel tangents with conjugate gradient projections is developed in order to efficiently find a good descent direction along which the value of objective function in problem (21)-(25) can be greatly improved. By conducting the HSR approach, a search direction of descent can be efficiently generated and a new iterate is created. The search process will be terminated at a Karush-Kuhn-Tucker (KKT) point where a good local optimal solution for problem (21)-(25) can be identified. In the following, a conjugate gradient projection method is firstly introduced to obtain a descent search direction. Applying the first order partial derivatives given in Chiou [6] for the performance index in (21), the gradients of (21) can be presented as follows at (y_0, Ψ_0, q_0).

$$\nabla_y P(y_0, \Psi_0) = \nabla_y P_0(y_0, \Psi_0, q_0) + \nabla_q P_0(y_0, \Psi_0, q_0) \nabla_y q(y_0, \Psi_0) \tag{26}$$

$$\nabla_{\Psi}P(y_0,\Psi_0)=\nabla_{\Psi}P_0(y_0,\Psi_0,q_0)+\nabla_q P_0(y_0,\Psi_0,q_0)\nabla_{\Psi}q(y_0,\Psi_0) \tag{27}$$

3.1. Conjugate Gradient Projections

Lemma 1 <Conjugate gradient directions> Consider a sequence of iterates $\left\{\beta_k=\begin{pmatrix}y_k\\ \Psi_k\end{pmatrix}\right\}$, $k=1,2,3,...$, solving a unconstrained problem $\Theta(\beta)$ according to

$$\beta_{k+1}=\beta_k+l_k d_k \tag{28}$$

where d_k is a search direction and l_k is the step length which minimizes the objective function $\Theta(\beta)$ along d_k from β_k. Let $d_1=-\nabla\Theta(\beta_1)$, d_k can be decided by

$$d_k=-\nabla\Theta(\beta_k)+\frac{\nabla\Theta^t(\beta_k)\nabla\Theta(\beta_k)}{\nabla\Theta^t(\beta_{k-1})\nabla\Theta(\beta_{k-1})}d_{k-1},\qquad k=2,3,..... \tag{29}$$

Then for the sequence of points $\{\beta_k\}$ generated by the conjugate gradient directions,

$$\Theta(\beta_k)>\Theta(\beta_{k+1}),\qquad k=1,2,3,....... \tag{30}$$

whenever $\nabla\Theta(\beta_k)\neq 0$.

Proof. Multiply equation (29) by $\nabla\Theta(\beta_k)\neq 0$, it becomes

$$\nabla\Theta^t(\beta_k)d_k=-\nabla\Theta^t(\beta_k)\nabla\Theta(\beta_k)+\frac{\nabla\Theta^t(\beta_k)\nabla\Theta(\beta_k)}{\nabla\Theta^t(\beta_{k-1})\nabla\Theta(\beta_{k-1})}\nabla\Theta^t(\beta_k)d_{k-1} \tag{31}$$

Since $\nabla\Theta^t(\beta_k)d_{k-1}=0$ by choosing the step length value l_k minimizing $\Theta(\beta_k)$ along the search direction d_k, the equation (31) becomes

$$\nabla\Theta^t(\beta_k)d_k=-\|\nabla\Theta(\beta_k)\|^2<0,\qquad k=1,2,3,.... \tag{32}$$

Thus for sufficiently small ε, $\varepsilon>0$, we have

$$\Theta(\beta_k)>\Theta(\beta_k+\varepsilon d_k) \tag{33}$$

Because by definition l_k is the step length which minimizes Θ along d_k from β_k, it implies

$$\Theta(\beta_k) > \Theta(\beta_k + \varepsilon d_k) \geq \Theta(\beta_k + l_k d_k) = \Theta(\beta_{k+1}) \tag{34}$$

which completes this proof. □

The direction generated by (29) for an unconstrained nonlinear problem is a descent direction which strictly decreases the objective function $\Theta(\beta)$ provided that the corresponding gradient value is not zero. In order to efficiently identify the feasible points for problem (21)-(25), we apply the conjugate gradient directions to a linear constraint set of problem (21)-(25) where a projected matrix is employed in the following manner.

Theorem 2. <A Conjugate gradient projection approach> In problem (21)-(25), a sequence of feasible iterates $\{\beta_k\}$ can be generated according to

$$\beta_{k+1} = \beta_k + l_k H_k d_k \tag{35}$$

where d_k is the conjugate gradient direction determined by (29) and l_k is the step length minimizing P along d_k for which the decision variable β_k is within the feasible region. Suppose that M_k has full rank at β_k, which is the gradient of active constraints in problem (21)-(25) and the projection matrix H_k is of the following form. Let I denote identity matrix,

$$H_k = I - M_k^t (M_k M_k^t)^{-1} M_k \tag{36}$$

A new search direction $\widetilde{d}_k$ can be determined in the following form.

$$\widetilde{d}_k = H_k d_k \tag{37}$$

Then the sequence of feasible points β_k generated by the conjugate gradient projection approach monotonically decreases the performance value,

$$P(\beta_k) > P(\beta_{k+1}), \qquad k = 1,2,3,... \tag{38}$$

whenever $H_k \nabla P(\beta_k) \neq 0$ and $\nabla P(\beta_k)$ is from (26)-(27) where $\beta_k = \begin{pmatrix} y_k \\ \Psi_k \end{pmatrix}$.

Proof. Following the results of *Lemma 1,* we have

$$\nabla P^t(\beta_k) d_k = -\nabla P^t(\beta_k) \nabla P(\beta_k) < 0, \qquad k = 1,2,3,.... \tag{39}$$

Multiply equation (39) by the projection matrix H_k, it becomes

$$\begin{aligned}&\nabla P^t(\beta_k)H_k d_k \\ &= -\nabla P^t(\beta_k)H_k \nabla P(\beta_k) \\ &= -\nabla P^t(\beta_k)H_k^t H_k \nabla P(\beta_k) \\ &= -\left\|H_k \nabla P(\beta_k)\right\|^2 < 0\end{aligned} \tag{40}$$

Thus for sufficiently small ε, $\varepsilon > 0$, we have

$$P(\beta_k) > P(\beta_k + \varepsilon\tilde{d}_k) \tag{41}$$

Because by definition l_k is the step length which minimize P along $\tilde{d}_k$ from β_k, it implies

$$P(\beta_k) > P(\beta_k + \varepsilon\tilde{d}_k) \geq P(\beta_k + l_k\tilde{d}_k) = P(\beta_{k+1}) \tag{42}$$

which completes this proof. □

Theorem 3. In *Theorem 2* when $H_k \nabla P(\beta_k) = 0$, if all the Lagrange multipliers corresponding to the active constraint gradients in problem (21)-(25) are positive or zeros, it implies the current β_k is a KKT point. Otherwise choose one negative Lagrange multiplier, say ς_j, and construct a new $\overline{M}_k$ of the active constraint gradients by deleting the *jth* row of M_k, which corresponds to the negative component ς_j, and make the projection matrix of the following form

$$H_k = I - \overline{M}_k^t(\overline{M}_k\overline{M}_k^t)^{-1}\overline{M}_k \tag{43}$$

The search direction then is determined by (37) and the results of *Theorem 2* hold.

Proof. Let ς_j be a negative component of the Lagrange multiplier and H_k defined in (36), we show $H_k \nabla P(\beta_k) \neq 0$. By contradiction, suppose

$$H_k \nabla P(\beta_k) = 0 \tag{44}$$

and let $\omega_k = -\left(\overline{M}_k\overline{M}_k^t\right)^{-1}\overline{M}_k \nabla P(\beta_k)$, then (44) can be rewritten as

$$0 = \nabla P(\beta_k) + \overline{M}_k^t\omega_k \tag{45}$$

For any $\varsigma_j < 0$, there exists a corresponding *j*th row, r_j of the active constraint in problem (21)-(25) and $\hat{\omega}_k$ such that

$$0 = H_k \nabla P(\beta_k)$$
$$= \nabla P(\beta_k) + \overline{M}_k^t \hat{\omega}_k + \varsigma_j r_j^t \tag{46}$$

We subtract (46) from (45) and it follows

$$0 = \overline{M}_k^t (\omega_k - \hat{\omega}_k) - \varsigma_j r_j^t \tag{47}$$

since $\varsigma_j \neq 0$ which contradicts the assumption that $\overline{M}_k$ has full rank. Thus $H_k \nabla P(\beta_k) \neq 0$.

□

Corollary 4 <Stopping condition> If $\beta_k = \begin{pmatrix} y_k \\ \Psi_k \end{pmatrix}$ is a KKT point for problem (21)-(25) then the search process may stop; otherwise a new descent direction at β_k can be generated according to *Theorems 2-3*. □

3.2. A Hybrid Search Heuristic (HSR)

In order to identify feasible points for problem (21)-(25) efficiently, a hybrid search heuristic combining the technique of parallel tangents with conjugate gradient projections is proposed as follows.

Step 1. (Initials) Start with $\beta_k = \begin{pmatrix} y_k \\ \Psi_k \end{pmatrix}$, set index $k = 1$.

Step 2. (Find equilibrium flows and conduct sensitivity analysis) Solve a traffic assignment problem (3) or (4) with specified decision variable β_k. Obtain the first order partial derivatives via equation (14).

Step 3. (Perform the Conjugate Gradient Projection approach) Calculate the projection matrix and decide the search direction $\tilde{d}_k$ via (37). If $\tilde{d}_k = 0$, go to Step 7. If $k = 1$ go to Step 4; otherwise go to Step 5.

Step 4. Find optimal step length α_{opt} in search direction $\tilde{d}_k$ along which the objective function value is minimized. Let $\beta_{k+1} = \beta_k + \alpha_{opt}\tilde{d}_k$, and set $k \leftarrow k+1$ return to Step 2.

Step 5. Find optimal step length α_{opt} in search direction $\tilde{d}_k$ along which the objective function value is minimized. Set $\hat{\beta}_{k+1} = \beta_k + \alpha_{opt}\tilde{d}_k$. Go to Step 6.

Step 6. (Conduct parallel tangents line search) Let ς_j be the most negative component of the Lagrange multiplier ς, and let $\tilde{d}_{jk}$ and β_{jk} be respectively be the *j*th corresponding component of $\tilde{d}_k$ and β_k. Introduce

$$\underline{\lambda} = \min\left\{\frac{\beta_{jk}}{\left|\tilde{d}_{jk}\right|}, if \quad \tilde{d}_{jk} < 0, \forall j \in L\right\} \tag{48}$$

and

$$\bar{\lambda} = \min\left\{\frac{\varsigma_j - \beta_{jk}}{\tilde{d}_{jk}}, if \quad \tilde{d}_{jk} > 0, \forall j \in L\right\} \tag{49}$$

Let

$$\lambda_{\max} = \min\{\underline{\lambda}, \bar{\lambda}\} \tag{50}$$

and

$$\beta_{k+1} = \beta_{k-1} + \lambda^{*}(\hat{\beta}_{k+1} - \beta_{k-1}) \tag{51}$$

Find λ^{*} such that

$$P(\beta_{k+1}) = \underset{0 \le \lambda \le \lambda_{\max}}{Min}\{P(\beta_{k-1} + \lambda(\hat{\beta}_{k+1} - \beta_{k-1}))\} \tag{52}$$

Go to Step 7.

Step 7. If $H_k \nabla P(\beta_k) \neq 0$, let $k \leftarrow k+1$ and return to Step 2. Otherwise, check Lagrange multiplier ς vector. If $\varsigma \geq 0$ then β_k is KKT point and stop. Otherwise find ς_j the most negative component of vector ς. Set $k \leftarrow k+1$ and return to Step 2.

4. Numerical Calculations

In this section, numerical experiments are conducted on a real data Sioux Falls network with selected signalized junctions as shown in Figure 1. The proposed *HSR*, the non-optimal *iterative optimization assignment (IOA)* and the *sensitivity analysis based (SAB)* methods are conducted with two distinct sets of initials for the signalized road network design problem subject to link capacity expansions. The non-signalized link travel time and link investment cost functions used are adopted from Suwansirikul *et al.* [22, p 261-262] where the convex investment function form is adopted from Abdulaal and LeBlanc [1, p 28], together with the data input details. Computational results are shown in Table 1 where the results for *system optimal solutions (SO for short)* are computed on the basis of system optimizations (see Chiou [7]). As shown in Figure 1 10 sets of signalized junctions and candidate links for capacity expansions are taken into account in the following illustration. Performance index is expressed in terms of vehicles while the conversion value is referred to that used in the work of Suwansirikul *et al.* As it is observed in Table 1, the multiple local optima exist due to the non-convexity of the area traffic control network and evidently each method leads to a different solution.

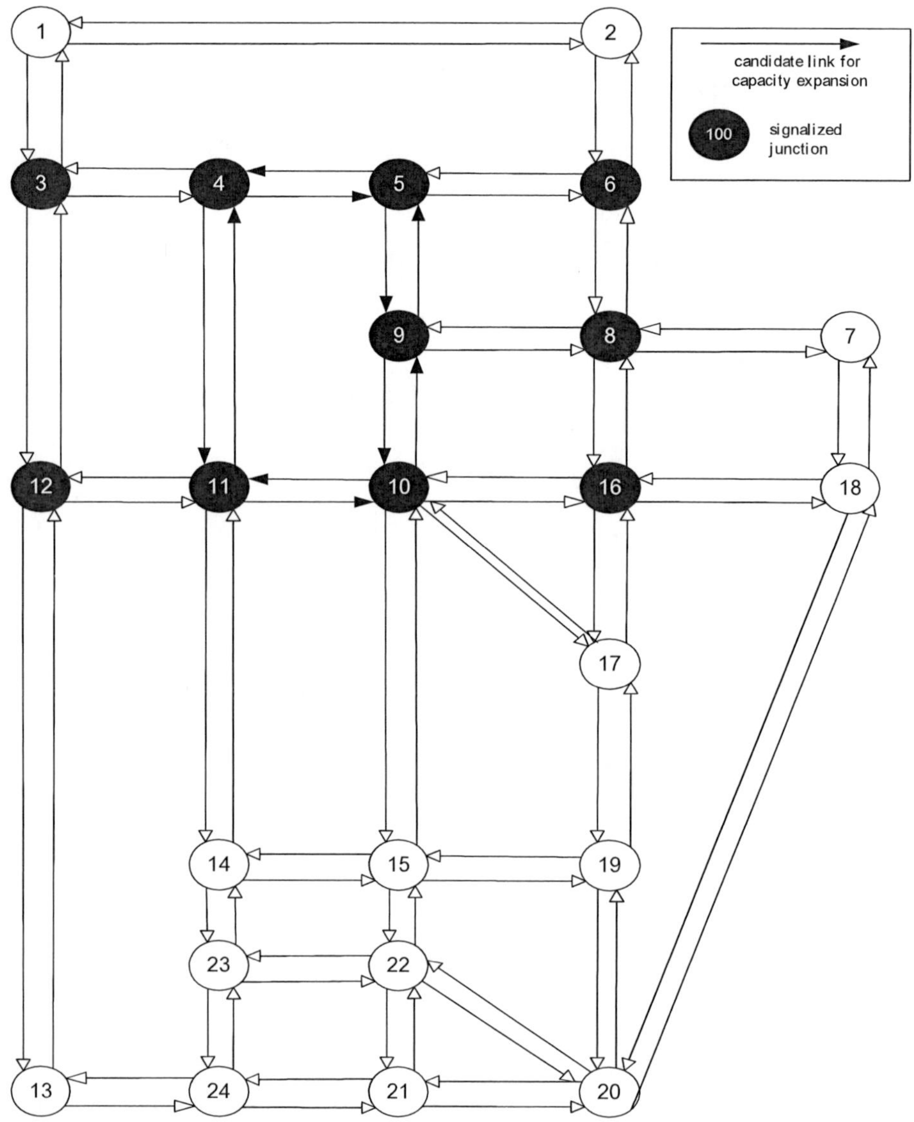

Figure 1. Sioux Falls network.

As it is seen in Table 1, the proposed heuristic *HSR* achieved the best performance with the value of 82.1 and 82 veh for the two different initials while the non-optimal solutions – *IOA*, achieved the worst performance with the highest values of 99.2 and 108.7 veh. The relative differences between the values achieved by the proposed *HSR* and *SO* are less than 1% for the two distinct sets of initials while those did by the *SAB* are over 15% and those did by the *IOA* are up to 30%. Regarding the computational efforts required in the computation processes for the area traffic control problem subject to link capacity expansion, as it seen from Table 1, the corresponding number of equilibrium assignments solved by the proposed *HSR* is far less than other traditional methods.

Table 1. Computational results for Sioux Falls network

variable/ algorithm	1st set				2nd set			
	IOA	SAB	SO	HSR	IOA	SAB	SO	HSR
Initial value of $1/\zeta$ (in sec)	72	72	72	72	118	118	118	118
Initial value of ϕ_{jm}/ζ (in sec)	28	28	28	28	50	50	50	50
Initial value of y_a	6.5	6.5	6.5	6.5	12	12	12	12
$1/\zeta$	110	110	110	110	132	132	132	132
ϕ_{13}/ζ	43	51	37	55	42	47	52	55
ϕ_{14}/ζ	38	50	39	51	40	49	50	52
ϕ_{15}/ζ	40	44	46	57	41	46	49	57
ϕ_{16}/ζ	39	49	43	52	44	41	48	50
ϕ_{18}/ζ	57	50	44	48	59	56	53	58
ϕ_{19}/ζ	52	49	51	50	55	54	57	55
$\phi_{1,10}/\zeta$	49	48	52	51	49	48	51	52
$\phi_{1,11}/\zeta$	45	51	53	55	53	54	55	55
$\phi_{1,12}/\zeta$	37	50	39	59	44	45	52	56
$\phi_{1,16}/\zeta$	47	50	49	52	48	49	51	52
$y_{(4,11)}$	4.6	3.9	5.4	4.9	5.7	3.9	5.4	4.9
$y_{(11,4)}$	3.9	3.8	5.3	5.2	1.6	3.8	5.3	5.2
$y_{(4,5)}$	1.2	2.6	1.6	1.9	5.6	2.6	1.8	1.9
$y_{(5,4)}$	1.5	2.7	1.7	2.6	1.6	2.7	1.7	2.1
$y_{(5,9)}$	2.3	2.6	2.8	2.2	3.1	2.9	2.8	2.3
$y_{(9,5)}$	2.4	2.9	2.7	2.8	3.8	2.9	2.7	2.7
$y_{(9,10)}$	5.6	4.3	5.7	5.7	7.6	4.3	5.7	5.7
$y_{(10,9)}$	4.8	4.4	4.9	5.5	3.8	4.4	4.9	5.5
$y_{(10,11)}$	5.4	4.7	4.3	4.7	7.3	4.7	4.8	4.5
$y_{(11,10)}$	6.2	4.8	4.3	4.1	3.6	4.8	4.3	4.7
PI (in veh)	99.2	95.34	81.7	82.1	108.7	94.4	81.5	82
#	78	69	50	24	80	70	53	27

where ϕ_{jm}/ζ denotes the duration of greens for signal group j at junction m measured in sec and $1/\zeta$ denotes the common cycle time measured in sec. PI denotes the performance index value measured in veh-h/h, and # denotes the number of equilibrium traffic assignment solved.

5. Conclusions and Further Issues

In this paper, we presented an improved heuristic *HSR* for a signalized road network design problem with link capacity expansions. The proposed method *HSR* has been illustrated successfully with promising results on the real data Sioux Falls road network as compared to traditional methods by consistently yielding better performance and less computational efforts. Consider further issues like solving the generalized user equilibrium assignments with asymmetric cost mappings are being taken into account and empirically computational experiments are being carried out.

Acknowledgements

The author is greatly appreciated for Taiwan National Science Council via grant NSC-96-2416-H-259-010-MY2.

References

[1] Abdulaal, M. & LeBlanc, L.J. (1979). Continuous equilibrium network design models. *Transportation Research,* 13B, 19-32.

[2] Allsop, R. E. & Charlesworth, J. A. (1977). Traffic in a signal-controlled road network: an example of different signal timings inducing different routeings. *Traffic Engineering Control*, 18, 262-264.

[3] Ban, J. X., Liu, H.X., Ferris, M. C. & Ran, B. (2006). A general MPCC model and its solution algorithm for continuous network design problem, *Mathematical and Computer Modelling*, 43, 493-505.

[4] Cascetta, E. Gallo, M. & Montella, B. (2006). Models and algorithms for the optimization of signal settings on urban networks with stochastic assignment models, *Annals of Operations Research,* 144, 301-328.

[5] Ceylan, H. & Bell, M.G.H. (2004). Traffic signal timing optimization based on genetic algorithm approach, including drivers routing, *Transportation Research Part B*, 38, 329-342.

[6] Chiou, S-W. (2003). TRANSYT derivatives for area traffic control optimization with network equilibrium flows. *Transportation Research Part B*, 37, 263-290.

[7] Chiou, S-W. (2005). Bilevel programming for the continuous transport network design problem, *Transportation Research Part B*, 39, 361-383.

[8] Chiou, S-W. (2007), Optimal design of signal-controlled road network, *Applied Mathematics and Computation*, 189(1), 1-8.

[9] Chiou, S-W. (2008), A hybrid approach for optimal design of signalized road network, *Applied Mathematical Modelling*, 32(2), 195-207.

[10] Cipriani, E. & Fusco, G. (2004) Combined signal setting design and traffic assignment problem, *European Journal of Operational Research* 155, 569-583.

[11] Clegg, J. Smith, M. J., Xiang, Y. & Yarrow, R. (2001). Bilevel programming applied to optimizing urban transportation, *Transportation Research Part B*, 35, 41-70.

[12] Dickson, T. J. (1981). A note on traffic assignment and signal timings in a signal-controlled road network, *Transportation Research- B* 15, 267-271.

[13] Friesz, T. L., Tobin, R. L., Cho, H-L & Mehta, N.J. (1990). Sensitivity analysis based algorithms for mathematical programs with variational inequality constraints. *Mathematical Programming*, 48, 265-284.

[14] Gershwin, G. B. & Tan, H. N. (1979). Hybrid optimization: optimal static traffic control constrained by drivers' route choice behavior, Massachusetts Institute of Technology, laboratory for Information and Decision System Report LIDS-p-870.

[15] Heydecker, B. G. & Khoo, T. K. (1990). The equilibrium network design problem. Proceedings of AIRO'90 Conference on Models and Methods for Decision Support, Sorrento, pp. 587-602.

[16] Luo, Z. Q., Pang, J-S. & Ralph, D. (1996) Mathematical Programs with Equilibrium Constraints. Cambridge University Press, Cambridge, U.K..

[17] Meng, Q. Yang, H. & Bell, M.G.H. (2001). An equivalent continuously differentiable model and a locally convergent algorithm for the continuous network design problem, *Transportation Research Part B* 35, 83-105.

[18] Outrata, J. Kocvara, M. & Zowe, J. (1998) Nonsmooth Approach to Optimization Problems with Equilibrium Constraints: Theory, Applications and Numerical Results. Kluwer Academic Publishers, Boston.

[19] Shimizu, K., Ishizuka, Y. & Bard, J. (1997). Nondifferentiable and Two-level Mathematical Programming. Kluwer Academic Publishers, Boston.

[20] Smith, M. J. (1980). A local traffic control policy which automatically maximizes the overall travel capacity of an urban road network. *Traffic Engineering and Control*, 21, 298-302.

[21] Suh, S. & Kim, T. J. (1992). Solving nonlinear bilevel programming models of the equilibrium network design problem: a comparative review. *Annals of Operations Research*, 34, 203-218.

[22] Suwansirikul, C., Friesz, T. L. & Tobin, R. L. (1987) Equilibrium decomposed optimization: a heuristic for continuous equilibrium network design problem, *Transportation Science* 21, 254-263.

[23] Tobin, R. L. & Friesz, T. L. (1988) Sensitivity analysis for equilibrium network flow, *Transportation Science* 22, 242-250.

[24] Vincent, R. A., Mitchell, A. I. & Robertson, D. I. (1980). *User guide to TRANSYT*, version 8. TRRL Report, LR888, Crowthorne, Transport and Road Research Laboratory.

[25] Yang, H. & Bell, M. G. H. (1998). Models and algorithms for road network design: a review and some new developments. *Transport Reviews,* 18, 257-278.

[26] Yang, H. & Yagar, S. (1995). Traffic assignment and signal control in saturated road networks. *Transportation Research Part A*, 29, 125-139.

In: Road Traffic: Safety, Modeling, and Impacts
Editors: S. E. Paterson and L. K. Allan, pp. 385-401
ISBN 978-1-60456-884-4

Chapter 9

SOCIAL COGNITIVE HUMAN FACTORS OF AUTOMOBILE DRIVING

Robert D. Mather
University of Central Oklahoma, USA

ABSTRACT

Social cognitive factors involved in automobile driving are generally understudied. Typically, human factors researchers who conduct research on driving examine the interaction of the driver and the vehicle, with emphasis on either the solitary driver or vehicle design. Social psychologists rarely look at the human factors of driving. However, as people must drive on roads together, social cognition plays a role in how people interact. Consequently, social cognition plays a central role in driving. Mather and DeLucia (2007) recently examined the interaction between social psychology and human factors of driving. Following their empirical study of implicit attitudes and pedestrian-vehicle collisions, the current chapter proposes various potential research topics on social cognitive human factors of automobile traffic safety. Basic and applied research in social psychology have much to contribute to research on the human factors of driving. Social interaction inside of the car can lead to distraction (e.g., cell phones; passenger interactions). Social interaction outside of the car can lead to death (e.g., teenagers tossing an item from one moving vehicle to another), injury (e.g., waving a car through when the other car is not clear), and saving lives (e.g., pointing to another driver's flat tire). Some possible areas of social psychological research that could contribute to research on the human factors of automobile driving include: motivation (e.g., need for closure), expectancies (e.g., second guessing another driver at a four-way stop), aggression (e.g., road rage), social facilitation (e.g., speeding up to pass another car or slowing down to keep from passing another car; general driving performance), attitudes and persuasion (e.g., increasing compliance with seat belt laws), and implicit racial attitudes (e.g., pedestrian-vehicle collisions). Potential contributions from these areas can be helpful to a single driver trying to drive safely and to anticipate danger from the road, obstacles, and other drivers. In summary, as automobile driving itself is an inherently social phenomenon, social psychological research is centrally relevant to research on driving. The current chapter examines in detail various social psychological research that is relevant to the human factors of automobile driving and traffic safety.

INTRODUCTION

Humans, because they live in groups and influence each other, are inherently social beings (Mather & Romo, 2007). We even drive on roads and crash into each other. The social nature of automobile driving is illustrated by Don Cheadle's character in the opening dialogue of the movie "Crash." Cheadle's character is a passenger in a vehicle that was just involved a rear-end collision initiated by another vehicle's driver. In reference to urban life, he states that, "...nobody touches you. We're always behind this metal and glass. I think we miss that touch so much that we crash into each other just so we can feel something" (Danbury & Haggis, 2004). While it is not likely that most traffic accidents are intentional (indeed, by definition an "accident" is not "intentional"), it does emphasize the fact that human beings are social and that automobile driving is not devoid of social interaction.

Social psychology is "an attempt to understand and explain how the thought, feeling, and behavior of individuals are influenced by the actual, imagined, or implied presence of other human beings" (Allport, 1954, p. 5)[11]. Social cognition is a social psychological perspective that draws upon the application of the methods and research of cognitive psychology to examine social psychological questions about how people make sense of people (Fiske & Taylor, 2008) within the context of cognition, motivation, and affect (emotion) (Kunda, 1999). The applied area of human factors "discovers and applies information about human behavior, abilities, limitations, and other characteristics to the design of tools, machines, systems, tasks, jobs, and environments for productive, safe, comfortable, and effective human use" (Sanders & McCormick, 1993, p. 5). While social psychology and human factors are very separate fields of study, they are each relevant to the task of automobile driving.

The logic of the argument in this chapter is relatively straightforward: 1) People drive automobiles on roads together, 2) social cognition plays a role in how people interact, and thus 3) social cognition plays a role in the human factors of automobile driving. Generally, the social cognitive factors involved in automobile driving are understudied. Typically, human factors researchers examine the interaction of the driver and the vehicle, with emphasis on either the driver alone or vehicle design. Social psychologists rarely look at the human factors of automobile driving, although a number of social psychology studies investigating other social phenomena have used driving situations (e.g., Doob & Gross, 1968; Kenrick & MacFarlane, 1986). One study that specifically examined social cognitive human factors of driving was conducted by Mather and DeLucia (2007), who examined the interaction between social psychology and the human factors of automobile driving. They examined the influence of the implicit racial attitudes of drivers on reaction times to pedestrians in driving simulations. This type of research illustrates that both basic and applied social psychology research can contribute to research on human factors in automobile driving.

[1] There is a distinction between "sociological social psychology," which draws upon literature and training in sociology, and "psychological social psychology," which draws upon literature and training in psychology. This distinction will be briefly explicated later in this chapter. This chapter uses "social psychology" to refer to "psychological social psychology."

Although people participate in many social interactions, social interaction that occurs inside of a car can be distracting to the automobile driver. For example, cell phone conversations that a driver engages in while driving divert attention away from the driving task (Strayer & Johnston, 2001). Conversations that a driver has with vehicle passengers also can divert attention away from the driving task (Strayer & Drews, 2007). Even children interacting with each other in the back seat of the vehicle can be distracting to a driver. These social interactions are important because distraction increases the risk of driver error.

Social interaction outside of the car can also contribute to driver error. Consider teenagers driving separate vehicles filled with passengers on an empty highway, playfully tossing an item from one moving vehicle to another. Such distractions could prove deadly. However, even well-meaning social interaction with other drivers can cause an accident. Consider a well-meaning driver stopped behind traffic on a two-lane, one-way access road. Let us suppose that the well-meaning driver leaves a space open so that an unsuspecting driver pulling out of a parking lot can turn right onto the access road (in front of the well-meaning driver). However, also consider that the unsuspecting driver in the parking lot is only turning right so he or she can get to the left turn lane ahead, meaning that he or she must turn into the far left lane when leaving the parking lot. Next, the unsuspecting car is waved through by the well-meaning car. However, if the unsuspecting car is not clear of traffic in the left lane, an accident might occur as the result of miscommunication during a social interaction between drivers. Despite this example, it is important to note that social interactions between drivers can have positive results as well. For example, police officers use their sirens and flashing lights to communicate to drivers that they are exceeding posted speed limits, and to inform drivers that they should vacate the roadway and to pull over to the shoulder of the road. This communication occurs in an effort to change the behavior of a driver who ideally will not exceed the posted speed limit in the future. Also, drivers will often point to other drivers to indicate that a driver has a flat tire of which he or she is not aware. This type of communication increases the safety of all drivers on the road by attempting to avoid the dangers of driving with a flat tire.

On the Concurrent Influence of Social Cognition and Human Factors

As referenced earlier, one of the most recognizable social interactions in the United States is the use of a cell phone while driving. Cell phone use has been shown to disrupt driving performance (Strayer & Johnston, 2001) and has even been shown to do so to a degree comparable to driving while intoxicated (Strayer, Drews, & Crouch, 2006). Using an eyetracker to record the movements and fixation times of the eyes, Strayer and Drews (2007) examined the use by participants of a hands-free cell phone while the participants drove in a simulated motor vehicle. They found evidence for inattention blindness, in which participants could not recall objects they had looked at while driving as effectively if they had been engaged in a conversation on the hands-free cell phone compared to those who were not engaged in a conversation on a hands-free cell phone. This occurred even when fixation time was controlled, meaning that degraded memory still existed for participants engaged in a conversation on a hands-free cell phone, even when they had looked at an object for the same amount of time as those who were not engaged in a conversation on a hands-free cell phone. They also found that participants did not reallocate their attentional resources based on an object's relevance to the safety of the driver. This failure to reallocate attention to safety is important because knowing where to look while driving helps the automobile driver to avoid risks (Pollatsek, Fisher, & Pradhan, 2006; Pollatsek, Narayanaan, Pradhan, & Fisher, 2006; Pradhan et al., 2005). Strayer and Drews also found psychophysiological evidence indicating that engaging in a conversation on hands-free cell phone degraded the encoding of the participants, as indicated by a decreased amplitude of the P300 event-related brain potential (ERP). Other evidence demonstrated that being engaged in a conversation with a passenger in the front seat leads to more effective driving behavior than being engaged in a conversation on a hands-free cell phone. The reason for this is the passenger can alter the conversation to indicate danger, approaching turns, etc. Thus, we can infer that a driver engaged in a social interaction with a passenger in his or her own car is in less danger than a driver engaged in a social interaction on a cell phone with a person who is a passenger in the front seat of another car. Such a scenario would put the driver on the cell phone at risk and the driver/passenger of the other car at risk, due to the passenger's not attending to the road for the other driver. Certainly, social interactions are quite complex, and so it follows that their influence on automobile driving is appropriately complex.

Since driving is a social phenomenon (e.g., we wouldn't need traffic lights if we were the only drivers), social psychological research is relevant to driving research. Some possible areas of social psychological research that could contribute to research on human factors of driving include motivation, expectancies, aggression, social facilitation, attitudes and persuasion, and implicit racial attitudes. This chapter will briefly examine some of the social psychology research that is relevant to human factors of automobile driving. Much of this research is related to the social cognitive perspective on social psychology.

Motivation

Motivation involves defining a goal, choosing a course of action to achieve the defined goal, and carrying out the course of action in pursuit of achieving the defined goal (Geen, 1995). In other words, motivation is what orients us to do something. For instance, a person's regulatory focus helps to dictate how they are strategically inclined, with a promotion focus orienting an individual towards positive things and a prevention focus orienting an individual away from negative things (Higgins, 1998). Werth and Forster (2007) found that prevention focus (both as a personality trait and as a manipulated variable) enhanced braking speed. Thus, regulatory focus is a potentially important motivational factor that should be examined in the study of the human factors of automobile driving.

Why do people do things? People behave as they do for many different reasons. Some of the internal motivating factors that have been described by social psychologists as unique individual differences include the need for cognition (Cacioppo, Petty, Kao, & Rodriguez, 1986), the need to belong (Baumeister & Leary, 1995), the need for cognitive closure (Kruglanski, Webster, & Klem, 1993), and the need to be accurate (Chen & Chaiken, 1999; Petty & Cacioppo, 1986). In particular, the need for cognitive closure seems like it might be directly relevant to driving behaviors.

The need for cognitive closure (Kruglanski, Webster, & Klem, 1993; Kruglanski & Webster, 1996) is the level to which an individual feels that they must find a definitive answer to a question. Thus, if a person is high in need for cognitive closure, they will not tolerate uncertainty—they simply must come to a solution, regardless of whether it is right or wrong. The need for cognitive closure can be examined as an individual difference measure (people naturally vary in their stable levels of need for cognitive closure) or a manipulated variable (people with experimentally imposed deadlines effectively become high in need for cognitive closure). A person who is high in need for cognitive closure views the world as black or white and doesn't handle uncertainty well. A person who is low in need for cognitive closure can handle uncertainty better.

Since a person who is high in need for cognitive closure is willing to be decisive regardless of whether or not the decision choice is correct, perhaps this motivation has an influence on driving behavior. A driver who is high in need for cognitive closure may feel the need to complete a merge into traffic without exploring the option of slowing down and waiting, simply for the sake of being decisive and eradicating the uncertainty that accompanies exploring other potential options. It is also possible that when a driver is in close proximity to several other moving vehicles on the highway at a high rate of speed, the dynamic nature of traffic necessitates a shift to need for cognitive closure—when traveling at a high rate of speed, a driver doesn't have the luxury of being able to quietly sit down and weigh the options for their vehicular movements. Thus, the specific driving situation could induce a need for cognitive closure. In such an instance, individual differences in need for cognitive closure would not influence performance on the task, although a mild induction of cognitive closure might augment existing tendencies. It is worth examining motivation from both the perspective of individual differences, as well as the role of the driving situation, in order to determine its influence on automobile driver behavior.

EXPECTANCIES

Our social behaviors do not occur in the isolation of our minds, and they certainly are not characterized by objective, rational thought. As humans, we bring our expectancies along to help make sense of our social world. That is, we interpret information based on the previous experiences and intuitive theories that we have developed over the course of our lives.

Sometimes a driver's memory is faulty due to expectations. Loftus and Palmer (1974) demonstrated that estimates of the speed of an automobile in an accident were influenced by inquiries that used different verbs. For example, after watching a film of an accident, the question "How fast were the cars going when they *smashed* into each other?" elicited higher estimates of automobile speed than the question "How fast were the cars going when they *hit* each other?" Additionally, subjects tested a week after viewing the accident were more likely to misremember broken glass as having been present at the accident when they had been asked the question with the verb of "smashed" a week earlier rather than the verb "hit".

Loftus, Miller, and Burns (1978) conducted a study examining eyewitness memory for an automobile accident involving a pedestrian. Participants viewed a series of 30 slides at 3 second a piece. The slides showed the collision occurring after the car either drove through a stop sign or a yield sign, and participants subsequently answered questions and selected the slides that they had seen from various pairs. Half of the participants were asked the question of, "Did another car pass the red Datsun while it was stopped at the stop sign?" while the other half of the participants were asked, "Did another car pass the red Datsun while it was stopped at the yield sign?" Participants were nearly twice as accurate in selecting the correct slides when the question embedded among other questions in the task between the presentation and paired-choice was congruent with what they had really seen. That is, many participants had memories that had been rewritten by the seemingly inconsequential question. This indicates that memory for what happens in a driving encounter can be malleable.

While expectancies can influence our perceptions, conformity can lead those who misperceive to the same erroneous conclusions based on their informational reliance on another observer's misperceptions. For instance, the autokinetic effect is a perceptual illusion that occurs when an individual stares at a spot of light in a dark room and sees it moving, though the spot of light does not actually move. Sherif's (1966) studies on the autokinetic effect found that when the autokinetic effect was tested in groups, the groups of people had different norms of actually reporting the existence of light movement. Additionally, Asch's (1952) studies showed that people are more likely to conform when faced with a unanimous majority, but will break from the group when they have a fellow dissenter.

There are several implications of conformity for driving behavior. First, if everyone else is exceeding the posted speed limit, a driver may conform to the norm of excessive speeding and will thus be more likely to speed. Second, if everyone else misperceives who is at fault in an accident (because of their expectations), it can be speculated that a person with the correct information (that is contrary to what the other witnesses believe that they perceived) will be unlikely to volunteer the information. Since people are very poor at understanding and reporting their own cognitive processes (Nisbett & Wilson, 1977), it is unlikely that most people would be aware of the biasing influences on their perceptions (Wilson & Brekke, 1994) in reporting information about an accident.

Additionally, expectations influence what people generally think about other people's interpersonal behaviors, intentions, characteristics, capabilities, and outcomes (Reich, Casa de Calvo, & Mather, 2008). For instance, when a driver reaches a four-way stop around the same time as other drivers, they must negotiate who will proceed through the intersection and in what order the vehicles will do so. The rule to yield to the car on the right is pointless if they all arrive simultaneously, or if the interpretations of when the vehicles actually arrive create differential perceptions within each of the drivers as to what the order should be. Necessarily, each individual driver is left to try and guess the intentions of the other drivers at the four-way stop. That is, each driver must attempt to predict the behavior of the other drivers. A driver might not trust other people in general, so he or she might wave the other driver through the intersection. Conversely, a driver might be quite trusting of other people in general, and wave the other drivers through the intersection. Finally, a driver with high generalized interpersonal expectations might decide to be assertive and proceed through the intersection, assuming that the others will yield to him or her. What motivates a driver in this decision? It is quite possible that such decisions are motivated by their expectancies of others, particularly their interpersonal expectancies. Future research is needed to examine the role of interpersonal expectancies in automobile driving behavior.

AGGRESSION

One driving behavior that has garnered much attention in the past few years but which has been around as long as drivers have shared roads is aggression. Aggression is defined by Geen (1990) as having three main elements: the delivery of a noxious stimulus to a victim, the intent to harm the victim, and the expectation that the behavior will harm the victim. Frustration, defined as "interference with the occurrence of goal responses at their accustomed time in the response sequence" (Berkowitz, 1969, p. 67) can lead to aggression.

As long as drivers compete for resources such as space, and do so as a function of time, aggressive drivers will stalk the roads. Indeed, Parry (1968) attempted to examine aggression in drivers, indicating that the issue itself is not a new problem. Some forms of aggression are relatively harmless, such as honking a horn to prompt a fellow driver or angrily muttering to a passenger about another driver. Other forms are more insidious, such as physical altercations resulting from undesired driver interactions on the road. Two high profile examples of road rage in the United States include Mike Tyson and Bob Gibson. Tyson, the former Heavyweight Boxing Champion, was convicted in 1999 of two counts of misdemeanor assault after a road rage incident (Branch, 1999). Gibson, a member of the National Baseball Hall of Fame, was involved in a road rage incident in 2002 ("Hall of Fame," 2002).

But what is road rage? Galovski, Malta, and Blanchard (2006) closely examined road rage and aggression in drivers. They suggested that the most common forms of aggressive driving include both verbal aggression (e.g., shouting insults) and gestural aggression (e.g., honking, making obscene gestures). Vehicular aggression (e.g., a driver using his or her vehicle to block other vehicles, following too closely or tailgating) is less common. Physical aggression (e.g., throwing objects, shooting another driver) occurs even less frequently. Galovski et al. found evidence that internationally, aggressive driving has a prevalence of fewer than 25% of all drivers. They suggested that longitudinal research is needed to answer the question of whether or not aggressive driving is increasing in prevalence. Why does it matter if drivers are aggressive or not? The answer is that aggressive driving contributes to motor vehicle fatalities.

How do the uses of the terms hostility, anger, and aggression differ in the research literature? Galovski et al. (2006) found that these terms were not consistently defined in their usage. Hostility generally referred to negative cognitive components that are associated with preemptive acts of interpersonal aggression, while anger generally dealt with the emotional component. For example, two individual difference measures can be used to examine anger toward other drivers. The Driving Anger Scale (DAS; Deffenbacher, Oetting, & Lynch, 1994) measures the amount of anger experienced while driving, and the Driving Vengeance Questionnaire (Wiesenthal, Hennesy, & Gibson, 2000) measures a driver's use of vengeance (revenge) to perceived threats in driving situations.

Deffenbacher, Deffenbacher, Lynch, and Richards (2003) used computer simulations to examine individual differences among drivers in terms of anger and aggression. They found that high anger drivers became angry more frequently than low anger drivers in a driving simulation, as well as in real life driving situations. Additionally, high anger drivers reported riskier driving behavior and more frequent loss of concentration, close calls, and moving violations compared to low anger drivers. Thus, anger as an individual difference measure was related to cognitive, affective and behavioral measures of driving behavior. Deffenbacher

(2008) found no evidence that driving anger differs for urban and rural drivers, but did find evidence that different driving situations evoked more anger, such as rush hour traffic situations.

Road rage is an obvious escalation of aggression in drivers. But what leads to road rage? Smaller aggressive acts of driving may escalate into road rage. One aggressive act that is common in driver interactions is the honking of an automobile horn. Additionally, temperature has been previously demonstrated to be related to aggression (Anderson, Anderson, Dorr, DeNeve, & Flanagan, 2000; Baron & Bell, 1976; Reifman, Larrick, & Fein, 1991). Kenrick and MacFarlane (1986) conducted a study of horn honking behavior in which the experimenters positioned a female confederate in a car at a one-lane exit of an intersection in a Phoenix, AZ, USA residential area, at which the traffic light had a 12 second green light. The study was conducted in the spring and summer. The confederate pulled the car to the intersection and remained there in front of exiting cars for 12 seconds while the observer recorded the honking behavior of the unsuspecting driver of the automobile stuck behind the confederate. The latency of the honk, number of honks, and total time honking were recorded by the observer. The researchers examined drivers who had their windows rolled down, with the rationale that they were not using their air conditioner during the trial, as well as drivers who had their windows rolled up. The researchers found that as the temperature outside of the automobile increased, horn honking also increased. This was only found to be the case for participants who had their windows rolled down. That is, temperature was positively correlated with horn honking.

Doob and Gross (1968) conducted a similar field study in which they varied the status of the car and driver that served as frustrating objects on the road. Using a high status car with high status driver attire and a low status car with low status driver attire, they found that drivers were more likely to honk at the low status car than they were to honk at the high status car, and men honked faster than women. Interestingly, two participants who found themselves behind the low status car at the intersection did not have their trials counted as aggressive because Doob and Gross operationally defined aggression as horn honking. These two automobiles actually hit the bumper of the low status car while attempting to prompt the low status car to proceed through the intersection. Thus, the low status car elicited extreme forms of aggression that were not included in the study!

In Salt Lake City, Utah, USA, Turner, Layton, and Simons (1975) conducted three field studies of honking behavior based on the paradigm used by Doob and Gross (1968). In their first study, Turner et al. found that women (92%) and men (58%) answered affirmatively to the statement, "If someone suddenly turns without signaling, I get annoyed." Additionally, men (77%) and women (56%) both answered affirmatively to the statement "I swear under my breath at other drivers." Thus, the majority of drivers who were sampled admitted that other drivers have the power to annoy them and that they swear (albeit privately) at other drivers. This research provides further evidence that driving is indeed a social event.

In their second study, Turner et al. (1975) replicated the Doob and Gross (1968) study with an additional manipulation in which a rifle was placed in a gun rack of a purportedly stalled older model pickup truck along with a bumper sticker with an aggressive label ("Vengeance") or nonaggressive label ("Friend"). They examined the reactions of 92 male drivers of relatively newer cars that were less than six years of age. Additionally, a "victim visibility" manipulation existed, in which a curtain in the back of the pickup truck either obstructed the view of the driver of the pickup truck (but not the gun rack) or did not obstruct the view of the driver. Results indicated that the closed curtain increased honking, and the rifle and vengeance bumper sticker combination increased honking when the curtain was closed, but not when it was open. Using a newer vehicle that was purportedly stalled, a third study examining both men and women separated the effects of the rifle from the bumper sticker. This study found that male drivers of new vehicles (high status) were more likely to honk with the rifle and vengeance bumper sticker combination than any of the other conditions, while this same condition produced lower rates of honking for male drivers of old vehicles (low status). Turner et al. speculated that lower status drivers might have inhibited their aggressive impulses in the presence of a high status vehicle decorated with aggressive stimuli.

Social Facilitation

The first social psychology study was conducted by Norman Triplett in 1897 (Coats & Feldman, 2001). Triplett's work on social facilitation indicated that the presence of others increased performance (Triplett, 1897). Social facilitation occurs when a person performs better on easy tasks and worse on difficult tasks in the presence of other people. Research has indicated that social facilitation occurs due to arousal (Bond & Titus, 1983) and also due to the opportunity for evaluation by other people (Aiello & Douthitt, 2001). Social facilitation has been shown to occur for impression formation (Thomas, Skitka, Christen, & Jurgena, 2002), and has even been demonstrated in species other than humans, such as cockroaches (Zajonc, Heingartner, & Herman, 1969).

There are several conditions under which social facilitation could potentially play a role in automobile driving behavior. For instance, traffic consists of other drivers, creating a situation in which other people are present. Compared to a driver in little or no traffic, a driver who merges into heavy traffic might speed up more than necessary to pass another car, or decelerate more than necessary to keep from passing another car. The driver's performance might be influenced by the mere presence of other drivers. Both of these cases would present potential dangers to the driver and to other drivers. If the merging lane ends abruptly, mistaking the magnitude of necessary acceleration and deceleration could result in collision with a wall or another car. Naturally, the complexity of the task increases with the addition of other cars in traffic. Each car in traffic is another time-to-contact (TTC) that must be estimated by the driver to successfully execute the merge. TTC refers to the rate of optical expansion of the other cars (DeLucia, Kaiser, Bush, Meyer, & Sweet, 2003). However, it is possible that the mere presence of other drivers can contribute to the driver's ability to execute the complex task.

Another straightforward hypothesis in which social facilitation might influence driving behavior is the possibility that social facilitation should degrade the performance of new drivers that have a passenger in the car who acts as a "backseat driver." A backseat driver would increase both the arousal of the novice driver, as well as create a situation in which the novice driver's performance is evaluated. In fact, social facilitation has been shown to be related to rated performance on driver's licensure tests (Rosenbloom, Shahar, Perlman, Estreich, & Kirzner, 2007). Additionally, the presence of a backseat driver should improve the performance of experienced drivers, assuming that the backseat drivers are not overly distracting.

Attitudes and Persuasion

An attitude is "a psychological tendency that is expressed by evaluating a particular entity with some degree of favor or disfavor" (Eagly & Chaiken, 1993, p. 1). An attitude is essentially an evaluation of something—Does a person like something or does he or she dislike it? An attitude is an evaluation with cognitive, emotional, and behavioral components (Mather & Romo, 2007). Persuasion is the act of changing an attitude (Petty, 1995).

Basic attitude and persuasion research from social psychology is used in applications of consumer psychology (Czellar, 2006) to advertising (Lucas & Britt, 1963), both for profit and for public service announcements (e.g., litter prevention; Cialdini, 2003). Indeed, Petty and Cacioppo (1996) discuss altruistic marketing as the use of basic behavioral research to contribute to society. As one of the most influential models of persuasion, the Elaboration Likelihood Model of Persuasion (Petty & Cacioppo, 1986) has found support for a central route to persuasion (persuasive processes requiring a great deal of thought) and a peripheral route to persuasion (persuasive processes requiring very little thought). When using the central route, people are influenced more by strong and high quality arguments. While using the peripheral route, people are more influenced by superficial cues such as the attractiveness of the messenger. The ELM has been used in AIDS prevention (Petty, Gleicher, & Jarvis, 1993) and drug abuse prevention (Petty, Baker, & Gleicher, 1991). Similarly, attitude and persuasion research is relevant to enforcing seat belt laws (Durbin, Smith, Kallan, Elliott, & Winston, 2007; Kim & Yamashita, 2007; Reinfurt, Williams, Wells, & Rodgman, 1996; Shin, Hong, & Waldron, 1999; Stasson & Fishbein, 1990; Trafimow & Fishbein, 1994; Ulmer, Preusser, Preusser, & Cosgrove, 1995), reducing driving under the influence of alcohol (Dula, Dwyer, & LeVerne, 2007), increasing driver acceptance of distraction mitigation strategies (Donmez, Boyle, Lee, & McGehee, 2006), reducing personal car use (Eriksson, Garvill, & Nordlund, 2008), and increasing airbag safety (Nelson, Sussman, & Graham, 1999). The ELM is useful in crafting effectively persuasive messages for many driver safety issues.

Leon Festinger (1957) proposed the theory of cognitive dissonance in which people strive to have consistency (consonance) among their attitudes, beliefs, and behaviors. When these elements are not consistent with each other, an aversive state of dissonance occurs. Changing one of the elements is one way to reduce the dissonance, such as changing an attitude to match a behavior. Stone et al. (1997) examined cognitive dissonance in a study on HIV prevention and condom usage. Results indicated that participants were more likely to use condoms after being induced with a hypocrisy in which they indicated personal reasons for which they had failed to use condoms in the past and then subsequently recorded a videotaped speech promoting the use of condoms for safe sex. The dissonance was aroused due to the hypocrisy and a behavior (condom purchase) was influenced. In an attempt to generalize the findings to aid in the reduction of road rage incidents, Takaku (2006) also applied a hypocrisy manipulation to examine the influence of dissonance arousal on the reduction of negative emotions. Similarly, such a paradigm could be used to examine hypocrisy conditions for individuals who believe that cell phone use while driving is dangerous, but who continue to engage in the risky behavior nonetheless.

Implicit Racial Attitudes

In presenting the argument for implicit social cognition, Greenwald and Banaji (1995) proposed the concept of an implicit attitude. Implicit attitudes are essentially evaluative associations that are based on past experience and are not open to our awareness, while explicit attitudes are the attitudes of which we are aware. Implicit attitudes are measured with implicit measures, such as reaction times, that are relatively impervious to social desirability effects. Explicit measures are measured with explicit measures, such as self-report, that are subject to social desirability effects. Implicit measures predict nonverbal behavior while explicit attitudes predict verbal behavior (Dovidio, Kawakami, Johnson, Johnson, & Howard, 1997; Fiske & Taylor, 2008).

One implicit measure that is widely used is the Implicit Association Test (IAT) (Greenwald, McGhee, & Schwartz, 1998), which is based on the fact that people respond faster to concepts that are more highly associated with each other than to those that are less associated. One of the first attempts to use social cognitive methods to assess implicit attitudes in drivers was conducted by Harre and Sibley (2007) using the IAT. Harre and Sibley used both implicit and explicit measures to assess driver's attitudes towards their own driving abilities relative to those of other drivers. They found that both implicit and explicit measures of attitudes regarding their own driving abilities predicted a driver's optimism about being in a crash.

Another recent application of social cognitive methods and theory to driving performance was conducted by Mather and DeLucia (2007). Mather and DeLucia proposed that racial differences in pedestrian-vehicle collisions may be due, in part, to contributions of implicit associations to reaction times as well as to effects of visual contrast that may differ due to skin color. Previous research had shown support for visual contrast as a factor in pedestrian conspicuity (Sleight, 1972), but no research had previously examined the effects of a driver's implicit racial attitudes on reaction times to pedestrians in a driving simulation. In driving simulations, Mather and DeLucia found evidence for visual contrast contributing to detection of pedestrian stimuli that had previously been shown to activate racial attitudes, even when the stimuli were presented to the participant below the threshold of awareness. Although they did not find evidence to support their hypothesis that implicit racial attitudes influenced reaction times in driving simulations, other methods of measurement may still find an effect of implicit racial attitudes. Another way of testing this hypothesis would be to use IAT scores to predict a participant's reaction times to pedestrians of different ethnicities and skin tones in a driving simulation, or to correlate explicit racial attitudes with driving performance. Regardless of measurement, the series of studies by Mather and DeLucia represented a scientific step towards integrating traditional social cognition research with human factors research on automobile driving.

CONCLUSION

Although each of the social psychological areas discussed in this chapter can be examined from the social cognition perspective, some are more frequently examined from such a perspective than others. Specifically, motivation, expectancies, attitudes/persuasion, and implicit racial attitudes are often examined from the social cognitive perspective within traditional social psychological literature. Potential contributions from each of these areas in social psychology can be helpful to automobile drivers trying to drive safely and anticipate danger from the road, obstacles, and other drivers. Particularly, social cognition research has the potential to contribute to the area of research concerned with human factors of automobile driving.

There are, however, publication difficulties of such interdisciplinary research. In this chapter, many of the specific social psychology studies that made interesting examinations of the interface between social psychology and automobile driving behavior were not published in the top mainstream social psychology journals (e.g., *Journal of Personality and Social Psychology*, *Personality and Social Psychology Bulletin*, *Journal of Experimental Social Psychology*), but rather were published in other outlets (e.g., *Journal of Applied Social Psychology*, *Environment and Behaviour*, *Psychological Reports*, *The Journal of Social Psychology*). The lack of such studies in the top mainstream publication outlets of social psychology indicate that the interdisciplinary application of basic social psychology findings to research on human factors of automobile driving is not of much importance to the field of social psychology. In this chapter, many of the applied psychology studies that made interesting examinations of the interface between social psychology and automobile driving behavior were found in *Accident Analysis and Prevention* and *Transportation Research Part F*. This indicates that the social aspect of human factors research of automobile driving is an important issue to the transportation research community. However, applied experimental psychology and research on the human factors of transportation safety are necessarily interdisciplinary research ventures, as they focus on improving the complex task of automobile driving. Thus, it is important for the social element to be considered in research on automobile driving. Since automobile driving is a complex task composed of complex driver behaviors that include social interactions both within and between vehicles, it is important to take many elements of the automobile driving task into account and to develop theories that explore this complex task at different levels of analysis.

As an example of the importance of such theory development, Factor, Mahalel, and Yair (2007) recently made the argument that sociological explanation is necessary to appropriately understand automobile driver behavior. In their "Social Accident" model, they emphasized the study of the manner in which group factors such as culture and society influence the interaction between automobile drivers. Their premise was that many traffic accidents occur when drivers bring unique understandings of the rules of automobile driving to the social context of the road. They stated that, "the interaction between two or more drivers could be examined as a function of the reciprocal relationship between society and culture at the macro level and attitudes and behaviors of drivers at the micro level" (p. 915). Thus, their sociological model of traffic safety differs from the current social cognitive perspective in that their model examines the macro level of social interaction, while the social cognitive perspective proposed in this chapter examines the micro level of social interaction. Such a distinction in the unit of measurement is fundamental to understanding the divergent perspectives of "social psychology" as conceptualized by sociologists and psychologists. Sociologists emphasize theories about society, while social psychologists emphasize theories addressing individual susceptibility to social influence (Aronson, Wilson, & Akert, 2007). This chapter outlined a social cognitive perspective that emphasized the manner in which the social world of an individual automobile driver influences how a driver navigates a social interaction with other drivers and pedestrians. The logic of this chapter is based on the idea that people drive automobiles on roads together, and that social cognition plays a role in how people interact. Consequently, social cognition plays a central role in automobile driving. Social cognitive factors involved in automobile driving have not been studied sufficiently and have not been integrated into current discussions of transportation safety with regard to the human factors research of automobile driving. This chapter suggests that social cognition research is important to the understanding of the complex task of automobile driving and can subsequently contribute to traffic safety.

Acknowledgments

Preparation of this chapter was supported in part by a grant from the Joe C. Jackson College of Graduate Studies and Research at the University of Central Oklahoma. I also thank Jamie Gill for her assistance in the preparation of this chapter.

References

Aiello, J. R., & Douthitt, E. A. (2001). Social facilitation from Triplett to electronic performance monitoring. *Group Dynamics: Theory, Research, and Practice, 5*, 163-180.

Allport, G. W. (1954). The historical background of modern social psychology. In G. Lindzey (Ed.), *Handbook of social psychology* (pp. 3-56). Reading, MA: Addison Wesley.

Anderson, C. A., Anderson, K. B., Dorr, N., DeNeve, K. M., & Flanagan, M. (2000). Temperature and aggression. In M. P. Zanna (Ed.) *Advances in experimental social psychology* (Vol. 32, pp. 63-133). San Diego: Academic Press.

Aronson, E., Wilson, T. D., & Akert, R. M. (2007). *Social psychology* (6th ed.). Upper Saddle River, NJ: Pearson Prentice Hall

Asch, S. E. (1952). *Social psychology*. New York: Prentice Hall.

Baron, R. A., & Bell, P. A. (1976). Aggression and heat: The influence of ambient temperature, negative affect, and a cooling drink on physical aggression. *Journal of Personality and Social Psychology, 33*, 245-255.

Baumeister, R. F., & Leary, M. R. (1995). The need to belong: Desire for interpersonal attachments as a fundamental human motivation. *Psychological Bulletin, 117*, 497-529.

Berkowitz, L. (1969). Social motivation. In G. Lindzey & E. Aronson (Eds.), *The handbook of social psychology* (2nd ed., Vol. 3, pp. 50-135). Reading, MA: Addison Wesley.

Bond, C. F., Jr., & Titus, L. J. (1983). Social facilitation: A meta-analysis of 241 studies. *Psychological Bulletin, 94*, 264-292.

Branch, G. E. III. (1999, February 8). Tyson considers whether to appeal case, sentence. *USA Today*, p. 11C

Cacioppo, J. T., Petty, R. E., Kao, C. F., & Rodriguez, R. (1986). Central and peripheral routes to persuasion: An individual difference perspective. *Journal of Personality and Social Psychology, 51*, 1032-1043.

Chen, S., & Chaiken, S. (1999). The heuristic-systematic model in its broader context. In S. Chaiken & Y. Trope (Eds.), *Dual-process theories in social psychology* (pp. 73-96). New York: Guilford.

Cialdini, R. B. (2003). Crafting normative messages to protect the environment. *Current Directions in Psychological Science, 12,* 105-109.

Coats, E. J., & Feldman, R. S. (Eds.) (2001). *Classic and contemporary readings in social psychology* (3rd. ed.). Upper Saddle River, NJ: Prentice Hall.

Czellar, S. (2006). Self-presentational effects in the Implicit Association Test. *Journal of Consumer Psychology, 16*, 92-100.

Danbury, B. (Producer), & Haggis, P. (Director). (2004). *Crash* [Motion picture]. United States: Lion's Gate Entertainment.

Deffenbacher, J. L. (2008). Anger, aggression, and risky behavior on the road: A preliminary study of urban and rural differences. *Journal of Applied Social Psychology, 38*, 22-36.

Deffenbacher, J. L., Deffenbacher, D. M., Lynch, R. S., & Richards, T. L. (2003). Anger, aggression, and risky behavior: A comparison of high and low anger drivers. *Behavior Research and Therapy, 41*, 701-718.

Deffenbacher, J. L., Oetting, E. R., & Lynch, R. S. (1994). Development of a driving anger scale. *Psychological Reports, 74*, 83-91.

DeLucia, P. R., Kaiser, M. K., Bush, J. M., Meyer, L. E., & Sweet, B. T. (2003). Information integration in judgments of time to contact. *The Quarterly Journal of Experimental Psychology, 56A(7)*, 1165-1189.

Donmez, B., Boyle, L. N., Lee, J. D., & McGehee, D. V. (2006). Drivers' attitudes toward imperfect distraction mitigation strategies. *Transportation Research Part F, 9*, 387-398.

Doob, A. N., & Gross, A. E. (1968). Status of frustrator as an inhibitor of horn-honking responses. *The Journal of Social Psychology, 78*, 213-218.

Dovidio, J. F., Kawakami, K., Johnson, C., Johnson, B., & Howard, A. (1997). On the nature of prejudice: Automatic and controlled processes. *Journal of Experimental Social Psychology, 33*, 510-540.

Dula, C. S., Dwyer, W. O., & LeVerne, G. (2007). Policing the drunk driver: Measuring law enforcement involvement in reducing alcohol-impaired driving. *Journal of Safety Research, 38*, 267-272.

Durbin, D. R., Smith, R., Kallan, M. J., Elliott, M. R., & Winston, F. K. (2007). Seat belt use among 13-15 year olds in primary and secondary enforcement law states. *Accident Analysis and Prevention, 39*, 524-529.

Eagly, A. E., & Chaiken, S. (1993). *The psychology of attitudes.* Fort Worth, TX: Harcourt Brace.

Eriksson, L., Garvill, J., & Nordlund, A. M. (2008). Interrupting habitual car use: The importance of car habit strength and moral motivation for personal car use reduction. *Transportation Research Part F, 11*, 10-23.

Factor, R., Mahalel, D., & Yair, G. (2007). The social accident: A theoretical model and a research agenda for studying the influence of social and cultural characteristics on motor vehicle accidents. *Accident Analysis and Prevention, 39*, 914-921.

Festinger, L. (1957). *A theory of cognitive dissonance*. Evanston, IL: Row, Peterson, & Company.

Fiske, S. T., & Taylor, S. E. (2008). *Social cognition: From brains to culture*. New York: McGraw-Hill.

Galovski, T. E., Malta, L. S., & Blanchard, E. B. (2006). *Road rage: Assessment and treatment of the angry, aggressive driver*. Washington, D.C.: American Psychological Association.

Geen, R. G. (1990). *Human aggression*. Pacific Grove, CA: Brooks/Cole.

Geen, R. G. (1995). *Human motivation: A social psychological approach*. Pacific Grove, CA: Brooks/Cole.

Greenwald, A. G., & Banaji, M. R. (1995). Implicit social cognition: Attitudes, self-esteem, and stereotypes. *Psychological Review, 102*, 4-27.

Greenwald, A. G., McGhee, D. E., & Schwartz, J. L. K. (1998). Measuring individual differences in implicit cognition: The Implicit Association Test. *Journal of Personality and Social Psychology, 74*, 1464-1480.

Hall of Fame pitcher Gibson won't face charges in fight. (2002, March 9). *St. Petersburg Times*, p. 5C.

Harre, N., & Sibley, C. G. (2007). Explicit and implicit self-enhancement biases in drivers and their relationship to driving violations and crash-risk optimism. *Accident Analysis and Prevention, 39*, 1155-1161.

Higgins, E. T. (1998). Promotion and prevention: Regulatory focus as a motivational principle. In M. P. Zanna (Ed.) *Advances in experimental social psychology* (Vol. 30, pp. 1-46). San Diego: Academic Press.

Kenrick, D. T., & MacFarlane, S. W. (1986). Ambient temperature and horn honking: A field study of the heat/aggression relationship. *Environment and Behaviour, 18*, 179-191.

Kim, K., & Yamashita, E. Y. (2007). Attitudes of commercial motor vehicle drivers towards safety belts. *Accident Analysis and Prevention, 39*, 1097-1106.

Kruglanski, A. W., & Webster, D. M. (1996). Motivated closing of the mind: "Seizing" and "freezing". *Psychological Review, 103*, 263-283.

Kruglanski, A. W., Webster, D. M, & Klem, A. (1993). Motivated resistance and openness to persuasion in the presence or absence of prior information. *Journal of Personality and Social Psychology, 65*, 861-876.

Kunda, Z. (1999). *Social cognition: Making sense of people*. Cambridge, MA: MIT Press.

Loftus, E. F., Miller, D. G., & Burns, H. J. (1978). Semantic integration of verbal information into visual memory. *Journal of Experimental Psychology: Human Learning and Memory, 4*, 19-31.

Loftus, E. F., & Palmer, J. C. (1974). Reconstruction of automobile destruction: An example of the interface between language and memory. *Journal of Verbal Learning and Verbal Behavior, 13*, 585-589.

Lucas, D. B., & Britt, S. H. (1963). *Measuring advertising effectiveness*. New York: McGraw-Hill.

Mather, R. D., & DeLucia, P. R. (2007). Testing for effects of racial attitudes and visual contrast on the speed of a driver's response to a pedestrian. *Transportation Research Part F, 10*, 437-446.

Mather, R. D., & Romo, A. (2007). *Automaticity and cognitive control in social behavior.* Southlake, TX: Fountainhead.

Nelson, T. F., Sussman, D., & Graham, J. D. (1999). Airbags: An exploratory survey of public knowledge and attitudes. *Accident Analysis and Prevention, 31*, 371-379.

Nisbett, R. E., & Wilson, T. D. (1977). Telling more than we know: Verbal reports on mental processes. *Psychological Review, 84*, 231-259.

Parry, M. H. (1968). *Aggression on the road.* New York: Tavistock.

Petty, R. E. (1995). Attitude change. In A. Tesser (Ed.), *Advanced social psychology* (pp. 194-255). New York: McGraw-Hill.

Petty, R. E., Baker, S. M., & Gleicher, F. (1991). Attitudes and drug abuse prevention: Implications of the Elaboration Likelihood Model of persuasion. In L. Donohew, H. E. Sypher, & W. J. Bukoski (Eds.), *Persuasive communication and drug abuse prevention* (pp. 71-90). Hillsdale, NJ: Lawrence Erlbaum.

Petty, R. E., & Cacioppo, J. T. (1986). The elaboration likelihood model of persuasion. In L. Berkowitz (Ed.), *Advances in Experimental Social Psychology* (Vol. 19, pp. 123-205). New York: Academic Press.

Petty, R. E., & Cacioppo, J. T. (1996, February). Addressing disturbing and disturbed consumer behavior: Is it necessary to change the way we conduct behavioral science? *Journal of Marketing Research,* 1-11.

Petty, R. E., Gleicher, F., & Jarvis, W. B. G. (1993). Persuasion and AIDS prevention. In J. B. Pryor & G. D. Reeder (Eds.), *The social psychology of HIV infection* (pp. 155-182). Hillsdale, NJ: Lawrence Erlbaum.

Pollatsek, A., Fisher, D. L., & Pradhan, A. (2006). Identifying and remedying failures of selective attention in younger drivers. *Current Directions in Psychological Science, 15*, 255-259.

Pollatsek, A., Narayanaan, V., Pradhan, A., & Fisher, D. L. (2006). Using eye movements to evaluate a PC-based risk-awareness and perception training program on a driving simulator. *Human Factors, 48*, 447-464.

Pradhan, A. K., Hammel, K. R., DeRamus, R., Pollatsek, A., Noyce, D. A., & Fisher, D. L. (2005). Using eye movements to evaluate effects of driver age on risk perception in a driving simulator. *Human Factors, 47*, 840-852.

Reich, D. A., Casa de Calvo, M. P., & Mather, R. D. (2008). *The role of interpersonal expectancies in predicting social judgments: The Interpersonal Expectancy Scale.* Manuscript in preparation, Texas Tech University.

Reifman, A. S., Larrick, R. P., & Fein, S. (1991). Temper and temperature on the diamond: The heat-aggression relationship in major league baseball. *Personality and Social Psychology Bulletin, 17*, 580-585.

Reinfurt, D., Williams, A., Wells, J., & Rodgman, E. (1996). Characteristics of drivers not using seat belts in a high belt use state. *Journal of Safety Research, 27*, 209-215.

Rosenbloom, T., Shahar, A., Perlman, A., Estreich, D., & Kirzner, E. (2007). Success on a practical driver's license test with and without the presence of another testee. *Accident and Analysis Prevention, 39*, 1296-1301.

Sanders, M. S., & McCormick, E. J. (1993). *Human factors in engineering and design* (7th ed.). New York: McGraw Hill.

Sherif, M. (1966). *The psychology of social norms*. New York: Harper & Row.

Shin, D., Hong, L., & Waldron, I. (1999). Possible causes of socioeconomic and ethnic differences in seat belt use among high school students. *Accident and Analysis Prevention, 31*, 485-496.

Stasson, M., & Fishbein, M. (1990). The relation between perceived risk and preventive action: A within-subject analysis of perceived driving risk and intentions to wear seatbelts. *Journal of Applied Social Psychology, 20*, 1541-1557.

Stone, J., Wiegand, A. W., Cooper, J., & Aronson, E. (1997). When exemplification fails: Hypocrisy and the motive for self-integrity. *Journal of Personality and Social Psychology, 72,* 54-65.

Strayer, D. L., & Drews, F. A. (2007). Cell-phone--induced driver distraction. *Current Directions in Psychological Science, 16*, 128-131.

Strayer, D. L., Drews, F. A., & Crouch, D. J. (2006). A comparison of the cell phone driver and the drunk driver. *Human Factors, 48*, 381-391.

Strayer, D. L., & Johnston, W. A. (2001). Driven to distraction: Dual-task studies of simulated driving and conversing on a cellular telephone. *Psychological Science, 12*, 462-466.

Takaku, S. (2006). Reducing road rage: An application of the dissonance-attribution model of interpersonal forgiveness. *Journal of Applied Social Psychology, 36*, 2362-2378.

Thomas, S. L., Skitka, L. J., Christen, S., & Jurgena, M. (2002). Social facilitation and impression formation. *Basic and Applied Social Psychology, 242*, 67-70.

Trafimow, D., & Fishbein, M. (1994). The importance of risk in determining the extent to which attitudes affect intentions to wear seatbelts. *Journal of Applied Social Psychology, 24*, 1-11.

Triplett, N. (1897). The dynamogenic factors in pacemaking and competition. *American Journal of Psychology, 9*, 507-533.

Turner, C. W., Layton, J. F., & Simons, L. S. (1975). Naturalistic studies of aggressive behavior: Aggressive stimuli, victim visibility, and horn honking. *Journal of Personality and Social Psychology, 31*, 1098-1107.

Ulmer, R. G., Preusser, C. W., Preusser, D. F., & Cosgrove, L. A. (1995). Evaluation of California's safety belt law change from secondary to primary enforcement. *Journal of Safety Research, 26*, 213-220.

Werth, L., & Forster, J. (2007). The effects of regulatory focus on braking speed. *Journal of Applied Social Psychology, 37*, 2764-2787.

Wiesenthal, D. L., Hennesy, D. A., & Gibson, P. M. (2000). The Driving Vengeance Questionnaire (DVQ): The development of a scale to measure deviant drivers' attitudes. *Violence and Victims, 15*, 115-136.

Wilson, T. D., & Brekke, N. (1994). Mental contamination and mental correction: Unwanted influences on judgments and evaluations. *Psychological Bulletin, 116*, 117-142.

Zajonc, R. B., Heingartner, A., & Herman, E. M. (1969). Social enhancement and impairment of performance in the cockroach. *Journal of Personality and Social Psychology, 13*, 83-92.

Reviewed by Mario Casa de Calvo, Texas A&M University - Kingsville

In: Road Traffic: Safety, Modeling and Impacts
Editors: S. E. Paterson and L. K. Allan, pp. 403-416
ISBN 978-1-60456-884-4

Chapter 10

DRIVER RESPONSES TO SPEED CAMERA ENFORCEMENT

Andrew P. Jones[1,*], Robin M. Haynes[1], Kate M. Blincoe[1] and Violet Sauerzapf[1]
[1]School of Environmental Sciences, University of East Anglia, Norwich, Norfolk, NR4 7TJ, United Kingdom

ABSTRACT

Road traffic accidents are a major source of mortality and morbidity in the United Kingdom. Speed is a known risk factor for accident risk, and speed cameras are being increasingly used to control vehicle speeds. However, if cameras are to be effective in modifying behaviour, it is important to understand how drivers respond to them. This research was undertaken using a postal questionnaire of drivers observed passing a speed camera in the county of Norfolk, England. Respondents were classified into a four category typology of conformers, deterred drivers, manipulators or defiers. Views and attitudes towards speed cameras and speed related behaviour were compared between groups. Differing perceptions and knowledge of the four types partially explained driving styles. Conformers and deterred drivers were least likely to exceed speed limits and were most favourable towards camera enforcement. Manipulators and defiers were younger, less experienced motorists who felt that they were unlikely to be prosecuted. This view means that manipulators and defiers are most difficult drivers to target for behavioural modification. In conclusion, the driver groups studied clearly showed divergent views and opinions. In order to improve the deterrent effect of speed cameras, specifically designed strategies should be used for the different groups.

* Tel: 00 44 1603 593127; Fax: 00 44 1603 591327;Email: a.p.jones@uea.ac.uk

INTRODUCTION

Fatalities from road traffic accidents (RTAs) in Britain have averaged over 3,400 per annum since 1998, and in 2003 a further 33,000 people were seriously injured on the nation's roads (Department for Transport, 2004). Although RTAs account for less than 1% of all deaths, a feature of mortality from this cause is that the highest incidence is amongst young adults, particularly those aged between 15 to 29 years. In this group, almost 17% of deaths occur as a result of collisions on the road. Because of this distinctive age distribution, RTAs are the cause of 3.6% of the total years of life lost to trauma and disease in Britain (Office of National Statistics, 2004).

Achieving reductions in levels of mortality and morbidity from RTAs is a national priority, and RTAs formed one of the five key areas in the Health of the National White Paper (Department of Health, 1992), and the more recent Saving Lives: Our Healthier Nation (Department of Health, 1999). It has long been recognised that there is a link between vehicle speed and accident risk, whereby vehicles travelling faster are at higher risk (Quimby et al, 1999). It is therefore unsurprising that improving levels of driver compliance with speed limits is seen as one way of reducing the burden of mortality and morbidity from this course. It is, however, noteworthy that driving faster than the speed limit permits is considered acceptable behaviour by many, and exceeding the speed limits on the United Kingdom's (UK) roads is endemic (DETR, 2000). Hence, if compliance is to be improved, it is important to understand attitudes to speeding and drivers' motivations to modify their behaviour.

Traffic systems have been described as complex 'social environments' in which self-perception is a key determinant of driving behaviour (Haglund and Aberg, 2000). Self-perception can be influenced by many things and may also be biased; the tendency for motorists to perceive their own driving differently to that of their fellow road users has been well documented. Svenson (1981), Groeger and Brown (1989) and McKenna (1993) are amongst the authors who have concluded that most drivers considered themselves to be more skilful than average, irrespective of their accident record. Walton and Bathurst (1998) argued that this 'self-enhancement bias' leads to road users believing that traffic rules, such as speed limits, only apply to others less able than themselves. Accordingly, self-perception, attitudes and beliefs are closely linked to drivers' choice of speed (Kimura, 1993).

The self-enhancement bias operates within the context of a culture that values speed and where speed on the road is desired and alluring (Corbett and Simon, 1999). Speeding is perceived as normal, and many drivers perceive speeding as one of the least serious traffic offences (Rothengatter, 1991). Indeed, a considerable proportion of road users believe that exceeding the limits will not lead to prosecution or a collision (Holland and Conner, 1996). Although the relationship between speed and the occurrence of road traffic accidents is well documented, a meta-analysis of published studies undertaken by Morrison et al (2003) found that it is insufficient to attempt to achieve reductions in collisions by simply lowering the statutory speed limit because drivers tend to respond poorly to such interventions. This may be caused by the combined effects of self-enhancement bias, drivers' perception of little danger associated with high speeds, and the infrequency of detection (Corbett, 2000). It is evident that novel interventions to enable the more effective enforcement of existing speed limits must be considered.

Speed cameras are an intervention which identifies speeding drivers for prosecution. The role of speed cameras is to decrease vehicle speeds by the deterrent effect of a fine, endorsement points on the motorist's driving licence, removal of a driving licence or, with the most serious offences, imprisonment. Speed cameras operate by taking photographic images of vehicles which pass them quicker than a set trigger speed, allowing the owner of the vehicle to be contacted using the vehicle registration details read from the images. Speed cameras may be sited in permanent, kerbside housings (fixed cameras), or operated on a transient basis from suitably marked vans located at the roadside (mobile cameras).

Studies in Australia and New Zealand have shown that speed cameras may significantly reduce collision and associated casualty numbers by lowering vehicle speeds. For example, Keall et al (2001) found that camera enforcement integrated with more general road safety campaigns resulted in a decrease in personal injury collisions of up to 32% in urban areas and 14% in rural areas. In Canada, the deployment of speed cameras was associated with a 2.8 kilometres per hour lowering of mean speeds and a 9% decrease in collisions at locations where the units were situated (Chen et al, 2002). Until recently, speed camera use in the UK has been sporadic because of the costs of purchasing and maintaining the devices. However, a pilot speed camera scheme that was trialled in eight police force areas in the year 2000 demonstrated encouraging results; in the first 2 years of operation, speeds fell at camera sites by an average of 3.7mph and, compared to the long term trend, there were 47% less fatalities and serious injuries (Gains et al, 2003).

Speed cameras may be an effective method of reducing vehicle speeds and hence decreasing collision frequency and severity. However, understanding motorist reactions to speed cameras in the social environments of traffic systems is difficult because responses to camera enforcement are often complex and multifaceted. An analysis of driver response to cameras was undertaken by Corbett (1995) and resulted in the development of a driver typology. The typology categorised drivers into four groups; *conformers* who always or nearly always comply with speed limits, the *deterred* who have lowered their speeds since the deployment of roadside cameras, *manipulators* who reduce speed for cameras but exceed the limit when not in a camera zone and, finally, *defiers* who speed regardless of camera enforcement. Additionally, the conformer group may be subdivided into true conformers who abide by all limits, and self-deceiving conformers who speed but *believe* they are obeying the law. This latter type of conformer may, for example, consider that it is legally acceptable to drive at speeds above the statutory limit but lower than the speed at which a camera would detect them. Corbett and Simon (1999) further studied the typology and found that there are distinct behavioural and personal differences between the groups.

This aim of the research presented here is to use an adapted version of the Corbett and Simon (1999) typology to present an updated assessment of motorists' perceptions of speeding, speed behaviour trends and general awareness. This is important because there have been considerable increases in levels of speed camera deployment in the UK since the original work by Corbett. A possible consequence of this is increased familiarity, and thus diminished impact on driver behaviour (Keall et al, 2001). This may have been exacerbated by media reports that cameras did not always contain photographic film.

In this chapter, we investigate drivers' attitudes, perceptions and behaviour in a rural area. The new context is important because exceeding the speed limit is more likely to have serious consequences in rural than urban environments. In comparison with urban areas, over four times the proportion of rural traffic collisions involve vehicles travelling at excessive speeds (DETR, 2000). Furthermore; road traffic collisions are more likely to result in killed or seriously injured casualties in rural areas compared with urban areas (Sharples and Fletcher, 2001). Hence this research was conducted on a sample of drivers in the county of Norfolk in Eastern England, which is one of the largest but most sparsely populated counties in England. It has a transport network of approximately 10,000 km of roads, with a per capita rate of fatalities and serious injuries caused by collisions which is higher than the national average (DfT, 2004).

METHODS

Over a four hour period, vehicles were photographed by a fixed-unit speed camera as they drove along a stretch of the A146 road at Loddon in Norfolk, a stretch of road that has a 60mph hour limit. The camera was set up for this research so that all vehicles in free flowing traffic that were travelling at over 50mph were photographed, not just those that were exceeding the speed limit. Vehicle registration details were read from the photographic images, allowing registered owners to be identified and sent a postal, self-completion questionnaire. The questionnaire was sent to 964 motorists with a covering letter which explained the purpose of the research, emphasised that all responses were anonymous, and requested that if the recipient of the questionnaire was not the driver of the car when it was photographed, then the questionnaire be passed on to that person.

The questionnaire asked drivers to report how they drive in general and in the vicinity of a camera on the survey road. Motorists were also asked to state their perception of the likelihood of prosecution for speeding, how in favour they were of camera enforcement, their familiarity with camera site locations, estimates of the speed at which a camera would photograph them, and questions about their beliefs on the relationship between speed and collisions. Additionally, drivers were questioned on their personal characteristics including age, sex, postcode, number of years of driving experience, traffic accident history and previous endorsement points. Postcodes were used to calculate the distance of each motorist's residence from the A146 Loddon camera site using a Geographical Information System software package (ArcGIS). Drivers' approval of cameras was ascertained by their strength of agreement with the statements shown in Table 1. Their responses to these statements ranged in five categories from 'strongly disagree' to 'strongly agree'. These responses were coded so that 1 indicated strong disapproval of cameras and 5 indicated strong approval. To calculate the camera favour score the coded responses over the seven statements were averaged for each respondent to give a score reflecting their overall attitude to cameras, whereby 1 demonstrated a strong negative reaction and 5 indicated a strong favourable response.

A number of questions were also included to allow motorists to identify themselves as a conformer, deterred driver, manipulator or defier. The methodology used to classify these driver types is outlined in Figure 1. In contrast to the Corbett and Simon (1999) methodology, we asked drivers to describe their camera related behaviour and to quantify the actual speeds they drove at, in order to minimise the possibility of subjective responses. Firstly, respondents were asked how often they drive above the limit in 60mph (rural) zones to separate those who 'usually' of 'nearly always' exceeded the limits from those who 'never', 'only rarely' or 'sometimes' did so. The second question asked how the introduction of speed cameras and signs in Norfolk had affected respondents' driving speeds and camera related behaviour. This allowed us to separate the speeding types (identified in question one) into manipulators or defiers, and the non-speeding types into conformers or deterred drivers. The questionnaire was designed to minimise the problems that can be associated with self-report data, and no driving behaviour was portrayed as negative or undesirable.

The responses from the questionnaires were entered into SPSS for Windows and differences in characteristics between the groups were tested. Where results are statistically significant at the 5% level, probabilities and test statistics are given in the text.

RESULTS

Of the 964 questionnaires despatched, 480 were returned, a response rate of 50%. The majority of respondents were male and the mean age was 52 years with 3% of respondents younger than 25 years and 4% older than 75 years. The total sample had a mean of 30.6 years of driving experience and drove on average 12,322 miles per year. A vehicle with an engine size greater than 2000cc was driven by 24% of respondents and a high performance vehicle was driven by 3%. Significant differences in certain characteristics were found between the males and the females. Females' mean age was 46 years, whilst males' mean age was 54 ($t = 32.56$, $p < 0.01$). Also, females had fewer years of driving experience, with a mean of 24 years driving compared to males' 34 years ($t = 55.12$, $p < 0.01$), they also drove fewer miles per year (mean 9264 miles compared to males' mean of 14059 miles) ($t = 7.02$, $p = 0.01$) and fewer females drove a car with an engine size larger than 2000cc (females 15%, males 28%) ($\chi^2 = 10.07$, $p < 0.01$).

Characteristics of the Driver Types

It was possible to categorise 467 (97%) of the respondents into one of the four typologies (13 drivers did not complete the relevant questions). In response to question one, 82% of motorists stated that they 'never', 'only rarely' or 'sometimes' exceeded the limit in a 60mph zone, whilst the remainder (18%) reported that they 'usually' or 'nearly always' exceeded the limit. Question two further categorised the drivers, resulting in 279 (60%) of the sample being defined as conformers, 121 (26%) as deterred drivers, 40 (9%) as manipulators and 27 (6%) as defiers.

Table 1. Statements used to ascertain the camera favour score

Cameras are an easy way to make money out of motorists
Cameras are meant to encourage drivers to keep to the limits, not punish them
Fewer accidents happen on roads where cameras are installed
Cameras mean that dangerous drivers are more likely to be caught
The use of speed cameras should be supported as a method of reducing casualties
The primary aim of speed cameras is to save lives
There are too many speed cameras in our local area

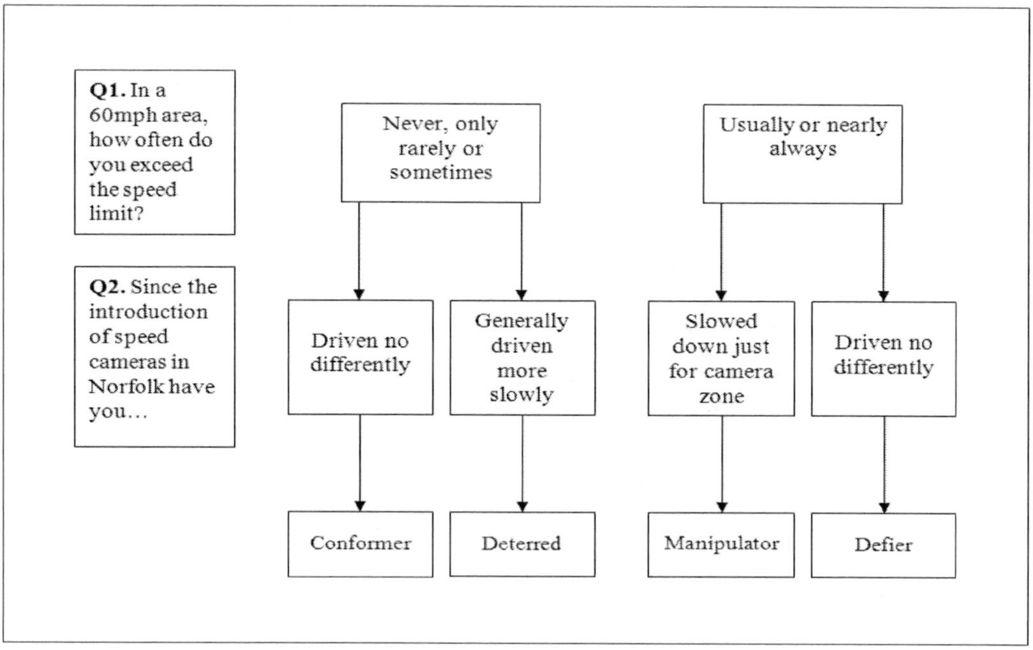

Figure 1. Determining the driver types based on driving in a 60mph zone.

The first stage of analysis focused on determining if drivers' characteristics varied between driver types. Table 2 shows selected characteristics of the sample stratified according to the four types. Conformers and deterred drivers were significantly older than manipulators and defiers ($t = 7.53$, $p < 0.01$) and were slightly more likely to be female. Unsurprisingly, age was positively correlated with years of driving experience ($t = 0.922$, $p < 0.01$) and accordingly conformers and the deterred had the greatest amount of driving experience and manipulators had the least of the four types. Whilst there was no statistically significant difference between the miles driven per year in the four groups, the manipulators did tend to report higher yearly mileages. Furthermore, the mean distance that the four driver types lived away from the camera site showed significant variation between groups; the conformers resided an average of 27km away, the deterred 23km, manipulators 40km and the defiers 60km ($F = 3.88$, $p = 0.01$).

Manipulators and the deterred were significantly more likely to drive a vehicle with high performance features compared to conformers or defiers ($\chi^2 = 8.95$, $p = 0.01$). Nevertheless, this trend was not reflected in the engine size of vehicles driven, where variations between groups were small. It appears that some deterred drivers may have become deterred because of their prior experiences; they were significantly more likely to have endorsement points on their licence than conformers and defiers ($\chi^2 = 20.12$, $p < 0.01$) and they were the second most likely group to have been involved in one or more collisions in the past three years. The conformers' driving history appears to reflect their cautious approach to driving; they were the least likely to have had a collision or to have endorsement points. The manipulators had significantly more endorsement points than the other drivers, suggesting that manipulative driving is not necessarily an effective way of avoiding prosecution. By contrast, the defiers were second only to the conformers in having avoided both collisions and endorsements.

Table 2. Characteristics of the sample

	Conformers	Deterred	Manipulators	Defiers	Total sample
Mean age (years)	52.6	53.0	42.3	45.9	51.6
Percentage male	65.1%	65.8%	67.5%	70.4%	65.8%
Mean years of driving experience	31.5	31.8	23.1	25.4	30.6
Mean miles driven per year	12,776	11,204	14,018	11,732	12,322
Mean distance of residences from A146 camera site (km)	27	23	40	60	30
Percentage with vehicle engine size > 2000cc	24.1%	22.2%	22.5%	18.5%	23.6%
Percentage who drive a high performance vehicle	1.1%	6.0%	7.5%	3.7%	3.4%
Percentage who have had 1or more collisions in past 3 years	10.3%	12.5%	12.8%	11.1%	11.5%
Percentage who currently have endorsement points on their licence	9.7%	24.0%	35.0%	11.1%	15.3%
Percentage who drive a company car	7.8%	5.2%	5.3%	7.7%	6.9%
N	279 (59.7%)	121 (25.9%)	40 (8.6%)	27 (5.8%)	480

Driving Behaviour

A comparison of drivers' characteristics provides an insight into the roles of prior experiences and personal qualities in defining different types of driver. However, whilst this profiling is useful, understanding how these characteristics impact on actual driving behaviour on the road is of more fundamental importance. Given that many studies have shown speeding is widespread, we considered the reasons behind why motorists speed and whether or not it is deliberate.

Of the total sample, 63% stated that many of their speeding incidents were likely or extremely likely to have been inadvertent. When asked for the reasons for 'deliberate' speeding, a third of the total sample reported that they were likely or extremely likely to exceed the limits because they felt pressured to do so by other road users (Table 3). Nevertheless, there was no significant difference between the proportions of each driver type who stated that they speed for this reason. Almost half of the sample stated that they were likely or extremely likely to speed because they were keeping up with the pace set by others, but again, no driver type was significantly more likely to report this than the others. However, compared to the conformers and deterred drivers, manipulators and defiers were significantly more likely say that they set their own 'safe speed' which was higher than the limit ($\chi^2 = 37.01$, $p < 0.01$) and they were also more likely to speed if they were in a hurry ($\chi^2 = 13.20$, $p = 0.01$).

We found that the causes of speeding show some variation between driver types, therefore it may be anticipated that different drivers' responses to speed cameras will vary with the type of camera used (fixed or mobile). Where mobile units are being deployed it is harder for drivers to predict whether a camera is present at a particular location. Our findings suggest that, compared to fixed camera units, mobile cameras seem to have an increased deterrent effect and may be particularly effective at tackling manipulative driving. Compared to the way that they drove in the vicinity of fixed cameras, 67% of manipulators stated that they drive more slowly where they thought there was a possibility of mobile speed camera enforcement. This percentage is significantly higher than the 55% of deterred, 31% of conformers and 30% of defiers who reported the same behaviour ($\chi^2 = 34.31$, $p < 0.01$).

An additional measure of the success of cameras is not just their capacity to reduce speeds at the sites where they are located, but also how much they create a general decrease in speeding. To investigate this we asked drivers about their behaviour on the A146 survey road, at both enforced sections (how they drive where they think there is a camera) and un-enforced areas (their general driving along the 60mph stretch of the road). Studying stated speeding behaviour on the un-enforced part of the survey road revealed a subset of 214 (47%) drivers who generally exceeded 60mph on the survey road (Table 4). This behaviour was reported by 52% of conformers, 38% of deterred drivers, 44% of manipulators and 48% of defiers. By contrast, only 30 (7%) of respondents stated that they generally drove above the limit in 60mph zones of the survey road where they thought a camera was present; a significantly lower proportion than when they believed there was not a camera present ($\chi^2 = 344.00$, $p < 0.01$). The conformers (2%) and the deterred (3%) were significantly less likely to state that they speed where they think that there is a camera present compared to the manipulators (36%) and the defiers (26%) ($\chi^2 = 88.29$, $p < 0.01$).

Table 3. Reasons for deliberate speeding

Percentage of drivers who stated they were 'likely' or 'extremely likely' to speed because:	Conformers	Deterred	Manipulators	Defiers	Total sample
Feel pressurised by other drivers to increase speed	34.7%	31.3%	30.8%	37.0%	33.6%
Simply keeping up with the speed set by other drivers	46.2%	42.6%	61.6%	51.8%	46.9%
Deliberately set own safe speed which is higher than the limit	12.6%	10.6%	46.2%	29.6%	16.0%
In a hurry and deliberately drive faster than the limit	24.9%	26.1%	46.1%	37.0%	27.7%
Number of respondents	279	121	40	27	480

Table 4. Exceeding the limits on the survey road

Percentage who stated that they generally exceed the speed limit	Conformers	Deterred	Manipulators	Defiers	Total sample
On the 60mph stretch of the survey road in free flowing traffic	51.7	37.9	43.6	48.1	47.2
On the 60mph stretch of the survey road in free flowing traffic where they think there is a camera	1.8	3.4	35.9	25.9	6.6

Knowledge and Perception

Although interventions such as mobile speed cameras may be able to adapt and influence driver actions, the question arises as to whether camera related behaviour is associated with the underlying cultural beliefs and perceptions that drivers may hold about the desirability and safety of speeding. In the long term, reducing the prevalence of such beliefs may have the greatest impact on reducing levels of speeding. To examine this, we investigated the sample's awareness and views of speed related issues such as the risks of speeding; the possibility of prosecution and the danger of being involved in a collision. Firstly we examined drivers' beliefs about whether higher speeds were linked with the likelihood of having collisions, both for their own driving and for that of other road users. The results are detailed in Table 5.

The proportion of drivers of all types who acknowledged that there was a link between speed and collision frequency on a personal level (37%) was significantly lower than when respondents were asked to consider drivers in general (56%) ($\chi^2 = 149.56$, $p < 0.01$). Nevertheless, the deterred were significantly more likely to acknowledge the link between speed and accidents, both specifically for themselves and also for all drivers. Manipulators and defiers were more likely to refute the general relationship between speed and collisions, suggesting that more of these drivers do not believe that their speeding behaviour is dangerous. It is evident that many drivers who chose to exceed the limits either did not believe, or refused to acknowledge, the dangers of speed.

The extent to which the decision to speed is based on assessing the risk of prosecution rather than fear of a collision may be of key interest for future road safety interventions. Of the whole sample, 19% thought it 'very likely' that there was a camera operating on a section of road with "police enforcement camera" signs, with just 4% of defiers giving this response, compared to 18% of conformers and manipulators, and 26% of deterred drivers ($\chi^2 = 26.40$, $p < 0.01$). This demonstrates the generally higher level of confidence amongst defiers that their speeding behaviour will not be detected. The deterred were significantly more likely to think that the camera would be functioning than the other driver groups ($\chi^2 = 10.66$, $p = 0.01$), suggesting that their cautious approach to speed may be a combination of awareness of the accident risk and fear of prosecution.

The view amongst some motorists that prosecution for speeding can be avoided was also demonstrated by the widespread belief that a certain level of transgression is allowed before speeding incurs punishment. Hence, all groups' average estimate for the threshold for prosecution on a 60mph road was above 60mph. The conformers' mean camera trigger speed estimate in a 60mph zone was the lowest at 63.4mph, followed by the deterred at 63.5mph and the manipulators at 64.9mph. The estimate for the defiers was significantly higher than the other groups at 65.1mph ($t = 2.31$, $p = 0.02$).

Although the speeding driver types (manipulators and defiers) share some similar beliefs about the risks of speeding, a difference between these motorists was the knowledge of where cameras were situated. Manipulators were most likely state to that they knew camera locations 'very well' or 'quite well' (71%) whilst the defiers were least likely to know (65%). Manipulators were also least likely to report that they were unaware of camera locations. Although none of these differences between the driver groups were found to be statistically significant at the 5% level, these findings reinforce the view that manipulative driving is in part enabled by awareness of camera location.

Whilst knowledge of camera trigger speeds and locations may facilitate certain driving behaviours, the way motorists choose to use or interpret such knowledge is likely to be determined by their attitudes and, crucially, whether they generally approve of camera enforcement. The mean camera favour score of the conformers (whereby 1 demonstrated strong disapproval of cameras and 5 indicated strong approval) was 2.5, whilst the deterred drivers' score was 2.7. These were significantly higher than the manipulators' and defiers' score of 2.3 ($\chi^2 = 7.64$, $p = 0.01$).

One of the principal methods used in attempts to modify driver attitudes and knowledge about speed cameras in the UK is the dissemination of road safety messages by local media coverage. We examined if levels of awareness of cost-recovery varied between the different driver groups in our sample and also whether the respondents had come into contact with any publicity regarding cameras. We expected that knowledge of cost-recovery may be associated with motorists' attitudes. However, only 18% of the total sample was aware that the fine monies contribute towards camera enforcement expenditure. In total, 31% stated they did not know what happened to the money, 27% of drivers thought that the income was used to increase Norfolk Constabulary funding, 16% believed it went to the Government and 8% thought that it paid for other road safety improvements. Interestingly, defiers and manipulators were significantly more likely to be aware of cost-recovery than the other driver types ($\chi^2 = 5.69$, $p = 0.02$). However, knowledge of this process was not found to be associated with respondents' camera favour scores, and hence it seems that publicity about such issues may not have influenced the attitudes of drivers who were aware of it.

Table 5. Views on the link between speed and accidents

Driver type	Is there a link between speed and accidents for drivers in general?				Is there a link between speed and accidents for you personally?			
	Driver responses:				Driver responses:			
	No	Don't know	It depends	Yes	No	Don't know	It depends	Yes
Conformers %	3.6%	1.4%	40.6%	54.3%	19.4%	4.3%	43.0%	33.3%
Deterred %	1.7%	0%	29.7%	68.6%	8.5%	2.5%	39.8%	49.2%
Manipulators %	15.0%	0%	45.0%	40.0%	25.0%	0%	47.5%	27.5%
Defiers %	7.4%	0%	48.1%	44.4%	18.5%	3.7%	51.9%	25.9%
Total sample %	4.2%	0.9%	39.0%	55.9%	16.6%	3.1%	43.7%	36.6%

When we examined general levels of awareness of publicity information, 32% of all respondents were aware of speed enforcement or road safety publicity, with the deterred drivers most likely to have heard of it (40%), followed by 33% of defiers, 28% of conformers, and 27% of manipulators. It was also noteworthy that those who were aware of publicity were slightly less likely to acknowledge the general risks of speed than those who were unaware (of those who were aware of publicity, 51% agreed with the general speed and accident link, whilst of those who were not aware of publicity, 58% agreed with the link). Interestingly, publicity awareness was not found to have any effect on drivers' camera favour score.

CONCLUSIONS

Our sample of drivers in the predominantly rural county of Norfolk had a complex response to speed cameras. Categorising the motorists into conformers, deterred drivers, manipulators and defiers has enabled the identification of key trends in attitudes and has also reinforced Haglund and Aberg's (2000) conclusion that many drivers operate as part of a broader more structured social environment. The results and their implications on the design of effective road safety initiatives are discussed here.

We were able to create profiles of the four types identified by adapting Corbett and Simon's (1999) typology. Whilst our method resulted in similar proportions of each driver type, unlike Corbett and Simon we did not deliberately over represent defiers. By reducing the subjectivity of what constitutes exceeding the limits we substantially increased our numbers of speeding drivers compared to if we had used the Corbett and Simon methodology.

The conformers and deterred drivers were the least likely to exceed the speed limits and were older, more likely to be female and had more years of driving experience than the defiers and manipulators. Many conformers appeared to be generally cautious drivers who were aware of the risks of accidents and prosecution and determined to avoid both. The deterred demonstrated a similar level of caution; however, this appeared to be partially caused by a higher frequency of both collisions and endorsements, illustrating how experience of the risks of speeding can have a deterrent effect (Corbett, 2000). It was evident that the conformers and the deterred drivers had more favourable attitudes toward camera enforcement so, in terms of enforcement, the key aim should be to maintain the level of support already felt by these driver types. This could be achieved by ensuring speed limit signage is perceived to be appropriate and consistent. Currently, local traffic authorities in the UK can change any speed limit on their roads and this can result in similar roads being given varying speed limits in different parts of the country (DETR, 2000). This is problematic because if drivers do not understand why roads are given a certain speed limit, then they are less respectful of limits and will cite inconsistency as a justification for speeding (Silcock et al, 2000).

As Corbett and Simon (1999) found, the speeding driver types were the younger, less experienced motorists with the most negative perception of camera enforcement and who lived the greatest distance away from the surveyed camera site. The manipulators drove most miles per year and were also more likely to have endorsement points on their licence, indicating either that manipulative driving is not necessarily an effective method of avoiding prosecution or that defiers, if apprehended, may become manipulators. Manipulators were also most likely to have had a collision in the past three years. The high level of prosecution accompanied by the worst collision record makes manipulators appear to be the primary group that should be targeted by road safety initiatives. However, the comparatively low level of collisions and endorsements that was reported by the defiers (they were second only to the conformers in having avoided both) contradicts Corbett and Simon's (1999) findings whereby the defiers were the most likely to have reported both collisions and endorsements. It is possible that some of these drivers currently speed because they have never experienced the negative consequences that can be associated with exceeding the limits and hence may not have seen their behaviour as risky.

The key differences between these driver types emerged when their beliefs and knowledge about cameras were examined. The defiers' mean trigger speed estimate was the highest, and they were least likely of all drivers to think that the cameras would be operating. Therefore, they perceived that the risk of prosecution as a result of speeding past a camera was low. Manipulators did not have this confidence, which could explain their braking behaviour. Unsurprisingly, manipulators also had the best knowledge of camera sites, and defiers, perhaps calculating that cameras would not be operating, had the poorest awareness of camera locations of the four cohorts.

The variations between these driver types suggest that specific enforcement strategies targeted at different groups may be most effective in reducing speeding behaviour. Our findings indicate that inconspicuous cameras may decrease manipulative driving behaviour. This is because, if road users are less certain of where cameras are located, they are forced to reduce speeds for the entire stretch of the road rather than just in close proximity to the camera (Keall et al, 2001). However, the concealment of cameras is not currently possible due to British Government guidelines which state that camera housings should be visible to road users (Gains et al, 2003). An additional problem with any concealment strategy, and one which has driven the Government guidelines on conspicuity, is that it could result in increased levels of anti-camera sentiments. To reduce defier-style speeding, different tactics are needed; an increased risk of prosecution, for example by reducing camera trigger speeds or increasing current levels of enforcement activity, would appear to be most effective.

Of course, there is the possibility that this survey has been affected by the problems of all self-report based analyses. Our response rate was high at 50%, but it is possible that those who provided a response are not fully representative of the wider population of drivers. Despite careful wording to ensure that no driving behaviour was portrayed as negative or inappropriate, there is the risk that some responses given by motorists were not representative of their true opinions or driving habits due to feelings of guilt about behaviours which they may perceive as being anti-social. As Corbett (2001) found, there is a tendency for faster road users to understate their normal speeds and for slower drivers to overstate theirs, which may also lead to some inconsistency of reported speeds. In applying the typology for this research, this may have decreased the differences between the cohorts, perhaps resulting in some defiers being categorised as conformers and vice versa. It is also noteworthy that the results presented here are from the early stages of the camera enforcement campaign in Norfolk, and hence drivers' views and reactions may evolve over time.

This chapter has analysed the driving characteristics of a sample of drivers in a rural area with the aim of providing a better understanding of the interactions that occur in the social environment of the traffic system. The chapter examined the demographic characteristics, attitudes and beliefs associated with different driving behaviours and determined the influence of drivers' perceptions of varying deployment methods on self-report driving habits. The driver groups studied clearly showed divergent views and opinions. Hence, in order to improve the deterrent effect of speed cameras on the UK's roads it is clear that specifically designed strategies should be used to target the different groups. To maintain the support of conformers and deterred drivers, speed limits should seem appropriate for the road type and be accompanied by frequent limit reminder signs to reduce accidental speeding. Manipulators could be tackled by the concealment of cameras. Alternatively, the use of mobile cameras may prove effective. For defiers, increasing the risk of prosecution is necessary and this could be achieved by the reduction of camera trigger speed thresholds or increasing the use of cameras.

ACKNOWLEDGEMENTS

We thank all the staff at the Norfolk Casualty Reduction Partnership who helped this research by the provision of funding, data and information, the drivers who completed the questionnaires, and Claire Corbett and Frances Simon, at Brunel University, for advice and assistance with the driver typology on which this research is based.

REFERENCES

Chen, G., Meckle, W., Wilson, J. (2002). Speed and safety effect of photoradar enforcement on a highway corridor in British Columbia, *Accident Analysis and Prevention*, 34, (2), 129-138.

Corbett, C. (1995). Road traffic offending and the introduction of speed cameras in England: the first self-report study, *Accident Analysis and Prevention*, 27, (3), 345-354.

Corbett, C. (2000). A typology of drivers' responses to speed cameras: implications for speed limit enforcement and road safety, *Psychology, Crime and Law*, 6, 305-330.

Corbett, C. (2001). Explanations for "understating" in self-reported speeding behaviour, *Transportation Research (Part F)*, 4, 133-150.

Corbett, C., Simon, F. (1999). *The Effects of Speed Cameras: How Drivers Respond, DTLR Report 11*. London: DTLR.

DETR: Department of Environment, Transport and the Regions (2000). *New Directions in Speed Management – A Review of Policy.* London: DETR

Department of Health (1992). *Health of the Nation.* London: Her Majesty's Stationary Office.

Department of Health (1999). *Saving Lives: Our Healthier Nation.* London: Her Majesty's Stationery Office.

Department for Transport (DfT) (2004). *Road Casualties Great Britain 2002, Annual Report.* London: Her Majesty's Stationary Office.

Gains, A., Humble, R., Heydecker, B., Robertson, S. (2003). *A Cost Recovery System for speed and red-light cameras: Two year pilot evaluation, Department for Transport, Road Safety Division.* London: Department for Transport.

Groeger, J.A., Brown, I.D. (1989). Assessing one's own and others' driving ability: Influences of sex, age and experience, *Accident Analysis and Prevention,* 21, (2), 155-168.

Haglund, M., Aberg, L. (2000). Speed choice in relation to speed limits and influences from other drivers, *Transportation Research (Part F),* 3, 39-51.

Holland, C.A., Conner, M.T. (1996). Exceeding the speed limit: An evaluation of the effectiveness of a police intervention, *Accident Analysis and Prevention*, 28, (5), 587-597.

Keall, M.D., Povey, L.J., Frith, W.J. (2001). The relative effectiveness of a hidden versus a visible speed camera programme, *Accident Analysis and Prevention*, 33, (2), 277-284.

Kimura, I. (1993). The relationship between drivers' attitudes to speeding and their speeding behaviour in hypothetical situations, *Hiroshima Forum for Psychology*, 15, 51-60.

McKenna, F.P. (1993). It won't happen to me: unrealistic optimism or illusion of control? *British Journal of Psychology*, 84, 39-50.

Morrison, D.S., Petticrew, M., Thomson, H. (2003). What are the most effective ways of improving population health through transport interventions? Evidence from systematic reviews, *Journal of Epidemiology and Community Health*, 57, (5): 327-333.

Office of National Statistics (2004). *Compendium of clinical and health indicators*. London: Her Majesties Stationery Office.

Quimby, A., Maycock, G., Palmer, C., Butteress, S. (1999). *The Factors that Influence a Driver's Choice of Speed – a Questionnaire Study, Report 325*. Crowthorne: Transport Research Laboratory.

Rothengatter, T. (1991). Automatic policing and information systems for increasing traffic law compliance, *Journal of Applied Behaviour*, 24, (1), 85-87.

Silcock, D., Smith, K., Knox, D., Beuret, K. (2000). *What limits speed? Factors that affect how fast we drive.* Basingstoke: AA Foundation for Road Safety Research.

Sharples, J.M. & Fletcher, J.P. (2001). *Tourist Road Accidents in Rural Scotland.* Edinburgh: Scottish Executive, Central Research Unit.

Svenson, O. (1981). Are We All Less Risky and More Skilful Than Our Fellow Drivers? *Acta Psychologica*, 47, 143-148.

Walton, D., Bathurst, J. (1998). An exploration of the perceptions of the average driver's speed compared to perceived driver safety and driving skill. *Accident Analysis and Prevention*, 30, (6), 821–830.

In: Road Traffic: Safety, Modeling and Impacts
Editors: S. E. Paterson and L. K. Allan, pp. 417-437
ISBN 978-1-60456-884-4

Chapter 11

METHODS AND ANALYSIS OF AVALANCHE RISK ASSESSMENT FOR AVALANCHE-PRONE ROADS: EXAMPLES AND COMPARISONS FOR THE MILFORD ROAD, NEW ZEALAND

Jordy Hendrikx[1] and Ian Owens[2]
[1]National Institute of Water and Atmospheric Research Ltd (NIWA), Christchurch, New Zealand
[2]Department of Geography, University of Canterbury, Christchurch, New Zealand

1. ABSTRACT

Avalanches pose a significant natural hazard in many parts of the world. Avalanche hazard is managed using a range of methods in different mountain areas. A first step for hazard management is a risk assessment to evaluate the hazard posed to a road. This assessment can then be used as a tool to determine the appropriate mitigation measures and techniques to be employed. Care must be taken when using any risk assessment equation as the sensitivity of the assumptions made are rarely fully understood. This chapter will examine the application of two key risk assessment methods to examine avalanche risk on highways, the Avalanche Hazard Index (AHI) and the probability of death to individuals (PDI). First, this chapter will review the historical development and modifications to the AHI and PDI methods. Second, this chapter will examine the sensitivity of these two methods to the various assumptions made in the analysis and results will be compared to other commonly accepted levels of risk for other hazards. Finally, using both methods, examples of risk assessments for several avalanche prone roads from around the world will be presented and compared to data from the Milford Road, New Zealand.

2. Introduction

In many parts of the world snow avalanches pose a significant natural hazard. Where this hazard interacts with people or their property it becomes a risk. Such is the case when a road passes through an area with an avalanche hazard. This scenario occurs in numerous locations in many countries around the world including; Argentina, Austria, Canada, Iceland, Norway, New Zealand, Pakistan, Switzerland and the USA to name a few. The risk is managed through a diverse range of methods, from active avalanche control (using explosives for artificial avalanche release) and large protective structures through to passive control by closing a road for the entire avalanche season. In many cases the avalanche risk posed to road traffic is unknown. However, as traffic volumes increase, or following an incident, there is often a desire to quantify the risk posed to road traffic from avalanches. Quantifying the risk is also the first step in a structured hazard management approach, as the assessment can then be used as a tool to determine the appropriate mitigation measures and techniques to be employed.

This chapter will provide a review of two main risk assessment methods for avalanche prone roads, the Avalanche Hazard Index (AHI) and the probability of death to individuals (PDI). Firstly, this chapter will review how the AHI has changed since its inception in 1974 (Avalanche Task Force, 1974) and since its application to the roads around the world (e.g. Fitzharris and Owens, 1980; Schaerer, 1989). The method for calculating PDI is also reviewed and recent modifications in the AHI are also presented (Hendrikx and Owens, 2008). Furthermore, the sensitivity of these methods to the various assumptions made in the analysis are examined and the results are compared to other commonly accepted levels of risk for other hazards, including avalanches, both in New Zealand, and elsewhere around the world. Finally, the level of risk on a selection of roads in New Zealand, Switzerland and Canada are compared in terms of collective risk and the AHI.

3. Methods of Avalanche Risk Assessment

3.1. The Avalanche Hazard Index (AHI)

The AHI was first developed in 1974 for use on highways in British Colombia, Canada (Avalanche Task Force, 1974). It was designed as a numerical expression of damage and loss as a result of the interaction between vehicles on a road and a snow avalanches (Schaerer, 1989). Since then it has been used on other roads around the world including; elsewhere in Canada (e.g. Schaerer, 1989; Stethem et al., 1995), the United States of America (e.g. Armstrong, 1981) and on the Milford Road, New Zealand (Fitzharris and Owens, 1980). The AHI is used to determine how serious avalanche problems are to allow comparisons of the hazard of different highways to establish priorities and determine the appropriate level of avalanche safety management and to show where control measures have the greatest effect (Schaerer, 1989). The AHI considers both moving and waiting traffic, and is a function of:

- The size and type of avalanche
- The frequency of avalanche occurrences

- The number of avalanche paths and the distance between them
- The total length of highway exposed
- The traffic volume
- The traffic speed
- The type of vehicle

Initial work on the Milford Road, New Zealand used the Avalanche Task Force(1974) approach (Fitzharris and Owens, 1980), to calculate the AHI using the encounter probability for moving traffic according to equation 1:

$$P_m = \frac{T(L+D)F}{V.3600.24} \tag{1}$$

where:

P_m = encounter probability for moving vehicles to be hit by avalanches (year^{-1})
T = average daily traffic volume per 24 hour counted in both directions (vehicles d^{-1})
V = average speed of the traffic (ms^{-1})
L = width of avalanche = average length of the road covered by the avalanche (m)
D = stopping distance on a snow-covered road for a vehicle with the speed V (m)
F = average frequency of avalanche occurrence (year^{-1})

This encounter probability is for an individual vehicle, regardless of whether it is a car, bus, truck, maintenance vehicle, travelling in a group or alone. Moving vehicles are exposed to the avalanches for only a few seconds, and the chances of being hit are low. However, the exposure time is much longer for vehicles that have stopped in the avalanche tracks. Waiting traffic has proven to be a significant cause of avalanche accidents (Schaerer, 1989). Vehicles may be stopped or waiting because of previous avalanche activity, fitting snow chains, or taking photographs (Fitzharris and Owens, 1980). The probability of encounter for waiting traffic depends on the probability of a subsequent avalanche in an adjacent or another part of the same path, frequency of avalanche occurrence and the number of vehicles waiting. Fitzharris and Owens(1980) calculated the AHI for the Milford Road using the encounter probability for waiting traffic according to equation 2:

$$P_w = P_s.F.N_w \tag{2}$$

where:

P_w = encounter probability for waiting traffic to be hit by avalanches (year^{-1})
F = average frequency of avalanche occurrence (year^{-1})
N_w = number of vehicles in the adjacent avalanche track (=L/15m)
L = width of avalanche = average length of road covered by the avalanche (m)
P_s = probability of a subsequent avalanche in an adjacent or another part of the same path. Fitzharris and Owens(1980) used 0.15.

Table 1. Avalanche type and weighting (Fitzharris and Owens, 1980, p.13)

Avalanche type	**Weighting**
k_1 Powder snow	1
k_2 Light snow	4
k_3 Deep snow	10
k_4 Plunging avalanche	12

As is common for initial risk assessments, Fitzharris and Owens(1980) undertook the assessment of the AHI on the Milford Road, by estimating the frequency and size of avalanches from historical information such as journal articles and newspapers (e.g. Smith, 1947), local knowledge as well as topographical and botanical field investigations. Estimates were made for the size and frequency of the three different types of avalanches, powder snow (k_1), light snow (k_2) and deep snow (k_3). Smith(1947) found that because of the very steep terrain certain snow conditions result in airborne avalanches. Therefore Fitzharris and Owens(1980) introduced an additional avalanche type, plunging avalanches (k_4). These different avalanche types have weightings that consider the relative cost and consequence of an encounter, as seen in Table 1.

Using these weightings, and the different frequencies and avalanche widths associated with each, Fitzharris and Owens(1980) calculated the AHI for the Milford Road using the encounter probability for moving and waiting traffic according to equation 3:

$$AHI = \sum_{k=1}^{4} W_k (P_m + P_w) \tag{3}$$

where:

$k =$ index for avalanche type k_1 to k_4
$W =$ weighting for avalanche type
$P_m =$ encounter probability for moving traffic
$P_w =$ encounter probability for waiting traffic

The 1980 AHI on the Milford Road was calculated at 46 for a winter traffic volume of 80 vehicles per day, where $P_m = 2$ and $P_w = 44$. According to the North American practice to group highways with respect to the avalanche hazard the Milford Road rated at a moderate hazard in 1980 (Table 2).

Moderate hazard was suggested to require avalanche control at selected sites, and be supplemented by warnings and occasional closures (Fitzharris and Owens, 1980). In the early 1980s this was the normal practice on the Milford Road. However, as the traffic volumes have increased (and the AHI hazard to high) the avalanche management has become more rigorous with the implementation of a full avalanche control programme, including the employment of experienced avalanche forecasters in 1984.

Table 2. Hazard level, AHI, and minimum avalanche management. Modified from Fitzharris and Owens (1980).

Hazard	AHI	Minimum management level
Very Low	< 1	Post signs in avalanche paths
Low	1 to 10	Warnings and road closures
Moderate	10 to 100	Avalanche control, selected sites
High	> 100	Full avalanche control, artificial release, structures

Schaerer(1989) modified the descriptions of the avalanche types and their weightings (Table 3) such that any subsequent calculations of AHI are about 0.7 to 0.9 times the indices calculated with the original weightings.

Schaerer(1989) also restated the method for calculating the AHI, provided some examples, and thereby attempted to make the process more repeatable for future risk assessments. In doing so he restated the equations for moving traffic as shown in equation 4:

$$P_{m_{ij}} = \frac{T(L_{ij} + D)}{R_{ij}.V.24000} \tag{4}$$

where:

$T =$ average daily traffic volume counted in both directions (vehicles d^{-1})
$L_{ij} =$ average length of road covered by avalanches of type j at avalanche path i (m)
$D =$ stopping distance for a vehicle with speed V on a snow covered road (m)
$R_{ij} =$ return period of occurrence of avalanches of class j at the avalanche path i (years)
$V =$ average speed of traffic (kmh^{-1})

Table 3. Avalanche type, description and weighting. Modified from Schaerer (1989).

Avalanche type	Avalanche description	Weighting
j_1 Powder snow	Less than 0.1m deep deposit and crosses the road at speeds less than $20ms^{-1}$. Conditions similar to those from blowing snow.	0
j_2 Slough	Less than 0.3m deep deposit, covers only part of the road, and often originates from a road cutting. Able to drive over or around the deposit.	0
j_3 Light snow	Between 0.3 and 1.0m deep deposit and flowing beyond the road. Cars would be pushed off the road but would not be buried	3
j_4 Deep snow	More than 1m deep deposit, and	10
j_5 Plunging avalanche	High speed dry snow falling long distances over cliffs or steep slopes. Extremely destructive due to their high wind speeds air blasts.	12

Compared to the previous equation of P_m, changes are observed in the units of traffic speed (from ms^{-1} to kmh^{-1}) and the use of return period of avalanche occurrences, the reciprocal of frequency.

Schaerer(1989) also restated the equations for waiting traffic as shown in equation 5:

$$P_{W_{i+1j}} = P_s . \frac{N_{wi+1j}}{R_{i+1j}} \tag{5}$$

where:

P_s =	probability that an avalanche will run along path i+1 while the traffic is waiting
N_{wi+1j} =	number of vehicles exposed to adjacent avalanche path, i+1
R_{i+1j} =	return period of occurrence of adjacent avalanches of type j at the avalanche path i (years)

P_s values ranging from 0.05 and 0.3 have been determined from observations at Rogers Pass, Canada (Schaerer, 1989). Fitzharris and Owens(1980), with no information about how an avalanche occurrence would be related to another avalanche at an adjacent site used a value of 0.15 for P_s. Armstrong(1981) suggested a lower value of P_s at 0.03-0.05 for Red Mountain Pass in Colorado. Schaerer(1989) related the P_s value to the characteristics of the avalanche starting zones, noting that a high P_s value would be appropriate for avalanche paths with similar aspects and terrain characteristics.

Compared to the previous equation of $P_{w,}$ changes can be observed with the use of the return period of avalanche occurrences, the reciprocal of frequency of avalanche occurrences per year. Otherwise the equation is the same; however a more thorough method of calculating the number of exposed vehicles (N_{wi+1j}), is used. To calculate the number of exposed vehicles Schaerer(1989) provided a series of equations that take into account the following factors:

- The average length of road each vehicle occupies (L_v) on a public road, normally 15m
- The waiting period (t), which is the time required for a road crew to react, normally 2 hours
- The safe waiting distance (S) between avalanche paths
- Average daily traffic volume (T)

Schaerer's(1989) equations work out the length of a queue of traffic (L_w), based on average traffic volume. The safe waiting distance between each avalanche path, combined with the queue length, will determine how many vehicles are then exposed in an adjacent avalanche path, in both directions. Recent work by Hendrikx and Owens (2008) has use the peak traffic volume from the low season, rather than the average volume as traffic flow was distinctly tidal in flow with two clear peaks. While this modification leads to an overestimation at times, there are also times when this method may in fact still be an underestimation e.g. the final two hours preceding road closure for avalanche hazard during the spring shoulder season.

Schaerer(1989) also considered the case of a second avalanche on a path which is already blocking the traffic. The probability that traffic is likely to be hit by another avalanche in path i is represented as shown in equation 6:

$$P'_{W_{ij}} = 0.5P'_s \frac{N_{wij}}{R_i}, N_{wij} \leq \frac{L_{ij}}{L_v} \tag{6}$$

where:

P'_s = probability of a second avalanche running along path i once one avalanche has already occurred.

Schaerer(1989) suggested that values for P'_s can range from 0 to 0.5 and must be chosen from a study of the terrain and for most avalanche paths with a single starting zone P'_s=0. In this analysis P'_s has been set at 0 as avalanches need to be substantial to reach the road, and will usually clear instabilities in a given starting zone.

Finally, the resultant AHI as proposed by Schaerer(1989) was expressed as the weighted expected frequency of encounters of moving (P_m) and waiting (P_w) vehicles with j type avalanches, summed over a road with n avalanche paths as shown in equation 7:

$$AHI = \sum_{i=1}^{n}\sum_{j=1}^{5} W_j (P_{m_{ij}} + P_{W_{ij}}) \tag{7}$$

where:

n = number of avalanche paths
j = index for avalanche type j_1 to j_5
W_j = weighting for type j avalanche
P_{mij} = probability of encounter for a moving vehicle in path i with avalanche type j.
P_{wij} = probability of encounter for a waiting vehicle in path i with avalanche type j.

Subsequently, the equations proposed by Schaerer(1989) have been used to calculate the AHI for many roads around the world. Examples include Rogers Pass, Canada (Stethem et al., 1995), Colorado Highways, USA (Mears, 1995) and the Milford Road, New Zealand (Hendrikx et al., 2006; Hendrikx and Owens 2008).

3.2. Probability of Death to an Individual

The PDI is a method used to express risk. It has been widely used for hazard assessment for a range of natural (e.g. landslides) and anthropogenic (e.g. dams) hazards. Weir(1998) undertook an assessment of the avalanche risk on the Milford Road using a PDI and Fatal Accident Rate (FAR) methods, where FAR is expressed as the probability of a fatality per 100 million person hours of exposure. However, recently there have been some further attempts to standardise a method to express the risk in terms of a PDI for a road with an avalanche hazard (e.g. Wilhelm, 1998; Kristensen et al., 2003; Margreth et al., 2003; Norwegian Geotechnical Institute (NGI), 2003). This section will describe the methods and subsequent modifications to the work by Wilhelm(1998) and Margreth et al.(2003). These methods were selected as they are based on fewer assumptions, the variables considered are more clearly explained, they have a large data set of road traffic avalanche fatalities and they have also been applied to several pass roads in Switzerland.

According to Wilhelm(1998) and Margreth et al.(2003), the collective risk (expressed as deaths per year) on a road crossed by n avalanche paths is given by equation 8:

$$CR = \frac{T.\beta}{24}\sum_{i=1}^{n}\frac{L_i}{R_i.V}\lambda_i \tag{8}$$

where:

$CR =$ collective risk (deaths year^{-1})
$T =$ average traffic volume for avalanche period (vehicles d^{-1})
$\beta =$ mean number of passengers per vehicle
$n =$ number of avalanche paths
$L_i =$ width of avalanche = average length of road covered by avalanche i (km)
$R_i =$ return period for avalanche i (years)
$V =$ vehicle speed in (kmh^{-1})
$\lambda_i =$ probability of death in a vehicle hit by an avalanche

The parameters, T, β, n, L_i, V and λ_i can be measured on site or determined from historical records. The return period , R_i, is more difficult to determine so Margreth et al.(2003) estimated it from slope angle in the starting zones and track. Slope angle and return period was then plotted for the available avalanche information from four pass roads and the R_i for the remaining paths having was determined from this relationship. Zisch et al.(2005) extended this approach to short term (day-to-day) avalanche risk using Monte Carlo analysis and return periods based on avalanche observations.

The risk, or probability of death to an individual (PDI) can be calculated according to Wilhelm (1998) and Margreth et al.(2003) using equation 9:

$$IR = \frac{z}{24}\sum_{i=1}^{n}\frac{L_i}{R_i.V}\lambda_i \tag{9}$$

Where:

$IR =$ individual probability of death (year^{-1})
$z =$ number of passages per day of that person (i.e. commuter passengers z = 2, road crew z = 6)

The number of passages for the road crew depends on the condition of the road, and Wilhelm(1998) suggested that it be set at 6. For a daily commuter on the road the number of passes is normally 2. In Switzerland it has been found that on pass roads the mean number of passengers per vehicle β is normally 1.6 (Margreth et al., 2003). In the Swiss Alps between 1946 and 1999, 167 passengers were buried by avalanches in their vehicles, of whom 30 persons or 18% died. Therefore λ_i = 0.18 in equations (8) and (9). Kristensen et al.(2003), despite lacking data, suggested a higher death rate of 40% or 0.40 because of Norway's topographic characteristics, remoteness and long rescue time. Schaerer(1989) however, provided a range of probabilities of death depending on avalanche type, from 0.05 for light snow to 0.25 for deep snow avalanches.

To better enumerate the PDI, Hendrikx et al.(2006) suggested the following improvements to the calculation to account for different avalanche types and resulting probabilities of death are suggested, for collective risk as shown in equation 10:

$$CR = \frac{T.\beta}{24}\sum_{i=1}^{n}\sum_{j=3}^{5}\frac{L_{ij}}{R_{ij}.V}\lambda_{ij} \tag{10}$$

and individual risk as shown in equation 11:

$$IR = \frac{z}{24}\sum_{i=1}^{n}\sum_{j=3}^{5}\frac{L_{ij}}{R_{ij}.V}\lambda_{ij} \tag{11}$$

where:

$j =$ avalanche type j_3 to j_5 (from Table 4)

Margreth et al.(2003), while not specifying different avalanche types, use 0.18 as the probability of death for an avalanche that can bury a vehicle, which may be approximately equal to the deep snow avalanche type of Schaerer(1989). However, Schaerer(1989) has a higher probability of death for this avalanche type at 0.25. In the absence of fatality data, plunging snow has been set at a high probability of death of 0.50 because of the continued evidence of the destructive nature of these avalanches on the Milford Road. Light snow avalanches were assigned a probability of death of 0.05 by Shearer(1989). As there is limited data available on the probability of death for each avalanche type, Hendrikx et al.(2006) used numbers on the upper limit for avalanche type j_3 and j_4 as suggested by Schaerer(1989) for the AHI and shown in Table 4.

When considering these modifications to the PDI equations it is interesting to note that the values in Table 4 indicate that the AHI weightings are not uniformly proportional to the PDI values. Most significant among these discrepancies is the difference between deep snow and plunging avalanches represented by an increase of 10 to 12 in the AHI, but 0.25 to 0.5 for the PDI. This could suggest that either the probability of death for a plunging avalanche may be too high when compared to the AHI weightings, or that the AHI weightings do not sufficiently account for the destructive nature of plunging avalanches.

While the modifications proposed by Hendrikx et al.,(2006) to the methods of Wilhelm(1998) and Margreth et al.(2003) add a variable probability of death, dependent on avalanche type, these equations unfortunately still do not consider the component of risk posed to waiting traffic. Waiting traffic has been shown to be extremely important in applications of the AHI (Fitzharris and Owens, 1980; Schaerer 1989). In the case of the Milford Road, waiting traffic contributed over 95% to the AHI (Fitzharris and Owens, 1980). Kristensen et al.(2003) and NGI(2003), have described a series of complex calculations which attempt to enumerate the increased risk for waiting traffic with increasing queues, while simultaneously decreasing the risk over time, as the probability of a second avalanche reduces. Unfortunately, the calculations were based on cumulative assumptions that were considered unrealistic for application on the Milford Road and other avalanche prone roads.

Hendrikx and Owens (2008) however, have further modified the equation by Wilhelm(1998) and Margreth et al.(2003) using the well established concepts within the AHI to allow for waiting traffic, by calculating the probability of encounter with a subsequent avalanche in an adjacent path as shown in equation 12 for collective risk:

$$CR_W = \frac{\beta}{24}\sum_{j=3}^{5}\lambda_j\sum_{i=1}^{n}\left(\frac{T.L_{ij}}{R_{ij}.V} + P_s\frac{N_{wi+1j}.d_{i+1j}}{R_{i+1j}}\right) \tag{12}$$

and equation 13 for individual risk:

$$IR_W = \frac{z}{24}\sum_{j=3}^{5}\lambda_j\sum_{i=1}^{n}\left(\frac{T.L_{ij}}{R_{ij}.V} + P_s\frac{N_{wi+1j}.d_{i+1j}}{R_{i+1j}}\right) \tag{13}$$

where:

$P_s =$ probability that an avalanche will run along path i+1 while the traffic is waiting

$N_{wi+1j} =$ number of vehicles exposed to adjacent avalanche path, i+1

$d_{i+1j} =$ length of time waiting vehicles are exposed to an avalanche in the adjacent path (hours)

$R_{i+1j} =$ return period of occurrence of adjacent avalanches of type j at the avalanche path i (years)

As the proposed equations by Hendrikx and Owens(2008) are only a recent addition to the literature on estimating the PDI for avalanche prone roads, the equations by Margreth et al.(2003) and Hendrikx et al.(2006) will be used in the following sections. Calculations of PDI using the Margreth et al.(2003) or Hendrikx et al.(2006) methods are likely to be significantly lower than the actual PDI would be during periods involving waiting traffic. This makes the direct comparison of the PDI for avalanche risk difficult to compare with other hazards. However, the Hendrikx et al.(2006) PDI method does still permit for easy comparisons between different roads, though this could be compromised if there were significant differences in the path configuration, e.g. single well spaced paths compared to clusters of paths.

Table 4. Avalanche type, probability of death and AHI weighting

Avalanche type	Probability of death	AHI weighting
j_1 Powder snow	0.00	0
j_2 Slough	0.00	0
j_3 Light snow	0.05	3
j_4 Deep snow	0.25	10
j_5 Plunging avalanche	0.50	12

4. SENSITIVITY OF THE METHODS

4.1. The Avalanche Hazard Index (AHI) Sensitivity

While the AHI equations have seen some modifications and additions since their inception in 1974 (Avalanche Task Force, 1974), there has been no assessment of their sensitivity to the various assumptions made. Hendrikx et al.(2006) used equation 7, the modified weightings proposed by Schaerer(1989) and a comprehensive data set from the Milford Road New Zealand to calculate the AHI of the controlled and theoretically uncontrolled avalanche regime, using the following parameters:

$T =$ 381 estimated average daily winter traffic volume in 2002 (URS, 2004)
$V =$ 11 ms^{-1} (Fitzharris and Owens, 1980)
$D =$ 20 m (Fitzharris and Owens, 1980)
$L_v =$ 15 m (Fitzharris and Owens, 1980)
$t =$ 2 h (Schaerer, 1989)
$L_w =$ 525m (Calculated using L_v and a peak winter flow of 35 vehicles h^{-1})
$P_s =$ 0.15 (Fitzharris and Owens, 1980)

The controlled avalanche regime was calculated to have an AHI of 2.76, comprised of a P_m of 0.57 and P_w of 2.19. The theoretically uncontrolled avalanche regime was calculated to have an AHI of 186.3, comprised of a P_m of 5.2 and P_w of 181.1. However, Hendrikx et al.(2006) asserted that these two calculations for the controlled and uncontrolled AHI on the Milford Road are both maximum and minimum values respectively due to the assumptions and constraints on the equations.

To further examine this, the following section will assess the sensitivity of AHI to the assumptions. Using the theoretically uncontrolled avalanche regime, the following parameters were varied: average speed V, stopping distance D, length of waiting queue L_w, probability of an adjacent avalanche path avalanching P_s, and length of road a vehicle occupies L_v. Only one parameter was changed at a time, leaving the remaining parameters unchanged, thereby allowing the examination of the sensitivity of the AHI to each parameter. T has expressly not been varied, as it is the effect of the other assumptions on the AHI with current traffic flow that is to be examined. The parameters were graphed against AHI as a ratio of the maximum value tested for each parameter (Figure 1), where:

$V =$ 6 ms^{-1} to 22 ms^{-1} (minimum and maximum road speed)
$D =$ 10 m to 80 m (dependent on speed and road snow cover)
$L_w =$ 100 m to 1500 m (based on an L_v of 15m and t of less than 1h to 3h)
$P_s =$ 0.05 and 0.3 (range as described for Rogers Pass (Schaerer, 1989)
$L_v =$ 5 m to 25 m (range dependent on ratio of cars to busses)

The AHI is strongly sensitive to Ps across the range, is strongly sensitive to both Lv and Lw in the low part of the range but insensitive towards the top of the range of variation while it is insensitive to variations in D and V. The length of waiting queue L_w increases the AHI at an approximately constant rate up to a threshold of 0.7 (1050m). In this example, when this threshold is reached, all of the safe waiting distances are utilized by the queue of waiting traffic and all possible queues extend into an avalanche path. Small increases thereafter are the result of the wider or adjoining avalanche paths being completely filled by waiting vehicles. The queue length is directly proportional to the time traffic is waiting, and the length of each vehicle on the road. In practical terms, queue length can therefore be decreased through regular road patrols which reduce the potential waiting time. There is a strong sensitivity of the AHI with P_s for the complete range of P_s values used. This supports the notion by Schearer(1989) that P_s must be determined from a local study of the avalanche terrain. As the length of road a vehicle occupies L_v, decreases the AHI increases sharply. This simply enumerates how many elements are at risk in any avalanche path as shorter vehicles will mean more vehicles will fit into an avalanche path. While this means that having only bus traffic will lower the calculated AHI, it may not lower the actual risk as the AHI does not account for the number of people in a vehicle. Reductions in risk could be expected if the same numbers of people were all travelling on buses, rather than individual cars, as the bus queue would be shorter, reducing the probability of the queue extending into an adjacent avalanche path. Changing values of average speed V, and stopping distance D, result in very slight changes of AHI. As average speed V increases, the AHI decreases very slightly, as moving vehicles are exposed to the hazard for a shorter period of time. As stopping distance D, increases the AHI increases slightly, as moving vehicles that cannot stop in time for an avalanche will be caught by that avalanche. Both variables only affect the P_m part of the AHI equation, not the P_w part, so are responsible for only very small changes in the overall AHI. The AHI is most strongly influenced by parameters that affect the P_w component. This analysis clearly shows the magnitude of the various assumptions in the AHI equations. Small changes in the L_w and the L_w parameters can lead to the doubling of the AHI.

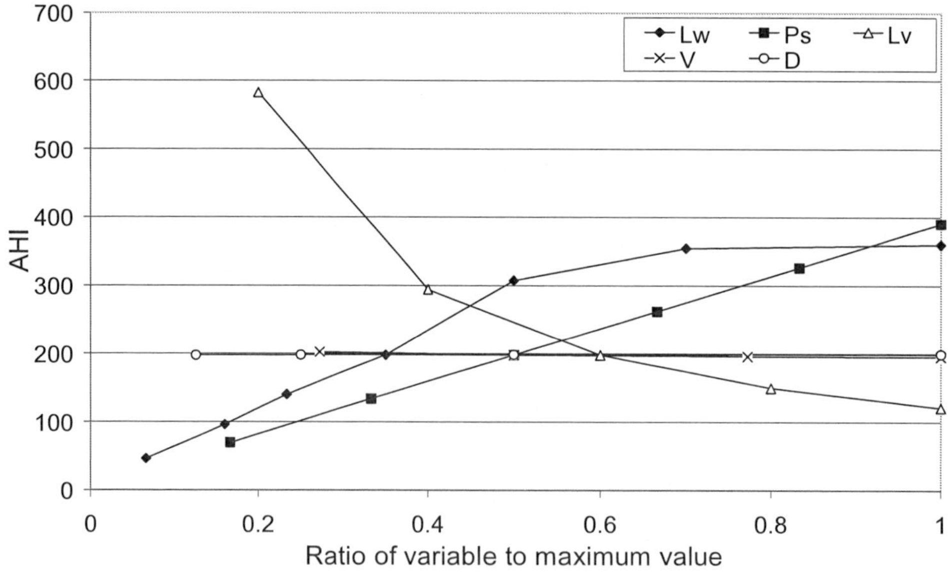

Figure 1. Sensitivity of AHI to variables in equations (4), (5) and (7), showing the ratio of variable to maximum value tested on the x axis and the AHI on the y axis.

4.2. Probability of Death to an Individual Sensitivity

While the PDI equations of Margreth et al.(2003) and Hendrikx et al.(2006) have seen some modifications and additions since, there has been no robust assessment of their sensitivity to the various assumptions made. Hendrikx et al.(2006) used equations 10 and 11 and a comprehensive data set from the Milford Road New Zealand to calculate the PDI of the controlled and theoretically uncontrolled avalanche regime, using the following parameters:

$C{:}B$ = 0.86 : 0.14 (URS, 2004)
β_B = 30 (Hendrikx et al., 2006)
β_C = 1.6 (Margreth et al., 2003)
V = 40 kmh^{-1} (Fitzharris and Owens, 1980)
λ_{i3} = 0.05 (Schaerer, 1989)
λ_{i4} = 0.25 (Schaerer, 1989)
λ_{i5} = 0.5

The controlled avalanche regime was calculated to have a PDI of 5.4 x 10^{-5} for a commuter with 2 passes per day and 1.6 x 10^{-4} for a member of the road crew with 6 passes per day. The collective risk was calculated at 0.058 deaths per year. The theoretically uncontrolled avalanche regime was calculated to have a PDI of 6.8 x 10^{-4} for a commuter with 2 passes per day and 2.1 x 10^{-3} for a member of the road crew with 6 passes per day. The collective risk was calculated at 0.726 deaths per year. Hendrikx et al.(2006) assert that these two calculations are both maximum and minimum values respectively due to the assumptions and constraints on the equations.

To further examine this, the following section will assess the sensitivity of PDI to the assumptions. Using the theoretically uncontrolled avalanche regime, the following parameters were varied: the ratio of cars to buses $C{:}B$, number of passengers per bus β_B number of passengers per car β_C, and speed of the vehicles V. Only one parameter was changed at a time, leaving the remaining parameters unchanged, thereby allowing the examination of the sensitivity of the PDI of a commuter. T has expressly not been varied, as it is the effect of the other assumptions on the PDI with present day traffic flow that is to be examined. The parameters were graphed against PDI as a ratio of the maximum parameter tested (Figure 2 and 3), where:

$C{:}B$ = 1.0 : 0.0 to 0.7 : 0.3 (no buses to over double the proportion of buses)
β_B = 20 to 40 (half full to almost completely full)
β_C = 1 to 4 (single occupancy to almost full)
V = 20 to 100 kmh^{-1} (minimum and maximum road speed)

The PDI is strongly sensitive to the variable speed across the range, however it is insensitive to all other parameters. The PDI is completely unaffected by the changes in the ratio of cars to buses $C{:}B$, number of passengers per bus β_B number of passengers per car β_C. This is because the risk to an individual remains the same regardless of whether they are in a bus or car, or the number of people in that vehicle with them. The speed of the vehicles V, does strongly affect the PDI. As the speed increases the PDI decreases, as faster moving vehicles are exposed to the hazard for a shorter period of time.

The collective risk shows a strong sensitivity to all of the parameters across the range. Only the variable speed, displays a negative relationship with the collective risk. As shown in Figure 3, the collective risk increases rapidly at a linear rate as the ratio of cars to buses *C:B*, reaches a minimum of 0.70 : 0.30 cars to buses. This is simply due to the higher number of people in a bus, and therefore greater consequences if hit by an avalanche. Likewise, as the number of people in each bus β_B, and car β_C, increases, so too does the collective risk. As with the PDI, the collective risk decreases as the speed of the vehicles *V*, increases. Increasing the speed of the vehicles will reduce both the PDI and collective risk deduced from avalanche risk, but will most likely increase the risks from other sources, e.g. road accidents.

As a final assessment of the sensitivity, the probability of death for each avalanche type λ_{ij} has also been varied in the range from:

λ_{i3} = 0.05 to 0.075
λ_{i4} = 0.17 to 0.42
λ_{i5} = 0.2 to 0.5

PDI for commuters and road crew, and collective risk did change as a result of these modifications, however they remained within the same orders of magnitude.

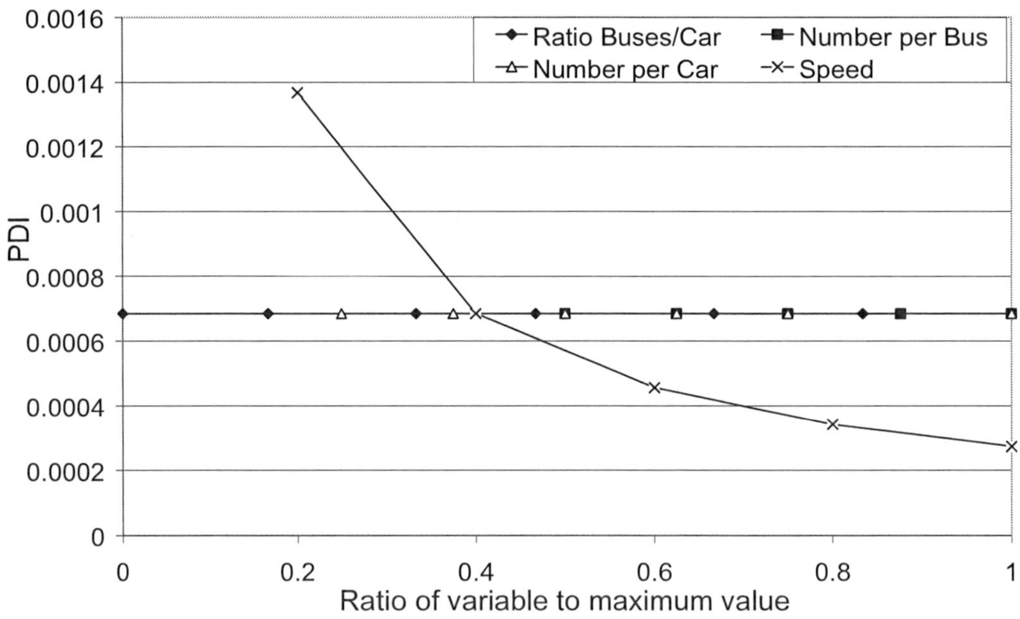

Figure 2. Commuter PDI sensitivity, with the ratio of variable to maximum value, on the x axis and PDI on the y axis.

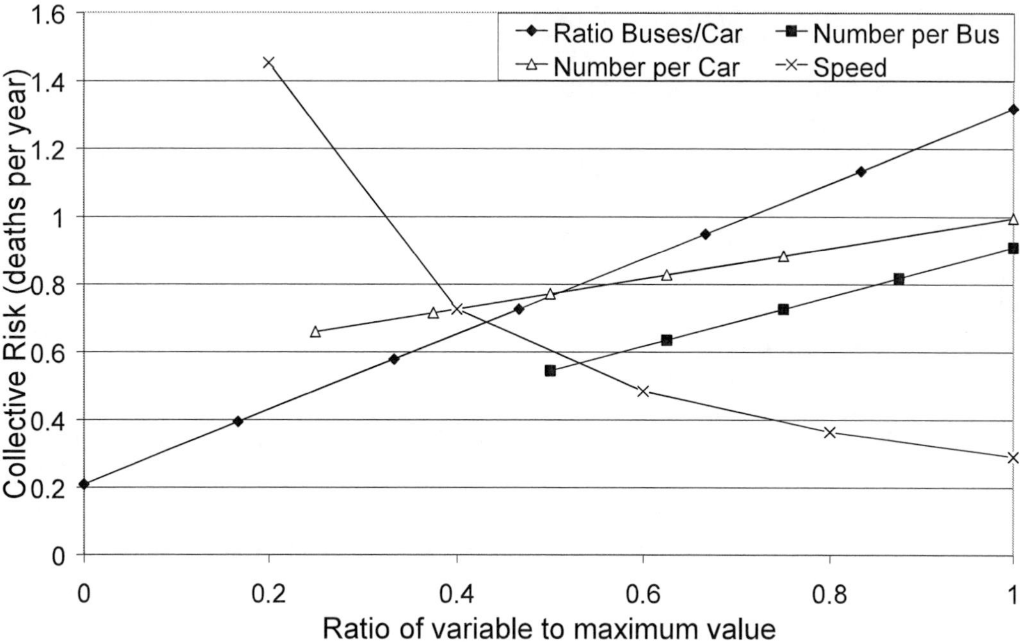

Figure 3. Collective risk sensitivity, with the ratio of variable to maximum value, on the x axis and collective risk on the y axis.

Table 5. AHI before and after control for Rogers Pass and the Milford Road

Road	Endangered road (km)	Number of avalanche paths	T (vehicle d^{-1})	AHI – with no control	AHI – after structural control	AHI – after structural control, standardised to T = 1000	AHI - residual, standardised to T = 1000	AHI - residual
Rogers Pass, 1974 (Avalanche Task Force, 1974)	36	65	905	335	174	192	?	?
Rogers Pass, 1987 (Schaerer, 1989)	36	65	1700	1004	235	138	8.8	<15
Rogers Pass, 1992 (Stethem et al., 1995)	36	130	2300	850	214	93	15.2	<35
Milford, 2002	29	50	381	186	186	488	7.3	<2.8

5. Discussion and Case Studies

Having highlighted how both the AHI and PDI methods have evolved and been modified, and how the assumptions made affect the final results, a comparison for several avalanche prone roads can now be presented. For both the AHI and PDI methods, the calculated level of risk experienced on the Milford Road, New Zealand will be compared to other roads around the world, and to other acceptable risks for natural and man made hazards.

5.1. Avalanche Hazard Index

Using equation 7 and the modified weightings proposed by Schaerer(1989) the AHI on the Milford Road can be compared to that of Rogers Pass, B.C., Canada. Table 5 shows the comparison of AHI before any control, after structural control (not including artillery), and the residual AHI for the Milford Road and Rogers Pass.

While the AHI has been modified over the period of comparison and the assumptions made by the various authors are not always clearly outlined, the values in Table 5 can still be used for the basis of a conservative comparison. T has been standardised to 1000 for all roads to permit comparison of the effect of the physical attributes of the avalanche paths, and therefore not just highlight the difference in traffic volumes. When the AHI after structural control is considered, based on a $T = 1000$, the Milford Road has a significantly higher AHI than the historical values for Rogers Pass (Table 5). It is interesting to note that per 1000 vehicles the AHI, after structural control, has decreased at Rogers Pass, clearly showing the effect of increasing the number of structures on the road. The Milford Road currently has no structures to mitigate the effect of an avalanche. When the residual AHI is considered, again based on $T = 1000$, the Milford Road has a much lower AHI than the historical values for Rodgers Pass.

Using Table 5, the effectiveness of the management and forecasting practices (following structural control) of each avalanche programme can also be compared. The AHI after structural control compared to the residual AHI is reduced by 179 or 84% on the 1992 Rogers Pass, compared to 183 or 98% on the Milford Road. However, when comparing the uncontrolled AHI, Rogers Pass clearly has a much more serious avalanche problem with a very high AHI, especially for the 1987 calculation.

Schaerer(1989) showed that people can accept an AHI of 1, and reducing this value would usually require control measures that are economically not justified, or would demand unacceptably long traffic delays. The calculations in Table 5 all show the residual AHI to exceed 1, suggesting that, according to the criterion of Schaerer(1989), the AHI is unacceptable on all these roads. However, the true AHI on these roads is lower than the calculated AHI, as the theoretical frequency of encounters has been found to be far greater than the observed number (Schaerer, 1989), and current road operators are aware of times of increasing hazard while the road is open, and will use mitigating strategies to reduce the hazard.

Table 6. Collective risk before and after control. Modified from Margreth et al.(2003)

Road	Affected road length (km)	Endangered road (km)	Number of avalanche paths	T (vehicle d^{-1})	Closed days (d $year^{-1}$)	Initial collective risk (deaths $year^{-1}$)	Initial collective risk, T = 1000	Residual collective risk (deaths $year^{-1}$)	Residual collective risk, T = 1000
Flüela	19.3	10.1	47	1000	> 95	0.70	0.70	0.13	0.13
Lukmanier	25	13.6	94	1000	> 68	0.51	0.51	0.07-0.13	0.07-0.13
Gotthard	24	15.7	55	6000	> 142	5.54	0.92	0.11	0.02
Milford	29	29	50	381	< 20	0.73	1.92	0.058	0.15

5.2. Probability of Death to an Individual

Using the equations of Margreth et al.(2003) and Hendrikx et al.(2006) the collective risk on the Milford Road can be compared to several pass roads in Switzerland (Table 6). While a direct comparison of the Milford Road will not be entirely equivalent (as the modified PDI calculations by Hendrikx et al.(2006) use higher probabilities of death for an avalanche), it will provide a basis for conservative comparison.

When the initial collective risk is considered, based on a $T = 1000$, the Milford Road is subject to significantly higher risk than three main pass roads in Switzerland; Flüela, Lukmanier and the Gotthard (Table 6). T has been standardised to 1000 for all roads to permit comparison of the effect of the physical attributes of the avalanche paths, rather than highlighting the difference in traffic volumes. When the residual risk is considered, again based on $T = 1000$, the Milford Road has only marginally higher risks than the Flüela and Lukmanier pass roads. When this is compared to the relative accessibility of the roads, the Milford Road is seldom closed more than 20 days $year^{-1}$, while the Flüela (between 1964 and 1971) was closed a minimum of 95 days $year^{-1}$, and the Lukmanier (between 1965 and 1997) was closed a minimum of 68 days $year^{-1}$ (Margreth et al., 2003). The residual collective risk on the Milford Road, as controlled by the avalanche programme, is only 8% of the initial collective risk. This is significantly less than 19% for the Flüela and 14-25% for the Lukmanier, which is controlled by artificial release on the northern side. The Gotthard however, has a very low residual collective risk, at only 2% of the initial collective risk, which reflects the long winter closures (> 142 days $year^{-1}$), thereby almost eliminating all risk.

Wilhelm(1999 [in] Margreth et al., 2003) suggests that a driver on a public road with a low probability of avoiding an avalanche should have a PDI lower than $1x10^{-5}$. This is compared to a PDI of $8.3x10^{-5}$ for a traffic accident in Switzerland (Margreth et al., 2003). NGI(2003) suggests that an avalanche encounter should be viewed as an 'obligatory' or involuntary risk, and as such should have a 1:10 ratio of the traffic accident rate. In Norway, the traffic accident rate is approximately 400 for a total population of 4 million, so the avalanche rate corresponds to approximately 1 per 100,000 inhabitants or $1x10^{-5}$. New Zealand in 2002, with a similar population of 3,939,100 and road traffic fatalities of 404 (Land Transport Safety Authority, 2003) had a PDI for road traffic fatalities of $1.03x10^{-4}$, similar to that of Norway. Applying the approach taken by NGI(2003) suggests that the Milford Road should have a PDI from avalanche risk no greater than $1.03x10^{-5}$. The risk analysis in Table 5 provides a higher number at 5.4×10^{-5} for a commuter with 2 passes, based on the assumptions listed in section 4.

To permit the comparison of avalanche risk in terms of PDI with risks from other hazards, the waiting traffic component that the equations of Wilhelm(1998), Margreth et al.(2003) and Hendrikx et al.(2006) neglect, must be considered. Using equation 12 and 13 we can account for moving and waiting traffic for collective and individual risk respectively. When using the previously stated parameters in section 4, and a value of 2 hour for d_{i+1j} is used, inclusion of the waiting traffic component increases the calculated individual and collective risk significantly as shown in Table 7

The values calculated in Table 7 must be viewed as an absolute maximum on the controlled Milford Road, as the speed of traffic is often greater than 40 kmh^{-1} and the current road operators are very aware of, and have measures in place to ensure safe practices during times of increasing hazard while the road is open, to minimise the length of time waiting traffic is exposed to an avalanche path. In this analysis two hours has been used as the exposure time for waiting traffic to highlight the impact of including waiting traffic into these equations. In reality, the whole queue will not be exposed for the same period of time, unless all the vehicles arrive simultaneously. In the opinion of the author, because of the control measures in place, the true risk on the controlled Milford Road is closer to 0.5 to 0.35 of the calculated value, which results in approximately 0.27 deaths $year^{-1}$ for the collective risk, $2.5x10^{-4}$ for individual risk for a commuter and $7.5x10^{-4}$ for individual risk for a member of the road crew.

Table 7. Collective risk and individual risk (deaths $year^{-1}$) including waiting traffic and not including waiting traffic, for a commuter and a member of the road crew on the controlled and theoretically uncontrolled Milford Road using 2002 traffic values. (Hendrikx and Owens, 2008)

	CR	CRw	IR Commuter	IRw Commuter	IR Road crew	IRw Road crew
Uncontrolled Milford Road	0.726	20.23	6.8×10^{-4}	1.9×10^{-2}	2.1×10^{-3}	5.7×10^{-2}
Controlled Milford Road	0.058	0.76	5.4×10^{-5}	7.1×10^{-4}	1.6×10^{-4}	2.1×10^{-3}

Suggested guidelines for individual risk in terms of PDI have been published by several organisations. In Iceland, new legislation for hazard zoning for residential areas has suggested that the upper limit of acceptable risk is $0.2x10^{-4}$ (Arnalds et al., 2004). In British Columbia, dam failure fatality probabilities must not exceed $1x10^{-4}$, which is based upon the concept that the risk to an individual from a dam failure should not exceed the individual natural death risk run by the safest population group (Hoek, 2000). The Australian Geomechanics Society (AGS) has suggested $1x10^{-4}$ for natural slopes and $1x10^{-5}$ for engineered slopes for landslides (AGS, 2000). Nielsen et al.(1994) suggested that the annual probability of fatality of $1x10^{-4}$ also defined the boundary between voluntary (restricted access to workers) and involuntary (general public) risk on dam sites. These results all suggest that when using the equations suggested by Hendrikx and Owens (2008), that the individual risk to a commuter on the controlled Milford Road, when accounting for waiting traffic, is unacceptably high when compared to the risk of other hazards. Therefore, to lower the level of risk on the Milford Road and any other avalanche prone road, it is imperative that stationary traffic should be avoided in the avalanche zone at all costs. In the case of the Milford Road, this is already common practice, with large areas of no stopping zones along the avalanche area, and large fines imposed on vehicles without snow chains (thereby reducing the likelihood of unintentional stops). Furthermore, while the individual and collective risk enumerated in deaths year^{-1} may still seem high, Schaerer(1989) noted that the theoretical frequency of encounters has been found to be far greater than the observed number, providing the example of Kootenay Pass, where the expected encounter frequency was six vehicles year^{-1} between 1965 and 1984, but on average only 1.9 vehicles year^{-1} were actually hit during this period.

Conclusion

This chapter has reviewed and examined the application of two key risk assessment methods to examine avalanche risk on highways, namely the Avalanche Hazard Index and the Probability of Death to Individuals. This review has shown how these equations have been modified and extended to further expand and refine their use. Sensitivity analysis of the results of the equations for a case study highlights the sensitivity of particular variables and therefore the need to carefully consider these variables when estimates must be made. Sensitivity analysis of the AHI shows that the length of a waiting queue, the probability of a second avalanche in an adjacent path, and the length of road a vehicle occupies to affect the AHI strongly. Sensitivity analysis of the PDI show that the speed of a vehicle is the only parameter that affects PDI, while all parameters tested affect the collective risk. Using both the AHI and PDI the Milford Road was compared to other roads around the world, highlighting how different management techniques such as structural control or passive control affect the residual AHI. A final assessment for the Milford Road, using modified PDI equations also permit the calculation of the waiting traffic component.

Acknowledgements

We would like to thank Transit New Zealand, Works Infrastructure Te Anau, and all the people that have worked on the Milford Road. Their observations and ongoing efforts have made this work possible.

References

Armstrong, B.R., 1981. A quantitative analysis of avalanche hazard on U.S. Highway 550, southwestern Colorado. Paper presented at the *Western Snow Conference*, 49, 95-104.

Arnalds, Þ, Jónasson, K. and Sirgurdsson, S., 2004. Avalanche hazard zoning in Iceland based on individual risk. *Annals of Glaciology*, 38, 285-290.

Australian Geomechanics Society Sub-committee on Landslide Risk Assessment, 2000. Landslide risk management concepts and guidelines. *Journal and News of the AGS*, 35, No. 1.

Avalanche Task Force, 1974. *Report on findings and recommendations, Appendix II*. British Columbia Department of Highways, Victoria, B.C., Canada.

Fitzharris, B.B. and Owens, I.F., 1980. *Avalanche atlas of the Milford Road; an Assessment of the Hazard to Traffic*. New Zealand Mountain Safety Council, Avalanche committee Report No. 4, 79 pp.

Hendrikx, J., Owens, I., Carran, W. and Carran, A., 2006. Avalanche risk evaluation with practical suggestions for risk minimization: A case study of the Milford Road, New Zealand. Proceedings of the *International Snow Science Workshop*, Telluride, Colorado, USA, 757-767.

Hendrikx, J. and Owens, I., 2008. Modified avalanche risk equations to account for waiting traffic on avalanche prone roads. *Cold Regions Science and Technology*, 51, 214–218.

Hoek, E., 2000. *Practical Rock Engineering – course notes by Evert Hoek*. A.A.Balkema Publishers, Rotterdam, Netherlands, 313 pp.

Kristensen, K., Harbitz, C.B. and Harbitz, A., 2003. Road Traffic and avalanches – methods for risk evaluation and risk management. *Surveys in Geophysics*, 24, 603-616.

Margreth, S., Stoffel, L. and Wilhelm, C., 2003. Winter opening of high alpine pass roads—analysis and case studies from the Swiss Alps. *Cold Regions Science and Technology*, 37, 467–482

Mears, A.I., 1995. Avalanche hazard Index for Colorado Highways. Report prepared for the Colorado Department of Transport. Report No. CDOT-DTD-R-95-17. 55p.

Nielsen, N.M., Hartford, D.N.D. and MacDonald, T.F., 1994. Selection of tolerable risk criteria for dam safety decision making. Paper presented at the *1994 Canadian Dam Safety Conference, Winnipeg, Manitoba*. BiTech Publishers, Vancouver, Canada, 355-369.

Norwegian Geotechnical Institute (NGI), 2003. *Road traffic and avalanches – methods for risk evaluation and risk management*. NGI report no. 20001289-4, Oslo, Norway, 33 pp.

Schaerer, P., 1989. The Avalanche Hazard Index. *Annals of Glaciology*, 13, 241-247.

Stethem, C., Schaerer, P., Jamieson, B. and Edworthy, J., 1995. Five mountain parks highway avalanche study. Paper presented at *International Snow Science Workshop*, Snowbird, Utah, USA, 72-79

URS, 2004. *Final draft scoping report for the SH94 Homer Tunnel east portal avalanche shed, Section 6.1 Traffic Statistics*. Unpublished draft report for Transit New Zealand, URS Report R001A3C, 6 pp.

Weir, P.L., 1998. Avalanche Risk Management – the Milford Road. [In] Elms, D. (ed.). *Owning the Future: Integrated Risk Management in Practice*. University of Canterbury Center for Advanced Engineering, Christchurch, New Zealand, pp. 275-292.

Wilhelm, C., 1998. *Quantitative risk analysis for evaluation of avalanche protection projects*. Norwegian Geotechnical Institute Publication, Oslo, Norway, vol. 203, pp. 288–293.

Zischg, A., S. Fuchs, M. Keiler, and J. Stötter, 2005. Temporal variability of damage potential on roads as a conceptual contribution towards a short-term avalanche risk simulation. *Natural Hazards and Earth System Science*, 5, 235-242.

In: Road Traffic: Safety, Modeling and Impacts
Editors: S. E. Paterson and L. K. Allan, pp. 439-453
ISBN 978-1-60456-884-4

Chapter 12

DRIVING BEHAVIOR AND COGNITIVE TASK PERFORMANCE OF FATIGUED DRIVERS: EFFECTS OF ROAD ENVIRONMENTS AND THEIR CHANGES

***Yung-Ching Liu*[*,1] and *Tsun-Ju Wu*[2]**
[1]Department of Industrial Engineering and Management, National Yunlin University of Science and Technology, Yunlin, Taiwan
[2]Phoenix Precision Technology Cooperation, Hsinchu, Taiwan

ABSTRACT

This study aims to explore the effects of different road environments and their changes on driving behaviors and cognitive task performance of fatigued drivers. Twenty-four participants volunteered in a 2 (road environment) x 3 (fatigue level) within-subject factorial design simulated driving experiment. Participants were asked to perform basic numerical calculation and distance estimation of traffic signs when driving normally, and provide answers to a questionnaire on fatigue rating. Results show that fatigued drivers faced greater attention demand, were less alert, and tended to overestimate the distance to roadside traffic signs. Fatigue caused by driving in complex road environment had the greatest negative impact on driving behavior and visual distance estimation, and the fatigue transfer effect worsened significantly both driving behavior and performance of fatigued drivers when switching from a complex to a monotonous road environment and vice versa. Notably, this study shows that fatigued drivers performed relatively better in arithmetic tasks than non-fatigued ones. In addition, when switching from a monotonous to a complex road environment, drivers' performance in visual distance estimation and arithmetic tasks improved though driving

[*] Corresponding author: Tel: +886-5-5342601 ext. 5124; fax: +886-5-5312073. E-mail address: Liuyc@ yuntech.edu.tw

behavior deteriorated, revealing that the fatigue effect upon drivers might be explained to some extent by the driver's alertness and arousal levels.

Key words: cognitive task, driver behavior, fatigue, fatigue-transfer effect, road environment

1. INTRODUCTION

Feeling fatigue is a common experience for most drivers. However, this usual experience could be a critical potential threat to road traffic safety. Driver fatigue has been shown to cause 1 to 3% of all crashes and 4% of fatalities in the United States (Lyznicki et al., 1998). Horne and Reyner (1995) reported that 16-20% of all motor vehicle crashes in the United Kingdom were related to sleep. From the self-reported factors of drivers, Maycock (1997) found that fatigue accounted for 9-10% of their recent accidents. In Australia, the figure has been reported to be 6% (Fell, 1994). There is a lot of evidence showing that fatigued drivers are a serious threat to road safety and prone to traffic accidents (Gander et al., 1993; Maycock, 1997; Williamson et al., 2001). Fatigue-related accidents have very serious consequences such as head-on collision (Pack et al., 1995; Horne & Reyner, 1999; Sagberg, 1999).

Most studies on fatigue driving focus on "endogenous" factors of drivers (Thiffault & Bergeron, 2003) such as sleep deprivation (Dinges, 1995), time of day (Horne & Reyner, 1999) and prolonged driving (Lisper et al., 1986), while research on "exogenous" factors such as road environment has been scarce. In countries like Taiwan and Japan with small total area but high population density and traffic load, cities lie in close proximity, separated by around an hour's drive, drivers thus experience within a short time widely different road environments shuttling between complex urban roadways and monotonous country paths.

Owing to the wide variations in road environment, driving in cities and country towns will require different amount of effort and shuttling between the two types of road environment having different complexities will also cause fatigue (Hancock & Warm, 1989; Hancock & Desmond, 2001).

For this reason, road geometry, roadside environment and other road factors that affect driving and make roads either complex or monotonous will directly affect the changes in the level of drivers' arousal and alertness (see Davis & Parasuraman, 1982), and this may be responsible for the occurrence of road accidents (Oron-Gilad & Hancock, 2005; Thiffault & Bergeron, 2003).

According to the attention resource theory, fatigue might reduce drivers' resource availability and deteriorate their subsidiary task performance (Harms, 1991). Complex cognitively demanding tasks are more vulnerable to the effect of fatigue (Pilcher & Huffcutt, 1996). Thus, complex traffic condition and road environment make driving task more demanding and affect drivers' performance.

Paradoxically, when the driving task is relatively difficult (e.g., on curved roads), fatigued drivers were able to navigate their vehicles and cope with the road curvatures better. While on straight roads, their performance tended to deteriorate, implying, to some extent, that fatigued drivers are not good in adjusting their effort effectively (Desmond & Matthews, 1997; Matthews & Desmond, 202). Desmond and Matthews (1997) described this as the effort-regulation hypothesis, an alternative perspective of the attention theory developed by Hancock and Warm (1989). Fatigue affects this effort-regulation most detrimentally in simulated driving under low-load condition (Matthews et al., 1996).

It is conceivable that monotony causes boredom and drowsiness, which is expressed in reduced effort on the task at hand (Craig & Cooper, 1992). On the other hand, more interesting and cognitively demanding tasks produce less fatigue effect (Kraemer et al., 2000). Thiffault and Bergeron (2003) found that in a monotonous driving situation, steering wheel movement of drivers is greater and occurs more often, showing that the effect of fatigue caused by a monotonous road environment on driver vigilance is relatively large. Driver fatigue is a physical and psychological state that lowers vigilance and alertness (Thiffault & Bergeron, 2003; Williamson et al. 1996). Dinges (1995) indicated that fatigued drivers show vigilance decrement in terms of slower visual reaction when driving on a monotonous road.

Fatigue has a negative effect on drivers' valid peripheral visual field (Rogé et al., 2003) and visual perception (Quant, 1992). Narrowing of visual field and poor visual acuity perception might cause roadside information (road signs, obstacles) to be undetected or distance to be misjudged, thus resulting in accidents.

The aforementioned fatigue effects caused by the changes in road environments on driving behavior, cognitive performance and visual perception ability of drivers are worth investigating. The present study will explore (1) the effect of fatigue produced by driving in different road environments, and (2) fatigue caused by changes in road environment, and how fatigue from the previous road stretch affects driving in the next road stretch of a different environment.

2. Methods

2.1. Participants

Twenty-four gender-balanced participants, aged between 24 and 30, holders of a valid driving license for over three years, driving on average twice a week and at least 5000 Km per year, were recruited. All participants' vision (corrected to over 0.9 and all passed Ishihara color test) and hearing (able to carry out a normal conversation with the experimenter when driving at 100 km/h using the simulator) were normal. Participants had no sleeping problems, no prior experience of using a driving simulator, and all performed the test at either 9:00-12:00 A.M. or 2:00-5:00 P.M. for keeping their circadian pattern in a similar condition (Craig et al., 2006). Each subject was paid US$30 for completing the test.

2.2. Apparatus

An interactive STI low-cost, fixed-base driving simulator developed by Systems Technology, Inc. Hawthorne, CA, USA was used in this study. The simulated vehicle cab, a VOLVO 340 DL, featured all normal automotive displays and controls (steering, brakes, and accelerator) found in an automatic vehicle. Different driving scenarios were projected onto a

120-inch screen with sound effects of the vehicles in motion broadcasted by two-channel amplifiers. Driving-related information such as speed and task instructions from the experimenter were projected on the head-up display (HUD) located 3.1 m in front of the driver. The vertical projection angle is between 6° and 12° below the driver's horizontal visual line, and the HUD area is about 32 (w) x 22 (h) cm^2 (~15 in^2). The display resolution is 800 x 600 dpi, and the presentation font (icon) size is 10 x 10 cm^2 (~1.8°).

2.3. Tasks

To understand the effect of fatigue induced by different road environments on driver's judgment, attention, comprehension, motor coordination, visual discrimination and reaction time, subjects were asked to perform the following tasks.

2.3.1. Driving

Subjects were asked to drive on the two types of motorway route on the driving simulator. These routes were designed to simulate the heavy traffic urban highway and monotonous rural roadway to generate workload and boredom, respectively for the driver as described above (Liu & Wen, 2004). Subjects had to observe the speed limit, comply with all traffic rules and safety regulations, and complete the required tasks as fast as possible making as few mistakes as possible.

2.3.2. Visual Traffic Sign Distance Estimation Task

In this task, subjects had to determine the distance of the vehicle from the "Beware of Pedestrian" sign standing 3 m away from the right-lane borderline. Fifteen seconds before the traffic sign appears, a verbal cue "Traffic Sign" was given to alert the subjects. When the subjects perceived that the traffic sign was at a 5-second distance away, they indicated by saying "Here" to complete the task and the experimenter started the stopwatch to record the exact time between the driver and the traffic sign. These distance estimation tasks were performed four times in each road stretch.

2.3.3. Arithmetic Task

Information related to the cognitive judgment task was displayed on the HUD for 5 seconds since it has been reported that 3-5 seconds are needed for the drivers to receive inputs from in-vehicle signing information systems (ISIS) (Collins et al. 1999). Five seconds before the task, the sign "Start" was shown as a pre-alert cue. Each sign will remain on the HUD for 5 seconds, followed by a 5-second interval and another sign. Subjects will have to complete the arithmetic tasks (addition or subtraction) and provide the answer orally. Arithmetic task assesses the information processing power and short-term memory of the subjects. Arithmetic tasks included two additions and two subtractions with no questions repeated and will be performed four times on each road stretch.

2.3.4. Psychological Fatigue Rating Task

The psychological fatigue rating questionnaire was adapted from Nilsson et al. (1997). The judgment scale started with zero: no fatigue, one: low fatigue and went to three: high fatigue. The questionnaire was divided into four dimensions, including perception, boredom, ankylosis and anxious fatigue characteristics. The related fatigue scales were given verbally by participants six times; that is, before and after driving on each road stretch and during switches of road environment.

2.4. Experimental Design and Procedures

To increase the driving burden, participants were required to drive at 100 km/h on busy city roads and at 70 km/h on monotonous country roads. Subjects drove in a non-fatigued state for 20 minutes on complex or monotonous road stretches and carried out related cognition/visual tasks mentioned above. The data from this stretch served as the baseline (the non-fatigued state) for comparison with the later fatigue-related results. A rest of 20 minutes followed allowing the drivers to return to a normal physical and mental state.

Published reports show that 60% of drivers become tired (Gainsky et al., 1993; Skipper & Wierwille, 1986) and 60% of fatal fatigue-related vehicle accidents occur within the first hour of driving (Summala & Mikkola, 1994). In view of these findings, all subjects in this study started, after resting, a 60-minute drive with no cognitive/visual works on a complex or monotonous road stretch to induce fatigue (fatigue induction road stretch), followed by 20 minutes of driving on a complex or monotonous road stretch and performing cognition/visual tasks (the fatigued state). Finally, there was 20 minutes of driving on a monotonous or complex road stretch (the fatigued-transfer state). The assessment items provided by this road design include: 20-minute driving on a complex or monotonous road stretch for inducing fatigue, and the fatigue-transfer effect within the last 20 minutes of driving which involves switching from a complex to a monotonous road stretch or vice versa. The road that starts with a monotonous environment is herein denoted as Experiment road 1 while Experiment road 2 refers to the road that starts with a complex environment. This study is a 2 (road environments, complex urban road vs. monotonous rural road) x 3 (fatigue states, non-fatigued vs. fatigued vs. fatigue-transfer) within-subjects factorial experimental design.

The aforementioned cognition/visual tasks appear at random every 120 seconds on the cognition/visual task road stretches four times each with no similar types of tasks appearing in succession. In addition, before and after driving on each road stretch and each switch of road environment, all participants were asked to verbally answer the related fatigue scale; six times in total, including when the non-fatigued state began and ended, at the beginning and end of the 60-min fatigue induction road stretch, at the end of the 20-minute workload-fatigued or monotonous state, and at the last 20 minutes.

Before the actual experiment, four volunteer drivers and the authors conducted a pilot study to ensure that this experimental design met the construct/content validities.

Each participant drove on each road stretch once using the simulator and each experiment lasted for about 140 minutes. In between the two road experiments, there was a one-day gap to allow full rest and to avoid any possible residual effect from the former road experiment. Which experiment road to be tested first is arranged according to the counter balance rule to avoid order-related effects.

2.5. Data Collection

Driving Behaviors

First, external driving behaviors were measured. They included (1) standard deviation of lateral lane position (feet), which is the position of the vehicle center with respect to the road's central dividing line; (2) frequency of major right-lane crossing, which occurs when the vehicle's center point crosses the road boundary; and (3) mean time per major right-lane crossing (sec). Second, attention-related driving behaviors were measured. They included: (1) frequency of single steering wheel reversal angle exceeding 6° since this difference indicates deterioration in steering performance or occurrence of shifted attention (Wierwille & Gutmann, 1978); (2) standard deviations in lateral acceleration, which are abrupt lateral maneuvers indicative of a vehicle that has come off the lane center track due to drivers' inattention (Dingus et al., 1997).

Traffic sign distance estimation performance: For this task, the experimenter recorded the time measure starting from the subject's indication to the traffic sign frame out as the distance estimation (in sec). The shorter the estimation time, the greater the overestimation of distance is made by drivers.

Arithmetic Task Performance

The response times (in sec) and accuracy rates (%) starting from the onset of the question displayed on the HUD to the time when subjects provided the correct answer verbally were collected.

Psychological fatigue ratings: The fatigue feeling felt by drivers was reported the six rating locations. These locations include start and end of the baseline road stretch, start and end of the 60-minute fatigue induction road stretch, the end location of 20-minute driving in the fatigued state, and the end of another 20-minute driving in the fatigue-transfer state.

3. RESULTS

3.1. Fatigue Ratings

Figure 1 displays drivers' feeling of fatigue at six locations when driving on each of the two experiment roads. An obvious increasing trend was found for both experiment roads, indicating that the longer the driver was on the road, the more fatigued they felt. Statistical analyses also revealed that drivers' fatigue feelings at locations 4, 5 and 6 were significantly higher than those felt at locations 1, 2 and 3 [for Experiment Road 1; $F(5, 115) = 30.208$, $p = .0001$; for Experiment Road 2: $F(5, 115) = 44.271$, $p = .0001$], while no significant differences were observed within each group.

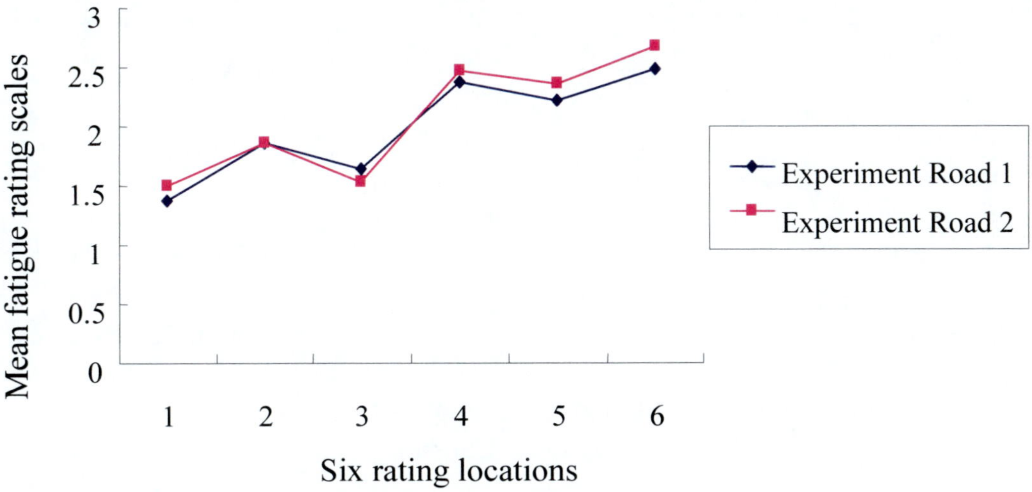

Locations 1 and 2: start/end places of the baseline road stretch.
Locations 3 and 4: start/end places of the 60-minute fatigue induction road stretch.
Location 5: end place of the fatigued road stretch
Location 6: end place of the fatigue-transfer road stretch.

Figure 1. Drivers' fatigue ratings conducted at six locations for Experiment Roads 1 and 2.

3.2. Driving Behavior

Steering Wheel Angle Behavior

Figure 2 depicts the frequency of driver's steering wheel reversal angle exceeding 6°. Statistical analysis of the data revealed significant difference between the two road environments [$F(1, 23) = 6.955$, $p = 0.015$]. The frequency of driver's steering wheel reversal angle exceeding 6° was higher (~20) in complex than in monotonous road environment (~14). On the other hand, three fatigue states and their interaction with road environments showed no significant difference. Notably, participant drivers in the fatigue-transfer state made more steering wheel angle reversals exceeding 6° when driving from a monotonous to a complex road stretch, but fewer reversals when driving from a complex to a monotonous road stretch.

Standard Deviation of Lateral Acceleration

Two main effects of road environment and fatigue state revealed significant differences in driver's lateral acceleration deviation measures [$F(1, 23) = 534.22$, $p= 0.0001$; $F(2, 46) = 7.183$, $p = 0.002$, respectively]. Standard deviations in lateral acceleration were larger in the complex than in the monotonous road environment. In addition, drivers in the fatigued state had larger standard deviations in lateral acceleration compared with those in both no-fatigue and fatigue-transfer states. A cross effect was noted when drivers in the fatigued state moved into the fatigue-transfer state. Fatigued drivers driving from the complex into the monotonous road stretch had significantly smaller standard deviation in lateral acceleration, compared with driving from the monotonous into the complex road stretch (Figure 3).

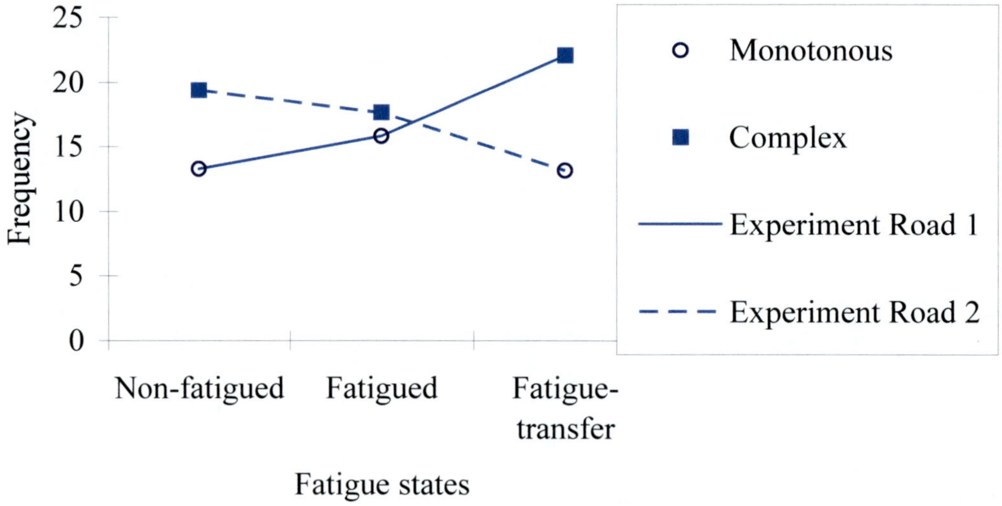

Figure 2. Frequency of driver's steering wheel reversal angle exceeding 6° for different experiment roads, road environments and fatigue states.

Lateral lane position measures: Figure 4 shows the standard deviation of driver's lateral lane position in different road environments. Drivers made larger lane deviation when driving on the complex than on the monotonous road stretch [F(1, 23) = 6.062, p = 0.022]. In addition, there were marked differences in lane position deviation under different fatigued driving situations [F(2, 46) = 11.059, p = 0.0001]. The variation in lane position was the smallest in the non-fatigued road stretches while no significant differences were found for the fatigued and fatigue-transfer road stretches.

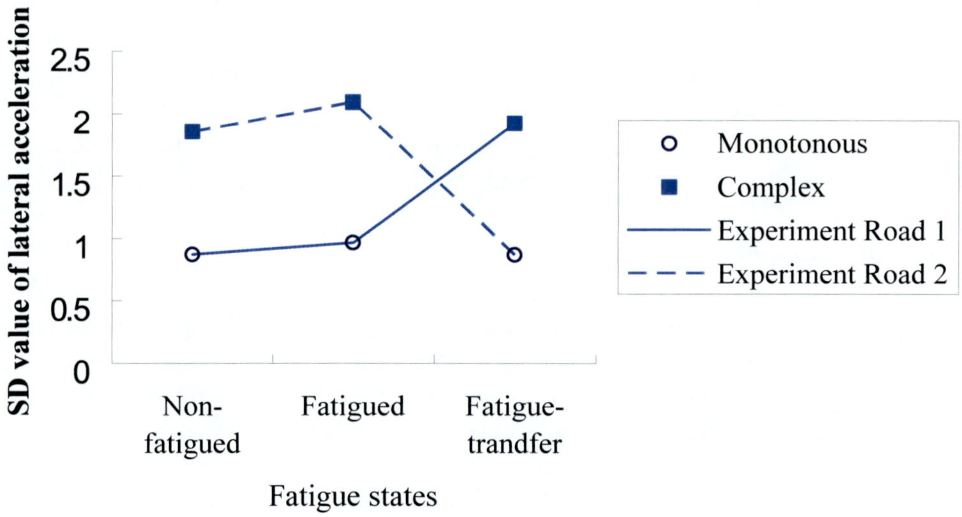

Figure 3. Driver's standard deviation of lateral acceleration for different experiment roads, road environments and fatigue states.

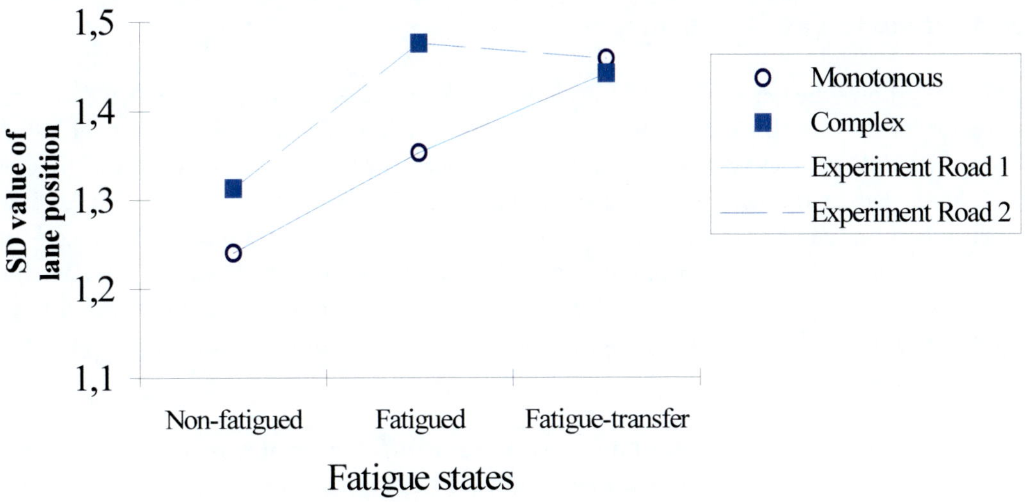

Figure 4. Standard deviation of lane position for different experiment roads, road environments and fatigue states.

Most notably, when switching from a monotonous to a complex road environment, the increase in road position deviation is the greatest though drivers made smaller lane position deviations. On the contrary, larger lane deviations were found among drivers switching into the monotonous from the complex road stretch than vice versa.

The frequency of vehicle major right-lane crossings and mean time per major right-lane crossing are shown in Table 1. As can be seen, participants made more major lane crossings in the complex than in the monotonous road environment (23.517 vs. 14.398) [$F(1, 23) = 109.859$, $p = 0.0001$]. The effect of different fatigue states were also statistically significant [$F(2, 46) = 3.557$, $p = 0.037$]. Post hoc mean comparisons indicated that drivers in fatigued and non-fatigued states (11.933 vs. 12.821) could be categorized as the same group, while those in non-fatigued and fatigue-transfer (13.161) states belonged to another group. Two-way interaction of road environments and fatigue states was found to be significant [$F(2, 46) = 5.212$, $p = 0.009$]. Drivers in the fatigue-transfer state made the largest number of major lane crossings (5.398) in the monotonous road environment than those in non-fatigued (4.369) and fatigued states (4.631). On the other hand, in the complex road environment, two groups were also found. One group comprised drivers in non-fatigued (8.452) and fatigue-transfer (7.763) states, and the other group was made up of those in fatigue-transfer and fatigued (7.302) states.

In terms of mean time per major right-lane crossing, the longest time (11.943 sec) was found among drivers in the fatigued state, followed by those in the fatigue-transfer state (10.640 sec), with the shortest time observed among non-fatigued drivers (7.214 sec) [$F(2, 46) = 33.252$, $p = .0001$].

3.3. Arithmetic Task Performance

Table 2 shows the reaction time of drivers in performing arithmetic tasks when driving on different road stretches [F(1, 23) = 6.241, p = 0.020] and in different fatigue states [F(2, 46) = 7.156, p = 0.002]. As can be seen, in the monotonous road environment, drivers' reaction was slower than that in the complex road environment (3.272 sec vs. 3.018 sec). In addition, drivers in both fatigue-transfer (3.442 sec) and non-fatigued (3.18 sec) states took longer time in responding to arithmetic tasks than those in the fatigued state (2.813 sec). When switching from a complex to a monotonous road environment, drivers took markedly longer time for calculation, thus causing a two-way interaction effect of road environments and fatigue states [F(2, 46) = 6.269, p = 0.004].

Table 1. Drivers' major right lane crossings for different experiment roads, road environments and fatigue states

Measures	Experiment Road 1			Experiment Road 2		
Number of lane crossings per minute	Mono. (non-fatigued)	Mono. (fatigued)	Mono. to Complex (fatigue-transfer)	Complex (non-fatigued)	Complex (fatigued)	Complex to Mono. (fatigue-transfer)
	4.369	4.631	7.763	8.452	7.302	5.398
Mean time per lane crossing	Mono. (non-fatigued)	Mono. (fatigued)	Mono. to Complex (fatigue-transfer)	Complex (non-fatigued)	Complex (fatigued)	Complex to Mono. (fatigue-transfer)
	6.3113	12.3194	10.2401	8.1171	11.5763	11.0396

Table 2. Drivers' performance measures in arithmetic and traffic sign distance estimation tasks

Measures	Experiment Road 1			Experiment Road 2		
Reaction time in arithmetic task (sec)	Mono. (non-fatigued)	Mono. (fatigued)	Mono. to Complex (fatigue-transfer)	Complex (non-fatigued)	Complex (fatigued)	Complex to Mono. (fatigue-transfer)
	3.154	2.672	2.893	3.207	2.954	3.991
Distance estimation (sec)	Mono. (non-fatigued)	Mono. (fatigued)	Mono. to Complex (fatigue-transfer)	Complex (non-fatigued)	Complex (fatigued)	Complex to Mono. (fatigue-transfer)
	4.5383	4.9208	4.7258	3.9679	3.4275	4.0629

3.4. Traffic Sign Distance Estimation

Table 2 also displays the distance estimation results among drivers in different road environment and fatigue states. As can be seen, the results were significantly different between the two road stretches [F(1, 23) = 24.453, p = 0.0001]. In the complex road environment, participant drivers tended to overestimate the distance between the traffic sign and themselves than in the monotonous road environment (3.819 sec vs. 4.728 sec). In addition, a two-way interaction between road environments and fatigue states is found [F(2, 46) = 4.593, p = 0.015] for the complex road experiment in different fatigue states (p = 0.322), but not for the monotonous road experiment. In the complex road environment (p = 0.0001), the time judged by participants driving in the fatigued state (3.4275 sec) was shorter than that in the non-fatigued (3.9679 sec) or fatigue-transfer state (4.063 sec), showing that drivers in the fatigue state tended to overestimate distance.

Switch in road environment switch also had a significant effect on distance judgment (p = 0.011). When switching from a monotonous to a complex road environment (4.7258 sec), drivers made more accurate estimation than when switching from a complex to a monotonous one (4.0629 sec).

4. DISCUSSION

Fatigue rating results validated that fatigue is produced after driving for 60 minutes (Figure 1). The overall trend is that fatigue increases with driving time; that is, the longer one drives, the more fatigued one gets. In addition, driving in the complex road environment (Experiment Road 2) causes drivers to feel fatigued earlier.

4.1. Effects of Road Environment on Fatigued Drivers' Behavior and Performance

We hypothesized that fatigued drivers who cannot concentrate well tend to spend more time correcting the direction of the vehicle and putting it back on the right path when it crosses the sideline. However, according to the results shown in Table 1, the number of lane crossings per minute was the highest among drivers in the non-fatigued state when driving in the complex road environment, though the mean time per lane crossing was the shortest. We attribute such results to the high oncoming traffic density that causes drivers to cling to the right side of the road to avoid oncoming vehicles. Being relatively more alert, they only cross the sideline occasionally and are quicker in correction their direction, thus remaining shorter over the sideline. Variations in driving direction also provide support for this phenomenon. There were relatively larger variations in lane position observed among drivers in both fatigued and fatigue-transfer states (Figure 4), meaning that even though non-fatigued drivers still crossed the right-hand sideline sometimes for a short duration, the overall driving direction was more stable compared with that in the two fatigued states.

Deviation in lateral acceleration (Figure 3), an effective measure for assessing changes in drivers' attention required, showed increased burden on attention resources of fatigued drivers in the complex road environment. In addition, fatigued drivers in the complex road environment overestimated the sign distance (judging subjectively an actual distance of 3.427 second as 5 seconds), evidencing that drivers' distance judgment ability declines with fatigue (Chi & Lin, 1998). When carrying out numerical calculations, fatigued drivers performed worse in the complex (reaction slowed) than in the monotonous road environment. The reaction time for fatigued drivers was not significantly slower than that for non-fatigued ones, suggesting that cognitive work is probably less affected by fatigue (Williamson et al., 2001). Alternatively, increased arousal among drivers may also lead to improved reaction times.

4.2. Effect of Road Environment Switch on Fatigued Drivers' Behavior and Performance

Fatigued drivers' behavior and performance results in the fatigue-transfer state are worth noting. The fatigue-transfer state comes right after the fatigued state, and it is expected that the already fatigued drivers will show inferior performance in both visual and numerical calculation tasks in this state, which may also vary with the driving time. However, no such inverse relationship between drivers' performance and driving time was observed. In other words, the performance of fatigued drivers did not decline progressively with increase in driving time.

Moreover, drivers behaved and performed differently when the road environment changed; that is, when switching from the complex to the monotonous road environment and vice versa. In this study, we hypothesized that fatigue from one road stretch will affect driving in the next road stretch, and named such effect as the "fatigue-transfer effect". More specifically, "monotonous (complex) fatigue-transfer effect" refers to how fatigue from driving on the monotonous (complex) road stretch affects drivers' behavior and performance when driving on the subsequent complex (monotonous) road stretch. Comparing the behavior and performance of drivers in the fatigue-transfer state on Experiment Road 1(2) with those of drivers in the fatigue state on Experiment Road 2(1) can shed light on the monotonous (complex) fatigue-transfer effect. Our experimental results show that although there was increase in frequency of steering wheel movements exceeding 6° (Figure 2), the monotonous fatigue-transfer effect did not lead to greater lateral acceleration deviation (Figure 3) or lateral lane position deviation (Figure 4). On the contrary, there was a reduction in both. Similarly, there was an increase in frequency of sideline crossings but a decrease in mean time per lane crossing. It is worth noting that drivers showed more accurate distance estimation judgment and did numerical calculations faster under the monotonous fatigue-transfer effect (Table 2). In short, there are no significant differences in driving behavior when switching from the monotonous to the complex road environment compared with fatigued driving in the complex road environment. On the other hand, switching from the monotonous to the complex road environment helped enhance visual and information processing capability of drivers. As for the complex fatigue-transfer effect, there was a decrease in frequency of steering wheel movements exceeding 6° (Figure 2). While lateral acceleration deviation (Figure 3) remained unaffected, there was greater variation in vehicle direction (Figure 4). In addition, there was an increase in frequency of sideline crossings but a decrease in mean time per lane crossing (Table 1). Of note is that drivers tended to overestimate road sign distance and do arithmetic more slowly under the complex fatigue-transfer effect (Table 2). According to the frequency of steering wheel reversal and lateral acceleration deviation, drivers' attention level did not change much, probably because of the road environment being straight and monotonous.

However, drivers' performance deteriorated when switching from the complex to the monotonous road environment, evidencing the negative impact of fatigue transfer on visual and information processing capability.

Our findings show that the driving behavior and performance of fatigued drivers are affected more by changes in road environment than by length of driving time. The switch from the complex to the monotonous road environment had a negative impact on both visual and numerical calculation capability, which was markedly greater than that when switching from the monotonous to the complex road environment. Drivers' subjective fatigue assessment results (Figure 2) and analysis of the fatigue-transfer effect show that the increase in fatigue under the monotonous fatigue-transfer effect (Location 6 on Experiment Road 1 vs. Location 5 on Experiment Road 2) was greater than that under the complex fatigue-transfer effect (Location 6 on Experiment Road 2 vs. Location 5 on Experiment Road 1). In other words, not only did switching from the complex to the monotonous road environment have a negative impact on driving behavior and performance, drivers also had a greater feeling of fatigue under such switch. Nevertheless, this phenomena could not be accounted for by the attention resource theory. Rather, changes in drivers' arousal and alertness level (Davis & Parasuraman, 1982) might offer a better explanation for our observations. When switching from the complex to the monotonous road environment, drivers may feel a greater sense of monotony, making them less alert, and thus reducing their visual and calculation capability. On the other hand, when switching from the monotonous to the complex road environment, drivers perceive greater complexity in the road environment, making them more alert, even though the increase in alertness is limited because they are already very fatigued. More interesting visual and cognition tasks may help enhance their arousal level (Kraemer et al., 2000), which will in turn improve their task performance.

Acknowledgements

The financial support from the National Science Council, Taiwan, under Contract No. NSC-93-2213-E-224-017 was gratefully acknowledged and sincere gratitude was also extended to all participants for their valuable efforts in completing this study.

References

Chi, C-F., & Lin, F-T. (1998). A comparison of seven visual fatigue assessment techniques in three data acquisition VDT tasks. *Human Factors, 40(4),* 577-590.

Collins, D.J., Biever, W.J., Dingus, T.A., Neale, V.L. (1999). An examination of driver performance under reduced visibility conditions when using an in-vehicle signing information system (isis). U.S. Department of Transportation Federal Highway Administration, Publication No. FHWA-RD- 99-130.

Craig, A., & Cooper, R.E. (1992). Symptoms of acute and chronic fatigue. In: A.P. Smith and D.M. Jones (Eds.), *Handbook of Human Performance, Vol. 3: State and Trait,* pp. 289-339. London: Academic Press.

Craig, A., Tran, Y., Wijesuriya, N., & Boord, P. (2006). A controlled investigation into the psychological determinants of fatigue. *Biological Psychology, 72,* 78-87.

Davies, D.R., & Parasuraman, R. (1982). *The Psychology of Vigilance.* New: York: Academic Press Publication.

Desmond, P.A., & Matthews, G. (1997). Implications of task-induced fatigue effects for in-vehicle countermeasures to driver fatigue. *Accident Analysis and Prevention, 4,* 515-523.

Dinges, D.F. (1995). An overview of sleepiness and accidents. *Journal of Sleep Research, 4,* 4-14.

Dingus, T.A., Hulse, M.C., Mollenhauer, M.A., Fleischman, R.N., McGehee, D.V., & Manakkal, N. (1997). Effects of age, system experience, and navigation technique on driving with an advanced traveler information system. *Human Factors, 39,* 177–199.

Fell, D. (1994). *Safety update: problem definition and countermeasure summary: Fatigue.* RUS No. 5, New South Wales Road Safety Bureau, Australia.

Gainsky T. L. (1993). Psychophysical determinants of stress in sustained attention. *Human Factors, 35,* 603-614.

Gander, P.H., Nguyen, D., Rosekind, M.R., & Connell, L.J. (1993). Age, circadian rhythms, and sleep loss in flight crews. *Aviation, Space Environment Medicine, 64*(3), 189-195.

Hancock, P.A., & Desmond, P.A. (2001). *Stress Workload and Fatigue.* Mahwah, NJ: Lawrence Erlbaum.

Hancock, P.A., & Warm, J.S. (1989). A dynamic model of stress and sustained attention. *Human Factors, 31,* 519-537.

Harms, L. (1991). Variation in drivers' cognitive load: Effects of driving through village areas and rural junctions. *Ergonomics, 34,* 151-160.

Horne, J. A., & Reyner L. A. (1999). Vehicle accidents related to sleep: a review. *Occupational and Environmental Medicine, 56*, 289-294.

Kraemer, S., Danker-Hopfe, H., Dorn, H., Schmidt, A., Ehlert, I., & Herrmannn, W.M. (2000). Time-of-day variations of indicators of attention: performance, physiologic parameters, and self-assessment of sleepiness. *Biological Psychology, 48,* 1069-1080.

Lisper, H-O., Laurell, H., & van Loon, J. (1986). Relation between time to falling asleep behind the wheel on a closed track and changes in subsidiary RT during prolonged driving on a motorway. *Ergonomics, 29,* 445-453.

Liu, Y.-C., & Wen, M.-H. (2004). Comparison of head-up display (hud) vs. head-down display (hdd): driving performance of commercial vehicle operators in Taiwan. *International Journal of Human-Computer Studies, 61,* 679-697.

Lyznicki, J.M., Doege, T.C., Davis, R.M., & Williams, W.A. (1998). Sleepiness, driving, and motor vehicle crashes. *JAMA, 279*(23), 1908-1913.

Matthews, G., & Desmond, P.A. (2002). Task-induced fatigue states and simulated driving performance. *The Quarterly Journal of Experimental Psychology, 55A*(2), 659-686.

Matthews, G., Sparkes, T.J., & Bygrave, H.M. (1996). Attentional overload, stress and simulated driving performance. *Human Performance, 9,* 77-101.

Maycock, G. (1997). Sleepiness and driving: the experience of U.K. drivers. *Accident Analysis and Prevention, 29,* 453-462.

Nilsson, T., Nelson, T. M., & Carlson, D. (1997). Development of fatigue symptoms during simulated driving. *Accident Analysis and Prevention, 29,* 479-488.

Oron-Gilad, T., & Hancock, P.A. (2005). Road environment and driver fatigue. *Proceedings of the Third International Driving Symposium on Human Factors in Driver Assessment, Training and Vehicle Design* (pp.318-324), June 27-30, Rockport, Maine.

Pack, A. I., Pack, A. M., Rodgman, E., Cucchiara, A., Dinges, D. F., & Schwab, C.W. (1995). Characteristics of crashes attributed to the driver having fallen asleep. *Accident Analysis and Prevention, 27,* 769-775.

Pilcher, J.J., & Huffcutt, A.I. (1996). Effects of sleep deprivation on performance: a meta-analysis. *Sleep, 19,* 318-326.

Quant, J.R. (1992). The effect deprivation and sustained military operations on near visual performance. *Aviation, Space, and Environmental Medicine, 63,* 172-176.

Rogé, J., Pébayle, T., Hannachi, S.EI., & Muzet, A. (2003). Effect of sleep deprivation and driving duration on the useful visual field in younger and older subjects during simulator driving. *Vision Research, 43,* 1465-1472.

Sagberg, F. (1999). Road accidents caused by drivers falling asleep. *Accident Analysis and Prevention, 31,* 639-649.

Skipper J. H., & Wierwille W. W. (1986). Drowsy driver detection using discriminate analysis. *Human Factors, 28,* 527-540.

Summala, H., & Mikkola, T. (1994). Fatal accidents among car and truck drivers: effects of fatigue, age and alcohol consumption. *Human Factors, 36,* 315-326.

Thiffault, P., & Bergeron, J. (2003). Fatigue and individual differences in monotonous simulated driving. *Personality and Individual Differences, 34,* 159-176.

Wierwille, W.W., & Gutmann, F. (1978). Comparison of primary and secondary task measures as a function of simulated vehicle dynamics and driving conditions. *Human Factors, 20 (2),* 233-244.

Williamson, A.M., Feyer A., L. A., & Friswell R. (1996). The impact of work practices on fatigue in long distance truck drivers. *Accident Analysis and Prevention 28*(6), 709-719.

Williamson, A.M., Feyer, A-M., Mattick, R., Friswell, R., & Finlay-Brown, S. (2001). Developing measures of fatigue using an alcohol comparison to validate the effects of fatigue on performance. *Accident Analysis and Prevention, 33(3),* pp. 313-326.

In: Road Traffic: Safety, Modeling and Impacts
Editors: S. E. Paterson and L. K. Allan, pp. 455-473
ISBN 978-1-60456-884-4

Chapter 13

PATHOMECHANISM OF HEAD INJURIES IN FATAL ROAD TRAFFIC ACCIDENTS

Klara Törő*[*], *Szilvia Fehér, Attila Dalos and György Dunay
Department of Forensic Medicine, Semmelweis University, Budapest, Hungary

ABSTRACT

Introduction

Injuries in traffic accidents are important causes of mortality in industrialized countries. Head trauma represents one of the most severe damages in road traffic accidents. Pedestrians, bicyclists and motor vehicle occupants have often suffered fatal intracranial injuries. The incidence of fatal head trauma depends on the role of victim, the speed, the protecting facilities of motor vehicles, and the available medical care. In this chapter the purpose of our examination was to calculate the rate of lethal intracranial injuries in different pathomechanisms of traffic accidents, and to evaluate the severity and outcome of head trauma.

Material and Methods

Cases of 1416 (1002 males, 414 females) fatal traffic accident were collected from 2000 to 2007 using the database of the Department of Forensic Medicine at Semmelweis University in Budapest. Autopsy reports were analyzed to determine the proportion of fatal head injuries. Types and location of injuries, blood alcohol concentrations were evaluated. Program Virtual crash version 2.2. was used for the simulation of vehicle accidents.

---

[*] Correspondence: Törő Klára, Department of Forensic Medicine, Semmelweis University, 1091-Hungary, Üllői út 93. Tel: 3612157300; Fax: 3612162676; E-mail: torok@igaz.sote.hu

Results

There were 329 (23.2%) cases with fatal head trauma suffered by 171 pedestrians, 34 bicyclists and 124 motor vehicle occupants. Among pedestrians and bicyclists there was a higher rate of head injuries, such as skull fractures, epidural hemorrhage, subdural hemorrhage, brain contusion. Thoracic damages, like traumatic aortic rupture, hemothorax, and abdominal damages, as liver rupture were dominant in motor vehicle occupants.

Conclusion

Head injuries play an important role in traffic accidental mortality. Our results underline the importance of preventive strategies in transportation, pointing out that different methods are necessary to reduce fatal injuries of various traffic participants. Traffic accidents usually carry legal consequences in proportion to the severity of the crash. The high incidence of accidents can be prevented by the collaboration between the representatives of several sciences, including forensic medicine. The elements of the prevention are the traffic rules and laws, the engineering, the traffic management and administration.

Key words: head injuries, traffic accidents, pedestrians, motor vehicle occupants

INTRODUCTION

The World Health Organization defines an accident as an unexpected, unplanned occurrence that may involve injury [1]. During 1990s, road traffic accidents ranked ninth among the leading causes of death in the World. Injuries in road traffic accidents are important causes of mortality in industrialized countries. The high incidence of accidents can be prevented by the collaboration between the representatives of several sciences, including forensic medicine. Road traffic participants are pedestrians, vehicle drivers, vehicle occupants, bicyclists, motorcyclists.

Definition of road traffic accident: Minimum one person and one vehicle are involved in an unexpected road traffic situation which may result in more or less severe injuries or damages to people and goods. (The crash of two pedestrians on the road is not a traffic accident.) The injuries can be characteristic for the types of collisions and may assist the reconstruction of the accidental mechanism.

Injuries in traffic accidents are important causes of mortality in industrialized countries [2,3]. These injuries represent a serious public health problem with high economic and social costs. The mortality rate of traffic accidents (13.7-15.8 death cases/100.000 citizens) is high in Hungary (Hungarian Statistical Annual Book 1999-2006). However, there was a decline of traffic accident mortality in the last decade, from 2001 on increasing rates are observed again. The previous reduction of death rate, attributable to motor vehicle crashes, represents the successful transport policy response to motorization challenges in Hungary [4,5,6]. Changes in driver and passenger behavior also have reduced motor vehicle collisions and injuries.

Enactment and enforcement of traffic safety laws, reinforced by public education, have led to safer behavior in driving. These include severe legal control of driving while under the influence of alcohol or drugs; force the use of safety belt, child-safety seat, and ware of helmet for motorcyclists. The reasons for the recently observed increased rates of fatal traffic accidents have not been understood yet. The urgent task is to collect more information about the nature and mechanisms of traffic accidents involving pedestrians, bicyclists and motor vehicle occupants to inform prevention policy makers [7].

Injury and death may occur in association with many forms of road transport. Road traffic accidents usually carry legal consequences in proportion to the severity of the crash. Nearly all common laws, jurisdictions impose some kind of requirement that parties involved in a collision (even with only stationary property) must stop at the scene, and exchange insurance or identification information. The elements of the prevention are the traffic rules and laws, the engineering, the traffic management and administration. Enactment and enforcement of traffic safety laws, reinforced by public education, have led to safer behavior in driving. These include severe legal control of driving while under the influence of alcohol or drugs; regulations as to the use of safety belt, child-safety seat, and wearing of helmet for motorcyclists.

Head trauma represents one of the most severe damages in road traffic accidents. Pedestrians, bicyclists and motor vehicle occupants have often suffered fatal intracranial injuries. The incidence of fatal head trauma depends on the role of victim, the speed, the protecting facilities of motor vehicles, and the available medical care.

In this chapter the purpose of our examination was to calculate the rate of lethal intracranial injuries in different pathomechanisms of traffic accidents, and to evaluate the severity and outcome of head trauma. We aimed to investigate the characteristic thread trauma of victims in a series of fatal road traffic accident. A comparison was made between pedestrians, bicyclists, and motor vehicle occupants (drivers and passengers). We assumed that traffic injuries would have various degree and distribution in the investigated groups.

Material and Methods

The survey target groups included victims of lethal traffic accidents. Cases of 1416 (1002 males, 414 females) fatal traffic accident were collected from 2000 to 2007 using the database of the Department of Forensic Medicine at Semmelweis University (Table 1). Figure 1 demonstrates the seasonal distribution of fatal traffic accidents in Budapest. Autopsy reports were analyzed to determine the proportion of fatal head injuries. Types and location of injuries, blood alcohol concentrations were evaluated. Information was collected from forensic autopsy records. Data were analyzed according to the roles of traffic participants (pedestrian, bicyclist, motor vehicle driver or passenger), age, and gender, duration of hospitalization, injured body region, and type of injury. Characteristic injuries of pedestrians, bicyclists and motor vehicle occupants were compared. Survival time and duration of medical treatment were investigated. Distribution of natural diseases of investigated groups was analyzed. The pattern of cranio-cerebral trauma was studied.

Table 1.

	Male No (%)	Female No (%)	All (100%)
Pedestrians	469 (63)	276 (37)	745
Bicyclists	86 (84.3)	16 (15.7)	102
Motorcyclists	127 (98.4)	2 (1.6)	129
Car (drivers, passengers)	277 (71.6)	110 (28.4)	387
Other	43 (81.1)	10 (18.9)	53
All	1002 (70.7)	414 (29.3)	1416

The material included subsets of 745 pedestrians, 102 bicyclists, and 569 (129 motorcycle and moped users, 387 car drivers and car passengers, 53 othres, as van driver or bus passengers) motor vehicle occupants. The average age of all injured person was 51.7 years.

We recorded the rate of alcohol-impaired drivers, and analyzed the levels of alcohol intoxication. The blood alcohol concentrations (BACs) were used only, if death occurred on the same day as the injury. Influence of alcohol was categorized as slight degree (BAC: 51-80 mg/100ml), mild degree (BAC: 81-150 mg/100ml), moderate (BAC: 151-250 mg/100 ml), severe (BAC: 251-350 mg/100ml) and very severe (BAC: above 351 mg/100ml). Alcohol-involved deaths were defined as those with detectable BAC of more than 50 mg/100ml.

Virtual crash version 2.2. was used for the simulation of vehicle crash with a complex real-time calculation about the mechanism of road traffic accidents.

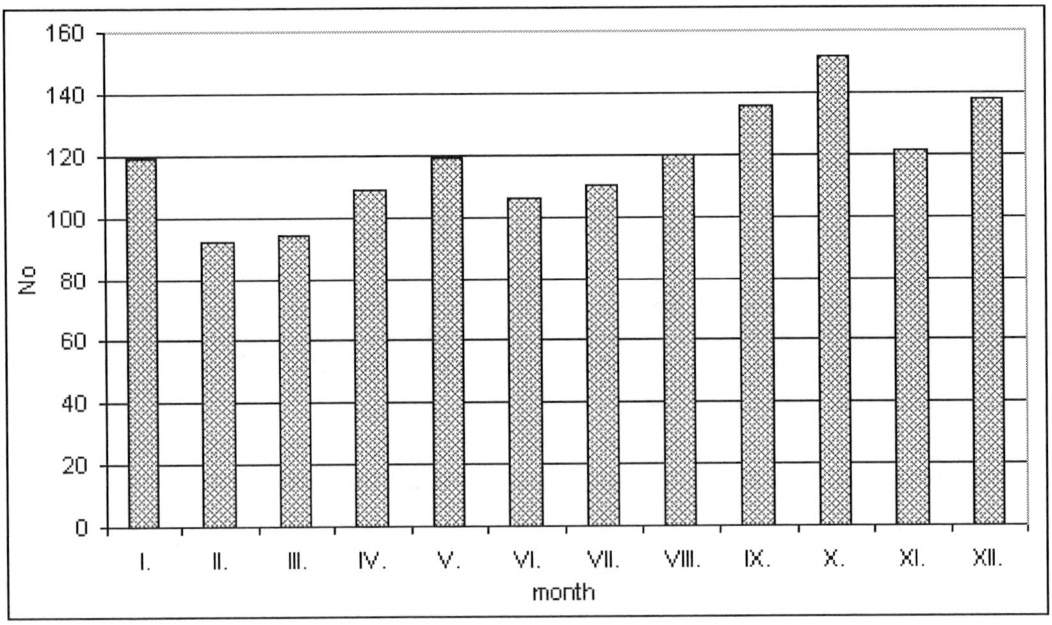

Figure 1. Seasonal distribution of fatal traffic accidents in Budapest.

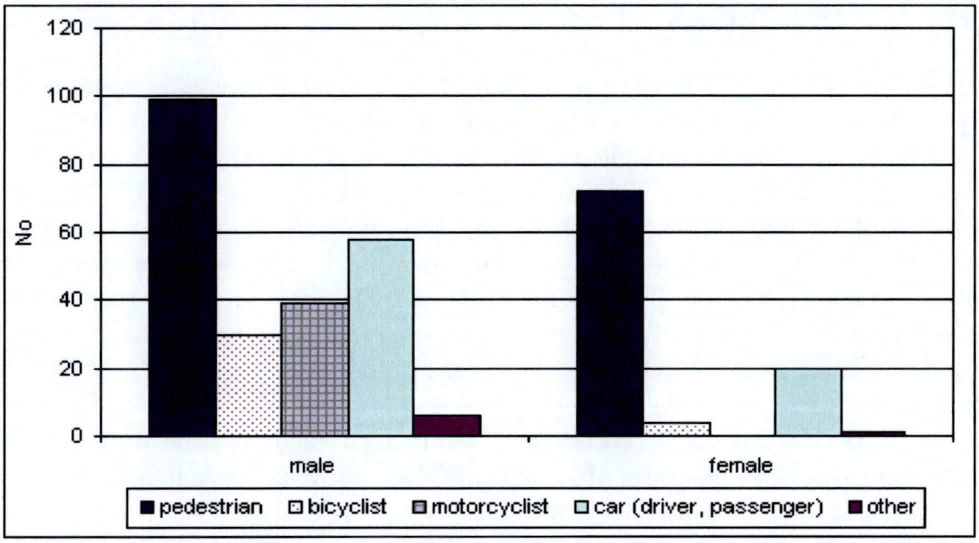

Figure 2. Head injuries among pedestrians, bicyclists and motor vehicle occupants.

Results

Sex and age distribution of fatal head injuries is presented in Table 2. The age groups of the victim were grouped into 8-year intervals ranging from 0 to 90 years. The youngest victim was a male child aged 1 year and the oldest was a 90 years-old male. There were 329 (23.2%) cases with fatal head trauma suffered by 171 pedestrians, 34 bicyclists and 124 motor vehicle occupants. Distribution of fatal head injuries is presented in Figure 2. Male victims exceeded females in every age group ($p<0.005$).

Table 2.

	Male			Female		
Age (year)	All	Fatal head injury No (%)	Other fatal injury No (%)	All	Fatal head injury No (%)	Other fatal injury No (%)
0-1	-	-	-	4	1 (25)	3 (75)
2-5	3	1 (33.3)	2 (66.7)	3	3 (100)	-
6-15	18	7 (38.9)	11 (61.1)	8	4 (50)	4 (50)
16-20	34	15 (44.1)	19 (55.9)	18	5 (27.8)	13 (72.2)
21-30	175	38 (21.7)	137 (78.3)	39	11 (28.2)	28 (71.8)
31-40	152	46 (30.3)	106 (69.7)	19	3 (15.8)	16 (84.2)
41-50	133	34 (25.6)	99 (74.4)	40	8 (20)	32 (80)
51-60	158	30 (19)	128 (81)	55	8 (14.5)	47 (85.5)
61-70	117	24 (20.5)	93 (79.5)	48	10 (20.8)	38 (79.2)
70<	204	37 (18.1)	167 (81.9)	176	44 (25)	132 (75)
Unknown	8	-	8 (100)	4	-	4 (100)
All	1002	232 (23.2)	770 (76.8)	414	97 (23.4)	317 (76.6)

Injuries of the Pedestrians

Primary injuries: due to the primary impact of the vehicle. This is the first phase of the accident. The injuries are resulted by the bumper or outstanding parts of the vehicle. It is typically seen on the legs (open, in-pieces fractures of the shins, a sack of hemorrhage between the layers of the skin, additional skinning). The primary injury can usually be found just below the knee level. *Second and third phase injuries*: due to striking the ground or other objects. These injuries are resulted by the falling down or scooping up of the body, and the car or another car may run over the body and cause severe injuries. Secondary injuries include sever or fatal head, chest and pelvis injuries. The most severe complications are head injuries among pedestrian victims. Figure 3 shows the simulation of the mechanism of pedestrian head injuries presenting by virtual crash program. *Running over injuries:* occur when a wheel passes over the body. Internal organs may be ruptured and large areas of skin and subcutaneous tissues may be ripped off. The tire of the vehicle may lead to a patterned blunt injury, which may be used for the identification of the vehicle. These injuries may occur in cases in which the car occupants are ejected from the car at the moment of the accident.

Among pedestrians and bicyclists there was a higher rate of head injuries, such as skull fractures, epidural hemorrhage, subdural hemorrhage, brain contusion. Thoracic damages, like traumatic aortic rupture, hemothorax, and abdominal damages, as liver rupture were dominant in motor vehicle occupants. Head injuries of bicyclists (Figure 4) are demonstrated by the complex real-time calculation.

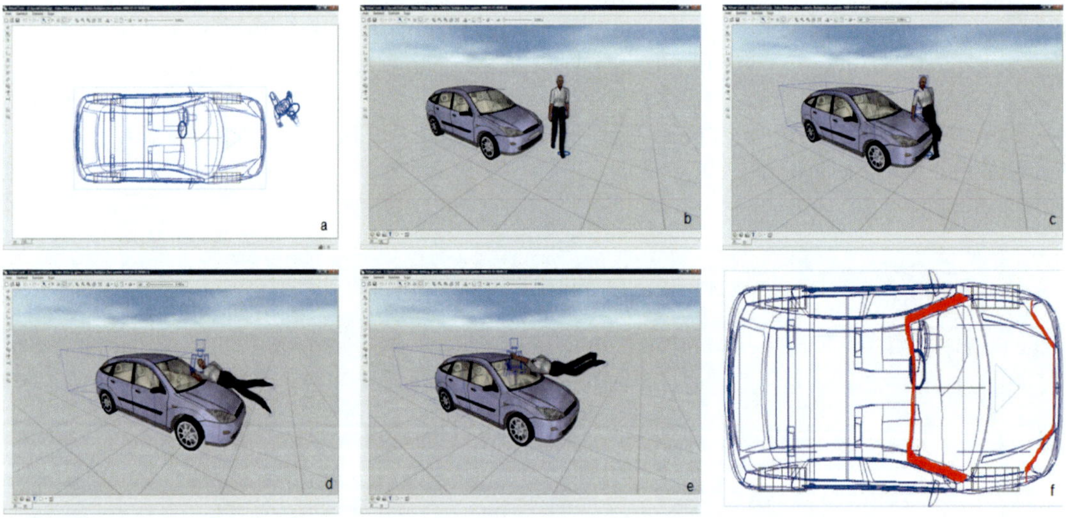

Figure 3. Simulation of the mechanism of pedestrian head injury (Virtual crash version 2.2.). *a/ Position of pedestrian and motor vehicle before the crash. b,c,d,e/ Process of the pedestrian damage with the primary and secondary injuries in case of 60 km/h vehicle speed. f/ Sources of head injuries of the pedestrian.*

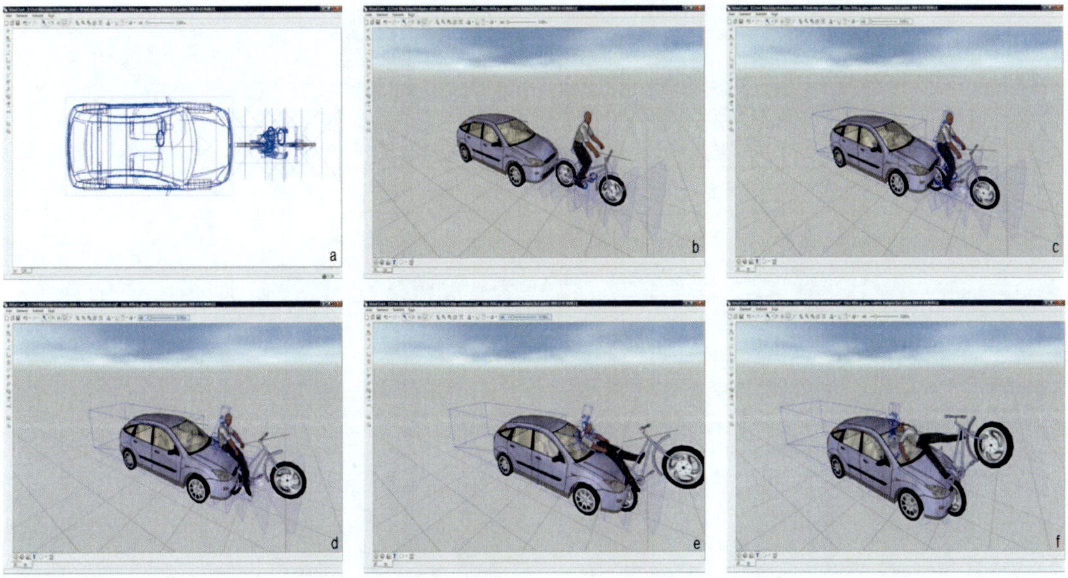

Figure 4. Simulation of the mechanism of bicyclist head injury (Virtual crash version 2.2.). *a/ Position of bicyclist and motor vehicle before the crash, v = 50km/h. b,c,d,e/ Process of the bicyclist damage in 0.2 sec time period. f/ Sources of head injuries of the bicyclist on the vehicle.*

Injuries of Motor Vehicle Occupants

Motor vehicle occupants suffer different injuries in connection with the position in the car and the used safety equipments. Mechanisms and characteristic of injuries among the front-seat occupants: the face and head hit the windscreen glass, frame or side-pillars, the occupants may smash the glass with their heads and may be ejected from the vehicle through the windscreen, the chest may be crushed against the fascia or steering wheel causing rib, sternum, heart or liver damage, the knees may stake the parcel shelf causing injuries and fractures, the lower limbs may be fractured by transmitted stress, which may also dislocate the hips and fracture the pelvis, the head swinging violently forward during deceleration may cause hyperflexion injury to the cervical or thoracic spine, if the head strikes a solid structure, the rebound may then cause hyperextension injury, if head restrains are not fitted to the seats (Figure 5), the status of the heart within the thorax, perhaps aided by hyperflexion, often tears the aorta at the termination of the descending arch, where the vessel meets the vertebral column.

Highest numbers of victims with fatal head injuries (Figure 6) were found in the 21–40 years group and least in the oldest age group. Age and sex distribution is presented in Figure 7 with the predominance of older victims.

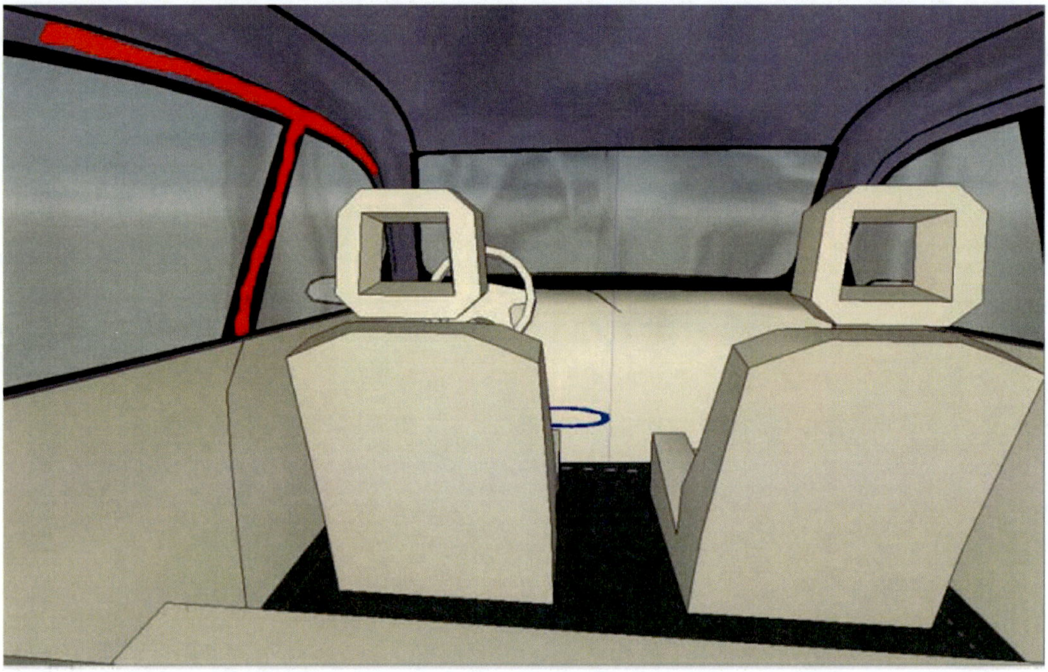

Figure 5. Simulation of the mechanism of head injury in motor vehicle (Virtual crash version 2.2.). *Sources of head injuries of victims sitting in the car.*

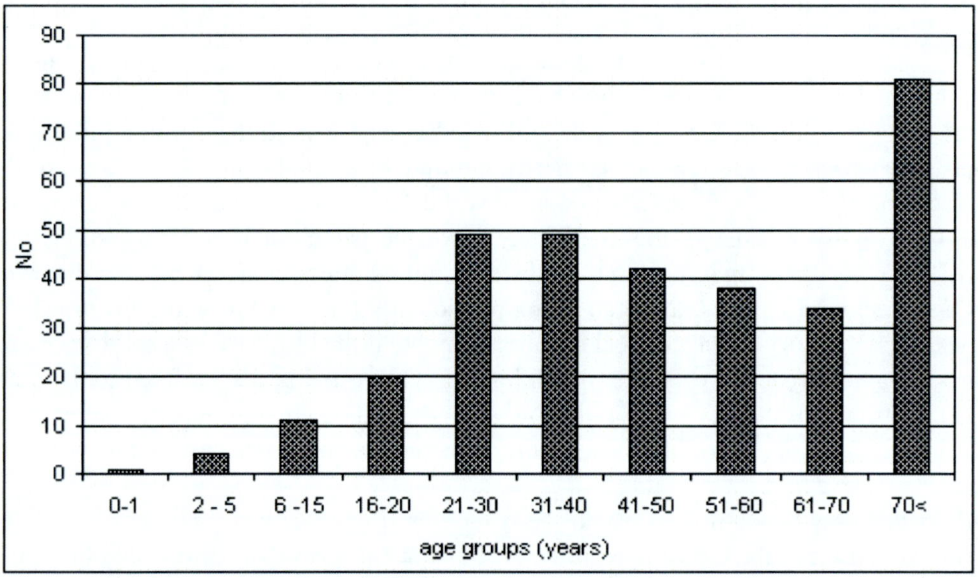

Figure 6. Fatal head injuries in different age groups of victims.

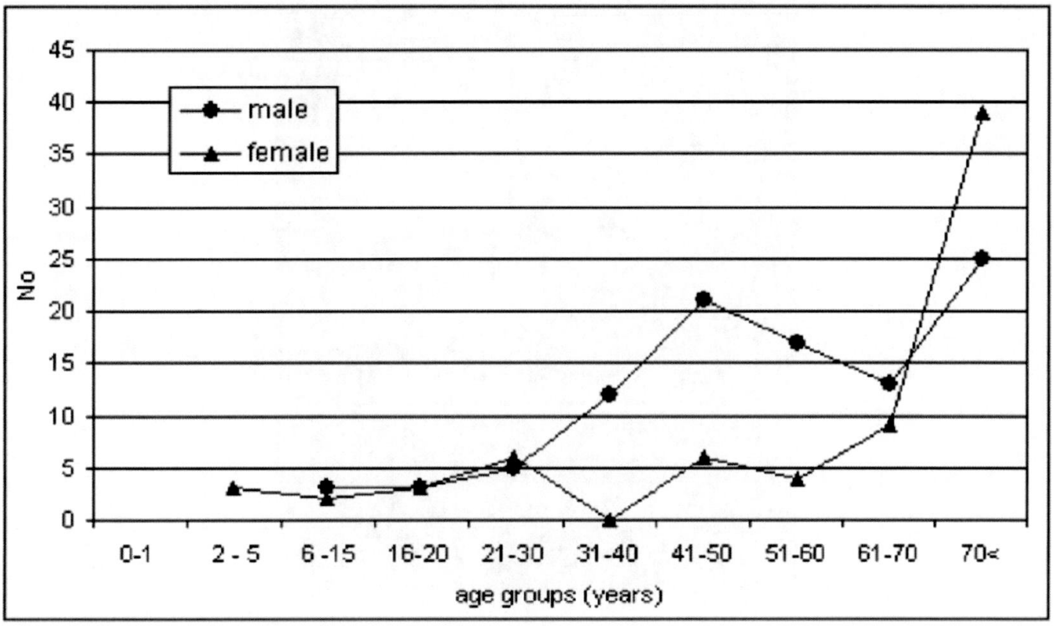

Figure 7. Age and sex distribution among pedestrians.

Pattern of Cranio-cerebral Trauma

External injury to the face and scalp were found in 78% of the victims. In 30% of cases, fractures of skull were found (Figure 8). Cranial vault was involved in 81%, base of skull in 34%, and both vault and base in 25% of cases. Anterior cranial fossa, parietal bone and temporal bone were the most commonly involved areas in fracture. Contusions and lacerations of brain (Figure 9) were found in equal number of cases (32%). Subdural hemorrhage (38%) (Figure 10) was observed most commonly, followed by subarachnoid hemorrhage (12%). Extradural hemorrhages were found in comparatively less number of cases (11%). Combinations of all hemorrhages were seen only in 4% of cases.

Post-mortem blood alcohol tests were done in 138 cases, with negative results in 19 (13.7%). Blood alcohol concentrations were evaluated such as slight degree in 77 (56%), mild degree in 2 (1.4%), moderate degree in 5 (3.6%), severe degree in 13 (9.4%), and very severe degree in 22 (15.9%) cases (Figure 11). No occasion with positive drug test was found. In this study material 94 (29 %) of victims died on the scene, while 235 (71%) person died on other places. During transportation to hospital 11 (3.3 %) persons died. Hospital treatment was received by 215 (65.3%) persons before they died. The mean survival period of victims was 6–24 h.

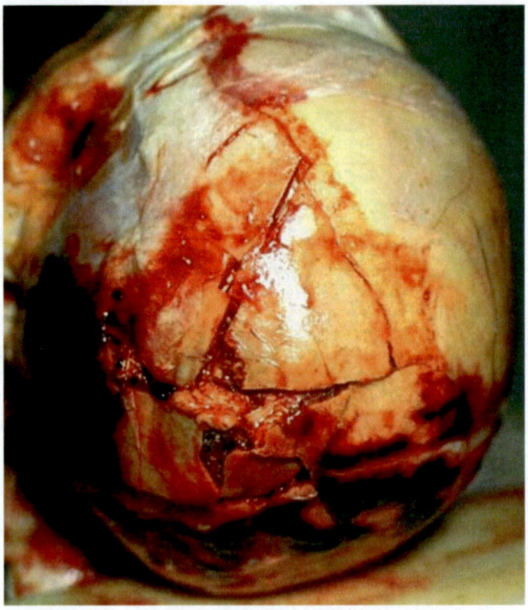

Figure 8. Skull fracture on a pedestrian victim.

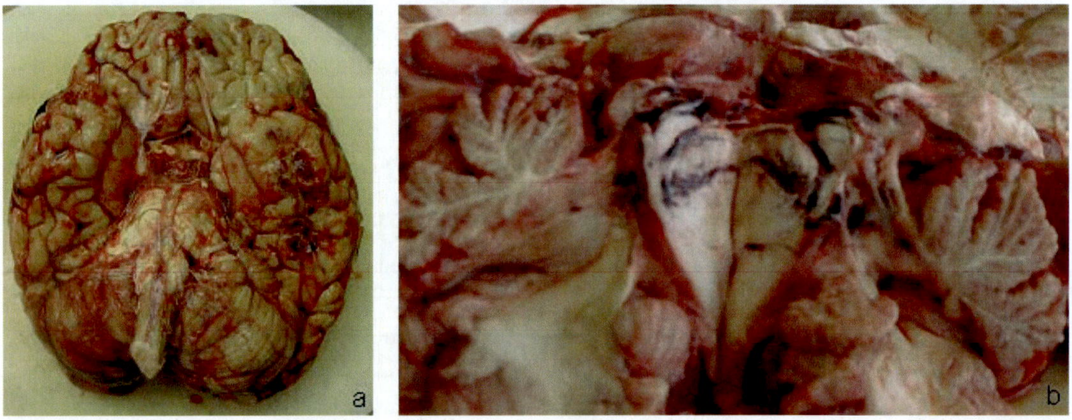

Figure 9. a/ Brain contusion of a traffic accident victim died on the scene, b/ Secondary hemorrhage in the pons tissue of a traffic accident victim died in hospital.

Seasonal occurrence of traffic accidents has not showed crucial differences. The rate of accidents on weekends was higher for motor vehicle occupants than for pedestrians and bicyclists. The average survival period was 4.8 days among pedestrians, bicyclist; and 10.8 days among motor vehicle occupants. In 45 (21%) cases death was due to trauma complications (pulmonary emboli, pneumonia, infections) developed during hospital treatments. The longer the period of hospital treatment was, the higher the rank of complications was.

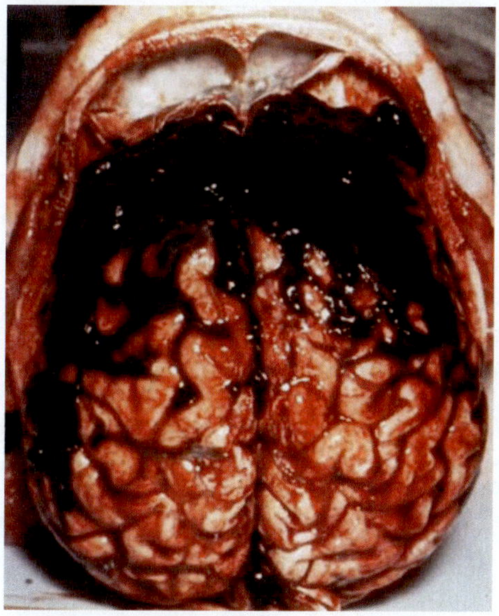

Figure 10. Subdural hemorrhage after traffic accident.

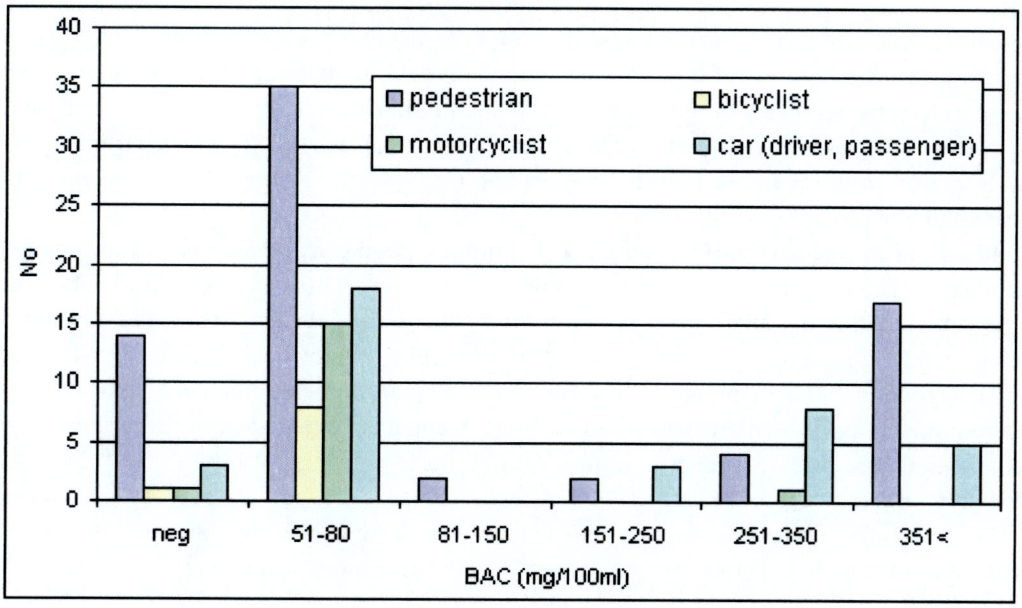

Figure 11. BACs among pedestrians, bicyclists and motor vehicle occupants.

CONCLUSION

In this chapter we demonstrate the occurrence and the possible pathomechanism of fatal head injuries during road traffic accidents. The population we studied included a spectrum of individuals died by significant head injuries followed traffic accidents. The results suggested the head injuries had a prior importance in fatal traffic accidents of pedestrians and bicyclists.

The current authors validated earlier reports that the most important predictor of mortality is the severity of head injury [8,9,10,11,12,13]. The incidence of head injuries and multiple contusions of lower limbs are frequent injuries in pedestrians and bicyclists [9,14,15]. Motor vehicle accidents are the most common causes of head injuries [16,17] or maxillofacial fractures [18]. The severity is the most important predictor of outcome in patients with head injuries [14,19,20]. However, severe head injuries resulting in brain death are the leading cause of mortality, a significant percentage of organ donations are obtained from these patients [21]. The demographic distribution was similar to those described in previous road traffic accident reports, and the majority of victims of traffic accidents were males [10,11,12,13,22,23,24,25,26,27]. In the present study in every age group male predominance was observed.

The distribution between motor and non-motor vehicle occupants does not reflect seasonal patterns. A previous study reported that peak incidence of pedestrian and bicyclist injuries occurred during the summer months and afternoon hours, whereas motor vehicle occupant injuries showed little seasonal variation and were more common during evening and night-time hours [16]. Most of the accidents in our study had occurred between afternoon and evening hours (2.00 p.m.–10.00 p.m.) followed by night and early morning [13,27]. Traffic during these hours is usually at their peak, and taking into account the increased pedestrian traffic.

Motor vehicle death rates among the elderly decreased substantially. The rate of bicyclist's injuries among 9- to 15-year-old children and of motorcycle injuries among adolescents between the ages of 12 and 16 years is high [16]. Young school-age children (5- to 9-years) [24] and elderly pedestrians (over 60 years of age) [28] are at particular risk. Motor vehicle injury patients admitted to hospital had high rate among persons younger than 30 years of age [26].

Road traffic accidents kill an estimated 1.2 million people worldwide each year, many of them has a fatal head injury. Current trends in population growth, industrialization and urbanization are putting heavy pressure on transport networks, particularly on the road. In the present study, the most common type of intra-cranial hemorrhage found was subdural hemorrhage and brain contusion. Injuries to the maxillofacial skeleton appear to be uncommon among pedestrians and motor vehicle occupants [29]. Based on experiences with medico-legal autopsies, traumatic injury of the vertebrobasilar artery (VBA) in traffic accident victims were found [30].

From the above facts, it is clear that the head is the most vulnerable part of the body involved in road traffic accidents, which alone accounts for most of the fatalities.

Pedestrians and Head Injuries

The pedestrians suffer most often from being struck by motor vehicles. Pedestrian crashes occur most frequently in urban areas where both pedestrian activity and traffic volumes are greater than in rural areas. In addition, many rural areas have no sidewalks, paths, or shoulders to serve as separated pedestrian facilities, and no lighting to increase the visibility of pedestrians at nighttimes. Road traffic pedestrian accidents demonstrate age distribution. While all age groups are more likely to be killed at no intersection locations, the numbers of pedestrians are higher for children and people in the older age groups. Moreover, some older pedestrians have physical, visual, and/or hearing impairments that place greater demand on intersection design.

It was concluded that vehicle body type characteristics play an important role in pediatric pedestrian injury severity and may offer engineering-based opportunities for injury control [31].

Bicyclist's, Motorcyclist's Accidents and Head Injuries

Bicycle riding is a popular form of recreation among persons of all ages, and related injuries cause significant morbidity and mortality. Most injuries occur in males and are associated with riding at high speed; most serious injuries and fatalities result from collisions with motor vehicles. The injuries are due to the projection from the unstable two-wheeled vehicle. The mechanism is a „flying – falling" one, when the outstanding parts of the body (face, hands, wrists, knees, elbows) suffer the impact of falling on the road. Other injuries are resulted by the parts of the vehicle such as abrasions and contusions resulted by the pedals or burns resulted by the exhaust pipe (motorcycle). The sitting-on position may be certified by the bruises on the buttock and parts of the leg due to the strike by the saddle. Running over injuries are also frequent.

Although superficial soft tissue injuries and musculoskeletal trauma are the most common injuries, head injuries are responsible for most fatalities and long-term disabilities. Overuse injuries may contribute to a variety of musculoskeletal complaints, compression neuropathies, perineal and genital complaints. Physicians treating such patients should consider medical factors, as well as suggest adjusting various components of the bicycle, such as the seat height and handlebars. Encouraging bicycle riders to wear helmets is a key to preventing injuries; protective clothing and equipment, and general safety advice also may offer some protection.

Fatal outcomes differ dramatically in the riding seasons, in accordance with the quality of roads, speed limits and availability of emergency medical service and trauma care. Thus, the most important comparison is before and after motorcycle helmet legislation within states. The enactment of comprehensive motorcycle helmet laws has consistently shown substantial decreases in motorcycle-related head injury death rates, and motorcycle fatalities increase 28-45% after repeal of mandatory helmet laws.

Motor Vehicle Occupants and Head Injuries

Whiplash is an injury mechanism to the neck caused by the neck bending forcibly forward and then backward, or vice versa. The injury usually involves the muscles, discs, nerves, and tendons in the neck. Most whiplash injuries are the result of a collision that includes sudden acceleration or deceleration. Many whiplash injuries occur when a person is involved in a car collision. Careful, defensive driving, wearing seatbelts, and using padded automobile headrests can lessen the likelihood of whiplash. Head restraints on car seats may prevent some whiplash injuries, by stopping the head from jolting backwards in a car crash [32].

A seat belt, sometimes called a safety belt, has been designed to secure the occupant of a car or other vehicle against harmful movement that may result from a collision or a sudden stop. As part of an overall occupant restraint system, seat belts are intended to reduce injuries by stopping the wearer from hitting hard interior elements of the vehicle or other passengers and by preventing the wearer from being thrown from the vehicle. The predominance of cervical injuries and the higher incidence of neck and facial injuries in car occupants may reflect thoracic-abdominal seat belt restraint [33,34].

Wrongful death can occur from various types of seat belt injuries [35]. For example, a passenger could be ejected upon a rear-end collision if the car was equipped with lap-only seat belt design. Other seat belt wrongful deaths can occur from a defect inertial release or a mechanical component failure. These failures can kill the occupant of the car when the seat belt releases during a car accident. Wrongful death in these situations often occurs when the passenger is ejected from the vehicle. In addition, in design defect cases where the seat belt anchor is improperly placed, wrongful death can occur when the passenger moves towards the roof during a rollover.

The unbelted occupants had a similar head injury severity pattern compared to belted occupants. The present study is another step to advance towards better understanding of the prevention, treatment and rehabilitation of side impact injuries [36].

Seat belts became available in new cars during the early 1970's, airbags during the 1980's. Since their introduction, airbags have become a controversial topic. Air bags supplement the safety belt by reducing the chance of the occupant's head and upper body striking some part of the vehicle's interior. They also help reduce the risk of serious injury by distributing crash forces more evenly across the occupant's body.

Airbags involve the extremely rapid deployment of a large cushion. While airbags can protect a person under the right circumstances, they can also injure or kill. Today all airbag control units recognize if a belt is used and set the trigger time accordingly. Newer airbags trigger at a lesser speed; nonetheless, passengers must remain at least 25 centimeters from the bag to avoid injury from the bag in a crash. Injuries such as abrasion of the skin, hearing damage (from the sound during deployment), head injuries, eye damage for spectacle wearers and breaking of the nose, fingers, hands or arms can occur as the airbag deploys. Airbags are designed to protect the head, neck and chest. Airbags combined with seat belts (lap and shoulder belt) do save lives and help prevent more serious injury. Researchers have found that airbags combined with seat belts effectively reduce spinal injuries from car accidents. Airbags were never designed to replace seat belts. That is why an airbag is called a 'supplemental restraint'. When the airbag deploys the seat belt helps protect the occupant by providing body support and preventing occupant ejection. The airbag creates a cushion between the occupant and the steering wheel, dashboard, windshield or other surfaces. Although many motor vehicle accidents cannot be prevented, drivers and passengers can reduce the risk of several injuries.

The installation of airbags in motor vehicles, in association with the use of seat belts, has reduced the incidence of head injuries, as well as significantly decreasing morbidity and mortality in motor vehicle accidents. Nevertheless, the literature on the subject increasingly refers to lesions related to airbag deployment. These are usually minor, but in certain circumstances, severe and fatal injuries can result [37].

It is essential for the forensic pathologists to have sufficient information about how the airbags work. For the medico-legal evaluation of injury pathomechanism forensic experts need data of the scene investigation and data about the types of cars in a case of fatal traffic accident. Timing is crucial in the airbag's ability to save lives in a head-on collision. An airbag must be able to deploy in a matter of milliseconds from the initial collision impact. It must also be prevented from deploying when there is no collision. Hence, the first component of the airbag system is a sensor that can detect head-on collisions and immediately trigger the airbag's deployment. However, when the car decelerates very quickly, as in a head-on crash, the ball suddenly moves forward and turns on an electrical circuit, initiating the process of inflating the airbag. Once the electrical circuit has been turned on by the sensor, a pellet of sodium aside (NaN_3) is ignited. A rapid reaction occurs, generating nitrogen gas (N_2). This process, from the initial impact of the crash to full inflation of the airbags, takes only about 40 milliseconds. Ideally, the body of the driver (or passenger) should not hit the airbag while it is still inflating. In order for the airbag to cushion the head and torso with air for maximum protection, the airbag must begin to deflate (*i.e.* decrease its internal pressure) by the time the body hits it. Otherwise, the high internal pressure of the airbag would create a surface as hard as stone.

Alcohol Influence

Drinking alcohol before driving is a main factor leading to lethal injuries linked to road traffic accidents [38]. Alcohol consumption patterns were significantly associated with car collisions between 15-24 years of age [16]. Male drivers who drove too fast and under the influence are involved more often in loss-of-control accidents than in other ones [39]. Alcoholic intoxication was frequent in drivers and also in pedestrians [5,38]. Concerning alcohol-related injury deaths, males had higher rates than females [40].

Head Complications, Clinical Management

Outcomes of patients with severe traumatic brain injury may potentially be improved by early targeting of the higher mean arterial pressure observed in survivors (mean arterial pressure 80 mm Hg), which may facilitate improved cerebral perfusion, and the immediate cause of death after head trauma could be the sudden hemodynamic failure [41,42].

Clinical investigations showed that serious injury rates were about ten times higher than mortality rates in motor vehicle accident [23]. Motor vehicle crashes were the greatest cause of potential years of life lost among the economically active male population [8]. Traffic crashes are the leading health threat to children in many countries [15,16,23,43].

Post-traumatic hypopituitarism (PTHP) is defined as pituitary hormonal deficiency occurring after head injury. PTHP has recently been increasingly recognized as an important medical condition among survivors of head trauma. It can be fatal if undiagnosed, and may contribute to the morbidity and mortality associated with traumatic brain injury (TBI). Although about 75% of PTHP cases are diagnosed within a year of TBI,1 delayed development of PTHP can also occur [44].

Hemorrhage under the internal limiting membrane or extending to the ora serrata were only seen in situations where crush injury was part of a fatal trauma scenario related to motor vehicles [45].

More road traffic injuries arrived in the emergency care units during summer without a specific distribution by weekdays, but with a higher probability of being a boy aged 5-9 years, presenting multiple trauma especially intracranial injury and requiring specialized hospital care [46].

Prevention

Cirera et al [26] reported that pedestrians and two-wheel motor vehicle occupants, besides accounting for two-thirds of motor vehicle injury cases, are the user groups with a greater risk of more severe injury, as well as higher chance of a hospital admission, independently of demographic and health care factors. Our data presented that every 5th person suffered traffic accident died of a complication after hospitalization. The occurrence of natural diseases suggests that coronary sclerosis is more frequent among motor vehicle occupants in spite of the younger average age of this group. A report on motor vehicle occupant versus pedestrian accidents has found that these events are common in the large urban traffic systems, and the mortality rate increases with age [47]. Factors such as the non-use of seat belts and helmets, being under the influence, and speeding are among the most important risk factors of traffic accidents. The high mortality rate of traffic accidents emphasizes the necessity of developing prevention strategies. Mortality rate is higher in the pedestrian subset than for motor vehicle victims [19].

Preventive strategies may include proper supervision and risk reduction with respect to use of alcohol or drugs. The data presented support the importance of primary prevention [48] in order to achieve the most effective, safe community interventions [20,49]. More emphasis should be placed on injury prevention and public education to prevent traffic injuries in young, vulnerable children. Head injuries appear to be a substantial factor affecting the outcome on injured pedestrians. Most of the fatal cases were associated with intracranial hemorrhages, skull fractures and brain damages. The high rate of severe head injuries in both of the motor and non-motor vehicle occupants suggests that the protection against head injury can be useful in fatal traffic accident prevention.

Effective road environmental interventions (eg, bicycle-friendly roadway design, intersections and crossings) along with efforts to promote safe personal behavior (e.g., helmet use and following rules of the road) are needed to help reduce injuries among cyclists while sharing the road [50]. Studies [4,51] have generally indicated that daytime use of running lights is associated with small to moderate reductions in multiple-vehicle daytime crashes, especially those involving vehicles approaching from the front or side. Road traffic policies such as pedestrian-friendly paths, separate lanes for light motor vehicles and heavy motor vehicles, and strict implementation of traffic rules and regulations may decrease the incidence of road traffic accidents and its fatalities.

REFERENCES

[1] Hogarth J. *Glossary of health care terminology,* WHO Publication, Copenhagen (1978).

[2] Macpherson A, Roberts I, Pless lB. Children's exposure to traffic and pedestrian injuries. *Am J Public Health* 1998;88:1840-3.

[3] The Centers for Disease Control and Prevention. Motor-vehicle safety: A 20th century public health achievement. *JAMA* 1999:281;2080-2.

[4] Williams AF, Lancaster KA. The prospects of daytime running lights for reducing vehicle crashes in the United States. *Public Health Reports* 1995;110:233-9.

[5] Varga T, Jeszenszky E, Simonka JA. Influence of alcohol: 114 hospitalized victims of traffic accidents. *For Sci Int* 1999;103:S25-S29.

[6] Varga T, Horvath J. Die Rolle der Stossgesschwindigkeit bei Verkehrsunfallen zwischen PKW's und Fussgangern. *Beitr Gerichtl Med* 1990;48:381-5.

[7] Wells S, Macdonald S. The relationship between alcohol consumption patterns and car, work, sport and home accidents for different age groups. *Accid Anal Prev* 1999;31:663-5.

[8] Barreto SM, Swerdlow AJ, Smith PG, Higgins CD, Andrade A. Mortality from injuries and other causes in a cohort of 21800 Brazilian steel workers. *Occup Environl Med* 1996;53:343-50.

[9] Ferrando J, Plasencia A, Ricart X, Segui-Gomez M. Motor vehicle injury patterns in emergency department patients in a South-European urban setting. *Proceedings: Annual Conference-Association for the advancement of automotive medicine* 2000;44:445-58.

[10] Akang EEU, Kuti MAO, Osunkoya AO. Pattern of fatal head injuries in Ibadan – a 10 year review. *Med Sci Law* 2002;42:160–6.

[11] Chandra J, Dogra TD, Dikshit PC. Pattern of cranio-intracranial injuries in fatal vehicular accidents in Delhi 1966–1976, *Med Sci Law 1979;*19:186–94.

[12] Gautam B, Verma SK, Sharma J J, Aggarwal NK. Pattern of road traffic accidents in North-East Delhi, *J For Med Toxicol* 2003;20:27–32.

[13] Patel NS. Traffic fatalities in Lusaka, Zambia, *Med Sci Law* 1979;19:61–5.

[14] Bhandari M, Guyatt G, Khera V, Kulkarni A, Sprague S, Schemitsch E. Operative management of lower extremity fractures in patients with head injuries. *Clin Orthop Rel Res* 2003;1:187-98.

[15] Eastridge B, Burgess AR. Pedestrian pelvic fractures: 5-year experience of a major urban trauma center. *J Trauma* 1997;42:695-700.

[16] Durkin, MS, LaraqueD, Lubman I, Barlow B. Epidemiology and prevention of traffic injuries to urban children and adolescents. *Pediatrics* 1999;103:1273-4.

[17] Kong LB, Lekawa M,Navarro RA, McGrath J, Cohen M, Margulies DR, Hiatt JR. Pedestrian-motor vehicle trauma: An analysis of injury profiles by age. *J Am Coll Surg* 1996;182:17-23.

[18] Alvi A, Doherty T, Lewen G. Facial fractures and concomitant injuries in trauma patients. *Laryngoscope* 2003;113:102-6.

[19] de Sousa RM, Regis FC, Koizumi Ms. Traumatic brain injury: differences among pedestrians and motor vehicle occupants. *Rev Saud Pub* 1999;33:85-94.

[20] Patrick DA, Bensard DO, Moore EE, Partington MD, Karrer FM. Driveway crush injuries in young children: a highly lethal, devastating, and potentially preventable event. *J Ped Surg* 1998;33:1712-5.

[21] Fisher B, Worthen M. Cardiac arrest induced by blunt trauma in children. *Ped Emerg Care* 1999;15:274-6.

[22] Wheatley J, Cass DT. Traumatic deaths in children: The importance of prevention. *Med J Australia* 1989;150:72-8.

[23] Barell V, Zadka P, Halperin B, Sidransky E. Childhood mortality from accidents in Israel. 1980-84. *IJMS* 1990;26:150-7.

[24] Rao R, Hawkins M, Guyer B. Children's exposure to traffic and risk of pedestrian injury in an urban setting. *J Urban Health: Bull N Y Acad Med* 1997;74:65-80.

[25] Namdaran F, Elton RA. A study of reported injury accidents among novice motorcycle riders in a Scottish region. *Accid Anal Prev* 1998;20:117-21.

[26] Cicera E, Plasencia A, Ferrando J, Segui-Gomez M. Factors associated with severity and hospital admission of motor vehicle injury cases in a southern European urban area. *Eur J Epidemiol* 2001;17:201-8.

[27] Jha Nilambar, D.K. Srinivasa, Roy Gautam and S. Jagdish, Epidemiological study of road traffic accident cases: a study from south India, *Indian J Commun Med* 2004;29:20–24.

[28] Fife D, Whitfield RA. Historical changes in motor vehicle death rates among the elderly. *Accid Anal Prev* 1988;20:393-8.

[29] Akama MK, Chindia ML, Macigo FG, Guthua SW. Pattern of maxillofacial and associated injuries in road traffic accidents. *East Afr Med J* 2007;84:287-95.

[30] Hiraiwa K, Sato T, Sasaki T, Mizusawa I, Nata M, Kodama N. Medico-legal aspects of traumatic injury of the vertebrobasilar artery. *Neurol Med Chir* (Tokyo) 2005;45:549-55.

[31] DiMaggio C, Durkin M, Richardson LD. The association of light trucks and vans with paediatric pedestrian deaths. *Int J Inj Contr Saf Promot.* 2006;13:95-9.

[32] Menon A, Pai VK, Rajeev A. Pattern of fatal head injuries due to vehicular accidents in Mangalore. *J Forensic Leg Med* 2008;15:75-7.

[33] Robertson A, Branfoot T, Barlow I, Giannoudis P. Spinal injury patterns resulting from car and motorcycle accidents. *Spine* 2002;27:2825-30.

[34] Glass RB, Sivit CJ, Sturm PF, Bulas 01, Eichelberger MR. Lumbar spine injury in a pediatric population: difficulties with computed tomographic diagnosis. *J Trauma* 1994;37:815-9.

[35] Bauzá G, Lamorte WW, Burke PA, Hirsch EF. High mortality in elderly drivers is associated with distinct injury patterns: analysis of 187,869 injured drivers. *J Trauma* 2008;64:304-10.

[36] Kumaresan S, Sances A Jr, Carlin F, Frieder R, Friedman K, Renfroe D. Biomechanics of side impact injuries: evaluation of seat belt restraint system, occupant kinematics and injury potential. *Conf Proc IEEE Eng Med Biol Soc.* 2006;1:87-90.

[37] Zaglia E, De Leo D, Lanzara G, Urbani U, Dolci M. Occipital condyle fracture: an unusual airbag injury. *J Forensic Leg Med* 2007;14:231-4.

[38] Reynaud M, Le Breton P, Gilot B, Vervialle F, Falissard B. Alcohol is the main factor in excess traffic accident fatalities in France. *Alcoholi* 2002;26:1833-9.

[39] Laapotti S, Keskinen E. Differences in fatal loss-of-control accidents between young male and female drivers. *Accid Anal Prev* 1998;30:435-42.

[40] Landen MG, Beller M, Funk E, Propst M, Middaugh J, Moolenaar RL. Alcohol-related injury death and alcohol availability in remote Alaska. *JAMA* 1997;278:1755-8.

[41] Henzler D, Cooper DJ, Tremayne AB, Rossaint R, Higgins A. Early modifiable factors associated with fatal outcome in patients with severe traumatic brain injury: a case control study. *Crit Care Med.* 2007;35:1196-7.

[42] Morentin B, Biritxinaga B. Massive pulmonary embolization by cerebral tissue after head trauma in an adult male. *Am J Forensic Med Pathol.* 2006;27:268-70.

[43] de Vries A, Kassam-Adams N, Cnaan A, Sherman-Slate E, Gallagher PR, Winston FK. Looking beyond the physical injury: Posttraumatic stress disorder in children and parents after pediatrics traffic injury. *Pediatrics* 1999;104:1293-9

[44] Nirmalaraj K, Holdaway I. Variable presentation of post-traumatic hypopituitarism. *N Z Med J* 2008;121:U2899.

[45] Gnanaraj L, Gilliland MG, Yahya RR, Rutka JT, Drake J, Dirks P, Levin AV. Ocular manifestations of crush head injury in children. *Eye.* 2007;21:3-4.

[46] Duma O, Roşu TS. Soliciting a model of a child emergency care unit in road accidents in the year 2005. *Rev Med Chir Soc Med Nat Iasi.* 2006;110:999-1003.

[47] Peng RY, Bongard FS. Pedestrian versus motor vehicle accidents: an analysis of 5000 patients. *J Am Coll Surg* 1999;189:343-8.

[48] Laraque 0, Barlow B, Durkin M. Prevention of youth injuries. *J Nat Med Ass* 1999;91:557-71.

[49] Agran PF, Anderson CL, Winn DG. Factors associated with restraint use of children in fatal crashes. *Pediatrics* 1998;2:630-1.

[50] Haileyesus T, Annest JL, Dellinger AM.Cyclists injured while sharing the road with motor vehicles. *Inj Prev.* 2007 Jun;13:202-6.

[51] Braver ER, Whitfield R, Ferguson SA. Seating positions and children's risk of dying in motor vehicle crashes. *Injury Prev* 1998;4:181-7.

In: Road Traffic: Safety, Modeling and Impacts
Editors: S. E. Paterson and L. K. Allan, pp. 475-486
ISBN 978-1-60456-884-4

Chapter 14

ROAD TRAUMA IN THE DEVELOPING WORLD

Sanket Srinivasa[1], Tarik Sammour[2] and Andrew G. Hill[2]
[1]Middlemore Hospital, Otahuhu, Auckland, NZ, USA
[2]South Auckland Clinical School, University of Auckland, USA

ABSTRACT

Road trauma is an important problem in the developing world, with serious implications particularly in resource poor areas. 90% of road traffic accident related deaths worldwide occur in low- and middle-income countries, accounting for approximately 1.1 million deaths each year. Road trauma mortality is expected to increase by 80% over the next twenty years in the developing world and become the second leading cause of death. This is in contrast to developed nations, where road traffic mortality has steadily declined.

The greater population density in developing countries means pedestrians are relatively more likely to be involved in road traffic accidents, with a lower incidence of multiple car involvement. Injury patterns reflect this. Aside from soft tissue injuries, the most frequent presentations involve limb fractures, followed by blunt abdominal trauma. Developing countries have lower rates of seatbelt usage causing different patterns of abdominal trauma, whilst a high number of two-wheelers also lead to patients presenting with more serious and varied patterns of injury. Oro-facial trauma is particularly common with two-wheelers.

The high incidence of road trauma in developing countries has been attributed to increasing number of vehicles and poorly developed infrastructure unable to accommodate different types of vehicles as well as large pedestrian populations. Studies have also demonstrated associations with speed and alcohol usage.

Trauma management in the clinical setting contrasts sharply to that in the developed world. This is mainly due to a lack of specialized care in some areas, with poor access to diagnostic radiology and operating theatre resources. In blunt abdominal trauma, for example, this necessitates a greater reliance on techniques like diagnostic peritoneal lavage for diagnosis, with a lower threshold for laparotomy. Conversely, the poorest nations can only manage patients conservatively due to a lack of facilities and expertise. Patient management is also compromised by late patient presentation caused by lack of education, geographic barriers, and transport constraints.

Improving trauma outcomes in the developing world remains a significant challenge. While the advent of specialist trauma centers and improved access to public hospitals is going some way to achieve this in certain parts of the world, the poorest nations still suffer. In order to effect change, advancements in infrastructure, and road legislation and its enforcement are required, as well as a commitment to public health and education projects. This must be instigated at a governmental and organizational level.

INTRODUCTION

Road trauma is an important problem in the developing world, with serious implications particularly in resource poor areas. 90% of road traffic accident related deaths worldwide occur in low and middle income countries, accounting for approximately 1.1 million deaths each year (see Figure 1). [1]

	Number	Rate per 100,000 population	Proportion of total (%)
Low and middle – income countries	**1,065,988**	**20.2**	**90**
High – income countries	**117,504**	**12.6**	**10**
Total	**1,183,492**	**19.0**	**100**

Figure 1. Estimated global road traffic injury-related deaths. [1]

Road trauma in the developing world is conservatively estimated to cost 65 billion dollars annually, which is significantly more than the total amount of foreign aid donated. [2] The burden of infectious diseases in the developing world, however, prevents a solitary focus on decreasing the incidence of road trauma and its consequences. Conversely, in the developed world, the road trauma mortality rate exceeded the mortality rate from infectious diseases by the year 1954, prompting medical organizations and political bodies to take a proactive approach by both proposing and endorsing legislation as well as evaluating clinical practice to optimize outcomes in trauma patients. [3]

The majority of research in this area is based in Western countries, leaving developing world road trauma relatively poorly explored. A large proportion of this research is based on car ergonomics, with safety devices such as airbags and seatbelt use extensively investigated. However, this is less relevant to the developing world, where the majority of victims are pedestrians. Some studies have estimated that as few as 5% of car occupants are injured in traffic accidents in the developing world. [4]

INCIDENCE

Road trauma mortality is expected to increase by 80% over the next twenty years in the developing world and become the second leading cause of death behind infectious diseases (see Figure 2). [1] This is in contrast to developed nations, where road traffic mortality has steadily declined. Road traffic accidents are the second leading cause of death for children aged 5 – 14 years and young adults aged 15 – 29 years. [1] Males are also more likely to be injured, which is economically disadvantageous. The true incidence of road trauma in developing countries is probably higher than reported as resource-poor nations lack the facilities to accurately record these data and are also faced with chronic under-reporting from victims, who may perceive the exercise of notification as being pointless. High pre-hospital mortality also contributes to inaccurate data.

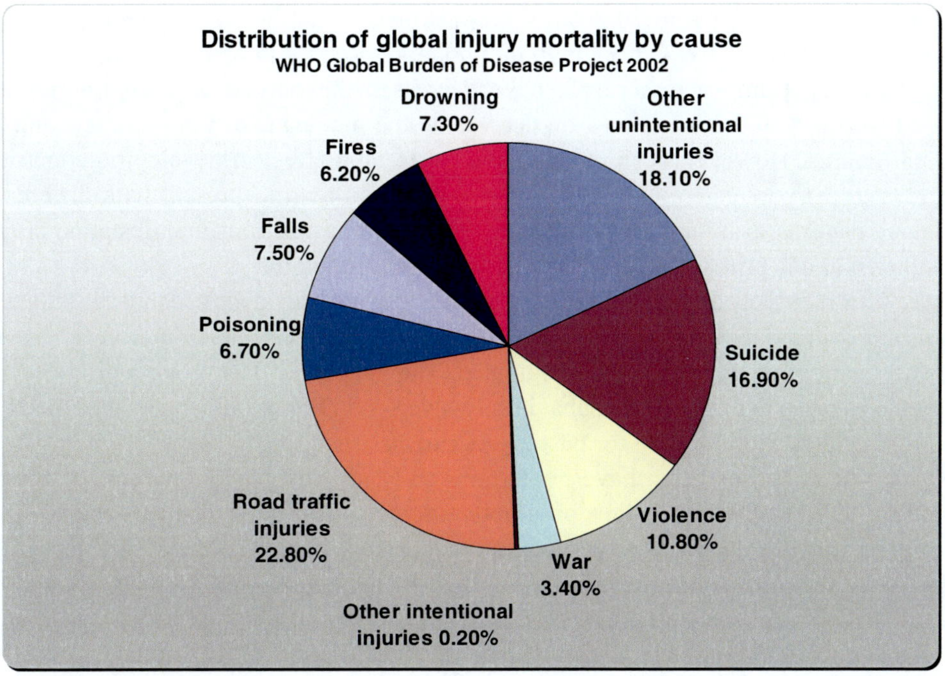

Figure 2. Causes of mortality. [1]

Developing countries show regional variance in the consequences of road trauma. As a generalization, there is a higher incidence of injury in urban areas whilst accidents in rural regions tend to have higher mortality. This may be because of less prompt care as well as higher speeds on less crowded rural roads.

ETIOLOGY

Rapidly industrialising countries are showing the highest incidence of road trauma. This is thought to be due to the recent phenomenal increase in the number of vehicles in an environment that is not adequately equipped to cope with the vehicular density. India, for example, is expected to have approximately 267 million vehicles, excluding two-wheelers, by the year 2050. [1] In Vietnam, as the number of motor vehicles increased by 14%, road deaths and injuries increased by 31% and 16% respectively. [5] In the last 30 years there has been a four to five-fold increase in road fatalities in Kenya. [6] Moreover, combined with high numbers of vehicles, other road-users in the form of pedestrians, cyclists and others also occupy the traffic scene forming a potentially dangerous combination of high and low speed traffic mix leading to irregular traffic flow.

The rise in vehicle numbers in developing countries has also been contributed to by increasing numbers of motorcycles, scooters and bicycles (all grouped under the category of two-wheelers). In conditions of poverty and traffic congestion, two-wheelers provide a suitable and economical alternative for individuals (see Figure 3). The high number of two-wheelers and their exposed occupants contribute to both the high incidence and mortality rate of road trauma, and different patterns of injury compared to the developed world. A study from India estimates that up to 70-80% of crashes involve 2-wheelers, though findings vary depending on numbers of two-wheelers in individual regions. [4] Laws regarding compulsory helmet use whilst riding a motorcycle have decreased the burden of head injuries in the developed world, with the focus now on decreasing the incidence of intra-thoracic and intra-abdominal trauma. However, such efforts have yet to take effect in developing countries. In comparison to car passengers, occupants of two-wheelers tend to present with different and more severe patterns of injury. Head injury is the commonest serious manifestation after soft tissue injuries, and is primarily caused by helmets not being worn.

There is a relatively higher reliance on public transport in poorer countries. Somewhat paradoxically, this also contributes to increased morbidity and mortality as these arrangements are often informal and unregulated. The matatus of Kenya are a prime example. These vehicles often exhibit unsafe behaviour by overcrowding of passengers and speeding to adhere to schedules. Competition for passengers can also lead to aggressive driving.

Developing countries often have poor roads and lack adequate resources to administer road safety and education campaigns or simply enforce traffic laws. The cars themselves are often in poor condition and their road-worthiness is rarely inspected. Widespread corruption in these countries allows unsafe vehicles to take to the roads. Whilst seat-belt usage has dramatically decreased the severity of road traffic injuries in developed nations, studies have revealed that only approximately 50% of cars in the developing world even have seatbelts. [7]

Figure 3. Two-wheelers in India (taken from Manfred´s Travel Pictures with permission, copyright 1997-2007 Manfred Leiter).

Failure to adhere to traffic rules also puts occupants at risk and often there are more than the recommended number of passengers travelling in one vehicle. Children are put at particular risk as they are often seated on an adult's lap with no other protection. Evidence from Saudi Arabia documented 97% of all paediatric oro-facial trauma to be due to this situation in two-wheeler accidents. [8]

Several studies have also shown speed to be a causative factor. [9,10,11] Whilst this is similar to developed countries, the synergistic effects of poorly developed roads as well as non-adherence to traffic rules and decrepit vehicles combined with suboptimal trauma management leads to significantly higher morbidity and mortality.

There are minimal data on the prevalence of alcohol- related crashes in resource-poor countries but conservative estimates have approximated alcohol to be a causative factor in between 35-50% of crashes. [1, 10] Drugs such as 'miraa' have also been implicated in contributing to road trauma incidence and together with alcohol, form a major part of the social landscape in these countries.

Developed countries have recently focused on the issue of driver distraction with cellular phones being extensively investigated. Certain places, such as Victoria in Australia have banned the use of cellular phones whilst driving and recent studies from Britain have suggested that a driver talking on a cellular phone exhibits lower levels of alertness and slower reflexes than drivers under the influence of alcohol. [12] This issue has not been explored in the developing world but may be increasingly important in countries such as India and China as cellular phone use dramatically increases.

Obviously, a major underlying issue that both precipitates and perpetuates these problems is that of poverty, both at a national and individual level. The poor are more likely to be pedestrians in unsafe environments with decreased access to healthcare.

TRAUMA MANAGEMENT

Studies have consistently shown trauma outcomes to be worse in developing countries, compared to developed countries. [13-15] This is probably due to a combination of factors. It is difficult to compare severity of injury between populations because prognostic scores such as TRISS, RTS and NISS have poor applicability in resource-poor settings. [13,16-18] Research is slowly emerging regarding prognostic scores which may have greater validity for developed nations [17].

What can be said is that in the developing world, patterns of injury tend to be different. As already mentioned, the higher rate of pedestrian and two-wheeler involvement has a definite impact on patient presentation. Car occupants also often have different patterns of injury, namely because of the disappointingly low rate of seatbelt availability and usage in developing countries. Seatbelts protect the passenger from impact within the car, and ejection through windows and doors. The lack of restraint results in more dramatic, severe, and ultimately fatal injuries in developing countries.

Pre-hospital trauma care is also suboptimal as there is seldom a formal ambulance or paramedic service. Studies comparing pre-hospital care between Seattle and Mexico have shown that patients in Mexico had more unsecured airways, inadequate fluid resuscitation and arrived at hospital later. [15] One innovative solution implemented in Ghana involved training commercial drivers in first-aid, leading to an apparently lower mortality rate. [19]

There are fewer hospitals per capita, meaning people often have to travel great distances for healthcare. For instance, the specialist maxillofacial unit of The University of Maiduguri Teaching Hospital in Northern Nigeria services a population of approximately 30 million people. The consequent delay in reaching hospital and delay prior to subsequent treatment is significant.

The hospitals that are available are resource constrained. No where is this more obvious than in the lack of radiological imaging facilities. Computed tomography (CT) scanners, for example, are an expense that simply cannot be afforded in many of these hospitals. In contrast, the developed world is better equipped to deal with trauma with the advent of specialized trauma centers. Ongoing research has dictated optimal management strategies for individual presentations, with diagnostic algorithms that are readily available to thoroughly investigate and manage different conditions. South Africa is generally recognized to be amongst the forefront in trauma management, largely due to the high incidence of trauma seen. Approximately 80% of all laparotomies in Groote Schuur hospital in Capetown are due to gun-shot wounds. [20] Whilst the mechanism of trauma is mostly due to interpersonal violence, the South African experience has provided much insight on the balance between conservative versus operative management of trauma.

Alongside a lack of auxiliary resources, there is a dearth of adequately trained personnel at the facilities providing trauma care. There are often no trained trauma surgeons and intensivists to deal with acute presentations. A potential solution to this problem would be to develop strong links between such facilities and the international surgical and nursing colleges responsible for specialty training, in an attempt to foster training posts in the developed world. If basic infrastructure is invested in, such posts can provide trainees with invaluable experience, allowing them to up-skill both themselves and local health professional and potentially improving outcomes in developing countries. Such fellowships could also be reciprocal, with doctors from developing countries accommodated and trained via placement in established international trauma centers. Such exchanges also have the potential to further increase awareness of the plight of patients in developing countries.

Cultural and religious beliefs also vary trauma management in resource-poor countries. In Morocco, traditional healers that specialize in the treatment of fractures are widely prevalent and are preferred by the local community. [21] They are also more accessible and economical than doctors. Their social integration has made them a service that the medical professionals must work alongside, despite the knowledge that they may provide potentially suboptimal outcomes.

Abdominal Trauma Management

Abdominal trauma is a common consequence of road traffic accidents, often in combination with other, more obvious complaints like limb fractures. In the developing world, it poses a particular challenge with regards to both diagnosis and management.

As is the case in developed countries, the commonest injuries are liver lacerations and splenic injuries, followed by other solid organ, bowel, and vascular injuries. The majority of victims are pedestrians or on two-wheelers. Within car occupants, the highest risk passengers are unrestrained victims on the 'near side' in lateral impact collisions. [22] Since seatbelts are not as commonly used, there is a higher incidence of ejection from the vehicle, and a relatively lower incidence of seatbelt associated injuries (e.g. "seatbelt syndrome").

There is increased emphasis on managing conditions conservatively and, whilst there are various criteria on deciding when such a strategy is best, undoubtedly the biggest influence has been the availability, or lack there of, of diagnostic radiology. CT scanning plays an important role in excluding intra-abdominal pathology in the absence of clinical signs and symptoms, and in some centers is used routinely in all serious trauma victims in the same manner as a skeletal survey. Some studies have indicated that 50% of trauma patients with no abdominal signs or symptoms have an intra-abdominal injury, which may be more reliably diagnosed by CT scanning. [23] Developing nations often lack radiology assistance and hence there is often a delay in diagnosis of silent pathology, decreasing successful outcomes. Ultrasound, however, is more readily available in developing world hospitals and may be used as part of trauma management in the same way that the FAST (Focused Abdominal Sonography for Trauma) system has been utilized. In the absence of sophisticated radiology, there is a greater reliance on diagnostic peritoneal lavage. Conversely, some hospitals choose to explore any potential abdominal injury with exploratory laparotomies. This can lead to a high rate of negative laparotomies and several studies from Africa have demonstrated a high rate of potentially avoidable post-operative complications, such as wound infection and / or sepsis. [24] Obviously, avoidable laparotomies contribute to increased morbidity for patients and are economically unsound in situations where resources are already scarce. Trauma management in developing countries can often be saddeningly simple with the poorest countries being able to offer little more than conservative management and rudimentary comfort cares.

Orthopaedic Trauma Mangagment

Since the incidence of pedestrian and two-wheeler injuries is highest, fractures are very common in road trauma with the long bones of the lower limb being particularly vulnerable. A Nigerian study has reported 79% of motorcycle accident victims to suffer from limb injuries, with tibial fractures being the most common. [25] There is often little in the way of specialist orthopaedic care and the majority of injuries are managed by splinting, closed reduction and sometimes external fixators. Whilst studies report good mortality outcomes with these methods, there is little research in developing world settings focusing on disability outcomes or functional recovery following these techniques. This is particularly important as having a consequent functional impairment can contribute to individual and family poverty, thus causing other potential problems downstream.

Neurological Trauma Management

Head trauma is a common consequence of road trauma, particularly in two-wheeler accidents. Poor compliance with helmet use predisposes the population to severe head injury. With neurosurgical services largely non-existent, head injuries are almost exclusively managed conservatively and their diagnosis is clinical. Hence there are poor outcomes in head trauma and survival rates show a strong correlation with the Glasgow Coma Score- higher scores show considerably higher mortality. Surviving victims often have considerable neurological morbidity, reducing their potential to contribute to society.

Paediatric Trauma Management

Data on paediatric trauma is limited. The most common serious injuries in children are head injury and followed by limb trauma. Chest and abdominal trauma represents a small subset of injuries. The majority of pediatric abdominal injuries involve splenic trauma.

Preventative Interventions

Whilst surgical management of trauma is important, the greatest benefit lies in implementing changes via public health means. Developed countries have used the Haddon Matrix as a framework to design effective interventions to lower the incidence of road trauma. However, most developing countries cannot institute such changes due to resource constraints. Indeed, government expenditure on road safety in Pakistan is $0.07 per capita. [26]

Potential successful interventions may include mandatory seatbelt usage as this has been shown to reduce mortality and morbidity in developed countries. [1, 27-29] However, cars in developing countries often don't have functional seatbelts and enforcing such legislation would be impossible in some countries due to a lack of law-enforcement personnel. As mentioned previously, vehicles that are not roadworthy are allowed to be driven due to corruption. Moreover, decrepit vehicles are the most affordable and are thus preferred.

Helmets for motorcycle and bicycle users have also been shown to decrease the incidence of head injuries and their clinical sequalae, when worn correctly. [30-34] This is a feasible intervention as those individuals that can afford a motorcycle or bicycle should be able to afford a helmet. Further education is necessary to ensure adherence to correctly wearing helmets, as observational studies have shown that many helmet users don't fix the chin strap, thereby nullifying the benefit of helmets in the event of a crash. A study from Indonesia showed that helmet use is seen to be a law enforcement issue rather than about safety. [32] Hence many users wear one for token compliance but do not fasten the chin straps.

The separation of pedestrians from vehicles reduces incidence of road trauma. [35] Many developing countries do not have footpaths for pedestrians (see Figure 4). Furthermore, footpaths are often used as work space by small shops and vendors forcing pedestrians to walk out onto the road. Simple interventions such as providing elevated curbs or fencing off pedestrian areas or the middle of the road to prevent crossing at certain points have been linked to significant decreases in road trauma incidence, morbidity and mortality. [36]

Figure 4. Busy road in Nigeria. [40]

Interventions to limit speed can include speed-bumps as these are low cost and low-maintenance. In Kolkata, India, the municipal corporation routinely builds a speed-bump at accident sites. Rumble strips at a common accident site in Ghana reduced road trauma mortality by 55% over a 16 month period. [37]

Since 1977, Sweden has enforced the use of headlights at all times for vehicles. They have demonstrated that this increases driver visibility and reduces mortality by 11%. [38] Despite its success, however, most Western countries have not adopted this idea.

There are numerous other reported success stories in the management of road trauma. Government-backed initiatives in Colombia to introduce mandatory traffic insurance with consequent safety and education campaigns have been very successful in reducing the impact of road trauma, demonstrating both the importance of government involvement and potential success if administered properly. The government of Taiwan has demonstrated that a proactive approach can both stagnate and reduce the incidence of road trauma even though vehicle numbers are increasing. Through enforced legislation and safety campaigns, the mortality rate decreased by 26.8% over a five-year-period up to 2001, even though the number of cars increased to over ten million. [39]

It appears that education and changing individual attitudes is key to bringing the scourge of road trauma under control. Road accidents have traditionally been seen as random chance events and therefore prevention is not regarded as plausible. In Ghana, accidents are perceived as an act of god. Such viewpoints foster resistance to viewing this problem as a potential system failure that can be solved, and need to be overcome.

Conclusion

Road traffic accidents represent a significant burden in the developing world and will continue to become more prominent without an effective response. Whilst poverty and competing health issues prevent optimal outcomes in regard to both incidence and mortality, action at the level of the government and institutions as well as personal and collective responsibilities assumed by communities may alter an otherwise bleak outlook. Implementation of interventions can reverse the forecasted trends.

References

[1] Peden M, Scurfield R, Sleet D. et al, eds. World report on road traffic injury prevention, *World Health Organization,* Geneva, 2004.

[2] Nantulya VM, Reich MR. Equity dimensions of road traffic injuries in low and middle-income countries. *Inj Control Saf Promot.* 2003; 10(1-2):13-20.

[3] Trinca GW et al. R*educing Traffic Injury - A Global Challenge.* 1988.

[4] Mohan D. Road traffic deaths and injuries in India: time for action. *NaW Med J India.* 2004;17(2):63-66.

[5] Nantulya VM, Reich MR. The neglected epidemic: road traffic injuries in developing countries. *BMJ.* 2002;324:1139-1141.

[6] Otieno T, Woodfield JC, Bird P, Hill AG. Trauma in rural Kenya. *Injury. 2004;35:1128-1233.*

[7] Forjuoh SN, Li G. A review of successful transport and home injury interventions to guide developing countries. *Soc Sci Med.* 1996;43(11) : 1551 -1560.

[8] Lawoyin TO. Lawoyin DO. Lawoyin JO. Factors associated with oro-facial injuries among children in Al-Baha, Saudi Arabia. *African J Med & Med Sci.* 31(1):37-40, 2002 Mar.

[9] Afukaar FK. Speed control in developing countries: issues, challenges and opportunities in reducing road traffic injuries. *Inj Control Saf Promot.* 2003;10(1-2):77-81

[10] Odero W, Meleckidzedeck K, Heda PM. Road traffic injuries in Kenya: magnitude, causes and status of intervention. *Inj Control Saf Promot.* 2003;10(1-2):53-61.

[11] Koushki PA, Bustan MA, Kartam N. Impact of safety belt use on road accident injury and injury type in Kuwait. *Accid Anal Prev.* 2003:35:237-241.

[12] Australian Transport Safety Bureau (ATSB) 2004 Road Safety in Australia – A publication commemorating. *World Health Day* 2004.

[13] Scarpelini S, De Andrede JI et al. The TRISS method applied to the victims of traffic accidents attended at a tertiary level emergency hospital in a developing country. *Injury.* 2006;37(1):72-7.

[14] London JA, Mock CN, Quanshah, RE et al. Priorities for Improving Hospital-Based Trauma Care in an African City. *J of Trauma.* 2001;51(4):747-753.

[15] Arreola-Risa C, Mock CN, Padilla D, Cavazos L, Maier RV, Jurkovich GJ. Trauma Care Systems in Urban Latin America: The Priorities Should Be Pre-hospital and Emergency Room Management. *J of Trauma.* 1995;39(3):457-462.

[16] Moini M, Rezaishiraz H, Zafarghandi MR. Characteristics and Outcome of Injured Patients Treated in Urban Trauma Centers in Iran. *J of Trauma.* 2000;48(3):503-507.

[17] Talwar S, Jain S, Porwal R, Laddha BL, Prasad P. Trauma Scoring in a Developing Country. *Sing Med J.* 1999;40(06).

[18] Zafar H, Rehmani R, Raja AJ, Ali A, Ahmed M. Registry based trauma outcome: perspective of a developing country. *Emerg Med J.* 2002;19: 391-394.

[19] Tiska MA, Adu-Ampofo M, Boakye G, Tuuli L, Mock CN. A model of pre-hospital trauma training for lay persons devised in Africa. *Emerg Med J.* 2004; 21:237-239.

[20] Bautz PC. Emergency Laparoscopy for Tangential Abdominal Gunshot Wounds: 5 Years Prospective Study. *ANZ J of Surg.* 2002;72 Supplement 1:A59.

[21] Choffat, F. Treatment of Fractures by Traditional Healers in Morocco. *Sozial-und Praventivmedizin.* 1979;24(2-3):172-178.

[22] Newgard, Craig D. Lewis, Roger J. Kraus, Jess F. McConnell, K John. Seat position and the risk of serious thoracoabdominal injury in lateral motor vehicle crashes. *Accident Analys & Prev.* 2005;37(4):668-74.

[23] Brown CK et al. Diagnostic evaluation of patients with blunt abdominal trauma: a decision analysis. *Acad Emerg Med.* 2000;7:385.

[24] Ayoade BA et al. Abdominal Injuries in Olabisi Onabanjo University Teaching Hospital Sagamu, Nigeria: Pattern and Outcome *Nig J of Orthopaedics and Trauma. 2006;5(2):45 – 49.*

[25] Oluwadiya KS et al. Motorcycle limb injuries in a developing country. *West Afr J of Med.* 2004;23(1): 42-7.

[26] Bishai D, Hyder AA, Ghaffar A, Morrow RH, Kobusingye O. Rates of public investment for road safety in developing countries: case studies of Uganda and Pakistan. *Health Policy and Planning.* 2004;18(2):232-235

[27] Mohan D. Road traffic injuries-a neglected pandemic. *Bull World Health Organ.* 2003:81 (9):684-685.

[28] Koushki PA et al. Impact of safety belt use on road accident injury and injury type in Kuwait. *Accid Anal Prev.* 2003:35:237-241.

[29] Cummings P, Wells JD, Rivara FP. Estimating seat belt effectiveness using matched-pair cohort methods. *Accid Anal Prev.* 2003:35:143-149.

[30] Liu B, Ivers R, Norton R, et al. Helmets for preventing injury in motorcycle riders. *The Cochrane Database Syst Rev.* 2003:(4).

[31] Tsai MC, Hemenway D. Effect of the mandatory helmet law in Taiwan. *Inj Prev.* 1999;5:290-291.

[32] Conrad P, Bradshaw YS, Lamsudin R, et al. Helmets, injuries and cultural definitions: motorcycle injury in urban Indonesia. *Accid Anal Prev.* 1996; 28(2):1

[33] Peek-Asa C et al. The prevalence of non-standard helmet use and head injuries among motorcycle riders. *Accid Anal Prev.* 1999;31:229-233.

[34] Thompson DC, Rivara FP, Thompson R. Helmets for preventing head and facial injuries in bicyclists. *The Cochrane Database Syst Rev.* 1999;(4).

[35] Forjuoh SN, Li G. A review of successful transport and home injury interventions to guide developing countries. *Soc Sci Med.* 1996;43(11) : 1551 -1560.

[36] Berhanu G. Models relating traffic safety with road environment and traffic flows on arterial roads in Addis Ababa. *Accid Anal Prev.* 2004;36:697-704.

[37] Afukaar FK. Speed control in developing countries: issues, challenges and opportunities in reducing road traffic injuries. *Inj Control Saf Promot.* 2003;10(1-2):77-81.

[38] Andersson, K.; Nilsson, G. *The effects on accidents of compulsory use of running lights during daylight in Sweden.* VT1 Swedish Road and Transport Research Institute, Report no. 208A; 1981.

[39] Chiu W, Ho Y, Lee Y Sharp Decline of Injury Mortality Rate in a Developing Country. *J Trauma.* 2003;55:391-392.

[40] Maier Karl, This House Has Fallen: Midnight in Nigería. PublicAffairs; 1st edition, July 2000.

In: Road Traffic: Safety, Modeling and Impacts
Editors: S. E. Paterson and L. K. Allan, pp. 487-502
ISBN 978-1-60456-884-4

Chapter 15

Motorcycle Accidents and Prevention Programs in Malaysia

Roszalina Ramli*

Department of Oral and Maxillofacial Surgery, Faculty of Dentistry, Universiti Kebangsaan Malaysia, Jalan Raja Muda Abdul Aziz 50300 Kuala Lumpur, Malaysia

Roslan Abdul Rahman

Department of Oral and Maxillofacial Surgery, Faculty of Dentistry, Universiti Kebangsaan Malaysia, Jalan Raja Muda Abdul Aziz 50300 Kuala Lumpur, Malaysia

Radin Umar Radin Sohadi

Malaysian Institute of Road Safety Research (MIROS) Lot 125-135, Jln TKS 1, Taman Kajang Sentral 43000 Kajang, Selangor Darul Ehsan, Malaysia

Abstract

Asian Development Bank (ADB) showed that road crashes killed more than 75,000 people in ASEAN in 2003 and it cost more than 2.2 percent of the region's annual gross domestic product (GDP).

Malaysia as one of the Association of South East Asian Nations (ASEAN) countries is a developing nation encompassing approximately 25 million population comprising Malays, Chinese, Indians and other races. Road traffic accident (RTA) is a major issue in this country as it is the predominant cause of mortality and morbidity in young people. In 2006, RTA resulted in 6287 fatalities and 58% or 3683 were fatalities involving motorcyclists and the leading cause of death was head injuries as a result of non-wearing or not securing the helmets properly.

Safety Planning Matrix which was based from the Haddon Matrix was introduced comprising pre-crash, crash and post-crash strategies involving human, vehicle and environment factors. Human and environmental safety programs are among the most extensively addressed and implemented. Human prevention program involves behavior

* Correspondence:Associate Professor Dr. Roszalina Ramli E-mail: r2tdh2004@yahoo.co.uk Fax:+0060326982944

modification and police enforcement program. Behavior modification program is directed largely at young road users especially motorcyclists and this mainly involves the media and television commercials. Safety and speed limit awareness are the key features in the police enforcement program.

Under the environmental safety program, exposure control program involves encouraging commuters to use other modes of transport to reduce them from being physically exposed to bodily impact during motorcycle rides; conspicuity program to increase visibility of a motorcyclist and road engineering program which involves setting-up exclusive motorcycle lanes, paved road shoulder and end treatment of non-exclusive motorcycle lanes.

It is targeted by the Malaysian Government that by the year 2010, the accident rate should reduce to more than 50 percent, i.e., deaths due to RTA should decline from the current 4.2 deaths per 10,000 vehicles to 2 deaths per 10,000 vehicles by 2010.

INTRODUCTION

Malaysia is a rapidly developing multi-racial nation encompassing approximately 25 million population comprising Malays, Chinese, Indians and other races. Over the last decade, Malaysia has experienced enormous economic expansion and growth in population (average of about 3% per year), industrialization and motorization. Vehicles have increased significantly from 7,210, 089 in 1994 to 13,878,000 in 2004 while ownership increased from 2.7 persons per vehicle in 1994 to 1.9 persons per vehicle in 2004, and almost half of this number were motorcyclists [1]. The increase in the total number of motorcycles is also seen in the neighboring countries, for example in Thailand, Singapore and Taiwan. Motorcycle is regarded as an important mode of daily transportation in many nations particularly in Asian countries. People commute on motorcycles to work and run daily errands as they could get to their destination faster and cheaper than in a car or a public transport. The number of motorcycle ownership is high because of factors such as low purchase price of the vehicle, low insurance surcharge rate and early age for motorcyclist licensing which is 16 years of age. A survey conducted by the Universiti Putra Malaysia on motorcyclists' receptiveness towards the change in various transport policies and vehicle ownership showed that many road users still favor motorcycles as a mode of transport and although the Malaysian Government gradually increases the motorcycle insurance premium, this does not discourage them from owning motorcycles.

Consequently, the number of fatalities related to motorcyclists has increased significantly since 1988. The number has increased alongside with the increase in population and ownership of vehicles, from 148,801 in 1994 to 326, 817 cases in 2004 and most of the casualties were motorcyclists [1]. Motorcyclist fatality carries an overall relative risk of about 17 to 20 times greater than that of a car occupant [2]. As a result of this, the Government of Malaysia has undertaken a number of efforts to reduce motorcycle accidents and fatalities such as improvement of road conditions, behavior modification and intensive traffic enforcement.

The use of helmet for riders and pillion riders is mandatory in Malaysia and has been strictly enforced since 1973.

The Cost of Lives Due to Road Traffic Accident

In 2003, it was estimated the value of a life lost was about Ringgit Malaysia (RM) 1.2 million [3]. With 6,282 fatalities, the country lost about RM 7.5 billion. Malaysia's 2003 gross domestic product (GDP) was over RM 385.7 billion. This means the cost of lives lost due to road traffic accident (RTA) was about 2% of the 2003 GDP. This was actually a minimum figure as it involved fatalities alone. The figures grew higher (2.4% of 2003 GDP) when the non-fatal injuries were taken into account [3].

Malaysian Institute of Road Safety Research (MIROS)

Malaysian Institute of Road Safety Research (MIROS) was established on the 3rd of January 2007 under the Ministry of Transport to serve as a one-stop centre for the generation and dissemination of road safety information through the print media and a concerted training program. MIROS carries out relevant research and establish cost-effective road safety interventions to enhance the current road safety level. Professor Ir. Dr. Radin Umar Radin Sohadi is the founding Director General of MIROS. There are various research centers in MIROS such as the road user behavioral change research center, road engineering and environment research center and vehicle safety and biomechanics research center. Under these research centers, various units are established such as human factor unit and design and instrumentation unit under the behavioral change research center; accident database system and analysis unit, highway planning and traffic engineering unit and exposure control and sociology of transport unit under the road engineering and environment research center and pre and post injury research unit, biomechanics and injury control unit and safety health environment unit under the vehicle safety and biomechanics research center.

Profile of a Typical Malaysian Motorcyclist with Poor Road Safety Attitude

RTA in Malaysia predominantly involved young people and the profile of these young casualties is studied.

Age

Majority of motorcyclists involved in RTA were young men, between the age of 16 to 25 years old and mostly possess driving license of less than 3 years [4]. They often owned motorcycles with engine capacity of more than 150cc engine displacement. These young riders have poor attitudes towards traffic rules and speeding compared to the older age groups.

From the authors' observation, young riders often break the traffic law by riding too fast, race among themselves and behave mischievously on the road. Other violation of traffic law includes running against the red lights, riding on the wrong direction of the road and improper or not wearing the helmets at all. Effectiveness of the use of a helmet is defined as when a helmet is worn with the strap securely fastened. Thus, helmet compliancy is observed when a motorcyclist has his helmet properly secured. It is observed that more than 50% of adult riders do not fasten their helmets properly [5].

In relation to child pillion riders, most of the children are likely not to be using safety helmets at all or wearing incorrect size helmets, usually larger than their heads' sizes or they could be wearing fake helmets.

Race

Indian riders were the least compliance in helmet wearing compared to Chinese and Malays riders [5].

Gender

Male motorcyclists were less compliance than the female riders in relation to helmet wearing [4,5].

Apart from compliance in helmet wearing, female motorcyclists in general have more positive attitudes towards road safety and more likely to comply with traffic laws [6].

Marital Status

There was no difference in proper safety helmet use in single or married riders [5].

Education Level

Riders with only primary level of education had least compliance in crash helmet wearing compared to those with secondary level education and the most compliance were shown in riders with tertiary level of education [5].

Level of License Type

Riders with no license had the least compliance of proper helmet usage compared to those riders with full, probation and learners' license holders [5].

Riding Experience

Young riders with less than a year riding experience were shown to have low compliance in helmet wearing [5].

Urban versus Non-Urban

Motorcyclists riding in town areas were more likely to comply with the proper usage of safety helmets compared with motorcyclists riding in outside town areas. This is most probably due to the motorcyclists' perception of lower enforcement activity in rural areas compared to urban sites. The enforcement activities are in fact lower in the non-urban areas due to limited resources and larger spatial coverage to be taken by the enforcement teams [6].

Distance Traveled

The compliance level of proper usage of helmets was lower for those riders who traveled short distances (below 2 kilometers) compared with those who traveled long distance [6]. This is observed in many community areas including residential area, working place *e.g.* within hospital, building construction and many other work areas.

Safety helmets are frequently worn during longer distance trips because of fear of RTA. This is because as the travel distance increases, the contact hours on the road increases and the motorcyclists are at an increase risk of exposing themselves to RTA.

Injuries in Malaysian Motorcyclists

Fatal Injuries

The highest fatality was observed in the 16 to 25 years males [7]. Fatal injuries commonly occur during Saturdays and Sundays between 1200 and 1859 hours. The time indicated that RTA commonly occurred during weekends when the traffic was not as congested as during working days.

The predominant cause of fatal injuries was due to two or more vehicle colliding with each other (usually motorcycle and car) in side and head-on collision. Head-on collisions are four times likely to cause death than serious injuries [4]. Other types of collision in multiple vehicle crashes include rear-end, right-angle, side swipe and loss of control.

In single vehicle crash, loss of control seems to be responsible for majority of fatalities. This happened when motorcyclists failed to negotiate bends or curves or unable to control vehicle while trying to avoid other vehicles.

In relation to types of fatal injuries, head injury or traumatic brain injury (TBI) was found to be the most predominant with high incidence of base of the skull fractures with fatal brain injuries, followed by chest, neck and abdominal injuries [7,8].

TBI was defined as any recorded change in neurologic status, visualized on computed tomography (CT) or magnetic resonance imaging (MRI). TBI can occur as a result of skull fracture with focal or diffuse brain injury. The Glasgow Coma Scale (GCS) is used to assess the severity of TBI. Classification of TBI includes massive destruction and penetrating injury; brain stem and cerebellum injury; parenchyma injury which involves contusion, diffuse axonal injury and intracerebral hematoma; cerebral oedema, extradural hematoma, subdural hematoma, intraventricular hemorrhage, pneumocephalus and subarachnoid hemorrhage; and skull fracture in the form of base or vault fractures [9].

Cervical spine injuries often occurred at the lower part of cervical spine, the C4,C5 and the intervertebral joint between C5-C6. The injury can be classified based on the vertebral kinematics *i.e.* flexion, extension, lateral flexion and compression. Frontal impact causes flexion kinematics at the lower vertebrae causing majority of minor injuries unless it occurs at a very high energy impact. The C2, C4, C5, C5-C6 and C6 are the most frequently injured sites observed.

In rear-end motorcycle crash, kinematics experienced by the motorcyclists at the initial stage is similar to the whiplash injury. The most frequent sites involved were the C5 and C5-C6. This type of injury often resulted in spinal cord injury (SCI) involving subluxation and contusion. The most significant cervical kinematics in side impact was the lateral flexion commonly seen at C5, C5-C6, C6-C7 and C7. Upper cervical injuries had been recorded to occur in the side impact crash and skidding [8,10].

It is not uncommon for the cervical injuries to occur in combination depending on the direction and load of impact energy *i.e.* flexion-compression and lateral flexion injuries was observed in a skidded motorcyclist.

Non-fatal Injuries

Motorcyclists are constantly physically exposed to bodily impact and in an event of a crash; a rider or its pillion is likely to sustain some form of injury even in a relatively low impact collision.

Non-fatal injuries in motorcyclists often result in prolonged hospitalization. Injuries sustained in RTA contribute significantly to morbidity, disability and disfigurement and all these cause a rise in the costs of medical care. In a study looking at non-fatal injuries sustained by Malaysian motorcyclists between the periods of 1 January 1998 to 31 December 1998, it was found that the most common injuries involved the lower and upper limbs [11]. The length of hospital stay varied from one day to one month (mean of five days) [11]. The most frequent type of collision causing limb injuries was the rear-end type [11].

Apart from limb injuries, other injuries involved the head, spine, chest and abdomen [11]. Thoracic and lumbar spinal injuries outnumbered cervical injuries in the non-fatal injuries [11].

Trauma to the head and neck commonly occur concurrently with facial injuries. Although most of facial injuries were not life threatening, but some form of deformity usually occurred as the result of trauma and the medical cost to treat deformity can be quite demanding. Facial disfigurement with or without functional disability requires complex medical and surgical intervention over a prolonged period of time. Our study on facial injuries revealed that the lower face (from upper lip to the chin) was the most frequent site involved in motorcyclists whether they were wearing helmets or not [12]. This is in agreement with another study which showed that the chin was the most susceptible location of injury in helmeted patients, whereas the scalp was the most common location in the non-helmeted motorcyclists [13].

Facial deformity includes gross facial asymmetry due to bone fractures causing depression or misalignment of fragments, alteration in normal positions of structures for examples the orbits in orbital dystopia and deformity involving the nose and the jaw bones.

The aftermath of RTA often resulted in a 'crippled' man with partially recovered head injury, some form of facial deformity and mentally depressed. He is usually home-bound, unable to work and dependant on other people, usually family members to do his daily chores. From the economic point of view, the family loses income from this young man and especially if this man is the breadwinner of the family, the situation is going to be made worse. Moreover, the total care and medical expenses has to be borne by the family members. To the country, it is a tremendous loss of resources and estimated to cost about 6 billion ringgit in lost of productivity and health costs [4].

CRASH HELMETS

It is mandatory to discuss helmets alongside with motorcycle accidents as crash helmets are designed to protect the head from injuries and majority of death is due to trauma to the head.

Helmets reduce mortality and morbidity, reduce hospitalization and Intensive Care Unit admission compared to non-helmet users, as well as reducing the financial burden created by motorcycle-related injuries [14,15,16]. Helmets significantly reduce the probability of head and neck injuries by 53% and lead to a 72% reduction in the probability of death due to head and neck injuries [17]

Unhelmeted motorcyclists sustained substantially higher rate of facial and brain injuries compared to helmeted riders [18].

In Malaysia, the first Malaysian Standard MS 1 draft for helmet was introduced in 1969 and legally enforced in 1973[19]. Since then, helmet usage has increased sharply and motorcycle fatalities were reduced to 30% [20]. The current standard follows the Standard Institution of Malaysia MS.1:1996 [21].

Types of Helmets

The helmets can be classified into the full-face or open-face type. An open-face helmet either covers only the top of the head or covers the top and sides of the head but not the face and chin and this is known as the 'jet-pilot' type. A chin strap is present to secure the helmet firmly in its position. A full-face helmet, on the other hand covers the face and the chin with a chin bar.

Helmet Main Components and Their Function

There are four basic components in a helmet:

- the outer shell
- an impact-absorbing liner
- the comfort padding or foam
- the retention system

The outer shell is usually made of fiber-reinforced composites or thermoplastics, for example polycarbonate. The basic function of the shell is to prevent penetration of any foreign object from touching the head and distributes the impact load on large area of the helmet material. Some types of shells delaminate on impact or others may crack or break if forced on high impact energy.

Inside the shell, the liner is made of expanded polystyrene. The thickness of the liner is determined by helmet weight, practicality and style. The helmet thickness is usually in the range between 20 and 30 mm. The material must be able to absorb most of the energy per unit volume while limiting the load on the human head to a less than damaging level.

The comfort padding is a soft foam-and-cloth layer that is located next to the head. Its function is to provide an area of comfort and at the same time helps the helmet fitting snugly. This foam padding can be taken out for cleaning in some helmets.

The retention system comprises the chin strap. A strap is connected to each side of the shell.

In a full-face helmet, the face shield may be moveable or non-moveable.

A helmet is also designed to provide a minimum peripheral vision of 105 degrees to each side.

Helmets and Protection

From forensic point of view, a study from Japan found that full-face helmets significantly decreased the incidence of brain contusion and slightly but not significantly decreased the incidence of subdural and subarachnoid hemorrhage [22]. Full-face helmets are more likely to decrease direct force at the point of impact and therefore better in preventing brain contusion than open-face helmets. However, cervical fractures were slightly but not significantly prevented in motorcyclists wearing full-face helmets as full-face helmets have little protective effect on neck injuries [22].

Prevention

An essential tool for effective road crash injury prevention is the adoption of a system approach to identify problems, formulate strategy, set targets and monitor performance [23]. The key problems that lead to RTA and high mortality and morbidity among Malaysian motorcyclist have been identified such as conspicuity due to the small size of the rider or not wearing proper lighted riding gear, vulnerability to head injuries as the result of improper helmet wearing or substandard helmet, erratic behavior of young riders as well as unforgiving road environments and lack of concern from other road users. Many strategies have been formulated to address the problems. The target is by the year 2010, the accident rate should reduce to more than 50 percent *i.e.* deaths due to RTA should decrease from the current 4.2 deaths per 10,000 vehicles to 2 deaths per 10,000 vehicles by 2010 [24].

Phase		**Factors**		
		Human	Vehicles & equipment	Environment
Pre-crash	Crash prevention	Information Attitudes Impairment Police enforcement	Roadworthiness Lighting Braking Handling Speed management	Road design and road layout Speed limits Pedestrian facilities
Crash	Injury prevention during crash	Use of restraints Impairment	Occupant restraints Other safety devices Crash protective design	Crash protective roadside objects
Post-crash	Life sustaining	First-aid skill Access to medics	Ease of access Fire risk	Rescue facilities Congestion

Figure 1. The Haddon Matrix.

Strategic Programs

Issues related to RTA received serious attention from the Malaysian government following a massive tragedy in Karak Highway in 1990 where a significant number of fatalities were involved in a single accident. Following that, a Cabinet Committee on Road Safety with the Prime Minister as the chairman was formed. In 1996, the Malaysian government set a 5-year national road safety target to reduce road traffic mortality by 30% by the year 2000 through a series of well planned strategic programs which was devised into accident prevention and reduction, injury control and post injury reduction. This was based from Haddon Matrix which was introduced by William Haddon Jr in the 1960s. Haddon defined three phases of the time sequence of a crash event; the pre-crash, crash and post-crash. He also introduced the concept of epidemiological triad of human, machine and environment that can interact during each phase of a crash. This resulted in a nine-cell Haddon Matrix, with each of the matrix allowing opportunities for intervention to reduce RTA [25].

Integrated strategic programs produced a marked decline in morbidity and mortality as seen in many developed countries [26,27]. The formula on continuing planned systemic improvements aiming at vehicles, roads and users proved to be highly effective as evidenced by reports from countries with lowest mortality rates namely Netherlands, Sweden and United Kingdom [28].

2006 -2010 Target

MIROS is undertaking researches to support the 5-year *Road Safety Plan of Malaysia (2006-2010)* that encompasses Education, Enforcement, Engineering and the Environment to benchmark road safety standards with that of other developed countries.

Phase		**Factors**		
		Human	Vehicles & equipment	Environment
Pre-crash	Crash prevention	Road safety education Driver training program and grading of driving institutes Automated enforcement system Community based programs (CBP)	Type of government body approval Rating of vehicle Rear seatbelts	
Crash	Injury prevention during crash	Compliance Correct use of active safety features	Passive safety system Crash compatibility	Clear zones Barrier systems Crash cushions
Post-crash	Life sustaining	Enhancing skills of paramedics and first respondents	Ease of evacuation Evacuation tools	Ease of access by first respondent teams Trauma centre

Figure 2. Malaysia road safety intervention (2006-2010).

The targets are:

- to reduce fatality rates per 10 000 registered vehicles from 4.2 in 2005 to 2.00 by 2010
- to reduce prevailing the death rate from 23 per 100 000 population to 10 per 100 000 population
- to reduce prevailing the fatality rate of 18 per billion vehicle kilometer traveled (VKT) to 10 per billion VKT

Haddon Matrix was adopted in the building of prevention measures based on the nine-cell matrix.

Apart from extensive road safety intervention program outlined as above, there were also three main strategic programs targeted for different types of road users. The programs were Motorcycle Safety Program (MSP), Pedestrians Safety Programs (PSP) and Car Occupants Safety Programs (CSP).

Specific Programs for Prevention of Motorcycle Accidents

In 2001, a Motorcycle Safety Program (MSP) was introduced to tackle motorcycle-related accidents. It was conducted by Road Safety Research Centre (RSRC), Faculty of Engineering, Universiti Putra Malaysia with the aim to formulate a short and long term plan for motorcycle safety as well as to draw a national safety policy related to motorcycle safety.

Based on research analysis of 6000 detailed crash data as well as 193 different collision mechanism of motorcyclists in Malaysia, multiple programs was formulated for the MSP. The key problems identified for motorcyclists include conspicuity (looked but not seen), head injuries (vulnerability), poor behavior (targeted for young riders), unforgiving road environment for motorcyclists and concern from other road users. Since 2007, all these programs are continued or new programs being introduced under the flagship of MIROS.

1) Conspicuity Program

This program was started after an earlier research by RSRC which revealed the phenomenon of 'looked but failed to see' as the motorcyclists are at the peripheral vision of other drivers. This is the one of the main cause of motorcycle accidents even during daytime. Initiative that was carried out within this program includes the running headlight intervention which was carried out since 1992, proper riding gear such as reflective strips and reflective vest, light-colored clothing campaign and highly visibility painted motorcycles and visibility enhancement sticker.

2) Behavior Modification Program

i) Media Medium

This program was targeted at young motorcyclists as they make up the highest number of Malaysian road users. The media is used to address the problem and the issues related to road safety. This included:

a) Identification of specific target segments (such as young, male riders)
b) Identification of strategic issues (such as head injuries due to not wearing safety helmets or insecured helmets)
c) Development, pre-testing, production and auditing, running and evaluating the effectiveness of campaign materials and media

The issues aired through television and radio included the conspicuity of motorcycle during daytime and at night, proper use of helmets, injuries sustained in RTA and the effects of speeding and weaving. These were 'direct' messages given to the audience. Besides that, issues related to family values and well-being of oneself was also included in the media preventive programs.

A study was conducted to assess the efficacy of this program. It was found that a significant ($p<0.05$) increase in the proper usage of the helmets from 44% before the campaign to 66% after the campaign [29]. The use of light-colored clothing also increased significantly ($p<0.05$) following the media intervention [29].

ii) Police Enforcement Programs

Many police enforcement programs were imposed at the national level. Among those were *Ops Sikap, Ops Nasihat* and *Ops Topi*.

a) *Ops Sikap* (Behavioral Operation)

This is a special program on safety and speed limit awareness particularly addressed during busy festive seasons and this is known as *'Ops Sikap'*. Malaysia being a multi-racial country has many festive celebrations throughout the year. During these periods, the number of RTA with fatalities overshot the figures of those during non-festive period. Majority of death and serious injuries involved motorcyclists. Among the activities carried out during these periods were:

- 24-hour observation at selected road stretch locations. Drivers driving above the speed limit or driving in dangerous manner would be notified immediately and summonses were issued to warn them about their attitude
- Monitoring traffic movement and their pattern at selected locations
- Installing temporary warning signage at locations considered as hazardous
- 24-hour monitoring at identified hazardous locations
- Installing road studs and blinkers to alarm road users before arriving at the identified dangerous areas
- Repair all pot-holes

b) *Ops Nasihat* (Advice Operation) and *Ops Topi* (Helmet Operation)

From February to March 2007, '*Ops Nasihat*' and *'Ops Topi'* were introduced. During these operations, motorcyclists were flagged down and they were given warning notice and advice if found not complying with the law. '*Ops Nasihat*' was introduced prior to enforcing the law and issuing fines. The rationale is to provide a period of adjustment to the public towards the acceptance of the law on helmet. '*Ops Topi*' was carried out concurrently with *'Ops Nasihat'*. Complimentary crash helmets were distributed to riders who did not wear the standard type of helmets and to child pillion riders. Helmets were also given away to schoolchildren following road safety talks.

Apart from physical police enforcement activities, enforcement is also carried out by means of technology such as speed cameras which have been proven to be most effective in reducing RTA and fatalities in developed countries.

iii) Community Based Program (CBP)

Activities such as talks and seminars are targeted towards selected communities with the participation of various government and non- government bodies. Apart from seminars, other activities include presentation of complimentary helmets to families who own motorcycles, crash prevention advocacy, injury control advocacy, youth activities and helmets campaigns. The recent CBP was carried out in Kuang, Selangor, one of the selected rural areas in Peninsular Malaysia in 2007. MIROS in collaboration with RSRC spearheaded evidence-based study to measure safety helmet compliance levels following completion of the program. Another rural area, Ijok which also located in the state of Selangor was selected as the control. It was found that the total number of RTA decreased by 64% while in Ijok, total number of accidents increased by 24%. Following 8 weeks post-CBP in Kuang, there was an increase of 14% in the use of safety helmets while Ijok remains at the baseline with no difference in the compliance of helmet wearing [30].

3) Road Engineering Program

In 1995, the Malaysian Government introduced the world's first exclusive motorcycle lane and this has successfully reduced motorcycle accidents along trunk roads. This innovation resulted in a great success with a marked reduction of fatality of 600% [29].

Other initiatives in the road engineering program include reconstruction of paved road shoulder, end treatment of non-exclusive motorcycle lane and junction accident model. The latter is a linear model that was able to suggest that the number of motorcycle accidents is proportional to the power of traffic flow. The increase in number of motorcycle accidents is associated with an increase in motorcycle and non-motorcycle flows into the junction and it is the non-motorcycle flow that has the highest effect on the liability of motorcycle accidents. Thus this model allows traffic engineers to determine appropriate junction treatment criteria for specially designed motorcycle lane facilities which include traffic speed, lane width, number of lanes, shoulder width and land use for the exclusive motorcycle lane when approaching a three legged, major-minor priority junctions.

Motorcycle guardrail is another road engineering invention towards prevention of injury. In Malaysia, most of the major roads were installed with the W-beam guardrail system. Safety performance on this existing guardrail was assessed using the whole-body motion crash test scenarios via a 3-dimensional Multi Body System and a Finite element models. The motorcycle used was a KRISS 110cc; a locally manufactured model and the motorcyclist was represented by a standard Hybrid III 50th percentile dummy. The impact was studied at different angles and the injury criteria were calculated in relation to the head, neck, thorax and lower limb areas. The result of this study showed that the impact of the motorcycle crash at high speed at angles of 90° and 45° to the guardrails were more severe and involved head injuries [31]. Based on this study, the design of an alternative safer guardrail was made and recommended for use in replacement of the existing guardrail.

4) Injury Control Program

This program was developed as a comprehensive program which look into all aspects of motorcycle accidents from safety policies to vehicle and road engineering approach and medical and trauma management. Among the programs planned are the revision of helmet standards, new helmet design for children and a campaign on the proper wear of helmet.

Child Helmet

One important study that was carried out under the injury control program was on the design and development of child helmet. The study showed that more than 70% of children riding pillion in Malaysia do not wear helmet or wear inappropriate size and standard helmet. As a result, head injuries in children involved in RTA are significantly higher compared to adults [32]. The finite element modeling technique was used to developed new material and design for Malaysia children helmet. Currently, the helmet for children has been launched and available in most of helmet shops.

5) Education

Education is an important part in the injury control program. Education on road traffic safety is carried out at various levels *e.g.* in primary and secondary schools, driving schools, workplaces and to the high-risk drivers. Road safety education is a life-long process and should be introduced early in life. The aim for early education is to inculcate in children and parents an understanding of the dangers within the traffic environment. The road safety chapter is also included in the primary schools module and teachers are trained to educate children using various methods *i.e.* the workbooks, teaching aids and education kits and many others.

The current training program for drivers whether they are new or professional drivers are being reviewed by the Malaysian Government to ensure that the output is compliant to international best practice. Road Safety Department of Ministry of Transport is involved in grading the driving institutions to ensure their standard and quality to best practices worldwide.

REFERENCES

[1] Radin Umar RS. Updates of road safety status in Malaysia. *IATSS Research* 2005; 29: 78-80

[2] Radin Umar RS, Mackay GM, Hills BL. Preliminary analysis of motorcycle accidents: short term impacts of the running headlights campaign and regulations. *J Traffic Med* 1995; 23: 17-28.

[3] Asian Development Bank-Association of Southeast Asian Nations. Regional Road Safety Programs. *Accident costing report AC5:* Malaysia, 2003.

[4] Pang TY, Radin Umar RS, Azhar AA, Megat Ahmad MH, Mohd Nasir MT, Harwant S. Accident characteristics of injured motorcyclists in Malaysia. *Med J Malaysia* 2000; 55:45-49.

[5] Kulanthayan S, Radin Umar RS, Ahmad Hariza H, Mohd Nasir MT, Harwant S Compliance of proper safety helmet usage in motorcyclists. *Med J Malaysia* 2000; 55(1): 40-44.

[6] Kulanthayan S, Radin Umar RS, Ahmad Hariza H, Mohd Nasir MT. Modeling of compliance behavior of motorcyclists to proper usage of safety helmets in Malaysia. *J Crash Prevention and Injury Control* 2001;2(3): 239-246.

[7] Pang TY, Radin Umar RS, Azhar AA, Harwant S, Shahrom AW, Abd Halim M et al. Fatal injuries in Malaysian motorcyclists. *Int Med Res J* 1999; 3(2): 115-119.

[8] Ooi SS, Wong SV, Radin Umar RS, Azhar AA, Yeap JS, Megat Ahmad MM. Mechanisms of cervical spine injuries for non-fatal motorcycle road crash. *Med J Malaysia* 2004; 59(2): 146-52

[9] Javouhey E, Guérin AC, Chiron M. Incidence and risk factors of severe traumatic brain injury resulting from road accidents: a population-based study. *Accid Anal Prev* 2006; 38(2): 225-33.

[10] Kasantikul V, Ouellet JV, Smith TA. Head and neck injuries in fatal motorcycle collisions as determined by detailed autopsy. *Annual Proc AAAM* 2002; pp125-137.

[11] Pang TY, Radin Umar RS, Azhar AA, Harwant S. Non-fatal injuries in Malaysian motorcyclists. *Int Med Res J* 1999; 3(2): 111-114.

[12] Ramli R, Abdul Rahman R, Abdul Rahman N, Abdul Karim F, Krsna Rajandram R, Mohamad MS, Mat Nor G, Sohadi RU. Pattern of maxillofacial injuries in motorcyclists in Malaysia. *J Craniofac Surg* 2008; 19(2): 316-321.

[13] Johnson MR, McCarthy MC, Miller SF, Peoples JB. Craniofacial trauma in injured motorcyclists: the impact of helmet usage. *J Trauma* 1995; 38(6): 876-878.

[14] Brandt MM, Ahrns KS, Corpron CA, Franklin GA, Wahl WL. Hospital cost is reduced by motorcycle helmet use. *J Trauma* 2002; 53(3):469-71.

[15] Poole GV, Ward EF. Causes of mortality in patients with pelvic injuries. *Injury* 1994; 25(4):223-5.

[16] Rowland J, Rivara F, Salzberg P, Soderberg R, Maier R, Koepsell T. Motorcycle helmet use and injury outcome and hospitalization costs from crashes in Washington state. *Am J Public Health* 1996; 86(1): 41-45.

[17] Keng S-H. Helmet use and motorcycle fatalities in Taiwan. *Accident Analysis and Prevention* 2005; 37: 349-355.

[18] Eastridge BJ, Shafi S, Minei JP, Culica D, McConnel C, Gentilello L. Economic impact of motorcycle helmets: from impact to discharge. *J Trauma* 2006; 60: 978-984.

[19] Standards and Industrial Research Institute of Malaysia (SIRIM) (1969) Malaysian Standard MS 1: 1969: *Specifications for Protective Helmets for Motorcyclists,* Kuala Lumpur.

[20] Supramaniam V, Bell V, Sung J. Fatal motorcycle accidents and helmet laws in Peninsular Malaysia. *Accident Analysis and Prevention.* 1984; 16(3): 157-162.

[21] SIRIM, 1996. *Malaysian Standards. MS1: 1996; ICS 13.340.20. Specification for protective helmets for vehicle users* (2nd rev). Kuala Lumpur, Malaysia: Standards & Industrial Research Institute of Malaysia.

[22] Hitosugi M, Shigeta A, Takatsu A, Yokoyama T, Tokudome S. Analysis of fatal injuries to motorcyclists by helmet type. *Am J Forensic Med Pathol* 2004; 25(2): 125-128.

[23] Rothe JP, ed. *Driving lessons: exploring systems that make traffic safer*. Edmonton, University of Alberta Press, 2002.

[24] Malaysia's Road Safety Research (MIROS).Body starts operation tomorrow. *Bernama.com. Januar*y 02, 2007.

[25] Haddon Jr W. The changing approach to the epidemiology, prevention and amelioration of trauma: the transition to approaches etiologically rather than descriptively. *Am J Public Health* 1968; 58: 1431-1438.

[26] Centers for Disease Control and Prevention. Motor vehicle safety: a 20th century public health achievement. *Morbidity and Mortality Weekly Report* 1999; 48: 369-374.

[27] Lonero L et al. Road safety as a social construct. Ottawa, Northport Associates, 2002 (Transport Canada Report No. 8080-00-1112.

[28] Koornstra M et al. Sunflower: a comparative study of the development of road safety in Sweden, the United Kingdom and the Netherlands. Leidschendam, Institute for Road Safety Research, 2002.

[29] Radin Umar RS. Motorcycle safety programmes in Malaysia: how effective are they? *International Journal of Injury Control and Safety Promotion* 2006; 13(2): 71-79.

[30] *Malaysian Institute of Road Safety Research* (MIROS). Community based safety helmet programme in Kuang 2007[MRR5].

[31] Billiyamin IA, Radin Umar RS, Wong SV, Hamoda AMS. The behaviour of motorcyclists in runoff accident with W-beam guardrails. *2nd National Conference on Transportation Management: Issues and Challenges.* 20-21 October 2003, Shah Alam, Malaysia 2003, pp96-106.

[32] Ong WY. Design of motorcycle safety helmet for children in Malaysia. *Master of Science Thesis,* Universiti Putra Malaysia 2001.

In: Road Traffic: Safety, Modeling and Impacts
Editors: S. E. Paterson and L. K. Allan, pp. 503-519
ISBN 978-1-60456-884-4

Chapter 16

A SIMULATION OF A ROAD TRAFFIC ACCIDENT REPORTING SYSTEM

Andrew Greasley*
Aston Business School, Aston University, Birmingham, B4 7ET, UK

ABSTRACT

This chapter provides a descriptive case study to demonstrate the use of a process-based approach to change regarding the implementation of an information system for road traffic accident reporting in a UK police force. The supporting tools of process mapping and business process simulation are used in the change process and assist in communicating the current process design and people's roles in the overall performance of that design. The simulation model is also used to predict the performance of new designs incorporating the use of information technology. The case study examines both the steps in developing the simulation model and the organisational context in which the simulation study takes place. Change in this public sector organisation is discussed in terms of process, policy, legal and staffing issues and the simulation method is assessed in relation to these factors. The benefits of the simulation method in the case study are also outlined. For example, the ability of simulation to proof new designs was seen as particularly important in a government agency were past failures of information technology investments had contributed to a risk averse approach to their implementation.

INTRODUCTION

Although there is no agreed boundary between public and private sectors in the UK it is estimated that employment in public sector organisations is approximately 6 million, of which 550,000 are employees working for direct government agencies (i.e. civil servants). One of

* Tel.: +44-121-359-3611; fax: +44-121-359-5271; *email address*: a.greasley@aston.ac.uk

the largest employees in the world is the state-run National Health Service (NHS) with some 1.16 million staff employed in England alone (Walker, 2003). In recent years, business process approaches to change have become increasingly popular in the public sector (MacIntosh, 2003; Gulledge and Sommer, 2002; Thong et al., 2000; McAdam amd Donaghy, 1999; McAdam and Mitchell, 1998; Harrington et al., 1997; Halachmi and Bovaird, 1997). The original concept of process change through the use of business process reengineering (BPR) projects was first described by Davenport and Short (1990) and Hammer (1990). These authors defined a 'first generation' re-engineering concept of a new, radical, IT-led, mechanistic and inspirational approach which has been evolving into a 'second generation' process management concept of a hybrid, contingent, IT-enabled, holistic and systematic approach (Cypress, 1994). This evolution results from the recognition of an overemphasis on reductionist and mechanistic aspects, in which crucial issues such as people and strategic issues were simply ignored (Melão and Pidd, 2000). However a feature of process management remains that of a focus on business processes, with the optimum arrangement of these processes and tasks within the organisation considered to be a critical design variable determining the efficiency of the resulting structure (Orman, 1995). The task of redesigning the process structure is however complex, as changes will be made that may affect different, but interacting and interrelated dimensions of an organisation such as its processes, people, strategy, environment and culture and the reported failure rate of projects is high (CSC Index, 1994).

It has been argued that one of the major issues that contributes to the high failure rate in business change projects is a lack of tools for evaluating the effects of design solutions before implementation (Paolucci et al, 1997; Tumay, 1996). Without this ability mistakes brought by change can only be recognised once the redesigned processes have been implemented, when it is usually difficult and costly to correct a wrong decision. The risk of this occurring may be leading to less than radical design changes, or even simply automation of current process designs which is unlikely to maximise the benefit of investments in information technology. In relation to business processes in local government Kloot and Martin (2000) state that the lack of information on processes and the effect of process change on costs means that managers are uncertain about how to proceed when the effects of change cannot be estimated. In order to reduce the risk in the implementation of change projects Meel and Sol (1996) advocate the development of computer-based models of business processes to support the process of experimentation with alternative business structures. Bal (1998) also highlights the support process analysis tools can provide in the analysis of the current process design. These include achieving a full understanding of the process to be redesigned and creating a shared vision and understanding of the current process design among the redesign team. Haque et al (2003) show the use of a process modelling tool for analysis at both a high level process and project management level and at an operational and detailed process level.

Business Process Simulation

In general business process simulation (BPS) is seen as a way of analysing dynamic systems that exhibit variability. Profozich (1998) outlines the disadvantages of using a static analysis tool such as a spreadsheet for analysing such systems, which requires the assumption that each process will operate on the average. There are however limitations to the method. One disadvantage is the relatively large amount of time and cost needed to develop a model (Pidd, 1998). Further limitations can be apparent in the interpretation of simulation study results. For instance Fathee et al (1998) state that as the complexity, randomness and variability within a business process increase the range of the prediction obtained from the simulation model becomes too wide for a decision base.

Business process simulation (Tumay, 1996) is seen as a way of analysing dynamic systems that exhibit variability and has been traditionally used in a manufacturing context, but it is increasingly used in service and public sector organisations. Some examples of use are Levine & Aurand (1994) who describe the use of simulation to analyse an automated workflow system of an administration process, Greasley (2000) who describes the use of simulation to analyse a custody of prisoner process, Verma et al (2000) who describe its use in redesigning check-processing operations and Hays and Bebbington (2000) who describe its use in a government agency overseeing film and literature classification. The use of simulation to assess the implementation of information systems (IS) in particular has been described by Giaglis (1999) and Giaglis et al (1999) who outline the need to support IS evaluation by developing techniques for generating estimates of the organisational value of IS. Experimental methods such as simulation are suggested as capable of providing such estimates. Warren and Crosslin (1995) suggest simulation as a method of providing the justification necessary to support business decisions to redesign and make what are often massive investments in IT systems. However despite this reported potential the use of BPS in general is low, with a UK survey of current or past users who had been trained in the technique giving a usage rate of 21% across industry and 22% in public sector organisations (Melao and Pidd, 2003).

METHODOLOGY

A case–based research methodology was chosen with the aim to provide an example of practice and test the proposition that the supporting tools of process mapping and business process simulation can improve the success of a process-based approach to change in the context of a public sector organisation. In this study the author acted in a consultancy role and undertook both the process of the simulation study and the construction of the simulation model itself. This might imply an action research methodology and indeed in practice action research and case study research are similar in many of their procedures (Collis and Hussey, 2003) and both are associated with the same research paradigm of phenomenology. However in this case the purposes of the intervention from the client perspective was to provide data from the simulation for an organisational decision and was not concerned with the author's research interest in simulation use. Thus there is a separation between the outcome of the consultancy and the outcome of the research analysis. Action research is associated with the mutual control by the researcher and client of the research and analysis of results. In addition the final action plan to be implemented is usually the client's responsibility, supported by the researcher and the research report is often published jointly by the researcher and client (Collis and Hussey, 2003). Bryman and Bell (2003) state that although there is no single type of action research it can be broadly defined as an approach in which the action researcher and a client collaborate in the diagnosis of a problem and in the development of a solution based on the diagnosis.

In terms of the case study approach Gummeson (2000) states that it has the advantage of providing an holistic view of a process. According to a holistic view the whole is not identical to the sum of its parts and so the whole can only be understood by treating it as the central object of study. Also case studies are of particular value in applied social sciences where research often aims to provide practitioners with tools.

Although a single-site study has obvious limitations with respect to the generalisability of the findings, the case is not aimed at being representative, but rather exemplary, thus the researcher does not need to assume that what is observed is truly representative of all similar situations (Stuart et al., 2002). In this example a regional police force in the UK shows many of the attributes of many public sector organisations in its structure and reporting relationships, and the involvement of a number of stakeholders in determining strategic targets and operational rules. Collection of data such as process durations, task allocations and process relationships for the study was undertaken by the author and employees of the organisation in order to construct the process maps and simulation used in the analysis of the process designs.

The Road Traffic Accident Case Study

A case study is presented of a process change initiative to a police road traffic accident (RTA) reporting system. Two aspects of performance of the RTA system require particular attention. The need to speed process execution is seen as essential to provide a faster and more efficient service to vehicle drivers. In particular there is a need to provide UK government agencies, such as the Department for Transport, with accident statistics within a 4-week time period. The second aspect of performance which requires improvement is the need to reduce the relatively high staffing cost associated with the process. The total cost of traffic police staff is relatively high as their on-costs need to include the purchase and maintenance of a police patrol vehicle. There is also a need for extensive administrative support at locations across the area covered by the Police Force.

Three options were originally considered for a revised road traffic accident reporting systems, these being an updation of current paper forms, a telephone-based system or computerisation using mobile technology. These options will be considered in turn.

1. Updation of Paper Forms

This proposal was to update current paper-based forms amended to include new local and national requirements such as new causation codes and a combining of the currently separated injury and non-injury accidents. The advantages of this approach include continuity of use of the present system with which officers are familiar, robustness against any failure of telecommunications equipment, ability to scan information for later electronic transmission and requires a minimal training need for police officers. The disadvantages of this approach include the slow entry of information into the current forms, the high cost of printing forms, the cost of re-input of data by data entry staff (and potential problems due to duplication of information), the delays caused by non-availability of data entry staff during out of hours working and the time-consuming administrative burden for police officers.

2. Telephone-based System

This proposal involves police officers, support staff and the general public speaking directly to bureau input staff. The advantages of this approach are that using specialised trained staff to gather and record information will lead to standardised procedures and less errors and higher quality data. This approach also has the advantage of reducing costs in terms of printing and requiring minimal training of police officers. The disadvantage of this approach is the cost of ensuring that the working patterns of bureau staff are sufficient to meet demand over a 24 hour period. There is also still a requirement for officers to record initial details of the RTA on paper.

3. Computerised System

In the proposed computerised RTA reporting system the attending officer completes paper-based forms as before but this information is promptly converted to digital form using a combination of image capture and data recognition. The advantages of this approach are that images of text such as officer written notes can be entered by input bureau staff, saving officer time. Once in digital format the documents can be delivered electronically preventing data duplication and enabling faster distribution. Physical documents can still be held in a central repository for reference if needed. The disadvantages of this approach are the risk of the use of new technology, the cost of technology and the need to train police officers to use the new technology.

Due to the potential gains in efficiency and effectiveness it was decided by the Police Force to investigate the computerised system option. The current and proposed computerised process designs for actions leading from a road traffic accident will now be discussed in more detail.

The Current Road Traffic Accident Reporting System

The current process at the selected constabulary for the reporting and recording of accidents involves a police officer completing a number of paper-based forms following the report of an incident by the public. These forms are distributed to the traffic administration and data HQ departments for processing. The traffic administration section oversees the submission of witness evidence, either by post or in person and the collation of an abstract containing officer and witness statements for use by interested parties such as insurance companies and court proceedings. The data HQ section oversees mapping of the geographical location of the accident that is used for road transport initiatives such as traffic calming and speed cameras. The main stages of the RTA process are now described.

Following the notification of a road traffic incident to the police by the public, a decision is made to attend the scene of the incident. It may be that for a minor incident the parties involved are instructed to pursue proceedings with their insurance companies and the police have no further involvement. If it is necessary to attend the RTA scene the officer travels to the location of the incident. After an assessment is made of the incident the officer returns to the station to complete and submit the appropriate paperwork. Three forms are used by a police officer attending a RTA. Form '54' is used for injury accidents and is triplicated on yellow, pink and white forms. The yellow and pink forms are forwarded by the officer to the witness proforma process and the white form is forwarded to the data HQ section for location mapping. A single form '55' is used for non-injury incidents which is filed unless further action is to be taken as a result of a dispute or claim, when it is then passed to the witness proforma process. In addition pocket book entries (PBE) are taken when no official record is required but provide data that could be retrieved at a later date and transferred to the appropriate form. Amendment of forms may take place at a later date. Form '54' amendment forms contain yellow, pink and white sections for distribution. Form '54' (yellow and pink) and Form '55' amendments are scrutinised to see if changes require further action such as new witness statements. Form '54' (white) amendments are communicated to data HQ.

A location mapping process collates information and passes it to a local Council which provide a location grid reference from sketches and location information provided by the officer who attended the accident scene. The data is collated, sorted and then forms are mapped in batches by entering location codes on new forms. If all the necessary information is not available a memo is sent to the officer for further information and the process is repeated. The information is then sent electronically to the local council who return the required geographical location details.

The witness proforma process obtains accident witness and driver information and places it on a proforma sheet. If a witness is identified their details are taken and a proforma is sent to them. If a fatal accident has occurred then the officer obtains further details in person at a later date. If the proforma has not been returned after three weeks a reminder letter is sent to the witness. If there is still no response from the witness and the information requested is required for further proceedings, then an officer will obtain the statement in person.

The abstract preparation process collates and checks documents associated with the RTA process to ensure all the data needed has been received. A decision is made at this stage if further action is required after reviewing the evidence collected. If further action is required a number of forms are collated. If at this stage no prosecution is to take place, a letter informing the driver of this decision is sent. If a prosecution is to take place the officer will write an abstract, summarising the details of the case. If a court case is scheduled and a not guilty plea has been entered then the officer will be required to attend the court proceedings in person. Otherwise this is the end of the involvement of the officer.

The Proposed Road Traffic Accident Reporting System

In the proposed computerised RTA reporting system the attending officer completes paper-based forms as before but this information is promptly converted to digital form using a document image processing (DIP) system. This is achieved by a combination of image capture and data recognition through a facsimile link. Data recognition systems, such as optical character recognition (OCR) are used to process information that is entered in a structured format, such as options selected using a ticked box format. Image capture is used in the following ways. Documents are stored as images to enable input bureau staff to validate the OCR scanned data. Images which cannot be interpreted by data recognition software, such as hand drawn sketches of the RTA scene are stored for later retrieval. Images of text such as officer written notes can be entered by input bureau staff, saving officer time. Once in digital format the documents can be delivered electronically preventing data duplication and enabling faster distribution. Physical documents are held in a central repository for reference if needed.

The proposed computerised system also incorporates a geographical information system (GIS) in order to improve the accuracy of the recorded location details of the RTA. Currently location details are currently based on a written description of the RTA by an officer which leads to inconsistent results. The current location description by the officer is usually acceptable for city incidents where nearby street intersections and other features can be used to pinpoint a location. However on long stretches of road it is often difficult to pinpoint an exact spot. This is important because of the need to accurately pinpoint areas with high accident rates for road safety measures (e.g. road humps) and speed camera placement. Further inaccuracies can also occur when the officer description is converted by the local council using an Ordnance Survey (OS) map grid reference which is only accurate to 200 yards. In this proposal each officer is issued with a portable digital map on which to indicate the RTA location. This information is transmitted by a mobile link to a geographical information system (GIS) (Wiley & Keyser, 1998) which provides accurate location analysis of both injury and non-injury incidents using the geocode system (Radcliffe, 2000). The geocode system is a network of grids covering the UK which allow a location to be assigned within a $10m^2$ area. The GIS system will combine the accident location analysis with data relating to the location of pelican crossings, traffic lights, street parking and anything else that might contribute to accidents or affect schemes being proposed. Along with data on details on road conditions at the time of the accident this information will help determine a prioritised list of road safety improvement measures.

Redesign of the Road Traffic Accident Reporting System

Table 1 provides an adaptation of the model of a process centred approach to change described by Aguilar et. al. (1999).

Table 1. Support provided by Business Process Simulation in process change

Phase	Step	Support provided by Business Process Simulation
Assess 'As-Is'	Build and Communicate Process Map	Additional Support
	Measure and Analyse Process Performance	Major Support
Build 'To-Be'	Develop Future Process Design	Major Support
	Enable and Implement Future Process Design	Additional Support

The model shows the use of business process simulation not only to predict the performance of the 'to-be' design before resources are committed, but also to use the technique to construct a model of the 'as-is' state in order to understand the process and measure the variation that takes place in key performance measures. Following this model the steps undertaken in the study are now outlined.

Step 1: Build and Communicate Process Map

Data is required in order to construct a process map which describes the logic of the model (i.e. how the process elements are connected) and decision points within the process. The technique of process mapping involves interviewing personnel and observation of the relevant process which provides information which is used to draw a process map. The analysis shows the interrelationships between activities and identifies the elements and roles involved in process execution. Harrington (1991) outlines a number of mapping and flowcharting techniques. Decision points can be modelled by a conditional rule based method or by a probability distribution. Probability distributions for decision points, such as the proportion of injury and non injury events are derived from the sample data and take the form of a percentage. A process map of the proposed RTA reporting system using Microsoft VISIO software is shown in Figure 1.

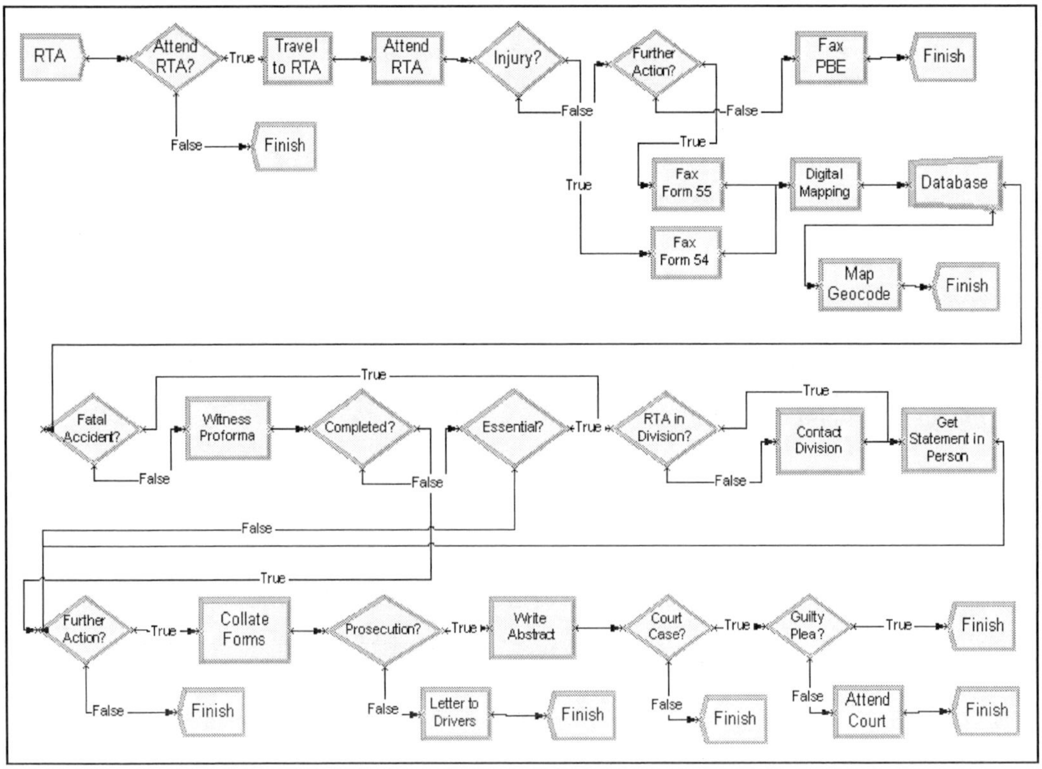

Figure 1. Process map of the proposed road traffic accident reporting system.

The second area of data required for the simulation model is for additional elements such as process durations, resource availability schedules and the timing of RTA occurrences. In this case probability distributions for process durations are derived from the sample data. In general a triangular distribution has been used for process durations which requires minimum, mean and maximum parameter values. Resource availability, in terms of a police officer attending the RTA, is assumed to be infinite as an RTA incident is treated as an 'emergency situation' and if the designated officer is unavailable an alternative officer is found. Over a period of 6 years there had not been an incident when no officer could be found, when required, to attend an RTA scene. In terms of the timing of RTA occurrences a three-month sample of road traffic accident (RTA) incidences was collected for the study. Although seasonal variations will occur in RTA patterns, the model is used to compare before and after redesign scenarios and not for forecasting purposes.

Step 2: Measure and Analyse Process Performance

The business process simulation was built using the ARENA™ simulation system (Kelton et al, 2001; Greasley, 2004) that incorporates a template of shapes called blocks that are placed on the computer screen and connected to represent the logic of the model. The user can then mouse click on each block and a pop-up menu allows the entry of parameters such as process durations. The main elements that make up a model are the CREATE, PROCESS and DECIDE blocks. The CREATE block generates the arrival of people (service applications), physical components (manufacturing applications) or information (information system applications) into the system. These are generically called entities in simulation terminology. It is necessary to define the rate of arrival of entities in the system by defining the time between arrivals (interarrival rate). This was achieved using sample data on the timing of the occurrence of RTA incidences. The PROCESS block is used to delay an entity for a predetermined time period (to represent the time taken to attend the RTA for example). It is also used to allocate resource time (e.g. traffic officer time) to a process. The DECIDE block is used to define decision points within the process. The entity can leave from two or more outputs from the block. For a two-output decision the routes are labelled true and false. These decision options can be implemented as either a percentage chance for each decision route (e.g. 70% entities follow the true route, 30% false) or by a rule-based decision (e.g. IF x THEN true route, else FALSE route).

Before experimental analysis of the model can begin it is necessary to ensure that the simulation provides a valid representation of the system. This process consists of verification and validation of the simulation model. Verification is analogous to the practice of 'debugging' a computer program. In this case the first check was to run the simulation model animation at a slow speed and observe the progress of the entities through the system to check for any logic errors. Then a number of test runs were undertaken and the results of the simulation noted and checked against real system performance and common-sense deductions. A verified model is a model which operates as intended by the modeller. However this does not necessarily mean that it is a satisfactory representation of the real system for the purposes of the study. This is the purpose of validation. In this case the results from the current RTA reporting system model could be compared against historical data of the actual system. A decision is made if model behaviour is close enough to the real system to meet the objectives of the study. Unlike verification, validation is a matter of judgement which involves a trade-off between the accuracy of measurement required and the amount of modelling effort required to achieve this.

Step 3: Develop Future Process Design

Because of the probability distributions used for RTA events, process times and decision points, the output measures of the simulation vary each time the simulation is run. Therefore it is necessary to run the simulation multiple times and form a confidence interval within which the average of the measure should lie. In this case the simulation was run 10 times for a simulated 28 days for the current and redesigned systems. The amount of road traffic officer time in hours was noted for each run for each scenario and a confidence interval calculated at a 95% level (Figure 2).

In this case the confidence intervals do not overlap and so there is a significant difference in the results (Robinson, 1994). In other words the difference in the result for the current and computerised systems is not due to random variation alone but due to an actual difference in performance. The results show that the mean officer hours required to undertake all the tasks associated with the RTA process is 2049 hours under the current system and 1750 hours under the new process design.

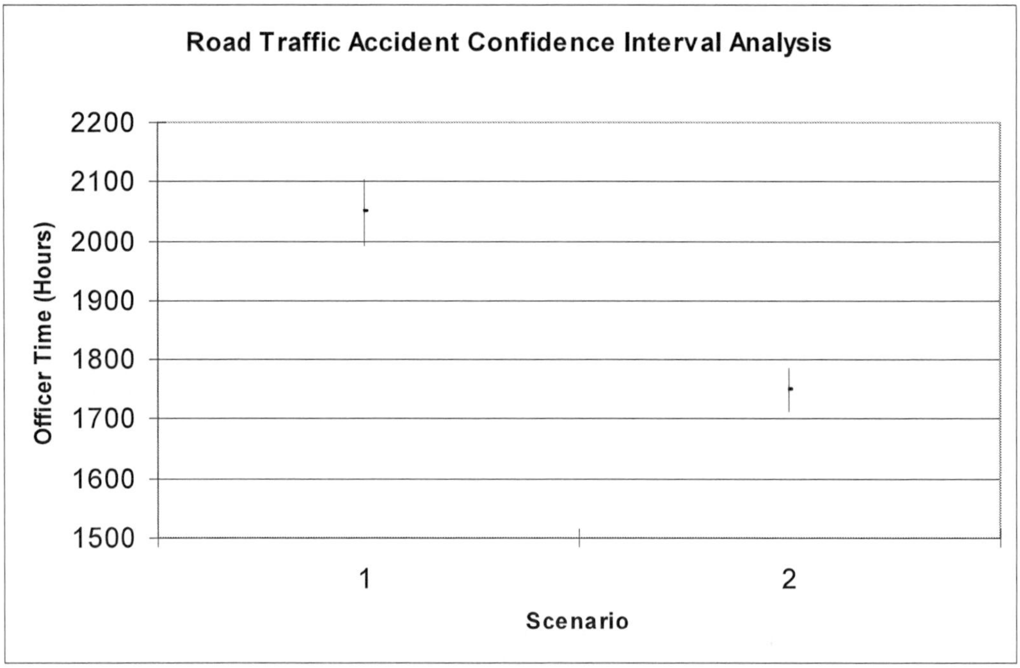

Figure 2. Road Traffic Accident Confidence Interval Analysis.
Scenario 1: Current Process Design .
Scenario 2: Redesigned System.

Step 4: Enable and Implement Future Process Design

The savings outlined in step 3 are dependent on a number of changes in process design. The need for officer time for transcribing and updating notes will be minimised by the use of Optical Character Recognition (OCR) transcription of officer notes by data centre personnel. A centralised data store will also save officer time by quicker storage and retrieval of information for the proforma and abstract preparation process. In addition the location mapping exercise will be simplified by the use of portable digital maps from which officers can indicate the RTA location. Process time will be speeded by workflow automation software which will prompt for timely response to requests for information in the witness proforma and abstract preparation processes. The use of a single point of contact for all data submissions and information requests will also reduce process execution time by eliminating search and delivery delays associated with paper records. The IS system will also have the benefit of improved data accuracy with a single database of all information and location analysis through the use of geocodes.

The simulation study focussed on savings made on the front-line road traffic officer staff but substantial savings can also be made by the centralisation of the traffic administration units. These units are currently located at a divisional level, which is a geographical subdivision of the Police Force area. This is necessary so that paperwork can be processed from officers returning to their local stations. However with the use of digital transmission of RTA information the geographical location of the administrative support can be centralised at a Force level. This can lead to less staff needs due to a centralised automation of processes through workflow and database technologies. The demand on separate divisions would also be aggregated at a corporate level leading to more efficient staff utilisation.

Discussion

A traditional view of a public sector organisation is that of an unchanging organisation structure where the principal linkages between parts are expressed in the form of reporting relationships. A process-based approach attempts to move to a situation where the focus is one concerning the flow of work, interactions of staff and the involvement of stakeholders such as customers, suppliers and government. For example there is a tendency for staff in different occupational groups, such as front-line police officers and administrative support staff, to concentrate on activities within the boundaries of that group. This tendency for each occupational group within a government agency to focus on only their particular area of operation is also apparent across different organisational types such as hospitals and universities. It could be argued that public sector organisations focus redesign efforts on front-line staff where failure to deliver service to the customer is highly visible (e.g. failure to attend a road traffic accident). There is less focus on 'back office' staff whose failure to complete a task may not be so visible in terms of their effect on performance measures. In this case study the Police categorised four main areas of change in regard to the RTA study of staffing, policy, legal and process issues.

In terms of staffing the major organisational change in terms of human resources initiated by the study was the centralisation of the support staffing structure. The process-based approach utilised in this study offered the opportunity for different staff groups to assess their role in the delivery of service in a number of ways. For example the actual procedure of building a process map helped people define roles and see who else does what. McIntosh (2003) reports that public sector organisations in particular spend a considerable amount of time modelling existing processes in order to build consensus on what currently happens, as a means of deciding how to proceed. The simulation tool was also able to investigate how local performance affected overall performance across organisational boundaries. For example the overall process speed is determined by the addition of multiple task durations and the sensitivity of the overall process speed to a change in the duration of a particular task can thus be determined by the simulation. In this case although the need to improve the efficiency and effectiveness of frontline officers provided the impetus for change, it became apparent during the process mapping stage that the support staff in the proposed centralised data entry facility would play a vital role in delivering the objectives of the study.

In terms of policy a feature of public sector agencies is the need to react to changing strategic objectives defined by external stakeholders such as the government. Hutton (1996) highlights the fact that sudden changes in policy in the public sector can take place which could have implications for BPR. It is proposed that the use of process mapping and simulation is able to quickly test the ability of the organisation to meet these challenges and thus increase its flexibility to do so.

In terms of legal issues when undertaking process management studies public sector organisations also need to take into consideration the need to meet regulatory or minimum service level requirements, the standardisation of processes across operating units (for example the definition of the role of a police officer is extensively documented and standardised across forces) and constraints imposed by external stakeholders who may not necessarily have an interest in the outcome of the process (for instance most people do not expect to be arrested but may try to influence police arrest procedures). This suggests a 'blank sheet' approach to process redesign is not realistic and in this case mapping the current process design was a valuable exercise in understanding actual and assumed constraints on the process design. It is proposed that the simulation tool is then able to facilitate a more creative approach to redesign, within the constraints placed on public sector organisations, by allowing experimentation without risk.

In terms of process issues the study has demonstrated the use of a process-based approach to change using the tools of process mapping and BPS to map out and simulate the processes involved in the RTA activity. The case study outcomes report that the proposed process design offers a number of potential benefits including faster delivery of road traffic accident information to government agencies through the use of digital mapping technology replacing location analysis by a third party. Faster process execution of document flows though workflow technology rather than the movement of paper records is also envisaged. Cost savings in terms of road traffic officer time are gained by eliminating the need to return to the police station to undertake administration duties. Savings are also predicted though a centralised administration facility, enabled through the digitisation of data, rather than paper records kept at a divisional level. The quality of the process is improved by the greater accuracy of road traffic accident location analysis through the use of digital mapping and geocodes. It is also anticipated that an improvement in the accuracy of RTA information will be gained though the use of a centralised database store replacing paper documentation.

A feature of this case study was that in this public sector organisation there was a lack of confidence in the introduction of information systems due to the fact that previous IS projects had often been delayed or over-budget. The publication of performance data by the UK government in order to provide information to consumers can make failures more visible than in a commercial organisation and so increase the risk of failure further. What the simulation was able to do was to both demonstrate how the new process would execute and quantify savings in officer time. Demonstration of the operation of the new design using the simulation animation display reduced uncertainty in how the new system would operate. Observation of performance measures provided by the model helped to secure an acceptance of the need for change by demonstrating the increased performance of the proposed system. In particular information provided by the model of staff cost savings quantified the benefits of change in terms of traffic officer on-costs. In general the need for public sector organisations to quantify outcomes to support business cases for action is likely to become prevalent. Shrader (2001) states that the need for a rigorous examination of strategy and a requirement to quantify project benefits will lead to the increasing use of simulation in US government agencies.

The limitations of the approach and tools used in the RTA case study are now discussed. For example the process mapping stage permitted understanding of the structure of current and proposed process designs, but only provided a static view and thus behaviour over time in response to fluctuating demand and resource availability cannot be predicted. Profozich (1998) outlines the disadvantages of using a static analysis tool such as a spreadsheet for analysing such systems, which requires the assumption that each process will operate on the average. In this study this limitation was overcome by the use of BPS which provides a dynamic view, but had the disadvantage of requiring a significant effort in terms of time, cost and modelling skills. This issue has led clients of simulation to report that they measure the success of a simulation project primarily in terms of delivery within budget (McHaney and Cronan, 2000) and require good project management skills from model builders, along with technical knowledge (Melão and Pidd, 2003).

In terms of staffing issues the approach assisted in communicating the real constraints to change, not just constraints determined by culture and working practices which would to a redesign that could be excessively conservative. The process mapping stage of the study provided a medium where new design ideas can be communicated between the redesign team. Later in the study the animation facilities of the BPS were also used to communicate new ideas. This feature of BPS is reported in Vreede and Verbraeck (1996) and Greasley (2004). There are limitations however to the ability of these tools to address the number of complex organisational issues, such as the management of a reduction of staffing levels and the movement of staff from divisional to a centralised traffic administration unit, involved in this study. Melão and Pidd (2000) report a potential danger of using simulation where the human aspect is only regarded as relevant as a resource for executing tasks and thus the socio-political considerations are neglected. The use of a multi-faceted approach to business process design using process mapping, business process simulation, system dynamics and soft approaches such as Soft Systems Methodology (SSM) (Checkland and Scholes, 1990) is suggested.

Further studies could extend the process design analysis with a study of the information infrastructure required. Warren and Crosslin (1995) suggest simulation as a method of providing the justification necessary to support business decisions to redesign and make what are often substantial investments in IT systems. Giaglis et al (1999) who outline the need to support IS evaluation by developing techniques for generating estimates of the organisational value of IS. Experimental methods such as simulation are suggested as capable of providing such estimates. A model of the information infrastructure would also have the advantage of assessing the feasibility of the project in terms of determining a specification for hardware such as communication devices and database systems. Changes to these aspects are usually undertaken using network simulation software, for example NetSim (www.tetcos.com). In support of this Painter et al (1996) argue that the process and infrastructure analysis can be integrated and present a methodology to achieve this.

CONCLUSION

A case study demonstrates the use of a process-based approach to change regarding the implementation of an information system for road traffic accident reporting in a UK police force. The supporting tools of process mapping and business process simulation are used in the change process and assist in communicating the current process design and people's roles in the overall performance of that design. The simulation model is also used to predict the performance of new designs incorporating the use of information technology. The approach is seen to have a number of advantages in the context of a public sector organisation. These include the ability for personnel to move from a traditional grouping of staff in occupational groups with relationships defined by reporting requirements to a view of their role in a process which delivers a performance to a customer. By running the simulation through time it is also possible to gauge how changes at an operational level can lead to the meeting of strategic targets over time. Also the ability of simulation to proof new designs was seen as particularly important in a government agency were past failures of information technology investments had been well publicised and had thus contributed to a more risk averse approach to their implementation. Limitations of the approach are also discussed. The process mapping stage permits understanding of the structure of current and proposed process designs, but is not able to predict behaviour over time. Business process simulation provides this facility but requires careful planning in order to ensure it is able to deliver results within cost and time targets. Finally the results of the simulation analysis must be considered within the constraints on public sector operations placed on them by internal and external stakeholders.

References

Aguilar, M., Rautert, T., Pater, A. (1999). Business process simulation: A fundamental step supporting process centred management", *Proceedings of the 1999 Winter Simulation Conference,* ed. by Farrington PA, Nembhard HB, Sturrock DT, Evans GW, SCS, pp. 1383-1392.

Bal, J. (1998). Process analysis tools for process improvement, *The TQM Magazine* 10 (5), 342-354.

CSC Index (1994). *State of re-engineering report,* CSC Index.

Checkland, P. and Scholes, J. (1990). *Soft Systems Methodology in Action,* John Wiley, Chichester.

Collis, J. and Hussey, R. (2003) *Business Research,* Second Edition, Palgrave Macmillan: Basingstoke.

Cypress, H. (1994). *Re-engineering,* OR/MS Today 21, 19-29.

Davenport, T.H. and Short, J.E. (1990). The new industrial engineering: Information technology and business process redesign, *Sloan Management Review 31,* Summer, 11-27.

Fathee, M.M.; Redd, R.; Gorgas, D.; Modarres, B. (1998) The Effects of Complexity on Business Process Reengineering: Values and Limitations of Modeling and Simulation Techniques, *Proceedings of the 1998 Winter Simulation Conference*, SCS, pp. 1339-1345.

Giaglis, G.M. (1999) On the Integrated Design and Evaluation of Business Processes and Information Systems, *Communication of the AIS*, 2, 5, July.

Giaglis, G.M., Mylonopoulos, N., Doukidis, G.I. (1999). The ISSUE methodology for quantifying benefits from information systems, Logistics Information Management 2, 50-62.

Greasley, A. (2000). A simulation analysis of arrest costs, J Opl Res Soc 51, 162-167.

Greasley, A. (2004). Simulation Modelling for Business, Ashgate Limited: Aldershot.

Gulledge, T.R. and Sommer, R.A. (2002). Business process management: public sector implications, Business Process Management Journal 8, 364-376.

Gummesson, E. (2000) *Qualitative Methods in Management Research,* Second Edition, Sage Publications, Thousand Oaks.

Haque, B., Pawar, K.S., Barson, R.J. (2003). The application of business process modelling to organisational analysis of concurrent engineering environments, *Technovation* 23, 147-162.

Halachmi, A. and Bovaird, T. (1997). Process reengineering in the public sector: learning some private sector lessons, *Technovation* 17 (5), 227-235.

Hammer, M. (1990). Re-engineering work: don't automate, obliterate, *Harvard Business Review,* 104-112.

Harrington, B., McLoughlin, K. and Riddell, D. (1997). Business process re-engineering in the public sector: a case study of the Contributions *Agency, New Technology, Work and Employment* 13 (1), 43-50.

Harrington, H.J. (1991). *Business Process Improvement: The breakthrough strategy for total quality, productivity, and competitiveness,* McGraw-Hill Inc.

Hays, M. and Bebbington, M. (2000). Simulation in public sector management: *A case study, Int Trans Opl Res* 7, 465-486.

Hutton, G. (1996). Business process re-engineering: A public sector view, in Armistead, C. and Rowland, P. (eds), *Managing Business Processes:* BPR and Beyond, John Wiley and Sons, Chichester.

Kelton, W.D., Sadowski, R.P., Sadowski, D.A. (2001). *Simulation with ARENA*, McGraw-Hill, Singapore.

Kloot, L. and Martin, J. (2000). Strategic performance management: A balanced approach to performance management issues in local government, *Management Accounting Research 11,* 231-251.

Levine, L.O. and Aurand, S.S. (1994). Evaluating Automated Work-Flow Systems for Administrative Processes, *Interfaces* 24, 141-151.

MacIntosh, R. (2003). BPR: alive and well in the public sector, *International Journal of Operations and Production Management* 23 (3), 327-344.

McAdam, R. and Donaghy, J. (1999). Business process re-engineering in the public sector: A study of staff perceptions and critical success factors, *Business Process Management Journal* 5 (1), 33-49.

McAdam, R. and Mitchell, N. (1998). Development of a business process re-engineering model applicable to the public sector, *Total Quality Managemen*t 9 (4), 160-163.

McHaney, R. and Cronan, T.P. (2000). Toward an empirical understanding of computer simulation implementation success, *Information and Management* 37, 135-151.

Meel, J.W. van and Sol, H.G. (1996). Business engineering: dynamic instruments for a dynamic world, *Simulation and Gaming* 27 (4), 440-661.

Melão, N. and Pidd, M., (2000). A conceptual framework for understanding business processes and business process modelling, *Information Systems Journal* 10, 105-129.

Melão, N. and Pidd, M. (2003). Use of business process simulation: A survey of practitioners, *Journal of the Operational Research Society* 54, 2-10.

Orman, L.V. (1995). A model management approach to business process engineering, *Proceedings of the American conference on information systems,* August, Pittsburgh, PA.

Painter, M.K., Fernandes, R., Padmanaban, N., Mayer, R.J. (1996). A methodology for integrating business process and infrastructure models, *Proceedings of the 1996 Winter Simulation Conference,* ed. by Charnes JM, Morrice DJ, Brunner DT, Swain JJ, SCS, pp. 1305-1312.

Paolucci, E., Bonci, F. and Russi, V. (1997). Redesigning organisations through business process re-engineering and object-orientation, *Proceedings of the 5th European conference on information systems,* pp. 587-601.

Pidd, M. (1998) *Computer Simulation in Management Science*, Fourth Edition, Wiley, Chicester.

Profozich, D. (1998). *Managing Change with Business Process Simulation,* Prentice Hall, New Jersey.

Radcliffe, J. (2000). Implementing and integrating crime mapping into a police intelligence environment, *International Journal of Police Science and Management* 2, 313-323.

Robinson, S. (1994). *Successful Simulation: A Practical Approach to Simulation Projects,* McGraw-Hill, Berkshire.

Shrader, S. (2001). Simulation in government: validating business strategy, Proceedings of the 2001 Winter Simulation Conference, *SCS,* 1259-1261.

Stuart, I., McCutcheon, D., Handfield, R., McLachlin, R., Samson, D. (2002). Effective case research in operations management: a process perspective, *Journal of Operations Management* 20, 419-433.

Thong, J.L. Yap, C.S. and Seah, K.L.. (2000). Business Process Reengineering in the Public Sector: The Case of the Housing Development Board in Singapore, *Journal of Management Information Systems* 17 (1), 245-270.

Tumay, K. (1996). Business Process Simulation, *Proceedings of the 1996 Winter Simulation Conference*, ed. by Charnes JM, Morrice DJ, Brunner DT, Swain JJ, SCS, pp. 93-98.

Verma, R., Gibbs, G.D. and Gilgan, R.J. (2000). Redesigning check-processing operations using animated computer simulation, *Business Process Management Journal* 6, 54-64.

Vreede, G.J. and Verbraeck A. (1996). Animating organizational processes: Insight eases change, *Simulation Practice and Theory* 4, 245-263.

Walker, D. (2003). New labour, *The Guardian,* 21 October 2003, pp. 25.

Warren, J.R. and Crosslin, R.L. (1995). Simulation Modeling for BPR, *Information Systems Management* 12, 32-43.

Wiley, R.B. and Keyser, T.K. (1998). Discrete event simulation experiments and geographic information systems in congestion management planning, *Proceedings of the 1998 Winter Simulation Conference*, ed. by Mederios DF, Watson EF, Carson JS, Manivannan MS, SCS, 1087-1093.

In: Road Traffic: Safety, Modeling and Impacts
Editors: S. E. Paterson and L. K. Allan, pp. 521-534
ISBN 978-1-60456-884-4

Chapter 17

SPATIAL MODELING OF AIR POLLUTION BASED ON TRAFFIC EMISSIONS IN URBAN AREAS

Lubos Matejicek
Institute for Environmental Studies, Charles University in Prague
Benatska 2, 128 01, Prague, Czech Republic
Zbynek Janour
Institute of Thermomechanics, Academy of Sciences
Dolejskova 5, 182 00, Prague, Czech Republic
Michal Strizik
Laboratory of Risk Research and Management, Technical University of Ostrava
Lumirova 13, 700 30, Ostrava, Czech Republic

ABSTRACT

Road traffic becomes a dominant source of air pollution in urban areas. The emissions of inorganic compounds and volatile organic compounds caused by motor vehicles have increased trends, particularly in urban areas, which together with other emitted compounds can cause human health problems in a long-time period. Procedures for estimating of road traffic emissions in the USA and in the European region are based on guidelines and recommendations. In order to implement the guidelines, decision-making tools are necessary to provide information in an easy understandable form. Thus, the spatial information system is needed to manage spatio-temporal data, provide analysis, solve numerical models and visualize the results. Generally, spatial data include street networks and monitoring networks. Temporal data are represented by changes of traffic intensity and by time series of measured pollutants. Geographic information systems (GISs) extended by environmental modeling tools offer advance estimates of traffic emissions improved by the digital terrain data and by the wind flow effects. The estimates of air pollution in the surrounded areas can be based on spatial interpolation. In a wide range of spatial interpolations, the deterministic techniques or the geostatistical methods are used in dependence on the input data and the results provided by the exploratory spatial data analysis (ESDA). Each spatial interpolation attached to the map layer defines concentration levels at a specified time period. The time series of map layers can demonstrate the variability of the air pollution distribution, which brings a new insight into this research.

In addition to mapping of the urban air pollution in dependence on the traffic intensity and the measured concentrations, numerical modeling and simulation in wind tunnels are presented as important tools for exploration of the dispersion and the transport of pollutants caused by wind flows and other effects. A number of software tools based on Gaussian dispersion principles are mentioned in the framework of US EPA guidelines.

As a case study, spatial modeling of air pollution is focused on the urban area of the city of Prague, because a number of motor vehicles registered on the Prague territory is growing. In addition to the registered vehicles, the specific phenomenon in the central Europe is represented by the abrupt increase in traffic of trucks. In spite of new methods for reduction of emissions from motor vehicles, urban development and successive reconstructions of existing roads cause other emissions. Thus, the street network also becomes a set of line sources of stirred up suspended particulates generated by the passing vehicles. These dust emissions, so-called secondary dust, are estimated and partially validated by measurements of the fraction PM_{10}. As an example, estimates of nitrogen oxides are carried out by spatial interpolation based on data from the surface monitoring network and the DIAL-LIDAR measurements.

Integration of spatio-temporal data together with environmental modeling tools brings new possibilities to compare spatial interpolations created for individual compounds at the high temporal and spatial resolution. In the framework of the GIS, mapping of air pollution and human exposures in streets, at workplace locations and residence addresses can help to reduce traffic emissions by optimization of transport scenarios.

Keywords: road traffic, air pollution, spatial modeling, GIS

1. Introduction

Transport-related emissions are rising and there is increasing consensus that the growth in motorized vehicles in urban areas is incompatible with health, and greenhouse-gas emissions. Traffic intensity and air pollution adverse health effects and overall environmental degradation. Mostly, transport is the only sector for which current emissions are higher than in the last seasons. In the USA, passenger-car emissions and emissions from trucks are increasing. In Europe, freight transport has more than doubled since 1970. In eastern parts of Europe, the 1990s are characterized by a major decline in rail transport and a concurrent increase in car and truck use. The escalation of urban transport is also significant in Chinese cities in the last decade. Thus, the transport, especially in urban areas, is widely recognized as a significant and increasing source of air pollution.

Many reviews are focused on regional and global impacts of transport emissions of air pollution, but mostly deal with the impact of emissions on local urban air quality, and consider only emissions related to the road transport. In addition to these reviews, some reports based on data from protection agencies (US EPA or European Environmental Agency) cover air and sea transport. There is also an emphasis on assessing possible solutions to environmental problems related to transport. But, too little action is taken on to reduce the environmental impacts. For all that the catalytic convertors are being fitted to every new gasoline-engine car, which is followed by similar developments in diesel emissions, impacts of road transport on urban air quality is still growing.

In spite of that the mode of transport is dependent on the selected area, which can include also ships and aircrafts, the attached case studies are focused on road traffic and its effects on air pollution in urban areas. Road transport is characterized from other sources of air pollution in that the pollutants are released in very close proximity to human receptors. It also reduces the opportunity for the atmosphere to dilute the pollutants. In addition to the close proximity, the roads in the city centre have buildings alongside that shelter the roads and reduce the wind speed at the sources of emissions to that on open roads. But this increase of transport emissions has not significant impact on damage human health because people usually spend larger amounts of time indoors [1]. In case of the buildings that line city streets, the indoor concentrations of pollutants are mainly dependent on outdoor concentrations from road traffic adjacent to windows and doors.

2. Numerical Modeling Focused on Air Pollution

In order to quantify impacts of transport emissions, dispersion of pollutants is studied by using of numerical modeling [2] or physical modeling [3]. The numerical dispersion models deal with simulation of increased concentrations near the major roads, compared to overall background levels of emitted compounds by motor vehicles. Due to the wide range of car classes, many compounds must be taken into account. Mostly, the dispersion modeling is focused on carbon monoxide (CO), nitrogen oxides (NO_x) ozone (O_3), coarse ($PM_{10\text{-}2.5}$), fine ($PM_{2.5}$) and ultrafine ($PM_{0.1}$) particles, black carbon (BC), polycyclic aromatic hydrocarbons (PAHs), and benzene. The numerical models are mostly based on Gaussian dispersion principles. Many state agencies and local authorities, especially in the USA, use the modeling program Industrial Source Complex-Short Term, Version 3 (ISCST3) approved and recommended by the US EPA. The ISCST3 implements a set of Gaussian plume-based models that are used to calculate downwind concentrations from point, line, and area sources. In dependence on a spatial scale of dispersion modeling, the roads are represented by line sources, and the crossroads, parking places, bus stops and other transport service places can be represented by point or area sources. While the ISCST3 is developed as the steady-state air emission modeling system, another dispersion model called CALPUFF represents a generalized non- steady-state air emission modeling system. The US EPA approves to use the CALPUFF as a guideline model for regulatory applications involving long range transport, and for near-field applications with non-steady-state effects. In case of local prediction of air pollution from road transport, many studies focused on the middle spatial scale tasks are based on these environmental modeling systems [4, 5, 6, 7]. The ISCST3 and the CALPUFF are complemented by other computer tools for processing of meteorological data, spatial modeling of buildings, and visualization of simulation results.

While the ISCST3 or the CALPUFF represent modeling guidelance, many dispersion models are directly focused on dispersion modeling of traffic emission sources. For example, the CALINE, based on Gaussian diffusion equations, employs a mixing zone concept for pollutant dispersion over the roadway. The prediction of pollutant concentrations is carried out for receptors located within 500 meters of the roadway. The inputs contain data about emissions sources, meteorology and site geometry. The special options are included for modeling of air quality near the intersections, street canyons and parking facilities. Nowadays, the CALINE together with the ISCST3 and other dispersion models are included in the US EPA model library, guidelines and recommendations.

3. Modeling in Wind Tunnels

In case of physical modeling, the relation between urban three-dimensional morphology and windiness is explored by wind tunnel experiments. Highly idealized city models consisting of blocks of buildings with street canyons are usually used for construction of the physical models. In addition to the physical models based on blocks of buildings, there are many complements for a more realistic simulation in the wind tunnels. Trees with various shapes of crowns are positioned inside the street canyons. Line sources are embedded in the bottom of the streets to release of the tracer gas for simulation of traffic emissions. Traffic-induced turbulence is simulated by rotating belts with thin plates etc. [8]. Physical modeling is used to be complemented by numerical modeling based on Navier-Stokes equations that are extended by tools for simulation of turbulent flows and calculation of boundary conditions. Many case oriented studies oriented on the urban environment are included in the books and papers focused on computational wind engineering [9]. The evaluation of surface flow fields in the urban environment can be extended by vertical flows caused by daytime heating processes in the summer periods and by other phenomena [10]. Thus, the wind tunnel experiments together with numerical simulation can provide the encouraging insight into air flows and offer to predict transport of air pollutants from traffic emission sources.

4. Methods Based on Exploratory Spatial Data Analysis

Due to required huge amounts of data and time consuming numerical calculations, another ways of environmental modeling are often used to predict concentration levels of pollutants in the neighborhood of emission sources. For example, the thematic maps focused on pollutant concentrations over urban areas are mostly based on spatial interpolation that can create the continuous surface of the pollutant concentration, standard errors of predictions and the probabilities that critical values are exceeded. In case of air pollution based on traffic emissions, the inputs into the interpolation tools are represented by estimated levels of concentration at the major emission sources formed by crossroads and road sections or by a surface monitoring network. The spatial interpolation by itself uses sample points at different locations. The sample points include estimates or measurements of selected pollutants.

In order to carry out the proper spatial interpolation, the exploratory spatial data analysis (ESDA) assists to gain a deeper understanding of the pollutant dispersion by creating histograms, Voronoi maps, quantile plots, trend analysis, semivariogram/covariance clouds and crosscovariance clouds. The ESDA is often integrated in the geographic information system (GIS), which allows highlighting the spatial objects on the map layers together with their attributes. Generally, two groups of interpolation techniques are provided by the GIS: deterministic and geostatistical. While the deterministic techniques use mathematical functions for spatial interpolation, geostatistical tools rely on both statistical and mathematical methods and, in addition to creating the surface, can assess the uncertainty of the predictions. The deterministic techniques include Inverse Distance Weighted (IDW) and polynomial interpolations. The geostatistical techniques are based on various kriging models [11]. Implementation of interpolation procedures is usually described in the GIS guidebooks that give more details about setting of the internal parameters [12].

5. Traffic Emission Inventories in Urban Areas

Inventories of atmospheric emissions from transport have been carried out in the framework of research projects and case oriented studies for decades. They are constructed on a range of scales from individual urban areas to countries or continents. The estimates of air pollution are structured in a different way depending on country legislation and on local rules that govern environmental protection. Thus, the dissimilarity between the inventories made in the USA and Europe reflect not only true differences in emissions, but also distinctions related to the way in which the inventory has been produced [13]. In the USA, many guidelines and traffic emission inventories are accessible through the web pages of the US EPA. The American urban areas are larger than the European urban areas that usually include historical centers linked by narrow streets, and the block of flats in the suburbs interconnected by the arterial roads. In the European Union, the emission inventory guidebooks are mostly prepared by the European Environmental Agency. Nearly each year, the updates are published and accessible on the Internet. For example, the ENEP/CORINAIR Emission Inventory Guidebook published in 2007 contains chapters focused on road transport [14]. The guidelines are discussed and applied in a number of studies [15, 16, 17]. Many modeling tools for estimation of traffic emissions in European cities are also based on CORINAIR emission inventories [18, 19, 20].

6. Spatial Estimation of Air Pollution by Laser Scanning

Remote sensing techniques, mostly used for detection and classification of the surface objects [21, 22], are also implemented in the systems for detection of atmospheric pollutants. Information about the pollutants is derived from spectroscopic interaction between radiation and atmospheric matter. The intensity of laser beams is decreased when passing through polluted areas of the atmosphere. In order to obtain the most precise measurements, the DIALs (Differential Absorption LIDARs) are employed in the mobile stations for surface measurements of the pollutants over the city road network. It represents active remote sensing methods where the pulse LIDAR techniques can estimate the distribution of the pollutants hundreds meters over the surface. The laser pulses are emitted in two different wavelengths, resonance and reference. The resonance wavelength is chosen within the absorption band of the measured air pollutant. Thus, the decrease of the intensity is higher than for the intensity of the reference wavelength that is outside of the absorption band. A portion of the laser light of both wavelengths is returned back to the DIAL system by molecules of gases (Rayleigh scattering) and by aerosols (Mie scattering). By comparing signal intensities on both wavelengths, the high resolution detector and processing unit can create the pollutant concentration profile along the trajectory of the laser pulses. The emitor and the detector are managed together by the rotary periscope that emits the laser beams and receives the backscarttered signals from any direction.

Applications of the LIDAR and the LIDAR-DIAL techniques show a growing capacity in environmental monitoring, especially, if spatial and temporal distributions are required [23]. A wide range of atmospheric contaminants, as for example volatile organic compounds, nitrogen oxides, ozone, sulphur dioxide and carbon monoxide, can be detected in the range of a few kilometers [24, 25, 26]. In addition to the measurement of individual pollutants over the urban areas, the horizontal and vertical concentration profiles allow estimating of the mixing height and the relationship between vertical mixing and ground-level pollutant concentrations. Thus, the ground-level concentrations of ozone and nitrogen oxides in urban canyons based on DIALs techniques can be compared with the conventional fixed point sampling of pollution gas concentrations obtained by conventional monitoring networks [27]. The measurements of the vertical and temporal variations based on a few observation periods bring other insight into air pollution transport and chemical interactions during a day period [28]. In order to analyze spatio-temporal measurements of air pollution together with surface spatial objects and traffic sources of pollution, the GISs are often used for the creation of map layers extended by concentration profiles [29]. In addition to the spatial data management and visualization, GISs can remarkably assist in spatial interpolation of the horizontal and vertical profiles derived from DIALs measurements [30].

7. GIS-Based Spatial Analysis and Visualization

GIS environment offers tools for exploratory spatial data analysis that can explore fixed point sampling and DIALs concentration profiles in dependency on surface emission sources, terrain models and other surface objects [31, 32]. Generally, GIS offers to integrate a wide range of spatial data, which can support spatial modeling and in some cases dynamic modeling. Various scenarios of data integration for environmental modeling are described in the book written by Goodchild et al. [33, 34]. In spite of a huge offer of the GIS analytical tools [35, 36, 37], there is still the need to develop new extensions that solve case oriented tasks in the framework of environmental analysis and modeling [38]. In some cases, the file exchange or sharing methods between the GIS and the standalone modeling tools must be used to solve large environmental projects that deal with decision-making processes.

The time series of pollutant concentrations from conventional fixed point sampling networks can be linked to the map layers together with the emission sources. The horizontal and vertical profiles from the DIALs offer to extend standard monitoring techniques by exploring pollutant concentrations over the surface. In case of numerical modeling and wind tunnel studies, the GIS extensions are needed to solve time consuming numerical calculations and to create the interconnection between the GIS and the standalone software tools based on US EPA modeling guidelines and other accessible resources. Nowadays, the RDBMS extended by modules for spatial data management, like ESRI's geodatabase, are used in order to support spatial analysis, numerical modeling and visualization in a more complex way that offers direct integration of spatial data formats, time series and analytical tools [39].

8. A Case Study for Spatio-Temporal Analysis of Air Pollution from Urban Traffic in Prague

In order to inform local authorities responsible for controlling road traffic in Prague, monitoring system of air pollution and other risks has been operating for a decade. This monitoring system was developed in the framework of the project, which is shared by the Department of Informatics at the Prague City Hall, by the Czech Hydrometeorological Institute and by the T-Soft company. Aggregated information about air pollution based on the surface monitoring network is presented on the Internet and accessible to public. The ground-level air pollution assessment shares data archived in the database of the Air Quality Information System of the Czech Republic. As an example, Figure 1 illustrates the temporal changes of nitrogen oxides during the time period of a few hours. The upper part contains buttons for interactive data management (functions: refresh, backward, forward, legend and report). The graph can display emission limit concentrations for emergency and regulation that are complemented by a list of selected monitoring stations (accessible on the right side).

Generally, the emission sources generating air pollutants are classified in the framework of the Air Pollution Source Register into a few classes in dependence on risk assessment. In case of the stationary emission sources, there are three classes in the register:

- extremely large and large source combustion processes with the heat-generating capacity above 5 MW and very significant technologies,
- middle-sized source combustion processes with the heat-generating capacity 0.2–5 MW and significant technologies,
- small source combustion processes with the heat-generating capacity under 0.2 MW and less significant technologies.

Monitoring of stationary emission sources is continuously updated and revalued in the framework of implementing of the new monitoring techniques and modeling procedures. The individual class is represented by the mobile emission sources, which are consecutively categorized in dependence on their shape to the point sources (crossroads, portals and tunnel ventilation exhausts, parking places) and the line sources (sections of the urban roads). In order to estimate the emissions from mobile sources in a more precise way, a few projects supported by the Ministry of the Environment of the Czech Republic have been created in the last decade. As an example, the map layer based on a model calculation of traffic emissions is presented in Figure 2. Model calculations take into account the type of vehicles, the composition of traffic flows, the speed, the slope and the estimates of increased emissions generated at a cold start of the vehicle engine. The following pollutants can be included into the tools: PM_{10}, sulphur dioxide, nitrogen oxides, carbon monoxide and hydrocarbons [40].

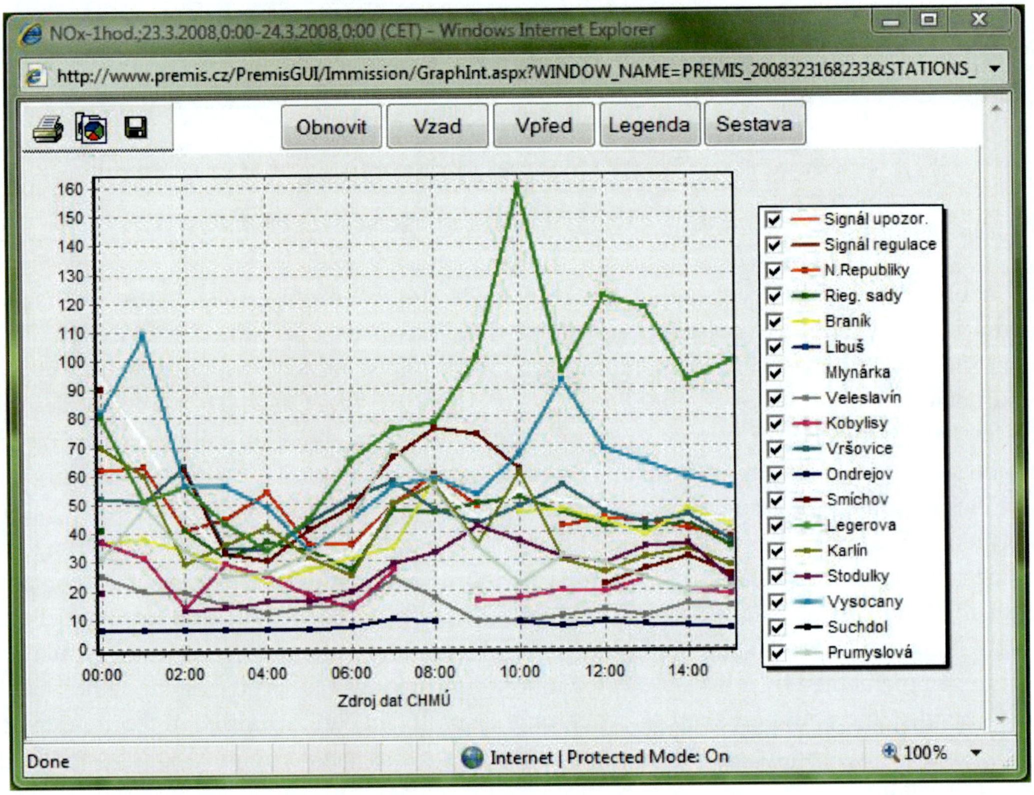

Figure 1. The actual information about air pollution in Prague accessible for public through the Internet (http://www.premis.cz, March-2008).

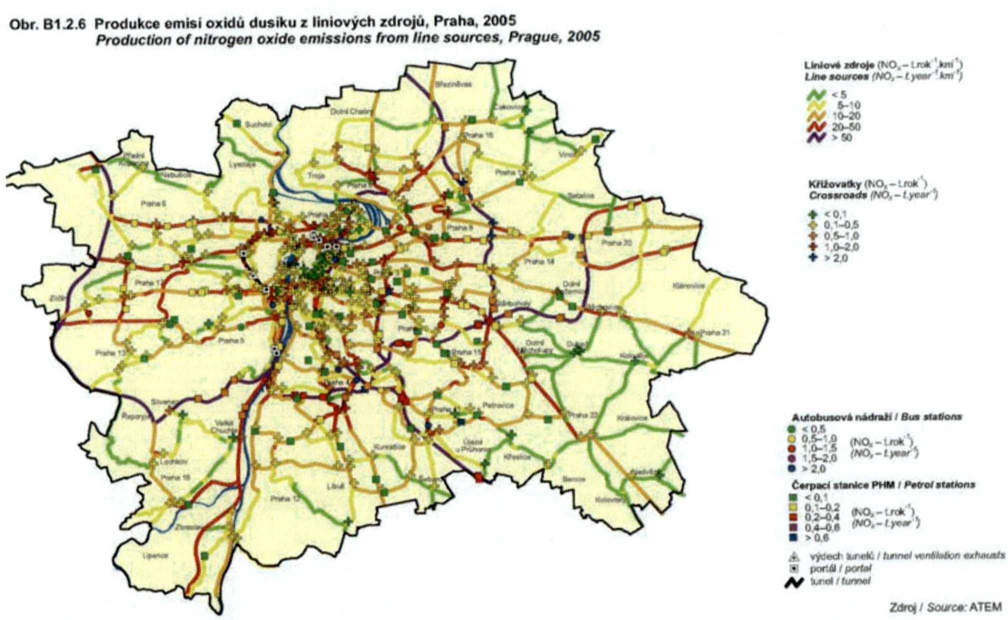

Figure 2. Map layers focused on modeling of nitrogen oxides from line sources (PRAGUE–ENVIRONMENT 2005, MHMP, 2006; http://envis.praha-mesto.cz/, March-2008).

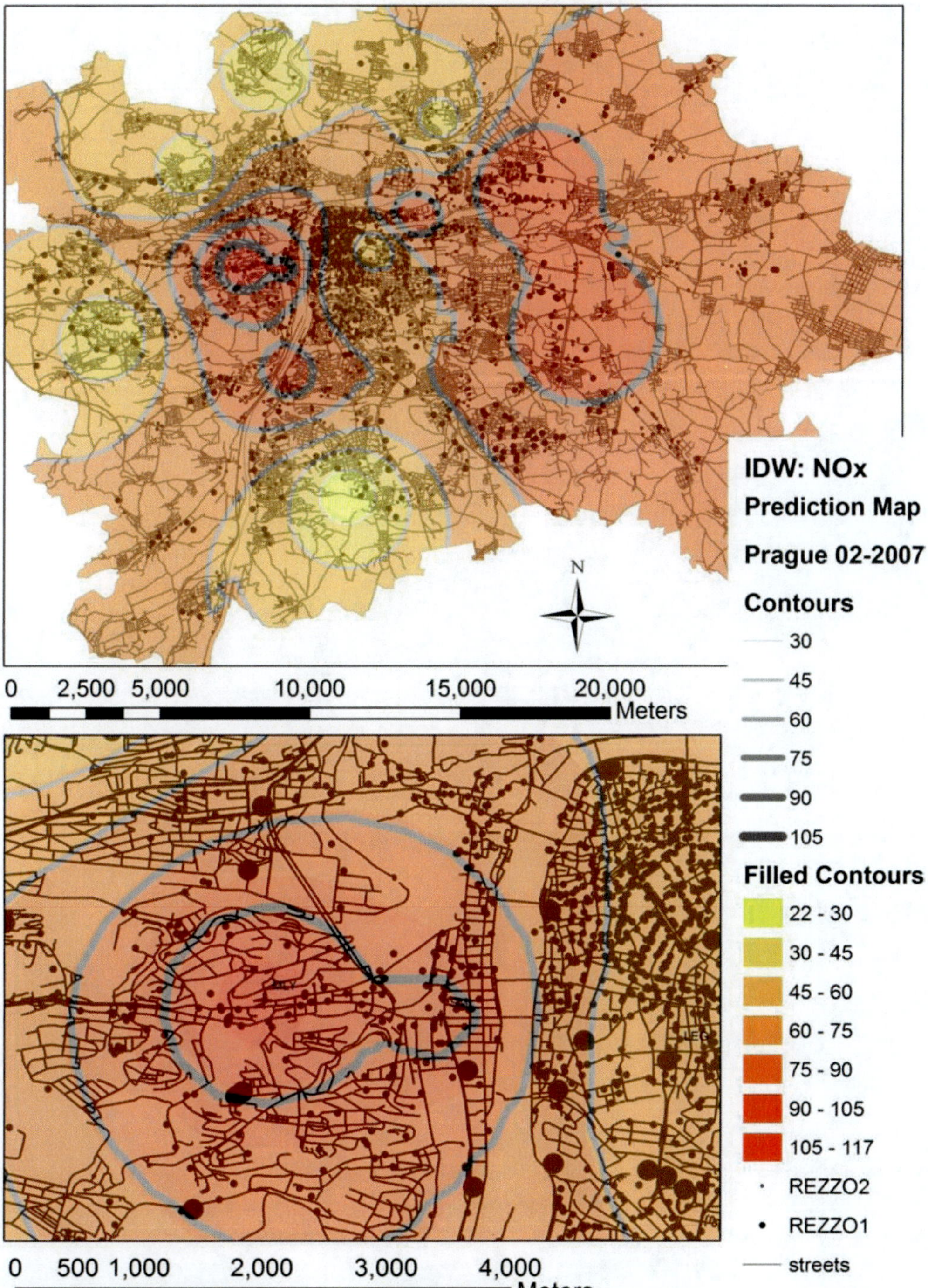

Figure 3. Map layers created in the GIS environment with spatial interpolations (IDW and ordinary kriging) based on data from surface monitoring network, the stationary sources of air emissions are marked in the map layer as points (REZZO1, REZZO2-the Czech guidelines).

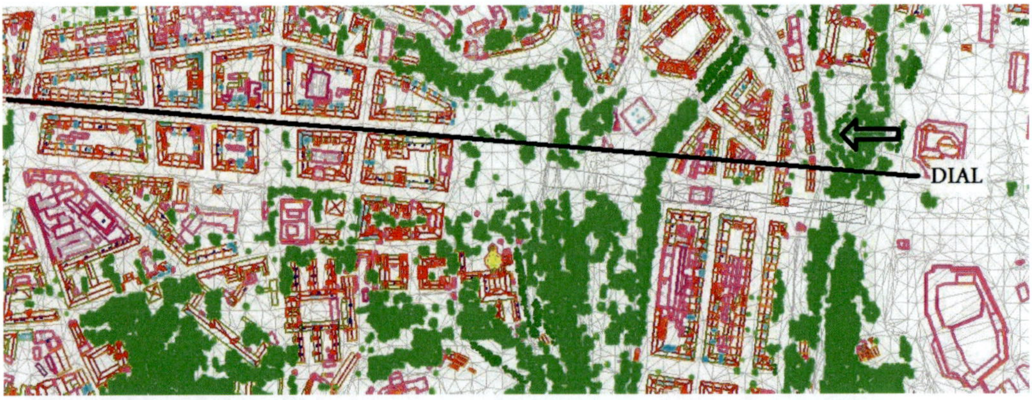

The location of the DIAL's vertical profile in the canyon of an arterial street in Prague.

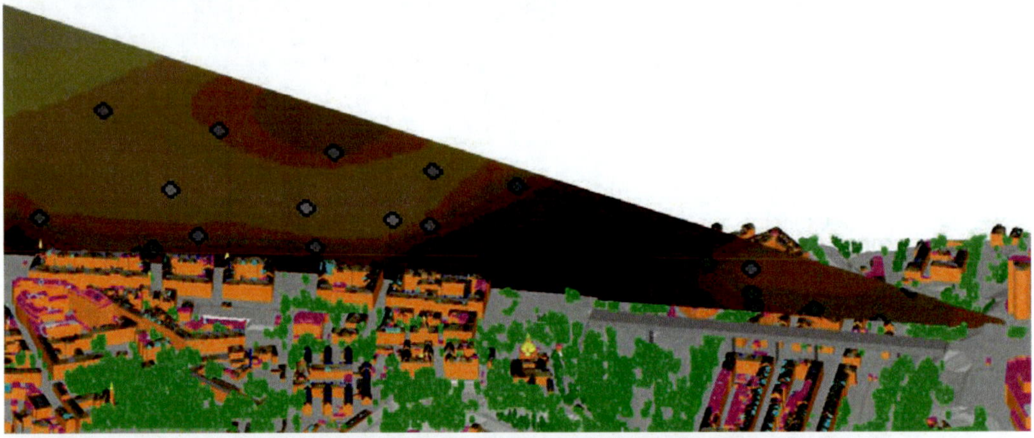

Visualization of the urban area complemented by the vertical image based on DIAL's data.

The detail view on the street canyon complemented by the image based on DIAL's data.

Figure 4. Processing of the DIAL-LIDAR data in the GIS and CAD environment.

The spatial interpolations based on data from the surface monitoring stations represent other source of information about living conditions in urban areas. The map layers focused on nitrogen oxides created in the GIS environment with the IDW and with the ordinary kriging are illustrated in Figure 3.

The distribution of the nitrogen dioxide in the street canyon located in the urban area of the city Prague is illustrated in Figure 4. The 3D scenes based on processing of the LIDAR-DIAL data are focused on the arterial street with heavy traffic. The measurements of the concentration located in spatial nodes on the laser lens are interpolated by the IDW and ordinary kriging. As an example, a few time periods are included into the output presentation.

CONCLUSION

A wide range of monitoring techniques and modeling procedures have been mentioned and shortly described in order to present the research activities focused on influence of the urban traffic on living conditions in urban areas. At present, the automotive traffic becomes the most important source of air pollution on the territory of large cities. Thus, the US EPA and the European Environmental Agency carry out many guidelines in the framework of various projects. The projects are shared by state agencies, research associations and universities from the whole world. Consecutively, the research outputs are implemented to the national legislation and guidelines [13, 14].

The environmental modeling supported by powerful computer tools and monitoring systems can predict dispersion and transport of pollutants. The environmental models used for regulatory applications are able to assist in decision-making processes of state authorities and local agencies. In order to improve the precision of predictions, there is still the need to develop new environmental models and to implement new monitoring techniques. A new research is represented by using data from high resolution satellite systems to identify surface objects and their properties in a more precise way. From this point of view, the remote sensing based on the DIAL-LIDARs represents the derived method that can bring a new insight into more detailed on-site air pollution estimates. The relationship between the urban morphology and the air pollution distribution can also help in the validation of numerical models and in a more realistic simulation in the wind tunnels [29, 30].

The attached case study focused on air pollution of the selected urban area in Prague introduces methodology implemented on the local level of the city Prague. In addition to the widely used techniques, spatial interpolations in the GIS environment based on data from the surface monitoring stations and the DIAL-LIDAR measurements are carried out to explore and visualize the distribution of the nitrogen oxides. However, detailed on-site air pollution estimates by the DIAL-LIDAR are not available for each road and time. But, in comparison with the estimates based on surface measurements and the standard modeling procedures, it can improve the prediction of air pollution in higher levels over the surface [41].

ACKNOWLEDGEMENTS

The attached case study is partly based on the data created in the framework of the project GACR 205050898, which was generally focused on measurement and modeling of air pollution in urban areas and flat landscape. The authors wish to thank the Lidar company managed by A. Cerny, P. Berger, M. Strizik and P. Engst.

The input spatial data for building of the 3D scenes are originally provided by the Institute of Municipal Informatics of Prague. The map layers of emission sources (registered as REZZO 1 and REZZO 2) and the temporal measurements of the surface air pollution transferred from the Internet are administrated by the IT Department of the Prague City Hall.

The spatio-temporal data were processed by ESRI-ArcGIS and Leica Geosystems-ERDAS Imagine in the GIS Laboratory, Faculty of Natural Science, Charles University in Prague in the framework of the project AVCR 1ET400760405 supported by the Information Society, Academy of Sciences in the Czech Republic.

REFERENCES

[1] Freijer, J.I., Bloemen, H.J.T., Loos, S., Marra, M., Rombout, P.J.A., Steentjes, G.M. and Veen, M.P. (1998). Modelling exposure of the Dutch population to air pollution. *Journal of Hazardous Materials 61*, 107-114.

[2] Venkatram, A., Isakov, V., Thoma, E. and Baldauf, R. (2007). Analysis of air quality data near roadways using a dispersion model. *Atmospheric Environment 41*, 9481-9497.

[3] Skote, M., Sandberg, M., Westerberg, U., Claesson, L. and Johansson A.V. (2005). Numerical and experimental studies of wind environment in an urban morphology. *Atmpospheric Environment 39*, 6147-6158.

[4] Wang, L., Parker, D.B., Parnell, C.B., Lacey, R.E. and Shaw, B.W. (2006). Comparison of CALPUFF and ISCST3 models for predicting downwind odor and source emission rates. *Atmospheric Environment 40*, 4663-4669.

[5] Song, Y., Zhang, M. and Cai, X. (2006). PM10 modeling of Beijing in the winter. *Atmospheric Environment 40*, 4126-4136.

[6] Souto, M.J., Souto, J.A., Munuzuri, V.P., Casares, J.J. and Bermudez, J.L. (2001). A comparision of operational Lagrangian particle and adaptive puff models for plume dispersion forecastting. *Atmospheric Environment 35*, 2349-2360.

[7] Hayes, E.T., Curran, T.P. and Dodd, V.A. (2006). A dispersion modeling approach to determine the odour impact of intensive poultry production units in Ireland. *Bioresource Technology 97*, 1773-1779.

[8] Gromke, C. and Ruck, B. (2007). Influence of trees on the dispersion of pollutants in an urban street canyon-Experimental investigation of the flow and concentration field. *Atmospheric Environment 41*, 3287-3302.

[9] Meroney, R.N. and Bienkiewicz, B. (1997). *Computational Wind Engineering 2*. Amsterdam: Elsevier.

[10] Fujino, T., Asaeda, T. and Ca, V.T. (1999). Numerical analyses of urban thermal environment in a basin climate-application of a k-ε model to complex terrain. *Journal of Wind Engineering and Industrial Aerodynamics 81*, 159-169.

[11] Cressie, N.A.C. (1993). *Statistics for Spatial Data*. New York: Wiley.

[12] Johnston, K., Ver Hoef, J.M., Krivoruchko, K. and Lucas, N. (2001). *Using ArcGIS Geostatistical Analyst*. Redlands, California: ESRI Press.

[13] Seika, M., Metz, N. and Harrison, R.M. (1996). Characteristics of urban and state emission inventories-comparison from Europe and the United States. *The Science of the Total Environment 189/190*, 221-234.

[14] EMEP/CORINAIR Emission Inventory Guidebook-2007. Technical report No 16/2007. Copenhagen: European Environmental Agency.

[15] Salt, J.E. and Moran, A. (1997). International greenhouse gas inventory systems: A comparison between CORINAIR and IPCC methodologies in the EU. *Global Environmental Change 7(4)*, 317-336.

[16] Borrego, C., Tchepel, O., Barros, N. and Miranda, A.I. (2000). Impact of road traffic emissions on air quality of the Lisbon region. *Atmospheric Environment 34*, 4683-4690.

[17] Thorpea, A.J., Harrisona, R.M., Boulterb, P.G. and McCrae, I.S. (2007). Estimation of particle resuspension source strength on a major London Road. *Atmospheric Environment 41*, 8007–8020.

[18] Monforti, F. and Pederzoli, A. (2005). THOSCANE: a tool to detail CORINAIR emission inventories. *Environmental Modelling & Software 20(5)*, 505-508.

[19] Costa, M. and Baldasano, J.M. (1995). Development of a source emission model for atmospheric pollutants in the Barcelona area. *Atmospheric Environment 30(2)*, 309-318.

[20] Klimont, Z., Amann, M., Cofala, J., Gyárfáš, F., Klaassen, G. and Schöpp, W. (1994). An emission inventory for the central European initiative 1988. *Atmospheric Environment 28(2)*, 235-246.

[21] Paine, D.P. and Kiser, J.D. (2003). *Aerial Photography and Image Interpretation*. New York: Wiley.

[22] Lassueur, T., Joost, S. and Randin, C.F. (2006). Very high resolution digital elevation models: Do they improve models of plant species distribution? *Ecological Modelling 198*, 139-153.

[23] Molero, F. and Jaque, F. (1999). The laser as a tool in environmental problems. *Optical Materials 13*, 167-173.

[24] Gondal, M.A. and Mastromarino, J. (2000). Lidar system for remote environmental studies. *Talanta 53*, 147-154.

[25] Dell'Aglio, M., Kholodnykh, A., Lassandro, R. and De Pascale, O. (2002). Development of a Ti: Sapphire DIAL system for pollutant monitoring and meteorological application. *Optics and Lasers in Engineering 37*, 233-244.

[26] Fiorani, L. and Durieux, E. (2001). Comparison among error calculation in differential absorption lidar measurements. *Optics & Laser Technology 33*, 371-377.

[27] Kim, S.W., Yoon, C.S., Won, J.G. and Choi, S.C. (2007). Geound-based remote sensing measurements of aerosol and ozone in an urban area: A case study of mixing height evolution and its effect on ground-level ozone concentrations. *Atmospheric Environment 41*, 7069-7081.

[28] Zwoździak, J., Zwoździak, A., Sówka, I., Ernst, K., Stacewicz, T., Szymański, A., Chudzyński, S., Czyżewski, A., Skubiszak, W. and Stelmaszczyk, K. (2001). Some results on ozone vertical distribution in atmospheric boundary layer from LIDAR and surface measurements over the Kamienczyk Valley, Poland. *Atmospheric Research 58*, 55-70.

[29] Matejicek, L. (2005). Spatial modelling of air pollution in urban areas with GIS: a case study on integrated database development. *Advances in Geosciences 4*, 63-68.

[30] Matejicek, L., Engst, P. and Janour, Z. (2006). A GIS-based approach to spatio-temporal analysis of environmental pollution in urban areas: A case study of Prague's environment extended by LIDAR data. *Ecological Modelling 199*, 261-277.

[31] Briggs, D.J., de Hoogh, C., Gulliver, J., Wills, J., Elliott, P., Kingham, S. and Smallbone, K. (2000). A regression-based method for mapping traffic-related air pollution: application and testing in four contrasting urban environments. *The Science of the Total Environment 253*, 151-167.

[32] Pujadas, M., Plaza, J., Terés, J., Artínano, B. and Millán, M. (2000). Passive remote sensing of nitrogen dioxide as a tool for tracking air pollution in urban areas: the Madrid urban plume, a case of study. *Atmospheric Environment 34*, 3041-3056.

[33] Maidment, D.R. (1996). Environmental Modeling within GIS. In M.F. Goodchild, L.T. Steyaert, B.O. Parks (Ed.), *GIS and Environmental Modeling: Progress and Research Issues* (pp. 315-323). Fort Collins: GIS World, Inc.

[34] Fedra, K. (1996). Distributed Models and Embedded GIS: Integration Strategies and Case Studies. In M.F. Goodchild, L.T. Steyaert, B.O. Parks (Ed.), *GIS and Environmental Modeling: Progress and Research Issues* (pp. 413-417). Fort Collins: GIS World, Inc.

[35] Zeiler, M. (1999). *Modeling Our World*. Redlands, California: ESRI Press.

[36] Mitchell, A. (2005). *The ESRI Guide to GIS Analysis. Volume 2: Spatial Measurements and Statistics*. Redlands, California: ESRI Press.

[37] Maguire, D.J., Batty, M., and Goodchild, M.F. (2005). *GIS, Spatial Analysis and Modeling*. Redlands, California: ESRI Press.

[38] Grünfeld, K. (2005). Integration spatio-temporal information in environmental monitoring data-a visualization approach applied to moss data. *Science of the Total Environment 347*, 1-20.

[39] MacDonald, A. (2001). *Building a Geodatabase*. Redlands, California: ESRI Press.

[40] Solc, J. and Pospisilova, J. (2007). *Prague-Environment* 2006. J.I.M. Liberec: Prague.

[41] Matejicek, L. (2006). Spatial estimations of air pollution in urban areas by LIDAR measurements. In: A., Voinov, A.J., Jakeman, A.E., Rizzoli (Ed.), *Proceedings of the iEMSs Third Biennial Meeting: "Summit on Environmental Modelling and Software"*. International Environmental Modelling and Software Society, Burlington, USA, July 2006.

INDEX

A

B

C

D

E

F

G

H

I

J

K

L

M

N

O

P

S

T

U

V

W

X

Y

Z